4.82, 4.93, 4.110, 4.113, 4.114, 4.115, 4.116, 4.117, 4.118, 4.119, 4.123, 4.137, I.1, I.2, I.3, 5.32, 5.71, 5.77, 5.100, 5.103, 5.123, 5.124, 5.125, 5.141, 6.3, 6.5, 6.9, 6.15, 6.16, 6.17, 6.18, 6.19, 6.30, 6.42, 6.43, 6.44, 6.45, 6.62, 6.63, 6.64, 6.65, 6.72, 6.74, 6.75, 7.7, 7.9, 7.12, 7.14, 7.15, 7.34, 7.38, 7.40, 7.41, 7.49, 7.50, 7.51, 7.53, 7.54, 7.57, 7.58, 7.59, 7.60, 7.61, 7.64, 7.65, 7.66, 7.67, 7.76, 7.77, 7.80, 7.81, 7.85, 7.88, 7.89, 7.90, 7.93, 7.94, 8.1, 8.3, 8.4, 8.7, 8.18, 8.19, 8.21, 8.24, 8.25, 8.38, 8.44, 8.46, 8.48, 8.50, 8.52, 8.53, 8.56, 9.6, 9.14, 9.17, 9.28, 9.29, 9.34, 10.15, 10.16, 10.17, 10.18, 11.13, 11.16, 11.17, 11.18, 11.19, 11.20, 11.21, 11.22, 12.14, 12.16, 12.18, 12.19, 12.22, 12.23, 12.24, 12.26, 12.33, 12.34, 12.39, 12.40, 12.41, 13.7, 13.8, 13.9, 13.10, 13.13, 13.14, 13.15, 13.23, 13.29, 14.3, 14.4, 14.5, 14.6, 14.7, 14.8, 14.11, 14.12, 14.18 14.19, 14.20, 14.21, 14.22, 14.29, 14.32, III.8, III.9, III.10, III.11, III.12, III.16, III.17, III.18, 15.1, 15.2, 15.3, 15.4, 15.5, 15.6, 15.7, 15.15, 15.16, 15.17,15.18, 15.20, 15.21, 15.33, 15.34, 16.1, 16.8, 16.9, 16.11, 16.15, 16.16, 16.17, 16.20, 16.22, 16.24, 16.27, 16.31, 16.33, 16.35, 16.36, 16.37, 16.38, 16.39, 16.43, 16.48, 16.49, 16.52, 16.61, 16.64. 16.65, 16.67, 16.70, 16.74, 16.75, 17.5, 17.6, 17.12, 17.20, 17.39, 17.55, 17.56, 17.57, 17.58, 17.59

Economics and Finance

2.65, 3.136, 3.137, 4.10, 4.11, 4.12, 4.34, 4.35, 4.50, 4.72, 10.12, 10.29, 16.41, 16.45, 16.55

Education and Child Development

1.2, 1.3, 1.4, 1.6, 1.8, 1.9, 1.17, 1.18, 1.45, 2.12, 2.13, 2.15, 2.19, 2.41, 2.75, 2.96, 3.2, 3.3, 3.4, 3.7, 3.10, 3.11, 3.12, 3.13, 3.30, 3.35, 3.41, 3.44, 3.45, 3.46, 3.48, 3.49, 3.50, 3.53, 3.86, 3.87, 3.88, 3.89, 3.90, 3.91, 3.92, 3.93, 3.98, 3.99, 3.113, 3.114, 3.115. 3.116, 3.117, 3.118, 3.119, 3.120, 3.121, 3.122, 3.123, 3.141, 3.150, 4.1, 4.24, 4.25, 4.26, 4.41, 4.42, 4.43, 4.44, 4.51, 4.68, 4.73, 4.85, 4.105, 4.106, 4.107, 4.120, 4.129, 4.131, 4.134, 4.136, 4.90, I.4, I.5, I.6, I.7, I.8, I.9, I.10, I.11, I.12, I.13, I.14, 5.2, 5.10, 5.27, 5.40, 5.67, 5.70, 5.108, 5.109, 5.17, 5.18, 5.19, 5.130, 5.131, 5.142, 5.145, 5.146, 6.71, 6.78, 7.37, 7.39, 7.69, 8.24, 9.9, 9.10, 9.11, 9.12, 9.21, 9.19, 9.20, 9.33, 9.37, 10.10, 10.11, 10.14, 10.28, 10.30, 10.31, 10.32, 11.1, 11.3, 11.5, 11.6, 11.23, 11.24, 11.25, 12.13, 12.14, 12.15, 12.16, 12.17, 12.28, 12.29, 12.30, 12.53, 12.54, 13.7, 13.8, 13.24, 13.26, 13.27, 13.28, 13.36, 13.37, 13.38, 13.39, 14.35, 14.36, 14.37, 14.38, III.4, III.5, III.6, III.13, III.14, III.15, III.26, 15.11, 15.12, 15.13, 15.14, 15.24, 16.6, 16.12, 16.18, 16.46, 16.50, 16.53

Environment

2.38, 2.42, 2.52, 2.59, 3.24, 3.25, 3.36, 3.38, 3.60, 3.66, 3.109, 3.132, 3.140, 4.29, 4.56, 4.57, 4.58, 4.59, 4.60, 6.80, 7.53, 7.54, 7.67, 7.83, 7.91, 8.23, 9.32, 9.43, 10.19, 10.22, 10.23, 10.24, 10.25, 10.26, 10.27, 10.36, 10.37, 11.14, 11.30, 11.31, 11.32, 11.33, 11.34, 11.35, 11.36, 11.37, 11.38, 11.39, 11.40, 12.8, 12.14, 12.16, 12.35, 12.37, 13.34, 13.35, 14.15, 14.16, 15.25, 16.5, 16.21, 16.54, 16.57, 16.68

Ethics

2.66, 2.67, 2.68, 2.69, 2.70, 2.71, 2.72, 2.73, 2.74, 2.75, 2.76, 2.77, 2.78, 2.79, 2.80, 2.81, 2.83, 2.84, 2.96, 2.98, 2.99, I.10, I.30, I.31, I.32, 5.113, 5.114, 6.34, 6.60, 6.70, 6.80, 7.34, 8.26, 8.28, 8.30, 9.7, 9.8, 9.16, 9.17, 9.28, 9.29, 9.35, III.19, III.20, III.21, 16.74, 16.75

Health and Nutrition

1.19, 1.20, 1.22, 1.59, 2.7, 2.10, 2.20, 2.22, 2.23, 2.24, 2.28, 2.30, 2.31, 2.33, 2.37, 2.39, 2.70, 2.71, 2.73, 2.74, 2.76, 2.78, 2.80, 2.81, 2.88, 2.94, 2.95, 3.8, 3.9, 3.33, 3.34, 3.37, 3.61, 3.62, 3.63, 3.70, 3.85, 3.104, 3.124, 3.125, 3.133, 3.134, 4.2, 4.31, 4.52, 4.53, 4.54, 4.69, 4.80, 4.81, 4.84, 4.87, 4.95, 4.96, 4.97, 4.98, 4.99, 4.100, 4.101, 4.102, 4.103, 4.104, 4.121, 4.122, 4.128, 4.139, I.22, I.23, I.24, I.25, I.26, I.27, I.28, 5.7, 5.23, 5.24, 5.29, 5.33, 5.36, 5.37, 5.38, 5.39, 5.50, 5.65, 5.73, 5.80, 5.111, 5.112, 5.115, 5.116, 5.148, 6.15, 6.16, 6.17, 6.18, 6.19, 6.23, 6.37, 6.38, 6.61, 7.18, 7.19, 7.30, 7.32, 7.33, 7.35, 7.36, 7.44, 7.52, 7.55, 7.56, 7.63, 7.68, 7.77, 7.78, 7.79, 7.82, 7.84, 7.86, 7.88, 7.89, 7.90, 8.1, 8.30, 8.48, 8.49, 8.54, 8.55, 8.57, II.11, II.12, II.13, II.23, II.26, II.27, 9.3, 9.4, 9.5, 9.21, 9.27, 9.36, 10.2, 10.5, 10.13, 10.38, 10.39, 11.15, 12.25, 12.36, 12.43, 12.44, 12.45, 12.46, 13.11, 13.16, 13.17, 13.18, 13.19, 13.20, 13.30, 13.31, 13.32, 14.13, 14.14, 14.23, 14.27, 14.31, 14.33, 14.34, 14.39, III.22, III.23, 15.23, 15.26, 15.36, 15.37, 15.38, 15.40, 16.58, 16.59, 16.60, 16.66, 17.24, 17.26, 17.27, 17.40, 17.50

Humanities and Social Sciences

2.8, 2.20, 2.22, 2.32, 2.61, 2.64, 2.66, 2.67, 2.68, 2.69, 2.79, 3.10, 3.32, 3.37, 3.64, 3.77, 3.96, 3.97, 4.20, 4.12, 4.22, 4.45, 4.74, 4.75, 4.79, 4.127, 4.132, 4.133, I.4, I.5, I.6, I.7, I.8, I.9, I.10, 5.9, 5.11, 5.12, 5.26, 5.57, 5.58, 5.68, 5.74, 5.113, 5.114, 5.122, 6.1, 9.16, 9.17, 9.18, 9.25, 9.26, 9.28, 9.29, 6.24, 6.31, 6.32, 6.33, 6.34, 6.35, 6.59, 6.60, 6.68, 6.69, 6.76, 6.77, 7.20, 7.27, 7.29, 7.37, 7.49, 7.50, 7.60, 7.64, 7.65, 7.87, II.17, II.18, 11.13, 11.27, 11.28, 11.29, 12.22, 12.31, 12.32, 12.38, 12.42, 12.55, 13.13, 13.14, 13.15, 13.21, 13.22, 13.29, 13.33, 14.24, 14.26, 14.28, III.1, III.2, III.3, III.19, III.20, III.21, 15.7, 15.8, 15.9, 15.14, 16.4, 16.28, 16.64, 16.65

International

2.41, 2.65, 2.76, 2.81, 3.28, 3.29, 3.36, 3.37, 3.40, 3.42, 3.61, 3.62, 3.63, 3.66, 3.97, 3.136, 3.137, 3.138, 3.139, 4.10, 4.11, 4.12, 4.20, 4.21, 4.22, 4.34, 4.39, 4.40, 4.128, 4.131, 5.24, 5.26, 5.94, 5.96, 5.140, 6.23, 6.24, 7.30, 8.33, II.1, II.2, II.11, II.12, II.13, 9.11, 9.12, 9.13, 10.12, 10.33, 10.34, 10.35, 11.27, 11.28, 11.29, 12.31, 12.32, 13.16, 13.17, 13.18, 13.19, 13.20, 14.19, 14.30, 15.19, 16.5, 16.41, 16.45, 16.55, 17.16

Law and Government Data

2.56, 2.77, 2.84, 3.6, 3.68, 6.80, 7.69, 9.7, 9.8, 9.35

Manufacturing, Products and Processes

4.110, 4.126, 5.77, 6.72, 7.28, 7.31, 7.35, 7.36, 8.22, 9.6, 9.34, 10.20, 10.21, 10.40, 13.7, 13.8, 17.9, 17.10, 17.11, 17.13, 17.14, 17.18, 17.19, 17.20, 17.21, 17.22, 17.23, 17.24, 17.25, 17.26, 17.27, 17.28, 17.29, 17.30, 17.31, 17.32, 17.33, 17.34, 17.35, 17.36, 17.37, 17.38, 17.39, 17.43, 17.44, 17.45, 17.46, 17.48, 17.49, 17.50, 17.51, 17.52, 17.55, 17.56, 17.57, 17.58, 17.59, 17.60

Science

3.74, 3.75, 3.83, 3.129, 3.143, 4.29, 4.30, 4.83, 4.88, 9.32, 10.33, 10.34, 10.35, 12.27, 12.47, 12.48, 12.49, 12.50, 12.51, 13.34, 15.25, 15.35, 15.39

Social Networking, Demographics, Characteristics of People

3.21, 3.30, 3.65, 3.146, 4.15, 4.16, 4.17, 4.70, 4.86, 4.93, 4.105, 4.106, 4.107, 4.111, 4.112, 5.46, 5.54, 5.80, 5.85, 5.99, 5.108, 5.109, 5.117, 5.118, 5.119, 5.120, 5.121, 5.124, 5.142, 5.143, 5.145, 5.146, 6.2, 6.35, 6.36, 7.26, 8.16, 8.17, II.1, II.2, II.8, II.9, II.10, II.16, II.28, II.29, 9.19, 9.20, 11.27, 11.28, 11.29, III.1, III.2, III.3, III.22, III.23, III.24, III.25, 16.56

Sports and Leisure

2.1, 2.26, 2.91, 3.1, 3.22, 3.23, 3.64, 3.142, 4.23, 4.24, 4.74, 4.89, 4.124, 4.125, 4.138, I.15, I.16, I.17, I.18, I.19, I.20, I.21, 5.3, 5.4, 5.5, 5.6, 5.15, 5.17, 5.21, 5.22, 5.28, 5.30, 5.31, 5.48, 5.51, 5.59, 5.66, 5.69, 5.76, 5.87, 5.94, 5.96, 5.104, 5.105, 5.106, 5.107, 5.110, 5.136, 5.137, 5.139, 5.140, 5.143, 5.147, 6.4, 6.10, 6.25, 6.26, 6.28, 6.51, 6.52, 6.54, 6.55, 6.62, 6.63, 6.65, 6.72, 6.74, 6.75, 7.1, 7.26, 7.80, 7.81, 7.84, 8.8, 8.14, 8.15, 8.33, II.11, II.12, II.13, II.14, II.15, II.23, 9.31, 10.15, 10.16, III.7, 15.1, 15.2, 15.3, 15.4, 15.5, 15.6, 15.15, 15.16, 15.17, 15.18, 15.20, 15.21, 16.6, 16.16.14, 16.33, 16.43, 16.56, 16.70, 16.71, 16.72, 16.73, 17.38, 17.51, 17.52, 17.60

Students

1.1, 1.2, 1.3, 1.4, 1.6, 1.8, 1.9, 1.10, 1.18, 1.45, 2.43, 2.44, 2.47, 2.51, 2.60, 2.93, 3.2, 3.3, 3.4, 3.5, 3.7, 3.11, 3.12, 3.13, 3.16, 3.17, 3.18, 3.39, 3.41, 3.44, 3.45, 3.46, 3.48, 3.49, 3.50, 3.53, 3.98, 3.99, 4.1, 4.24, 4.25, 4.26, 4.28, 4.41, 4.42, 4.43, 4.44, 4.51, 4.95, 4.96. 4.97, 4.98, 4.99, 4.100, 4.105, 4.106, 4.107, 4.129, 4.131, 4.138, I.4, I.5, I.6, I.7, I.8, I.9, I.10, I.11, I.12, I.13, I.14, 5.83, 5.108, 5.109, 5.117, 5.118, 5.119, 5.126, 5.127, 5.128, 5.129, 5.130, 5.131, 6.1, 6.5, 6.11, 6.12, 6.27, 6.66, 6.67, 6.81, 7.13, 7.18, 7.19, 7.26, 7.39, 7.69, 7.75, 7.84, 7.92, 8.2, 8.7, 8.9, 8.16, 8.17, 8.20, 8.24, 8.31, 8.47, 8.54, 8.55, II.8, II.9, II.10, II.16, II.17, II.18, II.28, II.29, 9.9, 9.10, 9.11, 9.12, 9.13, 9.27, 9.31, 9.33, 9.37, 10.5, 10.10, 10.11, 10.28, 11.1, 11.3, 12.13, 12.15, 12.17, 12.39, 12.42, 12.55, 14.33, 14.34, III.1, III.2, III.3, III.4, III.5, III.6, III.7, III.8, III.9, III.10, III.11, III.12, III.13, III.14, III.15, III.16, III.17, III.18, III.19, III.20, III.21, III.24, III.25, III.27, III.28, III.29, 17.4, 17.7, 17.8

Technology and the Internet

2.9, 2.12, 2.13, 2.15, 2.19, 2.43, 2.44, 2.52, 2.97, 3.1, 3.22, 3.23, 3.25, 3.27, 3.28, 3.29, 3.56, 3.57, 3.61, 3.64, 4.5, 4.6, 4.7, 4.8, 4.20, 4.21, 4.22, 4.32, 4.33, 4.39, 4.40, 4.78, I.1, I.2, I.3, 5.19, 5.20, 5.21, 5.22, 5.44, 5.45, 5.49, 5.83, 5.84, 5.93, 5.94, 5.95, 5.96, 5.97, 6.2, 6.32, 6.48, 6.49, 6.50, 6.57, 6.58, 6.64, 6.65, 6.68, 6.69, 6.70, 7.1, 7.6, 7.13, 7.20, 7.26, 7.40, 7.41, 7.57, 7.58, 7.59, 7.78, 8.8, 8.14, 8.15, 8.20, 9.31, 12.14, 12.16, 12.18, 12.42, 12.55, 14.30, 14.32, III.3, III.4, III.5, III.6, III.13, III.14, III.15, III.26, 16.6, 16.7, 16.14, 16.40, 16.44, 16.62, 16.63, 17.25

EXPLORING the PRACTICE of STATISTICS

David S. Moore *Purdue University*
George P. McCabe *Purdue University*
Bruce A. Craig *Purdue University*

W. H. Freeman and Company

Senior Publisher: Ruth Baruth
Acquisitions Editor: Karen Carson
Marketing Manager: Steve Thomas
Developmental Editor: Andrew Sylvester
Senior Media Editor: Laura Judge
Associate Media Editor: Catriona Kaplan
Editorial Assistant: Liam Ferguson
Marketing Assistant: Alissa Nigro
Photo Editor: Christine Buese
Cover/Text Designer: Blake Logan
Project Editor: Elizabeth Geller
Production Coordinator: Paul W. Rohloff
Composition and Illustrations: MPS Limited
Printing and Binding: Quad Graphics

Library of Congress Control Number: 2012952915

ISBN-13: 978-1-4641-0318-6
ISBN-10: 1-4641-0318-6

Printed in the United States of America

First printing

W. H. Freeman and Company
41 Madison Avenue
New York, NY 10010
Houndmills, Basingstoke RG21 6XS, England
www.whfreeman.com

BRIEF CONTENTS

DETAILED TABLE OF CONTENTS

TO TEACHERS: *About This Book*

Statistics is the science of data. *Exploring the Practice of Statistics* (*EPS*) is an introductory text based on this principle. We present methods of basic statistics in a way that emphasizes working with data and mastering statistical reasoning. *EPS* is elementary in mathematical level, but conceptually rich in statistical ideas. After completing a course based on our text, students will be able to think objectively about conclusions drawn from data and to use statistical methods in their own work.

In *EPS*, we combine attention to basic statistical concepts with a comprehensive presentation of the elementary statistical methods that students will find useful in their work. The philosophy behind *EPS* is the same as that for *Introduction to the Practice of Statistics* (*IPS*), now in its seventh edition. The level is suitable for beginners and there is an overall emphasis on *doing* statistics rather than just talking about the subject. This hands-on approach makes our course more similar to one on drawing or painting rather than art appreciation.

We believe that the best way for students to learn statistics is by doing statistics with real data. Our examples and exercises embody this belief. Many students enter a statistics course thinking that it is like a math course where every question has a single right answer that can be computed. Doing statistics is more than calculating a mean and a standard deviation. It involves the collection of data, the critical examination of data using graphical and numerical summaries, and the drawing of conclusions. In the exercises we frequently ask students to write a paragraph summarizing what they have discovered from their analysis and to state their conclusions. Other exercises ask them to compare approaches, to decide which they prefer, and to give reasons for their preference. Students are often uncomfortable with these higher level activities that involve synthesis and evaluation. Our examples are designed to help students with these tasks and illustrate the kinds of issues that are faced by people who analyze data.

The material in *EPS* is based on *IPS* but the approach is somewhat different. In *EPS*, we have shortened the probability and other introductory material so that we can more quickly move to the later chapters. In addition, we introduce the basics of inference with proportions rather than means. This choice allows us to begin our discussion of means with the t distributions and to eliminate the discussion of the Normal case with known standard deviation, a situation rarely encountered in statistical practice.

Statistics is a very exciting field. Two of us (McCabe and Craig) have spent over forty years directing the Statistical Consulting Service at Purdue University. Our experiences helping thousands of clients have given us valuable insights about how statistical methods should be used in practice. We hope that we have been able to share effectively our insights with you in this text and that after teaching from it, you will agree with us that statistics is a very exciting field.

GAISE

The College Report of the Guidelines for Assessment and Instruction in Statistics Education (GAISE) Project[1] was funded by the American Statistical Association to make recommendations for how introductory statistics courses should be taught. This report contains many interesting teaching suggestions and we strongly recommend that you read it. The philosophy and approach of *EPS* closely reflects the GAISE recommendations. Let's examine each of the recommendations in the context of *EPS*.

1. **Emphasize statistical literacy and develop statistical thinking**. Through our experiences as applied statisticians, we are very familiar with the components that are needed for the appropriate use of statistical methods. In Chapter 1 of *EPS*, we present the six steps in a statistical study: develop the topic of interest, develop research questions, collect and find data, evaluate the quality of data, perform statistical analysis, and draw conclusions. In examples and exercises throughout the text, we emphasize putting the analysis in the proper context and translating numerical and graphical summaries into conclusions.
2. **Use real data**. Many of the examples and exercises in *EPS* include data that that we have obtained from collaborators or consulting clients. Other data sets have come from research related to these activities. We have also used the Internet as a data source, particularly for data related to social media and other topics of interest to undergraduates. With our emphasis on real data, rather than artificial data chosen to illustrate a calculation, we frequently encounter interesting issues that we explore. These include outliers and nonlinear relationships. All data sets are available from the text Web site.
3. **Stress conceptual understanding, rather than mere knowledge of procedures**. With the software available today, it is very easy for almost anyone to apply a wide variety of statistical procedures, both simple and complex, to a set of data. Without a firm grasp of the concepts, such applications are frequently meaningless. In Chapter 1 of *EPS*, we give an overview of some of the basic concepts of our field. By using the methods that we present on real sets of data, we believe that the students will gain an excellent understanding of these concepts. Our emphasis is on the input (questions of interest, collecting or finding data, data) and the output (conclusions) for a statistical analysis. Formulas are given only where they will provide some insight into concepts.
4. **Foster active learning in the classroom**. As we mentioned above, we believe that statistics is exciting as something to do rather than something to talk about. Throughout the text we provide exercises in Use Your Knowledge sections that ask the students to perform some relatively simple tasks that reinforce the material just presented. Other exercises are particularly suited to be worked and discussed within a classroom setting.

5. **Use technology for developing concepts and analyzing data.** Technology has altered statistical practice in a fundamental way. In the past, some of the calculations that we performed were particularly difficult and tedious. In other words, they were not fun. Today, freed by software from the burden of computation, we can concentrate our efforts on the big picture: what questions are we trying to address with a study and what can we conclude from our analysis?

6. **Use assessments to improve and evaluate student learning.** Our goal for students who complete a course based on *EPS* is that they be able to design and carry out a statistical study for a project in their capstone course or other setting. Our exercises are oriented toward this goal. Many ask about the design of a statistical study and the collection of data. Others ask for a paragraph summarizing the results of an analysis. This recommendation includes the use of projects, oral presentations, article critiques, and written reports. We believe that students using this text will be well prepared to undertake these kinds of activities. Furthermore, we view these activities not only as assessments but also as valuable tools for learning statistics.

Teaching Recommendations

Although we have selected and ordered the material in this text based on our views of how statistics can be taught effectively, we have many respected colleagues who prefer to add a personal touch to their course by selecting and ordering material in a different way. If you prefer this approach, we'd like to present some options. Chapter 1 uses applets to introduce and explore some fundamental ideas in statistics. These applets and associated exercises can be used during instruction to promote class discussion and to initiate statistical thinking. However, because it is an overview and we expound on all of the material it contains in later chapters, it can be skipped or assigned as background reading instead. In Chapter 2, we explore the issues related to collecting data and in Chapters 3 and 4 we present methods for describing a single variable and relationships between two variables, respectively. Some instructors prefer to integrate some or all of this material with the later material on inference. In our experience, we have found that users of statistics are a bit too eager to perform inference and as a result neglect a careful description of their data. By separating the collection and exploration of the data from statistical inference in the way that we have done, students learn the importance of these steps and consider them essential parts of statistics.

We introduce statistical inference in Chapter 5. We have included only the parts of probability that we consider essential for a sound understanding of statistical inference. If you skip this chapter, you can integrate this material when the essentials of inference are presented in Chapters 6, 7, and 8. Unlike *IPS*, we introduce the basics of inference with proportions. This organization allows us to eliminate the discussion of the Normal case with known standard deviation, a situation rarely encountered in practice. We still begin our introduction of confidence intervals and significance tests using the Normal distribution and then switch to the t distributions when we discuss inference for means.

We believe that the most interesting and useful material in the text is in Part III, Chapters 9 through 14. We recommend that one of your goals as an instructor in a course for undergraduates from a variety of fields should be that your students would appropriately use statistical methods in a project for their capstone course. If you spend an excessive amount of your time on probability and the theoretical subtleties of inference, you are very likely to miss the big picture. When we make a technically correct statement about the meaning of a confidence interval or the result of a significance test, we often miss the real point of our analysis. For the practice of statistics, the emphasis should be on the interpretation of the result and the conclusions that can be drawn from an analysis of the data, with a minimum emphasis on the statistical subtleties. If you are teaching a course for students who are mathematically stronger and will major in statistics or actuarial science, the same general comments apply. They will learn probability and statistical theory in later courses. You have the opportunity to inspire them by explaining how useful statistics is in practice. For graduate students who will need to use statistics in their research, it is particularly important that you cover the chapters in Part III.

FEATURES OF *Exploring the Practice of Statistics*

EXAMPLES are provided throughout the text and help students develop their statistical understanding using real statistical problems encountered outside of the classroom setting.

EXAMPLE 6.8

College expectations. Each year the Cooperative Institutional Research Program (CIRP) Freshman Survey is administered to first-time incoming students at hundreds of colleges and universities. In 2011, 203,967 freshmen were polled. Of the respondents, 67.5% thought they had a very good chance of having at least a B average in college.[8] You decide to see if the view of incoming freshmen at your large university is more optimistic than this. You draw an SRS of $n = 200$ students and find $X = 147$ of them have this view. Can we conclude from this survey that incoming freshmen at your university have a more optimistic view of their grade point average than the general freshman population?

One way to answer this is to compute the probability of obtaining a sample proportion as large or larger than the observed $\hat{p} = 147/200 = 0.735$ assuming that, in fact, the population proportion at your university is 0.675. Software tells us that $P(X \geq 147) = 0.04$. Because this probability is relatively small, we conclude that observing a sample proportion of 0.735 is surprising when the true proportion is 0.675. The data provide evidence for us to conclude that the incoming freshmen at your university are more optimistic than the overall freshman population.

USE YOUR KNOWLEDGE Major concepts are immediately reinforced with problems that appear throughout the chapter (often following examples), allowing students to practice their skills as they work through the text.

USE YOUR KNOWLEDGE

3.6 Compare using a different type of rate. Refer to Example 3.6 on fatal workplace injuries.

(a) Find the rates per worker for the two groups.

(b) Find the rates per 10,000 workers for the two groups.

(c) Compare the rates that you calculated in (a) and (b) with the rates given in the example. Which do you prefer for effectively communicating the results to a general audience? Give reasons for your choice.

accurate estimate of σ whether or not the population has a Normal distribution. This fact is closely related to the law of large numbers.

Constructing a Normal quantile plot, stemplot, or boxplot to check for skewness and outliers is an important preliminary to the use of t procedures for small samples. For most purposes, the one-sample t procedures can be safely used when $n \geq 15$ unless an outlier or clearly marked skewness is present. CAUTION *Except in the case of small samples, the assumption that the data are an SRS from the population of interest is more crucial than the assumption that the population distribution is Normal.* Here are practical guidelines for inference on a single mean:[7]

LOOK BACK notes direct the reader to the first explanation of a topic, providing page numbers for easy reference. **CAUTION ICONS**, located throughout the chapters, provide signals to help students avoid common errors and misconceptions.

APPLET ICONS signal students to use related interactive statistical applets found on the *EPS* Web site.

5.98. APPLET **Use the *Probability* applet.** The *Probability* applet simulates tosses of a coin. You can choose the number of tosses n and the probability p of a head. You can therefore use the applet to simulate binomial random variables.

CHAPTER 3 EXERCISES

3.132. CHALLENGE **Fuel efficiency of hatchbacks and large sedans.** Let's compare the fuel efficiencies (miles per gallon) of model year 2009 hatchbacks and large sedans.[34] Here are the data:

Hatchbacks														
30	29	28	27	27	27	27	27	26	25	25	25	24	24	24
24	24	23	23	22	22	21	21	21	21	21	21	21	20	20
20	20	20	20	20	20	19	19	19	18	16	16			

Large sedans														
19	19	18	18	18	18	17	17	17	17	17	17	17	17	17
17	16	16	16	16	16	16	16	16	15	15	13	13		

Give graphical and numerical descriptions of the fuel efficiencies for these two types of vehicles. What are the main features of the distributions? Compare the two distributions and summarize your results in a short paragraph. MPGHATCHLARGE

SECTION AND CHAPTER EXERCISES Chapters are divided into sections, each of which is capped with a full set of exercises. A final set of culminating exercises appears at the end of each chapter. Exercises range from skill-building problems to more complex exercises that allow students to make judgments and draw conclusions based on real data and real scenarios. Special challenge exercises give students another type of exercise that may be more mathematically difficult or require deeper investigation or thought about a concept.

CHALLENGE ICONS indicate those exercises that may be more mathematical than others, require open-ended investigation, or require deeper thought about the basic concepts.

8.60. CHALLENGE **Sample size and margin of error.** In Section 8.1, we studied the effect of the sample size on the margin of error of the confidence interval for a single proportion. In this exercise we perform some calculations to observe this effect for the two-sample problem. Suppose that $\hat{p}_1 = 0.7$ and $\hat{p}_2 = 0.6$, and n represents the common value of n_1 and n_2. Compute the 95% margins of error for the difference in the two proportions for $n = 40, 50, 80, 100, 400, 500,$ and 1000. Present the results in a table and with a graph. Write a short summary of your findings.

PART I EXERCISES

I.1. Compare the smartphone apps. An experiment is designed to compare three different versions of an app for a smartphone. The apps will be given to 30 college-aged women, 30 college-aged men, 30 high-school-aged women, and 30 high-school-aged men. Within each group of 30, 10 will receive each version of the app. After using the app for a week, each subject will rate his or her satisfaction with the app on a scale from 1 to 10.

(a) What are the factors and levels in this experiment?

(b) Who are the experimental units? How many are there?

(c) Draw a sketch that describes the experiment.

PART REVIEWS *EPS* is divided into three parts: Looking at Data, Probability and Inference, and Topics in Inference. At the conclusion of each part, the authors provide a brief summary that puts the concepts covered in the chapters into the greater context of the course. The authors also provide a bulleted outline of the major concepts in each chapter. Part Reviews also include a full exercise set that asks students to apply concepts learned across all of the part's chapters.

ACKNOWLEDGMENTS

We are grateful to the many colleagues and students who have provided helpful comments and we hope that they will find this text another step forward. In particular, we would like to thank the following colleagues who offered specific comments on the manuscript:

Albert Bronstein, *University of Illinois at Chicago*
Shawn Chiappetta, *University of Sioux Falls*
Melanie Christian, *St. Lawrence College*
Greg Crow, *Point Loma Nazarene University*
Stephen Fox, *Hawaii Pacific University*
Kimberley Gilbert, *University of Georgia*
Lisa Green, *Middle Tennessee State University*
Brenda Gunderson, *University of Michigan*
Greg Henderson, *Hillsborough Community College*
Melinda Holt, *Sam Houston State University*
Patricia Humphrey, *Georgia Southern University*
Dick Jardine, *Keene State University*
Theodore Lai, *Hudson County Community College*
Catherine Matos, *Clayton State University*
Eric Matsuoka, *Leeward Community College*
Jackie Miller, *The Ohio State University*
Ron Miller, *Brigham Young University-Hawaii*
Mark Mills, *Central College*
Jason Molesky, *Lakeville Area Public Schools*
John Pfister, *Dartmouth College*
Lisa Rosenberg, *Elon University*
Thomas Songer, *Portland Community College, Sylvania*
David Unger, *University of Illinois at Urbana-Champaign*
David Vlieger, *Northwest Missouri State University*
Mark Wilson, *West Virginia University Institute of Technology*

The professionals at W. H. Freeman and Company, in particular Ruth Baruth, Karen Carson, Andrew Sylvester, Elizabeth Geller, Paul Rohloff, Blake Logan, and Christine Buese, have contributed greatly to the publication and future success of *EPS*. In addition, we would like to thank Don Gecewicz, Jackie Miller, and Julie Tesser for their valuable contributions. Most of all, we are grateful to the many friends and collaborators whose data and research questions have enabled us to gain a deeper understanding of the science of data. Finally, we would like to acknowledge the work of John W. Tukey whose contributions to data analysis have had such a great influence on us as well as a whole generation of applied statisticians.

For Students

PORTAL

www.yourstatsportal.com (Access code or online purchase required.) StatsPortal is the digital gateway to *Exploring the Practice of Statistics*, designed to enrich the course and enhance students' study skills through a collection of Web-based tools. StatsPortal integrates a rich suite of diagnostic, assessment, tutorial, and enrichment features, enabling students to master statistics at their own pace. StatsPortal is organized around three main teaching and learning components:

1. **Interactive e-Book** offers a complete and customizable online version of the text, fully integrated with all of the media resources available with *EPS*. The e-Book allows students to quickly search the text, highlight key areas, and add notes about what they're reading. Instructors can customize the e-Book to add, hide, and reorder content, add their own material, and highlight key text for students.
2. **Resources** organizes all of the resources for *EPS* into one location for ease of use. Resources include the following:
   - **NEW!** *LEARNINGCurve* is a formative quizzing system that offers immediate feedback at the question level to help students master course material.
   - **New! Stepped Tutorials** These new exercise tutorials (2–3 per chapter) feature algorithmically generated quizzing with step-by-step feedback and are easily assignable for homework.
   - **Statistical Video Series** consisting of StatClips, StatClips Examples, and Statistically Speaking "Snapshots." View animated lecture videos, whiteboard lessons, and documentary-style footage that illustrate key statistical concepts and help students visualize statistics in real-world scenarios.
   - **StatTutor Tutorials** offer audio-multimedia tutorials tied directly to the textbook, containing videos, applets, and animations.
   - **Statistical Applets** offer a series of interactive applets to help students master key statistical concepts and work exercises from the text.
   - **CrunchIt!® Statistical Software** allows users to analyze data from any online location. Designed with the beginner in mind, the software is not only easily accessible but also easy to use. CrunchIt!® offers all the basic statistical routines covered in introductory statistics courses and more.

- **Stats@Work Simulations** put students in the role of the statistical consultant, helping them better understand statistics interactively within the context of real-life scenarios.
- **EESEE Case Studies,** developed by The Ohio State University Statistics Department, teach students to apply their statistical skills by exploring actual case studies using real data.
- **Data sets** are available in ASCII, Excel, TI, Minitab, SPSS (an IBM Company*), S-PLUS, and JMP formats.
- **Student Solutions Manual** provides solutions to the odd-numbered exercises, with stepped-out solutions to select problems.
- **Statistical Software Manuals** for TI-83/84, Minitab, Excel, JMP, and SPSS provide instruction, examples, and exercises using specific statistical software packages.
- **Interactive Table Reader** allows students to use statistical tables interactively to seek the information they need.
- **Tables**

Resources (instructors only)

- **Instructor's Guide with Full Solutions** includes teaching suggestions, chapter comments, and detailed solutions to all exercises.
- **Test Bank** offers hundreds of multiple-choice questions.
- **Lecture PowerPoint slides** offers a detailed lecture presentation of statistical concepts covered in each chapter of *EPS*.
- **SolutionMaster** is a Web-based version of the solutions in the Instructor's Guide with Full Solutions. This easy-to-use tool allows instructors to generate a solution file for any set of homework exercises. Solutions can be downloaded in PDF format for convenient printing and posting. For more information or a demonstration, contact your local W. H. Freeman sales representative.

3. **Assignments** organizes assignments and guides instructors through an easy-to-create assignment process providing access to questions from the Test Bank and exercises from the text, including many algorithmic problems. The Assignment Center enables instructors to create their own assignments from a variety of question types for machine gradable assignments. This powerful assignment manager allows instructors to select their preferred policies in regard to scheduling, maximum attempts, time limitations, feedback, and more!

Companion Web site www.whfreeman.com/eps This open-access Web site includes statistical applets, data sets, supplementary exercises, statistical profiles, and self-quizzes. The Web site also offers four optional companion chapters covering bootstrap methods and permutation tests and statistics for quality control and capability.

---

*SPSS was acquired by IBM in October 2009.

Special Software Packages: Student versions of JMP, Minitab, S-PLUS, and SPSS are available on the student CD-ROM. This software is not sold separately and must be packaged with a text or a manual. Contact your W. H. Freeman representative for information or visit www.whfreeman.com.

Video Tool Kit (Access code or online purchase required.) This new Statistical Video Series consists of three types of videos aimed to illustrate key statistical concepts and help students visualize statistics in real-world scenarios:

- **StatClips** lecture videos, created and presented by Alan Dabney, Ph.D., Texas A&M University, are innovative visual tutorials that illustrate key statistical concepts. In 3–5 minutes, each StatClips video combines dynamic animation, data sets, and interesting scenarios to help students understand the concepts in an introductory statistics course.
- In **StatClips Examples,** Alan Dabney walks students through step-by-step examples related to the StatClips lecture videos to reinforce the concepts through problem solving.
- **SnapShots** videos, abbreviated, student-friendly versions of the **Statistically Speaking** video series, bring the world of statistics into the classroom. In the same vein as the successful PBS series **Against All Odds Statistics**, **Statistically Speaking** uses new and updated documentary footage and interviews that show real people using data analysis to make important decisions in their careers and in their daily lives. From business to medicine, from the environment to understanding the Census, **SnapShots** focus on why statistics is important for students' careers and how statistics can be a powerful tool to understand their world.

Student Solutions Manual This manual provides stepped-through solutions for all odd-numbered exercises in the text. Available electronically within the StatsPortal and the Online Study Center, as well as in print form.

ISBN: 1-4641-2682-8

Software Manuals Software manuals covering Minitab, Excel, SPSS, TI-83/84, and JMP are offered within StatsPortal and the Online Study Center. These manuals are also available in printed versions through custom publishing. They serve as basic introductions to popular statistical software options and guides to their use with *EPS*.

For Instructors

The **Instructor's Web site** (www.whfreeman.com/eps) requires user registration as an instructor and features all the student Web materials plus:

- Instructor version of **EESEE** (Electronic Encyclopedia of Statistical Examples and Exercises), with solutions to the exercises in the student version.
- **PowerPoint slides,** containing all textbook figures and tables.

- **Lecture PowerPoint slides,** offering a detailed lecture presentation of statistical concepts covered in each chapter of *EPS*.
- **Full answers** to the **Supplementary Exercises** on the student Web site.

Instructor's Solutions Manual This guide includes full solutions to all exercises and provides additional examples and data sets for class use, Internet resources, and sample examinations. It also contains brief discussions of the *EPS* approach for each chapter. Available electronically within the StatsPortal and IRCD, as well as in print form.

Test Bank The Test Bank contains hundreds of multiple-choice questions to generate quizzes and tests for each chapter of the text. Available on CD-ROM (for Windows and Mac), where questions can be downloaded, edited, and resequenced to suit each instructor's needs.

Enhanced Instructor's Resource CD-ROM: allows instructors to **search** and **export** (by key term or chapter) all the resources available on the companion Web site plus the following:

- All text images and tables
- Instructor's Guide with full solutions
- PowerPoint files and lecture slides
- Test Bank files

Course Management Systems W. H. Freeman and Company provides courses for Blackboard, WebCT (Campus Edition and Vista), Angel, Desire2 Learn, Moodle, and Sakai course management systems. These are completely integrated solutions that you can easily customize and adapt to meet your teaching goals and course objectives. Visit http://www.macmillanhighered.com/Catalog/other/Coursepack for more information.

i-clicker

i-clicker: i-clicker is a two-way radio-frequency classroom response solution developed by educators for educators. University of Illinois physicists Tim Stelzer, Gary Gladding, Mats Selen, and Benny Brown created the i-clicker system after using competing classroom response solutions and discovering they were neither classroom-appropriate nor student-friendly. Each step of i-clicker's development has been informed by teaching and learning. i-clicker is superior to other systems from both a pedagogical and technical standpoint. To learn more about packaging i-clicker with this textbook, please contact your local sales rep or visit www.iclicker.com.

TO STUDENTS: *What Is Statistics?*

Statistics is the science of collecting, organizing, and interpreting numerical facts, which we call *data*. We are bombarded by data in our everyday lives. The news mentions movie box-office sales, the latest poll of the president's popularity, and the average high temperature for today's date. Advertisements claim that data show the superiority of the advertiser's product. All sides in public debates about economics, education, and social policy argue from data. A knowledge of statistics helps separate sense from nonsense in this flood of data.

The study and collection of data are also important in the work of many professions, so training in the science of statistics is valuable preparation for a variety of careers. Each month, for example, government statistical offices release the latest numerical information on unemployment and inflation. Economists and financial advisors, as well as policymakers in government and business, study these data to make informed decisions. Doctors must understand the origin and trustworthiness of the data that appear in medical journals.

Politicians rely on data from polls of public opinion. Business decisions are based on market research data that reveal consumer tastes and preferences. Engineers gather data on the quality and reliability of manufactured products. Most areas of academic study make use of numbers and, therefore, also make use of the methods of statistics. This means it is extremely likely that your undergraduate research projects will involve, at some level, the use of statistics.

Learning from Data

The goal of statistics is to learn from data. To learn, we often perform calculations or make graphs based on a set of numbers. But to learn from data, we must do more than calculate and plot, because data are not just numbers; they are numbers that have some context that helps us learn from them.

Two-thirds of Americans are overweight or obese according to the Center for Disease Control and Prevention (CDC) Web site (www.cdc.gov/nchs/nhanes.htm). What does it mean to be obese or to be overweight? To answer this question we need to talk about body mass index (BMI). Your weight in kilograms divided by the square of your height in meters is your BMI. A person who is 6 feet tall (1.83 meters) and weighs 180 pounds (81.65 kilograms) will have a BMI of $81.65/(1.83)^2 = 24.4$ kg/m^2. How do we interpret this number? According to the CDC, a person is classified as overweight or obese if their BMI is 25 kg/m^2 or greater, and as obese if their BMI is 30 kg/m^2 or more. Therefore, two-thirds of Americans have a BMI of 25 kg/m^2 or more. The person who weighs 180 pounds and is 6 feet tall is not overweight or obese, but if he gains 5 pounds, his BMI would increase to 25.1 and he would be classified as overweight.

When you do statistical problems, even straightforward textbook problems, don't just graph or calculate. Think about the context and state your conclusions in the specific setting of the problem. As you are learning how to do statistical calculations and graphs, remember that the goal of statistics is not calculation for its own sake but gaining understanding from numbers. The calculations and graphs can be automated by a calculator or software, but you must supply the understanding. This book presents only the most common specific procedures for statistical analysis. A thorough grasp of the principles of statistics will enable you to quickly learn more advanced methods as needed. On the other hand, a fancy computer analysis carried out without attention to basic principles will often produce elaborate nonsense. As you read, seek to understand the principles as well as the necessary details of methods and recipes.

The Rise of Statistics

Historically, the ideas and methods of statistics developed gradually as society grew interested in collecting and using data for a variety of applications. The earliest origins of statistics lie in the desire of rulers to count the number of inhabitants or measure the value of taxable land in their domains. As the physical sciences developed in the seventeenth and eighteenth centuries, the importance of careful measurements of weights, distances, and other physical quantities grew. Astronomers and surveyors striving for exactness had to deal with variation in their measurements. Many measurements should be better than a single measurement, even though they vary among themselves. How can we best combine many varying observations? Statistical methods that are still important were invented to analyze scientific measurements.

By the nineteenth century, the agricultural, life, and behavioral sciences also began to rely on data to answer fundamental questions. How are the heights of parents and children related? Does a new variety of wheat produce higher yields than the old and under what conditions of rainfall and fertilizer? Can a person's mental ability and behavior be measured just as we measure height and reaction time? Effective methods for dealing with such questions developed slowly and with much debate.

As methods for producing and understanding data grew in number and sophistication, the new discipline of statistics took shape in the twentieth century. Ideas and techniques that originated in the collection of government data, in the study of astronomical or biological measurements, and in the attempt to understand heredity or intelligence came together to form a unified "science of data." That science of data—statistics—is the topic of this text.

The Organization of This Book

In the first chapter of this book, titled "Exploring Statistics," we ask you to use applets to explore some key ideas that are fundamental to your understanding of statistics. You will learn about sources of data, describing data, probability, and statistical inference. All of these topics will be discussed thoroughly in later chapters. In this chapter, we do not present the material for you to learn in the usual way. Rather, we provide a framework whereby you can explore these basic ideas and discover the fundamentals for yourself. For example, instead of describing what happens when you toss a fair coin many times, you will discover what happens by using an applet that allows you to simulate this process. As you explore statistics in this chapter, don't worry about the fine details. You will learn them later. Just sit back and enjoy your exploration.

Part I, "Looking at Data" has three chapters. The first of these, Chapter 2, discusses methods for producing data. Here you will learn basic principles of experiments and sample surveys. You can use these principles to design sampling schemes for collecting data that will enable you to examine questions that interest you. In addition, you will be able to evaluate critically the methods used by others to collect data. Chapters 2 and 3 follow a similar theme for looking at data: we use graphical and numerical summaries to describe data. In Chapter 3 we focus on a single variable, while in Chapter 4 we move to the more interesting case of relationships between a pair of variables. Chapter 4 contains a section on cautions concerning the use of the methods presented in the chapter, a theme repeated throughout the text.

Part II, "Probability and Inference" presents the probability foundation and the basics of statistical inference in four chapters. Chapter 5 provides the probability that is needed for an understanding of statistics. Probability is very interesting and some of you may undertake additional study in this area. Our focus, however, is to move quickly to the fundamentals of statistics. Chapter 6 provides these fundamentals for proportions, while Chapter 7 gives a parallel treatment for means. The basics here include confidence intervals and significance tests for one and two samples. In Chapter 8, we discuss some additional topics that are important for statistical inference such as how confidence intervals work, determining a sample size, and cautions regarding the use of significance tests.

Part III, "Topics in Inference," focuses on statistical inference for relationships between variables. The variables can be categorical or quantitative. In Chapter 9, we discuss relationships between pairs of categorical variables. The methods in Chapters 10 and 11 are called regression. In Chapter 10, there is an explanatory variable and a response variable and both are quantitative. In Chapter 11, we can have several explanatory variables. The methods in Chapters 12 and 13 are called analysis of variance. In Chapter 12, we have one quantitative response variable and one categorical explanatory variable. Chapter 13 generalizes this setting and allows two categorical explanatory variables. Finally, in Chapter 14, "Logistic Regression," we study the setting where the response variable is categorical with two possible values and there are one or more explanatory variables. We consider the methods in this part to be among the most useful in statistics. If you do not cover them in class, we urge you to study these chapters on your own. Armed with the background from the first eight chapters of the text, you will have no trouble mastering this material.

Supplementary chapters discuss additional topics and are available at the text Web site. These include Chapter 15, "Nonparametric Tests," Chapter 16, "Bootstrap Methods and Permutation Tests," and Chapter 17, "Statistics for Quality: Control and Capability."

What Lies Ahead

Exploring the Practice of Statistics is full of data from many different areas of life and study. Many exercises ask you to express briefly some understanding gained from the data. In practice, you would know much more about the background of the data you work with and about the questions you hope the data will answer. No textbook can be fully realistic. But it is important to form the habit of asking "What do the data tell me?" rather than just concentrating on making graphs and doing calculations.

You should have some help in automating many of the graphs and calculations. There are many kinds of statistical software, from spreadsheets to large programs for advanced users of statistics. The kind of computing available to learners varies a great deal from place to place—but the big ideas of statistics don't depend on any particular level of access to computing.

Because graphing and calculating are automated in statistical practice, the most important assets you can gain from the study of statistics are an understanding of the big ideas and the beginnings of good judgment in working with data. Ideas and judgment can't (at least yet) be automated. They guide you in telling the computer what to do and in interpreting its output. This book tries to explain the most important ideas of statistics, not just teach methods. Some examples of big ideas that you will meet are "always plot your data," "randomized comparative experiments," and "statistical significance."

You learn statistics by doing statistical problems. "Practice, practice, practice." Be prepared to work problems. The basic principle of learning is persistence. Being organized and persistent is more helpful in reading this book than knowing lots of math. The main ideas of statistics, like the main ideas of any important subject, took a long time to discover and take some time to master. The gain will be worth the pain.

ABOUT THE AUTHORS

David S. Moore is the Shanti S. Gupta Distinguished Professor of Statistics, Emeritus, at Purdue University and was 1998 president of the American Statistical Association. He received his A.B. from Princeton University and his Ph.D. from Cornell University, both in mathematics. He has written many research papers in statistical theory and served on the editorial boards of several major journals. Professor Moore is an elected fellow of the American Statistical Association and of the Institute of Mathematical Statistics and an elected member of the International Statistical Institute. He has served as program director for statistics and probability at the National Science Foundation.

In recent years, Professor Moore has devoted his attention to the teaching of statistics. He was the content developer for the Annenberg/Corporation for Public Broadcasting college-level telecourse **Against All Odds: Inside Statistics** and for the series of video modules **Statistics: Decisions through Data**, intended to aid in the teaching of statistics in schools. He is the author of influential articles on statistics education and of several leading texts. Professor Moore has served as president of the International Association for Statistical Education and has received the Mathematical Association of America's national award for distinguished college or university teaching of mathematics.

George P. McCabe is the Associate Dean for Academic Affairs in the College of Science and a Professor of Statistics at Purdue University. In 1966, he received a B.S. degree in mathematics from Providence College, and in 1970 a Ph.D. in mathematical statistics from Columbia University. His entire professional career has been spent at Purdue with sabbaticals at Princeton; the Commonwealth Scientific and Industrial Research Organization in Melbourne (Australia); the University of Berne (Switzerland); the National Institute of Standards and Technology (Boulder, Colorado); and the National University of Ireland in Galway. Professor McCabe is an elected fellow of the American Association for the Advancement of Science and of the American Statistical Association; he was 1998 Chair of its section on Statistical Consulting. From 2008 to 2010, he served on the Institute of Medicine Committee on Nutrition Standards for the National School Lunch and Breakfast Programs. He has served on the editorial boards of several statistics journals, has consulted with many major corporations, and has testified as an expert witness on the use of statistics. He received the 2012 Don Owen Award presented by the San Antonio Chapter of the American Statistical Association for contributions to statistical research, applications, and teaching.

Professor McCabe's research has focused on applications of statistics. Much of his recent work has been on problems of nutrition, including nutrient requirements, calcium metabolism, and bone health. He is author or coauthor of more than 160 publications in many different journals.

Bruce A. Craig is Professor of Statistics and Director of the Statistical Consulting Service at Purdue University. He received his B.S. in mathematics and economics from Washington University in St. Louis and his M.S. and Ph.D. in statistics from the University of Wisconsin–Madison. He is an active member of the American Statistical Association and was chair of its section on Statistical Consulting in 2009. He also is an active member of the Eastern North American Region of the International Biometrics Society and was elected by the voting membership to the Regional Committee from 2003 to 2006. Professor Craig has served on the editorial board of several statistical journals and has been a member of several data and safety monitoring boards, including Purdue's IRB. Professor Craig's research interests focus on the development of novel statistical methodology to address research questions in the life sciences. Areas of current interest are susceptibility testing of antimicrobials, protein structure determination, and animal abundance estimation. In 2005, he was named Purdue University Faculty Scholar.

DATA TABLE INDEX

(Chapters 16 and 17 are available online only.)

CHAPTER ONE

EXPLORING STATISTICS

1.1 Sources of Data
1.2 Describing Data
1.3 Probability and Statistical Inference

What percent of Facebook users are female? Which of two text message advertisements will lead to higher sales? Does the credit card debt of undergraduate students vary in different regions of the United States? There are many different ways to search for answers. In this text you will learn how to collect and use data to answer questions like these.

To start, we give an overview of some key ideas that are fundamental to your understanding and use of statistics. We assume that you are familiar with some of these ideas, such as computing a percent or an average, plotting data on a graph, and tossing a coin.

data

Data are numerical or qualitative descriptions of the objects that we want to study. Not all data are equally valuable. We will learn about collecting data ourselves and about using data that others have collected, sometimes to answer questions that are different from ours.

statistics

Statistics has traditionally focused on a collection of methods that translate data into answers to our questions. We prefer a broader view of the field that includes data collection and the effective presentation of the results as well as ethical considerations that arise in the use of statistics.

statistical study

A **statistical study** starts with a general topic of interest and ends with answers to specific research questions. Figure 1.1 illustrates the process. Let's look at the parts from start to finish.

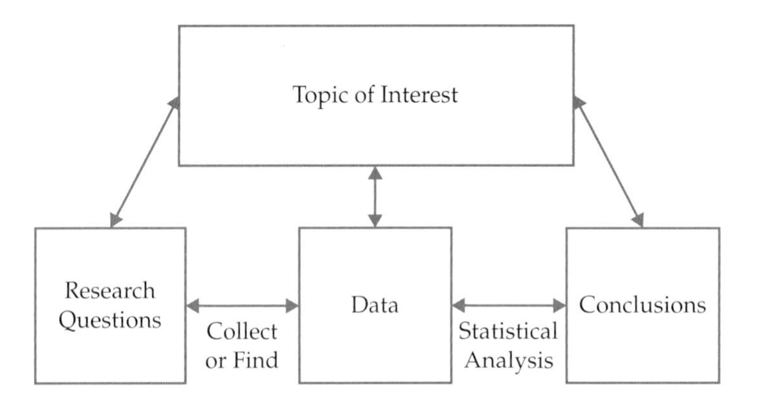

FIGURE 1.1 Outline of a statistical study.

THE SIX STEPS IN A STATISTICAL STUDY

1. **Topic of interest.** Here we develop a general idea of the area that we want to study. Teamwork can be very useful at this stage.
2. **Research questions.** We take our general ideas and translate them into research questions that can be answered with data. Note that we will often have more than one question.
3. **Collect or find data.** The research questions help us decide what kind of data we need to answer the questions. We may need to collect our own data, or data may be available from another source. If the data were collected by someone else, we need to evaluate the data collection process carefully to be sure that these data will meet our needs.
4. **Data.** Data are the central part of our process. Carefully collected data that address our topic of interest will allow us to obtain good answers to our research questions. If the data are flawed, the entire process falls apart.
5. **Analysis.** We use statistical methods to process the data and generate answers to our questions.
6. **Conclusions.** We take the results of our statistical analysis and present them in a form that we can use to describe to others what we have discovered. Graphical summaries can be very helpful here.

The topic of interest is the big-picture view of our process. Note that there are arrows from this box to research questions, data, and conclusions. The direct line to research questions is clear, but it is also important to keep the big picture in mind when we are dealing with the data and with the answers to our questions. The research questions, data, and conclusions tell us about the topic of interest. This is why the arrows also point back to the topic of interest from these boxes. It is easy to get so caught up with the details of the process that we lose sight of the big picture.

Most of the material in this text focuses on the topics that appear by the arrows between the boxes. How do we collect or find data? How should we analyze our data? To use statistics effectively, however, you need to keep in mind the entire process shown in Figure 1.1. Did your work provide answers to your questions that are based on a sound analysis of the data? Did you provide numerical summaries that communicate your findings effectively?

In the remainder of this chapter, we will explore three key ideas that are fundamental to your understanding of statistics. These are

1. **Sources of data.** We use the topic of interest and our research questions to design a plan for collecting data.
2. **Describing data.** With the data in hand, we use statistical methods to learn about our data.
3. **Probability and statistical inference.** Based on the foundation of probability theory, we draw conclusions regarding our research questions.

Discovery Learning

The topics covered in this chapter will all be discussed thoroughly in later chapters. Here, we explore these key ideas using applets that will help you to discover some basic concepts used in statistics. For example, you will use an applet that tosses a coin. In this chapter, we will not tell you what happens when you toss a coin many times. You will discover this by making many tosses with the applet. This approach is called **Discovery Learning.**[1] So, as you explore statistics in this chapter, relax and don't worry about the details. Focus on the concepts, patterns, and trends that arise from repeated use of the applets. We hope that you will find the experience to be fun.

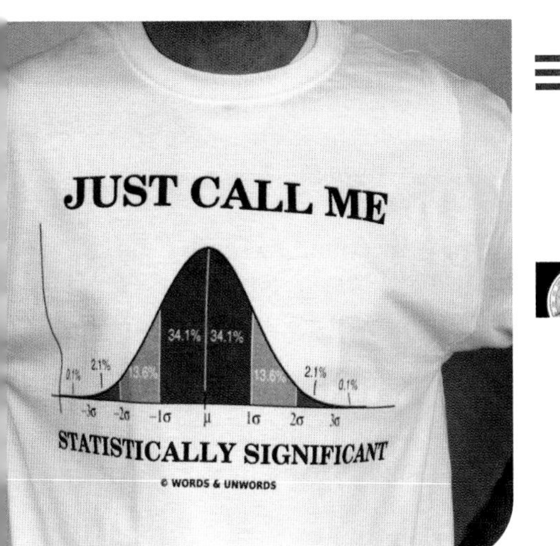

1.1 Sources of Data

Starting a statistical study

A statistical study starts with a topic of interest. Here is an example.

EXAMPLE 1.1

Should all undergraduate students take a statistics course? We are interested in attitudes about a proposed new undergraduate core curriculum requirement that all students take a statistics course. Do students at your university think this is a good idea?

Our topic of interest concerns student opinion about the new core requirement. This is a good place to start. But to decide what data to collect or find, we need to translate our topic of interest into something more specific: our research question (see Figure 1.1).

Research questions and data

To formulate our research question, we need to think about the data we will use to draw conclusions. We could ask a group of students to tell us what they think about this topic. Suppose, however, that we decide to ask some students to consider the statement "Requiring a statistics course in the new core curriculum is a good idea." We ask them to respond using a 1 to 5 scale with 1 = strongly disagree, 2 = disagree, 3 = not sure, 4 = agree, and 5 = strongly agree.

Use of the integers 1 to 5 in this setting is somewhat arbitrary. The responses strongly disagree to strongly agree could have been coded in many different ways. In addition, some researchers prefer to use a different number of possible responses, such as 7.

variable

We use the term **variable** to describe something that we measure or observe in a statistical study. In Example 1.1 the variable measures the opinion of each student studied using a 5-point scale. We will discuss different types of variables and their use in Chapter 3.

USE YOUR KNOWLEDGE

1.1 Ask different research questions. Give three other research questions that would provide data to address our general topic of interest about student opinions concerning the statistics requirement. State the possible responses to each question.

Sampling from a population

population

sample

Now that we have a question in a form that will provide data, we need to decide how to collect the data. Let's concentrate on fourth-year students. Their experience gives them a better perspective on the advantages and disadvantages of this requirement. We call this collection of students the **population** of interest. We will not ask all fourth-year students to answer the question. Instead, we will ask a **sample** of students to respond. If we do a good job of selecting the sample, we will be able to say something about the population. This is a fundamental concept in statistics: *we use data from a sample to say something about a population.*

USE YOUR KNOWLEDGE

1.2 Ask a few friends. Suppose that you ask a few of your friends who are in their fourth year whether or not they think that including statistics in the core curriculum is a good idea. Is this a good way to take a sample? Explain why or why not.

1.3 Ask the math majors. Suppose that you ask all math majors in their fourth year whether or not they think that including statistics in the core curriculum is a good idea. Is this a good way to take a sample? Explain why or why not.

The previous two exercises describe two ways to collect a sample that focus on particular subgroups of the population. Think again about the big picture. What is the topic of interest? What is the research question? What kind of conclusion will we be able to draw? If we sample just our friends or just the math majors, we will be able to conclude something about the opinions of these subgroups. But can we feel comfortable generalizing this conclusion to the population? In most cases, no. Here is another way to obtain a sample.

The *Simple Random Sample* applet on the text Web site gives a quick and easy way to select a sample in this setting. Suppose that there are 120 individuals in our population of fourth-year students and we want to select a sample of 12. Think about an ordered list of the population, numbered from 1 to 120. Here are the steps that we follow to obtain the sample.

1. Open the *Simple Random Sample* applet.
2. Set the population size by entering 120 in the box labeled "Population = 1 to."
3. Click on the "Reset" button and note that the "Population hopper" now contains the numbers 1, 2, . . . , 120.
4. Set the sample size by entering 12 in the box labeled "Select a sample of size."
5. Click on the "Sample" button.

Figure 1.2 shows the applet after completing these five steps. Notice that the "Sample bin" contains the numbers 39, 61, 41, 72, 81, 18, 36, 17, 60, 95, 40, and 63. We consult the list of students in our population, which is numbered from 1 to 120, and select the students with these numbers for our sample.

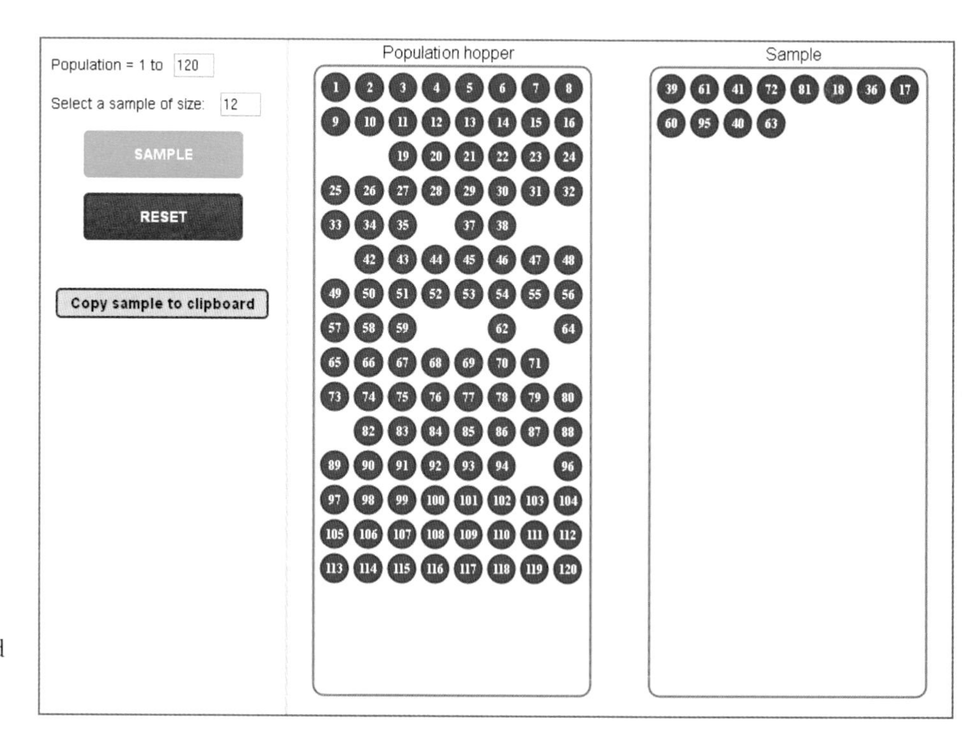

FIGURE 1.2 The *Simple Random Sample* applet selected a sample of size 12 from a population of size 120.

If we click on the "Reset" button and then the "Sample" button again, we will obtain a new sample. A second sample could be the numbers 119, 83, 5, 73, 11, 13, 113, 103, 20, 25, 97, and 27. After each sample, click on the "Reset" button to start over.

A key idea is that the applet is using a *random* method to select the sample. All possible subsets of 12 numbers are equally likely. We will learn more about populations and how to select different types of random samples in Chapter 2.

USE YOUR KNOWLEDGE

1.4 Consider a different population. Give two examples of different populations that would provide data to answer our general question about inclusion of a statistics course in the core curriculum. Do you think that changing the population would change the kind of conclusion that we could draw? Explain your answer.

1.5 Select a simple random sample. Use the *Simple Random Sample* applet to select a sample of 2 items from a population that contains the numbers 1, 2, 3, and 4. Repeat the process 20 times and record the number of times that a 3 is selected in the sample.

We now have most of the pieces that we need for our statistical study about including statistics in the core curriculum. Let's take a look at Figure 1.1 and check to see if anything is missing.

EXAMPLE 1.2

Including statistics in the core curriculum. We started with a general question about attitudes toward inclusion of a statistics course in the core curriculum. We then refined it by creating a specific question that we would ask. We also defined our population to be fourth-year students. We decided to take a sample from this population using the random sampling methods that we will discuss in Chapter 2.

The setting for many statistical studies is similar to our study of opinions about the core curriculum. We choose a set of research questions, define a population, select a sample, and then collect the data. Before we describe how to use the data to draw our conclusions, here is an example of a different but common setting for a statistical study.

A comparative study

EXAMPLE 1.3

Compare two strategies for learning statistics. Should you reread your statistics text, or is it better to spend the same amount of time working on the exercises that appear in the text? This is a general question and we need to translate it into a research question. Here is one version. After reading Chapter 6 and doing the exercises assigned by the instructor, should you spend an additional hour rereading the text or doing additional exercises? The outcome is the score on a quiz of the chapter administered immediately after the hour of additional study.

How should we collect the data? One possibility is to assign half the class to a group that rereads the text and the other half to a group that works on exercises.

USE YOUR KNOWLEDGE

1.6 Ask for volunteers. Suppose we ask the students to volunteer to be in the rereading group. We accept students into this group one at a time until we have the number needed. Evaluate this strategy for assigning students to the two groups.

We can use the *Simple Random Sample* applet to randomly assign students to the two groups. Suppose that we have 18 students and we want to assign 9 to each group. Similar to what we did in our previous use of this applet, we start with a numbered list of the 18 students. Here are the steps to obtain the two groups of students.

1. Open the *Simple Random Sample* applet.
2. Set the population size by entering 18 in the box labeled "Population = 1 to."
3. Click on the "Reset" button and note that the "Population hopper" now contains the numbers 1, 2, . . . , 18.
4. Set the sample size by entering 9 in the box labeled "Select a sample of size."
5. Click on the "Sample" button.

After completing these five steps, suppose that the "Sample bin" contains the numbers 13, 1, 18, 14, 9, 7, 5, 16, and 15. We consult the list of students in our population, which is numbered from 1 to 18, and select the students with these numbers for the group assigned to reread the text. The remaining students are assigned to the group that will do extra exercises.

Again, a key idea here is that we are using a *random* method to allocate subjects to groups in a comparative study. We will learn more about these studies in Chapter 2.

Through our examples, we have explored two key concepts in statistics. The first is **randomization.** *Random selection* of individuals from a population allows us to answer questions about the population from which the sample was drawn. This is the setting for our statistical study of the core curriculum. A random sample is more likely to be representative of the population than a convenience sample of friends or classmates.

randomization

Similarly, the *random assignment* of subjects to different treatments allows us to make a valid comparison of treatments. This is the setting for our statistical study comparing strategies for learning statistics. Random assignment allows us to avoid the bias that could result from other methods of assignment such as assigning the students who sit in the front half of the class to one treatment and those who sit in the back of the class to the other.

comparison

The second key concept is **comparison.** In Example 1.3 we wanted to draw a conclusion about effective ways to learn statistics. The two strategies for learning were similar in that both involved spending an additional hour learning material, followed by a quiz. The only difference was the activity used for learning: rereading the text or doing exercises.

Let's return to Figure 1.1. We have covered the topic of interest, research questions, and collecting or finding data. We are now at the point where we have our data. Next, we will use statistical analysis to describe our data and to draw conclusions. We will explore these ideas in the next section.

SECTION 1.1 SUMMARY

The **six steps in a statistical study** are

1. Topic of interest.
2. Research questions.
3. Collect or find data.
4. Data.
5. Analysis.
6. Conclusions.

For many statistical studies our focus of interest is a **population.** We take a **sample** of observations from the population for our study.

Taking a **simple random sample** is an easy and effective way to obtain a sample from a population.

Many statistical studies involve a **comparison** between treatments. For these studies, **randomization** is usually used to assign the subjects to the treatments.

SECTION 1.1 EXERCISES

For Exercises 1.1 to 1.3, see page 4; for Exercises 1.4 and 1.5, see page 6; and for Exercise 1.6, see page 7.

1.7. Statistical studies. Refer to Figure 1.1 (page 2). For each of the items below, explain what part of the figure they represent. Give reasons for your answers.

(a) Calculating the average quiz score of a sample of students.

(b) Who will win the next election for student body president?

(c) Is a new treatment for ACL knee surgery in college athletes better than the standard treatment with respect to a test of how well the knee works six months after the surgery?

(d) The study concluded that a flu vaccine can prevent adults from contracting the flu.

1.8. A statistics course. An instructor is interested in knowing whether a recent exam is viewed by the students as fair. There are 40 students in the class. The instructor received the opinions of 3 students who remained after class one day to complain about the exam.

(a) What is the population for this study?

(b) What is the sample?

(c) Is the sample a random sample? Explain why or why not.

(d) Do you think that the instructor should conclude that the opinions of the 3 students who remained after class represent those of the entire class? Explain why or why not.

1.9. An increase in tuition. Consider a possible increase in tuition as a topic of interest. Suggest three different research questions that could be used to learn something about this topic.

1.10. Which act to book. You are on the student concert committee. The choice of which act to book is the responsibility of your committee. What kind of data would you collect to help your committee make a good choice? Be specific and explain your answer.

1.11. APPLET **Select a sample.** Consider a population of 15 people. Use the *Simple Random Sample* applet to select a sample of size 5. Report the identification numbers of those selected in your sample.

1.12. APPLET **Select another sample.** Refer to the previous exercise. Click on the "Reset" button and then on the "Sample" button.

(a) Report the identification numbers for the new sample.

(b) How do these results compare with those that you found in the previous exercise?

(c) If you clicked "Reset" and then "Sample" again, what would you expect? Explain your answer.

1.13. APPLET **What do you think will happen?** Consider using the *Simple Random Sample* applet to select a sample of size 5 from a population of size 10.

(a) If you repeated the selection 20 times, about how many times do you think that the selected sample would include the individual labeled number 4?

(b) Use the *Simple Random Sample* applet to generate 20 samples, and report how many times the individual labeled 4 appeared in the sample.

(c) Compare this result with your answer to part (a).

(d) Repeat part (b) a few times and explain what you have learned by doing this exercise.

1.14. APPLET **Assign subjects to two treatments.** You are conducting a study to compare two treatments, A and B. Your study will have a total of 20 subjects, 10 for A and 10 for B. Use the *Simple Random Sample* applet to assign the subjects to treatments. Report the identification numbers of those assigned to each treatment.

1.15. APPLET **Select another sample.** Refer to the previous exercise. Click on the "Reset" button and then on the "Sample" button.

(a) Report the results of the new assignment.

(b) How do these results compare with those that you found in the previous exercise?

(c) If you clicked "Reset" and then "Sample" again, what would you expect? Explain your answer.

1.16. APPLET **What do you think will happen?** Consider using the *Simple Random Sample* applet to assign subjects to treatments as in the previous two exercises.

(a) If you repeated the assignment 20 times, about how many times do you think the individual labeled number 1 would be assigned to Treatment A?

(b) Use the *Simple Random Sample* applet to generate the assignments 20 times, and report how the results compare with your answer to part (a).

(c) Repeat parts (a) and (b) for the individual labeled 2.

(d) Explore the ideas presented in this exercise with additional work using the *Simple Random Sample* applet. Explain what you have done and summarize what you have learned in a short paragraph.

1.2 Describing Data

Recall that data are numerical or qualitative descriptions of the objects that we want to study. To learn from data, we need to summarize them. In this section we explore numerical and graphical ways to summarize data. We start with data for one variable.

Describing data for one variable

In Example 1.2, we asked a simple random sample of fourth-year students to reply to our question about including a statistics course in the core curriculum. Our data consist of the responses of the fourth-year students to our question. Each response is one of the numbers 1, 2, 3, 4, or 5. Suppose that we have responses from a random sample of 40 students in our population. Of these, 2 responded strongly disagree (1), 3 responded disagree (2), 5 responded not sure (3), 20 responded agree (4), and 10 responded strongly agree (5). How should we describe these data?

EXAMPLE 1.4

A numerical summary of the data. The variable in our statistical study is the response measured on the 5-point scale. Let's start with computing some percents. Two of the 40 students responded strongly disagree, response 1. The percent of students who gave this response is 2/40, or 5%. Similarly, the 3 students who responded disagree represent 3/40, or 7.5% of the sample. Here is a table giving the percents for all responses:

Response	Description	Count	Percent
1	Strongly disagree	2	5.0
2	Disagree	3	7.5
3	Not sure	5	12.5
4	Agree	20	50.0
5	Strongly agree	10	25.0
Total		40	100.0

Note that we have included a "Total" row in the table. It reminds us that the count total is 40 (the sample size) and that the percents sum to 100.

We can use the *One-Variable Statistical Calculator* applet to generate some summaries of our data. Here are the first steps in using the applet:

1. Open the *One-Variable Statistical Calculator* applet.
2. Click on the Data tab.
3. Enter the data into the display, one response (enter 1 twice, 2 three times, etc.) at a time.

Figure 1.3 shows the applet after completing these three steps. Notice that the first two entries contain the value 1; the next three contain the value 2; and

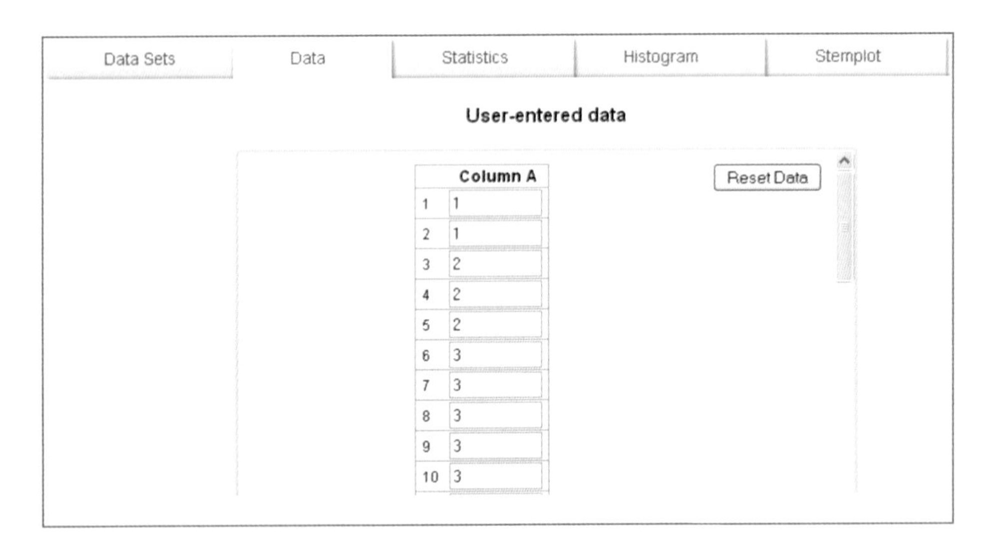

FIGURE 1.3 The *One-Variable Statistical Calculator* applet showing the Data tab with the first 10 sample values.

the next five contain the value 3. If we use the bar on the upper right of the display to scroll down the data list, we will see the entries for all 40 students in our sample.

EXAMPLE 1.5

A graphical summary of the data. The *One-Variable Statistical Calculator* applet provides several types of summaries of the data. We click on the Histogram tab, enter 0.5 in the "Lower value" box, enter 5.5 in the "Upper value" box, and enter 5 in the "Number of classes" box. Then, when we click on the "Update" button, we see the display given in Figure 1.4. Notice that the display gives a graphical view of our data. Each of the five values 1, 2, 3, 4, and 5 has a bar that shows the number of individuals having that value.

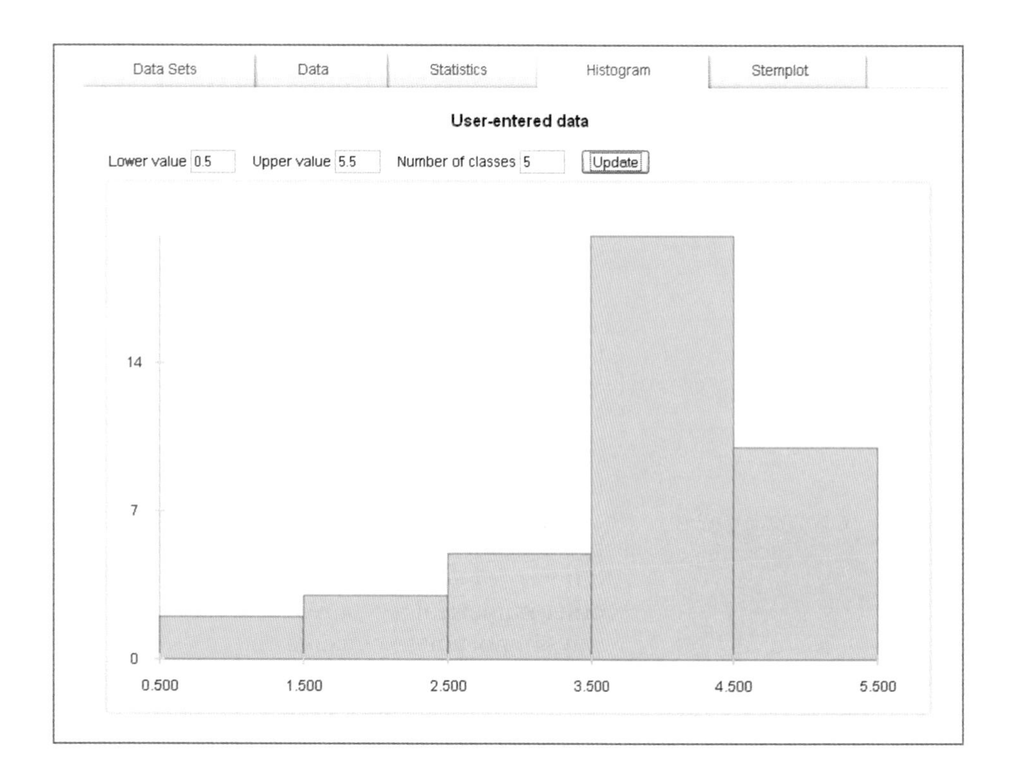

FIGURE 1.4 The *One-Variable Statistical Calculator* applet showing the Histogram tab, for Example 1.5.

The graphical summary shows that relatively few responses disagree with the proposed inclusion of a statistics course in the core curriculum. There were $2 + 3 = 5$ responses in these two categories (strongly disagree and disagree). Similarly, a large proportion of the sampled students indicated that they agree with the proposal. There were 20 students who agreed and 10 who strongly agreed.

EXAMPLE 1.6

Numerical summaries of the data. If we click on the Statistics tab of the *One-Variable Statistical Calculator,* we see several numerical summaries of our data. Figure 1.5 (page 12) shows these results. At the top of the display we see

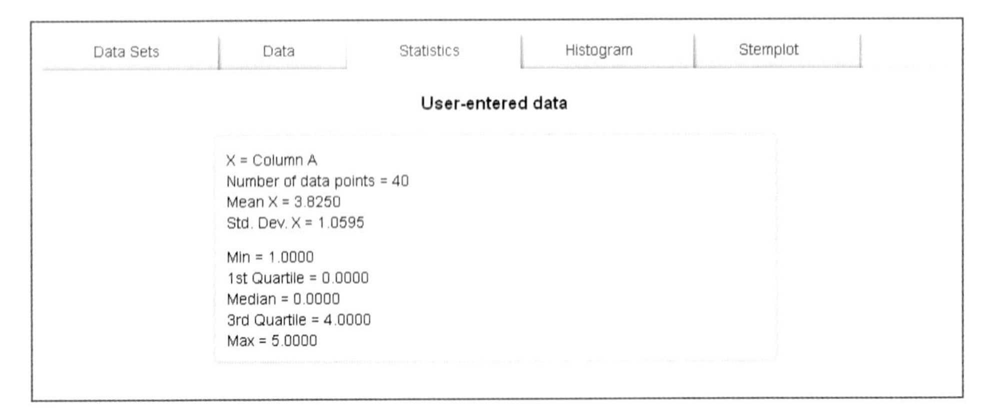

FIGURE 1.5 The *One-Variable Statistical Calculator* applet showing the Statistics tab, for Example 1.6.

$N = 40$. This tells us that our sample size is 40. It is easy to make a mistake when entering data into software. Checking the sample size on the output will help us to find and correct some kinds of mistakes.

mean

The display reports that the **mean** is 3.8250. We would report this as 3.8. You are probably familiar with this numerical summary of data. It is sometimes called the *average*. It is a measure of the center of a set of data. Sum the 40 values and divide by 40. The Min and Max are easy because these are the largest and smallest values of the variable in our sample.

The variable summarized in Example 1.4 is collected by first choosing descriptors: strongly agree, disagree, not sure, agree, strongly agree. We have coded these responses using the integers 1 to 5. Throughout this text, you will find examples and studies that use this sort of coding for opinions.

Some would argue that the mean is not a good summary for data like these because, for example, the difference between a 1 and a 2 is not the same as the difference between a 2 and a 3. On the other hand, we have found the mean and other numerical measures to be very effective summaries for such data. It is always important to consider the goals of the study when summarizing the data.

standard deviation

The other entries in the Statistics tab in Figure 1.5 are quantities that you may or may not have seen before. The **standard deviation** is a measure of how much a set of data is spread out. We will discuss all these numerical summaries in Chapter 3. For now, let's use the applet to explore how some of these summaries work.

USE YOUR KNOWLEDGE

1.17 Summarize data for one variable. The Data Sets tab of the *One-Variable Statistical Calculator* applet allows you to select a data set that contains the IQ scores for a sample of fifth-graders. Select the data and examine the summaries that the applet provides. Write a summary that includes only the ones that you have encountered previously and that you believe are effective in describing the data.

Whenever possible, graphical summaries should be used because they can visually display key characteristics of a set of data. The choice of a numerical statistical summary may depend on the intended audience for the answer to your question. In general, simpler is better. There are no strict rules to apply here but it is very important to consider this issue carefully.

EXAMPLE 1.7

A different numerical summary. For our core curriculum example, responses 4 and 5 correspond to agree and strongly agree. For a simple answer to our question, we could combine these two responses and summarize the data by reporting the proportion of responses in either of these two categories. In our sample of 40 students, 20 responded agree and 10 responded strongly agree. So the fraction of students who agree or strongly agree is $(20 + 10)/40 = 0.75$. Our simple summary here is *75% of the sampled students favor the inclusion of a statistics course in the core curriculum.*

Do we have an answer to our research question? Not quite. First, we need to decide what percent in favor in our population is enough to make a statement that the population of fourth-year students favor the inclusion of a statistics course in the core curriculum. It would be unreasonable to expect that everyone would agree. Would a simple majority of 50.1% be enough? What about two-thirds?

Another consideration is that we have obtained an estimate of this percent. If we took a different sample of 40 fourth-year students, we would expect to get a different percent in favor. This leads us to the idea of a margin of error, a plus or minus number that we attach to our estimate and that expresses the uncertainty in our reported value. We will discuss margins of error in detail in Chapter 6 and we will use them in most of the chapters that follow. We will discuss the margin of error in our opinion study after we first describe summaries for pairs of variables.

USE YOUR KNOWLEDGE

1.18 Which do you prefer? After Example 1.6 we summarized the results of our survey by giving the mean response for the five-point scale: 3.8. In Example 1.7 we reported that 75% of the students were in favor. Which summary do you prefer? Give reasons for your answer.

EXAMPLE 1.8

Summary for the statistics learning strategies study. In Example 1.3 we discussed a statistical study where we compare two strategies for learning statistics. The outcome was the score on a quiz. The usual numerical summaries for this situation are the means of the quiz scores for the two groups and the difference in these means.

In Chapter 3 we will learn more about graphical and numerical summaries that can be used to compare two groups.

Describing data for relationships

Many interesting statistical studies focus on relationships between two variables. Here is an example.

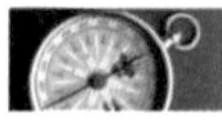

EXAMPLE 1.9

Eating fruits and vegetables and smoking. The Centers for Disease Control and Prevention (CDC) Behavior Risk Factor Surveillance System (BRFSS) collects data related to health conditions and risk behaviors.[2] The data are reported for each state. Let's look at the relationship between two of the variables in this data set. Fruits and Vegetables is the percent of adults in the state who report eating at least five servings of fruits and vegetables per day. Smoking is the percent who smoke every day. The data appear in Table 1.1.

TABLE 1.1

Fruit and vegetable consumption and smoking

State	Fruits and Vegetables (%)	Smoking (%)	State	Fruits and Vegetables (%)	Smoking (%)
Alabama	20.6	17.3	Montana	25.3	13.8
Alaska	24.2	15.1	Nebraska	24.1	13.4
Arizona	28.3	10.7	Nevada	21.9	16.4
Arkansas	21.8	17.0	New Hampshire	28.5	12.8
California	28.9	8.9	New Jersey	27.5	10.7
Colorado	25.8	12.3	New Mexico	22.4	12.8
Connecticut	28.5	11.6	New York	27.4	11.8
Delaware	21.4	13.4	North Carolina	21.6	15.5
Florida	26.2	12.8	North Dakota	21.9	13.3
Georgia	25.0	14.5	Ohio	20.8	15.5
Hawaii	28.7	11.8	Oklahoma	16.3	18.5
Idaho	22.3	12.5	Oregon	27.0	11.8
Illinois	24.6	15.1	Pennsylvania	25.4	15.6
Indiana	22.8	19.7	Rhode Island	25.6	13.4
Iowa	19.9	14.1	South Carolina	18.7	14.4
Kansas	18.8	13.2	South Dakota	18.6	12.0
Kentucky	18.4	20.5	Tennessee	26.4	18.0
Louisiana	19.6	16.0	Texas	25.2	11.7
Maine	28.6	14.0	Utah	22.8	6.8
Maryland	26.6	10.6	Vermont	30.0	12.2
Massachusetts	27.5	11.7	Virginia	26.3	12.4
Michigan	21.3	15.0	Washington	26.0	11.7
Minnesota	19.4	12.1	West Virginia	19.7	21.5
Mississippi	18.1	17.2	Wisconsin	24.4	14.1
Missouri	20.2	20.2	Wyoming	24.4	14.9

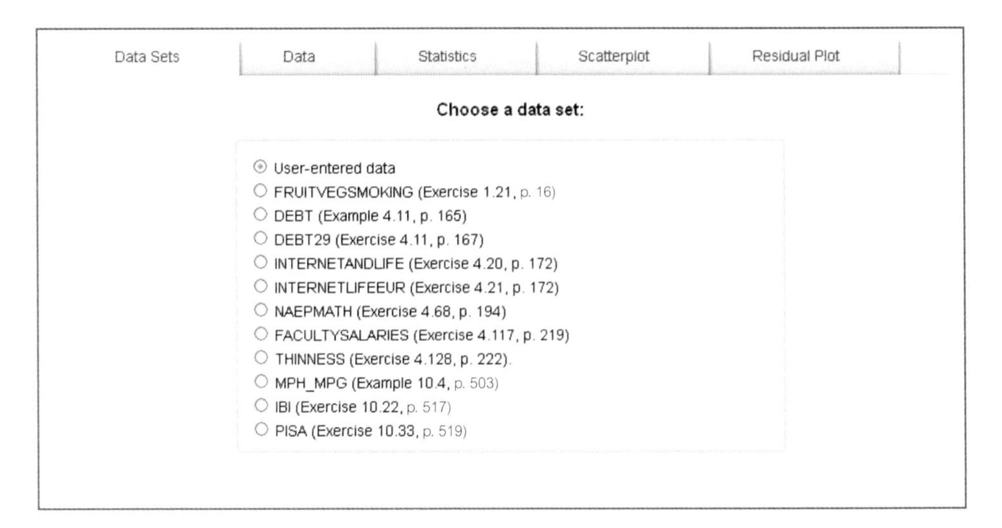

FIGURE 1.6 The *Two-Variable Statistical Calculator* applet as it appears when it is opened.

Let's explore the relationship between these two variables using the *Two-Variable Statistical Calculator* applet. Figure 1.6 shows the Data Sets tab that appears when we open the applet. We see several data sets that we can use with this applet. The one we want is titled "Fruit and vegetable consumption and smoking by state."

When we select this data set and then click on the Data tab, we see the data for the first 19 states. For Alabama, 20.6% of adults eat five or more servings of fruits and vegetables while 17.3% smoke every day; Alaskans eat more fruits and vegetables (24.2%) and they also smoke less.

Health professionals tell us that we should eat five or more servings of fruits and vegetables daily. They also tell us to avoid smoking. Alaska does better than Alabama on these two health behaviors. We can scroll down through the observations for the other states. Which state has the highest consumption of fruits and vegetables? The lowest smoking rate?

USE YOUR KNOWLEDGE

1.19 Is there a general pattern? Do you think that the comparison of Alabama and Alaska that we noticed could be part of a general pattern where eating fruits and vegetables is associated with smoking less? Give some ways that you would explore this idea.

scatterplot

We started our examination of a single variable by looking at a graphical summary, a histogram. Let's start our examination of a relationship using another graphical summary, a **scatterplot.** When we click on the Scatterplot tab in the *Two-Variable Statistical Calculator* applet, we see the display in Figure 1.7 (page 16). In this plot, the first variable, fruits and vegetables, is labeled "X", and the second variable, smoking, is labeled "Y."

You have probably seen graphs like this and have made some of your own. If not, don't worry, we will go through the details in Chapter 4. For now, let's just concentrate on the example.

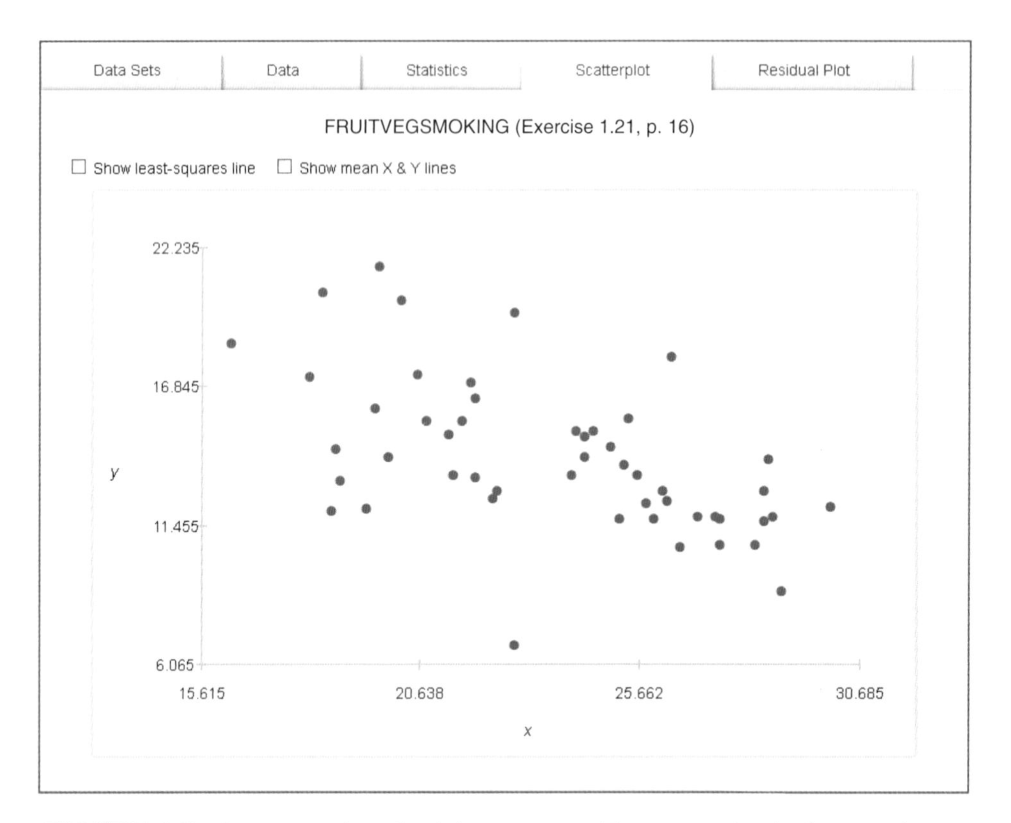

FIGURE 1.7 The Scatterplot tab of the *Two-Variable Statistical Calculator* applet.

USE YOUR KNOWLEDGE

1.20 Describe the plot. Examine the display in the Scatterplot tab carefully.

(a) Is the relationship that we observed for Alabama and Alaska true for all states? Give reasons for your answer.

(b) How would you describe the relationship between eating fruits and vegetables and smoking shown in this graph? Is it strong or weak? Is it positive or negative? (Don't worry about precise definitions of these terms; think about these things and explain your ideas.)

1.21 Sketch some scatterplots. Make some scatterplots similar to the one in Figure 1.7 that show the following kinds of relationships. You don't need to have 50 points; 10 or so will be plenty.

(a) A very strong positive relationship.

(b) A very weak negative relationship.

(c) No relationship.

When we explored data for one variable, we discussed the mean as a numerical summary. There is a similar summary that sometimes can be used to describe the relationship between two variables.

EXAMPLE 1.10

Add a line to the scatterplot. At the lower left of the display in the Scatterplot tab of the *Two-Variable Statistical Calculator* applet, there is a box that says "Show least-squares line." If we click this box, a line is added to the plot. The result is displayed in Figure 1.8. Notice that the equation of the line is given at the top of the scatterplot.

least-squares line

If you have made graphs of lines similar to the one in Figure 1.8, these ideas are somewhat familiar to you. Note that this line is called a **least-squares line.** We will study the details of this summary in Chapter 4.

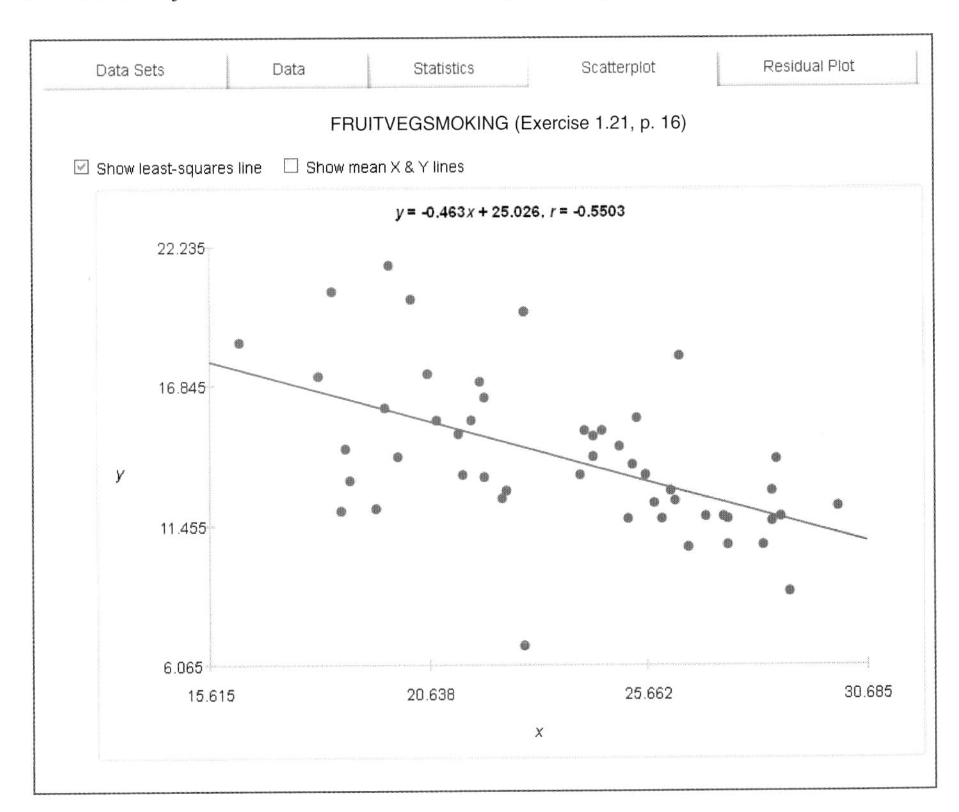

FIGURE 1.8 The Scatterplot tab of the *Two-Variable Statistical Calculator* applet with a line summarizing the relationship, for Example 1.10.

correlation

The Statistics tab in the *Two-Variable Statistical Calculator* applet also gives the equation of the least-squares line, as well as several other numerical summaries. The **correlation** is one of these summaries that is particularly useful. It measures the strength of a relationship. Section 4.2 discusses the correlation in detail.

USE YOUR KNOWLEDGE

1.22 Describe the plot with the line. Examine the display in Figure 1.8 carefully.

(a) What features of the plot does the line describe?

(b) What features of the plot does the line fail to describe?

SECTION 1.2 SUMMARY

We use **graphical** and **numerical** methods to summarize data.

Data for one variable can be summarized graphically using a **histogram.**

Data for one variable can be summarized numerically using **percents, means,** and **standard deviations.**

A relationship between two variables can be summarized graphically using a **scatterplot.**

A relationship between two variables can be summarized numerically using a **least-squares line** and the **correlation.**

SECTION 1.2 EXERCISES

For Exercise 1.17, see page 12; for Exercise 1.18, see page 13; for Exercise 1.19, see page 15; for Exercises 1.20 and 1.21, see page 16; and for Exercise 1.22, see page 17.

1.23. APPLET **Mean and standard deviation.** Enter the following data into the Data tab in the *One-Variable Statistical Calculator* applet: 8, 9, 10, 11, 12. Examine the Statistics tab and report the mean and the standard deviation.

1.24. APPLET **Mean for different data.** Refer to the previous exercise. Consider the data 8, 8, 9, 9, 10, 10, 11, 11, 12, 12.

(a) How would you expect the mean for these data to differ from what you found in the previous exercise?

(b) Check your answer to part (a) using the *One-Variable Statistical Calculator* applet.

(c) Repeat parts (a) and (b) of this exercise for the following data: 8, 8, 9, 9, 10, 10, 11, 11, 12, 13. Write a short explanation for what you have found.

1.25. APPLET **Standard deviation for another set of data.** Refer to Exercise 1.23. Consider the data 6, 8, 10, 12, 14.

(a) How would you expect the standard deviation for these data to differ from what you found in Exercise 1.23?

(b) Check your answer to part (a) using the *One-Variable Statistical Calculator* applet.

1.26. APPLET **The median.** Open the *Mean and Median* applet. Let's use this applet to explore the median, another measure of the center of a set of data. Click below the line to enter 3 observations anywhere on the line. Think about labeling the observations 1, 2, and 3 from left to right. In other words, the smallest observation is 1, the middle observation is 2, and the largest observation is 3.

(a) Which observation is the median: 1, 2, or 3?

(b) Move the observations around a bit. How does the median change?

(c) What do you conclude from this exercise?

1.27. APPLET **The median for different sample sizes.** Refer to the previous exercise. Use the *Mean and Median* applet to explore the median for 5, 7, 9, and 11 observations. Write a short paragraph explaining what you have learned from this exercise.

1.28. APPLET **The median for even-numbered sample sizes.** Let's use the *Mean and Median* applet to explore the median for an even (2, 4, 6, 8, etc.) number of observations.

(a) Describe a plan for using the applet to explore the behavior of the median of an even number of observations.

(b) Use your plan and summarize what you have learned.

1.29. APPLET **Mean and median.** Open the *Mean and Median* applet. Let's use this applet to explore these two measures of the center of a set of data. Click below the line to enter 5 observations on the left part of the line that are approximately equally spaced. The green arrow marks the location of the mean, and the red arrow marks the location of the median.

(a) Describe the location of the mean and the median.

(b) Drag the largest observation to the right and back again. How did the mean and median change? Summarize your findings.

1.30. APPLET **The least-squares line.** Open the *Correlation and Regression* applet. Let's use this applet to explore how the least-squares line changes when we move some observations. Start by putting 5 points on the plot that lie approximately on a straight line from the lower left to the upper right. Click on the box that says "Show least-squares line."

(a) Click on the point at the upper right and drag it straight down to the lower part of the display. Describe how the least-squares line changes when you do this.

(b) Drag the point that you moved back to its original position. Now, click on the middle point and drag it straight down to the lower part of the display. Describe how the least-squares line changes when you do this.

(c) Compare your results in parts (a) and (b), and summarize what you have learned about least-squares lines from this exercise.

1.31. 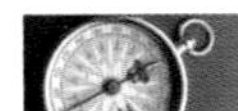 **Create your own exercise.** Use the *Correlation and Regression* applet to create your own exercise to explore how the least-squares line behaves when points are moved. Write a short report explaining what you have done and what you conclude.

1.3 Probability and Statistical Inference

One important consideration when we design a statistical study is the sample size. Intuition suggests that a larger sample size should provide more accurate information than a smaller sample size. In general, this intuition is correct. To make this idea more precise, we need some ideas from the theory of probability. Let's explore some of the fundamental concepts of probability that are useful for statistics.

EXAMPLE 1.11

Tossing coins. The simplest probability model is the repeated toss of a coin. We assume that the coin has no memory in the sense that the outcome of one toss does not influence the outcomes of any other tosses. "What happens if we toss a coin many times?" Let's answer this question in terms of the proportion of heads. We assume that the coin is fair. This means that heads and tails are equally likely.

Figure 1.9 shows the *Probability* applet as it appears when we open it. The probability of heads is set to 0.5, so we are tossing a fair coin. The number of

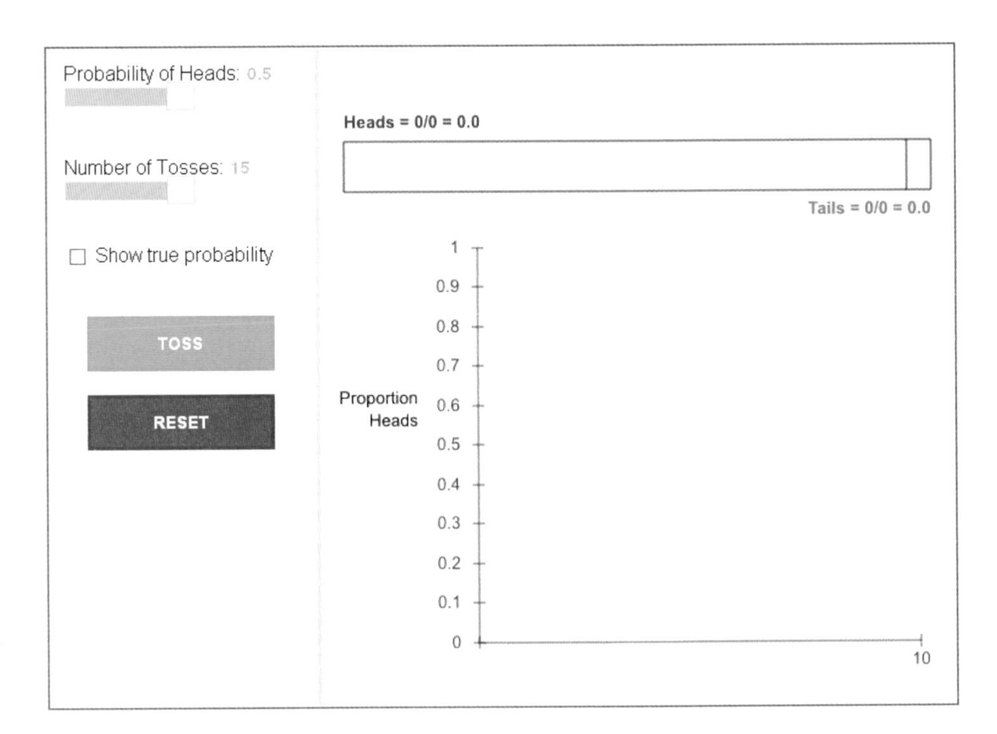

FIGURE 1.9 The *Probability* applet as it appears when it is opened, for Example 1.11.

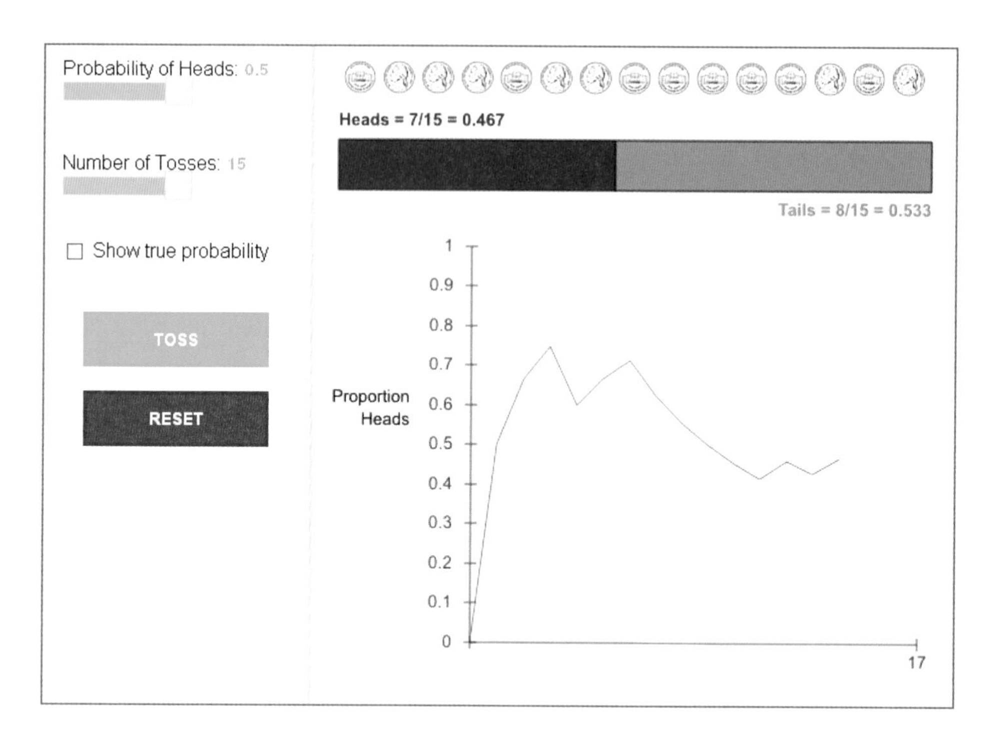

FIGURE 1.10 The *Probability* applet showing the results of tossing a coin 15 times, for Example 1.11.

tosses is set to 15. Then we click on the "Toss" button. The result is displayed in Figure 1.10. Notice that the proportion of heads is printed in the upper-left corner of the display. For our example, we have 7 heads in 15 tosses, so the proportion of heads is $7/15 = 0.467$.

Also notice that the proportion is 0 for the first toss (look at the graph at the right of the display). This means that the first toss was tails. As the number of tosses increases, the proportion of heads varies although it looks like the variation may be decreasing. Let's check this observation.

EXAMPLE 1.12

What happens if we toss 150 times? Click the "Toss" button 9 more times. The result is displayed in Figure 1.11. We clearly see that the variation decreases as the sample size increases. Also notice that the proportion of heads gets closer, in some sense, to the probability of heads, which is 0.5.

USE YOUR KNOWLEDGE

1.32 What if the probability of heads is 0.25? In the *Probability* applet, set the probability of heads to 0.25. Explore what happens for different numbers of tosses. Summarize what you have learned.

1.33 What if the probability of heads is 0.9? In the *Probability* applet, set the probability of heads to 0.9. Explore what happens for different numbers of tosses. Compare these results with those you found in Exercise 1.32 and summarize what you have learned.

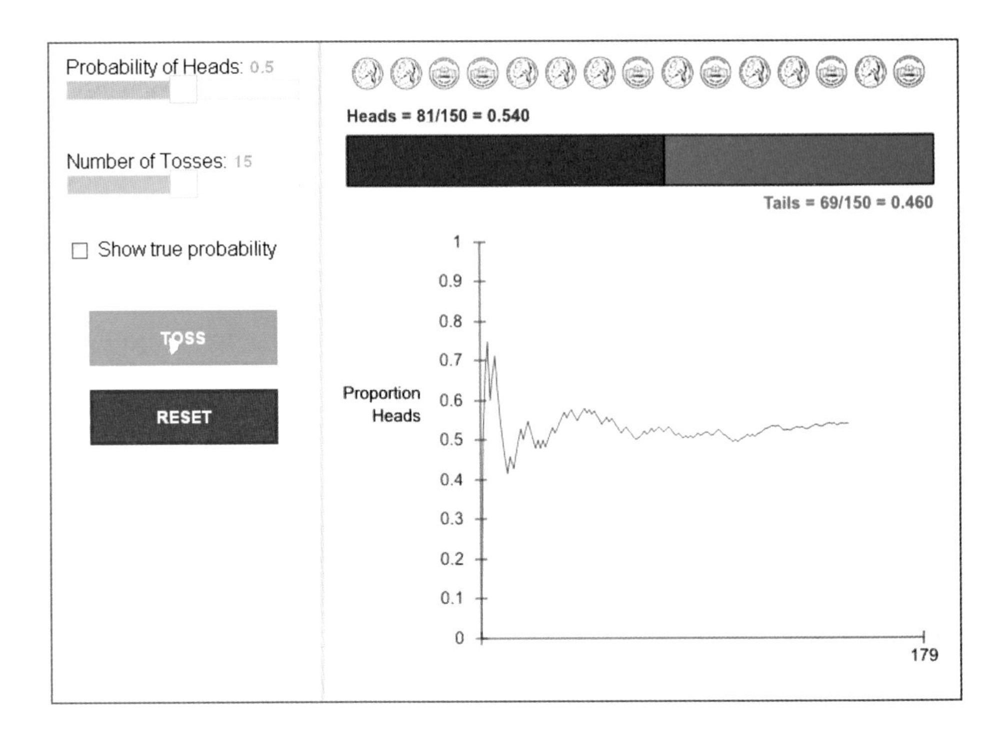

FIGURE 1.11 The *Probability* applet showing the results of tossing a coin 150 times, for Example 1.12.

So, for tossing coins, a larger sample will generally provide a sample proportion that is closer to the true probability of heads than will a smaller sample. This phenomenon is generally true: *larger samples provide better information than smaller samples.* In later chapters, we will explore how this phenomenon operates in many different situations.

Probability models

model

Tossing coins is something that is easy to understand and discuss. Most statistical studies, however, do not involve tossing coins. We discuss this setting because it gives us a **model** for many situations. The key characteristics here are that we have two possible outcomes for each trial (or toss) and that the outcome of one trial does not influence the outcome of any other trial. Examples of this setting include whether or not a randomly selected voter will say that he or she intends to vote for a particular candidate, whether or not a student athlete will graduate from college in six years, and whether or not a randomly selected student reports that he or she is a poor sleeper.

USE YOUR KNOWLEDGE

1.34 Think of some examples. Give three examples of situations where coin tossing provides a reasonable probability model. Note that the probability of heads can be any number between 0 and 1. Explain your answers.

1.35 Think of some bad examples. Give two examples of situations where there are two possible outcomes, as in the coin-tossing model, but the model is not appropriate. Explain your answers.

Coins and simple random samples

Consider a standard deck of 52 cards. You deal 5 cards to yourself. This scenario describes a process that is the basis for the *Simple Random Sample* applet. To make the connection, think about labeling the cards 1, 2, . . . , 52. This is the population and we would enter 52 in the box labeled "Population = 1 to" box. We want a sample of size 5, so we enter 5 in the box labeled "Select a sample of size."

Theoretical statisticians use probability models to derive statistical methods. The methods give correct results when the assumptions of the model hold.

Let's think about tossing coins and simple random sampling. With the coin toss, we really do not have a population that we are sampling. However, we can think of a very large population of items, half with a label of Heads and the other half with a label of Tails. As long as the population is large relative to the sample, the two models are very similar. We will use this framework for many of the methods that we will discuss in this text.

Unfortunately, in the real world, our assumed probability models are seldom completely accurate. Fortunately, situations in the real world are frequently close to our probability models, and the statistical methods that are derived from them are reasonably valid.

Other probability models

In Section 1.1, we briefly discussed comparative statistical studies. The probability model for this situation is the random assignment of subjects to two or more treatments. This model and the random sampling model are the two probability models that are the foundations for almost all the statistical methods described in this text.

Probability and the practice of statistics

When we perform a statistical study, the data we collect and analyze tell us something about the sample we used in our study. For most statistical studies, we are interested in more. For example, what can our data tell us about the population from which we drew the sample? What does the observed difference between the means of two groups tell us about the difference that we would see if we used very large samples for each group?

statistical inference

The tools of **statistical inference** allow us to answer questions like these. The two primary tools that we will study in this text are confidence intervals and significance testing. Here we will explore some of the basic ideas related to confidence intervals. In later chapters, we will study the details for both confidence intervals and significance testing in many different settings.

EXAMPLE 1.13

How do we interpret 75%? In Example 1.7 (page 13), we reported that 75% of sampled students favored including a statistics course in the core curriculum. We computed the 75% from our sample; 30 out of 40 students were in favor. Would we interpret this statistic differently if it was based on 3 out of 4 or if it was based on 300 out of 400?

USE YOUR KNOWLEDGE

1.36 What do you think? Recall the work that we did with coin tossing using the *Probability* applet in Examples 1.11 and 1.12. Think of the proportion of all fourth-year students as the target, much like the true probability of heads in our coin-tossing examples. Answer the question posed at the end of Example 1.13 using this framework.

Confidence intervals

Using probability theory, we can make these considerations more precise. When we have an estimate of some quantity from a sample, we give a measure of our uncertainty in our estimate. This value is called a **margin of error.**

margin of error

EXAMPLE 1.14

Cruelty on social-networking sites. A Pew Internet Poll asked teens aged 12 to 17 years about their experiences with social media.[3] Of the 623 teens who said that they used social media, 41% reported that they had witnessed online cruelty and meanness frequently or sometimes. The report states that the margin of error for the poll is plus or minus 6%.

The margin of error expresses the uncertainty in the estimates provided by the poll. In this case, we conclude that the true proportion of *all* teen users of social media who would say that they had witnessed online cruelty frequently or sometimes is 41% plus or minus 6%.

confidence interval

A **confidence interval** is another way to communicate the information provided by an estimate and a margin of error. We simply compute the estimate plus the margin of error ($41\% + 6\% = 47\%$) and the estimate minus the margin of error ($41\% - 6\% = 35\%$). So our confidence interval for the percent of teens who report witnessing online cruelty frequently or sometimes is 35% to 47%.

Confidence intervals are designed to have a high probability of containing the truth. Let's use an applet to explore the basic idea.

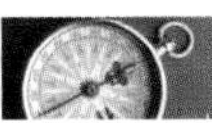

EXAMPLE 1.15

A confidence interval. When we open the *Confidence Interval* applet, we see the display given in Figure 1.12 (page 24). On the left is a panel that allows us to select a confidence level. We will leave this at the preselected value of 95%. The sample size is set to 20.

When we press the "Sample" button, we generate an estimate with its confidence interval. The display is given in Figure 1.13 (page 24). The black line added to the display is the confidence interval. The dot in the middle of the line is the estimate from sampled data, and the true value, given as μ, is represented by the green vertical line. Notice that the interval crosses the line that represents the true value. In other words, our interval *hit* the true value. This is indicated in the right panel, where we can see that we have one hit in a total of one try. The hit rate is 100%.

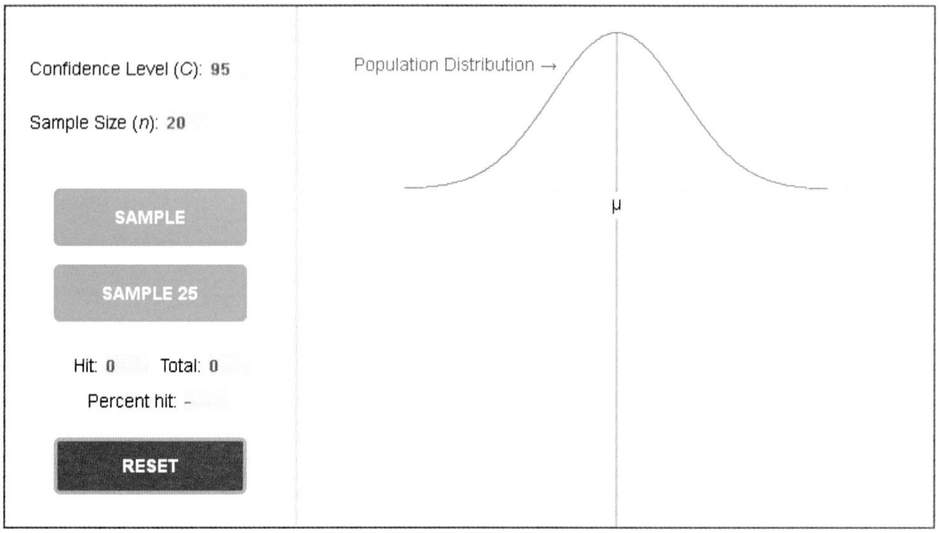

FIGURE 1.12 The *Confidence Interval* applet as it appears when it is opened, for Example 1.15.

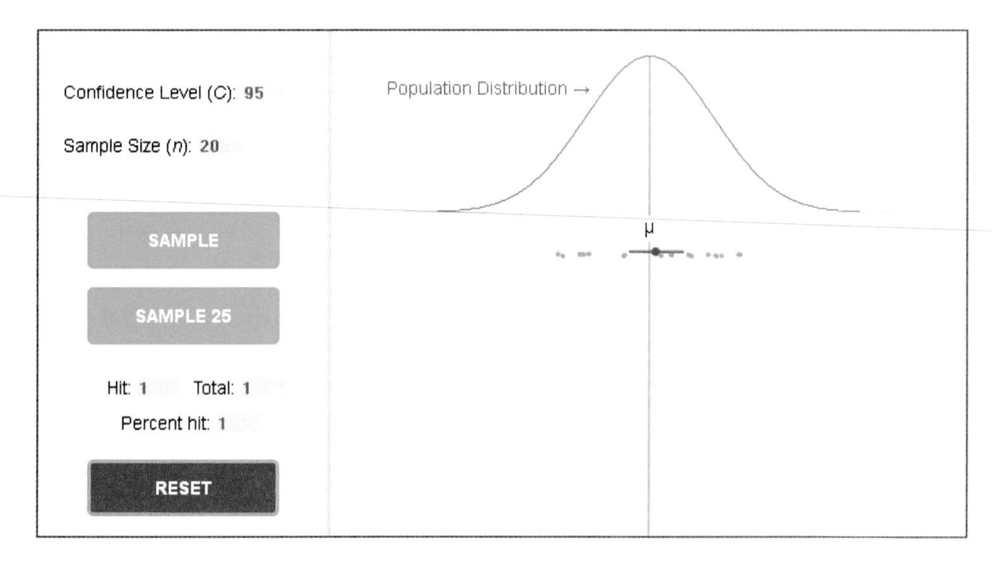

FIGURE 1.13 The *Confidence Interval* applet with one confidence interval, for Example 1.15.

Once again, don't be overly concerned about the details of how this interval was calculated. The key ideas that we are exploring with Figure 1.13 are

- We calculate a numerical summary using data from a sample.
- We calculate an interval that expresses the uncertainty in our summary.
- Our interval either contains or does not contain the true value.

We do not expect our interval to hit the true value every time. Let's explore what happens if we repeat this process many times.

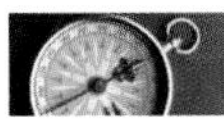

EXAMPLE 1.16

Look at many confidence intervals. Reset the *Confidence Interval* applet and click the "Sample 25" button. We see the 25 confidence intervals that result in Figure 1.14. Notice that 23 of the 25 intervals were hits. The 2 that missed are shown in red.

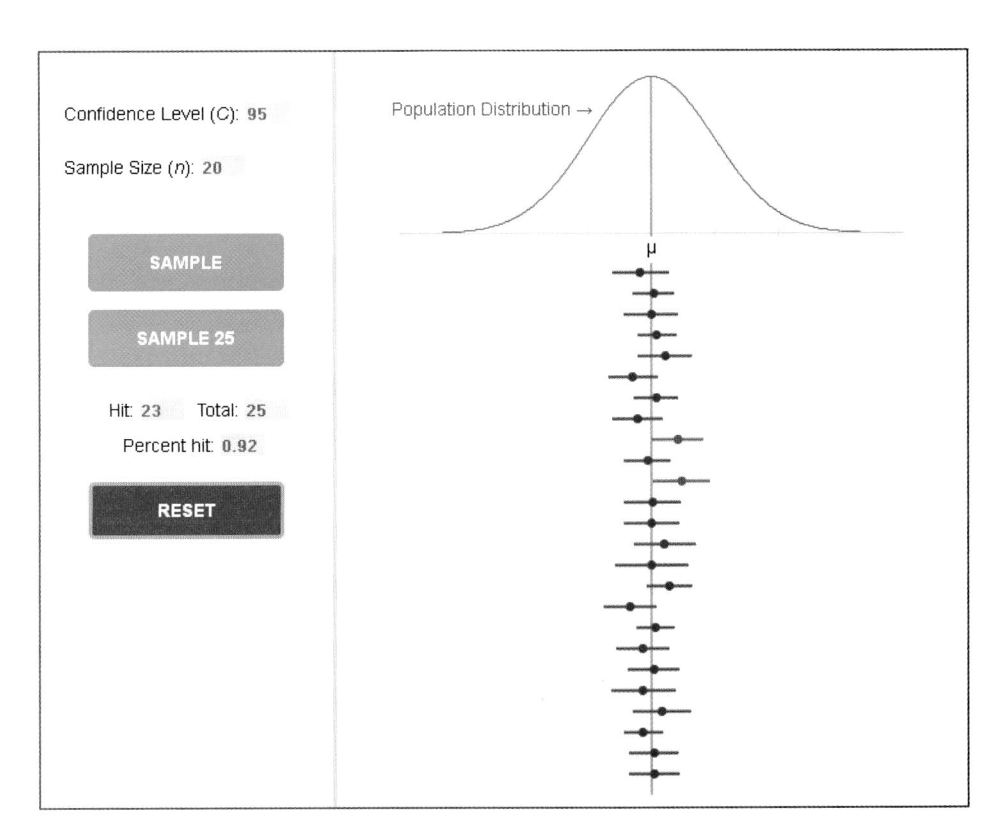

FIGURE 1.14 The *Confidence Interval* applet with 25 confidence intervals, for Example 1.16.

USE YOUR KNOWLEDGE

1.37 Generate some intervals. Use the *Confidence Interval* applet to generate 100 intervals by clicking the "Sample 25" button four times. Report the percent hit.

1.38 Repeat the process. Refer to the previous exercise.

(a) If you were to repeat the previous exercise, do you think you would get the same percent hit? Explain your answer.

(b) Check your answer by repeating the previous exercise 10 times. Report the percent hit values for each.

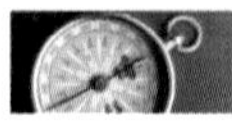

EXAMPLE 1.17

Confidence interval for statistics in the core curriculum. Refer to Example 1.4 (page 10), where we discussed the data for the study designed to address the issue of statistics in the core curriculum. In Example 1.6 we found that the mean is 3.8250. Using statistical software, we can find the 95% confidence interval. It is 3.47 to 4.16. We are 95% confident that the population mean is in this interval.

We have now finished our initial exploration of some key ideas in statistics using the applets. Here are some key points to keep in mind as we move forward:

- Randomization for surveys and experiments is a key for valid statistical inference.
- Many statistical methods are based on a consideration of what will happen in the long run.
- Keep in mind the framework presented in Figure 1.1 (page 2). Don't get lost in the many details that will follow.

SECTION 1.3 SUMMARY

Tossing coins provides a **probability model** that applies to many practical situations.

As we increase the number of tosses of a coin, the sample proportion of heads gets close to the true probability of heads.

We use **statistical inference** to generalize the results that we observe in a sample.

The **margin of error** is a measure of uncertainty in an estimate.

The two primary tools for statistical inference are **confidence intervals** and **significance testing.**

Probability models are the foundation for statistical inference.

SECTION 1.3 EXERCISES

For Exercises 1.32 and 1.33, see page 20; for Exercises 1.34 and 1.35, see page 21; for Exercise 1.36, see page 23; and for Exercises 1.37 and 1.38, see page 25.

1.39. True or false? Consider a coin that has probability of 0.5 of coming up heads. For each of the following statements, decide if it is true or false. Explain your answers.

(a) If you toss the coin 10 times, it will come up heads 5 times.

(b) If you toss the coin a very large number of times, the proportion of heads will be close to 0.5.

1.40. Margin of error. Do you think that the margin of error for a sample of 10 would be larger or smaller than the margin of error for a sample of size 100? Explain your answer.

1.41. APPLET **Generate 1000 confidence intervals.** In Example 1.15 we examined 50 confidence intervals using the *Confidence Interval* applet. Each time you click the "Sample 50" button, you generate an additional 50 confidence intervals. The number of intervals generated appears in the "Total" box. Repeatedly click the "Sample 50" button until you have 1000 intervals. Report the percent hit for your 1000 intervals.

1.42. APPLET **Change the confidence level.** Refer to the previous exercise. Change the confidence level to 90% by clicking this selection in the left panel. Then repeat the process of generating 1000 confidence intervals. Report the percent hit. How does this compare with the percent hit that you found in the previous exercise?

1.43. APPLET **Try 80% and 99%.** Refer to the previous two exercises. Repeat the generation of 1000 intervals for the two other choices of confidence level, 80% and 99%. Report the percent hit for the 1000 confidence intervals generated for these two selections, and compare these with those you found in the previous two exercises. Write a short explanation of the meaning of the confidence level based on the four possible selections available in the *Confidence Interval* applet.

1.44. Generalize the results. Based on what you have learned in the previous exercise, what do you expect to find for the percent hit with a large number of trials if the confidence level is 50%? Explain your answer.

CHAPTER 1 EXERCISES

1.45. Outline a statistical study. Outline a statistical study. Using Figure 1.1 as a guide, explain each part of your study.

1.46. Will attending statistics lectures improve your grade? The title of this exercise is a topic of interest. As a research question, it probably cannot be answered. Let's try to formulate a research question that *can* be answered with data. Here is one possibility: Do students who attend more than 80% of the statistics lectures in a particular statistics class receive higher grades than those who do not?

(a) Think about randomly assigning students to two groups: those who will attend at least 80% of the lectures and those who will not. Explain why this statistical study would be unethical.

(b) We can keep careful attendance records and classify the students at the end of the semester into the two groups. We can then compare the average grades. Explain the kind of conclusion you can draw from this statistical study. How would this kind of conclusion differ from the kind of conclusion that you could draw from the unethical study?

1.47. APPLET **Assign subjects to three treatments.** Use the *Simple Random Sample* applet to assign 30 subjects to three treatments, 10 to each. Explain all the steps you took in using the applet and report the results of your assignment.

1.48. APPLET **The mean.** For this exercise you will create data sets, enter them into the *One-Variable Statistical Calculator* applet, and examine the output in the Statistics tab. Each data set will contain 6 observations. Each observation can be 1, 2, 3, 4, or 5, as in our core curriculum example. For example, a data set could be 1, 2, 3, 3, 4, 5. You may need to explore several possibilities before finding the answers.

(a) Find a data set that maximizes the value of the mean.

(b) Find a data set that minimizes the value of the mean.

(c) Find two different data sets that have a mean equal to 3 but different standard deviations.

1.49. APPLET **The standard deviation.** Refer to the previous exercise.

(a) Find a data set that maximizes the value of the standard deviation.

(b) Find a data set that minimizes the value of the standard deviation.

1.50. APPLET **Summarize a relationship.** The Data Sets tab in the *Two-Variable Statistical Calculator* applet allows you to select a data set that contains the beer consumption and blood alcohol content for 16 volunteer students. Select the data and examine the summaries that the applet provides. Write a summary that includes only the ones that you have encountered previously and that you believe are effective in describing the relationship.

1.51. APPLET **Examine some coin-tossing trials.** Use the *Probability* applet to generate 25 tosses of a fair (set the probability of heads to 0.5) coin. Record the sample proportion (this number appears in the upper-left part of the display). Using the "Reset" button, repeat until you have 30 trials, each with 25 tosses.

(a) Open the *One-Variable Statistical Calculator* applet and enter the sample proportions for your 30 trials. Summarize the data using the information provided in the applet.

(b) Count the number of times the sample proportion is above 0.5. How does this number compare with what you would expect? Explain your answer.

PART ONE

LOOKING AT DATA

CHAPTER TWO

PRODUCING DATA

In this chapter, we will develop the skills needed to produce trustworthy data and to judge the quality of data produced by others. The techniques for producing data that we will study require no formulas, but they are among the most important ideas in statistics.

It is tempting to simply draw conclusions from our own experience, making no use of more broadly representative data. A magazine article about Pilates says that men need this form of exercise even more than women. The article, however, describes the benefits that two men received from taking Pilates classes and makes no reference to any women. A newspaper ad states that a particular brand of windows is "considered to be the best" and says that "now is the best time to replace your windows and doors." These types of stories, or *anecdotes,* sometimes provide topics of interest for future studies. However, this type of information does not give us a sound basis for drawing conclusions.

ANECDOTAL EVIDENCE

Anecdotal evidence is based on haphazardly selected individual cases, which often come to our attention because they are striking in some way. These cases may not be representative of any larger group of cases.

USE YOUR KNOWLEDGE

2.1 Wizards of Waverly Place. Your friends are big fans of *Wizards of Waverly Place,* a Disney Channel show about a family with three children who are training to be wizards. To what extent do you think you can generalize these preferences for this show to all students at your college?

2.2 Describe an anecdote. Find an example from some recent experience where anecdotal evidence is used to draw a conclusion that is not justified. Describe the example and explain why it cannot be used in this way.

2.3 Preference for Powerade. Ashley is a hard-core runner. She and all her friends prefer Powerade Ion[4] to Gatorade. Explain why Ashley's experience is not good evidence that most young people prefer Powerade to Gatorade.

2.4 Reliability of a product. A friend has driven a Toyota Camry for more than 200,000 miles with only the usual service maintenance expenses. Explain why not all Camry owners can expect this kind of performance.

Available data

Occasionally, data are collected for a particular purpose but can also serve as the basis for drawing sound conclusions about other research questions. We use the term *available data* for this type of data.

AVAILABLE DATA

Available data are data that were produced in the past for some other purpose but that may help answer a present question.

The library and the Internet can be good sources of available data. Because producing new data is expensive, we all use available data whenever possible. However, the clearest answers to present questions often require that data be produced to answer those specific questions. Here are two examples:

EXAMPLE 2.1

International consumer price indexes. If you visit the U.S. Bureau of Labor Statistics Web site, `bls.gov`, you can find many interesting sets of data and statistical summaries. One recent report compared the percent change in consumer price indexes of 9 countries for 2011. The 3.2% increase for the United States was the highest, while the lowest was Japan, with a 0.3% decrease.

FIGURE 2.1 Web sites of government statistical offices are prime sources of data. Here is a page about mathematics scores of children in grades 4 and 8 from the National Center for Education Statistics Web site.

EXAMPLE 2.2

Math skills of children. At the Web site of the National Center for Education Statistics, `nces.ed.gov`, you will find full details about the math skills of schoolchildren in the latest National Assessment of Educational Progress (Figure 2.1). Mathematics scores have slowly but steadily increased since 1990. All racial/ethnic groups, both men and women, and students in most states are getting better in math.

Many nations have a single national statistical office, such as Statistics Canada (`statcan.gc.ca`) or Mexico's INEGI (`inegi.mx`). More than 70 different U.S. agencies collect data. You can reach most of them through the government's FedStats site (`fedstats.gov`).

USE YOUR KNOWLEDGE

2.5 A popular product. A Web site claims that the Ikea Billy is a bookcase that is very popular among college students. What kind of information would you like to know about the background of this statement before you would consider it to be reliable?

A survey of college athletes is designed to estimate the percent who gamble. Do restaurant patrons give higher tips when their server repeats their order carefully? How much time do undergraduate college students spend on Facebook? Is the consumption of energy drinks increasing among high school students? The validity of our conclusions from the analysis of data collected to address issues like these rests on a foundation of carefully collected data. Statistical designs for producing these data rely on either *sampling* or *experiments*.

Sample surveys and experiments

sample surveys

How have the attitudes of Americans, on issues ranging from abortion to work, changed over time? **Sample surveys** are the usual tool for answering questions like these.

EXAMPLE 2.3

The General Social Survey. One of the most important sample surveys is the General Social Survey (GSS) conducted by the National Opinion Research Center, an organization for research and computing affiliated with the University of Chicago.[1] The GSS interviews about 3000 adult residents of the United States every second year.

sample
population

The GSS selects a **sample** of adults to represent the larger **population** of all English-speaking adults living in the United States. The idea of *sampling* is to study a part to gain information about the whole. Data are often produced by sampling a population of people or things. Opinion polls, for example, report the views of the entire country based on interviews with a sample of about 1000 people. Government reports on employment and unemployment are produced from a monthly sample of about 60,000 households. The quality of manufactured items is monitored by inspecting small samples each hour or each shift.

USE YOUR KNOWLEDGE

2.6 Find a sample survey. Use the Internet or some printed material to find an example of a sample survey that interests you. Describe the population, how the sample was collected, and some of the conclusions.

census

In all our examples, the expense of examining every item in the population makes sampling a practical necessity. Timeliness is another reason for preferring a sample to a **census,** which is an attempt to contact every individual in the entire population. We want information on current unemployment and public opinion next week, not next year. Moreover, a carefully conducted sample is often more accurate than a census. Accountants, for example, sample a firm's inventory to verify the accuracy of the records. Attempting to count every last item in the warehouse would be not only expensive but also inaccurate. Tired and bored people do not count carefully.

If conclusions based on a sample are to be valid for the entire population, a sound design for selecting the sample is required. Sampling designs are the topic of Section 2.2.

A sample survey collects information about a population by selecting and measuring a sample from the population. The goal is a picture of the population, disturbed as little as possible by the act of gathering information. Sample surveys are one kind of *observational study*.

An observational study, even one based on a statistical sample, is a poor way to determine what will happen if we change something. The best way to see the effects of a change is to do an **intervention**—where we actually impose the change. When our goal is to understand cause and effect, experiments are the only source of fully convincing data.

intervention

OBSERVATION VERSUS EXPERIMENT

In an **observational study** we observe individuals and measure variables of interest but do not attempt to influence the responses.

In an **experiment** we deliberately impose some treatment on individuals and we observe their responses.

USE YOUR KNOWLEDGE

2.7 H1N1 vaccines. A report issued by the Centers for Disease Control and Prevention stated that among 120 adults who received an injection of a monovalent H1N1 influenza A vaccine, 116, or 97%, had an effective response by three weeks after the vaccination. They also reported that the rates of adverse events such as headaches were not significantly different from those of a control group.[2] Is this an observational study or an experiment? Why?

2.8 Violent acts on prime-time TV. A typical hour of prime-time television shows three to five violent acts. Linking family interviews and police records shows a clear association between time spent watching TV as a child and later aggressive behavior.[3]
(a) Explain why this is an observational study rather than an experiment.
(b) Suggest several variables describing a child's home life that may be related to how much TV he or she watches. How would the effects of these variables make it difficult to conclude that more TV *causes* more aggressive behavior?

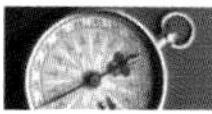

EXAMPLE 2.4

Child care and behavior. A study of child care enrolled 1364 infants in 1991 and planned to follow them through their sixth year in school. Twelve years later, the researchers published an article finding that "the more time children spent in child care from birth to age four-and-a-half, the more adults tended to rate them, both at age four-and-a-half and at kindergarten, as less likely to get along with others, as more assertive, as disobedient, and as aggressive."[4]

What can we conclude from this study? If parents choose to use child care, are they more likely to see these undesirable behaviors in their children?

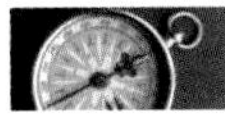

EXAMPLE 2.5

Is there a cause-and-effect relationship? Example 2.4 describes an observational study. Parents made all child care decisions and the study did not attempt to influence them. A summary of the study stated, "The study authors noted that their study was not designed to prove a cause and effect relationship. That is, the study cannot prove whether spending more time in child care causes children to have more problem behaviors."[5] Perhaps employed parents who use child care are under stress and the children react to their parents' stress. Perhaps single parents are more likely to use child care. Perhaps parents are more likely to place in child care children who already have behavior problems.

We can imagine an experiment that would remove these difficulties. From a large group of young children, choose some to be placed in child care and others to remain at home. This is an experiment because the treatment (child care or not) is imposed on the children. Of course, this particular experiment is neither practical nor ethical.

Should an experiment or sample survey that could possibly provide interesting and important information always be performed? How can we safeguard the privacy of subjects in a sample survey? What constitutes the mistreatment of people or animals who are studied in an experiment? These are questions of

ethics

ethics. In Section 2.3, we address ethical issues related to the design of studies and the analysis of data.

In Examples 2.4 and 2.5, we say that the effect of child care on behavior

confounded

is **confounded** with (mixed up with) other characteristics of families who use child care. Observational studies that examine the effect of a single variable on an outcome can be misleading when the effects of the explanatory variable are confounded with those of other variables. Because experiments allow us to isolate the effects of specific variables, we generally prefer them. Here is an example.

EXAMPLE 2.6

Which Web design sells more? A company that sells products on the Internet wants to decide which of two possible Web designs to use. During a two-week period they will use both designs and collect data on sales. They randomly select one of the designs to be used on the first day and then alternate the two designs on each of the following days. At the end of this period they compare the sales for the two designs.

randomization

Experiments usually require some sort of **randomization,** as in this example. We begin the discussion of statistical designs for data collection in Section 2.1 with the principles underlying the design of experiments.

USE YOUR KNOWLEDGE

2.9 Software for teaching creative writing. An educational software company wants to compare the effectiveness of its computer animation for teaching creative writing with that of a textbook presentation. The company tests the creative writing of each of a group of second-year college students and then randomly divides them into two groups. One group uses the animation, and the other studies the text. The company retests all the students and compares the increase in creative writing in the two groups. Is this an experiment? Why or why not?

2.10 Apples or apple juice. Food rheologists study different forms of foods and how the form of a food affects how full we feel when we eat it. One study prepared portions of apple juice and apples with the same number of calories. Half of the subjects were fed apples on one day followed by apple juice on a later day; the other half received the apple juice followed by the apples. After eating, the subjects were asked about how full they felt. Is this an experiment? Why or why not?

2.1 Design of Experiments

An experiment is a study where we actually do something to people, animals, or objects, and then we observe the response. Here is the basic vocabulary of experiments.

EXPERIMENTAL UNITS, SUBJECTS, TREATMENT

The individuals on which the experiment is done are the **experimental units.** When the units are human beings, they are called **subjects.** A specific experimental condition applied to the units is called a **treatment.**

factors

Because the purpose of an experiment is to reveal the response of one variable to changes in other variables, the distinction between explanatory and response variables is important. The explanatory variables in an experiment are often called **factors.**

level of a factor

Many experiments study the joint effects of several factors. In such an experiment, each treatment is formed by combining a specific value (often called a **level**) of each of the factors.

EXAMPLE 2.7

Are smaller class sizes better? Do smaller classes in elementary school really benefit students in areas such as scores on standard tests, staying in school, and going on to college? We might do an observational study that compares students who happened to be in smaller and larger classes in their early school years. Small classes are expensive, so they are more common in schools that serve richer communities. Students in small classes tend to also have other advantages: their schools have more resources, their parents are better educated, and so on. Confounding makes it impossible to isolate the effects of small classes.

The Tennessee STAR program was an experiment on the effects of class size. It has been called "one of the most important educational investigations ever carried out." The *subjects* were 6385 students who were beginning kindergarten. Each student was assigned to one of three *treatments:* regular class (22 to 25 students) with one teacher, regular class with a teacher and a full-time teacher's aide, and small class (13 to 17 students). These treatments are levels of a single *factor,* the type of class. The students stayed in the same type of class for four years, then all returned to regular classes. In later years, students from the small classes had higher scores on standard tests, were less likely to fail a grade, had better high school grades, and so on. The benefits of small classes were greatest for minority students.[6]

In principle, experiments can give good evidence for causation. Example 2.7 illustrates the big advantage of experiments over observational studies. All the students in the Tennessee STAR program followed the usual curriculum at their schools. Because students were assigned to different class types within their schools, school resources and family backgrounds were not confounded with class type. When students from the small classes did better than those in the other types, we can be confident that class size made the difference.

EXAMPLE 2.8

Repeated exposure to advertising. What are the effects of repeated exposure to an advertising message? The answer may depend both on the length of the ad and on how often it is repeated. An experiment investigated this question using undergraduate students as *subjects.* All subjects viewed a 40-minute television program that included ads for a digital camera. Some subjects saw a 30-second commercial; others, a 90-second version. The same commercial was shown either 1, 3, or 5 times during the program.

This experiment has two *factors:* length of the commercial, with 2 levels, and repetitions, with 3 levels. The 6 combinations of one level of each factor form 6 *treatments.* Figure 2.2 shows the layout of the treatments. After viewing, all the subjects answered questions about their recall of the ad, their attitude toward the camera, and their intention to purchase it. These are the *response variables.*[7]

Example 2.8 shows how experiments allow us to study the combined effects of several factors. The interaction of several factors can produce effects that could not be predicted from looking at the effects of each factor alone. Perhaps longer commercials increase interest in a product, and more commercials also

		Factor B Repetitions		
		1 time	3 times	5 times
Factor A Length	30 seconds	1	2	3
	90 seconds	4	5	6

FIGURE 2.2 The treatments in the study of advertising, for Example 2.8. Combining the levels of the two factors forms six treatments.

increase interest. Suppose, however, that we make a commercial longer *and* show it more often. In this case, viewers may get annoyed and their interest in the product could drop. The two-factor experiment in Example 2.8 will help us find out.

USE YOUR KNOWLEDGE

2.11 Food for a trip beyond the moon. Storing food for long periods of time is a major challenge for those planning for human space travel beyond the moon. One problem is that exposure to radiation decreases the length of time that food can be stored. One experiment examined the effects of nine different levels of radiation on a particular type of fat (lipid).[8] The amount of oxidation of the fat is the measure of the extent of the damage due to the radiation. Three samples are exposed to each radiation level. Give the experimental units, the treatment factor and its levels, and the response variable. There are many different types of lipids. To what extent do you think the results of this experiment can be generalized to other lipids?

2.12 Learning how to draw. A course in computer graphics technology requires students to learn drawing concepts. This topic is traditionally taught using supplementary material printed on paper. The instructor of the course believes that a Web-based interactive drawing program will be more effective in increasing the drawing skills of the students.[9] The 50 students who are enrolled in the course will be randomly assigned to either the paper-based instruction or the Web-based instruction. A standardized drawing test will be given before and after the instruction. Explain why this study is an experiment and give the experimental units, the treatments, and the response variable. Describe each factor and its levels. To what extent do you think the results of this experiment can be generalized to other settings?

Comparative experiments

Laboratory experiments in science and engineering often have a simple design with only a single treatment, which is applied to all experimental units. The design of such an experiment can be outlined as

Treatment ⟶ Observe response

For example, we may subject a beam to a load (treatment) and measure how much it bends (observation). When experiments are conducted in the field or with living subjects, such simple designs often yield invalid data. That is, we cannot tell whether the response was due to the treatment or to other variables.

EXAMPLE 2.9

T-shirt logos and sales. A student organization sells T-shirts to raise money to support its charitable activities. This year they decide to change the logo of the T-shirt in the hope that a more attractive logo will increase sales. They also organize a sales team to make personal contact with other students who might buy a T-shirt. Many people say that they like the new logo, and sales this year are higher than last year. The design of the experiment is

New logo ⟶ Observe sales

The T-shirt logo experiment was poorly designed to evaluate the effect of the new logo. Perhaps sales would have increased even with the old logo. Or the increase in sales could have been due to the personal contacts that were made by the sales team.

placebo effect

In medical settings this phenomenon is called a **placebo effect.** A placebo is a dummy treatment. People respond favorably to personal attention or to any treatment that they hope will help them.

For this experiment we don't know if the increase was due to the new logo, to the personal contacts with potential customers, or to a general environment where more students were inclined to buy T-shirts. The effect of the logo was confounded with other factors that could have had an effect on sales. The best way to avoid confounding is to do a comparative experiment. T-shirts with both logos could be sold and the sales of each could then be compared.

control group
treatment group

In medical settings, it is standard practice to randomly assign patients to either a **control group** or a **treatment group.** All patients are treated the same in every way except that the treatment group receives the treatment that is being evaluated.

"Uncontrolled" experiments in medicine and the behavioral sciences can be dominated by such influences as the details of the experimental arrangement, the selection of subjects, and the placebo effect. The result is often *bias.*

> **BIAS**
>
> A study has **bias** if it systematically favors certain outcomes.

An uncontrolled study of a new medical therapy, for example, is biased in favor of finding the treatment effective because of the placebo effect. Uncontrolled studies in medicine give new therapies a much higher success rate than proper comparative experiments do. Well-designed medical experiments usually compare several treatments.

USE YOUR KNOWLEDGE

2.13 Does using statistical software improve exam scores? An instructor in an elementary statistics course wants to know if using a new statistical software package will improve students' final-exam scores. He asks for volunteers and about half the class agrees to work with the new software. He compares the final-exam scores of the students who used the new software with the scores of those who did not. Discuss possible sources of bias in this study.

Randomization

experimental design

The **design of an experiment** first describes the response variable or variables, the factors (explanatory variables), and the layout of the treatments, with comparison as the leading principle. Figure 2.2 illustrates this aspect of the design of a study of response to advertising. The second aspect of design is the rule used to assign the experimental units to the treatments. Comparison of the effects of several treatments is valid only when all treatments are applied to similar groups of experimental units. If one corn variety is planted on more fertile ground, or if one cancer drug is given to more seriously ill patients, comparisons among treatments are meaningless.

If the groups assigned to treatments are quite different in a comparative experiment, we should be concerned that our experiment will be biased. How can we assign experimental units to treatments in a way that is fair to all treatments?

Experimenters often attempt to match groups by elaborate balancing acts. Medical researchers, for example, try to match the patients in a "new drug" experimental group and a "standard drug" control group by age, sex, physical condition, smoker or not, and so on. Matching is helpful but not always adequate.

Some important variables, such as how advanced a cancer patient's disease is, are so subjective that an experimenter might bias the study by, for example, assigning more advanced cancer cases to a promising new treatment in the hope that it will help them.

The statistician's remedy is to rely on *chance* to make an assignment that does not depend on any characteristic of the experimental units and that does not rely on the judgment of the experimenter in any way. The use of chance can be combined with matching, but the simplest design creates groups by chance alone. Here is an example.

EXAMPLE 2.10

Which new smartphone should be marketed? Two teams have prepared prototypes for a new smartphone. Before deciding which one will be marketed, they will be evaluated by college students. Forty students will receive a new phone. They will use it for two weeks and then answer some questions about how well they like the phone. The 40 students will be randomized, with 20 receiving each phone.

This experiment has a single factor (prototype) with two levels. The researchers must divide the 40 student subjects into two groups of 20. To do this in a completely unbiased fashion, they put the names of the 40 students in a hat, mix them up, and draw 20. These students will receive Phone 1, and the remaining 20 will receive Phone 2. Figure 2.3 outlines the design of this experiment.

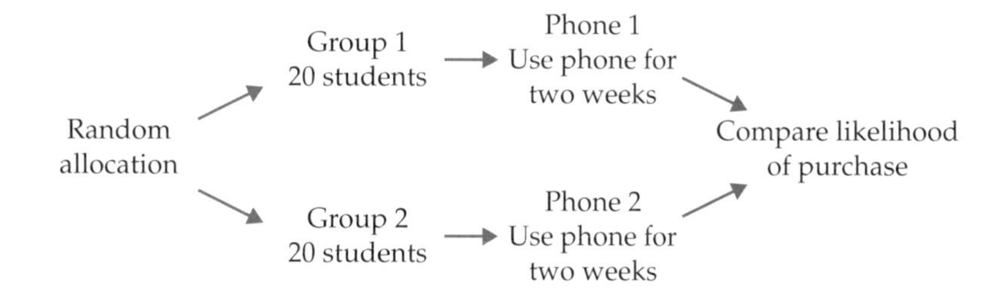

FIGURE 2.3 Outline of a randomized comparative experiment, for Example 2.10.

The use of chance to divide experimental units into groups is called randomization. The design in Figure 2.3 combines comparison and randomization to arrive at the simplest randomized comparative design. This "flowchart" outline presents all the essentials: randomization, the sizes of the groups and which treatment they receive, and the response variable. There are, as we will see later, statistical reasons for generally using treatment groups about equal in size.

USE YOUR KNOWLEDGE

2.14 Diagram the food for space travel experiment. Refer to Exercise 2.11 (page 39). Draw a diagram similar to Figure 2.3 that describes the food for space travel experiment.

2.15 Diagram the drawing experiment. Refer to Exercise 2.12 (page 39). Draw a diagram similar to Figure 2.3 that describes the computer graphics drawing experiment.

Randomized comparative experiments

The logic behind the randomized comparative design in Figure 2.3 is as follows:

- Randomization produces two groups of subjects that we expect to be similar in all respects before the treatments are applied.
- Comparative design helps ensure that influences other than the characteristics of the smartphone operate equally on both groups.
- Therefore, differences in the satisfaction with the phone must be due either to the characteristics of the smartphone or to the chance assignment of subjects to the two groups.

That "either-or" deserves more comment. We cannot say that *any* difference in the satisfaction with the two smartphones is caused by the characteristics of the smartphones. There would be some difference even if both groups used the same phone. Some students would be more likely to be highly favorable of any new smartphone. Chance can assign more of these students to one of the phones, so that there is a chance difference between the groups. We would not trust an experiment with just one subject in each group, for example. The results would depend too much on which phone got lucky and received the subject who was more likely to be highly satisfied. "Use enough subjects to reduce chance variation" is the third big idea of statistical design of experiments.

PRINCIPLES OF EXPERIMENTAL DESIGN

The basic principles of statistical design of experiments are

1. **Compare** two or more treatments.
2. **Randomize**—use impersonal chance to assign experimental units to treatments.
3. **Repeat** each treatment on many units to reduce chance variation in the results.

We hope to see a difference in the responses so large that it is unlikely to happen just because of chance variation. We can use the laws of probability (Chapter 5), which give a mathematical description of chance behavior, to learn if the treatment effects are larger than we would expect to see if only chance were operating. If they are, we call them *statistically significant*.

> **STATISTICALLY SIGNIFICANT**
> An observed effect so large that it would rarely occur by chance is called **statistically significant.**

You will often see the phrase "statistically significant" in reports of investigations in many fields of study. It tells you that the investigators found good evidence for the effect they were seeking.

How to randomize

The idea of randomization is to assign subjects to treatments by drawing names from a hat. In practice, experimenters use software to carry out randomization. Most statistical software will choose 20 out of a list of 40 at random, for example. The list might contain the names of 40 human subjects. The 20 chosen form one group, and the 20 that remain form the second group. The *Simple Random Sample* applet on the text Web site makes it particularly easy to choose treatment groups at random.

The Basics: Using a Table

You can randomize without software by using a *table of random digits*. Thinking about random digits helps you to understand randomization even if you will use software in practice. Table B at the back of the book is a table of random digits.

> **RANDOM DIGITS**
> A **table of random digits** is a list of the digits 0, 1, 2, 3, 4, 5, 6, 7, 8, 9 that has the following properties:
> 1. The digit in any position in the list has the same chance of being any one of 0, 1, 2, 3, 4, 5, 6, 7, 8, 9.
> 2. The digits in different positions are independent in the sense that the value of one has no influence on the value of any other.

You can think of Table B as the result of asking an assistant (or a computer) to mix the digits 0 to 9 in a hat, draw one, then replace the digit drawn, mix again, draw a second digit, and so on. The assistant's mixing and drawing save us the work of mixing and drawing when we need to randomize. Table B begins with the digits 19223950340575628713. To make the table easier to read, the digits appear in groups of five and in numbered rows. The groups and rows have no meaning—the table is just a long list of digits having both properties described above.

Our goal is to use random digits for the randomization required by our experiment. We need the following facts about random digits, which are consequences of the basic properties noted above:

- Any *pair* of random digits has the same chance of being any of the 100 possible pairs: 00, 01, 02, . . . , 98, 99.
- Any *triple* of random digits has the same chance of being any of the 1000 possible triples: 000, 001, 002, . . . , 998, 999.
- . . . and so on for groups of four or more random digits.

EXAMPLE 2.11

Randomize the students. In the smartphone experiment of Example 2.10, we must divide 40 students at random into two groups of 20 students each.

Step 1: *Label.* Give each student a numerical label, using as few digits as possible. Two digits are needed to label 40 students, so we use labels

01, 02, 03, . . . , 39, 40

It is also correct to use labels 00 to 39 or some other choice of 40 two-digit labels.

Step 2: *Table.* Start anywhere in Table B and read two-digit groups. Suppose we begin at line 130, which is

69051 64817 87174 09517 84534 06489 87201 97245

The first 10 two-digit groups in this line are

69 05 16 48 17 87 17 40 95 17

Each of these two-digit groups is a label. The labels 00 and 41 to 99 are not used in this example, so we ignore them. The first 20 labels between 01 and 40 that we encounter in the table choose students for the first phone. Of the first 10 labels in line 130, we ignore four because they are too high (over 40). The others are 05, 16, 17, 17, 40, and 17. The students labeled 05, 16, 17, and 40 will evaluate the first phone. Ignore the second and third 17s because that student is already in the group. Run your finger across line 130 (and continue to the following lines) until you have chosen 20 students. They are the students labeled

05, 16, 17, 40, 20, 19, 32, 04, 25, 29, 37, 39, 31, 18, 07, 13, 33, 02, 36, 23

You should check at least the first few of these. These students will receive the first phone. The remaining 20 will evaluate the second phone.

As Example 2.11 illustrates, randomization requires two steps: assign labels to the experimental units and then use Table B to select labels at random. Be sure that all labels are the same length so that all have the same chance to be chosen. Use the shortest possible labels—one digit for 10 or fewer individuals, two digits for 11 to 100 individuals, and so on. Don't try to scramble the labels as you assign them. Table B will do the required randomizing, so assign labels in any convenient manner, such as in alphabetical order for human subjects. You can read digits from Table B in any order—along a row, down a column, and so on—because the table has no order. As an easy standard practice, we recommend reading along rows.

Using Technology

It is easy to use statistical software or Excel to randomize. Here are the steps:

Step 1: *Label.* The first step in assigning labels to the experimental units is similar to the procedure we described previously. One difference, however, is that we are not restricted to using numerical labels. Any system where each experimental unit has a unique label identifier will work.

Step 2: *Use the computer.* Once we have the labels, we then create a data file with the labels and generate a random number for each label. In Excel, this can be done with the RAND() function. Finally, we sort the entire data set based on the random numbers. Groups are formed by selecting units in order from the sorted list.

This process is very much like writing the labels on a deck of cards, shuffling the cards, and dealing them out one at a time.

EXAMPLE 2.12

Using software for the randomization. Let's do a randomization similar to the one we did in Example 2.11, but this time using Excel. Here we will use 10 experimental units. We will assign 5 to the treatment group and 5 to the control group. We first create a data set with the numbers 1 to 10 in the first column. See Figure 2.4(a). Then we use RAND() to generate 10 random numbers in the second column. See Figure 2.4(b). Finally, we sort the data set based on the numbers in the second column. See Figure 2.4(c). The first 5 labels (7, 10, 9, 5, and 8) are assigned to the experimental group. The remaining 5 labels (4, 2, 3, 6, and 1) correspond to the control group.

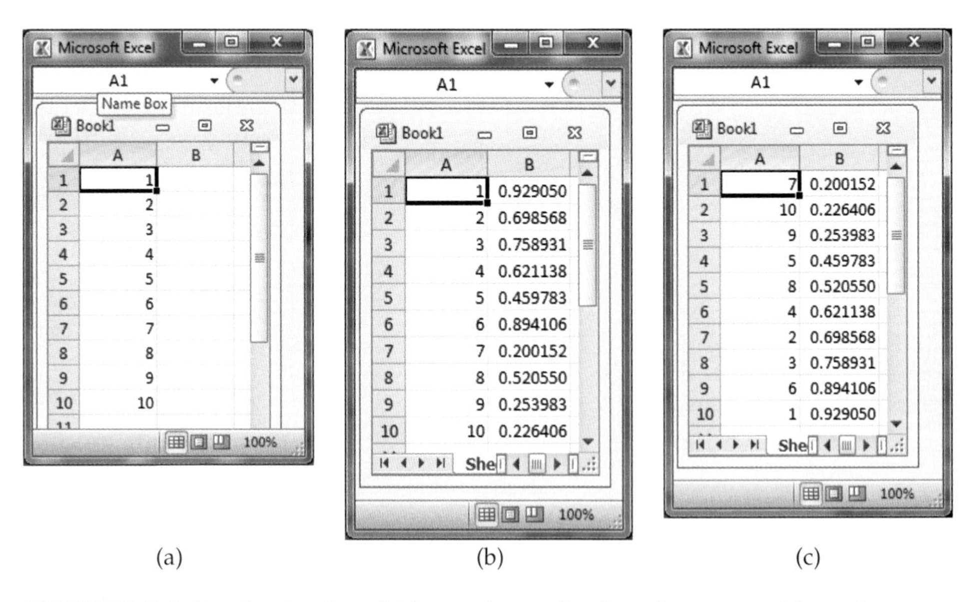

FIGURE 2.4 Randomization of 10 experimental units using a spreadsheet, for Example 2.12. (a) Labels. (b) Random numbers. (c) Sorted list of labels.

completely randomized design

When all experimental units are allocated at random among all treatments, as in Example 2.11, the experimental design is **completely randomized.** Completely randomized designs can compare any number of treatments. The treatments can be formed by levels of a single factor or by more than one factor.

EXAMPLE 2.13

Randomization for the TV commercial experiment. Figure 2.2 (page 38) displays six treatments formed by the two factors in an experiment on response to a TV commercial. Suppose that we have 150 students who are willing to serve as subjects. We must assign 25 students at random to each group. Figure 2.5 outlines the completely randomized design.

To carry out the random assignment, label the 150 students 001 to 150. (Three digits are needed to label 150 subjects.) Enter Table B and read three-digit groups until you have selected 25 students to receive Treatment 1 (a 30-second ad shown once). If you start at line 140, the first few labels for Treatment 1 subjects are 129, 048, and 003.

Continue in Table B to select 25 more students to receive Treatment 2 (a 30-second ad shown 3 times). Then select another 25 for Treatment 3 and so on until you have assigned 125 of the 150 students to Treatments 1 through 5. The 25 students who remain get Treatment 6. The randomization is straightforward but very tedious to do by hand. We recommend software or the *Simple Random Sample* applet. Exercise 2.35 shows how to use the applet to do the randomization for this example.

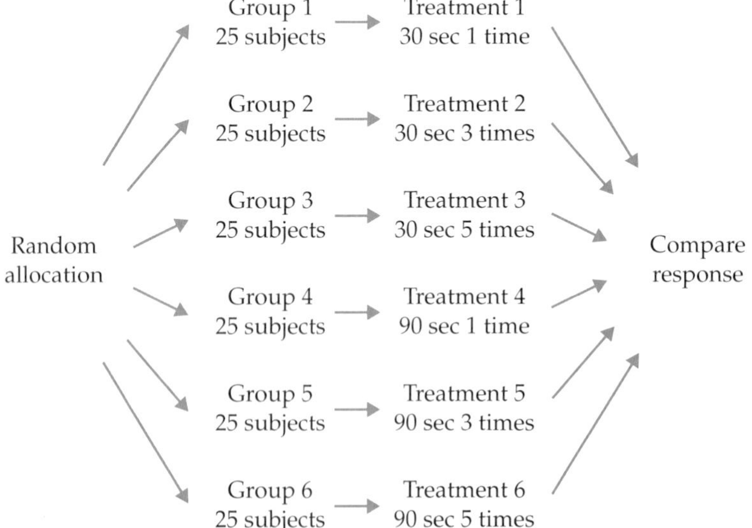

FIGURE 2.5 Outline of a completely randomized design comparing six treatments, for Example 2.13

USE YOUR KNOWLEDGE

2.16 Do the randomization. Use computer software to carry out the randomization for Example 2.13.

Cautions about experimentation

The logic of a randomized comparative experiment depends on our ability to treat all the experimental units identically in every way except for the actual treatments being compared. Good experiments therefore require careful attention to details. The ideal situation is where a study is **double-blind**—neither the subjects themselves nor the experimenters know which treatment any subject has received. The double-blind method avoids unconscious bias by, for example, a doctor who doesn't think that "just a placebo" can benefit a patient.

double-blind

Many—perhaps most—experiments have some weaknesses in detail. *The environment of an experiment can influence the outcomes in unexpected ways.* Although experiments are the gold standard for evidence of cause and effect, really convincing evidence usually requires that a number of studies in different places with different details produce similar results. Here are some brief examples of what can go wrong.

EXAMPLE 2.14

Placebo for a marijuana experiment. A study of the effects of marijuana recruited young men who used marijuana. Some were randomly assigned to smoke marijuana cigarettes, while others were given placebo cigarettes. This failed: the control group recognized that their cigarettes were phony and complained loudly. It may be quite common for blindness to fail because the subjects can tell which treatment they are receiving.[10]

lack of realism

The most serious potential weakness of experiments is **lack of realism.** The subjects or treatments or setting of an experiment may not realistically duplicate the conditions we really want to study. Here is an example.

EXAMPLE 2.15

Layoffs and feeling bad. How do layoffs at a workplace affect the workers who remain on the job? Psychologists asked student subjects to proofread text for extra course credit, then "let go" some of the workers (who were actually accomplices of the experimenters). Some subjects were told that those let go had performed poorly (Treatment 1). Others were told that not all could be kept and that it was just luck that they were kept and others let go (Treatment 2).

We can't be sure that the reactions of the students are the same as those of workers who survive a layoff in which other workers lose their jobs. Many behavioral science experiments use student subjects in a campus setting. Do the conclusions apply to the real world?

Lack of realism can limit our ability to apply the conclusions of an experiment to the settings of greatest interest. Most experimenters want to generalize their conclusions to some setting wider than that of the actual experiment. *Statistical analysis of an experiment cannot tell us how far the results will generalize to other settings.* Nonetheless, the randomized comparative experiment, because of its ability to give convincing evidence for causation, is one of the most important ideas in statistics.

Matched pairs designs

Completely randomized designs are the simplest statistical designs for experiments. They illustrate clearly the principles of control, randomization, and repetition. However, completely randomized designs are often inferior to more elaborate statistical designs. In particular, matching the subjects in various ways can produce more precise results than simple randomization.

matched pairs design

The simplest use of matching is a **matched pairs design,** which compares just two treatments. The subjects are matched in pairs. For example, an experiment to compare two advertisements for the same product might use pairs of subjects with the same age, sex, and income. The idea is that matched subjects are more alike than unmatched subjects, so that comparing responses within a number of pairs is more efficient than comparing the responses of groups of randomly assigned subjects. Randomization remains important: which one of a matched pair sees the first ad is decided at random. One common variation of the matched pairs design imposes both treatments on the same subjects, so that each subject serves as his or her own control. Here is an example.

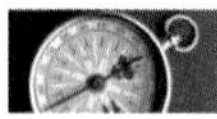

EXAMPLE 2.16

Matched pairs for the smartphone prototype experiment. Example 2.10 describes an experiment to compare two prototypes of a new smartphone. The experiment compared two treatments, Phone 1 and Phone 2. The response variable is the college student participants' satisfaction with the new smartphone. In Example 2.10, 40 student subjects were assigned at random, 20 students to each phone. This is a completely randomized design, outlined in Figure 2.3 (page 41). Subjects differ in how satisfied they are with smartphones in general. The completely randomized design relies on chance to create two similar groups of subjects.

If we wanted to do a matched pairs version of this experiment, we would have each college student use each phone for two weeks. An effective design would randomize the *order* in which the phones are evaluated by each student. This will eliminate bias due to the possibility that the first phone evaluated will be systematically evaluated higher or lower than the second phone evaluated.

The completely randomized design uses chance to decide which subjects will evaluate each smartphone prototype. The matched pairs design uses chance to decide which 20 subjects will evaluate Phone 1 first. The other 20 will evaluate Phone 2 first.

Block designs

The matched pairs design of Example 2.16 uses the principles of comparison of treatments, randomization, and repetition on several experimental units. However, the randomization is not complete (all subjects randomly assigned to treatment groups) but restricted to assigning the order of the treatments for each subject. *Block designs* extend the use of "similar subjects" from pairs to larger groups.

> **BLOCK DESIGN**
>
> A **block** is a group of experimental units or subjects that are known before the experiment to be similar in some way that is expected to affect the response to the treatments. In a **block design,** the random assignment of units to treatments is carried out separately within each block.

Block designs can have blocks of any size. A block design combines the idea of creating equivalent treatment groups by matching with the principle of forming treatment groups at random. Blocks are another form of *control.* They control the effects of some outside variables by bringing those variables into the experiment to form the blocks. Here are some typical examples of block designs.

EXAMPLE 2.17

Blocking in a cancer experiment. The progress of a type of cancer differs in women and men. A clinical experiment to compare three therapies for this cancer therefore treats sex as a blocking variable. Two separate randomizations are done, one assigning the female subjects to the treatments and the other assigning the male subjects. Figure 2.6 outlines the design of this experiment. Note that there is no randomization involved in making up the blocks. They are groups of subjects who differ in some way (sex in this case) that is apparent before the experiment begins.

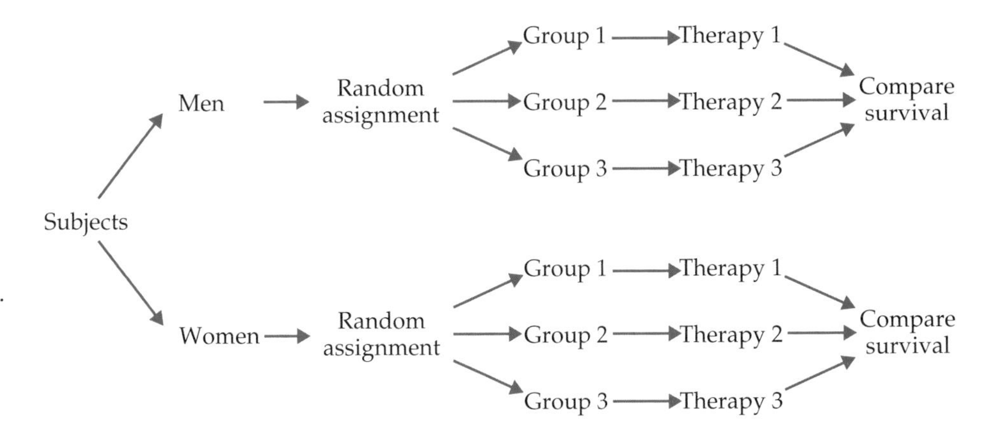

FIGURE 2.6 Outline of a block design, for Example 2.17. The blocks consist of male and female subjects. The treatments are the three therapies for cancer.

EXAMPLE 2.18

Blocking in an agriculture experiment. The soil type and fertility of farmland differ by location. Because of this variation, a test of the effect of how fields are tilled (two tillage types) and pesticide application (three application schedules) on soybean yields uses small fields as blocks. Each block is divided into six plots, and the six treatments are randomly assigned to plots separately within each block.

EXAMPLE 2.19

Blocking in an education experiment. The Tennessee STAR class size experiment (Example 2.7) used a block design. It was important to compare different class types in the same school because the children in a school come from the same neighborhood, follow the same curriculum, and have the same school environment outside class. In all, 79 schools across Tennessee participated in the program. That is, there were 79 blocks. New kindergarten students were randomly placed in the three types of class separately within each school.

Blocks allow us to draw separate conclusions about each block, for example, about men and women in the cancer study in Example 2.17. Blocking also allows more precise overall conclusions because the systematic differences between men and women can be removed when we study the overall effects of the three therapies. The idea of blocking is an important additional principle of statistical design of experiments. A wise experimenter will form blocks based on the most important unavoidable sources of variability among the experimental units. Randomization will then average out the effects of the remaining variation and allow an unbiased comparison of the treatments.

SECTION 2.1 SUMMARY

In an experiment, one or more **treatments** are imposed on the **experimental units** or **subjects.** Each treatment is a combination of **levels** of the explanatory variables, which we call **factors.**

A **placebo effect** is a positive response to a treatment that is not designed to have an effect.

The **design** of an experiment refers to the choice of treatments and the manner in which the experimental units or subjects are assigned to the treatments.

An observed effect is **statistically significant** if it is so large that it would rarely occur by chance.

The basic principles of statistical design of experiments are **control, randomization,** and **repetition.**

The simplest form of control is **comparison.** Experiments should compare two or more treatments to prevent **confounding** the effect of a treatment.

Randomization uses chance to assign subjects to the treatments. Randomization creates treatment groups that are similar (except for chance variation) before the treatments are applied. Randomization and comparison together prevent **bias,** or systematic favoritism, in experiments.

You can carry out randomization by giving numerical labels to the experimental units and using software or a **table of random digits** to choose treatment groups.

Repetition of the treatments on many units reduces the role of chance variation and makes the experiment more sensitive to differences among the treatments.

Good experiments require attention to detail as well as good statistical design. Many behavioral and medical experiments are **double-blind. Lack of realism** in an experiment can prevent us from generalizing its results.

In addition to comparison, a second form of control is to restrict randomization by forming **blocks** of experimental units that are similar in some way that is important to the response. Randomization is then carried out separately within each block.

Matched pairs are a common form of blocking for comparing just two treatments. In some matched pairs designs, each subject receives both treatments in a random order. In others, the subjects are matched in pairs as closely as possible, and one subject in each pair receives each treatment.

SECTION 2.1 EXERCISES

For Exercises 2.1 to 2.4, see page 32; for Exercise 2.5, see page 33; for Exercise 2.6, see page 34; for Exercises 2.7 and 2.8, see page 35; for Exercises 2.9 and 2.10, see pages 36–37; for Exercises 2.11 and 2.12, see page 39; for Exercise 2.13, see page 40; for Exercises 2.14 and 2.15, see page 42; and for Exercise 2.16, see page 46.

2.17. What is needed? Explain any flaws in each of the following proposed experiments and describe how you would improve the experiment.

(a) Two product promotion offers are to be compared. The first, which offers two items for $2, will be used in a store on Friday. The second, which offers three items for $3, will be used in the same store on Saturday.

(b) A study compares two marketing campaigns to encourage individuals to eat more fruits and vegetables. The first campaign is launched in Florida at the same time that the second campaign is launched in Minnesota.

(c) You want to evaluate the effectiveness of a new investment strategy. You try the strategy for one year and evaluate the performance of the strategy.

2.18. What's wrong? Explain what is wrong with each of the following randomization procedures and describe how you would do the randomization correctly.

(a) Twenty students are to be used to evaluate a new treatment. Ten men are assigned to receive the treatment, and 10 women are assigned to be the controls.

(b) Ten subjects are to be assigned to two treatments, 5 to each. For each subject, a coin is tossed. If the coin comes up heads, the subject is assigned to the first treatment; if the coin comes up tails, the subject is assigned to the second treatment.

(c) An experiment will assign 40 rats to four different treatment conditions. The rats arrive from the supplier in batches of 10 and the treatment lasts two weeks. The first batch of 10 rats is randomly assigned to one of the four treatments, and data for these rats are collected. After a one-week break, another batch of 10 rats arrives and is assigned to one of the three remaining treatments. The process continues until the last batch of rats is given the treatment that has not been assigned to the three previous batches.

2.19. Evaluate an online version of a course. An online version of a traditional course is to be evaluated by randomly assigning students to either the online version or the traditional course. The change in a standardized test score is the response variable. Explain why this experiment cannot be done in a double-blind fashion.

2.20. Can you change attitudes toward binge drinking? An experiment designed to change attitudes about binge drinking is to be performed using college students as subjects. Discuss some variables that you might use in a block design for this experiment.

2.21. Evaluate a new-employee orientation program. Your company runs a two-day orientation program Monday and Tuesday each week for new employees. A new program is to be compared with the current one. Set up an experiment to compare the new program with the old. Be sure to provide details regarding randomization and what outcome variables you will measure.

2.22. Medical magnets. Some claim that magnets can be used to reduce pain. Write a proposal requesting funding to test this claim. Give all the important details, including the number of subjects, issues concerning randomization, and how you will make the study double-blind.

2.23. Calcium and vitamin D. Vitamin D is needed for the body to use calcium. An experiment is designed to assess the effects of calcium and vitamin D supplements on the bones of first-year college students. The outcome measure is the total body bone mineral content (TBBMC), a measure of bone health. Three doses of calcium will be used: 0, 200, and 400 milligrams per day. The doses of vitamin D will be 0, 50, and 100 international units (IU) per day. The calcium and vitamin D will be given in a single tablet. All tablets, including those with no calcium and no vitamin D, will look identical. Subjects for the study will be 90 men and 90 women.

(a) What are the factors and the treatments for this experiment?

(b) Draw a picture explaining how you would randomize the 180 college students to the treatments.

(c) Use a spreadsheet to carry out the randomization.

2.24. CHALLENGE **The interaction of calcium and vitamin D.** Refer to the previous exercise. It is expected that the effect of vitamin D on TBBMC will be greatest when there is an adequate intake of calcium. In other words, the effect of vitamin D is expected to be greater with the higher doses of calcium. It is also expected that the men in the study will generally have adequate calcium and vitamin D in their diets, so the effect of the supplements will be very small at best. Draw pictures that illustrate these expected interactions.

2.25. CHALLENGE **Compare two versions of a product.** A coffeehouse wants to compare two new varieties of coffee.

(a) Describe an experiment where different customers evaluate each variety. Be sure to provide details, including how many customers you will use, issues related to randomization, and what evaluation data you will collect.

(b) Do the same for an experiment where each customer evaluates both varieties of coffee.

(c) Which experiment do you prefer? Give reasons for your answer.

2.26. The *Sports Illustrated* jinx. Some people believe that teams or individual athletes who appear on the cover of *Sports Illustrated* magazine will experience bad luck soon after they appear. Can you evaluate this belief with an experiment? Explain your answer.

For each of the experimental situations described in Exercises 2.27 to 2.29, identify the experimental units or subjects, the factors, the treatments, and the response variables.

2.27. How well do pine trees grow in shade? Ability to grow in shade may help pines in the dry forests of Arizona resist drought. How well do these pines grow in shade? Investigators planted pine seedlings in a greenhouse in either full light or light reduced to 5% of normal by shade cloth. At the end of the study, they dried the young trees and weighed them.

2.28. Will the students exercise more and eat better? Most American adolescents don't eat well and don't exercise enough. Can middle schools increase physical activity among their students? Can they persuade students to eat better? Investigators designed a "physical activity intervention" to increase activity in physical education classes and during leisure periods throughout the school day. They also designed a "nutrition intervention" that improved school lunches and offered ideas for healthy home-packed lunches. Each participating school was randomly assigned to one of the interventions, both interventions, or no intervention. The investigators observed physical activity and lunchtime consumption of fat.

2.29. Refusals in telephone surveys. How can we reduce the rate of refusals in telephone surveys? Most people who answer at all listen to the interviewer's introductory remarks and then decide whether to continue. One study made telephone calls to randomly selected households to ask opinions about the next election. In some calls, the interviewer gave her name, in others she identified the university she was representing, and in still others she identified both herself and the university. For each type of call, the interviewer either did or did not offer to send a copy of the final survey results to the person interviewed. Do these differences in the introduction affect whether the interview is completed?

2.30. Does aspirin prevent strokes and heart attacks? The Bayer Aspirin Web site claims, "Nearly five decades of research now link aspirin to the prevention of strokes and heart attacks." The most important evidence for this claim comes from the Physicians' Health Study, a large medical experiment involving 22,000 male physicians. One group of about 11,000 physicians took an aspirin every second day, while the rest took a placebo. After several years the study found that subjects in the aspirin group had significantly fewer heart attacks than subjects in the placebo group.

(a) Identify the experimental subjects, the factor and its levels, and the response variable in the Physicians' Health Study.

(b) Use a diagram to outline a completely randomized design for the Physicians' Health Study.

(c) What does it mean to say that the aspirin group had "significantly fewer heart attacks"?

2.31. Chronic tension headaches. Doctors identify "chronic tension-type headaches" as headaches that occur almost daily for at least six months. Can antidepressant medications or stress management training reduce the number and severity of these headaches? Are both together more effective than either alone? Investigators compared four treatments: antidepressant alone, placebo alone, antidepressant plus stress management, and placebo plus stress management. Outline the design of the experiment. The headache sufferers named in the table have agreed to participate in the study. Use software or Table B at line 151 to randomly assign the subjects to the treatments.

Anderson	Daye	Li	Park	Vassilev
Archberger	Engelbrecht	Lipka	Paul	Wang
Bezawada	Guha	Lu	Rau	Watkins
Cetin	Hatfield	Martin	Saygin	Xu
Cheng	Hua	Mehta	Shu	
Chronopoulou	Kim	Mi	Tang	
Codrington	Kumar	Nolan	Towers	
Daggy	Leaf	Olbricht	Tyner	

2.32. Guilt among workers who survive a layoff. Workers who survive a layoff of other employees at their location may suffer from "survivor guilt." A study of survivor guilt and its effects used as subjects 90 students who were offered an opportunity to earn extra course credit by doing proofreading. Each subject worked in the same cubicle as another student, who was an accomplice of the experimenters. At a break midway through the work, one of three things happened:

Treatment 1: The accomplice was told to leave; it was explained that this was because she performed poorly.

Treatment 2: It was explained that unforeseen circumstances meant there was only enough work for one person. By "chance," the accomplice was chosen to be laid off.

Treatment 3: Both students continued to work after the break.

The subjects' work performance after the break was compared with performance before the break.[11]

(a) Outline the design of this completely randomized experiment.

(b) If you are using software, randomly assign the 90 students to the treatments. If not, use Table B at line 153 to choose the first 4 subjects for Treatment 1.

2.33. Diagram the exercise and eating experiment. Twenty-four public middle schools agree to participate in the experiment described in Exercise 2.28. Use a diagram to outline a completely randomized design for this experiment. Then do the randomization required to assign schools to treatments. If you use Table B, start at line 160.

2.34. Price cuts on athletic shoes. Stores advertise price reductions to attract customers. What type of price cut is most attractive? Market researchers prepared ads for athletic shoes announcing different levels of discounts (20%, 40%, 60%, or 80%). The student subjects who read the ads were also given "inside information" about the fraction of shoes on sale (25%, 50%, 75%, or 100%). Each subject then rated the attractiveness of the sale on a scale of 1 to 7.[12]

(a) There are two factors. Make a sketch like Figure 2.2 (page 38) that displays the treatments formed by all combinations of levels of the factors.

(b) Outline a completely randomized design using 96 student subjects. Use software or Table B at line 111 to choose the subjects for the first treatment.

2.35. APPLET **Use the *Simple Random Sample* applet.** The *Simple Random Sample* applet allows you to randomly assign experimental units to more than two groups without difficulty. Example 2.13 (page 46) describes a randomized comparative experiment in which 150 students are randomly assigned to six groups of 25.

(a) Use the applet to randomly choose 25 out of 150 students to form the first group. Which students are in this group?

(b) The "population hopper" now contains the 125 students who were not chosen, in scrambled order. Click "Sample" again to choose 25 of these remaining students to make up the second group. Which students were chosen?

(c) Click "Sample" three more times to choose the third, fourth, and fifth groups. Don't take the time to write down these groups. Check that there are only 25 students remaining in the "population hopper." These subjects get Treatment 6. Which students are they?

2.36. APPLET **Use the *Simple Random Sample* applet.** You can use the *Simple Random Sample* applet to choose a treatment group at random once you have labeled the subjects. Example 2.11 (page 44) uses Table B to choose 20 students from a group of 40 in a study of smartphone preferences. Use the applet to choose the 20 students who will evaluate Phone 1. Which students did you choose? The remaining 20 students will evaluate Phone 2.

2.37. CHALLENGE **Health benefits of bee pollen.** "Bee pollen is effective for combating fatigue, depression, cancer, and colon disorders." So says a Web site that offers the pollen for sale. We wonder if bee pollen really does prevent colon disorders. Here are two ways to study this question. Explain why the first design will produce more trustworthy data.

(a) Find 400 women who do not have colon disorders. Randomly assign 200 to take bee pollen capsules and the other 200 to take placebo capsules that are identical in appearance. Follow both groups for 5 years.

(b) Find 200 women who take bee pollen regularly. Match each with a woman of the same age, race, and occupation who does not take bee pollen. Follow both groups for 5 years.

2.38. CHALLENGE **Measuring water quality in streams and lakes.** Water quality of streams and lakes is an issue of concern to the public. Although trained professionals typically are used to take reliable measurements, many volunteer groups are gathering and distributing information based on data that they collect.[13] You are part of a team to train volunteers to collect accurate water quality data. Design an experiment to evaluate the effectiveness of the training. Write a summary of your proposed design to present to your team. Be sure to include all the details that they will need to evaluate your proposal.

2.39. CHALLENGE **Calcium and the bones of young girls.** Calcium is important to the bone development of young girls. To study how the bodies of young girls process calcium, investigators used the setting of a summer camp. Calcium was given in punch at either a high or a low level. The camp diet was otherwise the same for all girls. Suppose that there are 50 campers.

(a) Outline a completely randomized design for this experiment.

(b) Describe a matched pairs design in which each girl receives both levels of calcium (with a "washout period," during which they eat their normal diet, between the times they receive each level of calcium). What is the advantage of the matched pairs design over the completely randomized design?

(c) The same randomization can be used in different ways for both designs. Label the subjects 01 to 50. You must choose 25 of the 50. Use Table B at line 110 to choose just the first 5 of the 25. How are the 25 subjects chosen treated in the completely randomized design? How are they treated in the matched pairs design?

2.40.  **Random digits.** Table B is a table of random digits. Which of the following statements are true of a table of random digits, and which are false? Explain your answers.

(a) There are exactly four 0s in each row of 40 digits.

(b) Each pair of digits has chance 1/100 of being 00.

(c) The digits 0000 can never appear as a group, because this pattern is not random.

2.2 Sampling Design

A political scientist wants to know what percent of college-age adults consider themselves conservatives. An automaker hires a market research firm to learn what percent of adults aged 18 to 35 recall seeing television advertisements for a hybrid vehicle. Government economists inquire about average household income.

In all these cases, we want to gather information about a large group of individuals. We will not, as in an experiment, impose a treatment to observe the response. Also, time, cost, and inconvenience forbid contacting every individual. In such cases, we gather information about only part of the group—a *sample*—to draw conclusions about the whole. **Sample surveys** are an important kind of observational study.

sample survey

POPULATION AND SAMPLE

The entire group of individuals that we want information about is called the **population.**

A **sample** is a part of the population that we actually examine to gather information.

Notice that "population" is defined in terms of our desire for knowledge. If we wish to draw conclusions about all U.S. college students, that group is our population even if only local students are available for questioning. The sample is the part from which we draw conclusions about the whole. The **design** of a sample survey refers to the method used to choose the sample from the population.

sample survey design

EXAMPLE 2.20

The Reading Recovery program. The Reading Recovery (RR) program has specially trained teachers work one-on-one with at-risk first-grade students to help them learn to read. A study was designed to examine the relationship between the RR teachers' beliefs about their ability to motivate students and the progress of the students whom they teach.[14] The Reading Recovery

International Data Evaluation Center Web site (www.idecweb.us) says that there are over 15,000 RR teachers. The researchers send a questionnaire to a random sample of 200 of these. The population consists of all RR teachers, and the sample is the 200 who were randomly selected.

Unfortunately, our idealized framework of population and sample does not exactly correspond to the situations that we face in many cases. In Example 2.20, the list of teachers was prepared at a particular time in the past. It is very likely that some of the teachers on the list are no longer working as RR teachers today. New teachers have been trained in RR methods and are not on the list. In spite of these difficulties, we still view the list as the population.

In reporting the results of a sample survey it is important to include all details regarding the procedures used. The proportion of the original sample who actually provide usable data is called the **response rate** and should be reported for all surveys. We may have out-of-date addresses for some who are still working as RR teachers. Some teachers may choose not to respond to our survey questions. If only 150 of the teachers who were sent questionnaires provided usable data, the response rate would be 150/200, or 75%. Follow-up mailings or phone calls to those who do not initially respond can help increase the response rate.

response rate

USE YOUR KNOWLEDGE

2.41 Job satisfaction in Mongolian universities. An educational research team wanted to examine the relationship between faculty participation in decision making and job satisfaction in Mongolian public universities. They are planning to randomly select 300 faculty members from a list of 2500 faculty members in these universities. The Job Descriptive Index will be used to measure job satisfaction, and the Conway Adaptation of the Alutto-Belasco Decisional Participation Scale will be used to measure decision participation. Describe the population and the sample for this study. Can you determine the response rate?

2.42 Taxes and forestland usage. A study was designed to assess the impact of taxes on forestland usage in part of the Upper Wabash River Watershed in Indiana.[15] A survey was sent to 772 forest owners from this region and 348 were returned. Consider the population, the sample, and the response rate for this study. Describe these based on the information given and indicate any additional information that you would need to give a complete answer.

Poor sample designs can produce misleading conclusions. Here is an example.

EXAMPLE 2.21

Sampling pieces of steel. A mill produces large coils of thin steel for use in manufacturing home appliances. The quality engineer wants to submit a sample of 5-centimeter squares to detailed laboratory examination. She asks a technician to cut a sample of 10 such squares. Wanting to provide "good" pieces of steel, the technician carefully avoids the visible defects in the coil material when cutting the sample. The laboratory results are wonderful, but the customers complain about the material they are receiving.

Online opinion polls are particularly vulnerable to bias because the sample who respond are not representative of the population at large. People who take the trouble to respond to an open invitation are not representative of the entire adult population.

In Example 2.21, the sample was selected in a manner that guaranteed that it would not be representative of the entire population. This sampling scheme displays *bias,* or systematic error, in favoring some parts of the population over others. Online polls use *voluntary response samples,* a particularly common form of biased sample.

VOLUNTARY RESPONSE SAMPLE

A **voluntary response sample** consists of people who choose themselves by responding to a general appeal. Voluntary response samples are biased because people with strong opinions, especially negative opinions, are most likely to respond.

The remedy for bias in choosing a sample is to allow impersonal chance to do the choosing, so that there is neither favoritism by the sampler nor voluntary response. Random selection of a sample eliminates bias by giving all individuals an equal chance to be chosen, just as randomization eliminates bias in assigning experimental subjects.

Simple random samples

The simplest sampling design amounts to placing names in a hat (the population) and drawing out a handful (the sample). This is *simple random sampling.*

SIMPLE RANDOM SAMPLE

A **simple random sample (SRS)** of size n consists of n individuals from the population chosen in such a way that every set of n individuals has an equal chance to be the sample actually selected.

Each treatment group in a completely randomized experimental design is an SRS drawn from the available experimental units. We select an SRS by labeling all the individuals in the population and using software or a table of random digits to select a sample of the desired size, just as in experimental randomization. Notice that an SRS gives each individual an equal chance to be chosen (thus avoiding bias in the choice). It also gives every possible sample an equal chance to be chosen. There are other random sampling designs that give each individual, but not each sample, an equal chance. One such design, systematic random sampling, is described in Exercise 2.56.

EXAMPLE 2.22

Spring break destinations. A campus newspaper plans a major article on spring break destinations. The authors intend to call a few randomly chosen resorts at each destination to ask about their attitudes toward groups of students as guests. Here are the resorts listed in one city. The first step is to label the members of this population as shown.

01	Aloha Kai	08	Captiva	15	Palm Tree	22	Sea Shell
02	Anchor Down	09	Casa del Mar	16	Radisson	23	Silver Beach
03	Banana Bay	10	Coconuts	17	Ramada	24	Sunset Beach
04	Banyan Tree	11	Diplomat	18	Sandpiper	25	Tradewinds
05	Beach Castle	12	Holiday Inn	19	Sea Castle	26	Tropical Breeze
06	Best Western	13	Lime Tree	20	Sea Club	27	Tropical Shores
07	Cabana	14	Outrigger	21	Sea Grape	28	Veranda

Now enter Table B, and read two-digit groups until you have chosen three resorts. If you enter at line 185, Banana Bay (03), Palm Tree (15), and Cabana (07) will be called.

Most statistical software will select an SRS for you, eliminating the need for Table B. The *Simple Random Sample* applet on the text Web site is a convenient way to automate this task.

Excel can do the job using the same ideas that we used when we randomized experimental units to treatments in designed experiments. There are four steps:

1. Create a data set with all the elements of the population in the first column.
2. Assign a random number to each element of the population; put these in the second column.
3. Sort the data set by the random number column.
4. The simple random sample is obtained by taking elements in the sorted list until the desired sample size is reached.

We illustrate the procedure with a simplified version of Example 2.22.

EXAMPLE 2.23

Select a random sample. Suppose that the population from Example 2.22 is only the first two rows of the display given there:

Aloha Kai	Captiva	Palm Tree	Sea Shell
Anchor Down	Casa del Mar	Radisson	Silver Beach

Note that we do not need the numerical labels to identify the individuals in the population. Suppose that we want to select a simple random sample of

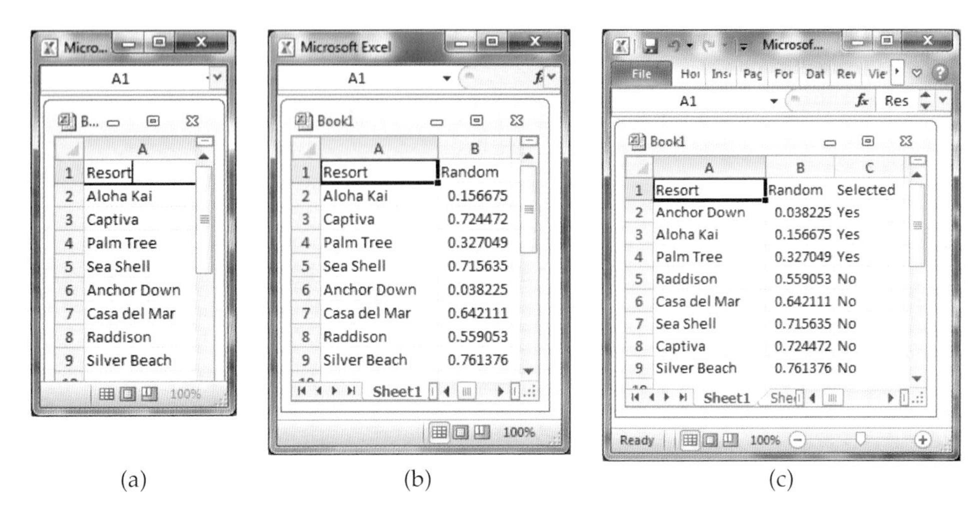

FIGURE 2.7 Selection of a simple random sample of resorts, for Example 2.23.

three resorts from this population. Figure 2.7(a) gives the spreadsheet with the population names. The random numbers generated by the RAND() function are given in the second column in Figure 2.7(b).[16] The sorted data set is given in Figure 2.7(c). We have added a third column to the spreadsheet to indicate which resorts were selected for our random sample. They are Anchor Down, Aloha Kai, and Palm Tree.

USE YOUR KNOWLEDGE

2.43 Ringtones for cell phones. You decide to change the ringtones for your cell phone by choosing 2 from a list of the 10 most popular ringtones.[17] Here is the list:

Sexy and I Know It	Red Solo Cup	Stronger	God Gave Me You
Mr. Wrong	Young, Wild, and Free	Dirt Road Anthem	Set Fire to the Rain
It Will Rain	Make Me Proud		

Select your two ringtones using a simple random sample.

2.44 Listen to three songs. The walk to your statistics class takes about 10 minutes, about the amount of time needed to listen to three songs on your iPod. You decide to take a simple random sample of top songs from a Billboard list of rock songs.[18] Here is the list:

Lonely Boy	These Days	Somebody That I Used to Know	Face to the Floor
Bully	The Sound of Winter	Walk	Paradise
Satellite	Shake It Out		

Select the three songs for your iPod using a simple random sample.

Stratified samples

The general framework for designs that use chance to choose a sample is a *probability sample*.

PROBABILITY SAMPLE

A **probability sample** is a sample chosen by chance. We must know what samples are possible and what chance, or probability, each possible sample has.

Some probability sampling designs (such as an SRS) give each member of the population an *equal* chance to be selected. This may not be true in more elaborate sampling designs. In every case, however, the use of chance to select the sample is the essential principle of statistical sampling.

Designs for sampling from large populations spread out over a wide area are usually more complex than an SRS. For example, it is common to sample important groups within the population separately, then combine these samples. This is the idea of a *stratified sample*.

STRATIFIED RANDOM SAMPLE

To select a **stratified random sample,** first divide the population into groups of similar individuals, called **strata.** Then choose a separate SRS in each stratum and combine these SRSs to form the full sample.

Choose the strata based on facts known before the sample is taken. For example, a population of election districts might be divided into urban, suburban, and rural strata.

A stratified design can produce more exact information than an SRS of the same size by taking advantage of the fact that individuals in the same stratum are similar to one another. Think of the extreme case in which all individuals in each stratum are identical: just one individual from each stratum is then enough to completely describe the population.

Strata for sampling are similar to blocks in experiments. We have two names because the idea of grouping similar units before randomizing arose separately in sampling and in experiments.

EXAMPLE 2.24

A stratified sample of dental claims. A dentist is suspected of defrauding insurance companies by describing some dental procedures incorrectly on claim forms and overcharging for them. An investigation begins by examining a sample of his bills for the past three years. Because there are five suspicious types of procedures, the investigators take a stratified sample. That is, they randomly select bills for each of the five types of procedures separately.

Multistage samples

Another common means of restricting random selection is to choose the sample in stages. This is common practice for national samples of households or people. For example, data on employment and unemployment are gathered by the government's Current Population Survey, which conducts interviews in about 60,000 households each month. The cost of sending interviewers to the widely scattered households in an SRS would be too high. Moreover, the government wants data broken down by states and large cities.

multistage sample

The Current Population Survey therefore uses a **multistage sampling design.** The final sample consists of clusters of nearby households that an interviewer can easily visit. Most opinion polls and other national samples are also multistage, though interviewing in most national samples today is done by telephone rather than in person, eliminating the economic need for clustering. The Current Population Survey sampling design is roughly as follows:[19]

Stage 1. Divide the United States into 2007 geographical areas called Primary Sampling Units, or PSUs. PSUs do not cross state lines. Select a sample of 754 PSUs. This sample includes the 428 PSUs with the largest population and a stratified sample of 326 of the others.

Stage 2. Divide each PSU selected into smaller areas called "blocks." Stratify the blocks using ethnic and other information and take a stratified sample of the blocks in each PSU.

Stage 3. Sort the housing units in each block into clusters of four nearby units. Interview the households in a probability sample of these clusters.

Analysis of data from sampling designs more complex than an SRS takes us beyond basic statistics. But the SRS is the building block of more elaborate designs, and analysis of other designs differs more in complexity of detail than in fundamental concepts.

Cautions about sample surveys

Random selection eliminates bias in the choice of a sample from a list of the population. Sample surveys of large human populations, however, require much more than a good sampling design.[20] To begin, we need an accurate and complete list of the population. Because such a list is rarely available, most samples suffer from some degree of *undercoverage.* A sample survey of households, for example, will miss not only homeless people but also prison inmates and students in dormitories. An opinion poll conducted by telephone will miss the large number of American households without residential phones. The results of national sample surveys therefore have some bias if the people not covered—who most often are poor people—differ from the rest of the population.

A more serious source of bias in most sample surveys is *nonresponse,* which occurs when a selected individual cannot be contacted or refuses to cooperate. Nonresponse to sample surveys often reaches 50% or more, even with careful planning and several callbacks. Because nonresponse is higher in urban areas, most sample surveys substitute other people in the same area to avoid favoring rural areas in the final sample. If the people contacted differ from those who are rarely at home or who refuse to answer questions, some bias remains.

> **UNDERCOVERAGE AND NONRESPONSE**
>
> **Undercoverage** occurs when some groups in the population are left out of the process of choosing the sample.
>
> **Nonresponse** occurs when an individual chosen for the sample can't be contacted or does not cooperate.

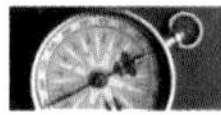

EXAMPLE 2.25

Nonresponse in the Current Population Survey. How bad is nonresponse? The Current Population Survey (CPS) has the lowest nonresponse rate of any poll we know: only about 4% of the households in the CPS sample refuse to take part, and another 3% or 4% can't be contacted. People are more likely to respond to a government survey such as the CPS, and the CPS contacts its sample in person before doing later interviews by phone.

The General Social Survey (Figure 2.8) is the nation's most important social science research survey. The GSS also contacts its sample in person, and it is run by a university. Despite these advantages, its most recent survey had a 30% rate of nonresponse.

FIGURE 2.8 Part of the home page for the General Social Survey (GSS). The GSS has assessed attitudes on a wide variety of topics since 1972. Its continuity over time makes the GSS a valuable source for studies of changing attitudes.

What about polls done by the media and by market research and opinion-polling firms? We usually don't know their rates of nonresponse, because they won't say. That itself is a bad sign. The Pew Research Center for the People and the Press designed a careful telephone survey and published the results: out of 2879 households called, 1658 were never at home, refused, or would not finish the interview. That's a nonresponse rate of 58%.[21]

Most sample surveys, and almost all opinion polls, are now carried out by telephone. This and other details of the interview method can affect the results. When presented with several options for a reply, such as "completely agree," "mostly agree," "mostly disagree," and "completely disagree," people tend to be a little more likely to respond to the first one or two options presented.

response bias

The behavior of the respondent or of the interviewer can cause **response bias** in sample results. Respondents may lie, especially if asked about illegal or unpopular behavior. The race or sex of the interviewer can influence responses to questions about race relations or attitudes toward feminism. Answers to questions that ask respondents to recall past events are often inaccurate because of faulty memory. For example, many people "telescope" events in the past, bringing them forward in memory to more recent time periods. "Have you visited a dentist in the last 6 months?" will often elicit a "Yes" from someone who last visited a dentist 8 months ago.[22]

EXAMPLE 2.26

Overreporting of voter behavior. "One of the most frequently observed survey measurement errors is the overreporting of voting behavior."[23] People know they should vote, so those who didn't vote tend to save face by saying that they did. Here are the data from a typical sample of 663 people after an election:

		What they said:	
		I voted	I didn't
What they did:	Voted	358	13
	Didn't vote	120	172

You can see that 478 people (72%) said that they voted, but only 371 people (56%) actually did vote.

wording of questions

The **wording of questions** is the most important influence on the answers given to a sample survey. Confusing or leading questions can introduce strong bias, and even minor changes in wording can change a survey's outcome. Here are some examples.

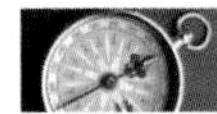

EXAMPLE 2.27

The form of the question is important. In response to the question "Are you heterosexual, homosexual, or bisexual?" in a social science research survey, one woman answered, "It's just me and my husband, so bisexual." The issue is serious, even if the example seems silly: reporting about sexual behavior is difficult because people understand and misunderstand sexual terms in many ways.

How do Americans feel about government help for the poor? Only 13% think we are spending too much on "assistance to the poor," but 44% think we are spending too much on "welfare." How do the Scots feel about the movement to become independent from England? Well, 51% would vote for "independence for Scotland," but only 34% support "an independent Scotland separate from the United Kingdom." It seems that "assistance to the poor" and "independence" are nice, hopeful words. "Welfare" and "separate" are negative words.[24]

The statistical design of sample surveys is a science, but this science is only part of the art of sampling. Because of nonresponse, response bias, and the difficulty of posing clear and neutral questions, you should hesitate to fully trust reports about complicated issues based on surveys of large human populations. *Insist on knowing the exact questions asked, the rate of nonresponse, and the date and method of the survey before you trust a poll result.*

SECTION 2.2 SUMMARY

A sample survey selects a **sample** from the **population** of all individuals about which we desire information. We base conclusions about the population on data about the sample.

The **design of a sample survey** refers to the method used to select the sample from the population. **Probability sampling designs** use impersonal chance to select a sample.

The basic probability sample is a **simple random sample (SRS).** An SRS gives every possible sample of a given size the same chance to be chosen.

Choose an SRS by labeling the members of the population and using a **table of random digits** to select the sample. Software can automate this process.

To choose a **stratified random sample,** divide the population into **strata,** groups of individuals that are similar in some way that is important to the response. Then choose a separate SRS from each stratum and combine them to form the full sample.

Multistage samples select successively smaller groups within the population in stages, resulting in a sample consisting of clusters of individuals. Each stage may employ an SRS, a stratified sample, or another type of sample.

Failure to use probability sampling often results in **bias,** or systematic errors in the way the sample represents the population. **Voluntary response** samples, in which the respondents choose themselves, are particularly prone to large bias.

In human populations, even probability samples can suffer from bias due to **undercoverage** or **nonresponse,** from **response bias** due to the behavior of the interviewer or the respondent, or from misleading results due to **poorly worded questions.**

SECTION 2.2 EXERCISES

For Exercises 2.41 and 2.42, see page 55; and for Exercises 2.43 and 2.44, see page 58.

2.45. What's wrong? Explain what is wrong with each of the following random selection procedures and decribe how you would do the randomization correctly.

(a) To determine the reading level of an introductory statistics text, you evaluate all the written material in the third chapter.

(b) You want to sample student opinions about a proposed change in procedures for changing majors. You hand out questionnaires to 100 students as they arrive for class at 7:30 A.M.

(c) A population of subjects is put in alphabetical order, and a simple random sample of size 10 is taken by selecting the first 10 subjects in the list.

2.46. What kind of sample? In each of the following situations, describe the sample as an SRS, a stratified random sample, a multistage sample, or a voluntary response sample. Explain your answers.

(a) There are 7 sections of an introductory statistics course. A random sample of 3 sections is chosen, and then random samples of 8 students from each of these sections are chosen.

(b) A student organization has 55 members. A table of random numbers is used to select a sample of 5.

(c) An online poll asks people who visit this site to choose their favorite television show.

(d) Separate random samples of male and female first-year college students in an introductory psychology course are selected to receive a one-week alternative instructional method.

2.47. Importance of students as customers. A committee on community relations in a college town plans to survey local businesses about the importance of students as customers. From telephone book listings, the committee chooses 150 businesses at random. Of these, 73 return the questionnaire mailed by the committee. What is the population for this sample survey? What is the sample? What is the rate (percent) of nonresponse?

2.48. CHALLENGE **Consumer spending.** A Gallup Poll used telephone interviews to collect data on consumer spending on different days of the week.[25] Here are the averages (in dollars) for each day of the week:

Monday	59	Friday	63
Tuesday	56	Saturday	73
Wednesday	55	Sunday	76
Thursday	59		

The data were collected between January 2 and October 21, 2009. Discuss how this choice may have affected the results.

2.49. CHALLENGE **Where do you get your news?** A Pew Research Center survey asked people about their main source for national and international news.[26] For one analysis Pew focused on comparing Fox with CNN. Here are the self-reported political affiliations (in percents) of the survey respondents and their main source of news:

Party	Fox	CNN	Other
Republican	34	12	54
Democrat	9	19	72
Independent	17	14	69

Write a report summarizing the results.

2.50. Identify the populations. For each of the following sampling situations, identify the population as exactly as possible. That is, say what kind of individuals the population consists of and say exactly which individuals fall into the population. If the information given is not complete, complete the description of the population in a reasonable way.

(a) A college has changed its core curriculum and wants to obtain detailed feedback from students during each of the first 12 weeks of the coming semester. Each week, a random sample of 5 students will be selected to be interviewed.

(b) The American Community Survey (ACS) replaced the census "long form" starting with the 2010 census. The main part of the ACS contacts 250,000 addresses by mail each month, with follow-up by phone and in person if there is no response. Each household answers questions about their housing, economic, and social status.

(c) An opinion poll contacts 1161 adults and asks them, "Which political party do you think has better ideas for leading the country in the twenty-first century?"

2.51. Interview residents of apartment complexes. You are planning a report on apartment living in a college town. You decide to select 5 apartment complexes at random for in-depth interviews with residents. Select a simple random sample of 5 of the following apartment complexes. If you use Table B, start at line 137.
DATA FILE RESIDENTS

Ashley Oaks	Country View	Mayfair Village
Bay Pointe	Country Villa	Nobb Hill
Beau Jardin	Crestview	Pemberly Courts
Bluffs	Del-Lynn	Peppermill
Brandon Place	Fairington	Pheasant Run
Briarwood	Fairway Knolls	Richfield
Brownstone	Fowler	Sagamore Ridge
Burberry	Franklin Park	Salem Courthouse
Cambridge	Georgetown	Village Manor
Chauncey Village	Greenacres	Waterford Court
Country Squire	Lahr House	Williamsburg

2.52. Using GIS to identify mint field conditions. A Geographic Information System (GIS) is to be used to distinguish different conditions in mint fields. Ground observations will be used to classify regions of each field as either healthy mint, diseased mint, or weed-infested mint. The GIS divides mint-growing areas into regions called pixels. An area contains 200 pixels. For a random sample of 25 pixels, ground measurements will be made to determine the status of the mint, and these observations will be compared with information obtained by the GIS. Select the random sample. If you use Table B, start at line 112 and choose only the first 5 pixels in the sample.

2.53. APPLET **Use the *Simple Random Sample* applet.** After you have labeled the individuals in a population, the *Simple Random Sample* applet automates the task of choosing an SRS. Use the applet to choose the sample in the previous exercise.

2.54. APPLET **Indiana counties.** There are 92 counties in Indiana. You want to choose an SRS of 10 of these counties for a study of county governments. Label the counties 01 to 92 and use the *Simple Random Sample* applet to choose your sample.

2.55. Repeated use of Table B. In using Table B repeatedly to choose samples or do randomization for experiments, you should not always begin at the same place, such as line 110. Why not?

2.56. Systematic random samples. Systematic random samples are often used at the last stage of a multistage sample to choose a sample of apartments in a large building or dwelling units in a block. An example will illustrate the idea of a systematic sample. Suppose that we must choose 4 addresses out of 100. Because $100/4 = 25$, we can think of the list as four lists of 25 addresses. Choose 1 of the first 25 at random, using Table B. The sample contains this address and the addresses 25, 50, and 75 places down the list from it. If 13 is chosen, for example, then the systematic random sample consists of the addresses numbered 13, 38, 63, and 88.

(a) A study of dating among college students wanted a sample of 200 of the 9000 single male students on campus. The sample consisted of every 45th name from a list of the 9000 students. Explain why the survey chooses every 45th name.

(b) Use Table B at line 115 to choose the starting point for this systematic sample.

2.57. CHALLENGE **Systematic random samples versus simple random samples.** The previous exercise introduces systematic random samples. Explain carefully why a systematic random sample *does* give every individual the same chance to be chosen but is *not* a simple random sample.

2.58. Random digit telephone dialing. An opinion poll in California uses random digit dialing to choose telephone numbers at random. Numbers are selected separately within each California area code. The size of the sample in each area code is proportional to the population living there. DATA FILE AREACODES

(a) What is the name for this kind of sampling design?

(b) The California area codes are

209	213	310	323	408	415	424	442	510	530
559	562	619	626	650	657	661	707	714	747
760	805	818	831	858	909	916	925	949	951

Another California survey does not call numbers in all area codes but starts with an SRS of 10 area codes. Choose such an SRS. If you use Table B, start at line 123.

2.59. Stratified sampling of forest areas. Stratified samples are widely used to study large areas of forest. Based on satellite images, a forest area in the Amazon basin is divided into 14 types. Foresters studied the four most commercially valuable types: alluvial climax forests of quality levels 1, 2, and 3, and mature secondary forest. They divided the area of each type into large parcels, chose parcels of each type at random, and counted tree species in a 20- by 25-meter rectangle chosen at random within each parcel selected. Here is some detail:

Forest type	Total parcels	Sample size
Climax 1	36	4
Climax 2	72	7
Climax 3	31	3
Secondary	42	4

Choose the stratified sample of 18 parcels. Be sure to explain how you assigned labels to parcels. If you use Table B, start at line 140.

2.60. Select club members to go to a convention. A club has 30 student members and 10 faculty members. The students are

Abel	Fisher	Huber	Moran	Reinmann
Carson	Golomb	Jimenez	Moskowitz	Santos
Chen	Griswold	Jones	Neyman	Shaw
David	Hein	Kiefer	O'Brien	Thompson
Deming	Hernandez	Klotz	Pearl	Utts
Elashoff	Holland	Liu	Potter	Vlasic

and the faculty members are

Andrews	Fernandez	Kim	Moore	Rabinowitz
Besicovitch	Gupta	Lightman	Phillips	Yang

The club can send 5 students and 3 faculty members to a convention and decides to choose those who will go by random selection. Select a stratified random sample of 5 students and 3 faculty members.

2.61. CHALLENGE **Stratified samples for alcohol attitudes.** At a party there are 30 students over age 21 and 20 students under age 21. You choose at random 3 of those over 21 and separately choose at random 2 of those under 21 to interview about attitudes toward alcohol. You have given every student at the party the same chance to be interviewed: what is that chance? Why is your sample not an SRS?

2.62. Stratified samples for accounting audits. Accountants use stratified samples during audits to verify a company's records of such things as accounts receivable. The stratification is based on the dollar amount of the item and often includes 100% sampling of the largest items. One company reports 5000 accounts receivable. Of these, 100 are in amounts over $50,000; 500 are in amounts between $1000 and $50,000; and the remaining 4400 are in amounts under $1000. Using these groups as strata, you decide to verify all the largest accounts and to sample 5% of the midsize accounts and 1% of the small accounts. How would you label the two strata from which you will sample? Use Table B, starting at line 115, to select the first 5 accounts from each of these strata.

2.63. The sampling frame. The list of individuals from which a sample is actually selected is called the **sampling frame.** Ideally, the frame should list every individual in the population, but in practice this is often difficult. A frame that leaves out part of the population is a common source of undercoverage.

(a) Suppose that a sample of households in a community is selected at random from the telephone directory. What households are omitted from this frame? What types of people do you think are likely to live in these households? These people will probably be underrepresented in the sample.

(b) It is usual in telephone surveys to use random digit dialing equipment that selects the last four digits of a telephone number at random after being given the area code and the exchange. The exchange is the first three digits of the telephone number. Which of the households that you mentioned in your answer to (a) will be included in the sampling frame by random digit dialing?

2.64. Survey questions. Comment on each of the following as a potential sample survey question. Is the question clear? Is it slanted toward a desired response?

(a) "Some cell phone users have developed brain cancer. Should all cell phones come with a warning label explaining the danger of using cell phones?"

(b) "Do you agree that a national system of health insurance should be favored because it would provide health insurance for everyone and would reduce administrative costs?"

(c) "In view of escalating environmental degradation and incipient resource depletion, would you favor economic incentives for recycling of resource-intensive consumer goods?"

2.65. CHALLENGE **Economic attitudes of Spaniards.** Spain's Centro de Investigaciones Sociológicos carried out a sample survey on the economic attitudes of Spaniards.[27] Of the 2496 adults interviewed, 72% agreed with the statement "Employees with higher performance must get higher pay." On the other hand, 71% agreed with the statement "Everything a society produces should be distributed among its members as equally as possible and there should be no major differences." Use these conflicting results as an example in a short explanation of why opinion polls often fail to reveal public attitudes clearly.

2.3 Ethics

The production and use of data, like all human endeavors, raise ethical questions. We won't discuss the telemarketer who begins a telephone sales pitch with "I'm conducting a survey." Such deception is clearly unethical. It enrages legitimate survey organizations, which find the public less willing to talk with them. Neither will we discuss those few researchers who, in the pursuit of professional advancement, publish fake data. There is no ethical question here—faking data to advance your career is just wrong. It will end your career when uncovered.

But just how honest must researchers be about real, unfaked data? Here is an example that suggests the answer is "More honest than they often are."

EXAMPLE 2.28

Provide all the critical information. Papers reporting scientific research are supposed to be short, with no extra baggage. Brevity can allow the researchers to avoid complete honesty about their data. Did they choose their subjects in a biased way? Did they report data on only some of their subjects? Did they try several statistical analyses and report only the ones that looked best? The statistician John Bailar screened more than 4000 medical papers in more than a decade as consultant to the *New England Journal of Medicine.* He says, "When it came to the statistical review, it was often clear that critical information was lacking, and the gaps nearly always had the practical effect of making the authors' conclusions look stronger than they should have."[28] The situation is no doubt worse in fields that screen published work less carefully.

The most complex issues of data ethics arise when we collect data from people. The ethical difficulties are more severe for experiments that impose some treatment on people than for sample surveys that simply gather information. Trials of new medical treatments, for example, can do harm as well as good to their subjects. Here are some basic standards of data ethics that must be obeyed by any study that gathers data from human subjects, whether sample survey or experiment.

BASIC DATA ETHICS

The organization that carries out the study must have an **institutional review board** that reviews all planned studies in advance to protect the subjects from possible harm.

All individuals who are subjects in a study must give their **informed consent** before data are collected.

All individual data must be kept **confidential.** Only statistical summaries for groups of subjects may be made public.

The law requires that studies funded by the federal government obey these principles. But neither the law nor the consensus of experts is completely clear about the details of their application.

Institutional review boards

The purpose of an institutional review board is not to decide whether a proposed study will produce valuable information. If a study is not designed well, subjects in it are put at risk for no gain. The board's purpose is, in the words of one university's board, "to protect the rights and welfare of human subjects (including patients) recruited to participate in research activities."

The board reviews the plan of the study and can require changes. It reviews the consent form to be sure that subjects are informed about the nature of the study and about any potential risks. Once research begins, the board monitors its progress at least once a year.

The most pressing issue concerning institutional review boards is whether their workload has become so large that their effectiveness in protecting subjects drops. When the government temporarily stopped human-subject research at Duke University Medical Center in 1999 due to inadequate protection of subjects, more than 2000 studies were going on. That's a lot of review work. There are shorter review procedures for projects that involve only minimal risks to subjects, such as most sample surveys. When a board is overloaded, there is a temptation to put more proposals in the minimal-risk category to speed the work.

USE YOUR KNOWLEDGE

The exercises in this section on ethics are designed to help you think about the issues that we are discussing and to formulate some opinions. In general there are no wrong or right answers, but you need to give reasons for your answers.

2.66 Do these proposals involve minimal risk? You are a member of your college's institutional review board. You must decide whether several research proposals qualify for lighter review because they involve only minimal risk to subjects. Federal regulations say that "minimal risk" means that the risks are no greater than "those ordinarily encountered in daily life or during the performance of routine physical or psychological examinations or tests." That's vague. Which of these do you think qualifies as "minimal risk"?
(a) Draw a drop of blood by pricking a finger to measure blood sugar.
(b) Draw blood from the arm for a full set of blood tests.
(c) Insert a tube that remains in the arm, so that blood can be drawn regularly.

2.67 Who should be on an institutional review board? Government regulations require that institutional review boards consist of at least five people, including at least one scientist, one nonscientist, and one person from outside the institution. Most boards are larger, but many contain just one outsider.
(a) Why should review boards contain people who are not scientists?
(b) Do you think that one outside member is enough? How would you choose that member? (For example, would you prefer a medical doctor? A member of the clergy? An activist for patients' rights?)

Informed consent

Both words in the phrase "informed consent" are important, and both can be controversial. Subjects must be *informed* in advance about the nature of a study and any risk of harm it may bring. In the case of a sample survey, physical harm is not possible. The subjects should be told what kinds of questions the survey will ask and about how much of their time it will take. Experimenters must tell subjects the nature and purpose of the study and outline possible risks. Subjects must then *consent* in writing.

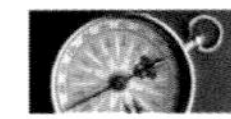

EXAMPLE 2.29

Who can give informed consent? Are there some subjects who can't give informed consent? It was once common, for example, to test new vaccines on prison inmates who gave their consent in return for good-behavior credit. Now we worry that prisoners are not really free to refuse, and the law forbids medical experiments in prisons.

Young children can't give fully informed consent, so the usual procedure is to ask their parents. A study of new ways to teach reading is about to start at a local elementary school, so the study team sends consent forms home to parents. Many parents don't return the forms. Can their children take part in the study because the parents did not say "No," or should we allow only children whose parents returned the form and said "Yes"?

What about research into new medical treatments for people with mental disorders? What about studies of new ways to help emergency room patients who may be unconscious or have suffered a stroke? In most cases, there is not time even to get the consent of the family. Does the principle of informed consent bar realistic trials of new treatments for unconscious patients?

These are questions without clear answers. Reasonable people differ strongly on all of them. There is nothing simple about informed consent.[29]

The difficulties of informed consent do not vanish even for capable subjects. Some researchers, especially in medical trials, regard consent as a barrier to getting patients to participate in research. They may not explain all possible risks; they may not point out that there are other therapies that might be better than those being studied; they may be too optimistic in talking with patients even when the consent form has all the right details.

On the other hand, mentioning every possible risk leads to very long consent forms that really are barriers. "They are like rental car contracts," one lawyer said. Some subjects don't read forms that run five or six printed pages. Others are frightened by the large number of possible (but unlikely) disasters that might happen and so refuse to participate. Of course, unlikely disasters sometimes happen. When they do, lawsuits follow and the consent forms become yet longer and more detailed.

Confidentiality

Ethical problems do not disappear once a study has been cleared by the review board, has obtained consent from its subjects, and has actually collected data about the subjects. It is important to protect the subjects' privacy by keeping all data about individuals confidential. The report of an opinion poll may say what percent of the 1500 respondents felt that legal immigration should be reduced. It may not report what *you* said about this or any other issue.

anonymity

Confidentiality is not the same as **anonymity.** Anonymity means that subjects are anonymous—their names are not known even to the director of the study. Anonymity is rare in statistical studies. Even where anonymity is possible (mainly in surveys conducted by mail), it prevents any follow-up to improve nonresponse or inform subjects of results.

Any breach of confidentiality is a serious violation of data ethics. The best practice is to separate the identity of the subjects from the rest of the data at once. Sample surveys, for example, use the identification only to check on who did or did not respond. In an era of advanced technology, however, it is no longer enough to be sure that each individual set of data protects people's privacy.

The government, for example, maintains a vast amount of information about citizens in many separate databases—census responses, tax returns, Social Security information, data from surveys such as the Current Population Survey, and so on. Many of these databases can be searched by computers for statistical studies.

A clever computer search of several databases might be able, by combining information, to identify you and learn a great deal about you even if your name and other identification have been removed from the data available for search. A colleague from Germany once remarked that "female full professor of statistics with a PhD from the United States" was enough to identify her among all the citizens of Germany. Privacy and confidentiality of data are hot issues among statisticians in the computer age.

EXAMPLE 2.30

Data collected by the government. Citizens are required to give information to the government. Think of tax returns and Social Security contributions. The government needs these data for administrative purposes—to see if we paid the right amount of tax and how large a Social Security benefit we are owed when we retire. Some people feel that individuals should be able to forbid any other use of their data, even with all identification removed. This would prevent using government records to study, say, the ages, incomes, and household sizes of Social Security recipients. Such a study could well be vital to debates on reforming Social Security.

USE YOUR KNOWLEDGE

2.68 How can we obtain informed consent? A researcher suspects that traditional religious beliefs tend to be associated with an authoritarian personality. She prepares a questionnaire that measures authoritarian tendencies and also asks many religious questions. Write a description of the purpose of this research to be read by subjects to obtain their informed consent. You must balance the conflicting goals of not deceiving the subjects as to what the questionnaire will tell about them and of not biasing the sample by scaring off religious people.

2.69 Should we allow this personal information to be collected? In which of the following circumstances would you allow collecting personal information without the subjects' consent?

(a) A government agency takes a random sample of income tax returns to obtain information on the average income of people in different occupations. Only the incomes and occupations are recorded from the returns, not the names.

(b) A social psychologist attends public meetings of a religious group to study the behavior patterns of members.

(c) A social psychologist pretends to be converted to membership in a religious group and attends private meetings to study the behavior patterns of members.

Clinical trials

Clinical trials are experiments that study the effectiveness of medical treatments on actual patients. Medical treatments can harm as well as heal, so clinical trials spotlight the ethical problems of experiments with human subjects. Here are the starting points for a discussion:

- Randomized comparative experiments are the only way to see the true effects of new treatments. Without them, risky treatments that are no better than placebos will become common.
- Clinical trials produce great benefits, but most of these benefits go to future patients. The trials also pose risks, and these risks are borne by the subjects of the trial. So we must balance future benefits against present risks.
- Both medical ethics and international human rights standards say that "the interests of the subject must always prevail over the interests of science and society."

The quoted words are from the 1964 Helsinki Declaration of the World Medical Association, the most respected international standard. The most outrageous examples of unethical experiments are those that ignore the interests of the subjects.

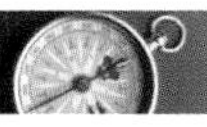

EXAMPLE 2.31

The Tuskegee study. In the 1930s, syphilis was common among black men in the rural South, a group that had almost no access to medical care. The Public Health Service Tuskegee study recruited 399 poor black sharecroppers with syphilis and 201 others without the disease to observe how syphilis progressed when no treatment was given. Beginning in 1943, penicillin became available to treat syphilis. The study subjects were not treated. In fact, the Public Health Service prevented any treatment until word leaked out and forced an end to the study in the 1970s.

The Tuskegee study is an extreme example of investigators following their own interests and ignoring the well-being of their subjects. A 1996 review said, "It has come to symbolize racism in medicine, ethical misconduct in human research, paternalism by physicians, and government abuse of vulnerable people." In 1997, President Clinton formally apologized to the surviving participants in a White House ceremony.[30]

Because "the interests of the subject must always prevail," medical treatments can be tested in clinical trials only when there is reason to hope that they will help the patients who are subjects in the trials. Future benefits aren't enough to justify experiments with human subjects. Of course, if there is already strong evidence that a treatment works and is safe, it is unethical *not* to give it. Here are the words of Dr. Charles Hennekens of the Harvard Medical School, who directed the large clinical trial that showed that aspirin reduces the risk of heart attacks:

> There's a delicate balance between when to do or not do a randomized trial. On the one hand, there must be sufficient belief in the agent's potential to justify exposing half the subjects to it. On the other hand, there must be sufficient doubt about its efficacy to justify withholding it from the other half of subjects who might be assigned to placebos.[31]

Why is it ethical to give a control group of patients a placebo? Well, we know that placebos often work. What is more, placebos have no harmful side effects. So in the state of balanced doubt described by Dr. Hennekens, the placebo group may be getting a better treatment than the drug group. If we *knew* which treatment was better, we would give it to everyone. When we don't know, it is ethical to try both and compare them.

The idea of using a control or placebo is a fundamental principle to be considered in designing experiments. In many situations, deciding what to use as an appropriate control requires some careful thought.

The choice of the control can have a substantial impact on how the results of an experiment are interpreted. Here is an example.

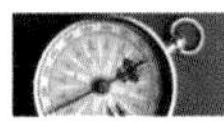

EXAMPLE 2.32

Attentiveness improves by nearly 20%. The manufacturer of a breakfast cereal designed for children claims that eating this cereal has been clinically shown to improve attentiveness by nearly 20%. The study used two groups of children who were tested before and after breakfast. One group received the cereal for breakfast, while breakfast for the control group was water. The results of the tests taken three hours after breakfast were used in the claim.

The Federal Trade Commission investigated the marketing of this product. They charged that the claim was false and violated federal law. The charges were settled and the company agreed to not use misleading claims in their advertising.[32]

It is not sufficient to obtain appropriate controls. The data must be collected from all groups in the same way. Here is an example of this type of flawed design.

EXAMPLE 2.33

Accurate identification of ovarian cancer. Two scientists published a paper claiming to have developed a very exciting new method to detect ovarian cancer using blood samples. When other scientists were unable to reproduce the results in different labs, the original work was examined more carefully. The original study looked at blood samples taken from women with ovarian cancer and from healthy controls. The blood samples were all analyzed using a mass spectrometer. The control samples were analyzed on one day, and the cancer samples were analyzed on the next day. This design was flawed because it could not control for changes over time in the measuring instrument.[33]

USE YOUR KNOWLEDGE

2.70 Is this study ethical? Researchers on aging proposed to investigate the effect of supplemental health services on the quality of life of older people. Eligible patients on the rolls of a large medical clinic were to be randomly assigned to treatment and control groups. The treatment group would be offered hearing aids, dentures, transportation, and other services not available without charge to the control group. The review board felt that providing these services to some but not other persons in the same institution raised ethical questions. Do you agree?

2.71 Should the treatments be given to everyone? Effective drugs for treating AIDS are very expensive, so most African nations cannot afford to give them to large numbers of people. Yet AIDS is more common in parts of Africa than anywhere else. Several clinical trials are looking at ways to prevent pregnant mothers infected with HIV from passing the infection to their unborn children, a major source of HIV infections in Africa. Some people say these trials are unethical because they do not give effective AIDS drugs to their subjects, as would be required in rich nations. Others reply that the trials are looking for treatments that can work in the real world in Africa and that they promise benefits at least to the children of their subjects. What do you think?

Behavioral and social science experiments

When we move from medicine to the behavioral and social sciences, the direct risks to experimental subjects are less acute, but so are the possible benefits to the subjects. Consider, for example, the experiments conducted by psychologists in their study of human behavior.

EXAMPLE 2.34

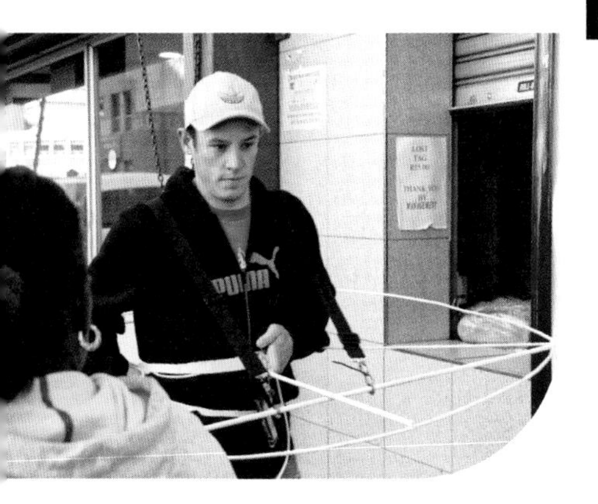

Personal space. Psychologists observe that people have a "personal space" and get annoyed if others come too close to them. We don't like strangers to sit at our table in a coffee shop if other tables are available, and we see people move apart in elevators if there is room to do so. Americans tend to require more personal space than people in most other cultures. Can violations of personal space have physical, as well as emotional, effects?

Investigators set up shop in a men's public restroom. They blocked off urinals to force men walking in to use either a urinal next to an experimenter (treatment group) or a urinal separated from the experimenter (control group). Another experimenter, using a periscope from a toilet stall, measured how long the subject took to start urinating and how long he kept at it.[34]

This personal space experiment illustrates the difficulties facing those who plan and review behavioral studies.

- There is no risk of harm to the subjects, although they would certainly object to being watched through a periscope. What should we protect subjects from when physical harm is unlikely? Possible emotional harm? Undignified situations? Invasion of privacy?
- What about informed consent? The subjects in Example 2.34 did not even know they were participating in an experiment. Many behavioral experiments rely on hiding the true purpose of the study. The subjects would change their behavior if told in advance what the investigators were looking for. Subjects are asked to consent on the basis of vague information. They receive full information only after the experiment.

The "Ethical Principles" of the American Psychological Association require consent unless a study merely observes behavior in a public place. They allow deception only when it is necessary to the study, does not hide information that might influence a subject's willingness to participate, and is explained to subjects as soon as possible. After the deception is explained, often it is required that subjects be asked again for their consent. The personal space study (from the 1970s) does not meet current ethical standards.

We see that the basic requirement for informed consent is understood differently in medicine and psychology. Here is an example of another setting with yet another interpretation of what is ethical. The subjects get no information and give no consent. They don't even know that an experiment may be sending them to jail for the night.

EXAMPLE 2.35

Domestic violence. How should police respond to domestic violence calls? In the past, the usual practice was to remove the offender and order him to stay out of the household overnight. Police were reluctant to make arrests because the victims rarely pressed charges. Women's groups argued that arresting offenders would help prevent future violence even if no charges were filed. Is there evidence that arrest will reduce future offenses? That's a question that experiments have tried to answer.

A typical domestic violence experiment compares two treatments: arrest the suspect and hold him overnight, or warn the suspect and release him. When police officers reach the scene of a domestic violence call, they calm the participants and investigate. Weapons or death threats require an arrest. If the facts permit an arrest but do not require it, an officer radios headquarters for instructions. The person on duty opens the next envelope in a file prepared in advance by a statistician. The envelopes contain the treatments in random order. The police either arrest the suspect or warn and release him, depending on the contents of the envelope. The researchers then watch police records and visit the victim to see if the domestic violence reoccurs.

The first such experiment appeared to show that arresting domestic violence suspects does reduce their future violent behavior. As a result of this evidence, arrest has become the common police response to domestic violence.

The domestic violence experiments shed light on an important issue of public policy. Because there is no informed consent, the ethical rules that govern clinical trials and most social science studies would forbid these experiments. They were cleared by review boards because, in the words of one domestic violence researcher, "These people became subjects by committing acts that allow the police to arrest them. You don't need consent to arrest someone."

SECTION 2.3 SUMMARY

Approval by an **institutional review board** is required for studies that involve humans or animals as subjects.

Human subjects must give **informed consent** if they are to participate in experiments.

Data on human subjects must be kept **confidential.**

Anonymity means that the names of the subjects are not known to anyone involved in the study.

Clinical trials are experiments that study the effectiveness of medical treatments on patients.

SECTION 2.3 EXERCISES

For Exercises 2.66 and 2.67, see page 68; for Exercises 2.68 and 2.69, see page 70–71; and for Exercises 2.70 and 2.71, see page 73.

2.72. What's wrong? Explain what is wrong in each of the following scenarios.

(a) Clinical trials are always ethical as long as they randomly assign patients to the treatments.

(b) The job of an institutional review board is complete when they decide to allow a study to be conducted.

(c) A treatment that has no risk of physical harm to subjects is always ethical.

2.73. How should the samples have been analyzed? Refer to the ovarian cancer diagnostic test study in Example 2.33 (pages 72–73). Describe how you would process the samples through the mass spectrometer.

2.74. The Vytorin controversy. Vytorin is a combination pill designed to lower cholesterol. It consists of a relatively inexpensive and widely used drug, Zocor, and a newer drug called Zetia. Early study results suggested that Vytorin was no more effective than Zetia. Critics claimed that the makers of the drugs tried to change the response variable for the study, and two congressional panels investigated why there was a two-year delay in the release of the results. Use the Web to search for more information about this controversy and write a report on what you find. Include an evaluation in the framework of ethical use of experiments and data. A good place to start your search would be to look for the phrase "Vytorin's Shortcomings."

2.75. Facebook and academic performance. *First Monday* is a peer-reviewed journal on the Internet. They published articles concerning Facebook and academic performance. Visit their Web site, `firstmonday.org`. Look at the first three articles in Volume 14, Number 5, May 4, 2009. Identify the key controversial issues involving the use of statistics that are addressed in these articles and write a report summarizing the facts as you see them. Be sure to include your opinions regarding ethical issues related to this work.

2.76. Informed consent to take blood samples. Researchers from Yale, working with medical teams in Tanzania, wanted to know how common infection with the AIDS virus is among pregnant women in that country. To do this, they planned to test blood samples drawn from pregnant women.

Yale's institutional review board insisted that the researchers get the informed consent of each woman and tell her the results of the test. This is the usual procedure in developed nations. The Tanzanian government did not want to tell the women why blood was drawn or tell them the test results. The government feared panic if many people turned out to have an incurable disease for which the country's medical system could not provide care. The study was canceled. Do you think that Yale was right to apply its usual standards for protecting subjects?

2.77. The General Social Survey. One of the most important nongovernment surveys in the United States is the National Opinion Research Center's General Social Survey. The GSS regularly monitors public opinion on a wide variety of political and social issues. Interviews are conducted in person in the subject's home. Are a subject's responses to GSS questions anonymous, confidential, or both? Explain your answer.

2.78. Anonymity and confidentiality in health screening. Texas A&M, like many universities, offers free screening for HIV, the virus that causes AIDS. The announcement says, "Persons who sign up for the HIV Screening will be assigned a number so that they do not have to give their name." They can learn the results of the test by telephone, still without giving their name. Does this practice offer anonymity or just confidentiality?

2.79. Anonymity and confidentiality in mail surveys. Some common practices may appear to offer anonymity while actually delivering only confidentiality. Market researchers often use mail surveys that do not ask the respondent's identity but contain hidden codes on the questionnaire that identify the respondent. A false claim of anonymity is clearly unethical. If only confidentiality is promised, is it also unethical to say nothing about the identifying code, perhaps causing respondents to believe their replies are anonymous?

2.80. Use of stored blood. Long ago, doctors drew a blood specimen from you as part of treating minor anemia. Unknown to you, the sample was stored. Now researchers plan to use stored samples from you and many other people to look for genetic factors that may influence anemia. It is no longer possible to ask your consent. Modern technology can read your entire genetic makeup from the blood sample.

(a) Do you think it violates the principle of informed consent to use your blood sample if your name is on it but you were not told that it might be saved and studied later?

(b) Suppose that your identity is not attached. The blood sample is known only to come from (say) "a 20-year-old white female being treated for anemia." Is it now OK to use the sample for research?

(c) Perhaps we should use biological materials such as blood samples only from patients who have agreed to allow the material to be stored for later use in research. It isn't possible to say in advance what kind of research, so this falls short of the usual standard for informed consent. Is it nonetheless acceptable, given complete confidentiality and the fact that using the sample can't physically harm the patient?

2.81. Testing vaccines. One of the most important goals of AIDS research is to find a vaccine that will protect against HIV. Because AIDS is so common in parts of Africa, that is the easiest place to test a vaccine. It is likely, however, that a vaccine would be so expensive that it could not (at least at first) be widely used in Africa. Is it ethical to test in Africa if the benefits go mainly to rich countries? The treatment group of subjects would get the vaccine, and the placebo group would later be given the vaccine if it proved effective. So the actual subjects would benefit—it is the future benefits that would go elsewhere. What do you think?

CHAPTER 2 EXERCISES

2.82. Experiments and surveys. Write a short report describing the differences and similarities between experiments and surveys. Include a discussion of the advantages and disadvantages of each.

2.83. Online behavioral advertising. The Federal Trade Commission (FTC) Staff Report "Self-Regulatory Principles for Online Behavioral Advertising" defines behavioral advertising as "the tracking of a consumer's online activities over time—including the searches the consumer has conducted, the Web pages visited and the content viewed—in order to deliver advertising targeted to the individual consumer's interests." The report suggests four governing concepts for their proposals. These are (1) transparency and control: when companies collect information from consumers for advertising, they should tell the consumer about how the data will be collected, and customers should be given a choice about whether to allow the data to be collected; (2) security and data retention: data should be kept secure and should be retained only as long as needed; (3) privacy: before data are used in a way that differs from what the companies initially promised, the companies should obtain consent from the consumer; and (4) sensitive data: affirmative express consent should be obtained before using any sensitive data.[35] Write a report discussing your opinions concerning online behavioral advertising and the four governing concepts. Pay particular attention to issues related to the ethical collection and use of statistical data.

2.84. Confidentiality at NORC. The National Opinion Research Center conducts a large number of surveys and has established procedures for protecting the confidentiality of their survey participants. For their Survey of Consumer Finances, they provide a pledge to participants regarding confidentiality. This pledge is available at `norc.org`. Review the pledge and summarize its key parts. Do you think that the pledge adequately addresses issues related to the ethical collection and use of data? Explain your answer.

2.85. Make it an experiment! In the following observational studies, describe changes that could be made to the data collection process that would result in an experiment rather than an observational study. Also, offer suggestions about unseen biases or lurking variables that may be present in the studies as they are described here.

(a) A friend of yours likes to play Texas hold 'em. Every time that he tells you about his playing, he says that he won.

(b) In an introductory statistics class you notice that the students who sit in the first two rows of seats had a higher score on the first exam than the other students in the class.

2.86. Name the designs. What is the name for each of these study designs?

(a) A study to compare two methods of preserving wood started with boards of Southern White Pine. Each board was ripped from end to end to form two edge-matched specimens. One was assigned to Method A; the other, to Method B.

(b) A survey on youth and smoking contacted by telephone 300 smokers and 300 nonsmokers, all 14 to 22 years of age.

(c) Does air pollution induce DNA mutations in mice? Starting with 40 male and 40 female mice, 20 of each sex were housed in a polluted industrial area downwind from a steel mill. The other 20 of each sex were housed at an unpolluted rural location 30 kilometers away.

2.87. Price promotions and consumer expectations. A researcher studying the effect of price promotions on consumer expectations makes up two different histories of the store price of a hypothetical brand of laundry detergent for the past year. Students in a marketing course view one or the other price history on a computer. Some students see a steady price, while others see regular promotions that temporarily cut the price. Then the students are asked what price they would expect to pay for the detergent. Is this study an experiment? Why? What are the explanatory and response variables?

2.88. Calcium and healthy bones. Adults need to eat foods or supplements that contain enough calcium to maintain healthy bones. Calcium intake is generally measured in milligrams per day (mg/d), and one measure of healthy bones is total body bone mineral density measured in grams per centimeter squared (TBBMD, g/cm^2). Suppose that you want to study the relationship between calcium intake and TBBMD.

(a) Design an observational study to study the relationship.

(b) Design an experiment to study the relationship.

(c) Compare the relative merits of your two designs. Which do you prefer? Give reasons for your answer.

2.89. Choose the type of study. Give an example of a question about pets and their owners, their behavior, or their opinions that would best be answered by

(a) a sample survey.

(b) an observational study that is not a sample survey.

(c) an experiment.

2.90. Compare the fries. Do consumers prefer the fries from Burger King or from McDonald's? Describe briefly the design of a blind matched pairs experiment to investigate this question. How will you use randomization?

2.91. Bicycle gears. How does the time it takes a bicycle rider to travel 100 meters depend on which gear is used and how steep the course is? It may be, for example, that higher gears are faster on the level but lower gears are faster on steep inclines. Discuss the design of a two-factor experiment to investigate this issue, using one bicycle with three gears and one rider. How will you use randomization?

2.92. CHALLENGE **Design an experiment.** The previous two exercises illustrate the use of statistically designed experiments to answer questions that arise in everyday life. Select a question of interest to you that an experiment might answer and carefully discuss the design of an appropriate experiment.

2.93. CHALLENGE **Design a survey.** You want to investigate the attitudes of students at your school about the faculty's commitment to teaching. The student government will pay the costs of contacting about 500 students.

(a) Specify the exact population for your study; for example, will you include part-time students?

(b) Describe your sample design. Will you use a stratified sample?

(c) Briefly discuss the practical difficulties that you anticipate; for example, how will you contact the students in your sample?

2.94. Compare two doses of a drug. A drug manufacturer is studying how a new drug behaves in patients. Investigators compare two doses: 5 milligrams (mg) and 10 mg. The drug can be administered by injection, by a skin patch, or by intravenous drip. Concentration in the blood after 30 minutes (the response variable) may depend both on the dose and on the method of administration.

(a) Make a sketch that describes the treatments formed by combining dosage and method. Then use a diagram to outline a completely randomized design for this two-factor experiment.

(b) "How many subjects?" is a tough issue. We will explain the basic ideas in Chapter 8. What can you say now about the advantage of using larger groups of subjects?

2.95. Would the results be different for men and women? The drug that is the subject of the experiment in the previous exercise may behave differently in men and women. How would you modify your experimental design to take this into account?

2.96. CHALLENGE **Informed consent.** The requirement that human subjects give their informed consent to participate in an experiment can greatly reduce the number of available subjects. For example, a study of new teaching methods asks the consent of parents for their children to be randomly assigned to be taught by either a new method or the standard method. Many parents do not return the forms, so their children must continue to be taught by the standard method. Why is it not correct to consider these children as part of the control group along with children who are randomly assigned to the standard method?

2.97. CHALLENGE **Two ways to ask sensitive questions.** Sample survey questions are usually read from a computer screen. In a Computer-Aided Personal Interview (CAPI), the interviewer reads the questions and enters the responses. In a Computer-Aided Self Interview (CASI), the interviewer stands aside and the respondent reads the questions and enters responses. One method almost always shows a higher percent of subjects admitting use of illegal drugs. Which method? Explain why.

2.98. Your institutional review board. Your college or university has an institutional review board that screens all studies that use human subjects. Get a copy of the document that describes this board (you can probably find it online).

(a) According to this document, what are the duties of the board?

(b) How are members of the board chosen? How many members are not scientists? How many members are not employees of the college? Do these members have some special expertise, or are they simply members of the "general public"?

2.99. Use of data produced by the government. Data produced by the government are often available free or at low cost to private users. For example, satellite weather data produced by the U.S. National Weather Service are available free to TV stations for their weather reports and to anyone on the Web. *Opinion 1:* Government data should be available to everyone at minimal cost. European governments, on the other hand, charge TV stations for weather data. *Opinion 2:* The satellites are expensive, and the TV stations are making a profit from their weather services, so they should share the cost. Which opinion do you support, and why?

CHAPTER THREE

LOOKING AT DATA—DISTRIBUTIONS

Statistics is the science of learning from data. Data are numerical or qualitative descriptions of the objects that we want to study. In this chapter, we will master the art of examining data.

A statistical analysis starts with a set of data. We construct a set of data by first deciding what *cases* we want to study. For each case, we record information about characteristics that we call *variables*.

CASES, LABELS, VARIABLES, AND VALUES

Cases are the objects described by a set of data. Cases may be customers, companies, subjects in a study, or other objects.

A **label** is a special variable used in some data sets to distinguish the different cases.

A **variable** is a characteristic of a case.

Different cases can have different **values** for the variables.

EXAMPLE 3.1

Over 16 billion sold. Apple's music-related products and services generated $1.5 billion in the fourth quarter of 2011. Since Apple started marketing iTunes in 2003, they have sold over 16 billion songs. Let's take a look at this remarkable product. Figure 3.1 is part of an iTunes playlist named EPS. The four songs shown are cases. They are numbered from 1 to 4 in the first column. These numbers are the labels that distinguish the four songs. The next five columns give other variables: the name (of the song), time (the length of time it takes to play the song), artist, album, and genre.

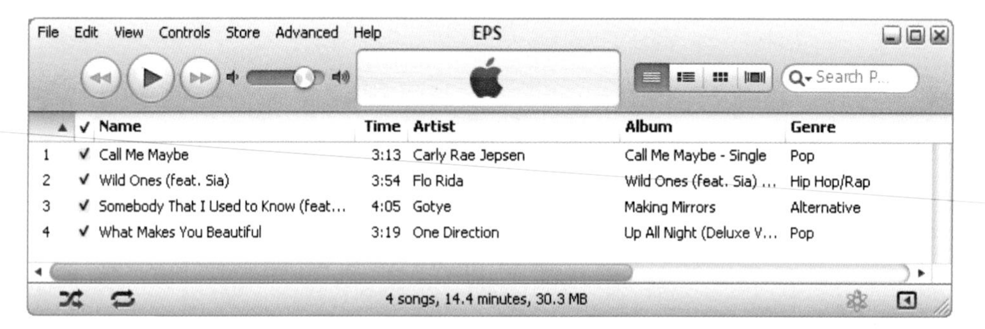

FIGURE 3.1 Part of an iTunes playlist, for Example 3.1.

Some variables, like the name of a song and the artist, simply place cases into categories. Others, like the length of a song, take numerical values for which we can do arithmetic. It makes sense to give an average length of time for a collection of songs, but it does not make sense to give an "average" album. We can, however, count the number of songs for different albums and do arithmetic with these counts.

CATEGORICAL AND QUANTITATIVE VARIABLES

A **categorical variable** places a case into one of several groups or categories.

A **quantitative variable** takes numerical values for which arithmetic operations such as adding and averaging make sense.

The **distribution** of a variable tells us what values it takes and how often it takes these values.

EXAMPLE 3.2

Categorical and quantitative variables in the iTunes playlist. The iTunes playlist contains five variables besides the label. These are the name, time, artist, album, and genre. The time is a quantitative variable. Name, artist, album, and genre are categorical variables.

An appropriate label for your cases should be chosen carefully. In our iTunes example, a natural choice of a label would be the name of the song. However, if you have more than one artist performing the same song, or the same artist performing the same song on different albums, then the name of the song would not uniquely label each of the songs in your playlist.

A quantitative variable such as the time in the iTunes playlist requires some special attention before we can do arithmetic with its values. The first song in the playlist has time equal to 3:29, that is, 3 minutes and 29 seconds. To do arithmetic with this variable, we should first convert all the values so that they have a single unit of measure. We could convert to seconds; 3 minutes is 180 seconds, so the total time is 180 + 29, or 209 seconds. An alternative would be to convert to minutes; 29 seconds is 0.483 seconds, so time in this way is 3.483 minutes.

USE YOUR KNOWLEDGE

3.1 Time in the iTunes playlist. In the iTunes playlist, do you prefer to convert the time to seconds or minutes? Give a reason for your answer.

In practice, any set of data is accompanied by background information that helps us understand the data. When you plan a statistical study or explore data from someone else's work, ask yourself the following questions:

1. **Who?** What **cases** do the data describe? **How many** cases appear in the data?
2. **What?** How many **variables** do the data contain? What are the **exact definitions** of these variables? In what **unit of measurement** is each variable recorded?
3. **Why? What purpose** do the data have? Do we hope to answer some specific questions? Do we want to draw conclusions about cases other than the ones we actually have data for? Are the variables that are recorded suitable for the intended purpose?

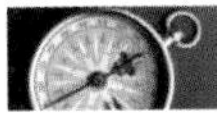

EXAMPLE 3.3

Data for students in a statistics class. Figure 3.2 (next page) shows part of a data set for students enrolled in an introductory statistics class. Each row gives the data for one student. The values for the different variables are in the columns. This data set has eight variables. ID is a label for each student. Exam1, Exam2, Homework, Final, and Project give the points earned, out of a possible total of 100, for each of these course requirements. Final grades are based on a possible 200 points for each exam and the final, 300 points for

Homework, and 100 points for Project. TotalPoints is the variable that gives the composite score. It is computed by adding 2 times Exam1, Exam2, and Final plus 3 times Homework plus 1 times Project. Grade is the grade earned in the course. This instructor used cutoffs of 900, 800, 700, etc. for the letter grades.

Microsoft Excel

	A	B	C	D	E	F	G	H
1	ID	Exam1	Exam2	Homework	Final	Project	TotalPoints	Grade
2	101	89	94	88	87	95	899	B
3	102	78	84	90	89	94	866	B
4	103	71	80	75	79	95	780	C
5	104	95	98	97	96	93	962	A
6	105	79	88	85	88	96	861	B

FIGURE 3.2 Spreadsheet for Example 3.3.

USE YOUR KNOWLEDGE

3.2 Who, what, and why for the statistics class data. Answer the who, what, and why questions for the statistics class data set.

3.3 Read the spreadsheet. Refer to Figure 3.2. Give the values of the variables Exam1, Exam2, and Final for the student with ID equal to 105.

3.4 Calculate the grade. A student whose data do not appear on the spreadsheet scored 86 on Exam1, 82 on Exam2, 77 for Homework, 90 on the Final, and 80 on the Project. Find TotalPoints for this student and give the grade earned.

spreadsheet

The display in Figure 3.2 is from an Excel **spreadsheet.** Spreadsheets are very useful for doing the kind of simple computations that you did in Exercise 3.4. You can type in a formula and have the same computation performed for each row.

Note that the names we have chosen for the variables in our spreadsheet do not have spaces. For example, we could have used the name "Exam 1" for the first exam score rather than Exam1. In some statistical software packages, however, spaces are not allowed in variable names. *For this reason, when creating spreadsheets for eventual use with statistical software, it is best to avoid spaces in variable names.* Another convention is to use an underscore (_) where you would normally use a space. For our data set, we could use Exam_1, Exam_2, and Final_Exam.

EXAMPLE 3.4

Cases and variables for the statistics class data. The data set in Figure 3.2 was constructed to keep track of the grades for students in an introductory statistics course. The cases are the students in the class. There are 8 variables in this data set. These include an identifier for each student and scores for the various course requirements. There are no units of measure for ID and grade. The other variables all have "points" as the unit.

EXAMPLE 3.5

Statistics class data for a different purpose. Suppose that the data for the students in the introductory statistics class were also to be used to study relationships between student characteristics and success in the course. For this purpose, we might want to use a data set like the spreadsheet in Figure 3.3. Here, we have decided to focus on the TotalPoints and Grade as the outcomes of interest. Other variables of interest have been included: Gender, PrevStat (whether or not the student has taken a statistics course previously), and Year (student classification as first, second, third, or fourth year). ID is a label, TotalPoints is a quantitative variable, and the remaining variables are all categorical.

Microsoft Excel

	A	B	C	D	E	F
1	ID	TotalPoints	Grade	Gender	PrevStat	Year
2	101	899	A	F	Yes	4
3	102	866	B	M	Yes	3
4	103	780	C	M	No	3
5	104	962	A	M	No	1
6	105	861	B	F	No	4

FIGURE 3.3 Spreadsheet for Example 3.5.

In our example, the possible values for the grade variable are A, B, C, D, and F. When computing grade point averages, many colleges and universities translate these letter grades into numbers using A = 4, B = 3, C = 2, D = 1, and F = 0. The transformed variable with numeric values is considered to be quantitative because we can average the numerical values across different courses to obtain a grade point average.

Sometimes, experts argue about numerical scales such as this. They ask whether or not the difference between an A and a B is the same as the difference between a D and an F. Similarly, many questionnaires ask people to respond on a 1 to 5 scale with 1 representing "strongly agree," 2 representing "agree," etc. Again we could ask about whether or not the five possible values for this scale are equally spaced in some sense. Nevertheless, from a practical point of view, the averages that can be computed when we convert categorical scales such as these to numerical values often provide a very useful way to summarize data.

USE YOUR KNOWLEDGE

3.5 Apartment rentals. A data set lists apartments available for students to rent. Information provided includes the monthly rent in dollars, whether or not cable is included free of charge, whether or not pets are allowed, the number of bedrooms, and the distance to the campus in miles. Describe the cases in the data set, give the number of variables, and specify whether each variable is categorical or quantitative.

Often the variables in a statistical study are easy to understand: height in centimeters, study time in minutes, and so on. But each area of work also has its own special variables. A psychologist uses the Minnesota Multiphasic Personality Inventory (MMPI), and a physical fitness expert measures "VO2 max," the volume of oxygen consumed per minute while exercising at maximum capacity. Both of these variables are measured with special **instruments.** VO2 max is measured by exercising while breathing into a mouthpiece connected to an apparatus that measures oxygen consumed. Scores on the MMPI are based on a long questionnaire, which is also an instrument.

instrument

Part of mastering your field of work is learning what variables are important and how they are best measured. Because details of particular measurements usually require knowledge of the particular field of study, we will say little about them.

Be sure that each variable really does measure what you want it to measure. A poor choice of variables can lead to misleading conclusions. Often, for example, the **rate** at which something occurs is a more meaningful measure than a simple count of occurrences.

rate

EXAMPLE 3.6

Injuries in the workplace. The Bureau of Labor Statistics keeps track of occupational injuries in various categories. In a recent year, there were 4613 fatal injuries among wage and salary workers but only 1044 fatal injuries among self-employed workers.[1]

Does this mean that the risk of an injury causing death is greater for wage and salary workers than for the self-employed? Not necessarily! A total of 136,670,000 persons were wage and salary workers, whereas only 10,544,000 persons were self-employed.

Let's compute the fatal injury rates for the two groups. Since the actual numbers of fatal injuries are small while the numbers of workers are large, rates such as these are expressed as fatal injuries per 100,000 workers. To calculate the rate, we take the ratio of fatal injuries to workers and multiply by 100,000. For wage and salary workers the rate is

$$\frac{4613}{136{,}670{,}000}100{,}000 = 3.4 \text{ per } 100{,}000 \text{ workers}$$

For self-employed workers the rate is

$$\frac{1044}{10{,}544{,}000}100{,}000 = 9.9 \text{ per } 100{,}000 \text{ workers}$$

When we compare the rates, we see that the self-employed are almost three times more likely than wage and salary workers to have a fatal injury at work.

USE YOUR KNOWLEDGE

3.6 Compare using a different type of rate. Refer to Example 3.6 on fatal workplace injuries.

(a) Find the rates per worker for the two groups.

(b) Find the rates per 10,000 workers for the two groups.

(c) Compare the rates that you calculated in (a) and (b) with the rates given in the example. Which do you prefer for effectively communicating the results to a general audience? Give reasons for your choice.

The preceding exercise illustrates an important point about presenting the results of your statistical calculations. *Always consider how to best communicate your results to a general audience.* For example, the numbers produced by your calculator or by statistical software frequently contain more digits than are needed. Be sure that you do not include extra information generated by software that will distract from a clear explanation of what you have found.

3.1 Displaying Distributions with Graphs

exploratory data analysis

Statistical tools and ideas help us examine data to describe their main features. This examination is called **exploratory data analysis.** Like an explorer crossing unknown lands, we want first to simply describe what we see. Here are two basic strategies that help us organize our exploration of a set of data:

- Begin by examining each variable by itself. Then move on to study the relationships among the variables.
- Begin with a graph or graphs. Then add numerical summaries of specific aspects of the data.

We will follow these principles in organizing our learning. This chapter presents methods for describing a single variable. We will study relationships among several variables in Chapter 4. Within each chapter, we will begin with graphical displays, then add numerical summaries for a more complete description.

Categorical variables: bar graphs and pie charts

distribution of a categorical variable

The values of a categorical variable are labels for the categories, such as "Yes" and "No." The **distribution of a categorical variable** lists the categories and gives either the **count** or the **percent** of cases that fall in each category.

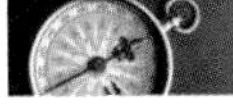

EXAMPLE 3.7

GPS

GPS market share. The Global Positioning System (GPS) uses satellites to transmit microwave signals that enable GPS receivers to determine the exact

location of the receiver. Here are the market shares for the major GPS receiver brands sold in the United States:[2]

Company	Percent (%)
Garmin	47
TomTom	19
Magellan	17
Mio	7
Other	10

Company is the categorical variable in this example, and its values are the names of the companies that provide GPS receivers in this market.

Note that the last value of the variable Company is "Other," which includes all receivers sold by companies other than the four listed by name. For data sets that have a large number of values for a categorical variable, we often create a category such as this that includes categories that have relatively small counts or percents. Careful judgment is needed when doing this. *You don't want to cover up some important piece of information contained in the data by combining data in this way.*

When we look at the GPS market share data set, we see that Garmin dominates the market with almost half of the sales. Using graphical methods to present the data will allow us to easily see this information and other characteristics of the data. We will now examine two graphical ways to present data.

EXAMPLE 3.8

DATA FILE GPS

bar graph

Bar graph for the GPS market share data. Figure 3.4 displays the GPS market share data using a **bar graph.** The heights of the five bars show the market shares for the four companies and the "Other" category.

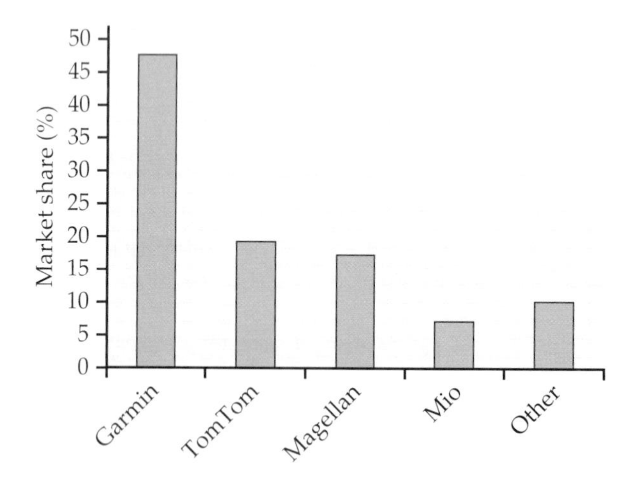

FIGURE 3.4 Bar graph for the GPS data, for Example 3.8.

The categories in a bar graph can be put in any order. In Figure 3.4, we ordered the companies based on their market share, with the "Other" category coming last. For other data sets, an alphabetical ordering or some other arrangement might produce a more useful graphical display.

You should always consider the best way to order the values of the categorical variable in a bar graph. Choose an ordering that will be useful to you. If you have difficulty, ask a friend if your choice communicates what you expect.

EXAMPLE 3.9

pie chart

Pie chart for the GPS market share data. The **pie chart** in Figure 3.5 helps us see what part of the whole each group forms. Here it is very easy to see that Garmin has about half of the market.

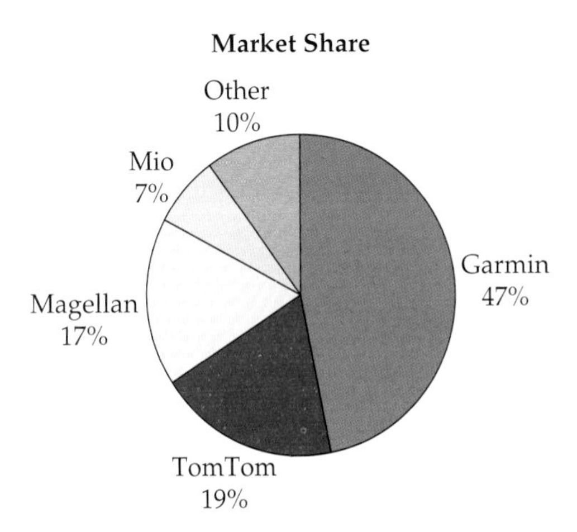

FIGURE 3.5 Pie chart for the GPS data, for Example 3.9.

To make a pie chart, you must include all the categories that make up a whole. A category such as "Other" in this example can be used, but the sum of the percents for all the categories should be 100%.

Bar graphs are more flexible. For example, you can use a bar graph to compare the numbers of students at your college majoring in biology, business, and political science. A pie chart cannot make this comparison because not all students fall into one of these three majors.

We now turn to the kinds of graphs that are used to describe the distribution of a quantitative variable. We will explain how to make the graphs by hand, because knowing this helps you understand what the graphs show. However, making graphs by hand is so tedious that software is almost essential for effective data analysis unless you have just a few observations.

Quantitative variables: stemplots

A *stemplot* (also called a stem-and-leaf plot) gives a quick picture of the shape of a distribution while including the actual numerical values in the graph. Stemplots work best for small numbers of observations that are all greater than 0.

STEMPLOT

To make a **stemplot,**

1. Separate each observation into a **stem,** consisting of all but the final (rightmost) digit, and a **leaf,** the final digit. Stems may have as many digits as needed, but each leaf contains only a single digit.
2. Write the stems in a vertical column with the smallest at the top, and draw a vertical line at the right of this column.
3. Write each leaf in the row to the right of its stem, in increasing order out from the stem.

EXAMPLE 3.10

Vitamin D. Your body needs vitamin D to use calcium when building bones. It is particularly important that young adolescents have adequate supplies of this vitamin because their bodies are growing rapidly. Vitamin D in the form 25-hydroxy vitamin D is measured in the blood and represents the stores available for the body to use. The units are nanograms per milliliter (ng/ml) of blood. Here are some values measured on a sample of 20 adolescent girls aged 11 to 14 years:[3]

16	43	38	48	42	23	36	35	37	34
25	28	26	43	51	33	40	35	41	42

To make a stemplot of these data, use the first digits as stems and the second digits as leaves. Figure 3.6 shows the steps in making the plot. The girl with a measured value of 16 ng/ml for vitamin D appears on the stem labeled 1 with a leaf of 6, while the girl with a measured value of 43 ng/ml appears on the stem labeled 4 with a leaf of 3.

The overall pattern of the data is fairly regular. The lowest value, 16 ng/ml, is somewhat lower than the next highest value, 23, but it is not particularly extreme.

(a)

Stem	
1	
2	
3	
4	
5	

(b)

Stem	Leaves
1	6
2	3586
3	8657435
4	3823012
5	1

(c)

Stem	Leaves
1	6
2	3568
3	3455678
4	0122338
5	1

FIGURE 3.6 Making a stemplot of the data in Example 3.10. (a) Write the stems. (b) Go through the data and write each leaf on the proper stem. For example, the values on the 2 stem are 23, 25, 28, and 26 in the order given in the display for the example. (c) Arrange the leaves on each stem in order out from the stem. The 2 stem now has leaves 3, 5, 6, and 8.

USE YOUR KNOWLEDGE

3.7 Make a stemplot. Here are the scores on the first exam in an introductory statistics course for 30 students in one section of the course:

80	73	92	85	75	98	93	55	80	90	92	80	87	90	72
65	70	85	83	60	70	90	75	75	58	68	85	78	80	93

Use these data to make a stemplot. Then use the stemplot to describe the distribution of the first-exam scores for this course. DATA FILE STATCOURSE

Making comparisons with stemplots

back-to-back stemplot

When you wish to compare two related distributions, a **back-to-back stemplot** with common stems is useful. The leaves on each side are ordered out from the common stem.

EXAMPLE 3.11

Vitamin D for boys. Here are the 25-hydroxy vitamin D values for a sample of 20 adolescent boys aged 11 to 14 years:

18	28	28	28	37	31	24	29	8	27
24	12	21	32	27	24	23	33	31	29

Figure 3.7 gives the back-to-back stemplot for the girls and the boys. The values on the left give the vitamin D measures for the girls, while the values on the right give the measures for the boys. The values of the boys tend to be lower than those for the girls.

Girls		Boys
	0	8
6	1	28
8653	2	134447788899
8765543	3	11237
8332210	4	
1	5	

FIGURE 3.7 A back-to-back stemplot to compare the distributions of vitamin D for samples of adolescent girls and boys, for Example 3.11.

splitting stems

trimming

Two modifications of the basic stemplot can be helpful in different situations. You can double the number of stems in a plot by **splitting each stem** into two: one with leaves 0 to 4 and the other with leaves 5 to 9. When the observed values have many digits, it is often best to **trim** the numbers by removing the last digit or digits before making a stemplot.

You must use your judgment in deciding whether to split stems and whether to trim, though statistical software will often make these choices for you. Remember that the purpose of a stemplot is to display the shape of a distribution. If there are many stems with no leaves or only one leaf, trimming will reduce the number of stems. Let's take a look at the effect of splitting the stems for our vitamin D data.

EXAMPLE 3.12

Back-to-back stemplot with split stems for vitamin D. Figure 3.8 gives the data from Examples 3.10 and 3.11 with split stems. Notice that we only needed one stem for 0 because there are no values between 0 and 4.

Girls		Boys
	0	8
	1	2
6	1	8
3	2	13444
865	2	7788899
43	3	1123
87655	3	7
332210	4	
8	4	
1	5	

FIGURE 3.8 A back-to-back stemplot with split stems to compare the distributions of vitamin D for samples of adolescent girls and boys, for Example 3.12.

USE YOUR KNOWLEDGE

3.8 Which stemplot do you prefer? Look carefully at the stemplots for the vitamin D data in Figures 3.7 and 3.8. Which do you prefer? Give reasons for your answer.

3.9 Why should you keep the space? Suppose that you had a data set for girls similar to the one given in Example 3.10, but that the observations of 33 ng/ml and 34 ng/ml were both changed to 35 ng/ml.
(a) Make a stemplot of these data for girls only using split stems.
(b) Should you use one stem or two stems for the 30s? Give a reason for your answer. (*Hint:* How would your choice reveal or conceal a potentially important characteristic of the data?)

Histograms

Stemplots display the actual values of the observations. This feature makes stemplots awkward for large data sets. Moreover, the picture presented by a stemplot divides the observations into groups (stems) determined by the number system rather than by judgment.

histogram

Histograms do not have these limitations. A **histogram** breaks the range of values of a variable into classes and displays only the count or percent of the observations that fall into each class. You can choose any convenient number of classes, but you should always choose classes of equal width.

Making a histogram by hand requires more work than a stemplot. Histograms do not display the actual values observed. For these reasons we prefer stemplots for small data sets.

The construction of a histogram is best shown by example. Most statistical software packages will make a histogram for you.

EXAMPLE 3.13

IQ

Distribution of IQ scores. You have probably heard that the distribution of scores on IQ tests is supposed to be roughly "bell-shaped." Let's look at some actual IQ scores. Table 3.1 displays the IQ scores of 60 fifth-grade students chosen at random from one school.

1. Divide the range of the data into classes of equal width. The scores in Table 3.1 range from 81 to 145, so we choose as our classes

$$75 \le \text{IQ score} < 85$$
$$85 \le \text{IQ score} < 95$$
$$\vdots$$
$$145 \le \text{IQ score} < 155$$

Be sure to specify the classes precisely so that each individual falls into exactly one class. A student with IQ 84 would fall into the first class, but a student with IQ 85 would fall into the second.

frequency
frequency table

2. Count the number of individuals in each class. These counts are called **frequencies,** and a table of frequencies for all classes is a **frequency table.**

Class	Count	Class	Count
$75 \le$ IQ score < 85	2	$115 \le$ IQ score < 125	13
$85 \le$ IQ score < 95	3	$125 \le$ IQ score < 135	10
$95 \le$ IQ score < 105	10	$135 \le$ IQ score < 145	5
$105 \le$ IQ score < 115	16	$145 \le$ IQ score < 155	1

3. Draw the histogram. First, on the horizontal axis mark the scale for the variable whose distribution you are displaying. That's the IQ score. The scale runs from 75 to 155 because that is the span of the classes we chose. The vertical axis contains the scale of counts. Each bar represents

TABLE 3.1

IQ test scores for 60 randomly chosen fifth-grade students

145	139	126	122	125	130	96	110	118	118
101	142	134	124	112	109	134	113	81	113
123	94	100	136	109	131	117	110	127	124
106	124	115	133	116	102	127	117	109	137
117	90	103	114	139	101	122	105	97	89
102	108	110	128	114	112	114	102	82	101

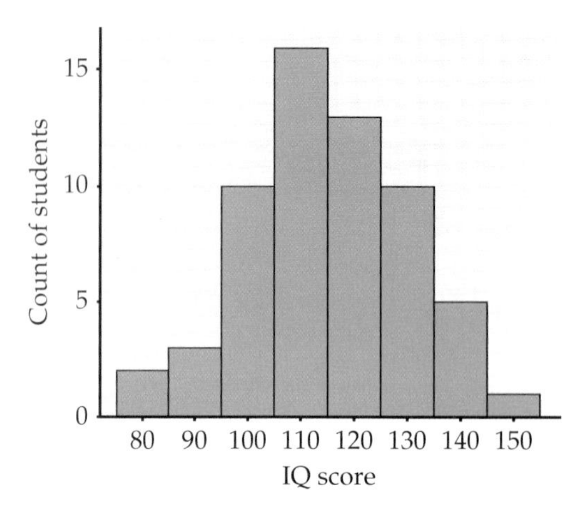

FIGURE 3.9 Histogram of the IQ scores of 60 fifth-grade students, for Example 3.13.

a class. The base of the bar covers the class, and the bar height is the class count. There is no horizontal space between the bars unless a class is empty, so that its bar has height zero. Figure 3.9 is our histogram. It does look roughly "bell-shaped."

Large sets of data are often reported in the form of frequency tables when it is not practical to publish the individual observations. In addition to the frequency (count) for each class, we may be interested in the fraction or percent of the observations that fall in each class. A histogram of percents looks just like a frequency histogram such as Figure 3.9. Simply relabel the vertical scale to read in percents. *Use histograms of percents for comparing several distributions that have different numbers of observations.*

USE YOUR KNOWLEDGE

3.10 Make a histogram. Refer to the first-exam scores from Exercise 3.7. Use these data to make a histogram using classes 50–59, 60–69, etc. Compare the histogram with the stemplot as a way of describing this distribution. Which do you prefer for these data?
STATCOURSE

Our eyes respond to the *area* of the bars in a histogram. Because the classes are all the same width, area is determined by height and all classes are fairly represented. There is no one right choice of the classes in a histogram. Too few classes will give a "skyscraper" graph, with all values in a few classes with tall bars. Too many will produce a "pancake" graph, with most classes having one or no observations. Neither choice will give a good picture of the shape of the distribution. You must use your judgment in choosing classes to display the shape. Statistical software will choose the classes for you. The software's choice is often a good one, but you can change it if you want.

You should be aware that the appearance of a histogram can change when you change the classes. The histogram function in the *One-Variable Statistical Calculator* applet on the text Web site allows you to choose the number of classes, so that it is easy to see how the choice of classes affects the histogram.

USE YOUR KNOWLEDGE

3.11 Change the classes in the histogram. Refer to the first-exam scores from Exercise 3.7 and the histogram you produced in Exercise 3.10. Now make a histogram for these data using classes 40–59, 60–79, and 80–100. Compare this histogram with the one that you produced in Exercise 3.10. Which do you prefer? Give a reason for your answer. STATCOURSE

3.12 Use smaller classes. Repeat the previous exercise using classes 55–59, 60–64, 65–69, etc.

Although histograms resemble bar graphs, their details and uses are distinct. A histogram shows the distribution of counts or percents among the values of a single variable. A bar graph compares the counts of different items. The horizontal axis of a bar graph need not have any measurement scale but simply identifies the items being compared. Draw bar graphs with blank space between the bars to separate the items being compared. Draw histograms with no space, to indicate that all values of the variable are covered.

Some spreadsheet programs, which are not primarily intended for statistics, will draw histograms as if they were bar graphs, with space between the bars. Often, you can tell the software to eliminate the space to produce a proper histogram.

Data analysis in action: don't hang up on me

Many businesses operate call centers to serve customers who want to place an order or make an inquiry. Customers want their requests handled thoroughly. Businesses want to treat customers well, but they also want to avoid wasted time on the phone. They therefore monitor the length of calls and encourage their representatives to keep calls short.

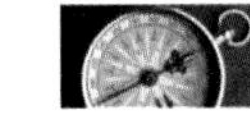

EXAMPLE 3.14

Calls to a customer service center. We have data on the length of all 31,492 calls made in a month to the customer service center of a small bank. Table 3.2 displays the lengths of the first 80 calls.[4]

TABLE 3.2

Service times (seconds) for calls to a customer service center

77	289	128	59	19	148	157	203
126	118	104	141	290	48	3	2
372	140	438	56	44	274	479	211
179	1	68	386	2631	90	30	57
89	116	225	700	40	73	75	51
148	9	115	19	76	138	178	76
67	102	35	80	143	951	106	55
4	54	137	367	277	201	52	9
700	182	73	199	325	75	103	64
121	11	9	88	1148	2	465	25

Take a look at the data in Table 3.2. In this data set the *cases* are calls made to the bank's call center. The *variable* recorded is the length of each call. The *units* are seconds. We see that the call lengths vary a great deal. The longest call lasted 2631 seconds, almost 44 minutes. More striking is that 8 of these 80 calls lasted less than 10 seconds. What's going on?

In Example 3.14 we started our study of the customer service center data by examining a few cases, the ones displayed in Table 3.2. It would be very difficult to examine all 31,492 cases in this way. How can we do this? Let's try a histogram.

EXAMPLE 3.15

CALLCENTER

Histogram for customer service center call lengths. Figure 3.10 is a histogram of the lengths (service time) of most of the 31,492 calls. We did not plot the few lengths greater than 1200 seconds (20 minutes). As expected, the graph shows that most calls last between about 1 minute and 5 minutes, with some lasting much longer when customers have complicated problems.

More striking is the fact that 7.6% of all calls are no more than 10 seconds long. It turned out that the bank penalized representatives whose average call length was too long—so some representatives just hung up on customers to bring their average length down. Neither the customers nor the bank were happy about this. The bank changed its policy, and later data showed that calls under 10 seconds had almost disappeared.

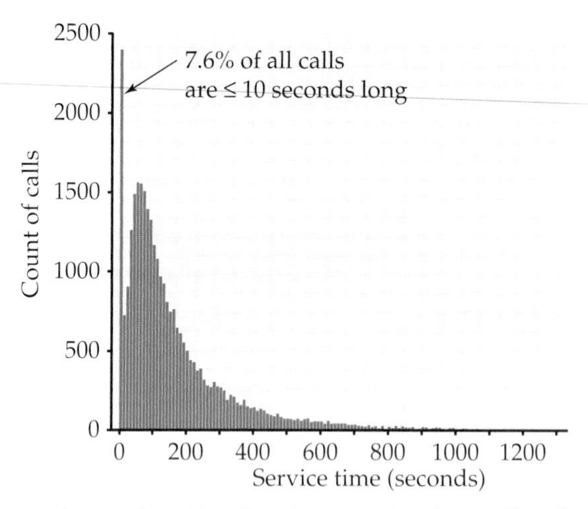

FIGURE 3.10 The distribution of call lengths for approximately 31,492 calls to a bank's customer service center, for Example 3.15. The data show a surprising number of very short calls. These are mostly due to representatives deliberately hanging up in order to bring down their average call length.

tails

The extreme values of a distribution are in the **tails** of the distribution. The high values are in the upper, or right, tail and the low values are in the lower, or left, tail. The overall pattern in Figure 3.10 is made up of the many moderate call lengths and the long right tail of more lengthy calls. The striking departure from the overall pattern is the surprising number of very short calls in the left tail.

Our examination of the call center data illustrates some important principles:

- After you understand the background of your data (cases, variables, units of measurement), the first thing to do is **plot** your data.
- When you look at a plot, look for an **overall pattern** and also for any **striking departures** from the pattern.

Examining distributions

Making a statistical graph is not an end in itself. The purpose of the graph is to help us understand the data. After you make a graph, always ask, "What do I see?" Once you have displayed a distribution, you can see its important features as follows.

> **EXAMINING A DISTRIBUTION**
>
> In any graph of data, look for the **overall pattern** and for striking **deviations** from that pattern.
>
> You can describe the overall pattern of a distribution by its **shape, center,** and **spread.**
>
> An important kind of deviation is an **outlier,** an individual value that falls outside the overall pattern.

In Section 3.2, we will learn how to describe center and spread numerically. For now, we can describe the center of a distribution by its *midpoint,* the value with roughly half the observations taking smaller values and half taking larger values. We can describe the spread of a distribution by giving the *smallest and largest values.* Stemplots and histograms display the shape of a distribution in the same way. Just imagine a stemplot turned on its side so that the larger values lie to the right. Some things to look for in describing shape are

modes
unimodal

- Does the distribution have one or several major peaks, called **modes**? A distribution with one major peak is called **unimodal.**

symmetric
skewed

- Is it approximately symmetric or is it skewed in one direction? A distribution is **symmetric** if the values smaller and larger than its midpoint are mirror images of each other. It is **skewed to the right** if the right tail (larger values) is much longer than the left tail (smaller values). It is **skewed to the left** if the left tail (smaller values) is much longer than the right tail (larger values).

Some variables commonly have distributions with predictable shapes. Many biological measurements on specimens from the same species and sex—lengths of bird bills, heights of young women—have symmetric distributions. Money amounts, on the other hand, usually have right-skewed distributions. There are many moderately priced houses, for example, but the few very expensive mansions give the distribution of house prices a strong right-skew.

EXAMPLE 3.16

Examine the histogram. What does the histogram of IQ scores (Figure 3.9, page 92) tell us?

Shape: The distribution is *roughly symmetric* with a *single peak* in the center. We don't expect real data to be perfectly symmetric, so we are satisfied if the two sides of the histogram are roughly similar in shape and extent.

Center: You can see from the histogram that the midpoint is not far from 110. Looking at the actual data shows that the midpoint is 114.

Spread: The spread is from 81 to 145. There are no outliers or other strong deviations from the symmetric, unimodal pattern.

The distribution of call lengths in Figure 3.10, on the other hand, is strongly *skewed to the right*. The midpoint, the length of a typical call, is about 115 seconds, or just under 2 minutes. The spread is very large, from 1 second to 28,739 seconds.

The longest few calls are *outliers*. They stand apart from the long right tail of the distribution, though we can't see this from Figure 3.10, which omits the largest observations. The longest call lasted almost 8 hours—that may well be due to equipment failure rather than an actual customer call.

USE YOUR KNOWLEDGE

3.13 Describe the first-exam scores. Refer to the first-exam scores from Exercise 3.7. Use your favorite graphical display to describe the shape, the center, and the spread of these data. Are there any outliers? STATCOURSE

Dealing with outliers

In data sets smaller than the service call data, you can spot outliers by looking for observations that stand apart (either high or low) from the overall pattern of a histogram or stemplot. *Identifying outliers is a matter for judgment. Look for points that are clearly apart from the body of the data, not just the most extreme observations in a distribution.*

You should search for an explanation for any outlier. Sometimes outliers point to errors made in recording the data. In other cases, the outlying observation may be caused by equipment failure or other unusual circumstances.

EXAMPLE 3.17

College students. How does the number of undergraduate college students vary by state? Figure 3.11 is a histogram of the numbers of undergraduate students in each of the 50 states.[5] Notice that about 65% of the states are

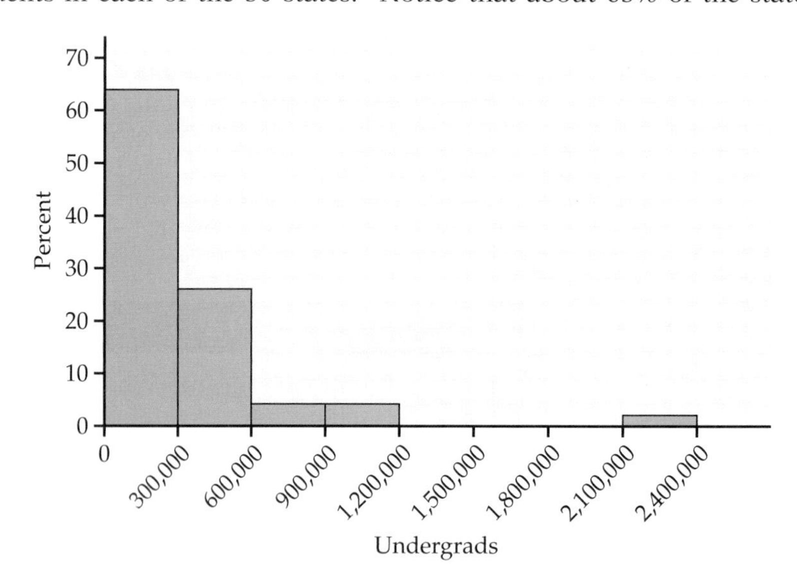

FIGURE 3.11 The distribution of the numbers of undergraduate college students for the 50 states, for Example 3.17.

included in the first bar of the histogram. These states have fewer than 300,000 undergraduates. The next bar includes another 36% of the states. These have between 300,000 and 600,000 students. The bar at the far right of the histogram corresponds to the state of California, which has over 2,000,000 undergraduates. In terms of this variable, California certainly stands apart from the other states. It is an outlier.

The state of California is an outlier in the Example 3.17 because it has a very large number of undergraduate students. California has the largest population of all the states, so we might expect it to have a large number of undergraduate college students. Let's look at these data in a different way.

EXAMPLE 3.18

College students per 1000. To account for the large variation in the populations of the states, we divide the number of undergraduate students by the state's population and then multiply by 1000. This gives the undergraduate college enrollment expressed as the number per 1000 people in each state. Here is a stemplot of the distribution:

```
3 | 7
4 | 1 1 1 1 1 3 3
4 | 5 5 5 5 6 6 6 7 7 7 7 7 8 8 8 8 9
5 | 0 0 1 1 1 2 2 2 4 4 4 4
5 | 5 6 6
6 | 0 0 0 0 1
6 | 7 9
6 | 1 2
7 | 7
```

California has 60 undergraduate students per 1000 people. This is one of the higher values in the distribution, but it is clearly not an outlier.

In Example 3.17 we looked at the distribution of the number of undergraduate students, while in Example 3.18 we adjusted these data by expressing the counts as number per 1000 people in each state. Which way is correct? The answer depends on why you are examining the data.

If you are interested in marketing a product to undergraduate students, the unadjusted numbers would be of interest. On the other hand, if you are interested in comparing states with respect to how well they provide opportunities for higher education to their residents, the population-adjusted values would be more suitable. *When presenting the results of any statistical analysis, always think about why you are doing the analysis, and this will guide you in choosing an appropriate analytic strategy.*

Here is an example with a different kind of outlier.

EXAMPLE 3.19

PTH

Healthy bones and PTH. Bones are constantly being built up (bone formation) and torn down (bone resorption). Young people who are growing have more formation than resorption. When we age, resorption increases to the

Solid bone matrix

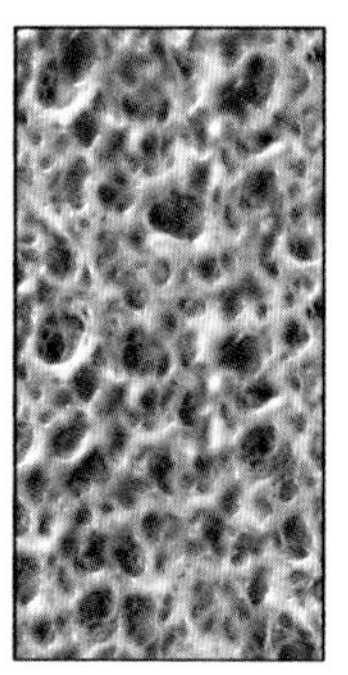

Weakened bone matrix

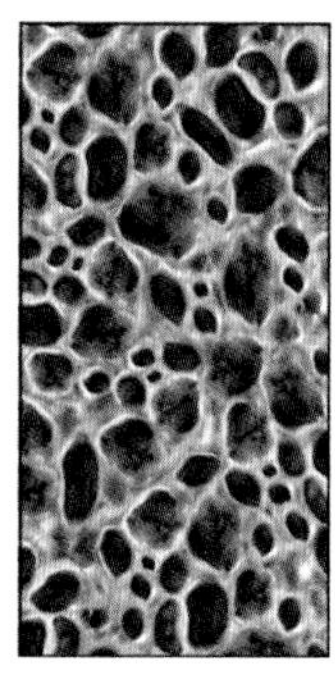

point where it exceeds formation. (The same phenomenon occurs when astronauts travel in space.) The result is osteoporosis, a disease associated with fragile bones that are more likely to break. The underlying mechanisms that control these processes are complex and involve a variety of substances. One of these is parathyroid hormone (PTH). Here are the values of PTH measured on a sample of 29 boys and girls aged 12 to 15 years:[6]

39	59	30	48	71	31	25	31	71	50	38	63	49	45	31
33	28	40	127	49	59	50	64	28	46	35	28	19	29	

The data are measured in picograms per milliliter of blood (pg/ml). The original data were recorded with one digit after the decimal point. They have been rounded to simplify our presentation here. Here is a stemplot of the data:

```
 1 | 9
 2 | 5 8 8 8 9
 3 | 0 1 1 1 3 5 8 9
 4 | 0 5 6 8 9 9
 5 | 0 0 9 9
 6 | 3 4
 7 | 1 1
 8 |
 9 |
10 |
11 |
12 | 7
```

The observation 127 clearly stands out from the rest of the distribution. A PTH measurement on this individual taken on a different day was similar to the rest of the values in the data set. We conclude that this outlier was caused by a laboratory error or a recording error, and we are confident in our decision to discard it for any additional analysis.

Time plots

Whenever data are collected over time, it is a good idea to plot the observations in time order. *Displays of the distribution of a variable that ignore time order, such as stemplots and histograms, can be misleading when there is systematic change over time.*

> **TIME PLOT**
>
> A **time plot** of a variable plots each observation against the time at which it was measured. Always put time on the horizontal scale of your plot and the variable you are measuring on the vertical scale.

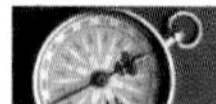

EXAMPLE 3.20

Seasonal variation in vitamin D. Although we get some of our vitamin D from food, most of us get about 75% of what we need from the sun. Cells in the skin make vitamin D in response to sunlight. If people do not get enough exposure to the sun, they can become deficient in vitamin D, resulting in weakened bones and other health problems. The elderly, who need more vitamin D than younger people, and people who live in northern areas where there is relatively little sunlight in the winter, are particularly vulnerable to these problems.

Figure 3.12 is a plot of the serum levels of vitamin D versus time of year for samples of subjects who visited a clinic in Switzerland.[7] Units for these measures are nanomoles per liter (nmol/l). The observations are grouped into periods of two months for the plot. The two-month averages are marked by dark circles and are connected by lines. The effect of the lack of sunlight in the winter months on vitamin D levels is clearly evident in the plot.

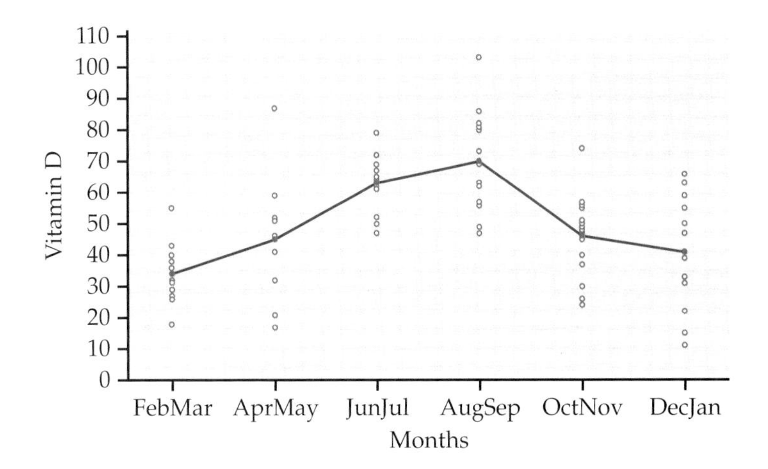

FIGURE 3.12 Plot of vitamin D levels versus months of the year, for Example 3.20.

The data described in Example 3.20 are based on a subset of the subjects in a study of 248 subjects. The investigators were particularly concerned about subjects whose levels were deficient, defined as a serum vitamin D level of less than 50 nmol/l. The investigators concluded that the deficiency rate was 3.8 times higher in February–March versus August–September: 91.2% versus 24.3%. To ensure that individuals from this population have adequate levels of vitamin D, some form of supplementation is needed, particularly during certain times of the year.

SECTION 3.1 SUMMARY

A data set contains information on a number of **cases.** Cases may be customers, companies, subjects in a study, or other objects. For each case, the data give values for one or more **variables.** A variable describes some characteristic of a case, such as a person's height, gender, or salary. Variables can have different **values** for different cases. A **label** is a variable that is used to identify different cases.

Some variables are **categorical** and others are **quantitative.** A categorical variable places each individual into a category, such as male or female. A quantitative variable has numerical values that measure some characteristic of each case, such as height in centimeters or annual salary in dollars.

Exploratory data analysis uses graphs and numerical summaries to describe the variables in a data set and the relations among them.

The **distribution** of a variable tells us what values it takes and how often it takes these values.

Bar graphs and **pie charts** display the distributions of categorical variables. These graphs use the counts or percents of the categories.

Stemplots and **histograms** display the distributions of quantitative variables. Stemplots separate each observation into a **stem** and a one-digit **leaf.** Histograms plot the **frequencies** (counts) or the percents of equal-width classes of values.

When examining a distribution, look for **shape, center,** and **spread** and for clear **deviations** from the overall shape.

Some distributions have simple shapes, such as **symmetric** or **skewed.** The number of **modes** (major peaks) is another aspect of overall shape. Not all distributions have a simple overall shape, especially when there are few observations.

Outliers are observations that lie outside the overall pattern of a distribution. Always look for outliers and try to explain them.

When observations on a variable are taken over time, make a **time plot** that graphs time horizontally and the values of the variable vertically.

SECTION 3.1 EXERCISES

For Exercise 3.1, see page 81; for Exercises 3.2 to 3.4, see page 82; for Exercise 3.5, see page 83; for Exercise 3.6, see page 85; for Exercise 3.7, see page 89; for Exercises 3.8 and 3.9, see page 90; for Exercise 3.10, see page 92; for Exercises 3.11 and 3.12, see page 93; and for Exercise 3.13, see page 96.

3.14. Employee data. The personnel department keeps records on all employees in a company. Here is the information that they keep in one of their data files: employee identification number, last name, first name, middle initial, department, number of years with the company, salary in dollars per year, education (coded as high school, some college, or college degree), and age in years.

(a) What are the cases for this data set?

(b) Describe each of these items as label, a quantitative variable, or a categorical variable.

(c) Set up a spreadsheet which could be used to record the data. Give appropriate column headings and five sample cases.

3.15. City rankings. A number of organizations rank cities based on various measures and produce lists of the top 10 or the top 100 best. Create a list of criteria that you would use to rank cities. Include at least eight variables and give reasons for your choices. Explain whether each variable is quantitative or categorical.

3.16. Survey of students. A survey of students in an introductory statistics class asked the following questions: (a) What is your age in years? (b) Do you like to dance (yes, no)? (c) Can you play a musical instrument (not at all, a little, pretty well)? (d) How much did you spend on food last week? (e) What is your height in inches? (f) Do you like broccoli (yes, no)? Classify each of these variables as categorical or quantitative and give reasons for your answers.

3.17. What questions would you ask? Refer to the previous exercise. Make up your own survey questions (at least six questions). Include at least two categorical variables and at least two quantitative variables. Tell which variables are categorical and which are quantitative. Give reasons for your answers.

3.18. Choosing a college or university. Popular magazines rank colleges and universities on their "academic quality" in serving undergraduate students. Describe five variables that you would like to see measured for each college if you were choosing where to study. Give reasons for each of your choices.

3.19. Favorite colors. What is your favorite color? One survey produced the following summary of responses to that question: blue, 42%; green, 14%; purple, 14%; red, 8%; black, 7%; orange, 5%; yellow, 3%; brown, 3%; gray, 2%; and white, 2%.[8] Make a bar graph of the percents and write a short summary of the major features of your graph. DATA FILE FAVORITECOLORS

3.20. Least-favorite colors. Refer to the previous exercise. The same study also asked people about their least-favorite color. Here are the results: orange, 30%; brown, 23%; purple, 13%; yellow, 13%; gray, 12%; green, 4%; white, 4%; red, 1%; black, 0%; and blue, 0%. Make a

bar graph of these percents and write a summary of the results. LEASTFAVCOLORS

3.21. Ages of survey respondents. The survey about color preferences reported the age distribution of the people who responded. Here are the results:

Age group (years)	1–18	19–24	25–35	36–50	51–69	70 and over
Count	10	97	70	36	14	5

(a) Add the counts and compute the percents for each age group.

(b) Make a bar graph of the percents.

(c) Describe the distribution.

(d) Explain why your bar graph is not a histogram.

3.22. Mobile browsing and iPhones. Users of iPhones were asked to respond to the statement "I do a lot more browsing on the iPhone than I did on my previous mobile phone" and responded as follows:[9] BROWSING

Response	Percent (%)
Strongly agree	54
Mildly agree	22
Mildly disagree	16
Strongly disagree	8

(a) Make a bar graph to display the distribution of the responses.

(b) Display the distribution with a pie chart.

(c) Summarize the information in these charts.

(d) Do you prefer the bar chart or the pie chart? Give a reason for your answer.

3.23. What did the iPhone replace? The survey in the previous exercise also asked iPhone users what phone, if any, the iPhone replaced. Here are the responses:

Response	Percent (%)	Response	Percent (%)
Motorola Razr	23.8	BlackBerry	13.0
Symbian	3.9	Windows Mobile	13.9
Sidekick	4.1	Replaced nothing	10.0
Palm	6.7	Other phone	24.5

Make a bar graph for these data. Carefully consider how you will order the responses. Explain why you chose the ordering that you did. PHONEREPLACE

3.24. Garbage. The formal name for garbage is "municipal solid waste." Here is a breakdown of the materials that made up American municipal solid waste:[10] GARBAGE

Material	Weight (million tons)	Percent of total (%)
Food scraps	31.7	12.5
Glass	13.6	5.3
Metals	20.8	8.2
Paper, paperboard	83.0	32.7
Plastics	30.7	12.1
Rubber, leather, textiles	19.4	7.6
Wood	14.2	5.6
Yard trimmings	32.6	12.8
Other	8.2	3.2
Total	254.1	100.0

(a) Add the weights and the percents for the nine materials given, including "Other." Each entry, including the total, is separately rounded to the nearest tenth. Therefore, the sum and the total may differ slightly because of **roundoff error.**

(b) Make a bar graph of the percents. The graph gives a clearer picture of the main contributors to garbage if you order the bars from tallest to shortest.

(c) If you use software, also make a pie chart of the percents. Comparing the two graphs, notice that it is easier to see the small differences among "Food scraps," "Plastics," and "Yard trimmings" in the bar graph.

3.25. Recycled garbage. Refer to the previous exercise. The following table gives the percent of the weight that was recycled for each of the categories. GARBAGE

Material	Weight (million tons)	Percent recycled (%)
Food scraps	31.7	2.6
Glass	13.6	23.7
Metals	20.8	34.8
Paper, paperboard	83.0	54.5
Plastics	30.7	6.8
Rubber, leather, textiles	19.4	14.7
Wood	14.2	9.3
Yard trimmings	32.6	64.1
Other	8.2	0.0
Total	254.1	

(a) Use a bar graph to display the percent recycled for these materials. Use the order of the materials given in the table above.

(b) Make another bar graph where the materials are ordered by the percent recycled, largest percent to smallest percent.

(c) Which bar graph, (a) or (b), do you prefer? Give a reason for your answer.

(d) Explain why it is inappropriate to use a pie chart to display these data. DATA FILE GARBAGE

3.26. Market share for search engines. The following table gives the market share for the major desktop search engines.[11] DATA FILE SEARCHENGINES

Search engine	Market share (%)	Search engine	Market share (%)
Google—Global	82.7	Ask—Global	0.6
Yahoo—Global	5.8	AOL—Global	0.4
Baidu	5.7	Other	0.7
Bing	3.9		

(a) Use a bar graph to display the market shares.

(b) Summarize what the graph tells you about market shares for search engines.

3.27. Spam. Email spam is the curse of the Internet. Here is a compilation of the most common types of spam:[12]

Type of spam	Percent (%)
Adult	14.5
Financial	16.2
Health	7.3
Leisure	7.8
Products	21.0
Scams	14.2

Make two bar graphs of these percents, one with bars ordered as in the table (alphabetical) and the other with bars in order from tallest to shortest. Comparisons are easier if you order the bars by height. A bar graph ordered from tallest to shortest bar is sometimes called a **Pareto chart,** after the Italian economist who recommended this procedure.

3.28. Facebook users by country. The following table gives the numbers of Facebook users by region of the world as of December 31, 2011:[13] DATA FILE FACEBOOKREGION

Region	Facebook users (in millions)
Asia	184
Africa	38
Europe	223
Latin America	142
North America	175
Middle East	18
The Caribbean	6
Oceania/Australia	13

(a) Use a bar graph to describe these data.

(b) Describe the major features of your graph in a short paragraph.

3.29. Facebook ratios. One way to compare the numbers of Facebook users for different regions of the world is to take into account the populations of these regions. Here are estimates of the populations for 2011 of the same geographic regions that we studied in the previous exercise:[14] DATA FILE FACEBOOKREGION

Region	Population (in millions)
Asia	3880
Africa	1038
Europe	816
Latin America	556
North America	347
Middle East	216
The Caribbean	41
Oceania/Australia	35

(a) Compute the ratios for each region by dividing the number of users from the previous exercise by the population size given in this exercise. Multiply these ratios by 100 to make the ratios similar to percents, and make a table of the results. Use the values in this table to answer the remaining parts of this exercise.

(b) Carefully examine the table, and summarize what it shows. Are there any extreme outliers? Which ones would you classify in this way?

(c) Use a stemplot to describe these data. You can list any extreme outliers separately from the plot.

(d) Describe the major features of these data using your plot and your list of outliers.

(e) How effective is the stemplot for summarizing these data? Give reasons for your answer.

(f) Explain why the values in the table that you constructed in part (a) are not the same as the percents of users who are from each region.

3.30. Women seeking graduate and professional degrees. Here are the percents of women among students seeking various graduate and professional degrees:[15] GRADDEGREES

Degree	Percent female (%)
Master's in business administration	39.8
Master's in education	76.2
Other master of arts	59.6
Other master of science	53.0
Doctorate in education	70.8
Other PhD degree	54.2
Medicine (MD)	44.0
Law	50.2
Theology	20.2

(a) Explain clearly why we cannot use a pie chart to display these data.

(b) Make a bar graph of the data. (Comparisons are easier if you order the bars by height.)

3.31. Vehicle colors. Vehicle colors differ among regions of the world. Here are data on the most popular colors for vehicles in North America and Europe:[16] VEHICLECOLORS

Color	North America percent (%)	Europe percent (%)
White	23	20
Black	18	25
Silver	16	15
Gray	13	18
Red	10	6
Blue	9	7
Brown/beige	5	5
Yellow/gold	3	1
Other	3	3

(a) Make a bar graph for the North America percents.

(b) Make a bar graph for the Europe percents.

(c) Now, be creative: make *one* bar graph that compares the two regions as well as the colors. Arrange your graph so that it is easy to compare the two regions.

3.32. Shakespeare's plays. Figure 3.13 is a histogram of the lengths of words used in Shakespeare's plays. Because there are so many words in the plays, we use a histogram of percents. What is the overall shape of this distribution? What does this shape say about word lengths in Shakespeare? Do you expect other authors to have word length distributions of the same general shape? Why?

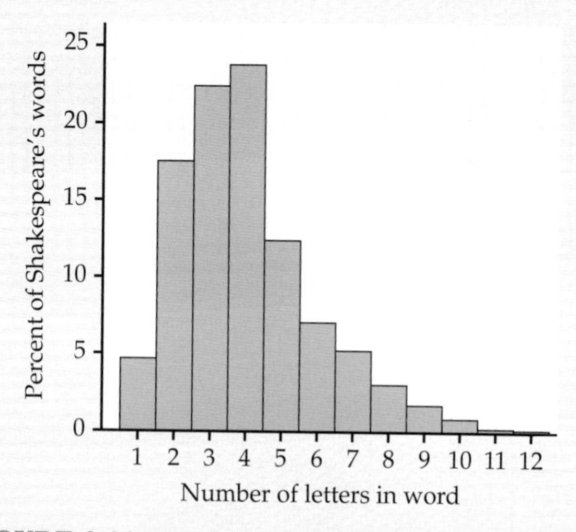

FIGURE 3.13 Histogram of the lengths of words used in Shakespeare's plays, for Exercise 3.32.

3.33. Diabetes and glucose. People with diabetes must monitor and control their blood glucose level. The goal is to maintain "fasting plasma glucose" between about 90 and 130 milligrams per deciliter (mg/dl). Here are the fasting plasma glucose levels for 18 diabetics enrolled in a diabetes control class, five months after the end of the class:[17]

141	158	112	153	134	95	96	78	148
172	200	271	103	172	359	145	147	255

Make a stemplot of these data and describe the main features of the distribution. (You will want to trim and also split stems.) Are there outliers? How well is the group as a whole achieving the goal for controlling glucose levels? GLUCOSE

3.34. Compare glucose in instruction and control groups. The study described in the previous exercise also measured the fasting plasma glucose of 16 diabetics who were given individual instruction on diabetes control. Here are the data:

128	195	188	158	227	198	163	164
159	128	283	226	223	221	220	160

Make a back-to-back stemplot to compare the class and individual instruction groups. How do the distribution shapes and success in achieving the glucose control goal compare? GLUCOSE

3.35. Vocabulary scores of seventh-grade students. Figure 3.14 displays the scores of all 947 seventh-grade students in the public schools of Gary, Indiana, on the vocabulary part of the Iowa Tests of Basic Skills.[18] Give a brief description of the overall pattern (shape, center, spread) of this distribution.

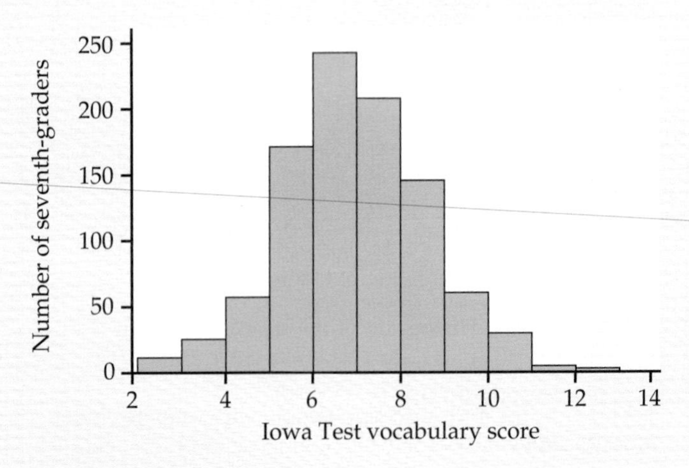

FIGURE 3.14 Histogram of the Iowa Tests of Basic Skills vocabulary scores of seventh-grade students in Gary, Indiana, for Exercise 3.35.

3.36. Carbon dioxide from burning fuels. Burning fuels in power plants or motor vehicles emits carbon dioxide (CO_2), which contributes to global warming. The data file CO2 gives CO_2 emissions per person from 134 countries.[19] CO2

(a) Why do you think we choose to measure emissions per person rather than total CO_2 emissions for each country?

(b) Display the CO_2 data in a graph. Describe the shape, center, and spread of the distribution. Which countries are outliers?

3.37. Thinness in Asia. Asian culture does not emphasize thinness, but young Asians are often influenced by Western culture. In a study of concerns about weight among young Korean women, researchers administered the Drive for Thinness scale (a questionnaire) to 264 female college students in Seoul, South Korea.[20] Drive for Thinness measures excessive concern with weight and dieting and fear of weight gain. Roughly speaking, a score of 15 is typical of Western women with eating disorders but is unusually high (90th percentile) for other Western women. Graph the data and describe the shape, center, and spread of the distribution of Drive for Thinness scores for these Korean students. Are there any outliers? THINNESS

3.38. CHALLENGE **Acidity of rainwater.** Changing the choice of classes can change the appearance of a histogram. Here is an example in which a small shift in the classes, with no change in the number of classes, has an important effect on the histogram. The data are the acidity levels (measured by pH) in 105 samples of rainwater. Distilled water has pH 7.00. As the water becomes more acidic, the pH goes down. The pH of rainwater is important to environmentalists because of the problem of acid rain.[21] ACIDRAIN

4.33	4.38	4.48	4.48	4.50	4.55	4.59	4.59	4.61	4.61
4.75	4.76	4.78	4.82	4.82	4.83	4.86	4.93	4.94	4.94
4.94	4.96	4.97	5.00	5.01	5.02	5.05	5.06	5.08	5.09
5.10	5.12	5.13	5.15	5.15	5.15	5.16	5.16	5.16	5.18
5.19	5.23	5.24	5.29	5.32	5.33	5.35	5.37	5.37	5.39
5.41	5.43	5.44	5.46	5.46	5.47	5.50	5.51	5.53	5.55
5.55	5.56	5.61	5.62	5.64	5.65	5.65	5.66	5.67	5.67
5.68	5.69	5.70	5.75	5.75	5.75	5.76	5.76	5.79	5.80
5.81	5.81	5.81	5.81	5.85	5.85	5.90	5.90	6.00	6.03
6.03	6.04	6.04	6.05	6.06	6.07	6.09	6.13	6.21	6.34
6.43	6.61	6.62	6.65	6.81					

(a) Make a histogram of pH with 14 classes, using class boundaries 4.2, 4.4, ..., 7.0. How many modes does your histogram show? More than one mode suggests that the data contain groups that have different distributions.

(b) Make a second histogram, also with 14 classes, using class boundaries 4.14, 4.34, ..., 6.94. The classes are those from (a) moved 0.06 to the left. How many modes does the new histogram show?

(c) Use your software's histogram function to make a histogram without specifying the number of classes or their boundaries. How does the software's default histogram compare with those in (a) and (b)?

3.39. CHALLENGE **Identify the histograms.** A survey of a large college class asked the following questions:

1. Are you female or male? (In the data, male = 0, female = 1.)

2. Are you right-handed or left-handed? (In the data, right = 0, left = 1.)

3. What is your height in inches?

4. How many minutes do you study on a typical weeknight?

Figure 3.15 shows histograms of the student responses, in scrambled order and without scale markings. Which histogram goes with each variable? Explain your reasoning.

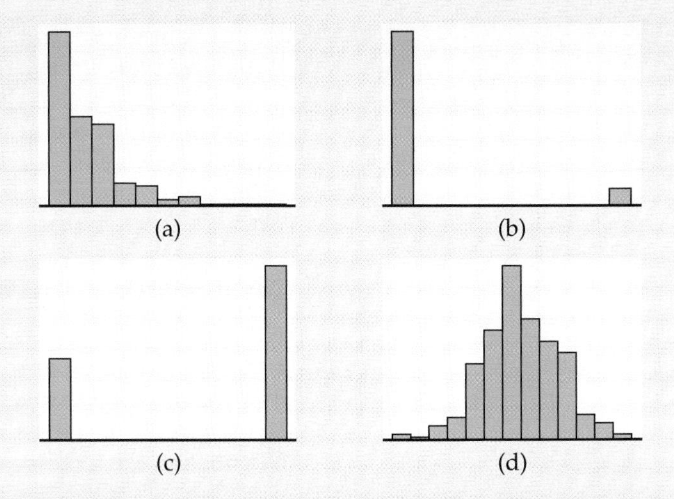

FIGURE 3.15 Match each histogram with its variable, for Exercise 3.39.

3.2 Describing Distributions with Numbers

We can begin describing distributions with graphs, but numerical summaries make comparisons between distributions more specific. A brief description of a distribution should include its *shape* and numbers describing its *center* and *spread.* We describe the shape of a distribution based on inspection of a histogram or a stemplot.

Now we will learn specific ways to use numbers to measure the center and spread of a distribution. We can calculate these numerical measures for any quantitative variable. But to interpret measures of center and spread, and to choose among the several measures we will learn, you must think about the shape of the distribution and the meaning of the data. The numbers, like graphs, are aids to understanding, not "the answer" in themselves.

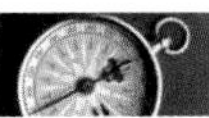

EXAMPLE 3.21

The distribution of business start times. An entrepreneur faces many bureaucratic and legal hurdles when starting a new business. The World Bank collects information about starting businesses throughout the world. They have determined the time, in days, to complete all of the procedures required to start a business.[22] Data for 195 countries are included in the data file (TimeToStart195). For this section we will examine data for a sample of 24 of these countries. Here are the data:

23	4	29	44	47	24	40	23	23	44	33	27
60	46	61	11	23	62	31	44	77	14	65	42

```
0 | 4
1 | 14
2 | 3333479
3 | 13
4 | 0244467
5 |
6 | 0125
7 | 7
```

FIGURE 3.16 Stemplot for sample of 24 business start times, for Example 3.21.

The stemplot in Figure 3.16 shows us the *shape, center,* and *spread* of the business start times. The stems are tens of days and the leaves are days. As is often the case when there are few observations, the shape of the distribution is irregular. There are peaks in the 20s and the 40s. The values range from 4 to 77 days with a center somewhere in the middle of these two extremes. There do not appear to be any outliers.

Measuring center: the mean

Numerical description of a distribution begins with a measure of its center or average. The two common measures of center are the *mean* and the *median.* The mean is the "average value" and the median is the "middle value." These are two different ideas for "center," and the two measures behave differently. We need precise recipes for the mean and the median.

THE MEAN $\bar{x}$

To find the **mean** $\bar{x}$ of a set of observations, add their values and divide by the number of observations. If the n observations are $x_1, x_2, \ldots, x_n$, their mean is

$$\bar{x} = \frac{x_1 + x_2 + \cdots + x_n}{n}$$

or, in more compact notation,

$$\bar{x} = \frac{1}{n}\sum x_i$$

The $\sum$ (capital Greek sigma) in the formula for the mean is short for "add them all up." The bar over the x indicates the mean of all the x-values. Pronounce the mean $\bar{x}$ as "x-bar." This notation is so common that writers who are discussing data use $\bar{x}$, $\bar{y}$, etc. without additional explanation. The subscripts on the observations x_i are just a way of keeping the n observations separate. They do not necessarily indicate order or any other special facts about the data.

EXAMPLE 3.22

Mean time to start a business. The mean time to start a business for the 24 countries in our data set is

$$\bar{x} = \frac{x_1 + x_2 + \cdots + x_n}{n}$$
$$= \frac{23 + 4 + \cdots + 42}{24}$$
$$= \frac{897}{24} = 37.375$$

The mean time to start a business is 37.4 days. Note that we have rounded the answer. Our goal is using the mean to describe the center of a distribution; it is not to demonstrate that we can compute with great accuracy. The additional digits do not provide any additional useful information. In fact, they distract our attention from the important digits that are meaningful. Do you think it would be better to report the mean as 37 days?

USE YOUR KNOWLEDGE

3.40 Include the outlier. The complete business start time data set with 195 countries has a few with very large start times. To construct the data set for Example 3.21, a random sample of 25 countries was selected. This sample included the South American country of Suriname, where the start time is 694 days. This country was deleted for Example 3.21. Reconstruct the original random sample by including Suriname. Show that the mean has increased to 64 days. (This is a rounded number. You should report the mean with two digits after the decimal.) DATA FILE TIMETOSTART25

3.41 Find the mean. Here are the scores on the first exam in an introductory statistics course for 10 students:

80 73 92 85 75 98 93 55 80 90

Find the mean first-exam score for these students. DATA FILE STATCOURSE

Exercise 3.40 illustrates an important weakness of the mean as a measure of center: *the mean is sensitive to the influence of a few extreme observations.* These may be outliers, but a skewed distribution that has no outliers will also pull the mean toward its long tail. Because the mean cannot resist the influence of extreme observations, we say that it is not a **resistant measure** of center.

resistant measure

A measure that is resistant does more than limit the influence of outliers. Its value does not respond strongly to changes in a few observations, no matter how large those changes may be. The mean fails this requirement because we can make the mean as large as we wish by making a large enough increase in just one observation. A resistant measure is sometimes called a **robust** measure.

robust

Measuring center: the median

We used the midpoint of a distribution as an informal measure of center in Section 3.1. The *median* is the formal version of the midpoint, with a specific rule for calculation.

THE MEDIAN M

The **median M** is the midpoint of a distribution. Half the observations are smaller than the median, and the other half are larger than the median. Here is a rule for finding the median:

1. Arrange all observations in order of size, from smallest to largest.
2. If the number of observations n is odd, the median M is the center observation in the ordered list. Find the location of the median by counting $(n + 1)/2$ observations up from the bottom of the list.
3. If the number of observations n is even, the median M is the mean of the two center observations in the ordered list. The location of the median is again $(n + 1)/2$ from the bottom of the list.

Note that the formula $(n + 1)/2$ does *not* give the median, just the location of the median in the ordered list. Medians require little arithmetic, so they are easy to find by hand for small sets of data. Arranging even a moderate number of observations in order is tedious, however, so that finding the median by hand for larger sets of data is unpleasant. Even simple calculators have an $\bar{x}$ button, but you will need computer software or a graphing calculator to automate finding the median.

EXAMPLE 3.23

Median time to start a business. To find the median time to start a business for our 24 countries, we first arrange the data in order from smallest to largest.

4	11	14	23	23	23	23	24	27	29	31	33
40	42	44	44	44	46	47	60	61	62	65	77

The count of observations $n = 24$ is even. The median, then, is the average of the two center observations in the ordered list. To find the location of the center observations, we first compute

$$\text{location of } M = \frac{n+1}{2} = \frac{25}{2} = 12.5$$

Therefore, the center observations are the 12th and 13th observations in the ordered list. The median is

$$M = \frac{33 + 40}{2} = 36.5$$

Note that you can use the stemplot directly to compute the median. In the stemplot the cases are already ordered, and you simply need to count from the top or the bottom to the desired location.

USE YOUR KNOWLEDGE

3.42 Include the outlier. Add Suriname, where the start time is 694 days, to the data set in Example 3.23 and show that the median is 40 days. Note that with this case included, the sample size is now 25, and the median is the 13th observation in the ordered list. Write out the ordered list and circle the outlier. Describe the effect of the outlier on the median for this set of data. TIMETOSTART25

3.43 Calls to a customer service center. The service times for 80 calls to a customer service center are given in Table 3.2 (page 93). Use these data to compute the median service time. CALLCENTER80

3.44 Find the median. Here are the scores on the first exam in an introductory statistics course for 10 students:

80 73 92 85 75 98 93 55 80 90

Find the median first-exam score for these students. STATCOURSE

Mean versus median

Exercises 3.40 and 3.42 illustrate an important difference between the mean and the median. Suriname is an outlier. It pulls the mean time to start a business up from 37 days to 64 days. The increase in the median is a lot less, from 36 days to 40 days.

The median is more *resistant* than the mean. If the largest starting time in the data set were 1200 days, the median for all 25 countries would still be 40 days. The largest observation just counts as one observation above the center, no matter how far above the center it lies. The mean uses the actual value of each observation and so will chase a single large observation upward.

The best way to compare the response of the mean and median to extreme observations is to use an interactive applet that allows you to place points on a line and then drag them with your computer's mouse. Exercises 3.71 to 3.73 use the *Mean and Median* applet for this book to compare mean and median.

The median and mean are the most common measures of the center of a distribution. The mean and median of a symmetric distribution are close together. If the distribution is exactly symmetric, the mean and median are exactly the same. In a skewed distribution, the mean is farther out in the long tail than is the median.

Let's examine endowments. The endowment for a college or university is money set aside and invested. The income from the endowment is usually used to support various programs. The distribution of the sizes of the endowments of colleges and universities is strongly skewed to the right. Most institutions have modest endowments, but a few are very wealthy. The median endowment of colleges and universities in a recent year was \$70 million—but the mean endowment was over \$320 million. The few wealthy institutions pulled the mean up but did not affect the median. *Don't confuse the "average" value of a variable (the mean) with its "typical" value, which we might describe by the median.*

We can now give a better answer to the question of how to deal with outliers in data. First, look at the data to identify outliers and investigate their causes. You can then correct outliers if they are wrongly recorded, delete them for good reason, or otherwise give them individual attention. The outlier in Example 3.19 (page 97) can be dropped from the data once we discover that it is an error. If you have no clear reason to drop outliers, you may want to use resistant methods, so that outliers have little influence over your conclusions. The choice is often a matter for judgment.

Measuring spread: the quartiles

A measure of center alone can be misleading. Two nations with the same median family income are very different if one has extremes of wealth and poverty and the other has little variation among families. A drug with the correct mean concentration of active ingredient is dangerous if some batches are much too high and others much too low.

We are interested in the *spread* or *variability* of incomes and drug potencies as well as their centers. The simplest useful numerical description of a distribution consists of both a measure of center and a measure of spread.

We can describe the spread or variability of a distribution by giving several percentiles. The median divides the data in two; half of the observations are above the median and half are below the median. We could call the median the 50th percentile. The upper **quartile** is the median of the upper half of the data. Similarly, the lower quartile is the median of the lower half of the data. With the median, the quartiles divide the data into four equal parts; 25% of the data are in each part.

quartile

percentile

We can do a similar calculation for any percent. The ***p*th percentile** of a distribution is the value that has p percent of the observations fall at or below it. To calculate a percentile, arrange the observations in increasing order and count up the required percent from the bottom of the list.

Our definition of percentiles is a bit inexact because there is not always a value with exactly p percent of the data at or below it. We will be content to take the nearest observation for most percentiles, but the quartiles are important enough to require an exact rule.

THE QUARTILES Q_1 AND Q_3

To calculate the quartiles:

1. Arrange the observations in increasing order and locate the median M in the ordered list of observations.
2. The **first quartile Q_1** is the median of the observations whose positions in the ordered list are to the left of the location of the overall median.
3. The **third quartile Q_3** is the median of the observations whose positions in the ordered list are to the right of the location of the overall median.

Here is an example.

EXAMPLE 3.24

TIMETOSTART24

Finding the quartiles. Here is the ordered list of the times to start a business in our sample of 24 countries:

4	11	14	23	23	23	23	24	27	29	31	33
40	42	44	44	44	46	47	60	61	62	65	77

The count of observations $n = 24$ is even, so the median is at position $(24 + 1)/2 = 12.5$, that is, between the 12th and the 13th observation in the ordered list. There are 12 cases above this position and 12 below it. The first quartile is the median of the first 12 observations, and the third quartile is the median of the last 12 observations. Check that $Q_1 = 23$ and $Q_3 = 46.5$.

Notice that the quartiles are resistant. For example, Q_3 would have the same value if the highest start time was 770 days rather than 77 days.

There are slight differences in the methods used by software to compute percentiles. However, the results will generally be quite similar, except in cases where the sample sizes are very small.

Be careful when several observations take the same numerical value. Write down all the observations and apply the rules just as if they all had distinct values.

USE YOUR KNOWLEDGE

3.45 Find the quartiles. Here are the scores on the first exam in an introductory statistics course for 10 students:

80 73 92 85 75 98 93 55 80 90

Find the quartiles for these first-exam scores.  STATCOURSE

EXAMPLE 3.25

Results from software. Statistical software often provides several numerical measures in response to a single command. Figure 3.17 displays such output from Minitab, JMP, and SPSS software for the data on the time to start a business. Examine the outputs carefully. Notice that they give different numbers of significant digits for some of these numerical summaries. Which output do you prefer?

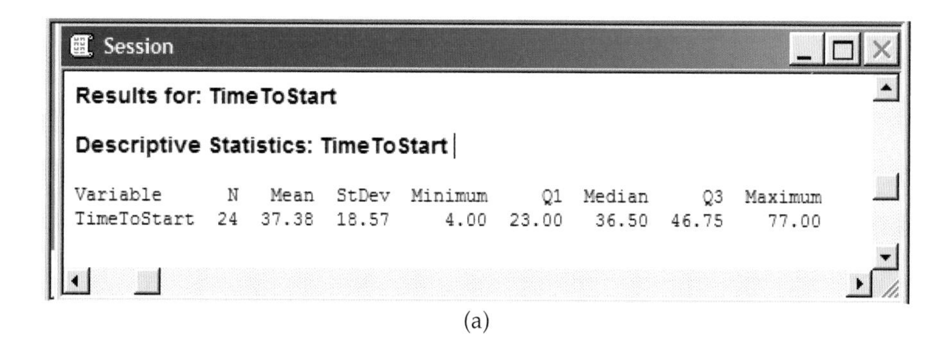

Session

Results for: TimeToStart

Descriptive Statistics: TimeToStart

Variable	N	Mean	StDev	Minimum	Q1	Median	Q3	Maximum
TimeToStart	24	37.38	18.57	4.00	23.00	36.50	46.75	77.00

(a)

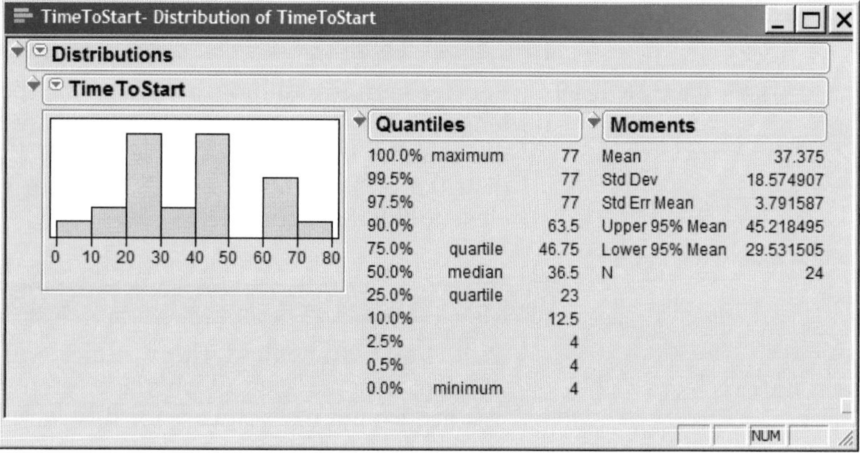

TimeToStart- Distribution of TimeToStart

Distributions

TimeToStart

Quantiles

100.0%	maximum	77
99.5%		77
97.5%		77
90.0%		63.5
75.0%	quartile	46.75
50.0%	median	36.5
25.0%	quartile	23
10.0%		12.5
2.5%		4
0.5%		4
0.0%	minimum	4

Moments

Mean	37.375
Std Dev	18.574907
Std Err Mean	3.791587
Upper 95% Mean	45.218495
Lower 95% Mean	29.531505
N	24

NUM

(b)

FIGURE 3.17 Descriptive statistics from (a) Minitab, (b) JMP, and (c) SPSS for the time to start a business, for Example 3.25. (*Continued on next page*)

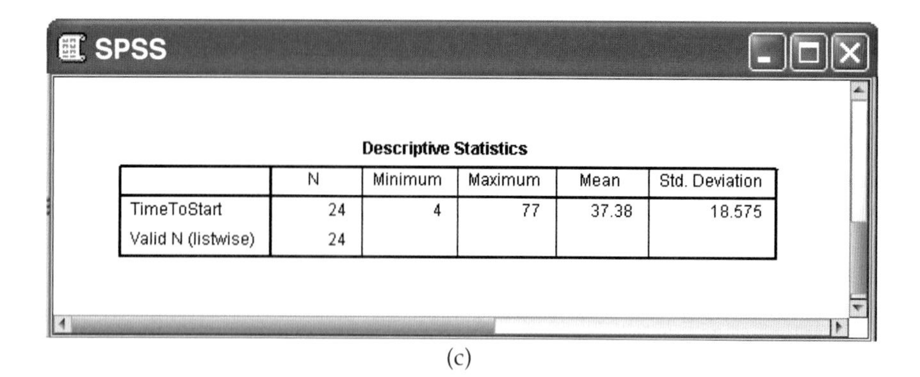
SPSS

Descriptive Statistics

	N	Minimum	Maximum	Mean	Std. Deviation
TimeToStart	24	4	77	37.38	18.575
Valid N (listwise)	24				

(c)

FIGURE 3.17 *Continued*

There are several rules for calculating quartiles, which often give slightly different values. The differences are always small. For describing data, just report the values that your software gives.

The five-number summary and boxplots

In Section 3.1, we used the smallest and largest observations to indicate the spread of a distribution. These single observations tell us little about the distribution as a whole, but they give information about the tails of the distribution that is missing if we know only Q_1, M, and Q_3. To get a quick summary of both center and spread, use all five numbers.

THE FIVE-NUMBER SUMMARY

The **five-number summary** of a set of observations consists of the smallest observation, the first quartile, the median, the third quartile, and the largest observation, written in order from smallest to largest. In symbols, the five-number summary is

$$\text{Minimum} \quad Q_1 \quad M \quad Q_3 \quad \text{Maximum}$$

EXAMPLE 3.26

Service center call lengths. Table 3.2 (page 93) gives the service center call lengths for the sample of 80 calls that we discussed in Example 3.14. The five-number summary for these data is 1.0, 54.5, 103.5, 200, and 2631. The distribution is highly skewed. The mean is 197 seconds, a value that is very close to the third quartile.

USE YOUR KNOWLEDGE

3.46 Verify the calculations. Refer to the five-number summary and the mean for service call lengths given in Example 3.26. Verify these results. Do not use software for this exercise and be sure to show all your work. CALLCENTER80

3.47 Find the five-number summary. Here are the scores on the first exam in an introductory statistics course for 10 students:

80 73 92 85 75 98 93 55 80 90

Find the five-number summary for these first-exam scores. STATCOURSE

The five-number summary leads to another visual representation of a distribution, the *boxplot.*

BOXPLOT

A **boxplot** is a graph of the five-number summary.

- A central box spans the quartiles Q_1 and Q_3.
- A line in the box marks the median M.
- Lines extend from the box out to the smallest and largest observations, or to a cutoff for suspected outliers.

When you look at a boxplot, first locate the median, which marks the center of the distribution. Then look at the spread. The quartiles show the spread of the middle half of the data, and the extremes (the smallest and largest observations) show the spread of the entire data set.

EXAMPLE 3.27

Service center call lengths. Table 3.2 (page 93) gives the call lengths for our sample of 80 service calls from our collection of 31,492 calls. In Exercise 3.46 you verified that the five-number summary for these data is 1.0, 54.5, 103.5, 200, and 2631. The boxplot is displayed in Figure 3.18. The skewness of the distribution is the major feature that we see in this plot. Note that the mean is marked with a "+" and appears very close to the third quartile at the upper edge of the box.

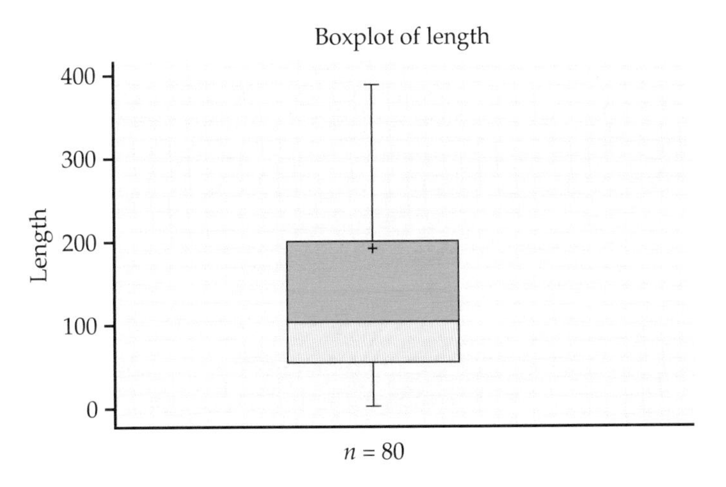

FIGURE 3.18 Modified boxplot for the sample of 80 service center call lengths, for Example 3.27.

USE YOUR KNOWLEDGE

3.48 Make a boxplot. Here are the scores on the first exam in an introductory statistics course for 10 students:

80 73 92 85 75 98 93 55 80 90

Make a boxplot for these first-exam scores. DATA FILE STATCOURSE

The 1.5 × *IQR* rule for suspected outliers

Look again at the boxplot for the 80 service center call lengths given in Figure 3.18 and the display of the data in Table 3.2 (page 93). There is a clear outlier, a call lasting 2631 seconds, more than twice the length of any other call.

How can we describe the spread of this distribution? The smallest and largest observations are extremes that do not describe the spread of the majority of the data. The distance between the quartiles (the range of the center half of the data) is a more resistant measure of spread. This distance is called the *interquartile range*.

THE INTERQUARTILE RANGE *IQR*

The **interquartile range *IQR*** is the distance between the first and third quartiles,

$$IQR = Q_3 - Q_1$$

For our data on service call lengths, $IQR = 200 - 54.5 = 145.5$. The quartiles and the *IQR* are not affected by changes in either tail of the distribution. They are therefore resistant, because changes in a few data points have no further effect once these points move outside the quartiles.

However, *no single numerical measure of spread, such as IQR, is very useful for describing skewed distributions.* The two sides of a skewed distribution have different spreads, so one number can't summarize them. We can often detect skewness from the five-number summary by comparing how far the first quartile and the minimum are from the median (left tail) with how far the third quartile and the maximum are from the median (right tail). The interquartile range is mainly used as the basis for a rule of thumb for identifying suspected outliers.

THE 1.5 × *IQR* RULE FOR OUTLIERS

Call an observation a suspected outlier if it falls more than $1.5 \times IQR$ above the third quartile or below the first quartile.

EXAMPLE 3.28

Outliers for call length data. For the call length data in Table 3.2,

$$1.5 \times IQR = 1.5 \times 145.5 = 218.25$$

Any values below $54.5 - 218.25 = -163.75$ or above $200 + 218.25 = 418.25$ are flagged as possible outliers. There are no low outliers, but the 8 longest calls are flagged as possible high outliers. Their lengths are

438 465 479 700 700 951 1148 2631

Statistical software often uses the $1.5 \times IQR$ rule. For example, the boxplot in Figure 3.18 plots the 8 call lengths identified by the $1.5 \times IQR$ rule as separate points. A plot such as this where suspected outliers are identified individually is called a **modified boxplot.**

modified boxplot

The distribution of call lengths is very strongly skewed. We may well decide that only the longest call is truly an outlier in the sense of deviating from the overall pattern of the distribution. The other 7 calls are just part of the long right tail. The $1.5 \times IQR$ rule does not remove the need to look at the distribution and use judgment. It is useful mainly to call our attention to unusual observations.

USE YOUR KNOWLEDGE

3.49 Find the *IQR*. Here are the scores on the first exam in an introductory statistics course for 10 students:

80 73 92 85 75 98 93 55 80 90

Find the interquartile range and use the $1.5 \times IQR$ rule to check for outliers. How low would the lowest score need to be for it to be an outlier according to this rule? DATA FILE STATCOURSE

The boxplot in Figure 3.18 tells us much more about the distribution of call lengths than the five-number summary or other numerical measures. The routine methods of statistics compute numerical measures and draw conclusions based on their values. These methods are very useful, and we will study them carefully in later chapters. But they cannot be applied blindly, by feeding data to a computer program, because *statistical measures and methods based on them are generally meaningful only for distributions of sufficiently regular shape.*

This principle will become clearer as we progress, but it is good to be aware at the beginning that quickly resorting to fancy calculations is the mark of a statistical amateur. Look, think, and choose your calculations selectively.

Measuring spread: the standard deviation

The five-number summary is not the most common numerical description of a distribution. That distinction belongs to the combination of the mean to measure center and the *standard deviation* to measure spread. The standard deviation measures spread by looking at how far the observations are from their mean.

THE STANDARD DEVIATION s

The **variance s^2** of a set of observations is the average of the squares of the deviations of the observations from their mean. In symbols, the variance of n observations $x_1, x_2, \ldots, x_n$ is

$$s^2 = \frac{(x_1 - \overline{x})^2 + (x_2 - \overline{x})^2 + \cdots + (x_n - \overline{x})^2}{n-1}$$

or, in more compact notation,

$$s^2 = \frac{1}{n-1}\sum(x_i - \overline{x})^2$$

The **standard deviation s** is the square root of the variance s^2:

$$s = \sqrt{\frac{1}{n-1}\sum(x_i - \overline{x})^2}$$

The idea behind the variance and the standard deviation as measures of spread is as follows: The deviations $x_i - \overline{x}$ display the spread of the values x_i about their mean $\overline{x}$. Some of these deviations will be positive and some negative because some of the observations fall on each side of the mean. In fact, *the sum of the deviations of the observations from their mean will always be zero.* Squaring the deviations makes them all positive, so that observations far from the mean in either direction have large positive squared deviations.

The variance is the average squared deviation. Therefore, s^2 and s will be large if the observations are widely spread about their mean, and small if the observations are all close to the mean.

EXAMPLE 3.29

Metabolic rate. A person's metabolic rate is the rate at which the body consumes energy. Metabolic rate is important in studies of weight gain, dieting, and exercise. Here are the metabolic rates of 7 men who took part in a study of dieting. (The units are calories per 24 hours. These are the same calories used to describe the energy content of foods.)

1792 1666 1362 1614 1460 1867 1439

Enter these data into your calculator or software and verify that

$$\overline{x} = 1600 \text{ calories} \qquad s = 189.24 \text{ calories}$$

Figure 3.19 plots these data as dots on the calorie scale, with their mean marked by an asterisk (*). The arrows mark two of the deviations from the mean. If you were calculating s by hand, you would find the first deviation as

$$x_1 - \bar{x} = 1792 - 1600 = 192$$

Exercise 3.70 asks you to calculate the seven deviations from Example 3.29, square them, and find s^2 and s directly from the deviations. Working one or two short examples by hand helps you understand how the standard deviation is obtained. In practice you will use either software or a calculator that will find s. The software outputs in Figure 3.17 give the standard deviation for the data on the time to start a business.

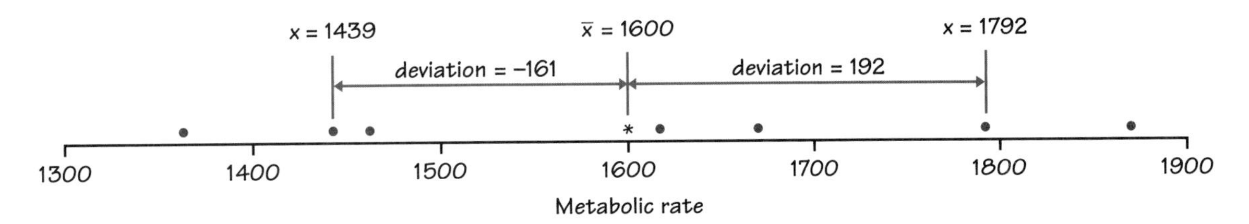

FIGURE 3.19 Metabolic rates for seven men, with the mean (*) and the deviations of two observations from the mean, for Example 3.29.

USE YOUR KNOWLEDGE

3.50 Find the variance and the standard deviation. Here are the scores on the first exam in an introductory statistics course for 10 students:

80 73 92 85 75 98 93 55 80 90

Find the variance and the standard deviation for these first-exam scores. STATCOURSE

The idea of the variance is straightforward: it is the average of the squares of the deviations of the observations from their mean. The details we have just presented, however, raise some questions.

Why do we square the deviations?

- First, the sum of the squared deviations of any set of observations from their mean is the smallest that the sum of squared deviations from any number can possibly be. This is not true of the unsquared distances. So squared deviations point to the mean as center in a way that distances do not.
- Second, the standard deviation turns out to be the natural measure of spread for a particularly important class of symmetric unimodal distributions, the *Normal distributions*. We will meet the Normal distributions in the next section.

Why do we emphasize the standard deviation rather than the variance?

- One reason is that s, not s^2, is the natural measure of spread for Normal distributions, which are introduced in the next section.
- There is also a more general reason to prefer s to s^2. Because the variance involves squaring the deviations, it does not have the same unit of measurement as the original observations. The variance of the metabolic rates, for example, is measured in squared calories. Taking the square root gives us a description of the spread of a distribution in the original measurement units.

Why do we average by dividing by $n - 1$ rather than n in calculating the variance?

- Because the sum of the deviations is always zero, the last deviation can be found once we know the other $n - 1$. So we are not averaging n unrelated numbers. Only $n - 1$ of the squared deviations can vary freely, and we average by dividing the total by $n - 1$.

degrees of freedom

- The number $n - 1$ is called the **degrees of freedom** of the variance or standard deviation. Many calculators offer a choice between dividing by n and dividing by $n - 1$, so be sure to use $n - 1$.

Properties of the standard deviation

Here are the basic properties of the standard deviation s as a measure of spread.

PROPERTIES OF THE STANDARD DEVIATION

- s measures spread about the mean and should be used only when the mean is chosen as the measure of center.
- $s = 0$ only when there is *no spread*. This happens only when all observations have the same value. Otherwise, $s > 0$. As the observations become more spread out about their mean, s gets larger.
- s, like the mean $\bar{x}$, is not resistant. A few outliers can make s very large.

USE YOUR KNOWLEDGE

3.51 A standard deviation of zero. Construct a data set with 5 cases that has a variable with $s = 0$.

The use of squared deviations renders s even more sensitive than $\bar{x}$ to a few extreme observations. For example, when we add Suriname to our sample of 24 countries for the analysis of the time to start a business (Exercises 3.40 and 3.42), we increase the standard deviation from 18.6 to 132.6! Distributions with outliers and strongly skewed distributions have large standard deviations. The number s does not give much helpful information about such distributions.

USE YOUR KNOWLEDGE

3.52 Effect of an outlier on the *IQR*. Find the *IQR* for the time to start a business with and without Suriname. What do you conclude about the sensitivity of this measure of spread to the inclusion of an outlier? DATA FILE TIMETOSTART24 DATA FILE TIMETOSTART25

Choosing measures of center and spread

How do we choose between the five-number summary and $\bar{x}$ and s to describe the center and spread of a distribution? Because the two sides of a strongly skewed distribution have different spreads, no single number such as s describes the spread well. The five-number summary, with its two quartiles and two extremes, does a better job.

CHOOSING A SUMMARY

The five-number summary is usually better than the mean and standard deviation for describing a skewed distribution or a distribution with strong outliers. Use $\bar{x}$ and s only for reasonably symmetric distributions that are free of outliers.

Remember that a graph gives the best overall picture of a distribution. Numerical measures of center and spread report specific facts about a distribution, but they do not describe its shape. Numerical summaries do not disclose the presence of multiple modes or gaps, for example. Always plot your data.

Changing the unit of measurement

The same variable can be recorded in different units of measurement. Americans commonly record distances in miles and temperatures in degrees Fahrenheit, while the rest of the world measures distances in kilometers and temperatures in degrees Celsius. Fortunately, it is easy to convert numerical descriptions of a distribution from one unit of measurement to another. This is true because a change in the measurement unit is a *linear transformation* of the measurements.

LINEAR TRANSFORMATIONS

A **linear transformation** changes the original variable x into the new variable x_{new} given by an equation of the form

$$x_{\text{new}} = a + bx$$

Adding the constant a shifts all values of x upward or downward by the same amount. In particular, such a shift changes the origin (zero point) of the variable. Multiplying by the positive constant b changes the size of the unit of measurement.

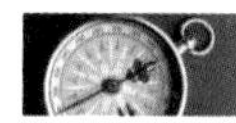

EXAMPLE 3.30

Change the units

(a) If a distance x is measured in kilometers, the same distance in miles is

$$x_{new} = 0.62x$$

For example, a 10-kilometer race covers 6.2 miles. This transformation changes the units without changing the origin—a distance of 0 kilometers is the same as a distance of 0 miles.

(b) A temperature x measured in degrees Fahrenheit must be reexpressed in degrees Celsius to be easily understood by the rest of the world. The transformation is

$$x_{new} = \frac{5}{9}(x - 32) = -\frac{160}{9} + \frac{5}{9}x$$

Thus, the high of 95°F on a hot American summer day translates into 35°C. In this case

$$a = -\frac{160}{9} \quad \text{and} \quad b = \frac{5}{9}$$

This linear transformation changes both the unit size and the origin of the measurements. The origin in the Celsius scale (0°C, the temperature at which water freezes) is 32° in the Fahrenheit scale.

Linear transformations do not change the shape of a distribution. If measurements on a variable x have a right-skewed distribution, any new variable x_{new} obtained by a linear transformation $x_{new} = a + bx$ (for $b > 0$) will also have a right-skewed distribution. If the distribution of x is symmetric and unimodal, the distribution of x_{new} remains symmetric and unimodal.

Although a linear transformation preserves the basic shape of a distribution, the center and spread will change. Because linear changes of measurement scale are common, we must be aware of their effect on numerical descriptive measures of center and spread. Fortunately, the changes follow a simple pattern.

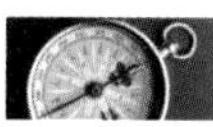

EXAMPLE 3.31

Use scores to find the points In an introductory statistics course, homework counts for 300 points out of a total of 1000 possible points for all course requirements. During the semester there were 12 homework assignments and each was given a grade on a scale of 0 to 100. The maximum total score for the 12 homework assignments is therefore 1200.

To convert the homework scores to final grade points, we need to convert the scale of 0 to 1200 to a scale of 0 to 300. We do this by multiplying the homework scores by 300/1200. In other words, we divide the homework scores by 4. Here are the homework scores and the corresponding final grade points for 5 students:

Student	1	2	3	4	5
Score	1056	1080	900	1164	1020
Points	264	270	225	291	255

These two sets of numbers measure the same performance on homework for the course. Since we obtained the points by dividing the scores by 4, the mean of the points will be the mean of the scores divided by 4. Similarly, the standard deviation of points will be the standard deviation of the scores divided by 4.

USE YOUR KNOWLEDGE

3.53 Calculate the points for a student. Use the setting of Example 3.31 to find the points for a student whose score is 950.

Here is a summary of the rules for linear transformations.

EFFECT OF A LINEAR TRANSFORMATION

To see the effect of a linear transformation on measures of center and spread, apply these rules:

- Multiplying each observation by a positive number b multiplies both measures of center (mean and median) and measures of spread (interquartile range and standard deviation) by b.
- Adding the same number a (either positive or negative) to each observation adds a to measures of center and to quartiles and other percentiles but does not change measures of spread.

In Example 3.31, when we converted from score to points, we described the transformation as dividing by 4. The multiplication part of the summary of the effect of a linear transformation applies to this case, because division by 4 is the same as multiplication by 0.25. Similarly, the second part of the summary applies to subtraction as well as addition, because subtraction is simply the addition of a negative number.

The measures of spread IQR and s do not change when we add the same number a to all the observations, because adding a constant changes the location of the distribution but leaves the spread unaltered. You can find the effect of a linear transformation $x_{new} = a + bx$ by combining these rules. For example, if x has mean $\bar{x}$, the transformed variable x_{new} has mean $a + b\bar{x}$.

SECTION 3.2 SUMMARY

A numerical summary of a distribution should report its **center** and its **spread** or **variability.**

The **mean** $\bar{x}$ and the **median** M describe the center of a distribution in different ways. The mean is the arithmetic average of the observations, and the median is their midpoint.

When you use the median to describe the center of the distribution, describe its spread by giving the **quartiles.** The **first quartile** Q_1 has one-fourth of the observations below it, and the **third quartile** Q_3 has three-fourths of the observations below it.

The **interquartile range** is the difference between the quartiles. It is the spread of the center half of the data. The **1.5 × *IQR* rule** flags observations more than 1.5 × IQR beyond the quartiles as possible outliers.

The **five-number summary** consisting of the median, the quartiles, and the smallest and largest individual observations provides a quick overall description of a distribution. The median describes the center, and the quartiles and extremes show the spread.

Boxplots based on the five-number summary are useful for examining distributions. The box spans the quartiles and shows the spread of the central half of the distribution. The median is marked within the box. Lines extend from the box to the extremes and show the full spread of the data. In a **modified boxplot,** points identified by the 1.5 × IQR rule are plotted individually.

The **variance** s^2 and especially its square root, the **standard deviation** s, are common measures of spread about the mean as center. The standard deviation s is zero when there is no spread and gets larger as the spread increases.

A **resistant measure** of any aspect of a distribution is relatively unaffected by changes in the numerical value of a small proportion of the total number of observations, no matter how large these changes are. The median and quartiles are resistant, but the mean and the standard deviation are not.

The mean and standard deviation are good descriptions for symmetric distributions without outliers. They are most useful for the Normal distributions introduced in the next section. The five-number summary is a better exploratory summary for skewed distributions.

Numerical measures of particular aspects of a distribution, such as center and spread, do not report the entire shape of most distributions. In some cases, particularly distributions with multiple peaks and gaps, these measures may not be very informative.

Linear transformations have the form $x_{new} = a + bx$. A linear transformation changes the origin if $a \neq 0$ and changes the size of the unit of measurement if $b > 0$. Linear transformations do not change the overall shape of a distribution. A linear transformation multiplies a measure of spread by b and changes a percentile or measure of center m into $a + bm$.

SECTION 3.2 EXERCISES

For Exercises 3.40 and 3.41, see page 107; for Exercises 3.42 to 3.44, see page 108; for Exercise 3.45, see page 111; for Exercises 3.46 and 3.47, see pages 112–113; for Exercise 3.48, see page 114; for Exercise 3.49, see page 115; for Exercise 3.50, see page 117; for Exercise 3.51, see page 118; for Exercise 3.52, see page 119; and for Exercise 3.53, see page 121.

3.54. CHALLENGE **Hurricanes and losses.** A discussion of extreme weather says, "In most states, hurricanes occur infrequently. Yet, when a hurricane hits, the losses can be catastrophic. Average annual losses are not a meaningful measure of damage from rare but potentially catastrophic events."[23] Why is this true?

3.55. The value of brands. A brand is a symbol or images that are associated with a company. An effective brand identifies the company and its products. Using a variety of measures, dollar values for brands can be calculated.[24] The most valuable brand is Coca-Cola, with a value of $71,861,000. Coke is followed by IBM at $69,905,000, Microsoft at $59,087,000, Google at $55,317,000, and GE at $42,808,000. For this exercise you will use the brand values, reported in millions of dollars, for the top 100 brands. DATA FILE BRANDS

(a) Graphically display the distribution of the values of these brands.

(b) Use numerical measures to summarize the distribution.

(c) Write a short paragraph discussing the dollar values of the top 100 brands. Include the results of your analysis.

3.56. Alcohol content of beer. Brewing beer involves a variety of steps that can affect the alcohol content. We have the percent alcohol for 86 domestic brands of beer.[25] DATA FILE BEER

(a) Use graphical and numerical summaries of your choice to describe these data. Give reasons for your choices.

(b) The data set contains an outlier. Explain why this particular beer is unusual and how its outlier status is related to how it is marketed.

3.57. An outlier for alcohol content of beer. Refer to the previous exercise. DATA FILE BEER

(a) Calculate the mean with and without the outlier. Do the same for the median. Explain how these statistics change when the outlier is excluded.

(b) Calculate the standard deviation with and without the outlier. Do the same for the quartiles. Explain how these statistics change when the outlier is excluded.

(c) Write a short paragraph summarizing what you have learned in this exercise.

3.58. Calories in beer. Refer to the previous two exercises. The data set also gives the calories per 12 ounces of beverage. DATA FILE BEER

(a) Analyze the data and summarize the distribution of calories for these 86 brands of beer.

(b) In the previous exercise you identified one brand of beer as an outlier. To what extent is this brand an outlier in the distribution of calories? Explain your answer.

(c) The distribution of calories suggests that there may be two groups of beers, which might be marketed differently. Examine the data carefully and describe the characteristics of the two groups.

3.59. Potatoes. A high-quality product is one that is consistent and has very little variability in its characteristics. Controlling variability can be more difficult with agricultural products than with those that are manufactured. The following table gives the weights, in ounces, of the 25 potatoes sold in a 10-pound bag. DATA FILE POTATOES

7.8	7.9	8.2	7.3	6.7	7.9	7.9	7.9	7.6	7.8	7.0	4.7	7.6
6.3	4.7	4.7	4.7	6.3	6.0	5.3	4.3	7.9	5.2	6.0	3.7	

(a) Summarize the data graphically and numerically. Give reasons for the methods you chose to use in your summaries.

(b) Do you think that your numerical summaries do an effective job of describing these data? Why or why not?

(c) There appear to be two distinct clusters of weights for these potatoes. Divide the sample into two subsamples based on the clustering. Give the mean and standard deviation for each subsample. Do you think that this way of summarizing these data is better than a numerical summary that uses all the data as a single sample? Give a reason for your answer.

3.60. Longleaf pine trees. The Wade Tract in Thomas County, Georgia, is an old-growth forest of longleaf pine trees (*Pinus palustris*) that has survived in a relatively undisturbed state since before the settlement of the area by Europeans. A study collected data about 584 of these trees.[26] One of the variables measured was the diameter at breast height (DBH). This is the diameter of the tree at 4.5 feet, and the units are centimeters (cm). Only trees with DBH greater than 1.5 cm were sampled. Here are the diameters of a random sample of 40 of these trees: DATA FILE LONGLEAF

10.5	13.3	26.0	18.3	52.2	9.2	26.1	17.6	40.5	31.8
47.2	11.4	2.7	69.3	44.4	16.9	35.7	5.4	44.2	2.2
4.3	7.8	38.1	2.2	11.4	51.5	4.9	39.7	32.6	51.8
43.6	2.3	44.6	31.5	40.3	22.3	43.3	37.5	29.1	27.9

(a) Find the five-number summary for these data.

(b) Make a boxplot.

(c) Make a histogram.

(d) Write a short summary of the major features of this distribution. Do you prefer the boxplot or the histogram for these data?

3.61. Blood proteins in children from Papua New Guinea. C-reactive protein (CRP) is a substance that can be measured in the blood. Values increase substantially within 6 hours of an infection and then reach a peak in 24 to 48 hours. In adults, chronically high values have been linked to an increased risk of cardiovascular disease. In a study of apparently healthy children aged 6 to 60 months in Papua New Guinea, CRP was measured in 90 children.[27] The units are milligrams per liter of blood (mg/l). Here are the data from a random sample of 40 of these children: DATA FILE CRP

0.00	3.90	5.64	8.22	0.00	5.62	3.92	6.81	30.61	0.00
73.20	0.00	46.70	0.00	0.00	26.41	22.82	0.00	0.00	3.49
0.00	0.00	4.81	9.57	5.36	0.00	5.66	0.00	59.76	12.38
15.74	0.00	0.00	0.00	0.00	9.37	20.78	7.10	7.89	5.53

(a) Find the five-number summary for these data.

(b) Make a boxplot.

(c) Make a histogram.

(d) Write a short summary of the major features of this distribution. Do you prefer the boxplot or the histogram for these data?

3.62. CHALLENGE **Transform blood protein values.** Refer to the previous exercise. With strongly skewed distributions such as this, we frequently reduce the skewness by taking a log transformation. We have a bit of a problem here, however, because some of the data are recorded as 0.00 and the logarithm of zero is not defined. For this variable, the value 0.00 is recorded whenever the amount of CRP in the blood is below the level that the measuring instrument is capable of detecting. The usual procedure in this circumstance is to add a small number to each observation before taking the logs. Transform these data by adding 1 to each observation and then taking the logarithm. Use the questions in the previous exercise as a guide to your analysis, and prepare a summary contrasting this analysis with the one that you performed in the previous exercise. DATA FILE CRP

3.63. CHALLENGE **Vitamin A deficiency in children from Papua New Guinea.** In the Papua New Guinea study that provided the data for the previous two exercises, the researchers also measured serum retinol. A low value of this variable can be an indicator of vitamin A deficiency. Here are the data on the same sample of 40 children from this study. The units are micromoles per liter of blood (μmol/l). DATA FILE VITAMINA

1.15	1.36	0.38	0.34	0.35	0.37	1.17	0.97	0.97	0.67
0.31	0.99	0.52	0.70	0.88	0.36	0.24	1.00	1.13	0.31
1.44	0.35	0.34	1.90	1.19	0.94	0.34	0.35	0.33	0.69
0.69	1.04	0.83	1.11	1.02	0.56	0.82	1.20	0.87	0.41

Analyze these data. Use the questions in the previous two exercises as a guide.

3.64. Luck and puzzle solving. Children in a psychology study were asked to solve some puzzles and were then given feedback on their performance. They then were asked to rate how luck played a role in determining their scores.[28] This variable was recorded on a 1 to 10 scale with 1 corresponding to very lucky and 10 corresponding to very unlucky. Here are the scores for 60 children:

1	10	1	10	1	1	10	5	1	1	8	1	10	2	1
9	5	2	1	8	10	5	9	10	10	9	6	10	1	5
1	9	2	1	7	10	9	5	10	10	10	1	8	1	6
10	1	6	10	10	8	10	3	10	8	1	8	10	4	2

Use numerical and graphical methods to describe these data. Write a short report summarizing your work. DATA FILE LUCK

3.65. Median versus mean for net worth. A report on the assets of American households says that the median net worth of U.S. families is $120,300. The mean net worth of these families is $556,300.[29] What explains the difference between these two measures of center?

3.66. Carbon dioxide emissions. The data file CO2 gives carbon dioxide (CO_2) emissions per person for 134 countries. The units are metric tons per person. The distribution is strongly skewed to the right. DATA FILE CO2

(a) Give the five-number summary. Explain why this summary suggests that the distribution is right-skewed.

(b) Which countries are outliers according to the $1.5 \times IQR$ rule? Make a stemplot or histogram of the data. Do you agree with the rule's suggestions about which countries are and are not outliers?

3.67. Mean versus median. A small accounting firm pays each of its six clerks $35,000, two junior accountants $80,000 each, and the firm's owner $320,000. What is the mean salary paid at this firm? How many of the employees earn less than the mean? What is the median salary?

3.68. Be careful how you treat the zeros. In computing the median income of any group, some federal agencies omit all members of the group who had no income. Give an example to show that the reported median income of a group can go down even though the group becomes economically better off. Is this also true of the mean income?

3.69. How does the median change? The firm in Exercise 3.67 gives no raises to the clerks and junior accountants, while the owner's salary increases to $455,000. How does this change affect the mean? How does it affect the median?

3.70. Metabolic rates. Calculate the mean and standard deviation of the metabolic rates in Example 3.29 (page 116), showing each step in detail. First find the mean $\bar{x}$ by summing the 7 observations and dividing by 7. Then find each of the deviations $x_i - \bar{x}$ and their squares. Check that the deviations have sum 0. Calculate the variance as an average of the squared deviations (remember to divide by $n - 1$). Finally, obtain s as the square root of the variance. DATA FILE METABOLIC

3.71. APPLET **Mean and median for two observations.** The *Mean and Median* applet allows you to place observations on a line and see their mean and median visually. Place two observations on the line by clicking below it. Why does only one arrow appear?

3.72. APPLET **Mean and median for three observations.** In the *Mean and Median* applet, place three observations on the line by clicking below it, two close together near the center of the line and one somewhat to the right of these two.

(a) Pull the single rightmost observation out to the right. (Place the cursor on the point, hold down a mouse button, and drag the point.) How does the mean behave? How does the median behave? Explain briefly why each measure acts as it does.

(b) Now drag the rightmost point to the left as far as you can. What happens to the mean? What happens to the median as you drag this point past the other two (watch carefully)?

3.73. APPLET **Mean and median for five observations.** Place five observations on the line in the *Mean and Median* applet by clicking below it.

(a) Add one additional observation *without changing the median.* Where is your new point?

(b) Use the applet to convince yourself that when you add yet another observation (there are now seven in all), the median does not change no matter where you put the seventh point. Explain why this must be true.

3.74. Hummingbirds and flowers. Different varieties of the tropical flower *Heliconia* are fertilized by different species of hummingbirds. Over time, the lengths of the flowers and the form of the hummingbirds' beaks have evolved to match each other. Here are data on the lengths in millimeters of three varieties of these flowers on the island of Dominica:[30]

H. bihai							
47.12	46.75	46.81	47.12	46.67	47.43	46.44	46.64
48.07	48.34	48.15	50.26	50.12	46.34	46.94	48.36
***H. caribaea* red**							
41.90	42.01	41.93	43.09	41.47	41.69	39.78	40.57
39.63	42.18	40.66	37.87	39.16	37.40	38.20	38.07
38.10	37.97	38.79	38.23	38.87	37.78	38.01	
***H. caribaea* yellow**							
36.78	37.02	36.52	36.11	36.03	35.45	38.13	37.1
35.17	36.82	36.66	35.68	36.03	34.57	34.63	

Make boxplots to compare the three distributions. Report the five-number summaries along with your graph. What are the most important differences among the three varieties of flower? DATA FILE HELICONIA

3.75. Compare the three varieties of flowers. The biologists who collected the flower length data in the previous exercise compared the three *Heliconia* varieties using statistical methods based on $\bar{x}$ and s. DATA FILE HELICONIA

(a) Find $\bar{x}$ and s for each variety.

(b) Make a stemplot of each set of flower lengths. Do the distributions appear suitable for use of $\bar{x}$ and s as summaries?

3.76. Imputation. Various problems with data collection can cause some observations to be missing. Suppose a data set has 20 cases. Here are the values of the variable x for 10 of these cases: DATA FILE IMPUTATION

17	6	12	14	20	23	9	12	16	21

The values for the other 10 cases are missing. One way to deal with missing data is called **imputation.** The basic idea is that missing values are replaced, or imputed, with values that are based on an analysis of the data that are not missing. For a data set with a single variable, the usual choice of a value for imputation is the mean of the values that are not missing. The mean for this data set is 15.

(a) Verify that the mean is 15 and find the standard deviation for the 10 cases for which x is not missing.

(b) Create a new data set with 20 cases by setting the values for the 10 missing cases to 15. Compute the mean and standard deviation for this data set.

(c) Summarize what you have learned about the possible effects of this type of imputation on the mean and the standard deviation.

3.77. CHALLENGE **Shakespeare's plays.** Look at the histogram of lengths of words in Shakespeare's plays, Figure 3.13 (page 103). The heights of the bars tell us what percent of words have each length. What is the median length of words used by Shakespeare? Similarly, what are the quartiles? Give the five-number summary for Shakespeare's word lengths. DATA FILE SHAKESPEARE

3.78. CHALLENGE **Create a data set.** Create a set of 5 positive numbers (repeats allowed) that have median 10 and mean 7. What thought process did you use to create your numbers?

3.79. CHALLENGE **Create another data set.** Give an example of a small set of data for which the mean is larger than the third quartile.

3.80. CHALLENGE **Deviations from the mean sum to zero.** Use the definition of the mean $\overline{x}$ to show that the sum of the deviations $x_i - \overline{x}$ of the observations from their mean is always zero. This is one reason why the variance and standard deviation use squared deviations.

3.81. CHALLENGE **A standard deviation contest.** This is a standard deviation contest. You must choose four numbers from the whole numbers 0 to 20, with repeats allowed.

(a) Choose four numbers that have the smallest possible standard deviation.

(b) Choose four numbers that have the largest possible standard deviation.

(c) Is more than one choice possible in either (a) or (b)? Explain.

3.82. Does your software give incorrect answers? This exercise requires a calculator with a standard deviation button or statistical software on a computer. The observations

20,001 20,002 20,003

have mean $\overline{x} = 20,002$ and standard deviation $s = 1$. Adding a 0 in the center of each number, the next set becomes

200,001 200,002 200,003

The standard deviation remains $s = 1$ as more 0s are added. Use your calculator or computer to calculate the standard deviation of these numbers, adding extra 0s until you get an incorrect answer. How soon did you go wrong? This demonstrates that calculators and computers cannot handle an arbitrary number of digits correctly.

3.83. Compare three varieties of flowers. Exercise 3.74 reports data on the lengths in millimeters of flowers of three varieties of *Heliconia*. In Exercise 3.75 you found the mean and standard deviation for each variety. Starting from the $\overline{x}$- and s-values in millimeters, find the means and standard deviations in inches. (A millimeter is 1/1000 of a meter. A meter is 39.37 inches.)

3.84. CHALLENGE **Changing units from inches to centimeters.** Changing the unit of length from inches to centimeters multiplies each length by 2.54 because there are 2.54 centimeters in an inch. This change of units multiplies our usual measures of spread by 2.54. This is true of IQR and the standard deviation. What happens to the variance when we change units in this way?

3.85. Weight gain. A study of diet and weight gain deliberately overfed 16 volunteers for eight weeks. The mean increase in fat was $\overline{x} = 2.42$ kilograms and the standard deviation was $s = 1.18$ kilograms. What are $\overline{x}$ and s in pounds? (A kilogram is 2.2 pounds.)

3.3 Density Curves and Normal Distributions

We now have a kit of graphical and numerical tools for describing distributions. What is more, we have a clear strategy for exploring data on a single quantitative variable:

1. Always plot your data: make a graph, usually a stemplot or a histogram.
2. Look for the overall pattern and for striking deviations such as outliers.
3. Calculate an appropriate numerical summary to briefly describe center and spread.

Technology has expanded the set of graphs that we can choose for Step 1. It is possible, though painful, to make histograms by hand. Using software, clever algorithms can describe a distribution in a way that is not feasible by hand, by fitting a smooth curve to the data in addition to or instead of a histogram. The curves used are called **density curves.** Before we examine density curves in detail, here is an example of what software can do.

EXAMPLE 3.32

Density curves for pH and numbers of undergraduate students. Figure 3.20 illustrates the use of density curves along with histograms to describe distributions. Figure 3.20(a) shows the distribution of the acidity (pH) of rainwater, from Exercise 3.38 (page 104). That exercise illustrates how the choice of classes can change the shape of a histogram. The density curve and the software's default histogram agree that the distribution has a single peak and is approximately symmetric.

Figure 3.20(b) shows a strongly skewed distribution, numbers of undergraduate students in the states from Example 3.17 (page 96). The histogram and density curve agree on the overall shape and on the long right tail.

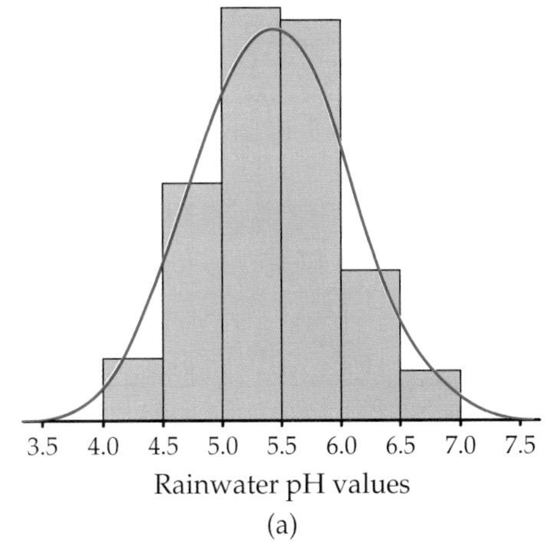

FIGURE 3.20 (a) The distribution of pH values measuring the acidity of 105 samples of rainwater, for Example 3.32. This roughly symmetric distribution is pictured with both a histogram and a density curve. (b) The distribution of the number of undergraduate students in the states, for Example 3.32. This right-skewed distribution is pictured with both a histogram and a density curve.

In general, software that draws density curves describes the data in a way that is less arbitrary than choosing classes for a histogram. A smooth density curve is, however, an idealization that pictures the overall pattern of the data but ignores minor irregularities as well as any outliers. We will concentrate, not on general density curves, but on a special class, the bell-shaped Normal curves.

Density curves

One way to think of a density curve is as a smooth approximation to the irregular bars of a histogram. Figure 3.21 shows a histogram of the scores of all 947 seventh-grade students in Gary, Indiana, on the vocabulary part of the Iowa Tests of Basic Skills. Scores of many students on this national test have a very regular distribution. The histogram is symmetric, and both tails fall off quite smoothly from a single center peak. There are no large gaps or obvious outliers. The curve drawn through the tops of the histogram bars in Figure 3.21 is a good description of the overall pattern of the data.

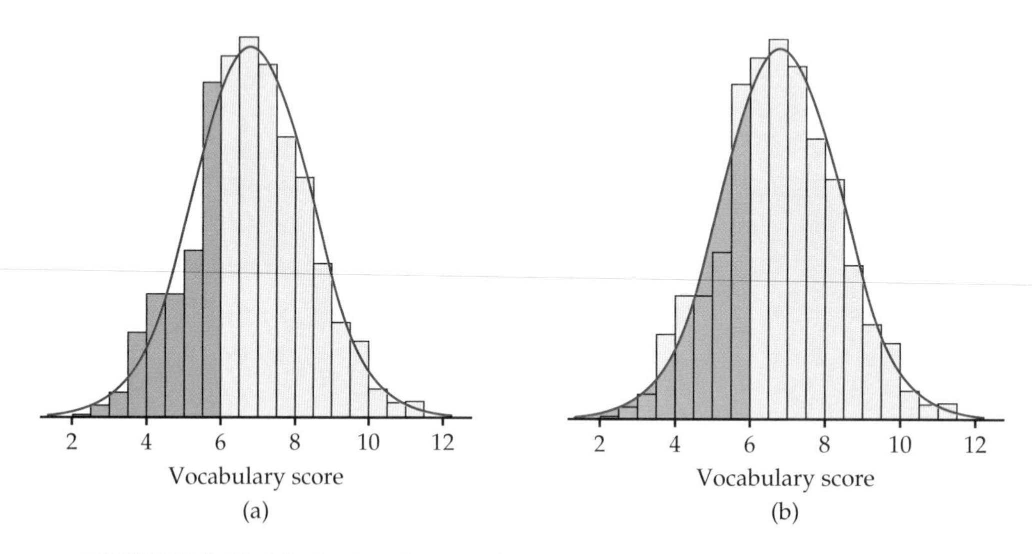

FIGURE 3.21 (a) The distribution of Iowa Test vocabulary scores for Gary, Indiana, seventh-graders. The shaded bars in the histogram represent scores less than or equal to 6.0. (b) The shaded area under the Normal density curve also represents scores less than or equal to 6.0. This area is 0.293, close to the true 0.303 for the actual data.

EXAMPLE 3.33

Vocabulary scores. In a histogram, the *areas* of the bars represent either counts or proportions of the observations. In Figure 3.21(a) we have shaded the bars that represent students with vocabulary scores 6.0 or lower. There are 287 such students, who make up the proportion $287/947 = 0.303$ of all Gary seventh-graders. The shaded bars in Figure 3.21(a) make up proportion 0.303 of the total area under all the bars. If we adjust the scale so that the total area of the bars is 1, the area of the shaded bars will be 0.303.

In Figure 3.21(b), we have shaded the *area under the curve* to the left of 6.0. Adjust the scale so that the total area under the curve is exactly 1. Areas under the curve then represent proportions of the observations. That is, *area = proportion.* The curve is then a density curve. The shaded area under the density curve in Figure 3.21(b) represents the proportion of students with score 6.0 or lower. This area is 0.293, only 0.010 away from the histogram result. You can see that areas under the density curve give quite good approximations of areas given by the histogram.

> **DENSITY CURVE**
>
> A **density curve** is a curve that
>
> - is always on or above the horizontal axis and
> - has area exactly 1 underneath it.
>
> A density curve describes the overall pattern of a distribution. The area under the curve and above any range of values is the proportion of all observations that fall in that range.

The density curve in Figure 3.21 is a *Normal curve.* Density curves, like distributions, come in many shapes. Figure 3.22 shows two density curves, a symmetric, Normal density curve and a right-skewed curve. A density curve of an appropriate shape is often an adequate description of the overall pattern of a distribution. Outliers, which are deviations from the overall pattern, are not described by the curve.

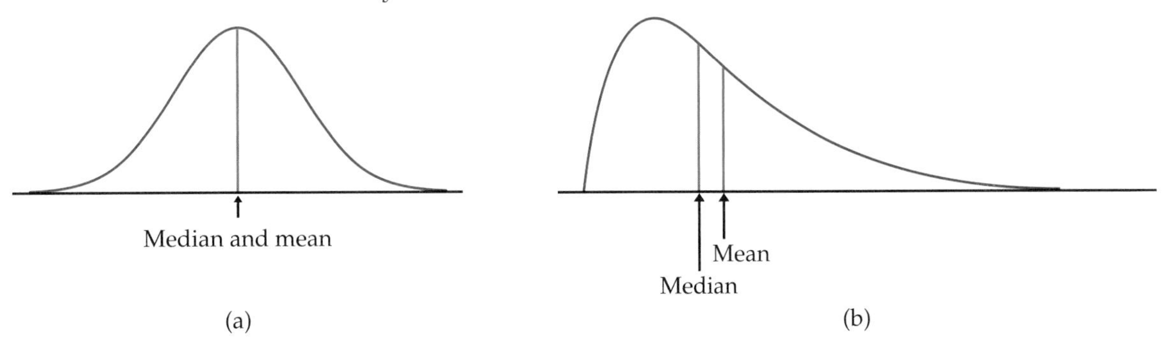

FIGURE 3.22 (a) A symmetric density curve with its mean and median marked. (b) A right-skewed density curve with its mean and median marked.

Measuring center and spread for density curves

Our measures of center and spread apply to density curves as well as to actual sets of observations, but only some of these measures are easily seen from the curve. A **mode** of a distribution described by a density curve is a peak point of the curve, the location where the curve is highest. Because areas under a density curve represent proportions of the observations, the **median** is the point with half the total area on each side. You can roughly locate the **quartiles** by dividing the area under the curve into quarters as accurately as possible by eye. The IQR is then the distance between the first and third quartiles. There are mathematical ways of calculating areas under curves. These allow us to locate the median and quartiles exactly on any density curve.

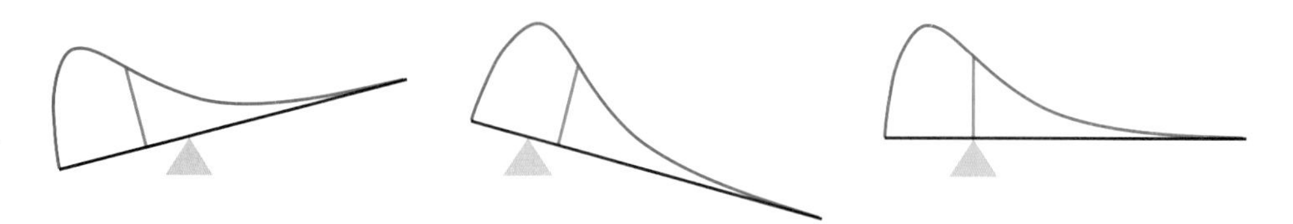

FIGURE 3.23 The mean of a density curve is the point at which it would balance.

What about the mean and standard deviation? The mean of a set of observations is their arithmetic average. If we think of the observations as weights strung out along a thin rod, the mean is the point at which the rod would balance. This fact is also true of density curves. The mean is the point at which the curve would balance if it were made out of solid material. Figure 3.23 illustrates this interpretation of the mean. We have marked the mean and median on the density curves in Figure 3.22.

A symmetric curve, such as the Normal curve in Figure 3.22(a), balances at its center of symmetry. Half the area under a symmetric curve lies on either side of its center, so this is also the median.

For a right-skewed curve, such as that shown in Figure 3.22(b), the small area in the long right tail tips the curve more than the same area near the center. The mean (the balance point) therefore lies to the right of the median. It is hard to locate the balance point by eye on a skewed curve. There are mathematical ways of calculating the mean for any density curve, so we are able to mark the mean as well as the median in Figure 3.22(b). The standard deviation can also be calculated mathematically, but it can't be located by eye on most density curves.

MEDIAN AND MEAN OF A DENSITY CURVE

The **median** of a density curve is the equal-areas point, the point that divides the area under the curve in half.

The **mean** of a density curve is the balance point, at which the curve would balance if made of solid material.

The median and mean are the same for a symmetric density curve. They both lie at the center of the curve. The mean of a skewed curve is pulled away from the median in the direction of the long tail.

A density curve is an idealized description of a distribution of data. For example, the symmetric density curve in Figure 3.21 is exactly symmetric, but the histogram of vocabulary scores is only approximately symmetric. We therefore need to distinguish between the mean and standard deviation of the density curve and the numbers $\bar{x}$ and s computed from the actual observations. The usual notation for the **mean** of an idealized distribution is μ (the Greek letter mu). We write the **standard deviation** of a density curve as σ (the Greek letter sigma).

mean μ
standard deviation σ

Normal distributions

One particularly important class of density curves has already appeared in Figures 3.21 and 3.22(a). These density curves are symmetric, unimodal, and bell-shaped. They are called **Normal curves,** and they describe *Normal distributions.* All Normal distributions have the same overall shape.

Normal curves

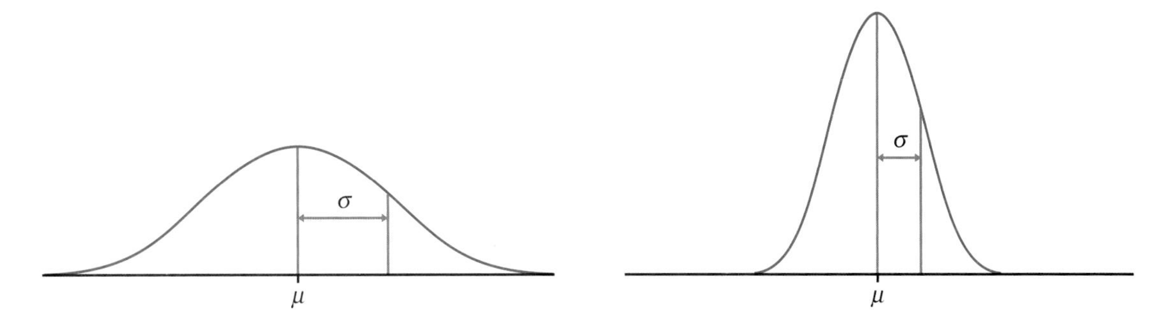

FIGURE 3.24 Two Normal curves, showing the mean μ and the standard deviation σ.

The exact density curve for a particular Normal distribution is specified by giving its mean μ and its standard deviation σ. The mean is located at the center of the symmetric curve and is the same as the median. Changing μ without changing σ moves the Normal curve along the horizontal axis without changing its spread. The standard deviation σ controls the spread of a Normal curve. Figure 3.24 shows two Normal curves with different values of σ. The curve with the larger standard deviation is more spread out.

The standard deviation σ is the natural measure of spread for Normal distributions. Not only do μ and σ completely determine the shape of a Normal curve, but we can locate σ by eye on the curve. Here's how. As we move out in either direction from the center μ, the curve changes from falling ever more steeply

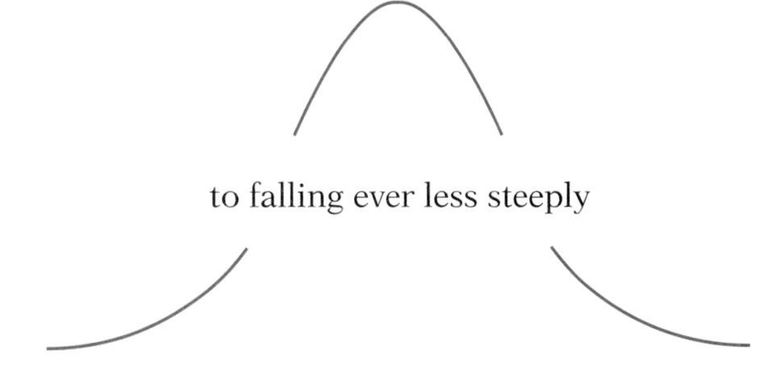

to falling ever less steeply

The points at which this change of curvature takes place are located at distance σ on either side of the mean μ. You can feel the change as you run your finger along a Normal curve, and so find the standard deviation. Remember that μ and σ alone do not specify the shape of most distributions, and that the shape of density curves in general does not reveal σ. These are special properties of Normal distributions.

There are other symmetric bell-shaped density curves that are not Normal. The Normal density curves are specified by a particular equation. The height of the density curve at any point x is given by

$$\frac{1}{\sigma\sqrt{2\pi}}e^{-\frac{1}{2}\left(\frac{x-\mu}{\sigma}\right)^2}$$

We will not make direct use of this fact, although it is the basis of mathematical work with Normal distributions. Notice that the equation of the curve is completely determined by the mean μ and the standard deviation σ.

Why are the Normal distributions important in statistics? Here are three reasons.

1. Normal distributions are good descriptions for some distributions of *real data*. Distributions that are often close to Normal include scores on tests taken by many people (such as the Iowa Test of Figure 3.21), repeated careful measurements of the same quantity, and characteristics of biological populations (such as lengths of baby pythons and yields of corn).
2. Normal distributions are good approximations to the results of many kinds of *chance outcomes*, such as tossing a coin many times.
3. Many *statistical inference* procedures based on Normal distributions work well for other roughly symmetric distributions.

However, *even though many sets of data follow a Normal distribution, many do not*. Most income distributions, for example, are skewed to the right and so are not Normal. Non-Normal data, like nonnormal people, not only are common but are also sometimes more interesting than their Normal counterparts.

The 68–95–99.7 rule

Although there are many Normal curves, they all have common properties. Here is one of the most important.

> **THE 68–95–99.7 RULE**
>
> In the Normal distribution with mean μ and standard deviation σ:
>
> - Approximately **68%** of the observations fall within σ of the mean μ.
> - Approximately **95%** of the observations fall within 2σ of μ.
> - Approximately **99.7%** of the observations fall within 3σ of μ.

Figure 3.25 illustrates the 68–95–99.7 rule. By remembering these three numbers, you can think about Normal distributions without constantly making detailed calculations.

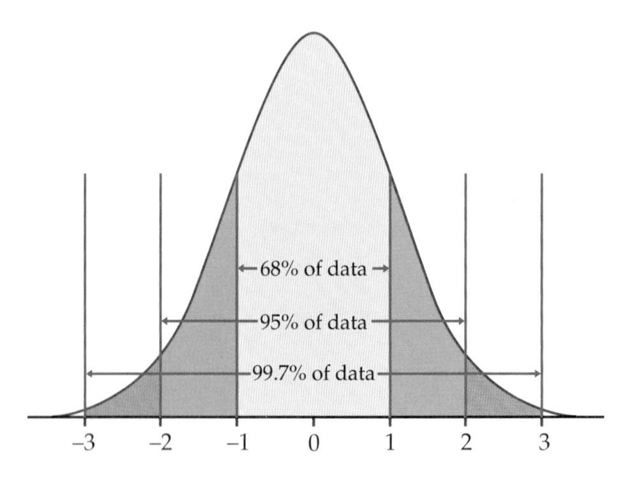

FIGURE 3.25 The 68–95–99.7 rule for Normal distributions.

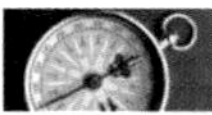

EXAMPLE 3.34

Heights of young women. The distribution of heights of young women aged 18 to 24 is approximately Normal with mean $\mu = 64.5$ inches and standard deviation $\sigma = 2.5$ inches. Figure 3.26 shows what the 68–95–99.7 rule says about this distribution.

Two standard deviations equals 5 inches for this distribution. The 95 part of the 68–95–99.7 rule says that the middle 95% of young women are between $64.5 - 5$ and $64.5 + 5$ inches tall, that is, between 59.5 inches and 69.5 inches. This fact is exactly true for an exactly Normal distribution. It is approximately true for the heights of young women because the distribution of heights is approximately Normal.

The other 5% of young women have heights outside the range from 59.5 to 69.5 inches. Because the Normal distributions are symmetric, half of these women are on the tall side. So the tallest 2.5% of young women are taller than 69.5 inches.

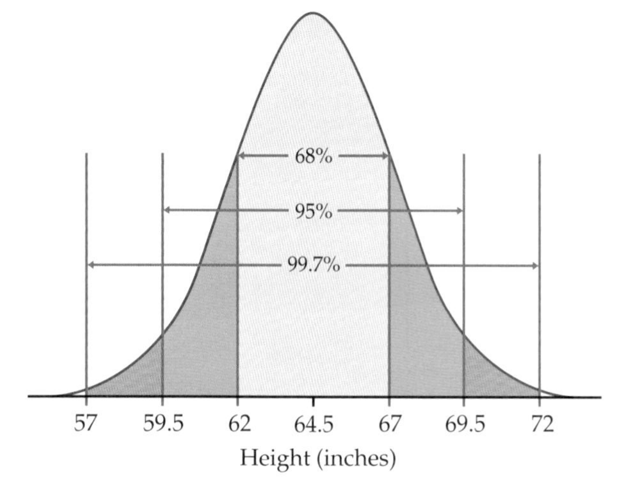

FIGURE 3.26 The 68–95–99.7 rule applied to the heights of young women, for Example 3.34.

Because we will mention Normal distributions often, a short notation is helpful. We abbreviate the Normal distribution with mean μ and standard deviation σ as $\mathbf{N(\mu, \sigma)}$. For example, the distribution of young women's heights is $N(64.5, 2.5)$.

$N(\mu, \sigma)$

USE YOUR KNOWLEDGE

3.86 Test scores. Many states have programs for assessing the skills of students in various grades. The Indiana Statewide Testing for Educational Progress (ISTEP) is one such program.[31] In a recent year 76,531 tenth-grade Indiana students took the English/language arts exam. The mean score was 572 and the standard deviation was 51. Assuming that these scores are approximately Normally distributed, $N(572, 51)$, use the 68–95–99.7 rule to give a range of scores that includes 95% of these students.

3.87 Use the 68–95–99.7 rule. Refer to the previous exercise. Use the 68–95–99.7 rule to give a range of scores that includes 99.7% of these students.

Standardizing observations

As the 68–95–99.7 rule suggests, all Normal distributions share many properties. In fact, all Normal distributions are the same if we measure in units of size σ about the mean μ as center. Changing to these units is called *standardizing*. To standardize a value, subtract the mean of the distribution and then divide by the standard deviation.

STANDARDIZING AND z-SCORES

If x is an observation from a distribution that has mean μ and standard deviation σ, the **standardized value** of x is

$$z = \frac{x - \mu}{\sigma}$$

A standardized value is often called a **z-score.**

A z-score tells us how many standard deviations the original observation falls away from the mean, and in which direction. Observations larger than the mean are positive when standardized, and observations smaller than the mean are negative.

EXAMPLE 3.35

Find some z-scores. The heights of young women are approximately Normal with $\mu = 64.5$ inches and $\sigma = 2.5$ inches. The z-score for height is

$$z = \frac{\text{height} - 64.5}{2.5}$$

A woman's standardized height is the number of standard deviations by which her height differs from the mean height of all young women. A woman 68 inches tall, for example, has z-score

$$z = \frac{68 - 64.5}{2.5} = 1.4$$

or 1.4 standard deviations above the mean. Similarly, a woman 5 feet (60 inches) tall has z-score

$$z = \frac{60 - 64.5}{2.5} = -1.8$$

or 1.8 standard deviations less than the mean height.

USE YOUR KNOWLEDGE

3.88 Find the z-score. Consider the ISTEP scores (see Exercise 3.86), which we can assume are approximately Normal, $N(572, 51)$. Give the z-score for a student who received a score of 620.

3.89 Find another z-score. Consider the ISTEP scores, which we can assume are approximately Normal, $N(572, 51)$. Give the z-score for a student who received a score of 510. Explain why your answer is negative even though all the test scores are positive.

We need a way to write variables, such as "height" in Example 3.34, that follow a theoretical distribution such as a Normal distribution. We use capital letters near the end of the alphabet for such variables. If X is the height of a young woman, we can then shorten "the height of a young woman is less than 68 inches" to "$X < 68$." We will use lowercase x to stand for any specific value of the variable X.

We often standardize observations from symmetric distributions to express them in a common scale. We might, for example, compare the heights of two children of different ages by calculating their z-scores. The standardized heights tell us where each child stands in the distribution for his or her age group.

Standardizing is a linear transformation that transforms the data into the standard scale of z-scores. We know that a linear transformation does not change the shape of a distribution, and that the mean and standard deviation change in a simple manner. In particular, *the standardized values for any distribution always have mean 0 and standard deviation 1.*

If the variable we standardize has a Normal distribution, standardizing does more than give a common scale. It makes all Normal distributions into a single distribution, and this distribution is still Normal. Standardizing a variable that has any Normal distribution produces a new variable that has the *standard Normal distribution.*

standard Normal distribution

THE STANDARD NORMAL DISTRIBUTION

The **standard Normal distribution** is the Normal distribution $N(0, 1)$ with mean 0 and standard deviation 1.

If a variable X has any Normal distribution $N(\mu, \sigma)$ with mean μ and standard deviation σ, then the standardized variable

$$Z = \frac{X - \mu}{\sigma}$$

has the standard Normal distribution.

Normal distribution calculations

Areas under a Normal curve represent proportions of observations from that Normal distribution. There is no formula for areas under a Normal curve. Calculations use either software that calculates areas or a table of areas. The table and most software calculate one kind of area: **cumulative proportions.** A cumulative proportion is the proportion of observations in a distribution that lie at or below a given value. When the distribution is given by a density curve, the cumulative proportion is the area under the curve to the left of a given value. Figure 3.27 (next page) shows the idea more clearly than words do.

cumulative proportion

The key to calculating Normal proportions is to match the area you want with areas that represent cumulative proportions. Then get areas for cumulative proportions either from software or (with an extra step) from a table. The following examples show the method in pictures.

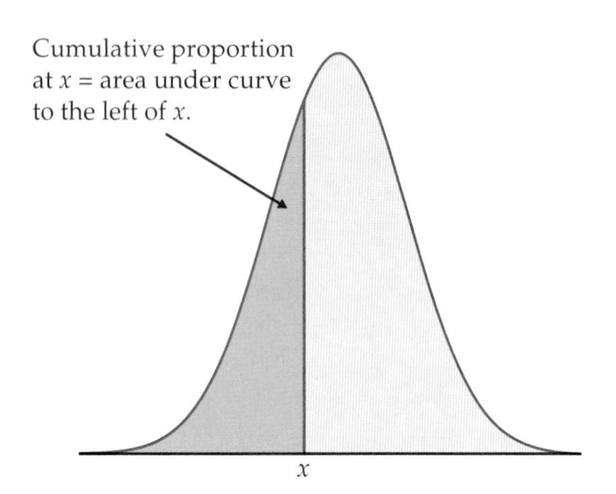

FIGURE 3.27 The *cumulative proportion* for a value x is the proportion of all observations from the distribution that are less than or equal to x. This is the area to the left of x under the Normal curve.

EXAMPLE 3.36

The NCAA standard for SAT scores. The National Collegiate Athletic Association (NCAA) requires Division I athletes to get a combined score of at least 820 on the SAT Mathematics and Verbal tests to compete in their first college year. (Higher scores are required for students with poor high school grades.) The scores of the 1.4 million students who took the SATs in a recent year were approximately Normal with mean 1026 and standard deviation 209. What proportion of all students had SAT scores of at least 820?

Here is the calculation in pictures: the proportion of scores above 820 is the area under the curve to the right of 820. That's the total area under the curve (which is always 1) minus the cumulative proportion up to 820.

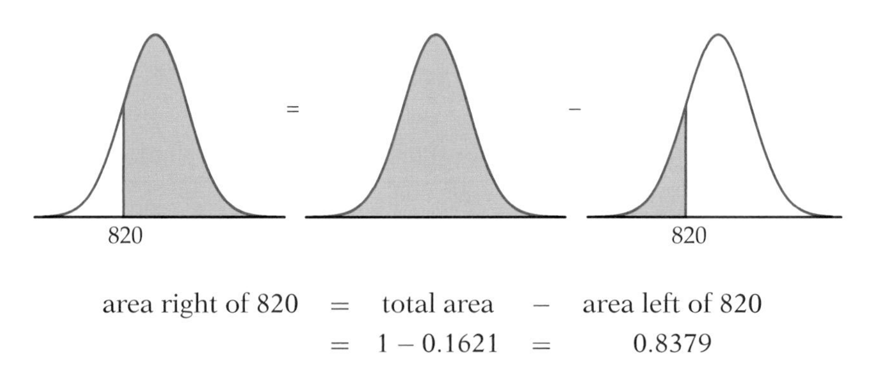

$$\begin{aligned} \text{area right of } 820 &= \text{total area} - \text{area left of } 820 \\ &= 1 - 0.1621 = 0.8379 \end{aligned}$$

That is, the proportion of all SAT takers who would be NCAA qualifiers is 0.8379, or about 84%.

There is *no* area under a smooth curve and exactly over the point 820. Consequently, the area to the right of 820 (the proportion of scores > 820) is the same as the area at or to the right of this point (the proportion of scores ≥ 820). The actual data may contain a student who scored exactly 820 on the SAT. That the proportion of scores exactly equal to 820 is 0 for a Normal distribution is a consequence of the idealized smoothing of Normal distributions for data.

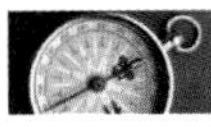

EXAMPLE 3.37

NCAA partial qualifiers. The NCAA considers a student a "partial qualifier" eligible to practice and receive an athletic scholarship, but not to compete, if the combined SAT score is at least 720. What proportion of all students who take the SAT are partial qualifiers? That is, what proportion have scores between 720 and 820? Here are the pictures:

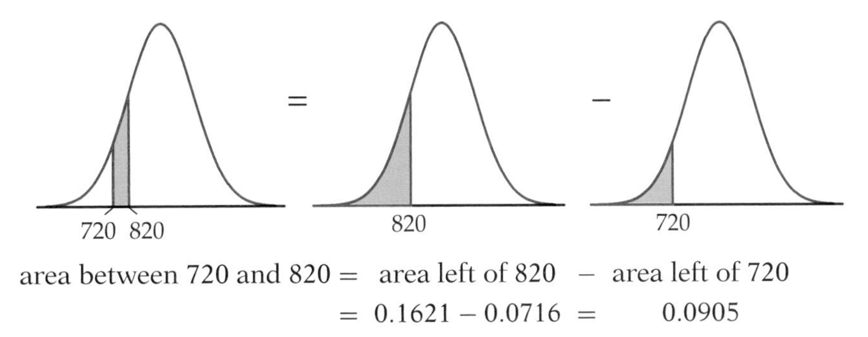

area between 720 and 820 = area left of 820 − area left of 720
= 0.1621 − 0.0716 = 0.0905

About 9% of all students who take the SAT have scores between 720 and 820.

How do we find the numerical values of the areas in Examples 3.36 and 3.37? If you use software, just plug in mean 1026 and standard deviation 209. Then ask for the cumulative proportions for 820 and for 720. (Your software will probably refer to these as "cumulative probabilities." We will learn in Chapter 5 why the language of probability fits.) If you make a sketch of the area you want, you will never go wrong.

You can use the *Normal Curve* applet on the text Web site to find Normal proportions. The applet is more flexible than most software—it will find any Normal proportion, not just cumulative proportions. The applet is an excellent way to understand Normal curves. But, because of the limitations of Web browsers, the applet is not as accurate as statistical software.

If you are not using software, you can find cumulative proportions for Normal curves from a table. That requires an extra step, as we now explain.

Using the standard Normal table

The extra step in finding cumulative proportions from a table is that we must first standardize to express the problem in the standard scale of z-scores. This allows us to get by with just one table, a table of *standard Normal cumulative proportions.* Table A in the back of the book gives cumulative proportions for the standard Normal distribution. The pictures at the top of the table remind us that the entries are cumulative proportions, areas under the curve to the left of a value z.

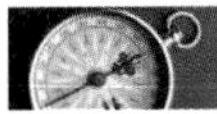

EXAMPLE 3.38

Find the proportion from z. What proportion of observations on a standard Normal variable Z take values less than 1.47?

Solution: To find the area to the left of 1.47, locate 1.4 in the left-hand column of Table A and then locate the remaining digit 7 as .07 in the top row. The entry opposite 1.4 and under .07 is 0.9292. This is the cumulative proportion we seek. Figure 3.28 illustrates this area.

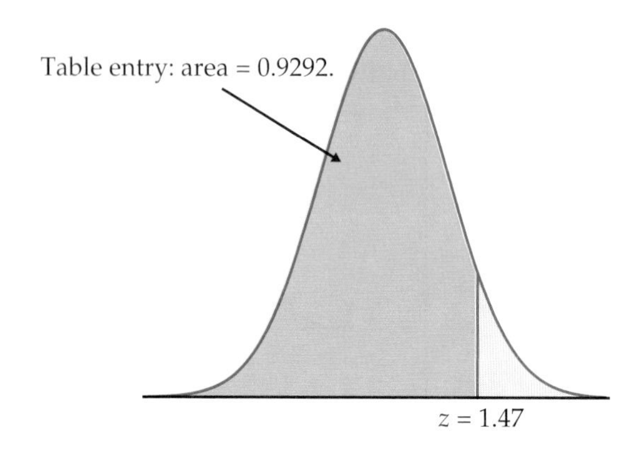

FIGURE 3.28 The area under a standard Normal curve to the left of the point $z = 1.47$ is 0.9292, for Example 3.38.

Now that you see how Table A works, let's redo the NCAA Examples 3.36 and 3.37 using the table.

EXAMPLE 3.39

Find the proportion from x. What proportion of all students who take the SAT have scores of at least 820? The picture that leads to the answer is exactly the same as in Example 3.36. The extra step is that we first standardize in order to read cumulative proportions from Table A. If X is SAT score, we want the proportion of students for which $X \geq 820$.

1. *Standardize.* Subtract the mean, then divide by the standard deviation, to transform the problem about X into a problem about a standard Normal Z:

$$X \geq 820$$

$$\frac{X - 1026}{209} \geq \frac{820 - 1026}{209}$$

$$Z \geq -0.99$$

2. *Use the table.* Look at the pictures in Example 3.36. From Table A, we see that the proportion of observations less than -0.99 is 0.1611. The area to the right of -0.99 is therefore $1 - 0.1611 = 0.8389$. This is about 84%.

The area from the table in Example 3.39 (0.8389) is slightly less accurate than the area from software in Example 3.36 (0.8379) because we must round z to two places when we use Table A. The difference is rarely important in practice.

EXAMPLE 3.40

Proportion of partial qualifiers. What proportion of all students who take the SAT would be partial qualifiers in the eyes of the NCAA? That is, what proportion of students have SAT scores between 720 and 820? First, sketch the areas, exactly as in Example 3.37. We again use X as shorthand for an SAT score.

1. *Standardize.*

$$\begin{array}{ccccc} 720 & \leq & X & < & 820 \\ \frac{720-1026}{209} & \leq & \frac{X-1026}{209} & < & \frac{820-1026}{209} \\ -1.46 & \leq & Z & < & -0.99 \end{array}$$

2. *Use the table.*

$$\begin{aligned} \text{area between } -1.46 \text{ and } -0.99 &= (\text{area left of } -0.99) - (\text{area left of } -1.46) \\ &= 0.1611 - 0.0721 = 0.0890 \end{aligned}$$

As in Example 3.37, about 9% of students would be partial qualifiers.

Sometimes we encounter a value of z more extreme than those appearing in Table A. For example, the area to the left of $z = -4$ is not given directly in the table. The z-values in Table A leave only area 0.0002 in each tail unaccounted for. For practical purposes, we can act as if there is zero area outside the range of Table A.

USE YOUR KNOWLEDGE

3.90 Find the proportion. Consider the ISTEP scores, which are approximately Normal, $N(572, 51)$. Find the proportion of students who have scores less than 620. Find the proportion of students who have scores greater than or equal to 620. Sketch the relationship between these two calculations using pictures of Normal curves similar to the ones given in Example 3.36.

3.91 Find another proportion. Consider the ISTEP scores, which are approximately Normal, $N(572, 51)$. Find the proportion of students who have scores between 620 and 660. Use pictures of Normal curves similar to the ones given in Example 3.37 to illustrate your calculations.

Inverse Normal calculations

Examples 3.34 to 3.40 illustrate the use of Normal distributions to find the proportion of observations in a given event, such as "SAT score between 720 and 820." We may instead want to find the observed value corresponding to a given proportion.

Statistical software will do this directly. Without software, use Table A backward, finding the desired proportion in the body of the table and then reading the corresponding z from the left column and top row.

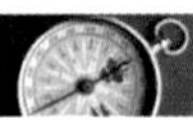

EXAMPLE 3.41

How high for the top 10%? Scores on the SAT Verbal test in recent years follow approximately the $N(505, 110)$ distribution. How high must a student score in order to place in the top 10% of all students taking the SAT?

Again, the key to the problem is to draw a picture. Figure 3.29 shows that we want the score x with area above it 0.10. That's the same as area below x equal to 0.90.

Statistical software has a function that will give you the x for any cumulative proportion you specify. The function often has a name such as "inverse cumulative probability." Plug in mean 505, standard deviation 110, and cumulative proportion 0.9. The software tells you that $x = 645.97$. We see that a student must score at least 646 to place in the highest 10%.

Without software, first find the standard score z with cumulative proportion 0.9, then "unstandardize" to find x. Here is the two-step process:

1. *Use the table.* Look in the body of Table A for the entry closest to 0.9. It is 0.8997. This is the entry corresponding to $z = 1.28$. So $z = 1.28$ is the standardized value with area 0.9 to its left.

2. *Unstandardize* to transform the solution from z back to the original x scale. We know that the standardized value of the unknown x is $z = 1.28$. So x itself satisfies

$$\frac{x - 505}{110} = 1.28$$

Solving this equation for x gives

$$x = 505 + (1.28)(110) = 645.8$$

This equation should make sense: it finds the x that lies 1.28 standard deviations above the mean on this particular Normal curve. That is the "unstandardized" meaning of $z = 1.28$. The general rule for unstandardizing a z-score is

$$x = \mu + z\sigma$$

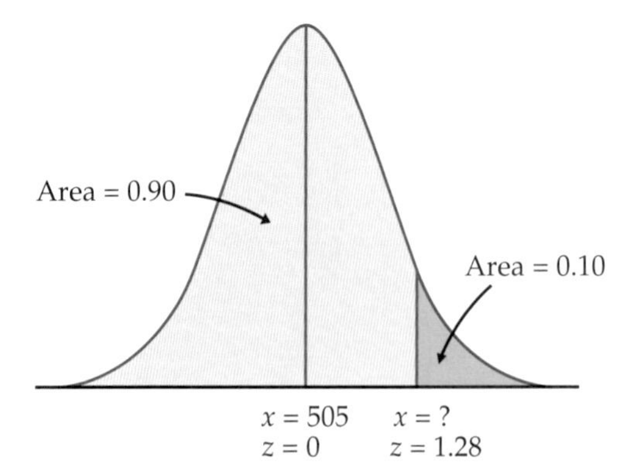

FIGURE 3.29 Locating the point on a Normal curve with area 0.10 to its right, for Example 3.41.

USE YOUR KNOWLEDGE

3.92 What score is needed to be in the top 25%? Consider the ISTEP scores, which are approximately Normal, $N(572, 51)$. How high a score is needed to be in the top 25% of students who take this exam?

3.93 Find the score that 80% of students will exceed. Consider the ISTEP scores, which are approximately Normal, $N(572, 51)$. Eighty percent of the students will score above x on this exam. Find x.

Normal quantile plots

The Normal distributions provide good descriptions of some distributions of real data, such as the Gary vocabulary scores. The distributions of some other common variables are usually skewed and therefore distinctly non-Normal. Examples include economic variables such as personal income and gross sales of business firms, the survival times of cancer patients after treatment, and the service lifetime of mechanical or electronic components. While experience can suggest whether or not a Normal distribution is plausible in a particular case, it is risky to assume that a distribution is Normal without actually inspecting the data.

A histogram or stemplot can reveal distinctly non-Normal features of a distribution, such as outliers, pronounced skewness, or gaps and clusters. If the stemplot or histogram appears roughly symmetric and unimodal, however, we need a more sensitive way to judge the adequacy of a Normal model. The most useful tool for assessing Normality is another graph, the **Normal quantile plot.**

Normal quantile plot

Here is the basic idea of a Normal quantile plot. The graphs produced by software use more sophisticated versions of this idea. It is not practical to make Normal quantile plots by hand.

1. Arrange the observed data values from smallest to largest. Record what percentile of the data each value occupies. For example, let's say that the smallest observation in a set of 20 is at the 5% point, the second smallest is at the 10% point, and so on.
2. Do Normal distribution calculations to find the values of z corresponding to these same percentiles. For example, $z = -1.645$ is the 5% point of the standard Normal distribution, and $z = -1.282$ is the 10% point. We call these values of Z **Normal scores.**
3. Plot each data point x against the corresponding Normal score. If the data distribution is close to any Normal distribution, the plotted points will lie close to a straight line.

Normal scores

Any Normal distribution produces a straight line on the plot because standardizing turns any Normal distribution into a standard Normal distribution. Standardizing is a linear transformation that can change the slope and intercept of the line in our plot but cannot turn a line into a curved pattern.

USE OF NORMAL QUANTILE PLOTS

If the points on a Normal quantile plot lie close to a straight line, the plot indicates that the data are Normal. Systematic deviations from a straight line indicate a non-Normal distribution. Outliers appear as points that are far away from the overall pattern of the plot.

Figures 3.30 to 3.32 are Normal quantile plots for data we have met earlier. The data x are plotted vertically against the corresponding standard Normal z-score plotted horizontally. The z-score scale generally extends from −3 to 3 because almost all of a standard Normal curve lies between these values. These figures show how Normal quantile plots behave.

EXAMPLE 3.42

Service center call lengths are not Normal. Figure 3.30 is a Normal quantile plot of the 80 call lengths in Table 3.2 (page 93). Because the plot is clearly curved, we conclude that these data are not Normally distributed. The shape of the curve is what we typically see with a distribution that is strongly skewed to the right.

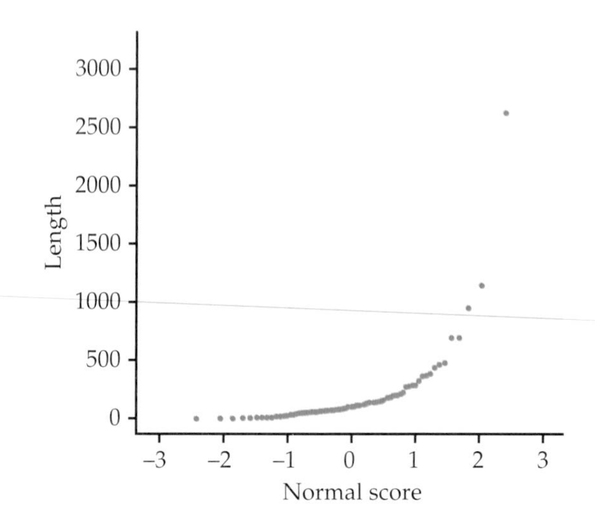

FIGURE 3.30 Normal quantile plot of the service center call lengths, for Example 3.42. This distribution is highly skewed.

EXAMPLE 3.43

Times to start a business are approximately Normal. Figure 3.31 (next page) is a Normal quantile plot of the data on times to start a business from Example 3.21. We have excluded Suriname, the outlier that you examined in Exercises 3.40 and 3.42. The plot is not particularly smooth but the overall pattern is approximately linear. Note that the sample size here is 24, much smaller than the 80 cases in the previous example. Because of the smaller sample size, the z-score scale extends from −2 to 2 rather than from −3 to 3. With smaller sample sizes such as this, we have less information about the true shape of the distribution. Nevertheless, our examination of the Normal quantile plot leads us to conclude that there is no clear deviation from Normality evident in the data.

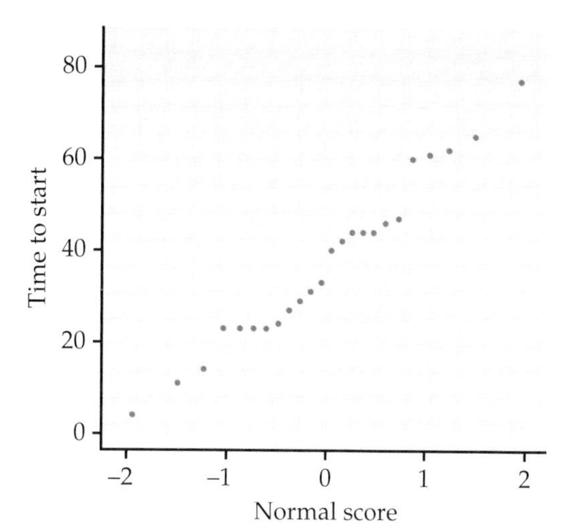

FIGURE 3.31 Normal quantile plot of 24 times to start a business, with the outlier, Suriname, excluded, for Example 3.43. This distribution is approximately Normal.

EXAMPLE 3.44

DATA FILE ACIDRAIN

Acidity of rainwater is approximately Normal. Figure 3.32 is a Normal quantile plot of the 105 acidity (pH) measurements of rainwater from Exercise 3.38 (page 104). Histograms don't settle the question of approximate Normality of these data, because their shape depends on the choice of classes. The Normal quantile plot makes it clear that a Normal distribution is a good description—there are only minor wiggles in a generally straight-line pattern.

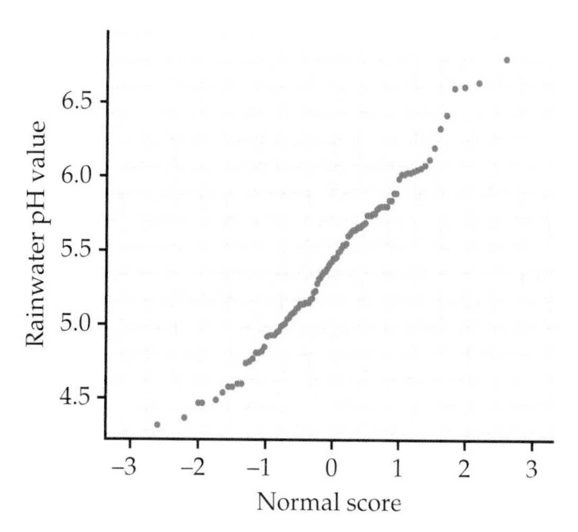

FIGURE 3.32 Normal quantile plot of the acidity (pH) values of 105 samples of rainwater, for Example 3.44.

CAUTION

As Figure 3.32 illustrates, real data almost always show some departure from the theoretical Normal model. *When you examine a Normal quantile plot, look for shapes that show clear departures from Normality. Don't overreact to minor wiggles in the plot.* When we discuss statistical methods that are based on the Normal model, we will pay attention to the sensitivity of each method to departures from Normality. Many common methods work well as long as the data are approximately Normal and outliers are not present.

SECTION 3.3 SUMMARY

The overall pattern of a distribution can often be described compactly by a **density curve.** A density curve has total area 1 underneath it. Areas under a density curve give proportions of observations for the distribution.

The **mean** μ (balance point), the **median** (equal-areas point), and the **quartiles** can be approximately located by eye on a density curve. The **standard deviation** σ cannot be located by eye on most density curves. The mean and median are equal for symmetric density curves, but the mean of a skewed curve is located farther toward the long tail than is the median.

The **Normal distributions** are described by bell-shaped, symmetric, unimodal density curves. The mean μ and standard deviation σ completely specify the Normal distribution $N(\mu, \sigma)$. The mean is the center of symmetry, and σ is the distance from μ to the change-of-curvature points on either side.

To **standardize** any observation x, subtract the mean of the distribution and then divide by the standard deviation. The resulting ***z*-score** $z = (x - \mu)/\sigma$ says how many standard deviations x lies from the distribution mean. All Normal distributions are the same when measurements are transformed to the standardized scale. In particular, all Normal distributions satisfy the **68–95–99.7 rule.**

If X has the $N(\mu, \sigma)$ distribution, then the standardized variable $Z = (X - \mu)/\sigma$ has the **standard Normal distribution** $N(0, 1)$. Proportions for any Normal distribution can be calculated by software or from the **standard Normal table** (Table A), which gives the **cumulative proportions** of $Z < z$ for many values of z.

The adequacy of a Normal model for describing a distribution of data is best assessed by a **Normal quantile plot,** which is available in most statistical software packages. A pattern on such a plot that deviates substantially from a straight line indicates that the data are not Normal.

SECTION 3.3 EXERCISES

For Exercises 3.86 and 3.87, see page 133; for Exercises 3.88 and 3.89, see page 134; for Exercises 3.90 and 3.91, see page 139; and for Exercises 3.92 and 3.93, see pages 140–141.

3.94. The effect of changing the standard deviation.

(a) Sketch a Normal curve that has mean 10 and standard deviation 3.

(b) On the same x axis, sketch a Normal curve that has mean 10 and standard deviation 1.

(c) How does the Normal curve change when the standard deviation is varied but the mean stays the same?

3.95. Know your density. Sketch density curves that might describe distributions with the following shapes:

(a) Symmetric, but with two peaks (that is, two strong clusters of observations).

(b) Single peak and skewed to the left.

3.96. Do women talk more? Conventional wisdom suggests that women are more talkative than men. One study designed to examine this stereotype collected data on the speech of 42 women and 37 men in the United States.[32] DATA FILE TALK

(a) The mean number of words spoken per day by the women was 14,297 with a standard deviation of 6441. Use the 68–95–99.7 rule to describe this distribution.

(b) Do you think that applying the rule in this situation is reasonable? Explain your answer.

(c) The men averaged 14,060 words per day with a standard deviation of 9056. Answer the questions in parts (a) and (b) for the men.

(d) Do you think that the data support the conventional wisdom? Explain your answer. Note that in Section 6.1 we will learn formal statistical methods to answer this type of question.

3.97. Data from Mexico. Refer to the previous exercise. A similar study in Mexico was conducted with 31 women and 20 men. The women averaged 14,704 words per day with a standard deviation of 6215. For men the mean was 15,022 and the standard deviation was 7864. DATA FILE TALKMEXICO

(a) Answer the questions from the previous exercise for the Mexican study.

(b) The means for both men and women are higher for the Mexican study than for the U.S. study. What conclusions can you draw from this observation.

3.98. Total scores. Here are the total scores of 10 students in an introductory statistics course: DATA FILE STATCOURSE

68 54 92 75 73 98 64 55 80 70

Previous experience with this course suggests that these scores should come from a distribution that is approximately Normal with mean 70 and standard deviation 10.

(a) Using these values for μ and σ, standardize the scores of these 10 students.

(b) If the grading policy is to give grades of A to the top 15% of scores based on the Normal distribution with mean 70 and standard deviation 10, what is the cutoff for an A in terms of a standardized score?

(c) Which students earned a grade of A in this course?

3.99. Assign more grades. Refer to the previous exercise. The grading policy says the cutoffs for the other grades correspond to the following: bottom 5% receive F, next 10% receive D, next 40% receive C, and next 30% receive B. These cutoffs are based on the $N(70, 10)$ distribution.

(a) Give the cutoffs for the grades in this course in terms of standardized scores.

(b) Give the cutoffs in terms of actual total scores.

(c) Do you think that this method of assigning grades is a good one? Give reasons for your answer.

3.100. A uniform distribution. If you ask a computer to generate "random numbers" between 0 and 1, you will get observations from a **uniform distribution.** Figure 3.33 graphs the density curve for a uniform distribution. Use areas under this density curve to answer the following questions.

(a) Why is the total area under this curve equal to 1?

(b) What proportion of the observations lie below 0.35?

(c) What proportion of the observations lie between 0.35 and 0.65?

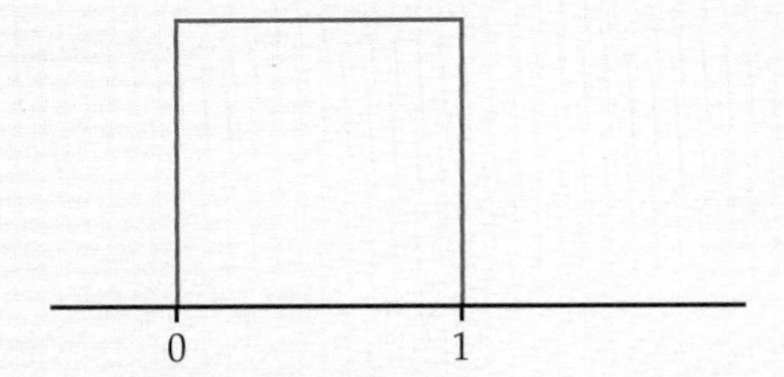

FIGURE 3.33 The density curve of a uniform distribution, for Exercise 3.100.

3.101. Use a different range for the uniform distribution. Many random number generators allow users to specify the range of the random numbers to be produced. Suppose that you specify that the outcomes are to be distributed uniformly between 0 and 4. Then the density curve of the outcomes has constant height between 0 and 4, and height 0 elsewhere.

(a) What is the height of the density curve between 0 and 4? Draw a graph of the density curve.

(b) Use your graph from (a) and the fact that areas under the curve are proportions of outcomes to find the proportion of outcomes that are less than 1.

(c) Find the proportion of outcomes that lie between 0.5 and 2.5.

3.102. Find the mean, the median, and the quartiles. What are the mean and the median of the uniform distribution in Figure 3.33? What are the quartiles?

3.103. Three density curves. Figure 3.34 displays three density curves, each with three points marked on it. At which of these points on each curve do the mean and the median fall?

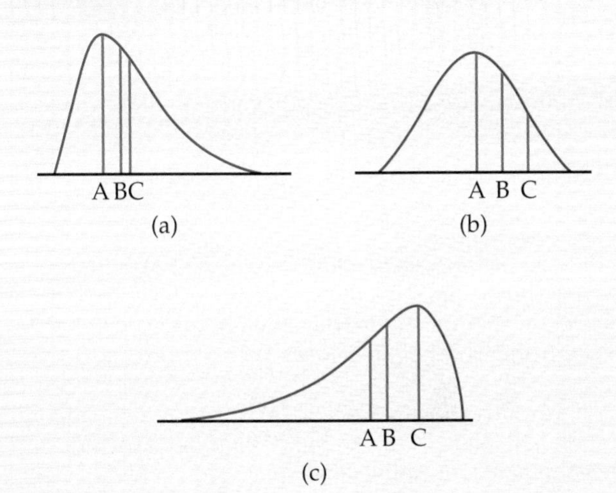

FIGURE 3.34 Three density curves, for Exercise 3.100.

3.104. Length of pregnancies. The length of human pregnancies from conception to birth varies according to a distribution that is approximately Normal with mean 266 days and standard deviation 16 days. Draw a density curve for this distribution on which the mean and standard deviation are correctly located.

3.105. APPLET **Use the *Normal Curve* applet.** The 68–95–99.7 rule for Normal distributions is a useful approximation. You can use the *Normal Curve* applet on the text Web site to see how accurate the rule is. Drag one flag across the other so that the applet shows the area under the curve between the two flags.

(a) Place the flags one standard deviation on either side of the mean. What is the area between these two values? What does the 68–95–99.7 rule say this area is?

(b) Repeat for locations two and three standard deviations on either side of the mean. Again compare the 68–95–99.7 rule with the area given by the applet.

3.106. Pregnancies and the 68–95–99.7 rule. The length of human pregnancies from conception to birth varies according to a distribution that is approximately Normal with mean 266 days and standard deviation 16 days. Use the 68–95–99.7 rule to answer the following questions.

(a) Between what values do the lengths of the middle 95% of all pregnancies fall?

(b) How short are the shortest 2.5% of all pregnancies? How long do the longest 2.5% last?

3.107. Horse pregnancies are longer. Bigger animals tend to carry their young longer before birth. The length of horse pregnancies from conception to birth varies according to a roughly Normal distribution with mean 336 days and standard deviation 3 days. Use the 68–95–99.7 rule to answer the following questions.

(a) Almost all (99.7%) horse pregnancies fall in what range of lengths?

(b) What percent of horse pregnancies are longer than 339 days?

3.108. APPLET **Use the *Normal Curve* applet.** Use the *Normal Curve* applet for the standard Normal distribution to say how many standard deviations above and below the mean the quartiles of any Normal distribution lie.

3.109. Acidity of rainwater. The Normal quantile plot in Figure 3.32 (page 143) shows that the acidity (pH) measurements for rainwater samples in Exercise 3.38 are approximately Normal. How well do these scores satisfy the 68–95–99.7 rule? To find out, calculate the mean $\bar{x}$ and standard deviation s of the observations. Then calculate the percent of the 105 measurements that fall between $\bar{x} - s$ and $\bar{x} + s$ and compare your result with 68%. Do the same for the intervals covering two and three standard deviations on either side of the mean. (The 68–95–99.7 rule is exact for any theoretical Normal distribution. It will hold only approximately for actual data.) DATA FILE ACIDRAIN

3.110. Find some proportions. Using either Table A or your calculator or software, find the proportion of observations from a standard Normal distribution that satisfies each of the following statements. In each case, sketch a standard Normal curve and shade the area under the curve that is the answer to the question.

(a) $Z > 1.65$

(b) $Z < 1.65$

(c) $Z > -0.76$

(d) $-0.76 < Z < 1.65$

3.111. Find more proportions. Using either Table A or your calculator or software, find the proportion of observations from a standard Normal distribution for each of the following events. In each case, sketch a standard Normal curve and shade the area representing the proportion.

(a) $Z \leq -1.8$

(b) $Z \geq -1.8$

(c) $Z > 1.6$

(d) $-1.8 < Z < 1.6$

3.112. Find some values of z. Find the value z of a standard Normal variable Z that satisfies each of the following conditions. (If you use Table A, report the value of z that comes closest to satisfying the condition.) In each case, sketch a standard Normal curve with your value of z marked on the axis.

(a) 22% of the observations fall below z.

(b) 40% of the observations fall above z.

3.113. High IQ scores. The Wechsler Adult Intelligence Scale (WAIS) is the most common IQ test. The scale of scores is set separately for each age group and is approximately Normal with mean 100 and standard deviation 15. The organization MENSA, which calls itself "the high IQ society," requires a WAIS score of 130 or higher for membership. What percent of adults would qualify for membership?

There are two major tests of readiness for college, the ACT and the SAT. ACT scores are reported on a scale from 1 to 36. The distribution of ACT scores is approximately Normal with mean $\mu = 21.5$ and standard deviation $\sigma = 5.4$. SAT scores are reported on a scale from 600 to 2400. SAT scores are approximately Normal with mean $\mu = 1509$ and standard deviation $\sigma = 321$. Exercises 3.114 to 3.123 are based on this information.

3.114. Compare an SAT score with an ACT score. Isabella scores 1820 on the SAT. Jermaine scores 29 on the ACT. Assuming that both tests measure the same thing, who has the higher score? Report the z-scores for both students.

3.115. Make another comparison. Jayden scores 16 on the ACT. Emily scores 1020 on the SAT. Assuming that both tests measure the same thing, who has the higher score? Report the z-scores for both students.

3.116. Find the ACT equivalent. Jose scores 2080 on the SAT. Assuming that both tests measure the same thing, what score on the ACT is equivalent to Jose's SAT score?

3.117. Find the SAT equivalent. Chloe scores 30 on the ACT. Assuming that both tests measure the same thing, what score on the SAT is equivalent to Chloe's ACT score?

3.118. Find an SAT percentile. Reports on a student's ACT or SAT usually give the percentile as well as the actual score. The percentile is just the cumulative proportion stated as a percent: the percent of all scores that were lower than this one. Maria scores 2090 on the SAT. What is her percentile?

3.119. Find an ACT percentile. Reports on a student's ACT or SAT usually give the percentile as well as the actual score. The percentile is just the cumulative proportion stated as a percent: the percent of all scores that were lower than this one. Jacob scores 19 on the ACT. What is his percentile?

3.120. How high is the top 10%? What SAT scores make up the top 10% of all scores?

3.121. How low is the bottom 20%? What SAT scores make up the bottom 20% of all scores?

3.122. Find the ACT quartiles. The quartiles of any distribution are the values with cumulative proportions 0.25 and 0.75. What are the quartiles of the distribution of ACT scores?

3.123. Find the SAT quintiles. The quintiles of any distribution are the values with cumulative proportions 0.20, 0.40, 0.60, and 0.80. What are the quintiles of the distribution of SAT scores?

3.124. Do you have enough "good cholesterol?" High-density lipoprotein (HDL) is sometimes called the "good cholesterol" because low values are associated with a higher risk of heart disease. According to the American Heart Association, people over the age of 20 years should have at least 40 milligrams per deciliter of blood (mg/dl) of HDL cholesterol.[33] U.S. women aged 20 and over have a mean HDL of 55 mg/dl with a standard deviation of 15.5 mg/dl. Assume that the distribution is Normal.

(a) What percent of women have low values of HDL (40 mg/dl or less)?

(b) HDL levels of 60 mg/dl are believed to protect people from heart disease. What percent of women have protective levels of HDL?

(c) Women with more than 40 mg/dl but less than 60 mg/dl of HDL are in the intermediate range, neither very good nor very bad. What proportion are in this category?

3.125. Men and HDL cholesterol. HDL cholesterol levels for men have a mean of 46 milligrams per deciliter of blood (mg/dl) with a standard deviation of 13.6 mg/dl. Answer the questions in the previous exercise for the population of men.

3.126. CHALLENGE **Quartiles for Normal distributions.** The quartiles of any distribution are the values with cumulative proportions 0.25 and 0.75.

(a) What are the quartiles of the standard Normal distribution?

(b) Using your numerical values from (a), write an equation that gives the quartiles of the $N(\mu, \sigma)$ distribution in terms of μ and σ.

(c) The length of human pregnancies from conception to birth varies according to a distribution that is approximately Normal with mean 266 days and standard deviation 16 days. Apply your result from (b): what are the quartiles of the distribution of lengths of human pregnancies?

3.127. CHALLENGE ***IQR* for Normal distributions.** Continue your work from the previous exercise. The interquartile range IQR is the distance between the first and third quartiles of a distribution.

(a) What is the value of the IQR for the standard Normal distribution?

(b) There is a constant c such that $IQR = c\sigma$ for any Normal distribution $N(\mu, \sigma)$. What is the value of c?

3.128. CHALLENGE **Outliers for Normal distributions.** Continue your work from the previous two exercises. The percent of the observations that are suspected outliers according to the $1.5 \times IQR$ rule is the same for any Normal distribution. What is this percent?

The remaining exercises for this section require the use of software that will make Normal quantile plots.

3.129. Three varieties of flowers. The study of tropical flowers and their hummingbird pollinators (Exercise 3.74, page 125) measured the lengths of three varieties of *Heliconia* flowers. We expect that such biological measurements will have roughly Normal distributions. DATA FILE HELICONIA

(a) Make Normal quantile plots for each of the three flower varieties. Which distribution is closest to Normal?

(b) The other two distributions show the same kind of mild deviation from Normality. In what way are these distributions non-Normal?

3.130. Use software to generate some data. Use software to generate 200 observations from the standard Normal distribution. Make a histogram of these observations. How does the shape of the histogram compare with a Normal density curve? Make a Normal quantile plot of the data. Does the plot suggest any important deviations from Normality? (Repeating this exercise several times is a good way to become familiar with how histograms and Normal quantile plots look when data actually are close to Normal.)

3.131. Use software to generate more data. Use software to generate 200 observations from the uniform distribution described in Exercise 3.100. Make a histogram of these observations. How does the histogram compare with the density curve in Figure 3.33? Make a Normal quantile plot of your data. According to this plot, how does the uniform distribution deviate from Normality?

CHAPTER 3 EXERCISES

3.132. CHALLENGE **Fuel efficiency of hatchbacks and large sedans.** Let's compare the fuel efficiencies (miles per gallon) of model year 2009 hatchbacks and large sedans.[34] Here are the data:

Hatchbacks

30 29 28 27 27 27 27 27 26 25 25 25 24 24 24
24 24 23 23 22 22 21 21 21 21 21 21 21 20 20
20 20 20 20 20 20 19 19 19 18 16 16

Large sedans

19 19 18 18 18 18 17 17 17 17 17 17 17 17 17
17 16 16 16 16 16 16 16 16 15 15 13 13

Give graphical and numerical descriptions of the fuel efficiencies for these two types of vehicles. What are the main features of the distributions? Compare the two distributions and summarize your results in a short paragraph. DATA FILE MPGHATCHLARGE

3.133. Binge drinking. The Behavioral Risk Factor Surveillance System (BRFSS) conducts a large survey of health conditions and risk behaviors in the United States.[35] The BRFSS data set contains data on 29 demographic and risk factors for each state. Use the percent of binge drinkers for this exercise. DATA FILE BRFSS

(a) Prepare a graphical display of the distribution and use your display to describe the major features of the distribution.

(b) Calculate numerical summaries. Give reasons for your choices.

(c) Write a short paragraph summarizing what the data tell us about binge drinking in the United States.

3.134. Eat your fruits and vegetables. Nutrition experts recommend that we eat five servings of fruits and vegetables each day. The BRFSS data set described in the previous exercise includes a variable that gives the percent of people who regularly eat five or more servings of fruits and vegetables. Answer the questions given in the previous exercise for this variable. DATA FILE BRFSS

3.135. CHALLENGE **Vehicle colors.** Vehicle color preferences differ in different regions. Here are data on the most popular colors in 2011 for several different regions of the world:[36]

Color	North America (%)	South America (%)	Europe (%)	China (%)	South Korea (%)	India (%)
White	23	17	20	15	25	28
Black	18	19	25	21	15	8
Silver	16	30	15	26	30	27
Gray	13	15	18	10	12	9
Red	10	11	6	7	4	6
Blue	9	1	7	9	4	5
Brown/beige	5	5	5	4	4	9
Yellow/gold	3	1	1	2	1	2
Other	3	1	3	6	5	6

Use the methods you have learned in this chapter to compare the vehicle color preferences in the different regions. Write a report summarizing your findings with an emphasis on similarities and differences across regions. Include recommendations related to marketing and advertising of vehicles in these regions. DATA FILE VCOLORSREGIONS

3.136. CHALLENGE **Balance of international payments for Canada.** Visit the Web page `statcan.gc.ca/tables-tableaux/sum-som/l01/cst01/econ01a-eng.htm`, which provides data on Canada's balance of international payments. Select some data from this Web page and use the methods that you learned in this chapter to create graphical and numerical summaries. Write a report summarizing your findings that includes supporting evidence from your analyses.

3.137. CHALLENGE **Canadian government revenue and expenditures by province and territory.** Visit the Web pages of Statistics Canada, `statcan.gc.ca/start-debut-eng.html`. Locate pages that give government revenue and expenditures for all provinces and territories. Select some data from these Web pages and use the methods that you learned in this chapter to create graphical and numerical summaries. Write a report summarizing your findings that includes supporting evidence from your analyses.

3.138. Internet use. The World Bank collects data on many variables related to development for countries throughout the world.[37] One of these is Internet use, expressed as the number of users per 100 people. The data file for this exercise gives this variable for 182 countries. Use graphical and numerical methods to describe this distribution. Write a short report summarizing what the data tell about worldwide Internet use. DATA FILE INTERNETUSE

3.139. Internet use in Europe. Refer to the previous exercise. Now examine the data only for the countries in Europe. Answer the questions in the previous exercise and compare your results with those that you found for all 182 countries. DATA FILE INTERNETEUROPE

3.140. Park space and population. Below are data on park and open space in several U.S. cities with high population density.[38] In this table, population is reported in thousands of people, and park and open space is called open space, with units of acres. DATA FILE PARKSPACE

City	Population	Open space
Baltimore	651	5,091
Boston	589	4,865
Chicago	2,896	11,645
Long Beach	462	2,887
Los Angeles	3,695	29,801
Miami	362	1,329
Minneapolis	383	5,694
New York	8,008	49,854
Oakland	399	3,712
Philadelphia	1,518	10,685
San Francisco	777	5,916
Washington, DC	572	7,504

(a) Make a bar graph for population. Describe what you see in the graph.

(b) Do the same for open space.

(c) For each city, divide the open space by population. This gives rates: acres of open space per thousand residents.

(d) Make a bar graph of the rates.

(e) Redo the bar graph that you made in part (d) by ordering the cities by their open space to population rate.

(f) Which of the two bar graphs in (d) and (e) do you prefer? Give reasons for your answer.

3.141. Compare two Normal curves. In Exercise 3.86, we worked with the distribution of ISTEP scores on the English/language arts portion of the exam for tenth-graders. We used the fact that the distribution of scores for the 76,531 students who took the exam was approximately $N(572, 51)$. These students were classified in a variety of ways, and summary statistics were reported for these different subgroups. When classified by gender, the scores for the women are approximately $N(579, 49)$, and the scores for the men are approximately $N(565, 55)$. Figure 3.35 gives the Normal density curves for these two distributions. Here is a possible description of these data: women score about 14 points higher than men on the ISTEP English/language arts exam. Critically evaluate this statement and then write your own summary based on the distributions displayed in Figure 3.35.

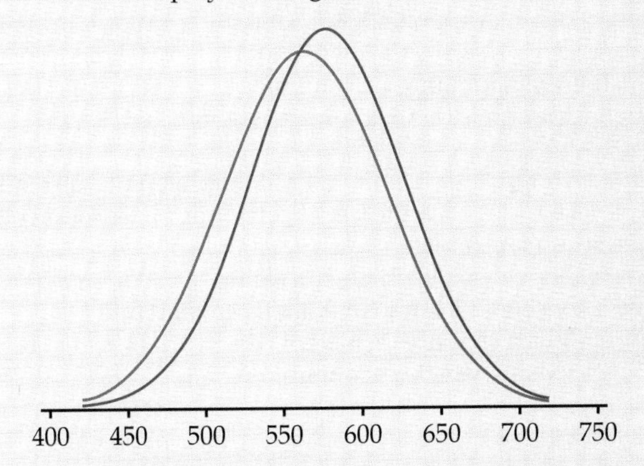

FIGURE 3.35 Normal density curves for ISTEP scores of women and men, for Exercise 3.141.

3.142. Leisure time for college students. You want to measure the amount of "leisure time" that college students enjoy. Write a brief discussion of two issues:

(a) How will you define "leisure time"?

(b) Once you have defined "leisure time," how will you measure Sally's leisure time this week?

3.143. Biological clocks. Many plants and animals have "biological clocks" that coordinate activities with the time

TABLE 3.3

Biological clock cycle lengths (hours) for a plant species in different locations

23.89	23.72	23.74	24.35	25.05	24.56	23.69	22.33	23.79	22.12
25.39	23.08	25.64	23.98	25.84	25.46	24.37	24.13	24.40	24.74
24.44	24.82	23.56	24.96	24.21	23.85	24.57	23.44	23.64	24.23
24.01	24.58	25.57	23.73	24.11	23.21	25.08	24.03	24.62	23.51
23.21	23.41	23.69	22.97	24.65	24.65	24.29	23.89	25.08	23.89
24.95	23.09	23.21	24.66	23.88	25.33	24.38	24.68	25.34	25.22
23.45	23.39	25.43	23.16	23.95	23.25	24.72	24.89	24.88	24.71
23.58	25.98	24.28	24.25	23.16	24.19	27.22	23.77	26.21	24.33
24.34	24.89	24.32	24.14	24.00	23.48	25.81	24.99	24.18	22.73
24.18	23.95	24.48	23.89	24.24	24.96	24.58	24.29	24.31	23.64
23.87	23.68	24.87	23.00	23.48	24.26	23.34	25.11	24.69	24.97
24.64	24.49	23.61	24.07	26.60	24.91	24.76	25.09	26.56	25.13
24.81	25.63	25.63	24.69	24.41	23.79	22.88	22.00	23.33	25.12
24.00	24.31	23.03	24.51	28.55	22.96	23.61	24.72	24.04	25.18
24.30	24.22	24.39	24.73	24.68	24.14	24.57	24.42	25.62	

of day. When researchers looked at the length of the biological cycle in the plant *Arabidopsis* by measuring leaf movements, they found that the length of the cycle is not always 24 hours. Further study discovered that cycle length changes systematically with north-south location. Table 3.3 contains cycle lengths for 149 locations around the world.[39] Describe the distribution of cycle lengths with a histogram and numerical summaries. In particular, how much variation is there among locations? DATA FILE BIOCLOCK

3.144. Product preference. Product preference depends in part on the age, income, and gender of the consumer. A market researcher selects a large sample of potential car buyers. For each consumer, she records gender, age in years, household income in dollars per year, and automobile preference. Which of these variables are categorical and which are quantitative?

3.145. Internet service. TopTenReviews publishes product reviews of consumer products in many different categories.[40] Here are the Internet service providers who made their top ten list, with the numbers of residential customers for each:

Service provider	Subscribers (millions)	Service provider	Subscribers (millions)
Comcast	17.0	Charter	5.5
Time Warner Cable	9.7	Verizon	4.3
AT&T	17.8	CenturyLink	6.4
Cox	3.9	SuddenLink	1.4
Optimum	3.3	EarthLink	1.6

Display these data in a graph. How many subscribers do the many smaller providers have? Add an "Other" entry in your graph. Business people looking at this graph see an industry that offers opportunities for larger companies to take over. DATA FILE INTERNETPROVIDE

3.146. Weights are not Normal. The heights of people of the same sex and similar ages follow Normal distributions reasonably closely. Weights, on the other hand, are not Normally distributed. The weights of women aged 20 to 29 have mean 141.7 pounds and median 133.2 pounds. The first and third quartiles are 118.3 pounds and 157.3 pounds. What can you say about the shape of the weight distribution? Explain your reasoning.

3.147. What graph would you use? What type of graph or graphs would you plan to make in a study of each of the following issues?

(a) What makes of cars do students drive? How old are their cars?

(b) How many hours per week do students study? How does the number of study hours change during a semester?

(c) Which radio stations are most popular with students?

(d) When many students measure the concentration of the same solution for a chemistry course laboratory assignment, do their measurements follow a Normal distribution?

3.148. Spam filters. A university department installed a spam filter on its computer system. During a 21-day period, 6693 messages were tagged as spam. How much spam you get depends on what your online habits are. Here are the counts for some students and faculty in this department (with log-in IDs changed, of course):

ID	Count	ID	Count	ID	Count	ID	Count
AA	1818	BB	1358	CC	442	DD	416
EE	399	FF	389	GG	304	HH	251
II	251	JJ	178	KK	158	LL	103

All other department members received fewer than 100 spam messages. How many did the others receive in total? Make a graph and comment on what you learn from these data. DATA FILE SPAMFILTERS

3.149. Two distributions. If two distributions have exactly the same mean and standard deviation, must their histograms have the same shape? If the distributions have the same five-number summary, must their histograms have the same shape? Explain.

3.150. CHALLENGE **Norms for reading scores.** Raw scores on behavioral tests are often transformed for easier comparison. A test of reading ability has mean 70 and standard deviation 10 when given to third-graders. Sixth-graders have mean score 80 and standard deviation 11 on the same test. To provide separate "norms" for each grade, we want scores in each grade to have mean 100 and standard deviation 20.

(a) What linear transformation will change third-grade scores x into new scores $x_{new} = a + bx$ that have the desired mean and standard deviation? (Use $b > 0$ to preserve the order of the scores.)

(b) Do the same for the sixth-grade scores.

(c) David is a third-grade student who scores 72 on the test. Find David's transformed score. Nancy is a sixth-grade student who scores 78. What is her transformed score? Who scores higher within his or her grade?

(d) Suppose that the distribution of scores in each grade is Normal. Then both sets of transformed scores have the $N(100, 20)$ distribution. What percent of third-graders have scores less than 75? What percent of sixth-graders have scores less than 75?

CHAPTER FOUR

LOOKING AT DATA—RELATIONSHIPS

In Chapter 3 we learned to use graphical and numerical methods to describe the distribution of a single variable. Many of the interesting examples of the use of statistics involve relationships between pairs of variables. Learning ways to describe relationships with graphical and numerical methods is the focus of this chapter.

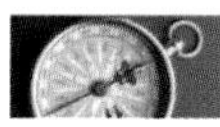

EXAMPLE 4.1

Stress and lack of sleep. Stress is a common problem for college students. Exploring factors that are associated with stress may lead to strategies that will help students to relieve some of the stress that they experience. Recent studies have suggested that a lack of sleep is associated with stress.[1] The two variables involved in the relationship here are lack of sleep and stress. The cases are the students who are the subjects for a particular study.

LOOK BACK
cases p. 80

When we study relationships between two variables, it is not sufficient to collect data on both variables. A key idea for this chapter is that *both variables must be measured on the same cases.* For Example 4.1, this means that each student must have a measured value for lack of sleep and for stress.

USE YOUR KNOWLEDGE

4.1 Relationship between first test and final exam. You want to study the relationship between the score on the first exam and the score on the final for the 30 students enrolled in an elementary statistics class. Who are the cases for your study?

We use the term *associated* to describe the relationship between two variables, such as stress and lack of sleep in Example 4.1. Here is another example where two variables are associated.

EXAMPLE 4.2

Size and price of a coffee beverage. You visit a local Starbucks to buy a Mocha Frappuccino. The barista explains that this blended coffee beverage comes in three sizes and asks if you want a Tall, a Grande, or a Venti. The prices are \$3.50, \$4.00, and \$4.50, respectively. There is a clear association between the size of the Mocha Frappuccino and its price.

ASSOCIATION BETWEEN VARIABLES

Two variables measured on the same cases are **associated** if knowing the values of one of the variables tells you something about the values of the other variable that you would not know without this information.

In the Mocha Frappuccino example, knowing the size tells you the exact price, so the association here is very strong. Many statistical associations, however, are simply overall tendencies that allow exceptions. Some people get adequate sleep and are highly stressed. Others get little sleep and do not experience much stress. The association here is much weaker than the one in the Mocha Frappuccino example.

Examining relationships

When you examine the relationship between two or more variables, first ask the preliminary questions that are familiar from Chapter 3:

- What *cases* do the data describe?
- What *variables* are present? How are they measured?
- Which variables are *quantitative* and which are *categorical?*

EXAMPLE 4.3

Stress and lack of sleep. A study of stress and lack of sleep collected data on 1125 students from an urban midwestern university. Two of the variables measured were the Pittsburgh Sleep Quality Index (PSQI) and the Subjective Units of Distress Scale (SUDS). In this study the cases are the 1125 students studied.[2] The PSQI is based on responses to a large number of questions that are summarized in a single variable that has a value between 0 and 21 for each subject. Therefore, we will treat the PSQI as a quantitative variable. SUDS is a similar scale with values between 0 and 100 for each subject. We will treat SUDS as a quantitative variable also.

In many situations, we measure a collection of categorical variables and then combine them in a scale that can be viewed as a quantitative variable. The PSQI is an example. We can also turn the tables in the other direction. Here is an example:

EXAMPLE 4.4

Hemoglobin and anemia. Hemoglobin is a measure of iron in the blood. The units are grams of hemoglobin per deciliter of blood (g/dl). Normal values depend on age and gender. Adult women typically have values between 12 and 16 g/dl.

Anemia (lack of iron in the blood) is a major problem in developing countries, and many studies have been designed to evaluate the problem. In these studies, computing the mean hemoglobin level of a sample of people is not particularly useful. For studies like this, using a definition of severe anemia as a hemoglobin level of less than 8 g/dl is more appropriate.

Now, researchers can administer different treatments and then compare the proportions of subjects who are severely anemic rather than compare the difference in the mean hemoglobin levels. The categorical variable, severely anemic or not, is much more useful than the quantitative variable, hemoglobin level.

When analyzing data to draw conclusions, it is important to carefully consider the best way to summarize the data. *Just because a variable is measured as a quantitative variable, it does not necessarily follow that the best summary is based on the mean (or the median).* As the previous example illustrates, converting a quantitative variable to a categorical variable is a very useful option to keep in mind.

USE YOUR KNOWLEDGE

4.2 Create a categorical variable from a quantitative variable. Consider the study described in Example 4.3. Some analyses compared three groups of students. The students were classified as having optimal sleep quality (a PSQI of 5 or less), borderline sleep quality (a PSQI of 6 and 7), or poor sleep quality (a PSQI of 8 or more). When the three groups of students are compared, is the PSQI being used as a quantitative variable or as a categorical variable? Explain your answer and describe some advantages to using the optimal, borderline, and poor categories in explaining the results of a study such as this.

4.3 Replace names by ounces. In the Mocha Frappuccino example, the variable size is categorical, with Tall, Grande, and Venti as the possible values. Suppose you converted these values to the number of ounces: Tall is 12 ounces, Grande is 16 ounces, and Venti is 24 ounces. For studying the relationship between ounces and price, describe the cases and the variables, and state whether each variable is quantitative or categorical.

A new question

When you examine the relationship between two variables, a new question becomes important:

- Is your purpose simply to explore the nature of the relationship, or do you hope to show that one of the variables can explain variation in the other? That is, are some of the variables *response variables* and others *explanatory variables?*

RESPONSE VARIABLE, EXPLANATORY VARIABLE

A **response variable** measures an outcome of a study. An **explanatory variable** explains or causes changes in the response variables.

EXAMPLE 4.5

Stress and lack of sleep. Refer to the study of stress and lack of sleep in Example 4.3. Here, the explanatory variable is the Pittsburgh Sleep Quality Index and the response variable is the Subjective Units of Distress Scale.

USE YOUR KNOWLEDGE

4.4 Sleep and stress or stress and sleep? Consider the scenario described in Example 4.3. Make an argument for treating the Subjective Units of Distress Scale as the explanatory variable and the Pittsburgh Sleep Quality Index as the response variable.

In some studies it is easy to identify explanatory and response variables. The following example illustrates one situation where this is true: when we actually set values of one variable to see how it affects another variable.

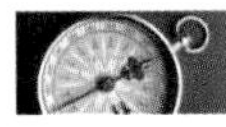

EXAMPLE 4.6

How much calcium do you need? Adolescence is a time when bones are growing very actively. If young people do not have enough calcium, their bones will not grow properly. How much calcium is enough?

Research designed to answer this question has been performed for many years at events called "Camp Calcium."[3] At these camps subjects eat a controlled diet that is identical except for the amount of calcium.

The amount of calcium retained by the body is the major response variable of interest. Since the amount of calcium consumed is controlled by the researchers, this variable is the explanatory variable.

When you don't set the values of either variable but just observe both variables, there may or may not be explanatory and response variables. Whether there are depends on how you plan to use the data.

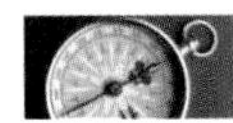

EXAMPLE 4.7

Student loans. A college student aid officer looks at the findings of the National Student Loan Survey. She notes data on the amount of debt of recent graduates, their current income, and how stressed they feel about college debt. She isn't interested in predictions but is simply trying to understand the situation of recent college graduates.

A sociologist looks at the same data with an eye to using amount of debt and income, along with other variables, to explain the stress caused by college debt. Amount of debt and income are explanatory variables, and stress level is the response variable.

In many studies, the goal is to show that changes in one or more explanatory variables actually *cause* changes in a response variable. But many explanatory-response relationships do not involve direct causation. The SAT scores of high school students help predict the students' future college grades, but high SAT scores certainly don't cause high college grades.

Some of the statistical techniques in this chapter require us to distinguish explanatory from response variables. Others make no use of this distinction. You will often see explanatory variables called **independent variables** and response variables called **dependent variables.** These are mathematical terms, not statistical terms. The idea behind this language is that response variables depend on explanatory variables. Because the words "independent" and "dependent" have other meanings in statistics that are unrelated to the explanatory-response distinction, we prefer to avoid those words.

independent variable
dependent variable

Most statistical studies examine data on more than one variable. Fortunately, statistical analysis of several-variable data builds on the tools used for examining individual variables. The principles that guide our work also remain the same:

- Start with a graphical display of the data.
- Look for overall patterns and deviations from those patterns.
- Based on what you see, use numerical summaries to describe specific aspects of the data.

4.1 Scatterplots

Let's start with an example. We will first look at a graphical display of the data. Later we will explore a numerical summary.

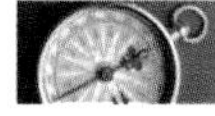

EXAMPLE 4.8

Spam botnets. A botnet is a remotely and silently controlled collection of networked computers. Botnets are illicitly created through the use of viruses, Trojans, and other malware to assimilate computers, or bots, into the botnet, generally without the knowledge of the computer owner. Some botnets can grow to many thousands of bots located all over the world.

A botnet that is used to send unwanted commercial emails, called spam, is called a spam botnet.[4] About 120 billion spam messages are sent per day, and the cost of dealing with spam messages is estimated to be $140 billion per year.[5] Here is some information about 10 large botnets:

Botnet	Bots (thousands)	Spams per day (billions)	Botnet	Bots (thousands)	Spams per day (billions)
Srizbi	315	60	Grum	50	2
Bobax	185	9	Ozdok	35	10
Rustock	150	30	Nucrypt	20	5
Cutwail	125	16	Wopla	20	0.6
Storm	85	3	Spamthru	12	0.35

The variables are the number of bots operated by the botnet and the number of spam messages per day produced by these bots. The first botnet listed is a botnet called Srizbi, which was discovered on June 30, 2007.[6] Srizbi has 315,000 bots that generate 60 billion spams per day.

USE YOUR KNOWLEDGE

4.5 Make a data set.

(a) Create a spreadsheet that contains the spam botnet data.

(b) How many cases are in your data set?

(c) Describe the labels, variables, and values that you used.

(d) Which columns give quantitative variables? SPAM

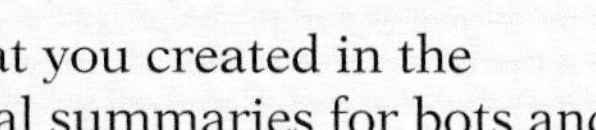

4.6 Use your data set. Using the data set that you created in the previous exercise, find graphical and numerical summaries for bots and spam messages per day. SPAM

The most common way to display the relation between two quantitative variables is a *scatterplot*. We explored the use of this tool in Chapter 1.

scatterplot p. 15

SCATTERPLOT

A **scatterplot** shows the relationship between two quantitative variables measured on the same individuals. The values of one variable appear on the horizontal axis, and the values of the other variable appear on the vertical axis. Each individual in the data appears as the point in the plot fixed by the values of both variables for that individual.

EXAMPLE 4.9

SPAM

Bots and spam messages. We think that a botnet that has a large number of bots would be capable of generating a large number of spam messages, relative to a botnet that has a smaller number of bots. Therefore, we think of the number of bots as an explanatory variable and the number of spam messages as a response variable. We begin our study of this relationship with a graphical display of the two variables.

Figure 4.1 gives a scatterplot that displays the relationship between the response variable, spam messages per day, and the explanatory variable, number of bots. Notice that 6 of the 10 botnets are clustered in the lower-left part of the plot with relatively low values for both bots and spam messages per day. On the other hand, the botnet Srizbi stands out with the highest values for both variables.

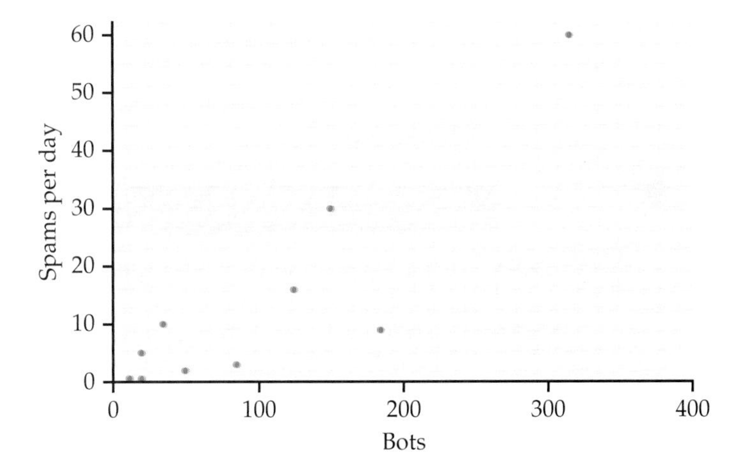

FIGURE 4.1 Scatterplot of bots (in thousands) versus spams per day (in billions), for Example 4.9.

Always plot the explanatory variable, if there is one, on the horizontal axis (the x axis) of a scatterplot. We usually call the explanatory variable x and the response variable y. If there is no explanatory-response distinction, either variable can go on the horizontal axis.

USE YOUR KNOWLEDGE

4.7 Make a scatterplot.

(a) Make a scatterplot similar to Figure 4.1 for the spam botnet data.

(b) Mark the location of the botnet Bobax on your plot. SPAM

4.8 Change the units.

(a) Create a spreadsheet with the spam botnet data using the actual values. In other words, for Srizbi use 315,000 for the number of bots and 60,000,000,000 for the number of spam messages per day.

(b) Make a scatterplot for the data coded in this way.

(c) Describe how this scatterplot differs from Figure 4.1. DATA FILE SPAM

Interpreting scatterplots

To look more closely at a scatterplot such as Figure 4.1, apply the strategies of exploratory analysis learned in Chapter 3. Here is a summary.

EXAMINING A SCATTERPLOT

In any graph of data, look for the **overall pattern** and for striking **deviations** from that pattern.

You can describe the overall pattern of a scatterplot by the **form, direction,** and **strength** of the relationship.

An important kind of deviation is an **outlier,** an individual value that falls outside the overall pattern of the relationship.

linear relationship

Figure 4.1 shows a clear *form:* the data lie in a roughly straight-line, or **linear,** pattern. To help us see this relationship, we can use software to put a straight line through the data. We will see more details about how this is done in Section 4.3.

EXAMPLE 4.10

SPAM

cluster

Scatterplot with a straight line. Figure 4.2 plots the botnet data with a fitted straight line. There is a **cluster** of points in the lower left. Although Srizbi appears to be an outlier, it lies roughly in the same linear pattern as the other botnets.

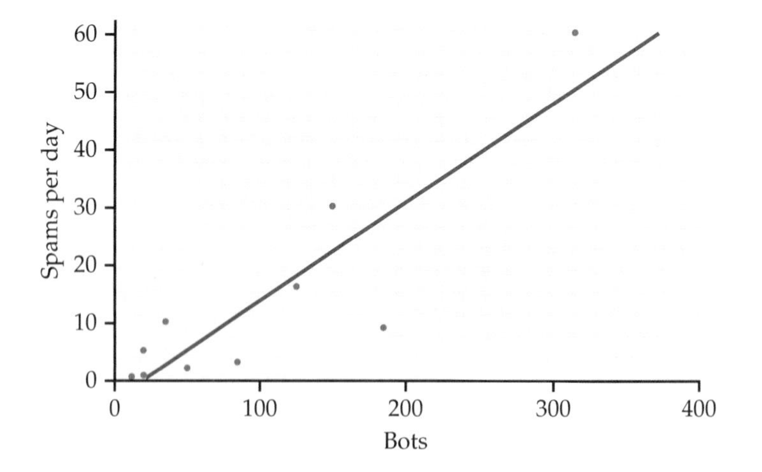

FIGURE 4.2 Scatterplot of bots (in thousands) versus spams per day (in billions) with a fitted straight line, for Example 4.10.

The relationship in Figure 4.2 also has a clear *direction:* botnets with more bots generate more spam messages than botnets that have fewer bots. This is a *positive association* between the two variables.

POSITIVE ASSOCIATION, NEGATIVE ASSOCIATION

Two variables are **positively associated** when above-average values of one tend to accompany above-average values of the other, and below-average values also tend to occur together.

Two variables are **negatively associated** when above-average values of one tend to accompany below-average values of the other, and vice versa.

strength of a relationship

The **strength of a relationship** in a scatterplot is determined by how closely the points follow a clear form. The overall relationship in Figure 4.1 is fairly moderate. Botnets with similar numbers of bots have a fair amount of scatter in the number of spam messages per day that they produce. Here is an example of a stronger linear relationship.

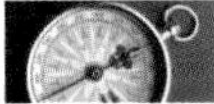

EXAMPLE 4.11

DEBT

Debt for 24 countries. The amount of debt owed by a country is a measure of its economic health. The Organisation for Economic Co-operation and Development (OECD) collects data on the central government debt for many countries. One of their tables gives the debt for 30 countries for the years 1998 to 2007.[7] Since there are a few countries with a very large amount of debt, let's concentrate on the 24 countries with debt less than US$ trillion in 2006. The 6 countries excluded are the United Kingdom, France, Germany, Italy, the United States, and Japan. The data are given in Figure 4.3.

Figure 4.4 is a scatterplot of the central government debt in 2007 versus the central government debt in 2006. The scatterplot shows a strong positive relationship between the debt in these two years.

USE YOUR KNOWLEDGE

4.9 Make a scatterplot. In our Mocha Frappuccino example, the 12-ounce drink costs $3.50, the 16-ounce drink costs $4.00, and the 24-ounce drink costs $4.50. Explain which variable should be used as the explanatory variable and make a scatterplot. Describe the scatterplot and the association between these two variables.

It is tempting to conclude that the strong linear relationship that we have found between the debt in 2006 and debt in 2007 for the 24 countries is evidence that the amount of debt for each country is approximately the same in the two years. The first exercise below asks you to explore this temptation.

Microsoft Excel - GovDebt200...

E25

	A	B	C
1	Country	Debt2006	Debt2007
2	Luxembourg	0.65	0.78
3	Iceland	4.04	4.92
4	Slovak Republic	18.44	22.79
5	New Zealand	21.44	27.82
6	Norway	43.03	49.05
7	Czech Republic	38.44	49.36
8	Australia	43.91	49.45
9	Ireland	47.30	55.29
10	Finland	77.58	82.54
11	Hungary	76.76	90.24
12	Denmark	94.29	92.74
13	Switzerland	101.28	107.49
14	Portugal	142.97	166.06
15	Sweden	185.01	182.11
16	Poland	164.41	205.97
17	Mexico	195.66	216.76
18	Austria	204.51	231.56
19	Turkey	244.89	286.96
20	Korea	295.07	309.34
21	Netherlands	278.67	316.07
22	Greece	297.93	352.80
23	Canada	347.68	389.83
24	Belgium	366.91	420.74
25	Spain	427.06	464.99
26			

FIGURE 4.3 Central government debt in 2006 and 2007 for 24 countries with debt in 2006 less than US$1 trillion, in US$ billions, for Example 4.11.

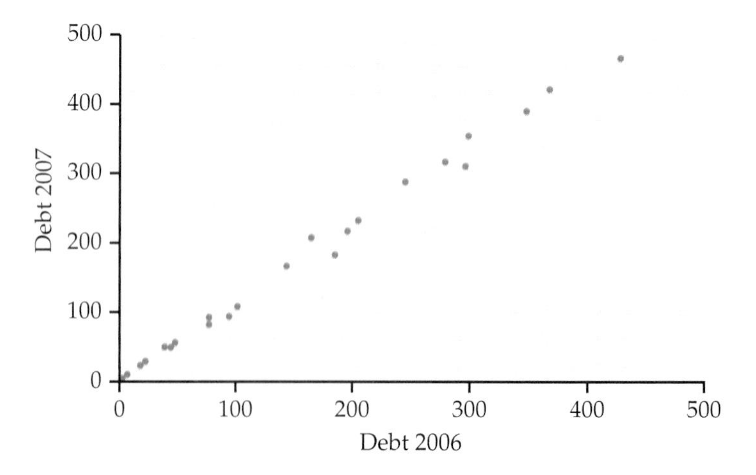

FIGURE 4.4 Scatterplot of debt in 2007 (US$ billions) versus debt in 2006 (US$ billions) for 24 countries with less than US$1 trillion debt in 2006, for Example 4.11.

USE YOUR KNOWLEDGE

4.10 Are the debts in 2006 and 2007 approximately the same? Use the methods you learned in Chapter 3 to examine whether or not the central government debts in 2006 and 2007 are approximately the same. (*Hint:* Think about creating a new variable that would help you to answer this question.) DEBT

4.11 What about the countries with very large debts? In Example 4.11 we excluded six countries. The original data set did not include a value for the debt in 2007 for Japan. Here are the debts, in US$ billions, for the other five countries:

Country	2006 debt	2007 debt
United Kingdom	1168	1231
France	1240	1454
Germany	1252	1409
Italy	1892	2167
United States	4848	5055

Add the data for these five countries to your data set, and make a scatterplot that includes the data for all 29 countries. Summarize the relationship. Do the additional data change the relationship? Explain your answer. DEBT29

Of course, not all relationships are linear. Here is an example where the relationship is described by a nonlinear curve.

EXAMPLE 4.12

Forbes.com Best Countries for Business. Forbes.com analyzes business climates in 120 countries and determines an ordered list of these countries called the Best Countries for Business.[8] Let's look at two of the variables that they use to determine the ranks in their list: gross domestic product per capita and unemployment rate. We exclude data from a few countries with very extreme values or missing values for one or more of the variables used in the rankings. Figure 4.5 is a scatterplot of gross domestic product per capita versus unemployment rate for 99 countries.

The scatterplot suggests that there is a negative relationship between these two variables. The relationship is approximately linear for small values of unemployment, particularly for values less than about 15%. However, after that point the curve decreases more slowly. Overall, we have a nonlinear relationship.

Look at Figure 4.5 carefully. Notice that the two variables that we are examining are both skewed toward large values. This shows up in the plot as the clustering of many of the data points in the left and in the lower part of the scatterplot. Skewed data are quite common in business applications of statistics,

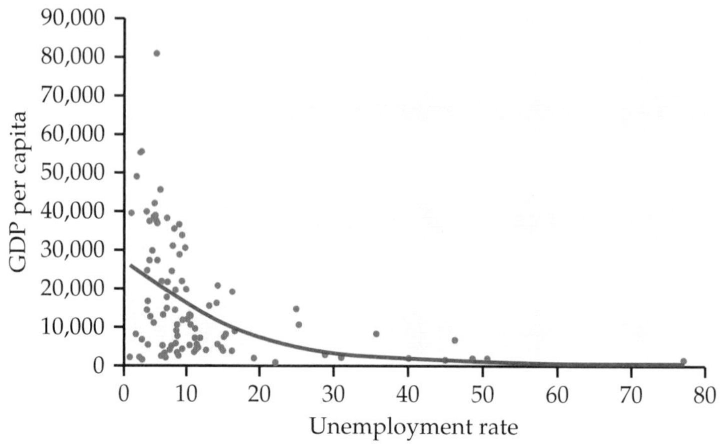

FIGURE 4.5 Scatterplot of gross domestic product per capita versus unemployment for 99 countries, for Example 4.12. There is a negative nonlinear relationship between these two variables.

transformation

particularly when the measured variable is some kind of count or amount. In these situations, we often apply a **transformation** to the data. This means that we replace the original values by the transformed values and then use the transformed values for our analysis.

Transforming data is common in statistical practice. There are systematic principles that describe how transformations behave and guide the search for transformations that will, for example, make a distribution more Normal or a curved relationship more linear.

The log transformation

log transformation

The most important transformation that we will use is the **log transformation.** Strictly speaking, this transformation is used for variables that have only positive values. Occasionally, we use it when there are zero values, but in this case we first replace the zero values by some small value, often one-half of the smallest positive value in the data set.

You have probably encountered logarithms in one of your high school mathematics courses as a way to do certain kinds of arithmetic. Logarithms are a lot more fun when used in statistical analyses. We will use natural logarithms. Statistical software and statistical calculators generally provide easy ways to perform this transformation.

EXAMPLE 4.13

Gross domestic product per capita and unemployment with logarithms. Figure 4.6 is a scatterplot of the log of the gross domestic product per capita versus the log of the unemployment rate. Notice how the data now fill up much of the central part of the scatterplot in contrast to the clustering that we noticed in Figure 4.5. Here we see that the relationship is essentially flat for values of log unemployment that are less than about 1.6. This point corresponds to an unemployment rate of about 5%. So for low unemployment rates, there appears to be little or no relationship between unemployment and gross domestic product per capita. On the other hand, for values of unemployment greater than 5%, we see an approximately linear negative relationship with gross domestic product per capita. High unemployment is associated with low gross domestic product per capita.

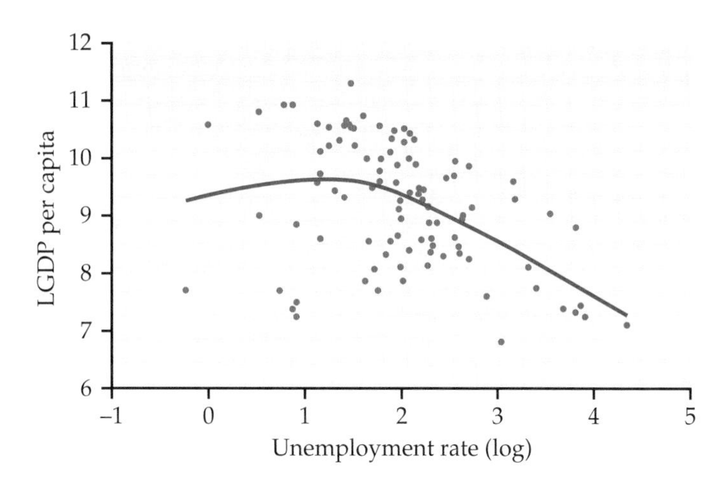

FIGURE 4.6 Scatterplot of log gross domestic product versus log unemployment for 99 countries, along with a smooth curve, for Example 4.13. There appears to be essentially no relationship between the variables for low unemployment values, up to about 5% (1.6 on the log scale), and a negative relationship for unemployment values greater than 5%.

Using transformations and interpreting scatterplots are arts that require judgment and knowledge about the variables that we are studying. *Always ask yourself if the relationship that you see makes sense.* If it does not, then additional analyses are needed to understand the data.

Adding categorical variables to scatterplots

In Examples 4.12 and 4.13, we looked at two of the variables used by Forbes .com to construct their Best Countries for Business list. They use several more variables, but do not give the details about exactly how their list is constructed. Let's take a look at how the two variables we examined relate to whether or not a country ranks high or low on the list.

> **CATEGORICAL VARIABLES IN SCATTERPLOTS**
>
> To add a categorical variable to a scatterplot, use a different plot color or symbol for each category.

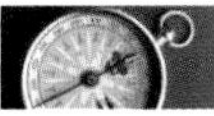

EXAMPLE 4.14

Gross domestic product per capita and the rankings. We start by creating a categorical variable that indicates whether or not a country ranks in the top half of the Best Countries for Business list. In Figure 4.7, we use the symbol H for countries that rank in the top half of the list and L for countries that rank in the bottom half of the list. Examine the scatterplot carefully. Notice that the countries in the top part of the plot, those with relatively high gross domestic product per capita, tend to rank high in the Best Countries for Business list. On the other hand, countries in the bottom part, those with relatively low gross domestic product per capita, tend to rank low. What about unemployment? No clear pattern is evident. Although the countries with the six highest unemployment rates are all ranked low, five of them also have very low values for gross domestic product per capita.

In this example, we used a quantitative variable, rank in the Best Countries for Business list, to create a categorical variable that indicated whether a

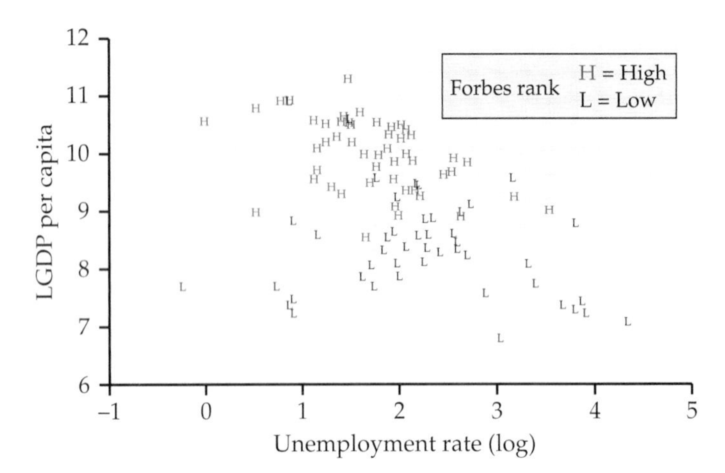

FIGURE 4.7 Scatterplot of log gross domestic product versus log unemployment for 106 countries, for Example 4.14. Countries in the top half of the Best Countries for Business ranking are plotted with the symbol H and those in the lower half of the ranking are plotted with the symbol L. Countries with high gross domestic product tend to be rated high, while the unemployment rate does not appear to be a major factor in distinguishing high versus low rankings.

country ranked high or low in the list. Of course, if the variable that you want to include in a scatterplot by means of a plotting symbol is already categorical, no conversion is required. *Careful judgment is needed in using this graphical method.* Don't be discouraged if your first attempt is not very successful. In performing a good data analysis, you will often produce several plots before you find the one that you believe to be most effective in describing the data.[9]

USE YOUR KNOWLEDGE

4.12 Change the plotting symbol. In Example 4.14 we used the plotting symbols H and L to distinguish countries that ranked high and low on the Best Countries for Business list. Let's see if we can learn anything more about these variables by refining our categorical variable further. Define a new categorical variable that has three distinct values corresponding to a rank in the top third (ranks 1 to 40), the middle third (ranks 41 to 80), and the bottom third (ranks 81 to 120). Choose appropriate plotting symbols and make a scatterplot similar to Figure 4.7 using this categorical variable. Describe your scatterplot and compare what you can learn from it with what we learned in Example 4.14. BESTCOUNTRIES

Categorical explanatory variables

Scatterplots display the association between two quantitative variables. To display a relationship between a categorical explanatory variable and a quantitative response variable, make a side-by-side comparison of the distribution of the quantitative variable for each value of the categorical explanatory variable. Back-to-back stemplots (page 89) and side-by-side boxplots (page 113) are useful tools for this purpose. We will study methods for describing the association between two categorical variables in Section 4.5 (page 202).

SECTION 4.1 SUMMARY

To study relationships between variables, we must measure the variables on the same cases.

If we think that a variable x may explain or even cause changes in another variable y, we call x an **explanatory variable** and y a **response variable.**

A **scatterplot** displays the relationship between two quantitative variables. Mark values of one variable on the horizontal axis (x axis) and values of the other variable on the vertical axis (y axis). Plot each individual's data as a point on the graph.

Always plot the explanatory variable, if there is one, on the x axis of a scatterplot. Plot the response variable on the y axis.

Plot points with different colors or symbols to see the effect of a categorical variable in a scatterplot.

In examining a scatterplot, look for an overall pattern showing the **form, direction,** and **strength** of the relationship, and then for **outliers** or other deviations from this pattern.

Form: Linear relationships, where the points show a straight-line pattern, are an important form of relationship between two variables. Curved relationships and **clusters** are other forms to watch for.

Direction: If the relationship has a clear direction, we speak of either **positive association** (high values of the two variables tend to occur together) or **negative association** (high values of one variable tend to occur with low values of the other variable).

Strength: The **strength** of a relationship is determined by how close the points in the scatterplot lie to a simple form such as a line.

To display the relationship between a categorical explanatory variable and a quantitative response variable, make a graph that compares the distributions of the response for each category of the explanatory variable.

SECTION 4.1 EXERCISES

For Exercise 4.1, see page 154; for Exercises 4.2, 4.3, and 4.4, see page 156; for Exercises 4.5 and 4.6, see page 158; for Exercises 4.7 and 4.8, see pages 159–160; for Exercise 4.9, see page 161; for Exercises 4.10 and 4.11, see page 163; and for Exercise 4.12, see page 166.

4.13. What's wrong? Explain what is wrong with each of the following:

(a) A boxplot can be used to examine the relationship between two variables.

(b) In a scatterplot we put the response variable on the y axis and the explanatory variable on the x axis.

(c) If two variables are positively associated, then high values of one variable are associated with low values of the other variable.

4.14. Make some sketches. For each of the following situations, make a scatterplot that illustrates the given relationship between two variables.

(a) A strong negative linear relationship.

(b) No apparent relationship.

(c) A weak positive relationship.

(d) A more complicated relationship. Explain the relationship.

4.15. Who does not have health insurance? The lack of adequate health insurance coverage is a major problem for many Americans. The Current Population Survey collected data on the characteristics of the uninsured.[10] The numbers of uninsured and the total number of people classified by age are as follows. The units are thousands of people. DATA FILE HEALTHINSURANCE

Age group	Number uninsured	Total number
Under 18 years	7,307	74,916
18 to 24 years	8,078	29,651
25 to 34 years	11,804	41,584
35 to 44 years	8,692	39,842
45 to 64 years	13,231	80,939
65 years and older	792	39,179

(a) Plot the number of uninsured versus age group.

(b) Find the total number of uninsured persons and use this total to compute the percent of the uninsured who are in each age group.

(c) Plot the percents versus age group.

(d) Explain how the plot you produced in part (c) differs from the plot that you made in part (a).

(e) Summarize what you can conclude from these plots.

4.16. Which age groups have the larger percent uninsured? Refer to the previous exercise. Let's take a look at the data from a different point of view. HEALTHINSURANCE

(a) For each age group calculate the percent who are uninsured using the number of uninsured persons and the total number of persons in each group.

(b) Make a plot of the percent uninsured versus age group.

(c) Summarize the information in your plot and write a short summary of what you conclude from your analysis.

4.17. Compare the two percents. In the previous two exercises, you computed percents in two different ways and generated plots of uninsured versus age group. Describe the difference between the two ways with an emphasis on what kinds of conclusions can be drawn from each. HEALTHINSURANCE

4.18. What's in the beer? The Web site Beer100.com advertises itself as "Your Place for All Things Beer." One of their things is a list of 86 domestic beer brands with the percent alcohol, calories per 12 ounces, and carbohydrates per 12 ounces (in grams).[11] BEER

(a) Figure 4.8 gives a scatterplot of carbohydrates versus percent alcohol. Give a short summary of what can be learned from the plot.

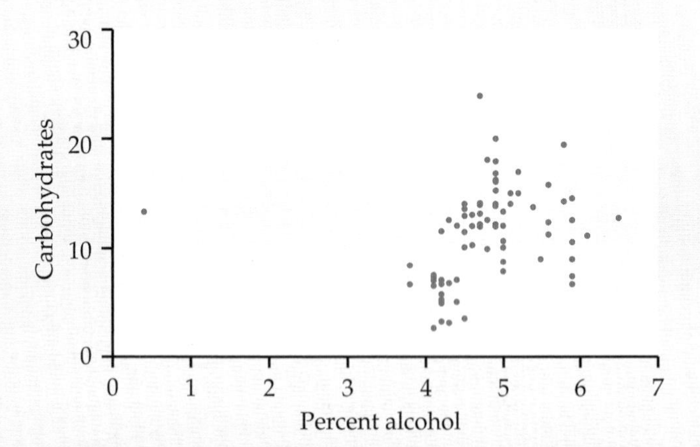

FIGURE 4.8 Scatterplot of carbohydrates (in grams per 12 ounces) versus percent alcohol for 86 brands of beer, for Exercise 4.18.

(b) One of the points is an outlier. Use the data file to find the outlier brand of beer. How is this brand of beer marketed as compared with the other brands?

(c) Remove the outlier from the data set and generate a scatterplot of the remaining data.

(d) Describe the relationship between carbohydrates and percent alcohol based on what you see in your scatterplot.

4.19. More beer. Refer to the previous exercise. BEER

(a) Make a scatterplot of calories versus percent alcohol using the data set without the outlier.

(b) Describe the relationship between these two variables.

4.20. Will you live longer if you use the Internet? The World Bank collects data on many variables related to world development for countries throughout the world. Two of these are Internet use, in number of users per 100 people, and life expectancy, in years.[12] Figure 4.9 is a scatterplot of life expectancy versus Internet users. INTERNETANDLIFE

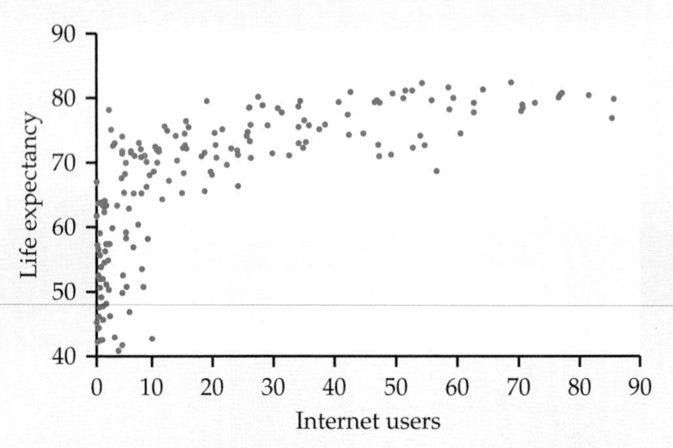

FIGURE 4.9 Scatterplot of life expectancy (in years) versus Internet users (per 100 people) for 181 countries, for Exercise 4.20.

(a) Describe the relationship between these two variables.

(b) A friend looks at this plot and concludes that using the Internet will increase the length of your life. Write a short paragraph explaining why the association seen in the scatterplot does not provide a reason to draw this conclusion.

4.21. Let's look at Europe. Refer to the previous exercise. Figure 4.10 gives a scatterplot for the same data for the 48 European countries in the data set. Compare this figure with Figure 4.9, which plots the data for all 181 countries in the data set. Write a paragraph summarizing the relationship between life expectancy and Internet use for European countries with an emphasis on how the European countries compare with the entire set of 181 countries.

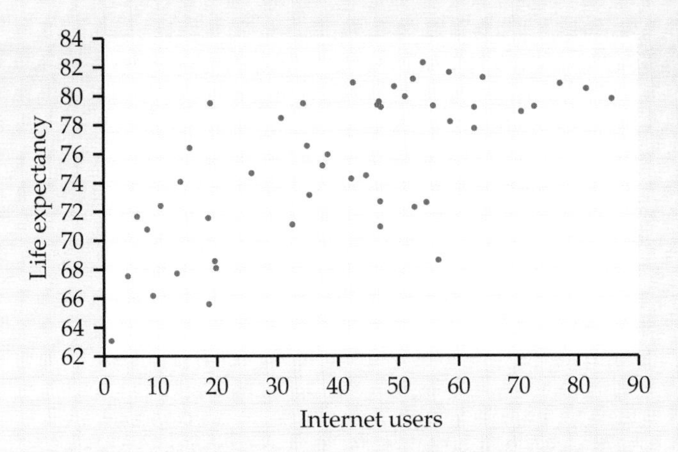

FIGURE 4.10 Scatterplot of life expectancy (in years) versus Internet users (per 100 people) for 48 European countries, for Exercise 4.21.

Be sure to take into account the fact that the software used to create these plots automatically chooses the range of values for each axis so that the space in the plot is used efficiently. In this case, the range of values for Internet use is the same for both scatterplots, but the range of values for life expectancy is quite different. INTERNETLIFEEUR

4.22. CHALLENGE **How would you make a better plot?** In the previous two exercises, we looked at the relationship between life expectancy and Internet use. First, we made a scatterplot for all 181 countries in the data set. Then we made one for the subset of 48 European countries. Explain how you would construct a single plot to make a comparison between the European countries and the other countries in the data set. (*Optional:* Make the plot if you have software that can do what you need.) INTERNETANDLIFE

4.23. Average temperatures. Here are the average temperatures in degrees for West Lafayette, Indiana, during the months of February through May: WLAFTEMPS

Month	February	March	April	May
Temperature (degrees F)	30	41	51	62

(a) Explain why month should be the explanatory variable for examining this relationship.

(b) Make a scatterplot and describe the relationship.

4.24. Relationship between first test and final exam. How strong is the relationship between the score on the first exam and the score on the final exam in an elementary statistics course? Here are data for eight students from such a course: STATCOURSE8

First-test score	153	144	162	149	127	118	158	153
Final-exam score	145	140	145	170	145	175	170	160

(a) Which variable should play the role of the explanatory variable in describing this relationship?

(b) Make a scatterplot and describe the relationship.

(c) Give some possible reasons why this relationship is so weak.

4.25. Relationship between second test and final exam. Refer to the previous exercise. Here are the data for the second test and the final exam for the same students: STATCOURSE8

Second-test score	158	162	144	162	136	158	175	153
Final-exam score	145	140	145	170	145	175	170	160

(a) Explain why you should use the second-test score as the explanatory variable.

(b) Make a scatterplot and describe the relationship.

(c) Why do you think the relationship between the second-test score and the final-exam score is stronger than the relationship between the first-test score and the final-exam score?

4.26. Add an outlier to the plot. Refer to the previous exercise. Add a ninth student whose scores on the second test and final exam would lead you to classify the additional data point as an outlier. Highlight the outlier on your scatterplot and describe the performance of the student on the second exam and final exam and why that leads to the conclusion that the result is an outlier. Give a possible reason for the performance of this student.

4.27. Explanatory and response variables. In each of the following situations, is it more reasonable to simply explore the relationship between the two variables or to view one of the variables as an explanatory variable and the other as a response variable? In the latter case, which is the explanatory variable and which is the response variable?

(a) The weight of a child and the age of the child from birth to 10 years.

(b) High school English grades and high school math grades.

(c) The rental price of apartments and the number of bedrooms in the apartment.

(d) The amount of sugar added to a cup of coffee and how sweet the coffee tastes.

(e) The student evaluation scores for an instructor and the student evaluation scores for the course.

4.28. Parents' income and student loans. How well does the income of a college student's parents predict how much the student will borrow to pay for college? We have data on parents' income and college debt for a sample of 1200 recent college graduates. What are the explanatory and response variables? Are these variables categorical or quantitative? Do you expect a positive or a negative association between these variables? Why?

4.29. Small falcons in Sweden. Often the percent of an animal species in the wild that survives to breed again is lower following a successful breeding season. This is part of nature's self-regulation, tending to keep population size stable. A study of merlins (small falcons) in northern Sweden observed the number of breeding pairs in an isolated area and the percent of males (banded for identification) who returned the next breeding season. Here are data for nine years:[13] DATA FILE FALCONS

Pairs	28	29	29	29	30	32	33	38	38
Percent	82	83	70	61	69	58	43	50	47

(a) Why is the response variable the *percent* of males that return rather than the *number* of males that return?

(b) Make a scatterplot. To emphasize the pattern, also plot the mean response for years with 29 and 38 breeding pairs and draw lines connecting the mean responses for the six values of the explanatory variable.

(c) Describe the pattern. Do the data support the theory that a smaller percent of birds survive following a successful breeding season?

4.30. Biological clocks. Many plants and animals have "biological clocks" that coordinate activities with the time of day. When researchers looked at the length of the biological cycles in the plant *Arabidopsis* by measuring leaf movements, they found that the length of the cycle is not always 24 hours. The researchers suspected that the plants adapt their clocks to their north-south position. Plants don't know geography, but they do respond to light, so the researchers looked at the relationship between the plants' cycle lengths and the length of the day on June 21 at their locations. The data file includes data on cycle length and day length, both in hours, for 146 plants.[14] Plot cycle length as the response variable against day length as the explanatory variable. Does there appear to be a positive association? Is it a strong association? Explain your answers. DATA FILE BIOCLOCKS

4.31. CHALLENGE **Body mass and metabolic rate.** Metabolic rate, the rate at which the body consumes energy, is important in studies of weight gain, dieting, and exercise. The following table gives data on the lean body mass and resting metabolic rate for 12 women and 7 men who are subjects in a study of dieting. Lean body mass, given in kilograms, is a person's weight leaving out all fat. Metabolic rate is measured in calories burned per 24 hours, the same calories used to describe the energy content of foods. The researchers believe that lean body mass is an important influence on metabolic rate. DATA FILE BODYMASS

Subject	Sex	Mass	Rate	Subject	Sex	Mass	Rate
1	M	62.0	1792	11	F	40.3	1189
2	M	62.9	1666	12	F	33.1	913
3	F	36.1	995	13	M	51.9	1460
4	F	54.6	1425	14	F	42.4	1124
5	F	48.5	1396	15	F	34.5	1052
6	F	42.0	1418	16	F	51.1	1347
7	M	47.4	1362	17	F	41.2	1204
8	F	50.6	1502	18	M	51.9	1867
9	F	42.0	1256	19	M	46.9	1439
10	M	48.7	1614				

(a) Make a scatterplot of the data, using different symbols or colors for men and women.

(b) Is the association between these variables positive or negative? What is the form of the relationship? How strong is the relationship? Does the pattern of the relationship differ for women and men? How do the male subjects as a group differ from the female subjects as a group?

4.2 Correlation

A scatterplot displays the form, direction, and strength of the relationship between two quantitative variables. Linear (straight-line) relations are particularly important because a straight line is a simple pattern that is quite common. We say a linear relationship is strong if the points lie close to a straight line, and weak if they are widely scattered about a line.

Our eyes are not good judges of how strong a relationship is. The two scatterplots in Figure 4.11 depict exactly the same data, but the plot on the right is drawn smaller in a large field. The plot on the right seems to show a stronger relationship. Our eyes can be fooled by changing the plotting scales or the amount of white space around the cloud of points in a scatterplot.[15]

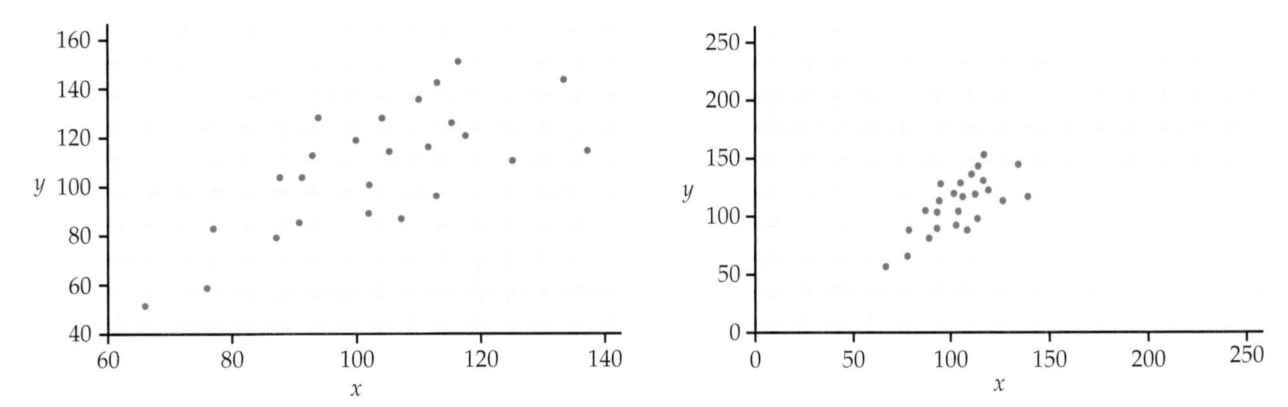

FIGURE 4.11 Two scatterplots of the same data. The linear pattern in the plot on the right appears stronger because of the surrounding space.

We need to follow our strategy for data analysis by using a numerical measure to supplement the graph. *Correlation* is the measure we use.

The correlation r

We have data on variables x and y for n individuals. Think, for example, of measuring height and weight for n people. Then x_1 and y_1 are your height and your weight, x_2 and y_2 are my height and my weight, and so on. For the ith individual, height x_i goes with weight y_i. Here is the definition of correlation.

CORRELATION

The **correlation** measures the direction and strength of the linear relationship between two quantitative variables. Correlation is usually written as r.

Suppose that we have data on variables x and y for n individuals. The means and standard deviations of the two variables are $\overline{x}$ and s_x for the x-values, and $\overline{y}$ and s_y for the y-values. The correlation r between x and y is

$$r = \frac{1}{n-1} \sum \left(\frac{x_i - \overline{x}}{s_x} \right) \left(\frac{y_i - \overline{y}}{s_y} \right)$$

As always, the summation sign $\sum$ means "add these terms for all the individuals."

The formula for the correlation r is a bit complex. It helps us see what correlation is but is not convenient for actually calculating r. In practice you should use software or a calculator that finds r from keyed-in values of two variables x and y.

The formula for r begins by standardizing the observations. Suppose, for example, that x is height in centimeters and y is weight in kilograms and that we have height and weight measurements for n people. Then $\bar{x}$ and s_x are the mean and standard deviation of the n heights, both in centimeters. The value

$$\frac{x_i - \bar{x}}{s_x}$$

is the standardized height of the ith person, familiar from Chapter 3. The standardized height says how many standard deviations above or below the mean a person's height lies. Standardized values have no units—in this example, they are no longer measured in centimeters. Standardize the weights also. The correlation r is an average of the products of the standardized height and the standardized weight for the n people.

USE YOUR KNOWLEDGE

4.32 Spam botnets. In Exercise 4.5 you made a data set for the botnet data. Use that data set to compute the correlation between the number of bots and the number of spam messages per day. DATA FILE SPAM

4.33 Change the units. In the previous exercise bots were given in thousands and spam messages per day were recorded in billions. In Exercise 4.8 you created a data set using the actual values. For example, Srizbi has 315,000 bots and generates 60,000,000,000 spam messages per day. DATA FILE SPAM

(a) Find the correlation between bots and spam messages using this data set.

(b) Compare this correlation with the one that you computed in the previous exercise.

(c) What can you say in general about the effect of changing units in this way on the size of the correlation?

Properties of correlation

The formula for correlation helps us see that r is positive when there is a positive association between the variables. Height and weight, for example, have a positive association. People who are above average in height tend to also be above average in weight. Both the standardized height and the standardized weight for such a person are positive. People who are below average in height tend also to have below-average weight. Then both standardized height and standardized weight are negative. In both cases, the products in the formula for r are mostly positive and so r is positive. In the same way, we can see that r is negative when the association between x and y is negative. More detailed study of the formula gives more detailed properties of r. Here is what you need to know to interpret correlation:

- Correlation makes no use of the distinction between explanatory and response variables. It makes no difference which variable you call x and which you call y in calculating the correlation.

- *Correlation requires that both variables be quantitative.* For example, we cannot calculate a correlation between the incomes of a group of people and what city they live in, because city is a categorical variable.
- Because r uses the standardized values of the observations, r does not change when we change the units of measurement of x, y, or both. Measuring height in inches rather than centimeters and weight in pounds rather than kilograms does not change the correlation between height and weight. The correlation r itself has no unit of measurement; it is just a number.
- Positive r indicates positive association between the variables, and negative r indicates negative association.
- The correlation r is always a number between -1 and 1. Values of r near 0 indicate a very weak linear relationship. The strength of the relationship increases as r moves away from 0 toward either -1 or 1. Values of r close to -1 or 1 indicate that the points lie close to a straight line. The extreme values $r = -1$ and $r = 1$ occur only when the points in a scatterplot lie exactly along a straight line.

- Correlation measures the strength of only the linear relationship between two variables. *Correlation does not describe curved relationships between variables, no matter how strong they are.*
- *Like the mean and standard deviation, the correlation is not resistant: r is strongly affected by a few outlying observations.* Use r with caution when outliers appear in the scatterplot.

The scatterplots in Figure 4.12 illustrate how values of r closer to 1 or -1 correspond to stronger linear relationships. To clarify the meaning of r, the standard deviations of both variables in these plots are equal and the horizontal and vertical scales are the same. In general, it is not so easy to guess the value of r from the appearance of a scatterplot.

Remember that changing the plotting scales in a scatterplot may mislead our eyes, but it does not change the standardized values of the variables and therefore cannot change the correlation. To explore how extreme observations can influence r, use the *Correlation and Regression* applet available on the text Web site.

Finally, remember that *correlation is not a complete description of two-variable data,* even when the relationship between the variables is linear. You should give the means and standard deviations of both x and y along with the correlation. (Because the formula for correlation uses the means and standard deviations, these measures are the proper choices to accompany a correlation.) Conclusions based on correlations alone may require rethinking in the light of a more complete description of the data.

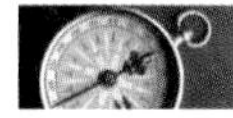

EXAMPLE 4.15

Scoring of figure skating in the Olympics. Until a scandal at the 2002 Olympics brought change, figure skating was scored by judges on a scale from

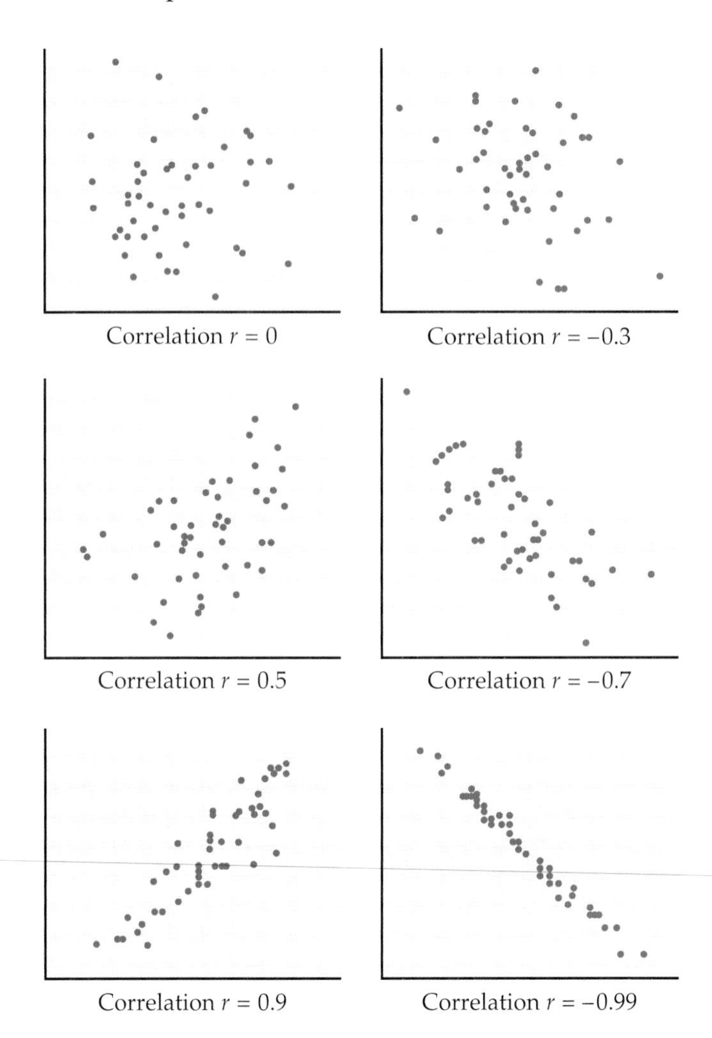

FIGURE 4.12 How the correlation r measures the direction and strength of a linear association.

0.0 to 6.0. The scores were often controversial. We have the scores awarded by two judges, Pierre and Elena, to many skaters. How well do they agree? We calculate that the correlation between their scores is $r = 0.9$. But the mean of Pierre's scores is 0.8 point lower than Elena's mean.

These facts in the example above do not contradict each other. They are simply different kinds of information. The mean scores show that Pierre awards lower scores than Elena. But because Pierre gives *every* skater a score about 0.8 point lower than Elena, the correlation remains high. Adding the same number to all values of either x or y does not change the correlation. If both judges score the same skaters, the competition is scored consistently because Pierre and Elena agree on which performances are better than others. The high r shows their agreement. But if Pierre scores some skaters and Elena others, we must add 0.8 points to Pierre's score to arrive at a fair comparison.

SECTION 4.2 SUMMARY

The **correlation** r measures the direction and strength of the linear (straight line) association between two quantitative variables x and y. Although you can calculate a correlation for any scatterplot, r measures only linear relationships.

Correlation indicates the direction of a linear relationship by its sign: $r > 0$ for a positive association and $r < 0$ for a negative association.

Correlation always satisfies $-1 \leq r \leq 1$ and indicates the strength of a relationship by how close it is to -1 or 1. Perfect correlation, $r = \pm 1$, occurs only when the points lie exactly on a straight line.

Correlation ignores the distinction between explanatory and response variables. The value of r is not affected by changes in the unit of measurement of either variable.

Correlation is not resistant, so outliers can greatly change the value of r.

SECTION 4.2 EXERCISES

For Exercises 4.32 and 4.33, see page 172.

4.34. Thinking about correlation. Figure 4.4 (page 162) is a scatterplot of 2007 debt versus 2006 debt for 24 countries. Is the correlation r for these data near -1, clearly negative but not near -1, near 0, clearly positive but not near 1, or near 1? Explain your answer. DATA FILE DEBT

4.35. Brand names and generic products.

(a) If a store always prices its generic "store brand" products at 90% of the brand name products' prices, what would be the correlation between the prices of the brand name products and the store brand products? (*Hint:* Draw a scatterplot for several prices.)

(b) If the store always prices its generic products $1 less than the corresponding brand name products, then what would be the correlation between the prices of the brand name products and the store brand products?

4.36. Strong association but no correlation. Here is a data set that illustrates an important point about correlation: DATA FILE CORRELATION

X	20	30	40	50	60
Y	10	30	50	30	10

(a) Make a scatterplot of Y versus X.

(b) Describe the relationship between Y and X. Is it weak or strong? Is it linear?

(c) Find the correlation between Y and X.

(d) What important point about correlation does this exercise illustrate?

4.37. Alcohol and carbohydrates in beer. Figure 4.8 (page 168) gives a scatterplot of carbohydrates versus percent alcohol in 86 brands of beer. Compute the correlation for these data. DATA FILE BEER

4.38. Alcohol and carbohydrates in beer revisited. Refer to the previous exercise. The data that you used to compute the correlation includes an outlier. DATA FILE BEER

(a) Remove the outlier and recompute the correlation.

(b) Write a short paragraph about the possible effects of outliers on a correlation using this example to illustrate your ideas.

4.39. Will you live longer if you use the Internet? Figure 4.9 (page 168) is a scatterplot of life expectancy versus Internet users per 100 people for 181 countries. In Exercise 4.20 you described this relationship. Make a plot of the data similar to Figure 4.9 and report the correlation. DATA FILE INTERNETANDLIFE

4.40. Let's look at Europe. Refer to the previous exercise. Figure 4.10 (page 169) gives a scatterplot of the same data for the 48 European countries in the data set. DATA FILE INTERNETLIFEEUR

(a) Make a plot of the data similar to Figure 4.10.

(b) Report the correlation.

(c) Summarize the differences and similarities between the relationship for all 181 countries and the results that you found in this exercise for the European countries only.

4.41. Second test and final exam. In Exercise 4.25 you looked at the relationship between the score on the second test and the score on the final exam in an elementary statistics course. Here are the data: DATA FILE STATCOURSE8

Second-test score	158	162	144	162	136	158	175	153
Final-exam score	145	140	145	170	145	175	170	160

(a) Find the correlation between these two variables.

(b) Do you think that the correlation between the first test and the final exam should be higher than, approximately equal to, or lower than the correlation between the second test and the final exam? Give a reason for your answer.

4.42. First test and final exam. Refer to the previous exercise. Here are the data for the first test and the final exam: DATA FILE STATCOURSE8

First-exam score	153	144	162	149	127	118	158	153
Final-exam score	145	140	145	170	145	175	170	160

(a) Find the correlation between these two variables.

(b) In Exercise 4.24 (page 169) we noted that the relationship between these two variables is weak. Does your calculation of the correlation support this statement? Explain your answer.

(c) Examine part (b) of the previous exercise. Does your calculation agree with your prediction?

4.43. The effect of a different point. Examine the data in Exercise 4.41 and add a ninth student who has low scores on the second test and the final exam and fits the overall pattern of the other scores in the data set. Calculate the correlation and compare it with the correlation that you calculated in Exercise 4.41. Write a short summary of your findings.

4.44. The effect of an outlier. Refer to Exercise 4.41. Add a ninth student whose scores on the second test and final exam would lead you to classify the additional data point as an outlier. Recalculate the correlation with this additional case and summarize the effect it has on the value of the correlation.

4.45. Heights of people who date each other. A student wonders if tall women tend to date taller men than do short women. She measures herself, her dormitory roommate, and the women in the adjoining rooms; then she measures the next man each woman dates. Here are the data (heights in inches): DATA FILE DATEHEIGHTS

Women (x)	66	64	66	65	70	65
Men (y)	72	68	70	68	71	65

(a) Make a scatterplot of these data. Based on the scatterplot, do you expect the correlation to be positive or negative? Near ±1 or not?

(b) Find the correlation r between the heights of the men and women.

(c) How would r change if all the men were 6 inches shorter than the heights given in the table? Does the correlation tell us whether women tend to date men taller than themselves?

(d) If heights were measured in centimeters rather than inches, how would the correlation change? (There are 2.54 centimeters in an inch.)

(e) If every woman dated a man exactly 3 inches taller than herself, what would be the correlation between male and female heights?

4.46. An interesting set of data. Make a scatterplot of the following data:

x	1	2	3	4	10	10
y	1	3	3	5	1	11

Use your calculator to show that the correlation is about 0.5. What feature of the data is responsible for reducing the correlation to this value despite a strong straight-line association between x and y in most of the observations? DATA FILE INTERESTING

4.47. APPLET **Use the applet.** You are going to use the *Correlation and Regression* applet to make different scatterplots with 10 points that have correlation close to 0.8. CAUTION *Many patterns can have the same correlation. Always plot your data before you trust a correlation.*

(a) Stop after adding the first 2 points. What is the value of the correlation? Why does it have this value no matter where the 2 points are located?

(b) Make a lower-left to upper-right pattern of 10 points with correlation about $r = 0.8$. Make a rough sketch of your scatterplot.

(c) Make another scatterplot, this time with 9 points in a vertical stack at the left of the plot. Add one point far to the right and move it until the correlation is close to 0.8. Make a rough sketch of your scatterplot.

(d) Make yet another scatterplot, this time with 10 points in a curved pattern that starts at the lower left, rises to the right, then falls again at the far right. Adjust the points up or down until you have a quite smooth curve with correlation close to 0.8. Make a rough sketch of this scatterplot also.

4.48. APPLET **Use the applet.** Go to the *Correlation and Regression* applet. Click on the scatterplot to create a group of 10 points in the lower-right corner of the scatterplot with a strong straight-line negative pattern (correlation about -0.9).

(a) Add one point at the upper left that is in line with the first 10. How does the correlation change?

(b) Drag this last point down until it is opposite the group of 10 points. How small can you make the correlation? Can you make the correlation positive? *A single outlier can greatly strengthen or weaken a correlation. Always plot your data to check for outlying points.*

4.49. What's wrong? Each of the following statements contains a blunder. Explain in each case what is wrong.

(a) "There is a high correlation between the age of American workers and their occupation."

(b) "We found a high correlation ($r = 1.19$) between students' ratings of faculty teaching and ratings made by other faculty members."

(c) "The correlation between the gender of a group of students and the color of their cell phones was $r = 0.23$."

4.50. High correlation does not mean that the values are the same. Investment reports often include correlations. Following a table of correlations among mutual funds, a report adds, "Two funds can have perfect correlation but different levels of risk. For example, Fund A and Fund B may be perfectly correlated, yet Fund A moves 20% whenever Fund B moves 10%." Write a brief explanation, for someone who knows no statistics, of how this can happen. Include a sketch to illustrate your explanation.

4.51. Student ratings of teachers. A college newspaper interviews a psychologist about student ratings of the teaching of faculty members. The psychologist says, "The evidence indicates that the correlation between the research productivity and teaching rating of faculty members is close to zero." The paper reports this as "Professor McDaniel said that good researchers tend to be poor teachers, and vice versa." Explain why the paper's report is wrong. Write a statement in plain language (don't use the word "correlation") to explain the psychologist's meaning.

4.3 Least-Squares Regression

Correlation measures the direction and strength of the linear (straight-line) relationship between two quantitative variables. If a scatterplot shows a linear relationship, we would like to summarize this overall pattern by drawing a line on the scatterplot. A *regression line* summarizes the relationship between two variables, but only in a specific setting: when one of the variables helps explain or predict the other. That is, regression describes a relationship between an explanatory variable and a response variable.

REGRESSION LINE

A **regression line** is a straight line that describes how a response variable y changes as an explanatory variable x changes. We often use a regression line to **predict** the value of y for a given value of x. Regression, unlike correlation, requires that we have an explanatory variable and a response variable.

EXAMPLE 4.16

Fidgeting and fat gain. Does fidgeting keep you slim? Some people don't gain weight even when they overeat. Perhaps fidgeting and other "nonexercise activity" (NEA) explains why—the body might spontaneously increase nonexercise activity when fed more. Researchers deliberately overfed 16 healthy young adults for 8 weeks. They measured fat gain (in kilograms) and, as an explanatory variable, increase in energy use (in calories) from activity other

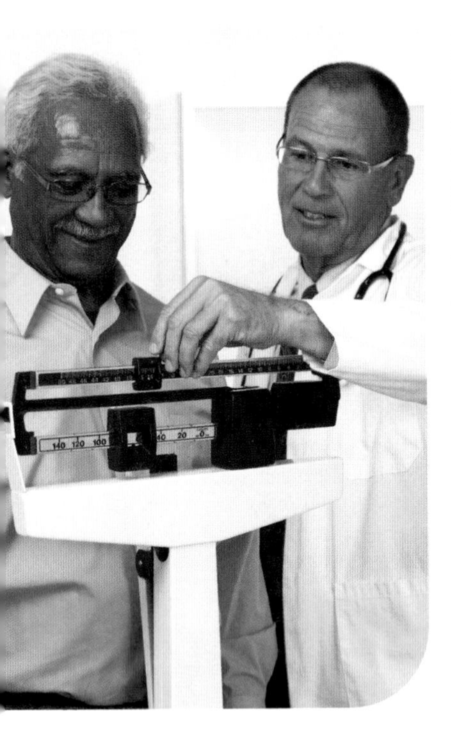

than deliberate exercise—fidgeting, daily living, and the like. Here are the data:[16]

NEA increase (cal)	−94	−57	−29	135	143	151	245	355
Fat gain (kg)	4.2	3.0	3.7	2.7	3.2	3.6	2.4	1.3
NEA increase (cal)	392	473	486	535	571	580	620	690
Fat gain (kg)	3.8	1.7	1.6	2.2	1.0	0.4	2.3	1.1

Figure 4.13 is a scatterplot of these data. The plot shows a moderately strong negative linear association with no outliers. The correlation is $r = -0.7786$. People with larger increases in nonexercise activity do indeed gain less fat. A line drawn through the points will describe the overall pattern well.

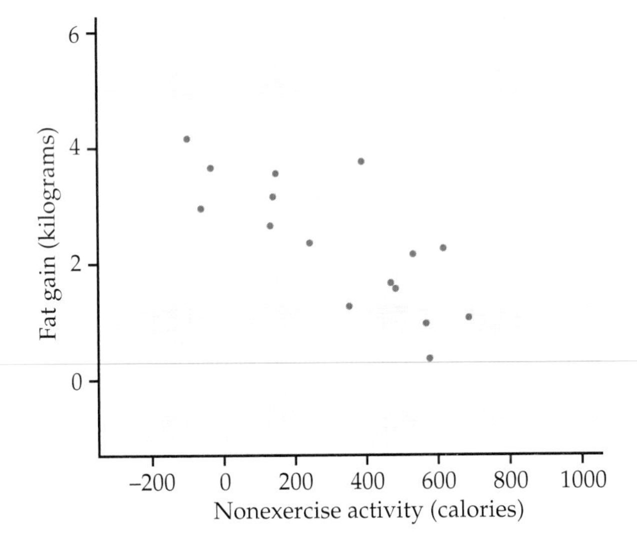

FIGURE 4.13 Fat gain after 8 weeks of overeating plotted against the increase in nonexercise activity over the same period, for Example 4.16.

Fitting a line to data

When a scatterplot displays a linear pattern, we can describe the overall pattern by drawing a straight line through the points. Of course, no straight line passes exactly through all the points. **Fitting a line** to data means drawing a line that comes as close as possible to the points. The equation of a line fitted to the data gives a concise description of the relationship between the response variable y and the explanatory variable x.

fitting a line

STRAIGHT LINES

Suppose that y is a response variable (plotted on the vertical axis) and x is an explanatory variable (plotted on the horizontal axis). A straight line relating y to x has an equation of the form

$$y = b_0 + b_1 x$$

In this equation, b_1 is the **slope,** the amount by which y changes when x increases by one unit. The number b_0 is the **intercept,** the value of y when $x = 0$.

In practice, we will use software to obtain values of b_0 and b_1 for a given set of data.

EXAMPLE 4.17

Regression line for fat gain. Any straight line describing the nonexercise activity data has the form

$$\text{fat gain} = b_0 + (b_1 \times \text{NEA increase})$$

In Figure 4.14 we have drawn the regression line with the equation

$$\text{fat gain} = 3.505 - (0.00344 \times \text{NEA increase})$$

The figure shows that this line fits the data well. The slope $b_1 = -0.00344$ tells us that fat gained goes down by 0.00344 kilogram for each added calorie of NEA increase.

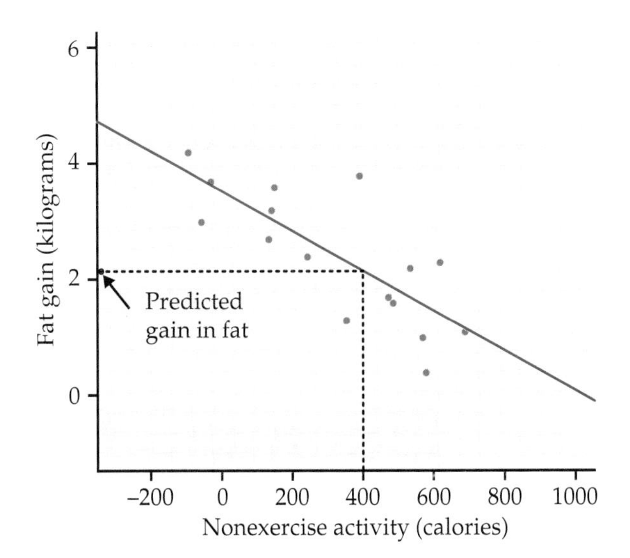

FIGURE 4.14 A regression line fitted to the nonexercise activity data and used to predict fat gain for an NEA increase of 400 calories, for Examples 4.17 and 4.18.

The slope b_1 of a line $y = b_0 + b_1x$ is the *rate of change* in the response y as the explanatory variable x changes. The slope of a regression line is an important numerical description of the relationship between the two variables. For Example 4.17, the intercept is $b_0 = 3.505$ kilograms. This value is the estimated fat gain if NEA does not change. When we substitute the value 0 for the NEA increase, the regression equation gives 3.505 (the intercept) as the predicted value of the fat gain.

USE YOUR KNOWLEDGE

4.52 Plot the data with the line. Make a sketch of the data in Example 4.16 and plot the line

$$\text{fat gain} = 4.505 - (0.00344 \times \text{NEA increase})$$

on your sketch. Explain why this line does not give a good fit to the data.

Prediction

prediction

We can use a regression line to **predict** the response y for a specific value of the explanatory variable x.

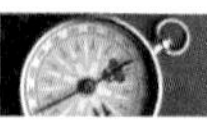

EXAMPLE 4.18

Prediction for fat gain. Based on the linear pattern, we want to predict the fat gain for an individual whose NEA increases by 400 calories when she overeats. To use the fitted line to predict fat gain, go "up and over" on the graph in Figure 4.14. From 400 calories on the x axis, go up to the fitted line and over to the y axis. The graph shows that the predicted gain in fat is a bit more than 2 kilograms.

If we have the equation of the line, it is faster and more accurate to substitute $x = 400$ in the equation. The predicted fat gain is

$$\text{fat gain} = 3.505 - (0.00344 \times 400) = 2.13 \text{ kilograms}$$

The accuracy of predictions from a regression line depends on how much scatter about the line the data show. In Figure 4.14, fat gains for similar increases in NEA show a spread of 1 or 2 kilograms. The regression line summarizes the pattern but gives only roughly accurate predictions.

USE YOUR KNOWLEDGE

4.53 Predict the fat gain. Use the regression equation in Example 4.17 to predict the fat gain for a person whose NEA increases by 600 calories.

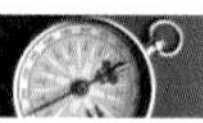

EXAMPLE 4.19

Is this prediction reasonable? Can we predict the fat gain for someone whose nonexercise activity increases by 1500 calories when she overeats? We can certainly substitute 1500 calories into the equation of the line. The prediction is

$$\text{fat gain} = 3.505 - (0.00344 \times 1500) = -1.66 \text{ kilograms}$$

That is, we predict that this individual loses fat when she overeats. This prediction is not trustworthy. Look again at Figure 4.14. An NEA increase of 1500 calories is far outside the range of our data. We can't say if increases this large ever occur, or if the relationship remains linear at such extreme values. Predicting fat gain when NEA increases by 1500 calories *extrapolates* the relationship beyond what the data show.

EXTRAPOLATION

Extrapolation is the use of a regression line for prediction far outside the range of values of the explanatory variable x used to obtain the line. Such predictions are often not accurate and should be avoided.

USE YOUR KNOWLEDGE

4.54 Would you use the regression equation to predict? Consider the following values for NEA increase: −400, 200, 500, 1000. For each, decide whether you would use the regression equation in Example 4.17 to predict fat gain or whether you would be concerned that the prediction would not be trustworthy because of extrapolation. Give reasons for your answers.

Least-squares regression

Different people might draw different lines by eye on a scatterplot. This is especially true when the points are widely scattered. We need a way to draw a regression line that doesn't depend on our guess as to where the line should go. No line will pass exactly through all the points, but we want one that is as close as possible. We will use the line to predict y from x, so we want a line that is as close as possible to the points in the *vertical* direction. That's because the prediction errors we make are errors in y, which is the vertical direction in the scatterplot.

The line in Figure 4.14 predicts 2.13 kilograms of fat gain for an increase in nonexercise activity of 400 calories. If the actual fat gain turns out to be 2.3 kilograms, the error is

$$\begin{aligned} \text{error} &= \text{observed gain} - \text{predicted gain} \\ &= 2.3 - 2.13 = 0.17 \text{ kilograms} \end{aligned}$$

Errors are positive if the observed response lies above the line, and negative if the response lies below the line. We want a regression line that makes these prediction errors as small as possible. Figure 4.15 illustrates the idea. For clarity, the plot shows only three of the points from Figure 4.14, along with the line, on an expanded scale. The line passes below two of the points and above one of them. The vertical distances of the data points from the line appear as

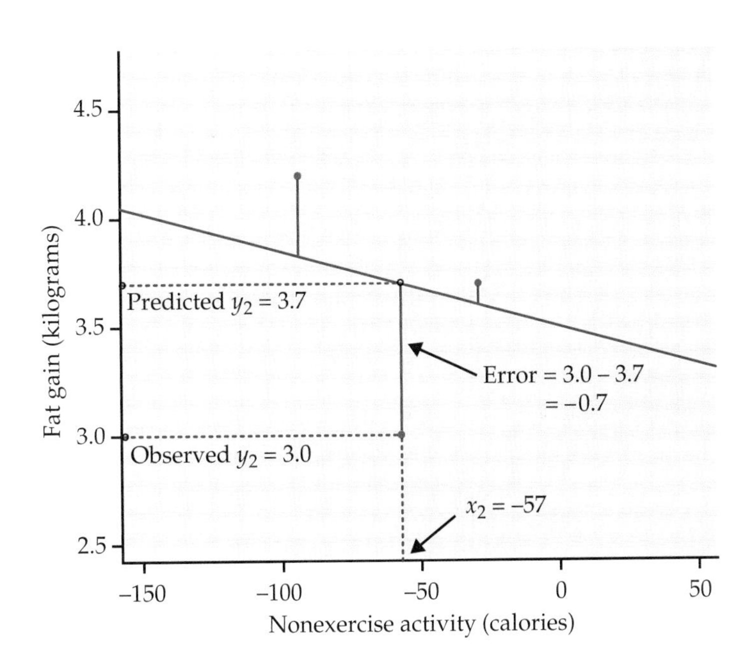

FIGURE 4.15 The least-squares idea: make the errors in predicting y as small as possible by minimizing the sum of their squares.

vertical line segments. A "good" regression line makes these distances as small as possible. There are many ways to make "as small as possible" precise. The most common is the *least-squares* idea. The line in Figures 4.14 and 4.15 is in fact the least-squares regression line.

> **LEAST-SQUARES REGRESSION LINE**
>
> The **least-squares regression line of y on x** is the line that makes the sum of the squares of the vertical distances of the data points from the line as small as possible.

Here is the least-squares idea expressed as a mathematical problem. We represent n observations on two variables x and y as

$$(x_1, y_1),\ (x_2, y_2),\ \ldots,\ (x_n, y_n)$$

If we draw a line $y = b_0 + b_1x$ through the scatterplot of these observations, the line predicts the value of y corresponding to x_i as $\hat{y}_i = b_0 + b_1x_i$. We write $\hat{y}$ (read "y-hat") in the equation of a regression line to emphasize that the line gives a *predicted* response $\hat{y}$ for any x. The predicted response will usually not be exactly the same as the actually *observed* response y. The method of least squares chooses the line that makes the sum of the squares of these errors as small as possible. To find this line, we must find the values of the intercept b_0 and the slope b_1 that minimize

$$\sum(\text{error})^2 = \sum(y_i - b_0 - b_1x_i)^2$$

for the given observations x_i and y_i. For the NEA data, for example, we must find the b_0 and b_1 that minimize

$$(4.2 - b_0 + 94b_1)^2 + (3.0 - b_0 + 57b_1)^2 + \cdots + (1.1 - b_0 - 690b_1)^2$$

where b_0 and b_1 are the intercept and slope of the least-squares line.

You will use software or a calculator with a regression function to find the equation of the least-squares regression line from data on x and y. We will therefore give the equation of the least-squares line in a form that helps our understanding but is not efficient for calculation.

> **EQUATION OF THE LEAST-SQUARES REGRESSION LINE**
>
> We have data on an explanatory variable x and a response variable y for n individuals. The means and standard deviations of the sample data are $\bar{x}$ and s_x for x and $\bar{y}$ and s_y for y, and the correlation between x and y is r. The equation of the least-squares regression line of y on x is
>
> $$\hat{y} = b_0 + b_1x$$
>
> with **slope**
>
> $$b_1 = r\frac{s_y}{s_x}$$
>
> and **intercept**
>
> $$b_0 = \bar{y} - b_1\bar{x}$$

EXAMPLE 4.20

Check the calculations. Verify from the data in Example 4.16 that the mean and standard deviation of the 16 increases in NEA are

$$\bar{x} = 324.8 \text{ calories} \quad \text{and} \quad s_x = 257.66 \text{ calories}$$

The mean and standard deviation of the 16 fat gains are

$$\bar{y} = 2.388 \text{ kg} \quad \text{and} \quad s_y = 1.1389 \text{ kg}$$

The correlation between fat gain and NEA increase is $r = -0.7786$. The least-squares regression line of fat gain y on NEA increase x therefore has slope

$$\begin{aligned} b_1 &= r\frac{s_y}{s_x} = -0.7786\left(\frac{1.1389}{257.66}\right) \\ &= -0.00344 \text{ kg per calorie} \end{aligned}$$

and intercept

$$\begin{aligned} b_0 &= \bar{y} - b_1\bar{x} = 2.388 - (-0.00344)(324.8) \\ &= 3.505 \text{ kg} \end{aligned}$$

The equation of the least-squares line is

$$\hat{y} = 3.505 - 0.00344x$$

When doing calculations like this by hand, you may need to carry extra decimal places in the preliminary calculations to get accurate values of the slope and intercept. Using software or a calculator with a regression function eliminates this worry.

Interpreting the regression line

The slope $b_1 = -0.00344$ kilograms per calorie in Example 4.20 is the change in fat gain as NEA increases. The units "kilograms of fat gained per calorie of NEA" come from the units of y (kilograms) and x (calories). Although the correlation does not change when we change the units of measurement, the equation of the least-squares line does change. The slope in grams per calorie would be 1000 times as large as the slope in kilograms per calorie, because there are 1000 grams in a kilogram. The small value of the slope, $b_1 = -0.00344$, does not mean that the effect of increased NEA on fat gain is small—it just reflects the choice of kilograms as the unit for fat gain. *The slope and intercept of the least-squares line depend on the units of measurement—you can't conclude anything from their size.*

EXAMPLE 4.21

Regression using software. Figure 4.16 displays the basic regression output for the nonexercise activity data from three statistical software packages. Other software produces very similar output. You can find the slope and intercept of the least-squares line, calculated to more decimal places than we need, in all three outputs. The software also provides information that we do not yet need, including some that we trimmed from Figure 4.16.

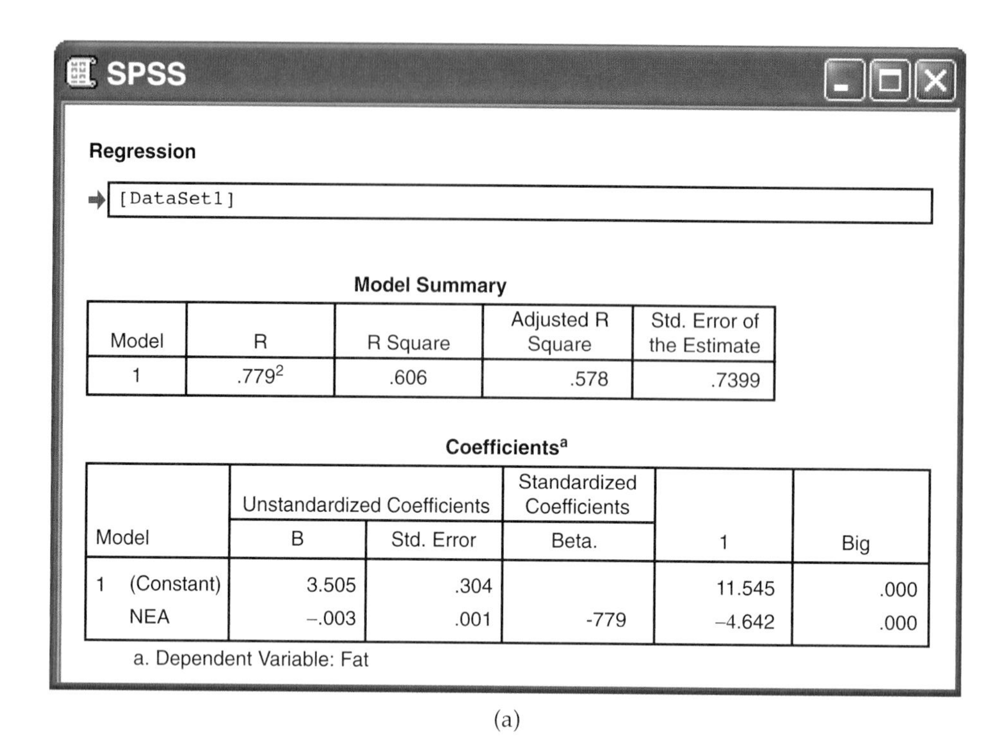

SPSS

Regression

[DataSet1]

Model Summary

Model	R	R Square	Adjusted R Square	Std. Error of the Estimate
1	.779[2]	.606	.578	.7399

Coefficients[a]

Model	Unstandardized Coefficients		Standardized Coefficients	1	Big
	B	Std. Error	Beta.		
1 (Constant)	3.505	.304		11.545	.000
NEA	–.003	.001	-779	–4.642	.000

a. Dependent Variable: Fat

(a)

FIGURE 4.16 Regression results for the nonexercise activity data from three statistical software packages: (a) SPSS, (b) Minitab, (c) JMP. Other software produces similar output. (*Continued*)

Part of the art of using software is to ignore the extra information that is almost always present. Look for the results that you need. Once you understand a statistical method, you can read output from almost any software.

Facts about least-squares regression

Regression is one of the most common statistical settings, and least squares is the most common method for fitting a regression line to data. Here are some facts about least-squares regression lines.

Fact 1. There is a close connection between correlation and the slope of the least-squares line. The slope is

$$b_1 = r\frac{s_y}{s_x}$$

This equation says that along the regression line, **a change of one standard deviation in x corresponds to a change of r standard deviations in y.** When the variables are perfectly correlated ($r = 1$ or $r = -1$), the change in the predicted response $\hat{y}$ is the same (in standard deviation units) as the change in x. Otherwise, because $-1 \leq r \leq 1$, the change in $\hat{y}$ is less than the change in x. As the correlation grows less strong, the prediction $\hat{y}$ moves less in response to changes in x.

Fact 2. The least-squares regression line always passes through the point $(\bar{x}, \bar{y})$ on the graph of y against x. So the least-squares regression line of

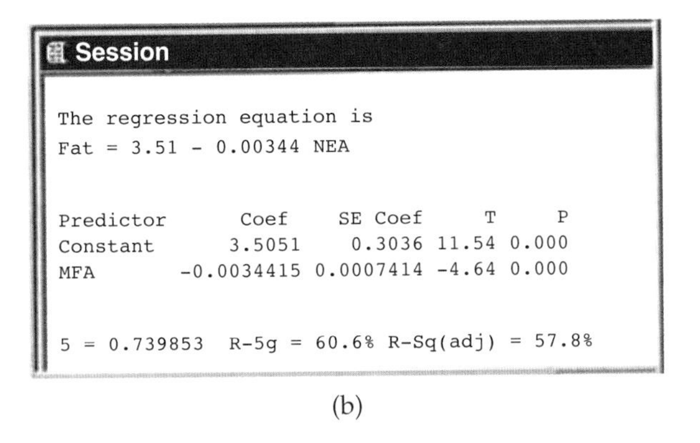

(b)

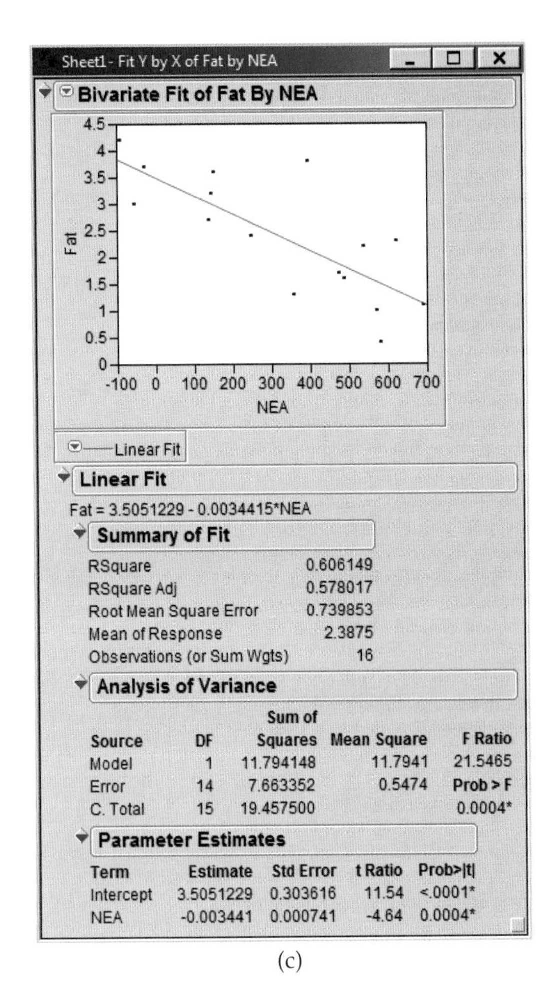

(c)

FIGURE 4.16 *(Continued)*

y on x is the line with slope rs_y/s_x that passes through the point $(\overline{x}, \overline{y})$. We can describe regression entirely in terms of the basic descriptive measures $\overline{x}$, s_x, $\overline{y}$, s_y, and r.

Fact 3. The distinction between explanatory and response variables is important in regression. Least-squares regression looks at the distances of the data points from the line only in the y direction. If we reverse the roles of the two variables, we get a different least-squares regression line.

Correlation and regression

Least-squares regression looks at the distances of the data points from the line only in the y direction. So the two variables x and y play different roles in regression.

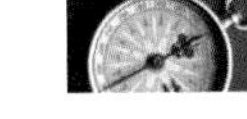

EXAMPLE 4.22

DATA FILE SPAM

Bots and spam messages per day. Figure 4.17 is a scatterplot of the spam botnet data described in Example 4.8 (page 158). There is a positive linear

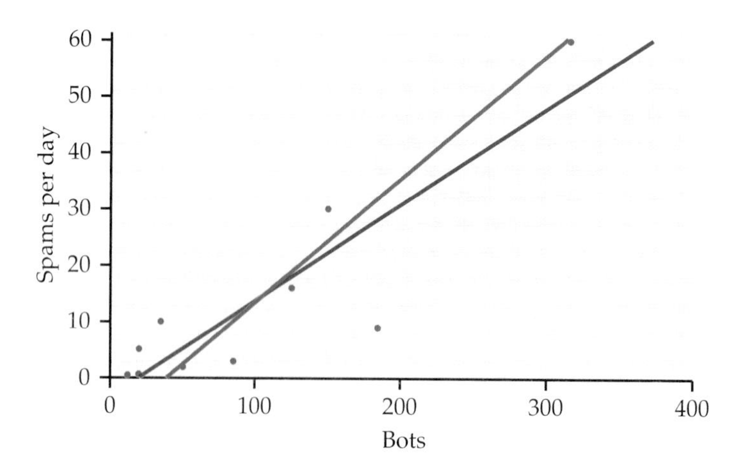

FIGURE 4.17 Scatterplot of spams per day versus the number of bots for 10 botnets, for Example 4.22. The two lines are the two least-squares regression lines: using bots to predict spams per day (red) and using spams per day to predict bots (blue).

relationship. The two lines on the plot are the two least-squares regression lines. The regression line using bots to predict spam messages per day is red. The regression line using spam messages per day to predict bots is blue.

Regression of spams per day on bots and regression of bots on spams per day give different lines. In the regression setting, you must decide which variable is explanatory.

Even though the correlation r ignores the distinction between explanatory and response variables, there is a close connection between correlation and regression. We saw that the slope of the least-squares line involves r. Another connection between correlation and regression is even more important. In fact, the numerical value of r as a measure of the strength of a linear relationship is best interpreted by thinking about regression. Here is the fact we need.

r^2 IN REGRESSION

The **square of the correlation, r^2,** is the fraction of the variation in the values of y that is explained by the least-squares regression of y on x.

The correlation between NEA increase and fat gain for the 16 subjects in Example 4.16 (page 177) is $r = -0.7786$. Because $r^2 = 0.606$, the straight-line relationship between NEA and fat gain explains about 61% of the vertical scatter in fat gains in Figure 4.14 (page 179).

When you report a regression, give r^2 as a measure of how successfully the regression explains the response. The software outputs in Figure 4.16 include r^2, either in decimal form or as a percent.

When you see a correlation, square it to get a better feel for the strength of the association. Perfect correlation ($r = -1$ or $r = 1$) means the points lie exactly on a line. Then $r^2 = 1$ and all the variation in one variable is accounted for by the linear relationship with the other variable. If $r = -0.7$ or $r = 0.7$, $r^2 = 0.49$ and about half the variation is accounted for by the linear relationship. In the r^2 scale, correlation ± 0.7 is about halfway between 0 and ± 1.

USE YOUR KNOWLEDGE

4.55 What fraction of the variation is explained? Consider the following correlations: -0.9, -0.5, -0.3, 0, 0.3, 0.5, and 0.9. For each, give the fraction of the variation in y that is explained by the least-squares regression of y on x. Summarize what you have found from performing these calculations.

Understanding r^2

Here is a more specific interpretation of r^2. The fat gains y in Figure 4.14 range from 0.4 kilograms to 4.2 kilograms. The variance of these responses, a measure of how variable they are, is

$$\text{variance of observed values } y = 1.297$$

Much of this variability is due to the fact that as x increases from -94 to 690 calories it pulls y along with it. If the only variability in the observed responses were due to the straight-line dependence of fat gain on NEA, the observed gains would lie exactly on the regression line. That is, they would be the same as the predicted gains $\hat{y}$. We can compute the predicted gains by substituting the NEA values for each subject into the equation of the least-squares line. Their variance describes the variability in the predicted responses. The result is

$$\text{variance of predicted values } \hat{y} = 0.786$$

This is what the variance would be if the responses fell exactly on the line, that is, if the linear relationship explained 100% of the observed variation in y. Because the responses don't fall exactly on the line, the variance of the predicted values is smaller than the variance of the observed values. Here is the fact we need:

$$\begin{aligned} r^2 &= \frac{\text{variance of predicted values } \hat{y}}{\text{variance of observed values } y} \\ &= \frac{0.786}{1.297} = 0.606 \end{aligned}$$

This fact is always true. The squared correlation gives the variance the responses would have if there were no scatter about the least-squares line as a fraction of the variance of the actual responses. This is the exact meaning of "fraction of variation explained" as an interpretation of r^2.

These connections with correlation are special properties of least-squares regression. They are not true for other methods of fitting a line to data. One reason that least squares is the most common method for fitting a regression line to data is that it has many convenient special properties.

SECTION 4.3 SUMMARY

A **regression line** is a straight line that describes how a response variable y changes as an explanatory variable x changes.

The most common method of fitting a line to a scatterplot is least squares. The **least-squares regression line** is the straight line $\hat{y} = b_0 + b_1x$ that minimizes the sum of the squares of the vertical distances of the observed y-values from the line.

You can use a regression line to **predict** the value of y for any value of x by substituting this x into the equation of the line. **Extrapolation** beyond the range of x-values spanned by the data is risky.

The **slope** b_1 of a regression line $\hat{y} = b_0 + b_1x$ is the rate at which the predicted response $\hat{y}$ changes along the line as the explanatory variable x changes. Specifically, b_1 is the change in $\hat{y}$ when x increases by 1. The numerical value of the slope depends on the units used to measure x and y.

The **intercept** b_0 of a regression line $\hat{y} = b_0 + b_1x$ is the predicted response $\hat{y}$ when the explanatory variable $x = 0$. This prediction is not particularly useful unless x can actually take values near 0.

The least-squares regression line of y on x is the line with slope $b_1 = rs_y/s_x$ and intercept $b_0 = \bar{y} - b_1\bar{x}$. This line always passes through the point $(\bar{x}, \bar{y})$.

Correlation and regression are closely connected. The correlation r is the slope of the least-squares regression line when we measure both x and y in standardized units. The square of the correlation r^2 is the fraction of the variance of one variable that is explained by least-squares regression on the other variable.

SECTION 4.3 EXERCISES

For Exercise 4.52, see page 179; for Exercise 4.53, see page 180; for Exercise 4.54, see page 181; and for Exercise 4.55, see page 187.

4.56. Open space and population. The New York City Open Accessible Space Information System Cooperative (OASIS) is an organization of public and private sector representatives that has developed an information system designed to enhance the stewardship of open space.[17] Data from the OASIS Web site for 12 large U.S. cities follow. The variables are population in thousands and total park or other open space within city limits in acres. DATA FILE OASIS

City	Population	Open space
Baltimore	651	5,091
Boston	589	4,865
Chicago	2,896	11,645
Long Beach	462	2,887
Los Angeles	3,695	29,801
Miami	362	1,329
Minneapolis	383	5,694
New York	8,008	49,854
Oakland	399	3,712
Philadelphia	1,518	10,685
San Francisco	777	5,916
Washington, DC	572	7,504

(a) Make a scatterplot of the data using population as the explanatory variable and open space as the response variable.

(b) Is it reasonable to fit a straight line to these data? Explain your answer.

(c) Find the least-squares regression line. Report the equation of the line and draw the line on your scatterplot.

(d) What proportion of the variation in open space is explained by population?

4.57. Prepare the report card. Refer to the previous exercise. One way to compare cities with respect to the amount of open space that they have is to compare the actual open space with the open space that is predicted by the least-squares regression line.

Using the data and the regression line from the previous exercise, compute the difference between the actual open space and the predicted open space for each city. Cities with positive differences are doing better than predicted, while those with negative differences are doing worse.

Make a table with the city name and the difference, ordered from best to worst by the size of the difference. (In the following section, we will learn more about how to use these differences, which we call "residuals.") DATA FILE OASIS

4.58. Is New York an outlier? Refer to Exercises 4.56 and 4.57. Write a short paragraph about the data point corresponding to New York City. Is this point an outlier? If it were deleted from the data set, would the least-squares regression line change very much? Compare the analysis results with and without this observation. DATA FILE OASIS

4.59. Open space per person. Refer to Exercises 4.56 to 4.58. Open space in acres per person is an alternative way to report open space. Divide open space by population to compute the value of this variable for each city. Using this new variable as the response variable and population as the explanatory variable, answer the questions given in Exercise 4.56. How do your new results compare with those that you found in that exercise? OASIS

4.60. A different report card. Refer to Exercise 4.57. Prepare a report card based on the analysis of open space per person that you performed in Exercise 4.59. Write a short paragraph comparing this report card with the one that you prepared in Exercise 4.57. Which do you prefer? Give reasons for your answer. OASIS

4.61. Alcohol and carbohydrates in beer. Figure 4.8 (page 168) gives a scatterplot of carbohydrates versus percent alcohol in 86 brands of beer. In Exercise 4.37 you calculated the correlation between these two variables. Find the equation of the least-squares regression line for these data. BEER

4.62. Alcohol and carbohydrates in beer revisited. Refer to the previous exercise. The data that you used to compute the least-squares regression line include an outlier. BEER

(a) Remove the outlier and recompute the least-squares regression line.

(b) Write a short paragraph about the possible effects of outliers on a least-squares regression line using this example to illustrate your ideas.

4.63. Always plot your data! Table 4.1 presents four sets of data prepared by the statistician Frank Anscombe to illustrate the dangers of calculating without first plotting the data.[18] ANSCOMBE

(a) Without making scatterplots, find the correlation and the least-squares regression line for all four data sets. What do you notice? Use the regression line to predict y for $x = 10$.

(b) Make a scatterplot for each of the data sets and add the regression line to each plot.

(c) In which of the four cases would you be willing to use the regression line to describe the dependence of y on x? Explain your answer in each case.

4.64. Data generated by software. The following 20 observations on Y and X were generated by a computer program. GENERATEDDATA

Y	*X*	*Y*	*X*
34.38	22.06	27.07	17.75
30.38	19.88	31.17	19.96
26.13	18.83	27.74	17.87
31.85	22.09	30.01	20.20
26.77	17.19	29.61	20.65
29.00	20.72	31.78	20.32
28.92	18.10	32.93	21.37
26.30	18.01	30.29	17.31
29.49	18.69	28.57	23.50
31.36	18.05	29.80	22.02

(a) Make a scatterplot and describe the relationship between Y and X.

TABLE 4.1

Four data sets for exploring correlation and regression

					Data Set A						
x	10	8	13	9	11	14	6	4	12	7	5
y	8.04	6.95	7.58	8.81	8.33	9.96	7.24	4.26	10.84	4.82	5.68
					Data Set B						
x	10	8	13	9	11	14	6	4	12	7	5
y	9.14	8.14	8.74	8.77	9.26	8.10	6.13	3.10	9.13	7.26	4.74
					Data Set C						
x	10	8	13	9	11	14	6	4	12	7	5
y	7.46	6.77	12.74	7.11	7.81	8.84	6.08	5.39	8.15	6.42	5.73
					Data Set D						
x	8	8	8	8	8	8	8	8	8	8	19
y	6.58	5.76	7.71	8.84	8.47	7.04	5.25	5.56	7.91	6.89	12.50

(b) Find the equation of the least-squares regression line and add the line to your plot.

(c) What percent of the variability in Y is explained by X?

(d) Summarize your analysis of these data in a short paragraph.

4.65. Add an outlier. Refer to the previous exercise. Add an additional observation with $Y = 50$ and $X = 30$ to the data set. Repeat the analysis that you performed in the previous exercise and summarize your results, paying particular attention to the effect of this outlier. DATA FILE GENDATA21A

4.66. Add a different outlier. Refer to the previous two exercises. Add an additional observation with $Y = 29$ and $X = 50$ to the original data set. DATA FILE GENDATA21B

(a) Repeat the analysis that you performed in the first exercise and summarize your results, paying particular attention to the effect of this outlier.

(b) In this exercise and in the previous one, you added an outlier to the original data set and reanalyzed the data. Write a short summary of the changes in correlations that can result from different kinds of outliers.

4.67. The regression equation. The equation of a least-squares regression line is $y = 12 + 6x$.

(a) What is the value of y for $x = 5$?

(b) If x increases by one unit, what is the corresponding increase in y?

(c) What is the intercept for this equation?

4.68. Progress in math scores. Every few years, the National Assessment of Educational Progress asks a national sample of eighth-graders to perform the same math tasks. The goal is to get an honest picture of progress in math. Here are the last few national mean scores, on a scale of 0 to 500:[19] DATA FILE NAEPMATH

Year	1990	1992	1996	2000	2003	2005	2008	2011
Score	263	268	272	273	278	279	281	283

(a) Make a time plot of the mean scores, by hand. This is just a scatterplot of score against year. There is a slow linear increasing trend.

(b) Find the regression line of mean score on year step-by-step. First calculate the mean and standard deviation of each variable and their correlation (use a calculator with these functions). Then find the equation of the least-squares line from these. Draw the line on your scatterplot. What percent of the year-to-year variation in scores is explained by the linear trend?

(c) Now use software or the regression function on your calculator to verify your regression line.

4.69. Metabolic rate and lean body mass. Compute the mean and the standard deviation of the metabolic rates and lean body masses in Exercise 4.31 (page 170) and the correlation between these two variables. Use these values to find the slope of the regression line of metabolic rate on lean body mass. Also find the slope of the regression line of lean body mass on metabolic rate. What are the units for each of the two slopes? DATA FILE BODYMASS

4.70. Heights of husbands and wives. The mean height of American women in their early twenties is about 64.5 inches, and the standard deviation is about 2.5 inches. The mean height of men the same age is about 68.5 inches, with standard deviation about 2.7 inches. If the correlation between the heights of husbands and wives is about $r = 0.5$, what is the equation of the regression line of the husband's height on the wife's height in young couples? Draw a graph of this regression line. Predict the height of the husband of a woman who is 67 inches tall.

4.71. CHALLENGE **A property of the least-squares regression line.** Use the equation for the least-squares regression line to show that this line always passes through the point $(\bar{x}, \bar{y})$.

4.72. CHALLENGE **Best countries for business.** Figure 4.5 (page 164) gives a scatterplot of the gross domestic product per capita versus the unemployment rate for 99 countries. DATA FILE BESTCOUNTRIES

(a) Plot the data and add the least-squares regression line to the plot.

(b) Is it appropriate to use this least-squares regression line to describe the relationship shown in your plot? Explain your answer.

4.73. CHALLENGE **Class attendance and grades.** A study of class attendance and grades among first-year students at a state university showed that in general students who attended a higher percent of their classes earned higher grades. Class attendance explained 16% of the variation in grade index among the students. What is the numerical value of the correlation between percent of classes attended and grade index?

4.4 Cautions about Correlation and Regression

Correlation and regression are among the most common statistical tools. They are used in more elaborate form to study relationships among many variables, a situation in which we cannot see everything we need by studying a single scatterplot. We need a firm grasp of the use and limitations of these tools, both now and as a foundation for more advanced statistical methods.

Residuals

A regression line describes the overall pattern of a linear relationship between an explanatory variable and a response variable. Deviations from the overall pattern are also important. In the regression setting, we see deviations by looking at the scatter of the data points about the regression line. The vertical distances from the points to the least-squares regression line are as small as possible in the sense that they have the smallest possible sum of squares. Because they represent "leftover" variation in the response after fitting the regression line, these distances are called *residuals*.

> **RESIDUALS**
>
> A **residual** is the difference between an observed value of the response variable and the value predicted by the regression line. That is,
>
> $$\text{residual} = \text{observed } y - \text{predicted } y$$
> $$= y - \hat{y}$$

EXAMPLE 4.23

Residuals for fat gain. Example 4.16 (page 177) describes measurements on 16 young people who volunteered to overeat for 8 weeks. Figure 4.18(a) is a scatterplot of these data. The data show a strong negative linear association. The least-squares line is

$$\text{fat gain} = 3.505 - (0.00344 \times \text{NEA increase})$$

One subject's NEA rose by 135 calories. That subject gained 2.7 kilograms of fat. The predicted gain for 135 calories is

$$\hat{y} = 3.505 - (0.00344 \times 135) = 3.04 \text{ kg}$$

The residual for this subject is therefore

$$\begin{aligned} \text{residual} &= \text{observed } y - \text{predicted } y \\ &= y - \hat{y} \\ &= 2.7 - 3.04 = -0.34 \text{ kg} \end{aligned}$$

Most regression software will calculate and store residuals for you.

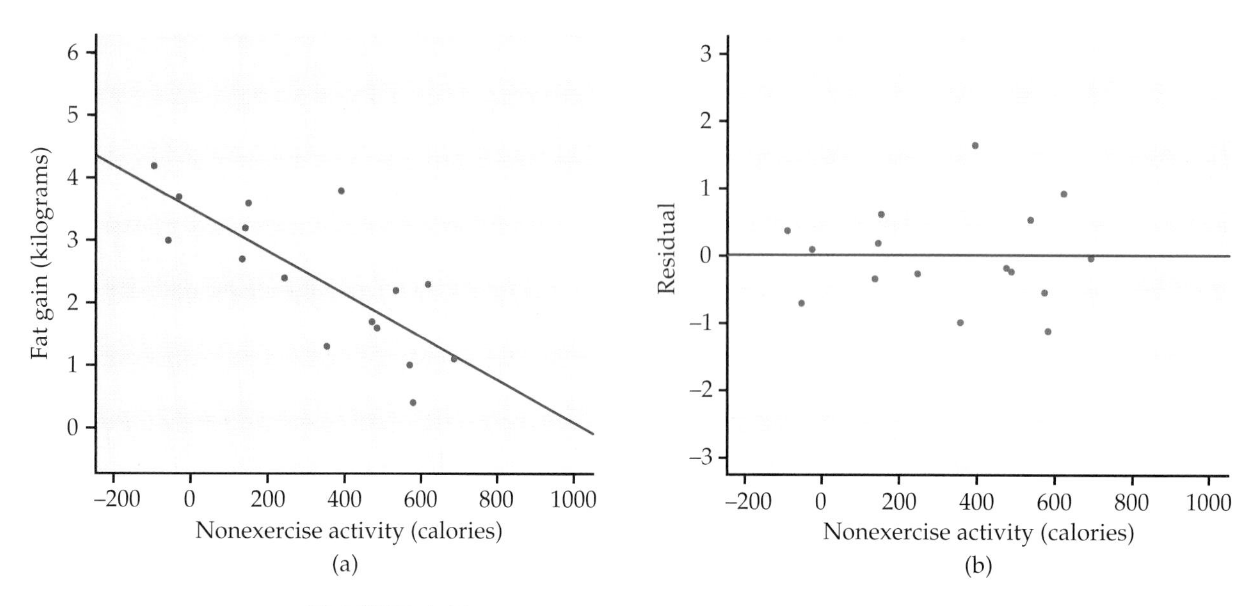

FIGURE 4.18 (a) Scatterplot of fat gain versus increase in NEA, with the least-squares regression line, for Example 4.23. (b) Residual plot for the regression displayed in Figure 4.18(a). The line at $y = 0$ marks the mean of the residuals.

USE YOUR KNOWLEDGE

4.74 Find the predicted value and the residual. Another individual in the NEA data set has NEA increase equal to 143 calories and fat gain equal to 3.2 kg. Find the predicted value of fat gain for this individual and then calculate the residual. Explain why this residual is positive.

Because the residuals show how far the data fall from our regression line, examining the residuals helps assess how well the line describes the data. Although residuals can be calculated from any model fitted to the data, the residuals from the least-squares line have a special property: **the mean of the least-squares residuals is always zero.**

USE YOUR KNOWLEDGE

4.75 Find the sum of the residuals. Here are the 16 residuals for the NEA data rounded to two decimal places:

0.37	−0.70	0.10	−0.34	0.19	0.61	−0.26	−0.98
1.64	−0.18	−0.23	0.54	−0.54	−1.11	0.93	−0.03

Find the sum of these residuals. Note that the sum is not exactly zero because of roundoff error. DATA FILE FIDGET

Plotting the residuals

You can see the residuals in the scatterplot of Figure 4.18(a) by looking at the vertical deviations of the points from the line. The *residual plot* in Figure 4.18(b) makes it easier to study the residuals by plotting them against the explanatory variable, increase in NEA.

> **RESIDUAL PLOTS**
>
> A **residual plot** is a scatterplot of the regression residuals against the explanatory variable. Residual plots help us assess the fit of a regression line.

Because the mean of the residuals is always zero, the horizontal line at zero in Figure 4.18(b) helps orient us. This line (residual = 0) corresponds to the fitted line in Figure 4.18(a). The residual plot magnifies the deviations from the line to make patterns easier to see. If the regression line catches the overall pattern of the data, there should be *no pattern* in the residuals. That is, the residual plot should show an unstructured horizontal band centered at zero. The residuals in Figure 4.18(b) do have this irregular scatter.

You can see the same thing in the scatterplot of Figure 4.18(a) and the residual plot of Figure 4.18(b). It's just a bit easier in the residual plot. Deviations from an irregular horizontal pattern point out ways in which the regression line fails to catch the overall pattern. Here is an example:

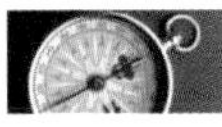

EXAMPLE 4.24

Patterns in the life expectancy and Internet use residuals. Figure 4.9 (page 168) gives a scatterplot of life expectancy versus Internet use for 181 countries. The first thing that we see in the plot is the large number of countries with low Internet use. In addition, the relationship between life expectancy and Internet use appears to be curved. For low values of Internet use, there is a clear relationship, while for higher values, the curve becomes fairly flat.

Figure 4.19 gives the residuals for the regression when we predict life expectancy from number of Internet users. Look at the left part of the plot,

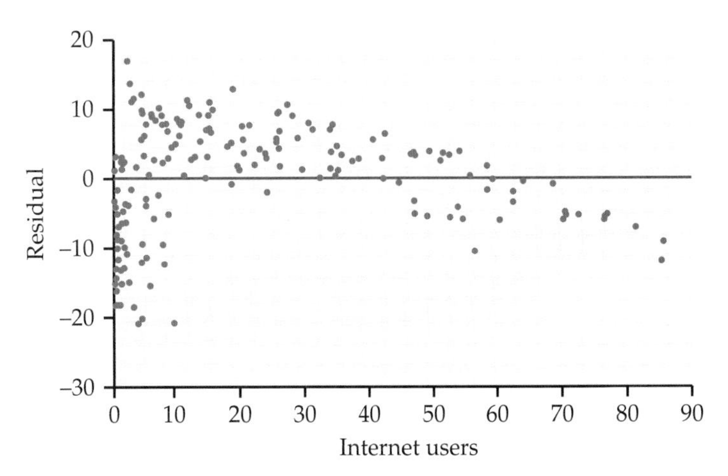

FIGURE 4.19 Residual plot for the regression of life expectancy on Internet users, for Example 4.24.

where the values of Internet use are low. Here we see that there are many more negative residuals than positive ones. Now look at the right part of the plot, where the values of Internet use are high. Here we see that the residuals also tend to be negative.

The residual pattern in Figure 4.19 is characteristic of a simple curved relationship. *There are many ways in which a relationship can deviate from a linear pattern.* We now have an important tool for examining these deviations. Use it often and carefully when you study relationships.

Outliers and influential observations

When you look at scatterplots and residual plots, look for striking individual points as well as for an overall pattern. Here is an example of data that contain some unusual cases.

EXAMPLE 4.25

Diabetes and blood sugar. People with diabetes must manage their blood sugar levels carefully. They measure their fasting plasma glucose (FPG; in milligrams per deciliter of blood) several times a day with a glucose meter. Another measurement, made at regular medical checkups, is called HbA1c. This is roughly the percent of red blood cells that have a glucose molecule attached. It measures average exposure to glucose over a period of several months. Table 4.2 gives data on both HbA1c and FPG for 18 diabetics five months after they had completed a diabetes education class.[20]

Because both FPG and HbA1c measure blood glucose, we expect a positive association. The scatterplot in Figure 4.20 shows a surprisingly weak relationship, with correlation $r = 0.4819$. The line on the plot is the least-squares regression line for predicting FPG from HbA1c. Its equation is

$$\hat{y} = 66.4 + 10.41x$$

It appears that one-time measurements of FPG can vary quite a bit among people with similar long-term levels of blood glucose, as measured by HbA1c.

TABLE 4.2

Two measures of glucose level in diabetics

Subject	HbA1c (%)	FPG (mg/dl)	Subject	HbA1c (%)	FPG (mg/dl)	Subject	HbA1c (%)	FPG (mg/dl)
1	6.1	141	7	7.5	96	13	10.6	103
2	6.3	158	8	7.7	78	14	10.7	172
3	6.4	112	9	7.9	148	15	10.7	359
4	6.8	153	10	8.7	172	16	11.2	145
5	7.0	134	11	9.4	200	17	13.7	147
6	7.1	95	12	10.4	271	18	19.3	255

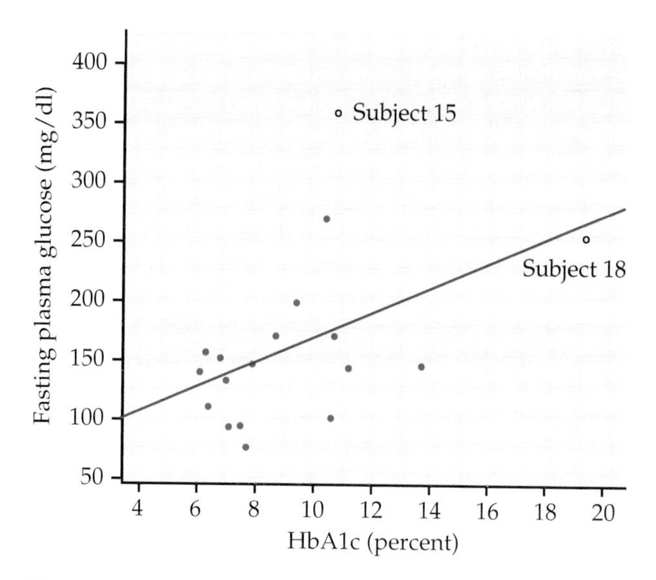

FIGURE 4.20 Scatterplot of fasting plasma glucose against HbA1c (which measures long-term blood glucose), with the least-squares line, for Example 4.25.

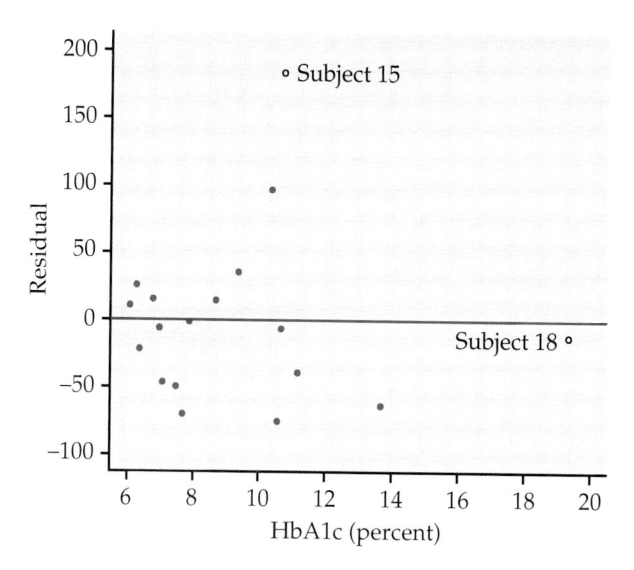

FIGURE 4.21 Residual plot for the regression of fasting plasma glucose on HbA1c. Subject 15 is an outlier in fasting plasma glucose. Subject 18 is an outlier in HbA1c that may be influential but does not have a large residual.

Two unusual cases are marked in Figure 4.20. Subjects 15 and 18 are unusual in different ways. Subject 15 has dangerously high FPG and lies far from the regression line in the y direction. Subject 18 is close to the line but far out in the x direction. The residual plot in Figure 4.21 confirms that Subject 15 has a large residual and that Subject 18 does not.

Points that are outliers in the x direction, like Subject 18, can have a strong influence on the position of the regression line. Least-squares lines make the sum of squares of the vertical distances to the points as small as possible. A point that is extreme in the x direction with no other points near it pulls the line toward itself.

OUTLIERS AND INFLUENTIAL OBSERVATIONS IN REGRESSION

An **outlier** is an observation that lies outside the overall pattern of the other observations. Points that are outliers in the y direction of a scatterplot have large regression residuals, but other outliers need not have large residuals.

An observation is **influential** for a statistical calculation if removing it would markedly change the result of the calculation. Points that are outliers in the x direction of a scatterplot are often influential for the least-squares regression line.

Influence is a matter of degree—how much does a calculation change when we remove an observation? It is difficult to assess influence on a regression line without actually doing the regression both with and without the suspicious observation. A point that is an outlier in x is often influential. But if the point happens to lie close to the regression line calculated from the other observations, then its presence will move the line only a little and the point will not be influential.

The influence of a point that is an outlier in y depends on whether there are many other points with similar values of x that hold the line in place. Figures 4.20 and 4.21 identify two unusual observations. How influential are they?

EXAMPLE 4.26

Influential observations. Subjects 15 and 18 both influence the correlation between FPG and HbA1c, in opposite directions. Subject 15 weakens the linear pattern. If we drop this point, the correlation increases from $r = 0.4819$ to $r = 0.5684$. Subject 18 extends the linear pattern. If we omit this subject, the correlation drops from $r = 0.4819$ to $r = 0.3837$.

To assess influence on the least-squares line, we recalculate the line leaving out a suspicious point. Figure 4.22 shows three least-squares lines. The solid line is the regression line of FPG on HbA1c based on all 18 subjects. This is the same line that appears in Figure 4.20. The dotted line is calculated from all subjects except Subject 18. You see that point 18 does pull the line down toward itself. But the influence of Subject 18 is not very large—the dotted and solid lines are close together for HbA1c values between 6 and 14, the range of all except Subject 18.

The dashed line omits Subject 15, the outlier in y. Comparing the solid and dashed lines, we see that Subject 15 pulls the regression line up. The influence is again not large, but it exceeds the influence of Subject 18.

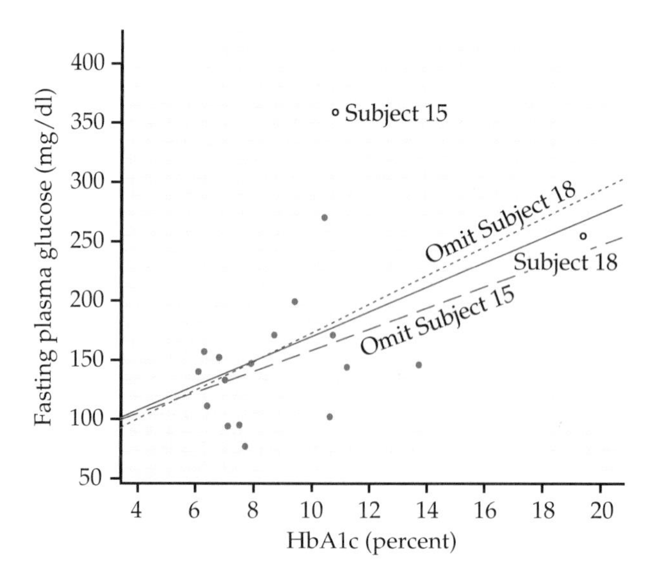

FIGURE 4.22 Three regression lines for predicting fasting plasma glucose from HbA1c, for Example 4.26. The solid line uses all 18 subjects. The dotted line leaves out Subject 18. The dashed line leaves out Subject 15. "Leaving one out" calculations are the surest way to assess influence.

The best way to see how points that are extreme in x can influence the regression line is to use the *Correlation and Regression* applet on the text Web site. As Exercise 4.92 (page 202) demonstrates, moving one point can pull the line to almost any position on the graph.

We did not need the distinction between outliers and influential observations in Chapter 3. A single large salary that pulls up the mean salary $\overline{x}$ for a group of workers is an outlier because it lies far above the other salaries. It is also influential because the mean changes when it is removed. In the regression setting, however, not all outliers are influential. Because influential observations draw the regression line toward themselves, we may not be able to spot them by looking for large residuals.

Beware the lurking variable

Correlation and regression are powerful tools for measuring the association between two variables and for expressing the dependence of one variable on the other. These tools must be used with an awareness of their limitations. We have seen that

- Correlation measures *only linear association,* and fitting a straight line makes sense only when the overall pattern of the relationship is linear. Always plot your data before calculating.
- *Extrapolation* (using a fitted model far outside the range of the data that we used to fit it) often produces unreliable predictions.
- Correlation and least-squares regression are *not resistant.* Always plot your data and look for potentially influential points.

Another caution is even more important: the relationship between two variables can often be understood only by taking other variables into account. *Lurking variables* can make a correlation or regression misleading.

LURKING VARIABLE

A **lurking variable** is a variable that is not among the explanatory or response variables in a study and yet may influence the interpretation of relationships among those variables.

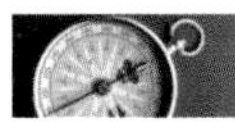

EXAMPLE 4.27

Discrimination in medical treatment? Studies show that men who complain of chest pain are more likely to get detailed tests and aggressive treatment such as bypass surgery than are women with similar complaints. Is this association between gender and treatment due to discrimination?

Perhaps not. Men and women develop heart problems at different ages—women with heart problems are, on the average, between 10 and 15 years older than men. Aggressive treatments are more risky for older patients, so doctors may hesitate to recommend them. Lurking variables—the patient's age and condition—may explain the relationship between gender and doctors' decisions.

Here is an example of a different type of lurking variable.

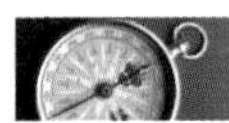

EXAMPLE 4.28

Gas and electricity bills. A single-family household receives bills for gas and electricity each month. The 12 observations for a recent year are plotted with the least-squares regression line in Figure 4.23. We have arbitrarily chosen to put the electricity bill on the x axis and the gas bill on the y axis. There is a clear negative association. Does this mean that a high electricity bill causes the gas bill to be low and vice versa?

To understand the association in this example, we need to know a little more about the two variables. In this household, heating is done by gas and cooling is done by electricity. Therefore, in the winter months the gas bill will be relatively high and the electricity bill will be relatively low. The pattern is reversed in the summer months. The association that we see in this example is due to a lurking variable: time of year.

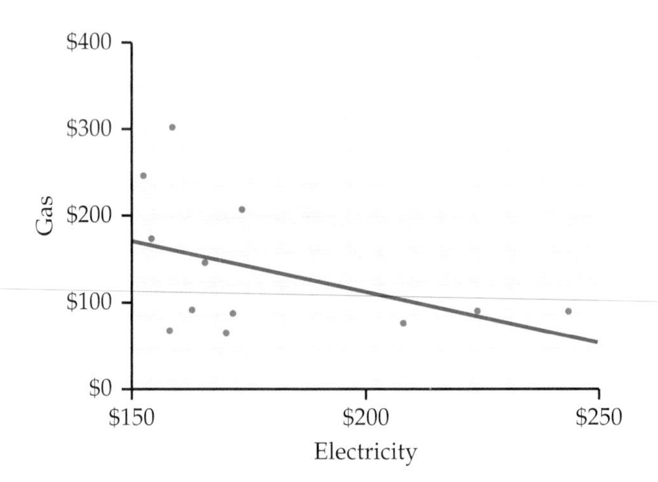

FIGURE 4.23 Scatterplot with least-squares regression line for predicting monthly charges for gas using monthly charges for electricity for a household, for Example 4.28.

Correlations that are due to lurking variables are sometimes called "nonsense correlations." The correlation is real. What is nonsense is the suggestion that the variables are directly related so that changing one of the variables *causes* changes in the other. Remember that an association between two variables x and y can reflect many types of relationship among x, y, and one or more lurking variables.

ASSOCIATION DOES NOT IMPLY CAUSATION

An association between an explanatory variable x and a response variable y, even if it is very strong, is not by itself good evidence that changes in x actually cause changes in y.

Lurking variables sometimes create a correlation between x and y, as in Examples 4.27 and 4.28. When you observe an association between two variables, always ask yourself if the relationship that you see might be due to a lurking variable. As in Example 4.28, time is often a likely candidate.

SECTION 4.4 SUMMARY

You can examine the fit of a regression line by plotting the **residuals,** which are the differences between the observed and predicted values of y. Be on the lookout for points with unusually large residuals and also for nonlinear patterns and uneven variation about the line.

Also look for **influential observations,** individual points that substantially change the regression line. Influential observations are often outliers in the x direction, but they need not have large residuals.

Correlation and regression must be **interpreted with caution.** Plot the data to be sure that the relationship is roughly linear and to detect outliers and influential observations.

Lurking variables may explain the relationship between the explanatory and response variables. Correlation and regression can be misleading if you ignore important lurking variables.

We cannot conclude that there is a cause-and-effect relationship between two variables just because they are strongly associated. **High correlation does not imply causation.**

SECTION 4.4 EXERCISES

For Exercises 4.74 and 4.75, see page 192.

4.76. What's wrong? Each of the following statements contains an error. Describe each error and explain why the statement is wrong.

(a) If the residuals are all positive, this implies that there is a positive relationship between the response variable and the explanatory variable.

(b) A negative relationship can never be due to causation.

(c) A lurking variable is always a response variable.

4.77. What's wrong? Each of the following statements contains an error. Describe each error and explain why the statement is wrong.

(a) High correlation implies causation.

(b) An outlier will always have a small residual.

(c) If we have data at values of x equal to 1, 2, 3, 4, and 5, and we try to predict the value of y using a least-squares regression line, we are doing an extrapolation.

4.78. Use of the Internet and a long life. Exercise 4.20 (page 168) asks the question "Will you live longer if you use the Internet?" Figure 4.9 (page 168) is a scatterplot of life expectancy in years versus number of Internet users for 181 countries. The scatterplot shows a positive association between these two variables. Do you think that this plot indicates that Internet use causes people to live longer? Give another possible explanation for why these two variables are positively associated. DATA FILE INTERNETANDLIFE

4.79. How's your self-esteem? People who do well tend to feel good about themselves. Perhaps helping people feel good about themselves will help them do better in their jobs and in life. For a time, raising self-esteem became a goal in many schools and companies. Can you think of explanations for the association between high self-esteem and good performance other than "Self-esteem causes better work"?

4.80. Are big hospitals bad for you? A study shows that there is a positive correlation between the size of a hospital (measured by its number of beds x) and the median number of days y that patients remain in the hospital. Does this mean that you can shorten a hospital stay by choosing a small hospital? Why?

4.81. Does herbal tea help nursing-home residents? A group of college students believes that herbal tea has remarkable powers. To test this belief, they make weekly visits to a local nursing home, where they visit with the residents and serve them herbal tea. The nursing-home staff reports that after several months many of the residents are healthier and more cheerful. We should commend the students for their good deeds but doubt that herbal tea helped the residents. Identify the explanatory and response variables in this informal study. Then explain what lurking variables account for the observed association.

4.82. Price and ounces. In Example 4.2 (page 154) and Exercise 4.3 (page 156) we examined the relationship between the price and the size of a Mocha Frappuccino. The 12-ounce Tall drink costs \$3.50, the 16-ounce Grande is \$4.00, and the 24-ounce Venti is \$4.50.

(a) Plot the data and describe the relationship. (Explain why you should plot size in ounces on the x axis.)

(b) Find the least-squares regression line for predicting the price using size. Add the line to your plot.

(c) Draw a vertical line from the least-squares line to each data point. This gives a graphical picture of the residuals.

(d) Find the residuals and verify that they sum to zero.

(e) Plot the residuals versus size. Interpret this plot.

4.83. Average monthly temperatures. Here are the average monthly temperatures for Chicago, Illinois: CHICAGOTEMPS

Month	1	2	3	4	5	6
Temperature (°F)	21.0	25.4	37.2	48.6	58.9	68.6

Month	7	8	9	10	11	12
Temperature (°F)	73.2	71.7	64.4	52.8	40.0	26.6

In this table, months are coded as integers, with January corresponding to 1 and December corresponding to 12.

(a) Plot the data with month on the x axis and temperature on the y axis. Describe the relationship.

(b) Find the least-squares regression line and add it to the plot. Does the line give a good fit to the data? Explain your answer.

(c) Calculate the residuals and plot them versus month. Describe the pattern and explain what the residual plot tells you about the relationship between temperature and month in Chicago.

(d) Do you think you would find a similar pattern if you plotted the same kind of data for another city?

(e) Would your answer to part (d) change if the other city was Melbourne, Australia? Explain why or why not.

4.84. Growth of infants in Egypt. A study of nutrition in developing countries collected data from the Egyptian village of Nahya. Here are the mean weights (in kilograms) for 170 infants in Nahya who were weighed each month during their first year of life:[21] INFANTGROWTH

Age (months)	1	2	3	4	5	6	7	8	9	10	11	12
Weight (kg)	4.3	5.1	5.7	6.3	6.8	7.1	7.2	7.2	7.2	7.2	7.5	7.8

(a) Plot weight against time.

(b) A hasty user of statistics enters the data into software and computes the least-squares line without plotting the data. The result is

```
The regression equation is
weight = 4.88 + 0.267 age
```

Plot this line on your graph. Is it an acceptable summary of the overall pattern of growth? Remember that you can calculate the least-squares line for *any* set of two-variable data. It's up to you to decide if it makes sense to fit a line.

(c) Fortunately, the software also prints out the residuals from the least-squares line. In order of age along the rows, they are

−0.85	−0.31	0.02	0.35	0.58	0.62
0.45	0.18	−0.08	−0.35	−0.32	−0.28

Verify that the residuals have sum zero (except for roundoff error). Plot the residuals against age and add a horizontal line at zero. Describe carefully the pattern that you see.

4.85. CHALLENGE **A lurking variable.** The effect of a lurking variable can be surprising when individuals are divided into groups. In recent years, the mean SAT score of all high school seniors has increased. But the mean SAT score has decreased for students at each level of high school grades (A, B, C, and so on). Explain how grade inflation in high school (the lurking variable) can account for this pattern. CAUTION *A relationship that holds for each group within a population need not hold for the population as a whole. In fact, the relationship can even change direction.*

4.86. CHALLENGE **Another example.** Here is another example of the group effect cautioned about in the previous exercise. Explain how as a nation's population grows older mean income can go down for workers in each age group, yet still go up for all workers.

4.87. Basal metabolic rate. Careful statistical studies often include examination of potential lurking variables. This was true of the study of the effect of nonexercise activity (NEA) on fat gain (Example 4.16, page 177), our lead example in Section 4.3. Overeating may lead our bodies to spontaneously increase NEA (fidgeting and the like). Our bodies might also spontaneously increase their basal metabolic rate (BMR), which is a measure of energy use while resting. If both energy uses increase, regressing fat gain on NEA alone would be misleading. Here are data on BMR and fat gain for the same 16 subjects whose NEA we examined earlier: FIDGET

BMR increase (cal)	117	352	244	−42	−3	134	136	−32
Fat gain (kg)	4.2	3.0	3.7	2.7	3.2	3.6	2.4	1.3

BMR increase (cal)	−99	9	−15	−70	165	172	100	35
Fat gain (kg)	3.8	1.7	1.6	2.2	1.0	0.4	2.3	1.1

The correlation between NEA and fat gain is $r = -0.7786$. The slope of the regression line for predicting fat gain from NEA is $b_1 = -0.00344$ kilogram per calorie. What are the correlation and slope for BMR and fat gain? Explain why these values show that BMR has much less effect on fat gain than does NEA.

4.88. Gas chromatography. Gas chromatography is a technique used to detect very small amounts of a substance, for example, a contaminant in drinking water. Laboratories use regression to calibrate such techniques. The following data show the results of five measurements for each of four amounts of the substance being investigated.[22] The explanatory variable x is the amount of substance in the specimen, measured in nanograms (ng), units of 10^{-9} gram. The response variable y is the reading from the gas chromatograph. GASCHROMO

Amount (ng)	Response				
0.25	6.55	7.98	6.54	6.37	7.96
1.00	29.7	30.0	30.1	29.5	29.1
5.00	211	204	212	213	205
20.00	929	905	922	928	919

(a) Make a scatterplot of these data. The relationship appears to be approximately linear, but the wide variation in the response values makes it hard to interpret.

(b) Compute the least-squares regression line of y on x, and plot this line on your graph.

(c) Now compute the residuals and make a plot of the residuals against x. It is much easier to see deviations from linearity in the residual plot. Describe carefully the pattern displayed by the residuals.

4.89. Golf scores. Here are the golf scores of 6 members of a women's golf team in the first two rounds of the National Collegiate Athletic Association tournament:[23] GOLFNCAA

Player	Gulyanamitta	Hernandez	Hoffmeister
Round 1	80	74	76
Round 2	76	72	76

Player	LeBlanc	Mess	Sinha
Round 1	76	85	88
Round 2	77	83	83

(a) Plot the data with the Round 1 scores on the x axis and the Round 2 scores on the y axis.

(b) Describe the relationship.

(c) Calculate the least-squares regression line and add it to your plot.

(d) Circle the observation for Maria Hernandez. She was the NCAA champion in this tournament.

4.90. Climate change. Drilling down beneath a lake in Alaska yields chemical evidence of past changes in climate. Biological silicon (in milligrams of silicon per gram of water), left by the skeletons of single-celled creatures called diatoms, measures the abundance of life in the lake. A rather complex variable based on the ratio of certain isotopes relative to ocean water gives an indirect measure of moisture, mostly from snow. As we drill down, we look farther into the past. Here are data from 2300 to 12,000 years ago:[24] SILICONISOTOPE

Isotope (%)	Silicon (mg/g)	Isotope (%)	Silicon (mg/g)	Isotope (%)	Silicon (mg/g)
−19.90	97	−20.71	154	−21.63	224
−19.84	106	−20.80	265	−21.63	237
−19.46	118	−20.86	267	−21.19	188
−20.20	141	−21.28	296	−19.37	337

(a) Make a scatterplot of silicon (response) against isotope (explanatory). Ignoring the outlier, describe the form, direction, and strength of the relationship. The researchers say that this relationship and relationships among other variables they measured are evidence for cyclic changes in climate that are linked to changes in the sun's activity.

(b) The researchers single out one point in their plot, which they say "is an outlier that was excluded in the correlation analysis." Circle this outlier on your graph. What is the correlation with and without this point? The point strongly influences the correlation.

(c) Is the outlier also strongly influential for the regression line? Calculate and draw on your graph the regression lines with and without the outlier, and discuss what you see.

4.91. APPLET **Use the applet.** It isn't easy to guess the position of the least-squares line by eye. Use the *Correlation and Regression* applet to compare a line you draw with the least-squares line. Click on the scatterplot to create a group of 15 to 20 points from lower left to upper right with a clear positive straight-line pattern (correlation around 0.7). Click the "Draw line" button and use the mouse to draw a line through the middle of the cloud of points from lower left to upper right.

Note the "thermometer" that appears above the plot. The red portion is the sum of the squared vertical distances from the points in the plot to the least-squares line. The green portion is the "extra" sum of squares for your line—it shows by how much your line misses the smallest possible sum of squares.

(a) You drew a line by eye through the middle of the pattern. Yet the right-hand part of the bar is probably almost entirely green. What does that tell you?

(b) Now click the "Show least-squares line" box. Is the slope of the least-squares line smaller (the new line is less steep) or larger (line is steeper) than that of your line? If you repeat this exercise several times, you will consistently get the same result.

CAUTION *The least-squares line minimizes the vertical distances of the points from the line. It is not the line through the "middle" of the cloud of points.* This is one reason why it is hard to draw a good regression line by eye.

4.92. APPLET **Use the applet.** Go to the *Correlation and Regression* applet. Click on the scatterplot to create a group of 10 points in the lower-right corner of the scatterplot with a strong straight-line pattern (correlation about −0.9). In Exercise 4.48 (page 176) you started to see that correlation r is not resistant. Now click the "Show least-squares line" box to display the regression line.

(a) Add one point at the upper left that is far from the other 10 points but exactly on the regression line. Why does this outlier have no effect on the line even though it changes the correlation?

(b) Now move this last point down until it is opposite the group of 10 points. You see that one end of the least-squares line chases this single point, while the other end remains near the middle of the original group of 10. What makes the last point so influential?

4.93. Education and income. There is a strong positive correlation between years of education and income for economists employed by business firms. (In particular, economists with doctorates earn more than economists with only a bachelor's degree.) There is also a strong positive correlation between years of education and income for economists employed by colleges and universities.

But when all economists are considered, there is a *negative* correlation between education and income. The explanation for this is that business pays high salaries and employs mostly economists with bachelor's degrees, while colleges pay lower salaries and employ mostly economists with doctorates.

Sketch a scatterplot with two groups of cases (business and academic) that illustrates how a strong positive correlation within each group and a negative overall correlation can occur together.

4.94. Dangers of not looking at a plot. Table 4.1 (page 189) presents four sets of data prepared by the statistician Frank Anscombe to illustrate the dangers of calculating without first plotting the data.[25] DATA FILE ANSCOMBE

(a) Use x to predict y for each of the four data sets. Find the predicted values and residuals for each of the four regression equations.

(b) Plot the residuals versus x for each of the four data sets.

(c) Write a summary of what the residuals tell you for each data set and explain how the residuals help you to understand these data.

4.5 Data Analysis for Two-Way Tables

LOOK BACK quantitative and categorical variables p. 80

When we study relationships between two variables, one of the first questions we ask is whether each variable is quantitative or categorical. For two quantitative variables, we use a scatterplot to examine the relationship, and we fit a line to the data if the relationship is approximately linear. If one of the variables is quantitative and the other is categorical, we can use the methods in Chapter 3 to describe the distribution of the quantitative variable for each value of the categorical variable. This leaves us with the situation where both variables are categorical. In this section we discuss methods for studying these relationships.

Some variables—such as gender, race, and occupation—are inherently categorical. Other categorical variables are created by grouping values of a quantitative variable into classes. Published data are often reported in grouped form to save space. To describe categorical data, we use the *counts* (frequencies) or *percents* (relative frequencies) of individuals that fall into various categories.

The two-way table

two-way table

A key idea in studying relationships between two variables is that both variables must be measured on the same individuals or cases. When both variables are categorical, the raw data are summarized in a **two-way table** that gives counts of observations for each combination of values of the two categorical variables. Here is an example.

EXAMPLE 4.29

DATA FILE BINGEGENDER

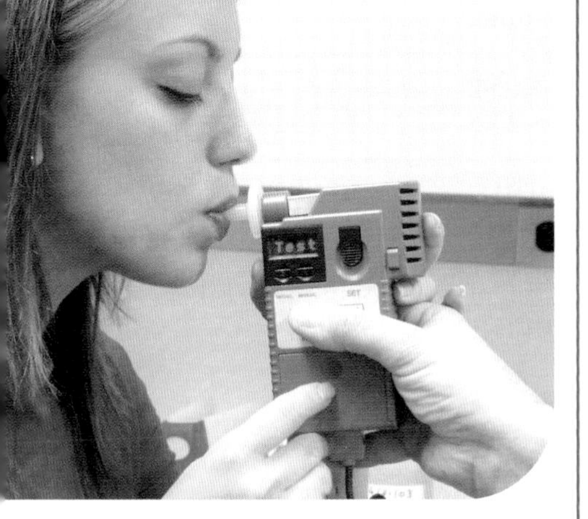

Binge drinking by college students. Alcohol abuse has been described by college presidents as the number one problem on campus, and it is an important cause of death in young adults. How common is it? A survey of 17,096 students in U.S. four-year colleges collected information on drinking behavior and alcohol-related problems.[26] The researchers defined "frequent binge drinking" as having five or more drinks in a row three or more times in the past two weeks. Here is the two-way table classifying students by gender and whether or not they are frequent binge drinkers:

Two-way table for frequent binge drinking and gender

	Gender	
Frequent binge drinker	**Men**	**Women**
Yes	1630	1684
No	5550	8232

We see that there are 1630 male students who are frequent binge drinkers and 5550 male students who are not.

USE YOUR KNOWLEDGE

4.95 Read the table. How many female students are binge drinkers? How many are not? DATA FILE BINGEGENDER

row and column variables

cell

For the binge-drinking example, we could view gender as an explanatory variable and frequent binge drinking as a response variable. This is why we put gender in the columns (like the x axis in a regression) and frequent binge drinking in the rows (like the y axis in a regression). We call binge drinking the **row variable** because each horizontal row in the table describes the drinking behavior. Gender is the **column variable** because each vertical column describes one gender group. Each combination of values for these two variables appears in a **cell** in the table. For example, the cell corresponding to women who are not frequent binge drinkers contains the number 8232. This table is called a 2×2 table because there are 2 rows and 2 columns.

To describe relationships between two categorical variables, we compute different types of percents. Our job is easier if we expand the basic two-way table by adding various totals. We illustrate the idea with our binge-drinking example.

EXAMPLE 4.30

Add the margins to the table. We expand the table in Example 4.29 by adding the totals for each row, for each column, and the total number of all the observations. Here is the result:

Two-way table for frequent binge drinking and gender

	Gender		
Frequent binge drinker	**Men**	**Women**	**Total**
Yes	1,630	1,684	3,314
No	5,550	8,232	13,782
Total	7,180	9,916	17,096

In this study there are 7180 male students. The total number of binge drinkers is 3314, and the total number of individuals in the study is 17,096.

USE YOUR KNOWLEDGE

4.96 Read the margins of the table. How many women are subjects in the binge-drinking study? What is the total number of students who are not binge drinkers? BINGEGENDER

In this example, be sure that you understand how the table is obtained from the raw data. Think about a data file with one line per subject. There would be 17,096 lines, or records, in this data set. In the two-way table, each individual is counted once and only once. As a result, the sum of the counts in the table is the total number of individuals in the data set. *Most errors in the use of categorical-data methods come from a misunderstanding of how these tables are constructed.*

Joint distribution

We are now ready to compute some proportions that help us understand the data in a two-way table. Suppose that we are interested in the men who are binge drinkers. The proportion of these is simply 1630 divided by 17,096, or 0.095. We would estimate that 9.5% of college students are male frequent binge drinkers. For each cell, we can compute a proportion by dividing the cell entry by the total sample size. The collection of these proportions is the **joint distribution** of the two categorical variables.

joint distribution

EXAMPLE 4.31

The joint distribution. For the binge-drinking example, the joint distribution of binge drinking and gender is

Joint distribution of frequent binge drinking and gender

	Gender	
Frequent binge drinker	**Men**	**Women**
Yes	0.095	0.099
No	0.325	0.482

Because this is a distribution, the sum of the proportions should be 1. For this example the sum is 1.001. The difference is due to roundoff error.

USE YOUR KNOWLEDGE

4.97 Explain the computation. Explain how the entry for the women who are not binge drinkers in Example 4.31 is computed from the table in Example 4.30. BINGEGENDER

From the joint distribution we see that the percents of men and women who are frequent binge drinkers are similar in the population of college students. For the men 9.5% are frequent binge drinkers; for women the percent is slightly higher, at 9.9%.

Note, however, that the proportion of women who are not frequent binge drinkers is also higher than the proportion of men. One reason for this is that there are more women in the sample than men. To understand this set of data we will need to do some additional calculations. Let's look at the distribution of gender.

Marginal distributions

marginal distribution

When we examine the distribution of a single variable in a two-way table, we are looking at a **marginal distribution.** There are two marginal distributions, one for each categorical variable in the two-way table. They are very easy to compute.

EXAMPLE 4.32

The marginal distribution of gender. Look at the table in Example 4.30. The total numbers of men and women are given in the bottom row, labeled "Total." Our sample has 7180 men and 9916 women. To find the marginal distribution of gender we simply divide these numbers by the total sample size, 17,096. The marginal distribution of gender is

Marginal distribution of gender		
	Men	**Women**
Proportion	0.420	0.580

Note that the proportions sum to 1. There is no roundoff error.

Often we prefer to use percents rather than proportions. Here is the marginal distribution of gender described with percents:

Marginal distribution of gender		
	Men	**Women**
Percent	42.0%	58.0%

Which form do you prefer?

The other marginal distribution for this example is the distribution of binge drinking.

EXAMPLE 4.33

The marginal distribution of binge drinking in percents. Here is the marginal distribution of the binge-drinking variable (in percents):

Marginal distribution of frequent binge drinking		
	Yes	**No**
Percent	19.4%	80.6%

USE YOUR KNOWLEDGE

4.98 Explain the marginal distribution. Explain how the marginal distribution of frequent binge drinking given in Example 4.33 is computed from the entries in the table given in Example 4.30. BINGEGENDER

LOOK BACK
bar graphs and pie charts p. 85

Each marginal distribution from a two-way table is a distribution for a single categorical variable. We can use a bar graph or a pie chart to display such a distribution. For our two-way table, we will be content with numerical summaries: for example, 58% of these college students are women, and 19.4% of the students are frequent binge drinkers. When we have more rows or columns, the graphical displays are particularly useful.

Describing relations in two-way tables

The table in Example 4.30 contains much more information than the two marginal distributions of gender alone and frequent binge drinking alone. We need to do a little more work to examine the relationship. Relationships among categorical variables are described by calculating appropriate percents from the counts given. What percents do you think we should use to describe the relationship between gender and frequent binge drinking?

EXAMPLE 4.34

Women who are frequent binge drinkers. What percent of the women in our sample are frequent binge drinkers? This is the count of the women who are frequent binge drinkers as a percent of the number of women in the sample:

$$\frac{1684}{9916} = 0.170 = 17.0\%$$

USE YOUR KNOWLEDGE

4.99 Find the percent. Show that the percent of men who are frequent binge drinkers is 22.7%. 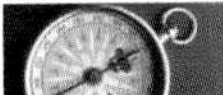

Recall that when we looked at the joint distribution of gender and binge drinking, we found that among all college students in the sample, 9.5% were male frequent binge drinkers and 9.9% were female frequent binge drinkers. The percents are fairly similar because the counts for these two groups, 1630 and 1684, are close. The calculations that we just performed, however, give us a different view. When we look separately at women and men, we see that the proportions of frequent binge drinkers are somewhat different, 17.0% for women versus 22.7% for men.

Conditional distributions

conditional distribution

In Example 4.34 we looked at the women alone and examined the distribution of the other categorical variable, frequent binge drinking. Another way to say this is that we conditioned on the value of gender being female. Similarly, we can condition on the value of gender being male. When we condition on the value of one variable and calculate the distribution of the other variable, we obtain a **conditional distribution.** Note that in Example 4.34 we calculated only the percent for frequent binge drinking. The complete conditional distribution gives the proportions or percents for all possible values of the conditioning variable.

EXAMPLE 4.35

Conditional distribution of binge drinking for women. For women, the conditional distribution of the binge-drinking variable in terms of percents is

Conditional distribution of binge drinking for women		
	Yes	**No**
Percent	17.0%	83.0%

Note that we have included the percents for the two possible values, Yes and No, of the binge-drinking variable. These percents sum to 100%.

USE YOUR KNOWLEDGE

4.100 A conditional distribution. Perform the calculations to show that the conditional distribution of binge drinking for men is BINGEGENDER

Conditional distribution of binge drinking for men		
	Yes	**No**
Percent	22.7%	77.3%

Comparing the conditional distributions (Example 4.35 and Exercise 4.100) reveals the nature of the association between gender and frequent binge drinking. In this set of data the men are more likely to be frequent binge drinkers than the women.

Bar graphs can help us to see relationships between two categorical variables. No single graph (such as a scatterplot) portrays the form of the relationship between categorical variables, and no single numerical measure (such as the correlation) summarizes the strength of an association. Bar graphs are flexible enough to be helpful, but you must think about what comparisons you want to display. For numerical measures, we must rely on well-chosen percents or on more advanced statistical methods.[27]

Of course, we prefer to use software to compute the joint, marginal, and conditional distributions. A two-way table contains a great deal of information in compact form. Making that information clear almost always requires using percents. *You must decide which percents you need.*

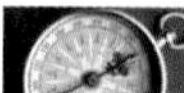

EXAMPLE 4.36

Software output. Figure 4.24 gives computer output for the data in Example 4.29 using SPSS, Minitab, and JMP. There are minor variations among software packages, but these are typical of what is usually produced. Each cell in the 2×2 table has four entries. These are the count (the number of observations in the cell), the conditional distributions for rows and columns, and the joint distribution. Note that all of these are expressed as percents rather than proportions. Marginal totals and distributions are given in the rightmost column and the bottom row.

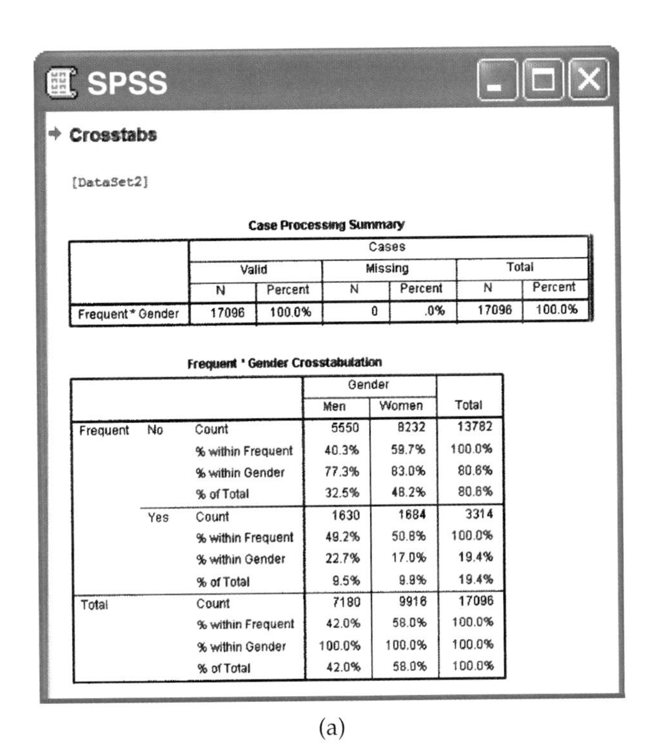

SPSS

Crosstabs

[DataSet2]

Case Processing Summary

	Cases					
	Valid		Missing		Total	
	N	Percent	N	Percent	N	Percent
Frequent * Gender	17096	100.0%	0	.0%	17096	100.0%

Frequent * Gender Crosstabulation

			Gender		
			Men	Women	Total
Frequent	No	Count	5550	8232	13782
		% within Frequent	40.3%	59.7%	100.0%
		% within Gender	77.3%	83.0%	80.6%
		% of Total	32.5%	48.2%	80.6%
	Yes	Count	1630	1684	3314
		% within Frequent	49.2%	50.8%	100.0%
		% within Gender	22.7%	17.0%	19.4%
		% of Total	9.5%	9.9%	19.4%
Total		Count	7180	9916	17096
		% within Frequent	42.0%	58.0%	100.0%
		% within Gender	100.0%	100.0%	100.0%
		% of Total	42.0%	58.0%	100.0%

(a)

Session

Tabulated statistics: Frequent, Gender

```
Using frequencies in Count

Rows: Frequent   Columns: Gender

            Men   Women     All

No         5550    8232   13782
          40.27   59.73  100.00
          77.30   83.02   80.62
          32.46   48.15   80.62

Yes        1630    1684    3314
          49.19   50.81  100.00
          22.70   16.98   19.38
           9.53    9.85   19.38

All        7180    9916   17096
          42.00   58.00  100.00
         100.00  100.00  100.00
          42.00   58.00  100.00

Cell Contents:      Count
                    % of Row
                    % of Column
                    % of Total
```

(b)

Sheet1-Fit Y by X of Gender by Frequent

Contingency Analysis of Gender By Frequent

Weight: Count

Contingency Table

Frequent	Count Total % Col % Row %	Gender: Men	Women	
	No	5550 32.46 77.30 40.27	8232 48.15 83.02 59.73	13782 80.62
	Yes	1630 9.53 22.70 49.19	1684 9.85 16.98 50.81	3314 19.38
		7180 42.00	9916 58.00	17096

(c)

FIGURE 4.24 Computer output from three statistical software packages for the binge-drinking study in Example 4.36: (a) SPSS, (b) Minitab, and (c) JMP. Other software produces similar output.

Most software packages order the row and column labels numerically or alphabetically. In general, it is better to use words rather than numbers for the column labels. This sometimes involves some additional work, but it avoids the kind of confusion that can result when you forget the real values associated with each numerical value.

You should verify that the entries in Figure 4.24 correspond to the calculations that we performed in Examples 4.31 to 4.35. In addition, verify the calculations for the conditional distributions of gender for each value of the binge-drinking variable.

Simpson's paradox

As is the case with quantitative variables, the effects of lurking variables can strongly influence relationships between two categorical variables. Here is an example that demonstrates the surprises that can await the unsuspecting consumer of data.

EXAMPLE 4.37

DATA FILE CUSTSERVICEREP

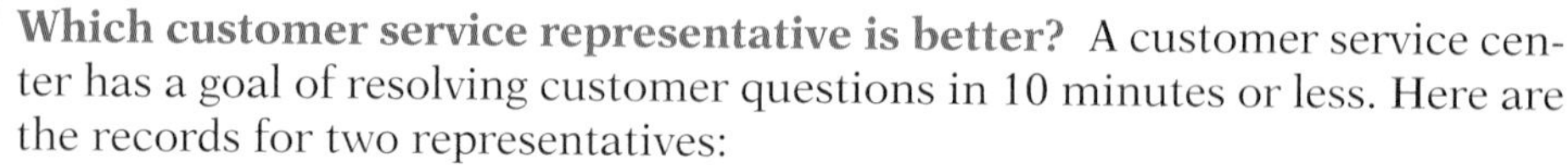

Which customer service representative is better? A customer service center has a goal of resolving customer questions in 10 minutes or less. Here are the records for two representatives:

	Representative	
Goal met	**Alexis**	**Peyton**
Yes	172	118
No	28	82
Total	200	200

Alexis has met the goal 172 times out of 200, a success rate of 86%. For Peyton, the success rate is 118 out of 200, or 59%. Alexis clearly has the better success rate.

Let's look at the data in a little more detail. The data summarized come from two different weeks in the year.

EXAMPLE 4.38

DATA FILE CUSTSERVICEREP

Let's look at the data more carefully. Here are the counts broken down by week:

	Week 1		Week 2	
Goal met	**Alexis**	**Peyton**	**Alexis**	**Peyton**
Yes	162	19	10	99
No	18	1	10	81
Total	180	20	20	180

For Week 1, Alexis met the goal 90% of the time (162/180) while Peyton met the goal 95% of the time (19/20). Peyton had the better performance in Week 1. What about Week 2? Here Alexis met the goal 50% of the time (10/20) while the success rate for Peyton was 55% (99/180). Peyton again had the better performance. How does this analysis compare with the analysis that combined the counts for the two weeks? That analysis clearly showed that Alexis had the better performance, 86% versus 59%.

These results can be explained by a lurking variable related to week. The first week was during a period when the product had been in use for several months. Most of the calls to the customer service center concerned problems that had been encountered before. The representatives were trained to answer these questions and usually had no trouble in meeting the goal of resolving the problems quickly. On the other hand, the second week occurred shortly after the release of a new version of the product. Most of the calls during this week concerned new problems that the representatives had not yet encountered. Many more of these questions took longer than the 10-minute goal to resolve.

Look at the total in the bottom row of the detailed table. During the first week, when calls were easy to resolve, Alexis handled 180 calls and Peyton handled 20. The situation was exactly the opposite during the second week, when the calls were difficult to resolve. There were 20 calls for Alexis and 180 for Peyton.

The original two-way table, which did not take account of week, was misleading. This example illustrates *Simpson's paradox*.

SIMPSON'S PARADOX

An association or comparison that holds for all of several groups can reverse direction when the data are combined to form a single group. This reversal is called **Simpson's paradox.**

The lurking variables in our Simpson's paradox example are categorical. That is, they break the observations into groups, workweek. *Simpson's paradox is an extreme form of the fact that observed associations can be misleading when there are lurking variables.*

three-way table

The data in Example 4.38 are given in a **three-way table** that reports counts for each combination of three categorical variables: week, representative, and whether or not the goal was met. In our example, we looked at the three-way table by constructing two two-way tables for representative by goal, one for each week. The original table can be obtained by adding the corresponding counts for these two tables. This process is called **aggregating** the data. When we aggregated data in our example, we ignored the variable week, which then became a lurking variable. *Conclusions that seem obvious when we look only at aggregated data can become quite different when the data are examined in more detail.*

aggregation

SECTION 4.5 SUMMARY

A **two-way table** of counts organizes data about two categorical variables. Values of the **row variable** label the rows that run across the table, and values of the **column variable** label the columns that run down the table. Two-way tables are often used to summarize large amounts of data by grouping outcomes into categories.

The **joint distribution** of the row and column variables is found by dividing the count in each cell by the total number of observations.

The **row totals** and **column totals** in a two-way table give the **marginal distributions** of the two variables separately. It is clearer to present these distributions as percents of the table total. Marginal distributions do not give any information about the relationship between the variables.

To find the **conditional distribution** of the row variable for one specific value of the column variable, look only at that one column in the table. Find each entry in the column as a percent of the column total.

There is a conditional distribution of the row variable for each column in the table. **Comparing conditional distributions is one way to describe the association between the row and the column variables.** It is particularly useful when the column variable is the explanatory variable. When the row variable is explanatory, find the conditional distribution of the column variable for each row and compare these distributions.

Bar graphs are a flexible means of presenting categorical data. There is no single best way to describe an association between two categorical variables.

We present data on three categorical variables in a **three-way table,** printed as separate two-way tables for each level of the third variable. A comparison between two variables that holds for each level of a third variable can be changed or even reversed when the data are **aggregated** by summing over all levels of the third variable. **Simpson's paradox** refers to the reversal of a comparison by aggregation. It is an example of the potential effect of lurking variables on an observed association.

SECTION 4.5 EXERCISES

For Exercise 4.95, see page 203; for Exercise 4.96, see page 204; for Exercise 4.97, see page 205; for Exercise 4.98, see page 206; for Exercise 4.99, see page 207; and for Exercise 4.100, see page 208.

4.101. Exercise and adequate sleep. A survey of 656 boys and girls who were 13 to 18 years old asked about adequate sleep and other health-related behaviors. The recommended amount of sleep is six to eight hours per night.[28] In the survey 59.4% of the respondents reported that they got less than this amount of sleep on school nights. An exercise scale was developed and was used to classify the students as above or below the median in this domain. Here is the 2 × 2 table of counts with students classified as getting or not getting adequate sleep and by the exercise variable: DATA FILE SLEEP

	Exercise	
Enough sleep	**High**	**Low**
Yes	151	115
No	148	242

(a) Find the distribution of adequate sleep for the high exercisers.

(b) Do the same for the low exercisers.

(c) Summarize the relationship between adequate sleep and exercise using the results of parts (a) and (b).

4.102. Adequate sleep and exercise. Refer to the previous exercise. DATA FILE SLEEP

(a) Find the distribution of exercise for those who get adequate sleep.

(b) Do the same for those who do not get adequate sleep.

(c) Write a short summary of the relationship between adequate sleep and exercise using the results of parts (a) and (b).

(d) Compare this summary with the summary that you obtained in part (c) of the previous exercise. Which do you prefer? Give a reason for your answer.

4.103. Which hospital is safer? Insurance companies and consumers are interested in the performance of hospitals. The government releases data about patient outcomes in hospitals that can be useful in making informed health care decisions. Here is a two-way table of data on the survival of patients after surgery in two hospitals. All patients undergoing surgery in a recent time period are included. "Survived" means that the patient lived at least 6 weeks following surgery. HOSPITALS

	Hospital A	Hospital B
Died	63	16
Survived	2037	784
Total	2100	800

What percent of Hospital A patients died? What percent of Hospital B patients died? These are the numbers one might see reported in the media.

4.104. Patients in "poor" or "good" condition. Refer to the previous exercise. Not all surgery cases are equally serious, however. Patients are classified as being in either "poor" or "good" condition before surgery. Here are the data broken down by patient condition. The entries in the original two-way table are just the sums of the "poor" and "good" entries in this pair of tables. HOSPITALS

Good Condition	Hospital A	Hospital B
Died	6	8
Survived	594	592
Total	600	600

Poor Condition	Hospital A	Hospital B
Died	57	8
Survived	1443	192
Total	1500	200

(a) Find the death rate for Hospital A patients who were classified as "poor" before surgery. Do the same for Hospital B. In which hospital do "poor" patients fare better?

(b) Repeat (a) for patients classified as "good" before surgery.

(c) What is your recommendation to someone facing surgery and choosing between these two hospitals?

(d) How can Hospital A do better in both groups, yet do worse overall? Look at the data and carefully explain how this can happen.

4.105. Full-time and part-time college students. The Census Bureau provides estimates of numbers of people in the United States classified in various ways.[29] Let's look at college students. The following table gives us data to examine the relation between age and full-time or part-time status. The numbers in the table are expressed as thousands of U.S. college students. USCOLSTUDENTS

U.S. college students by age and status

	Status	
Age (years)	Full-time	Part-time
15–19	3388	389
20–24	5238	1164
25–34	1703	1699
35 and over	762	2045

(a) What is the U.S. Census Bureau estimate of the number of full-time college students aged 15 to 19?

(b) Give the joint distribution of age and status for this table.

(c) What is the marginal distribution of age? Display the results graphically.

(d) What is the marginal distribution of status? Display the results graphically.

4.106. Condition on age. Refer to the previous exercise. Find the conditional distribution of status for each of the four age categories. Display the distributions graphically and summarize their differences and similarities. USCOLSTUDENTS

4.107. Condition on status. Refer to the previous two exercises. Compute the conditional distribution of age for each of the two status categories. Display the distributions graphically and write a short paragraph describing the distributions and how they differ. USCOLSTUDENTS

4.108. Complete the table. Here are the row and column totals for a two-way table with two rows and two columns:

a	b	200
c	d	200
200	200	400

Find *two different* sets of counts a, b, c, and d for the body of the table that give these same totals. This shows that the relationship between two variables cannot be obtained from the two individual distributions of the variables.

4.109. Construct a table with no association. Construct a 3×3 table of counts where there is no apparent association between the row and column variables.

4.110. Survey response rates. A market research firm conducted a survey of companies in its state. They mailed a questionnaire to 300 small companies, 300 medium-sized companies, and 300 large companies. The rate of nonresponse is important in deciding how reliable survey results are. Here are the data on response to this survey: RESPONSERATES

	Response		
Size of company	Yes	No	Total
Small	175	125	300
Medium	145	155	300
Large	120	180	300

(a) What was the overall percent of nonresponse?

(b) Describe how nonresponse is related to the size of the business. (Use percents to make your statements precise.)

(c) Draw a bar graph to compare the nonresponse percents for the three size categories.

(d) Using the total number of responses as a base, compute the percent of responses that come from each of small, medium, and large businesses.

(e) The sampling plan was designed to obtain equal numbers of responses from small, medium, and large companies. In preparing an analysis of the survey results, do you think it would be reasonable to proceed as if the responses represented companies of each size equally?

CHAPTER 4 EXERCISES

4.111. Popularity of a first name. The Social Security Administration maintains lists of the top 1000 names for boys and girls born each year since 1879.[30] The name "Atticus" made the list in seven recent years. Here are the ranks for those years: ATTICUS

Year	2004	2005	2006	2007	2008	2009	2010
Rank	935	792	767	682	689	609	564

(a) Plot rank versus year.

(b) Find the equation of the least-squares regression line and add it to your plot.

(c) Do these data suggest that the name "Atticus" is becoming more popular, less popular, or staying the same in popularity over this period of time? Give reasons for your answer.

4.112. You select the name. Refer to the previous exercise. Choose a first name and find the rank of this name for the past several years from the Social Security Web site, `ssa.gov/OACT/babynames`. Answer the questions from the previous exercise for this name.

4.113. Salaries and raises. For this exercise we consider a hypothetical employee who starts working in Year 1 with a salary of $50,000. Each year, her salary increases by approximately 5%. By Year 20, she is earning $126,000. The following table gives her salary for each year (in thousands of dollars): RAISES

Year	Salary	Year	Salary	Year	Salary	Year	Salary
1	50	6	63	11	81	16	104
2	53	7	67	12	85	17	109
3	56	8	70	13	90	18	114
4	58	9	74	14	93	19	120
5	61	10	78	15	99	20	126

(a) Figure 4.25 is a scatterplot of salary versus year with the least-squares regression line. Describe the relationship between salary and year for this person.

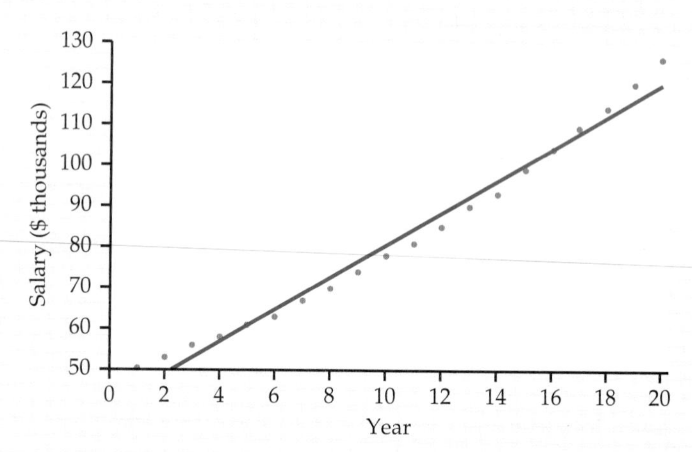

FIGURE 4.25 Plot of salary versus year, with the least-squares regression line, for an individual who receives approximately a 5% raise each year for 20 years, for Exercise 4.113.

(b) The value of r^2 for these data is 0.9832. What percent of the variation in salary is explained by year? Would you say that this is an indication of a strong linear relationship? Explain your answer.

4.114. Look at the residuals. Refer to the previous exercise. Figure 4.26 is a plot of the residuals versus year. RAISES

(a) Interpret the residual plot.

(b) Explain how this plot highlights the deviations from the least-squares regression line that you can see in Figure 4.25.

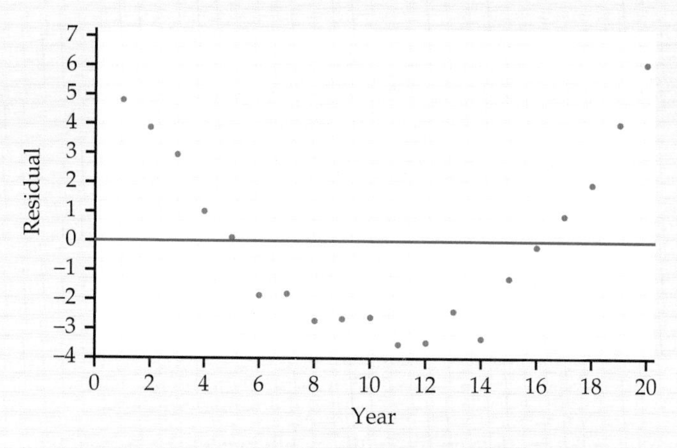

FIGURE 4.26 Plot of residuals versus year for an individual who receives approximately a 5% raise each year for 20 years, for Exercise 4.114.

4.115. Try logs. Refer to the previous two exercises. Figure 4.27 is a scatterplot with the least-squares regression line for log salary versus year. For this model, $r^2 = 0.9995$. RAISES

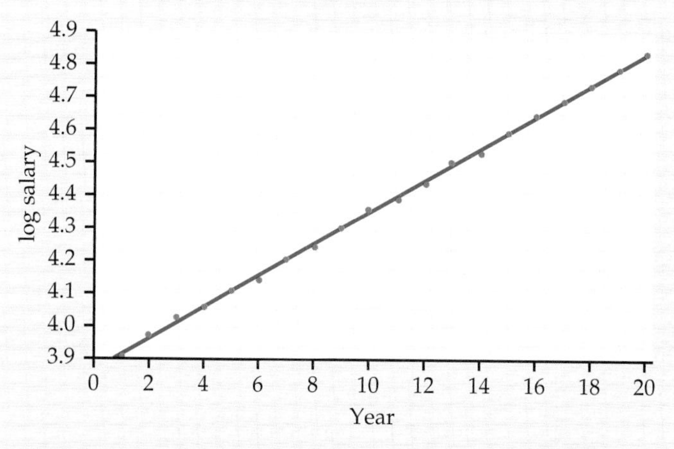

FIGURE 4.27 Plot of log salary versus year, with the least-squares regression line, for an individual who receives approximately a 5% raise each year for 20 years, for Exercise 4.115.

(a) Compare this plot with Figure 4.25. Write a short summary of the similarities and the differences.

(b) Figure 4.28 is a plot of the residuals for the model using year to predict log salary. Compare this plot with Figure 4.26 and summarize your findings.

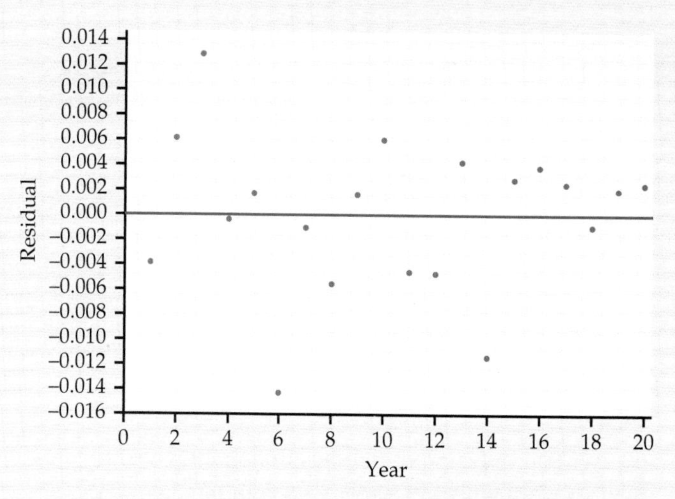

FIGURE 4.28 Plot of residuals, based on log salary, versus year for an individual who receives approximately a 5% raise each year for 20 years, for Exercise 4.115.

4.116. Do some predictions. The individual whose salary we have been studying wants to do some financial planning. Specifically, she would like to predict her salary 5 years into the future, that is, for Year 25. She is willing to assume that her employment situation will be stable for the next 5 years and that it will be similar to the last 20 years. RAISES

(a) Predict her salary for Year 25 using the least-squares regression equation constructed to predict salary from year.

(b) Predict her salary for Year 25 using the least-squares regression equation constructed to predict log salary from year. Note that you will need to take the predicted log salary and convert this value back to the predicted salary. Many calculators have a function that will perform this operation.

(c) Which prediction do you prefer? Explain your answer.

(d) Someone looking at the numerical summaries and not the plots for these analyses says that because both models have very high values of r^2, they should perform equally well in doing this prediction. Write a response to this comment.

(e) Write a short summary about the value of graphical summaries and the problems of extrapolation using what you have learned in studying the salary data.

4.117. Faculty salaries. Here are the salaries for a sample of professors in a mathematics department at a large midwestern university for the academic years 2010–2011 and 2011–2012. FACULTYSALARIES

(a) Construct a scatterplot with the 2011–2012 salaries on the vertical axis and the 2010–2011 salaries on the horizontal axis.

(b) Comment on the form, direction, and strength of the relationship in your scatterplot.

2010–2011 salary ($)	2011–2012 salary ($)	2010–2011 salary ($)	2011–2012 salary ($)
145,800	146,900	137,650	140,350
114,800	117,800	133,160	136,485
111,000	114,600	75,972	80,472
99,700	103,900	77,000	80,500
113,000	115,200	83,500	87,000
112,790	115,250	142,850	145,830
104,500	109,250	123,506	126,906
150,000	154,080	116,100	119,400

(c) What proportion of the variation in 2011–2012 salaries is explained by 2010–2011 salaries?

4.118. Find the line and examine the residuals. Refer to the previous exercise. FACULTYSALARIES

(a) Find the least-squares regression line for predicting 2011–2012 salaries from 2010–2011 salaries.

(b) Analyze the residuals, paying attention to any outliers or influential observations. Write a summary of your findings.

4.119. Bigger raises for those earning less. Refer to the previous two exercises. The 2010–2011 salaries do an excellent job of predicting the 2011–2012 salaries. Is there anything more that we can learn from these data? In this department there is a tradition of giving higher-than-average raise percent to those whose salaries are lower. Let's see if we can find evidence to support this idea in the data. FACULTYSALARIES

(a) Compute the raise percent for each faculty member. Take the difference between the 2011–2012 salary and the 2010–2011 salary, divide by the 2010–2011 salary, and then multiply by 100. Make a scatterplot with raise as the response variable and the 2010–2011 salary as the explanatory variable. Describe the relationship that you see in your plot.

(b) Find the least-squares regression line and add it to your plot.

(c) Analyze the residuals. Are there any outliers or influential cases? Make a graphical display and include this in a short summary of what you conclude.

(d) Is there evidence in the data to support the idea that greater raise percent are given to those with lower salaries? In your answer, include numerical and graphical summaries to support your conclusion.

4.120. Graduation rates. One of the factors used to evaluate undergraduate programs is the proportion of incoming students who graduate. This quantity, called the graduation rate, can be predicted by variables such as the SAT or ACT scores and the high school records of the incoming students. One of the components that *U.S. News & World Report* uses when evaluating colleges is the difference between the actual graduation rate and the rate predicted by a regression equation.[31] In this chapter, we call this quantity the residual. Explain why the residual is a better measure to evaluate college graduation rates than the raw graduation rate.

4.121. CHALLENGE **Eating fruits and vegetables and smoking.** The Centers for Disease Control and Prevention (CDC) Behavior Risk Factor Surveillance System (BRFSS) collects data related to health conditions and risk behaviors.[32] Data in the BRFSS are aggregated by state. Figure 4.29 is a plot of two of the BRFSS variables. Fruits and Vegetables is the percent of adults in the state who report eating at least five servings of fruits and vegetables per day. Smoking is the percent who smoke every day. BRFSS

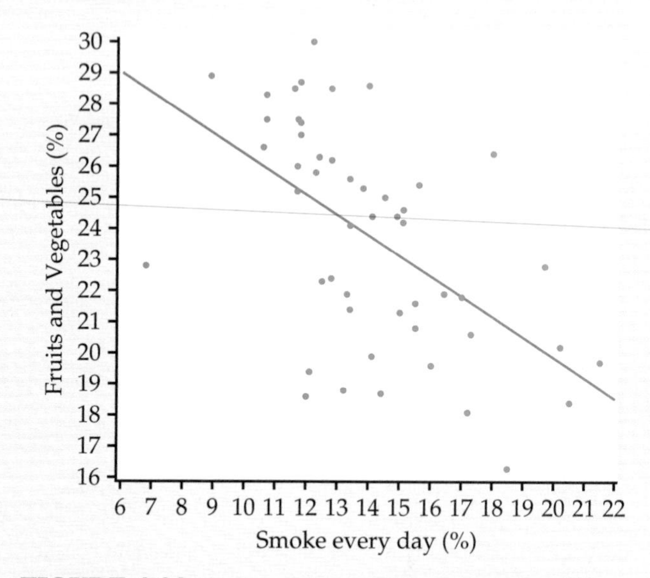

FIGURE 4.29 Fruits and Vegetables versus Smoking with least-squares regression line, for Exercise 4.121.

(a) Describe the relationship between Fruits and Vegetables and Smoking. Explain why you might expect this type of association.

(b) Find the correlation between the two variables.

(c) For Utah, 22.8% eat at least five servings of fruits and vegetables per day and 6.8% smoke every day. Find Utah on the plot and describe its position relative to the other states.

TABLE 4.3

Fruit and vegetable consumption and smoking

State	Fruits and Vegetables (%)	Smoking (%)	State	Fruits and Vegetables (%)	Smoking (%)
Alabama	20.6	17.3	Montana	25.3	13.8
Alaska	24.2	15.1	Nebraska	24.1	13.4
Arizona	28.3	10.7	Nevada	21.9	16.4
Arkansas	21.8	17.0	New Hampshire	28.5	12.8
California	28.9	8.9	New Jersey	27.5	10.7
Colorado	25.8	12.3	New Mexico	22.4	12.8
Connecticut	28.5	11.6	New York	27.4	11.8
Delaware	21.4	13.4	North Carolina	21.6	15.5
Florida	26.2	12.8	North Dakota	21.9	13.3
Georgia	25.0	14.5	Ohio	20.8	15.5
Hawaii	28.7	11.8	Oklahoma	16.3	18.5
Idaho	22.3	12.5	Oregon	27.0	11.8
Illinois	24.6	15.1	Pennsylvania	25.4	15.6
Indiana	22.8	19.7	Rhode Island	25.6	13.4
Iowa	19.9	14.1	South Carolina	18.7	14.4
Kansas	18.8	13.2	South Dakota	18.6	12.0
Kentucky	18.4	20.5	Tennessee	26.4	18.0
Louisiana	19.6	16.0	Texas	25.2	11.7
Maine	28.6	14.0	Utah	22.8	6.8
Maryland	26.6	10.6	Vermont	30.0	12.2
Massachusetts	27.5	11.7	Virginia	26.3	12.4
Michigan	21.3	15.0	Washington	26.0	11.7
Minnesota	19.4	12.1	West Virginia	19.7	21.5
Mississippi	18.1	17.2	Wisconsin	24.4	14.1
Missouri	20.2	20.2	Wyoming	24.4	14.9

(d) For California, the percents are 28.9% for Fruits and Vegetables and 8.9% for Smoking. Find California on the plot and describe its position relative to the other states.

(e) Pick your favorite state and write a short summary in which you estimate its position relative to states that you consider to be similar. Then use Table 4.3 to determine if your guess is supported by the data. Summarize your results.

4.122. CHALLENGE **Education and eating fruits and vegetables.** Refer to the previous exercise. The BRFSS data file contains a variable called EdCollege, the proportion of adults who have completed college. DATA FILE BRFSS

(a) Plot the data with Fruits and Vegetables on the x axis and EdCollege on the y axis. Describe the overall pattern of the data.

(b) Add the least-squares regression line to your plot. Does the line give a summary of the overall pattern? Explain your answer.

(c) Pick out a few states and use their position in the graph to write a short summary of how they compare with other states.

(d) Can you conclude that earning a college degree will cause you to eat five servings of fruits and vegetables per day? Explain your answer.

4.123. Predicting text pages. The editor of a statistics text would like to plan for the next edition. A key variable is the number of pages that will be in the final version. Text files are prepared by the authors using a word processor called LaTeX, and separate files contain figures and tables. For the previous edition of the text, the number of pages in the LaTeX files can easily be determined, as well as the number of pages in the final version of the text. Here are the data: DATA FILE TEXTPAGES

Chapter	1	2	3	4	5	6	7	8	9	10	11	12	13
LaTeX pages	77	73	59	80	45	66	81	45	47	43	31	46	26
Text pages	99	89	61	82	47	68	87	45	53	50	36	52	19

(a) Plot the data and describe the overall pattern.

(b) Find the equation of the least-squares regression line and add the line to your plot.

(c) Find the predicted number of pages for the next edition if the number of LaTeX pages is 62.

(d) Write a short report for the editor explaining to her how you constructed the regression equation and how she could use it to estimate the number of pages in the next edition of the text.

4.124. CHALLENGE **Points scored in women's basketball games.** Use the Internet to find the scores for the past season's women's basketball team at a college of your choice. Is there a relationship between the points scored by your chosen team and the points scored by the opponent? Summarize the data and write a report on your findings.

4.125. CHALLENGE **Look at the data for men.** Refer to the previous exercise. Analyze the data for the men's team from the same college, and compare your results with those for the women.

4.126. CHALLENGE **Plywood strength.** How strong is a building material such as plywood? To be specific, support a 24-inch by 2-inch strip of plywood at both ends and apply force in the middle until the strip breaks. The modulus of rupture (MOR) is the force needed to break the strip. We would like to be able to predict MOR without actually breaking the wood. The modulus of elasticity (MOE) is found by bending the wood without breaking it. Both MOE and MOR are measured in pounds per square inch. Here are data for 32 specimens of the same type of plywood:[33] DATA FILE MOEMOR

MOE	MOR	MOE	MOR
2,005,400	11,591	2,181,910	12,702
1,166,360	8,542	1,559,700	11,209
1,842,180	12,750	2,372,660	12,799
2,088,370	14,512	1,580,930	12,062
1,615,070	9,244	1,879,900	11,357
1,938,440	11,904	1,594,750	8,889
2,047,700	11,208	1,558,770	11,565
2,037,520	12,004	2,212,310	15,317
1,774,850	10,541	1,747,010	11,794
1,457,020	10,314	1,791,150	11,413
1,959,590	11,983	2,535,170	13,920
1,720,930	10,232	1,355,720	9,286
1,355,960	8,395	1,646,010	8,814
1,411,210	10,654	1,472,310	6,326
1,842,630	10,223	1,488,440	9,214
1,984,690	13,499	2,349,090	13,645

Can we use MOE to predict MOR accurately? Use the data to write a discussion of this question.

4.127. Distribution of the residuals. Some statistical methods require that the residuals from a regression line have a Normal distribution. The residuals for the nonexercise activity example are given in Exercise 4.75 (page 192). Is their distribution close to Normal? Make a Normal quantile plot to find out.

4.128. Asian culture and thinness. Asian culture does not emphasize thinness, but young Asians are often influenced by Western culture. In a study of concerns about weight among young Korean women, researchers administered the Drive for Thinness scale (a questionnaire) to 264 female college students in Seoul, South Korea. This scale measures excessive concern with weight and dieting and fear of weight gain. In Exercise 3.37 (page 104), you examined the distribution of Drive for Thinness scores among these college students. The study looked at several explanatory variables. One was Body Dissatisfaction, also measured by a questionnaire. DATA FILE THINNESS

(a) Make a scatterplot of Drive for Thinness (response) against Body Dissatisfaction. The appearance of the plot is a result of the fact that both variables take only whole-number values. Such variables are common in the social and behavioral sciences.

(b) Add the least-squares line to your plot. The line shows a linear relationship. How strong is this relationship? Body Dissatisfaction was more strongly correlated with Drive for Thinness than any of the other explanatory variables examined. Rather weak relationships are common in social and behavioral sciences because individuals vary a great deal. Using several explanatory variables together improves prediction of the response. This is *multiple regression,* discussed in Chapter 11.

4.129. CHALLENGE **Midterm-exam scores and final-exam scores.** We expect that students who do well on the midterm exam in a course will usually also do well on the final exam. Gary Smith of Pomona College looked at the exam scores of all 346 students who took his statistics class over a 10-year period.[34] The least-squares line for predicting final-exam score from midterm-exam score was $\hat{y} = 46.6 + 0.41x$.

Octavio scores 10 points above the class mean on the midterm. How many points above the class mean do you predict that he will score on the final? (*Hint:* What is the predicted final-exam score for the class mean midterm score $\bar{x}$?) This is an example of *regression to the mean,* the phenomenon that gave "regression" its name: students who do well on the midterm will, on the average, do less well on the final, but still above the class mean.)

4.130. Firefighters and fire damage. Someone says, "There is a strong positive correlation between the number of firefighters at a fire and the amount of damage the fire does. So sending lots of firefighters just causes more damage." Explain why this reasoning is wrong.

4.131. University degrees in Asia. Asia has become a major competitor of the United States and Western Europe in education as well as economics. Here are counts of first university degrees in science and engineering in the three regions:[35] DATA FILE UNIVDEGREES

	Region		
Field	**United States**	**Western Europe**	**Asia**
Engineering	61,941	158,931	280,772
Natural science	111,158	140,126	242,879
Social science	182,166	116,353	236,018

Direct comparison of counts of degrees would require us to take into account Asia's much larger population. We can, however, compare the distribution of degrees by field of study in the three regions. Do this using calculations and graphs, and write a brief summary of your findings.

4.132. CHALLENGE **Motivation to participate in volunteer service.** A study examined patterns and characteristics of volunteer service for young people from high school through early adulthood.[36] Here are some data that can be used to compare males and females on participation in unpaid volunteer service or community service and motivation for participation:

	Participants			
	Motivation			
Gender	**Strictly voluntary**	**Court-ordered**	**Other**	**Nonparticipants**
Men	31.9%	2.1%	6.3%	59.7%
Women	43.7%	1.1%	6.5%	48.7%

Note that the percents in each row sum to 100%. Graphically compare the volunteer service profiles for men and women. Describe any differences that are striking.

4.133. CHALLENGE **Look at volunteers only.** Refer to the previous exercise. Recompute the table for volunteers only. To do this, take the entries for each motivation and divide by the percent of volunteers. Do this separately for each gender. Verify that the percents sum to 100% for each gender. Give a graphical summary to compare the motivation of men and women who are volunteers. Compare this with your summary in the previous exercise, and write a short paragraph describing similarities and differences in these two views of the data.

4.134. An example of Simpson's paradox. Mountain View University has professional schools in business and law. Here is a three-way table of applicants to these professional schools, categorized by gender, school, and admission decision:[37] DATA FILE ADMISSIONS

Business			Law		
	Admit			Admit	
Gender	**Yes**	**No**	**Gender**	**Yes**	**No**
Male	400	200	Male	90	110
Female	200	100	Female	200	200

(a) Make a two-way table of gender by admission decision for the combined professional schools by summing entries in the three-way table.

(b) From your two-way table, compute separately the percents of male and female applicants admitted. Male applicants are admitted to Mountain View's professional schools at a higher rate than female applicants.

(c) Now compute separately the percents of male and female applicants admitted by the business school and by the law school.

(d) Explain carefully, as if speaking to a skeptical reporter, how it can happen that Mountain View appears to favor males when this is not true within each of the professional schools.

4.135. Construct an example with four schools. Refer to the previous exercise. Make up a similar table for a hypothetical university having four different schools that illustrates the same point. Carefully summarize your table with the appropriate percents.

4.136. CHALLENGE **Class size and class level.** A university classifies its classes as either "small" (fewer than 40 students) or "large." A dean sees that 62% of Department A's classes are small, while Department B has only 40% small classes. She wonders if she should cut Department A's budget and insist on larger classes. Department A responds to the dean by pointing out that classes for third- and fourth-year students tend to be smaller than classes for first- and second-year students. The following three-way table gives the counts of classes by department, size, and student audience. Write a short report for the dean that summarizes these data. Start by computing the percents of small classes in the two departments and include other numerical and graphical comparisons as needed. Here are the numbers of classes to be analyzed: DATA FILE CLASSSIZE

	Department A			Department B		
Year	**Large**	**Small**	**Total**	**Large**	**Small**	**Total**
First	2	0	2	18	2	20
Second	9	1	10	40	10	50
Third	5	15	20	4	16	20
Fourth	4	16	20	2	14	16

4.137. Identity theft. A study of identity theft looked at how well consumers protect themselves from this increasingly prevalent crime. The behaviors of 61 college students were compared with the behaviors of 59 nonstudents.[38] Participants in the study were asked to respond "Yes" or "No" to the following statement: "When asked to create a password, I have used either my mother's maiden name, or my pet's name, or my birth date, or the last four digits of my social security number, or a series of consecutive numbers." For the students, 22 agreed with this statement while 30 of the nonstudents agreed.

(a) Display the data in a two-way table and analyze the data. Write a short summary of your results.

(b) The students in this study were junior and senior college students from two sections of a course in Internet marketing at a large northeastern university. The nonstudents were a group of individuals who were recruited to attend commercial focus groups on the West Coast conducted by a lifestyle marketing organization. Discuss how the method of selecting the subjects in this study relates to the conclusions that can be drawn from it.

4.138. CHALLENGE **Athletes and gambling.** A survey of student athletes that asked questions about gambling behavior classified students according to the National Collegiate Athletic Association (NCAA) division.[39] For male student athletes, the percents who reported wagering on collegiate sports are given here along with the numbers of respondents in each division:

Division	**I**	**II**	**III**
Percent	17.2%	21.0%	24.4%
Number	5619	2957	4089

(a) Analyze the data. Give details and a short summary of your conclusion.

(b) The percents in the preceding table are given in the NCAA report, but the numbers of male student athletes in each division who responded to the survey question are estimated based on other information in the report. To what extent do you think this has an effect on the results?

(c) Some student athletes may be reluctant to provide this kind of information, even in a survey where there is no possibility that they can be identified. Discuss how this fact may affect your conclusions.

4.139. CHALLENGE **Health conditions and risk behaviors.** The data file BRFSS gives several variables related to health conditions and risk behaviors as well as demographic information for the 50 states and the District of Columbia. Pick at least three pairs of variables to analyze. Write a short report on your findings. DATA FILE BRFSS

PART ONE REVIEW

LOOKING AT DATA

In Chapter 1 we explored the key ideas that are fundamental to your understanding of statistics. We started with an overview of the basic elements of a statistical study: the topic of interest, the research questions, collecting or finding data, the data, the analysis, and the conclusions. Through the use of applets, we explored random sampling, describing data graphically and numerically for one and two variables, probability, and confidence intervals. All the topics explored in Chapter 1 are discussed thoroughly in later chapters. Here, we used the applets to discover the basics.

In Chapter 2 we explored issues related to the collection of data. The first section focused on the design of experiments, where treatments are imposed on the experimental units and the response is measured. The principles of comparison, randomization, and including sufficient numbers of repetitions were explored in detail. Methods for randomly assigning experimental units to treatments were also discussed. The second section of Chapter 2 concerned sampling design, where data are collected from samples that are drawn from a population. The goal is to use the sample data to make an inference about the population. Methods for selecting random samples were also discussed. The final section in Chapter 2 discussed ethical issues related to the collection and analysis of data.

In Chapter 3 we explored methods for examining data for a single variable. In the first section graphical methods for displaying the distribution of categorical and quantitative variables were described. These included bar graphs and pie charts for categorical variables, and stemplots and histograms for quantitative variables. Numerical summaries were discussed in the second section. These included the mean and the standard deviation, which are appropriate for distributions that are approximately symmetric with no large outliers, and the median and interquartile range for distributions that are skewed or that contain outliers. The boxplot was introduced as a graphical summary of the median, quartiles, minimum, and maximum. The final section of Chapter 3 introduced the idea of a density curve as a model for continuous data. The most useful density curve for statistics, the Normal distribution, was discussed in detail.

In Chapter 4 we explored methods for examining relationships between pairs of variables. As we did in Chapter 3 for a single variable, we began with a graphical summary. We used a scatterplot to display the relationship between two quantitative variables. We looked for the overall pattern in the display and for striking deviations from the overall pattern. We described the form, the direction, and the strength of the relationship. We noted outliers that fell outside the overall pattern. A brief second section discussed the correlation, a measure of the strength of a linear relationship. In the third section we discussed using a least-squares regression line to summarize a linear relationship and the use of the square of the correlation as a measure of the proportion of the variability in the response variable that is explained by the explanatory variable. A fourth section discussed cautions needed for the appropriate use of correlation and regression. A final section explored ways of examining the relationship between a pair of categorical variables.

Chapter 2: Producing Data

- In an experiment, one or more **treatments** are imposed on the **experimental units** or **subjects.** Each treatment is a combination of **levels** of the explanatory variables, which we call **factors.** (page 37)
- A **placebo effect** is a positive response to a treatment that is not designed to have an effect. (page 40)
- The **design** of an experiment refers to the choice of treatments and the manner in which the experimental units or subjects are assigned to the treatments. (page 40)
- The basic principles of statistical design of experiments are **control, randomization,** and **repetition.** (page 42)
- An observed effect is **statistically significant** if it is so large that it would rarely occur by chance. (page 43)
- The simplest form of control is **comparison.** Experiments should compare two or more treatments to prevent **confounding** the effect of a treatment with other influences, such as lurking variables. (page 40)
- **Randomization** uses chance to assign subjects to the treatments. Randomization creates treatment groups that are similar (except for chance variation) before the treatments are applied. Randomization and comparison together prevent **bias,** or systematic favoritism, in experiments. (page 40)
- You can carry out randomization by giving numerical labels to the experimental units and using **software** or a **table of random digits** to choose treatment groups. (page 43)
- **Repetition** of the treatments on many units reduces the role of chance variation and makes the experiment more sensitive to differences among the treatments. (page 42)

- Good experiments require attention to detail as well as good statistical design. Many behavioral and medical experiments are **double-blind. Lack of realism** in an experiment can prevent us from generalizing its results. (page 47)
- In addition to comparison, a second form of control is to restrict randomization by forming **blocks** of experimental units that are similar in some way that is important to the response. Randomization is then carried out separately within each block. (page 48)
- **Matched pairs** are a common form of blocking for comparing just two treatments. In some matched pairs designs, each subject receives both treatments in a random order. In others, the subjects are matched in pairs as closely as possible, and one subject in each pair receives each treatment. (page 48)
- A sample survey selects a **sample** from the **population** of all individuals about which we desire information. We base conclusions about the population on data about the sample. (page 54)
- The **design of a sample survey** refers to the method used to select the sample from the population. **Probability sampling designs** use impersonal chance to select a sample. (page 59)
- The basic probability sample is a **simple random sample (SRS).** An SRS gives every possible sample of a given size the same chance to be chosen. (page 56)
- Choose an SRS by labeling the members of the population and using a **table of random digits** to select the sample. **Software** can automate this process. (page 57)
- To choose a **stratified random sample,** divide the population into **strata,** groups of individuals that are similar in some way that is important to the response. Then choose a separate SRS from each stratum and combine them to form the full sample. (page 59)
- **Multistage samples** select successively smaller groups within the population in stages, resulting in a sample consisting of clusters of individuals. Each stage may employ an SRS, a stratified sample, or another type of sample. (page 60)
- Failure to use probability sampling often results in **bias,** or systematic errors in the way the sample represents the population. **Voluntary response** samples, in which the respondents choose themselves, are particularly prone to large bias. (page 56)
- In human populations, even probability samples can suffer from bias due to **undercoverage** or **nonresponse,** from **response bias** due to the behavior of the interviewer or the respondent, or from misleading results due to **poorly worded questions.** (page 61)
- Approval by an **institutional review board** is required for studies that involve humans or animals as subjects. (page 67)
- Human subjects must give **informed consent** if they are to participate in experiments. (page 68)
- Data on human subjects must be kept **confidential.** (page 69)

- **Anonymity** means that the names of the subjects are not known to anyone involved in the study. (page 69)
- **Clinical trials** are experiments that study the effectiveness of medical treatments on patients. (page 71)

Chapter 3: Looking at Data—Distributions

- A data set contains information on a number of **cases.** Cases may be people, animals, or things. For each case, the data give values for one or more **variables.** A variable describes some characteristic of a case, such as a person's height, gender, or salary. Variables can have different **values** for different cases. A **label** is a variable that is used to identify different cases. (page 80)
- Some variables are **categorical** and others are **quantitative.** A categorical variable places each individual into a category, such as male or female. A quantitative variable has numerical values that measure some characteristic of each case, such as height in centimeters or annual salary in dollars. (page 80)
- **Exploratory data analysis** uses graphs and numerical summaries to describe the variables in a data set and the relations among them. (page 85)
- The **distribution** of a variable tells us what values it takes and how often it takes these values.(page 85)
- **Bar graphs** and **pie charts** display the distributions of categorical variables. These graphs use the counts or percents of the categories. (page 85)
- **Stemplots** and **histograms** display the distributions of quantitative variables. Stemplots separate each observation into a **stem** and a one-digit **leaf.** (page 87)
- Histograms plot the **frequencies** (counts) or the percents of equal-width classes of values. (page 90)
- When examining a distribution, look for **shape, center,** and **spread** and for clear **deviations** from the overall shape. (page 95)
- Some distributions have simple shapes, such as **symmetric** or **skewed.** The number of **modes** (major peaks) is another aspect of overall shape. Not all distributions have a simple overall shape, especially when there are few observations. (page 95)
- **Outliers** are observations that lie outside the overall pattern of a distribution. Always look for outliers and try to explain them. (page 96)
- When observations on a variable are taken over time, make a **time plot** that graphs time horizontally and the values of the variable vertically. (page 98)
- A numerical summary of a distribution should report its **center** and its **spread** or **variability.** (page 105)
- The **mean** $\bar{x}$ and the **median** M describe the center of a distribution in different ways. The mean is the arithmetic average of the observations, and the median is their midpoint. (page 106)

- When you use the median to describe the center of the distribution, describe its spread by giving the **quartiles.** The **first quartile Q_1** has one-fourth of the observations below it, and the **third quartile Q_3** has three-fourths of the observations below it. (page 109)
- The **interquartile range** is the difference between the quartiles. It is the spread of the center half of the data. The **1.5 × IQR rule** flags observations more than $1.5 \times IQR$ beyond the quartiles as possible outliers. (page 114)
- The **five-number summary** consisting of the median, the quartiles, and the smallest and largest individual observations provides a quick overall description of a distribution. The median describes the center, and the quartiles and extremes show the spread. (page 112)
- **Boxplots** based on the five-number summary are useful for examining several distributions. The box spans the quartiles and shows the spread of the central half of the distribution. The median is marked within the box. Lines extend from the box to the extremes and show the full spread of the data. In a **modified boxplot,** points identified by the $1.5 \times IQR$ rule are plotted individually. (page 113)
- The **variance s^2** and especially its square root, the **standard deviation s,** are common measures of spread about the mean as center. The standard deviation s is zero when there is no spread and gets larger as the spread increases. (page 116)
- A **resistant measure** of any aspect of a distribution is relatively unaffected by changes in the numerical value of a small proportion of the total number of observations, no matter how large these changes are. The median and quartiles are resistant, but the mean and the standard deviation are not. (page 107)
- The mean and standard deviation are good descriptions for symmetric distributions without outliers. They are most useful for the Normal distributions. The five-number summary is a better exploratory summary for skewed distributions. (page 119)
- Numerical measures of particular aspects of a distribution, such as center and spread, do not report the entire shape of most distributions. In some cases, particularly distributions with multiple peaks and gaps, these measures may not be very informative. (page 119)
- **Linear transformations** have the form $x_{new} = a + bx$. A linear transformation changes the origin if $a \neq 0$ and changes the size of the unit of measurement if $b > 0$. Linear transformations do not change the overall shape of a distribution. A linear transformation multiplies a measure of spread by b and changes a percentile or measure of center m into $a + bm$. (page 119)
- The overall pattern of a distribution can often be described compactly by a **density curve.** A density curve has total area 1 underneath it. Areas under a density curve give proportions of observations for the distribution. (page 128)
- The **mean** μ (balance point), the **median** (equal-areas point), and the **quartiles** can be approximately located by eye on a density curve. The

standard deviation σ cannot be located by eye on most density curves. The mean and median are equal for symmetric density curves, but the mean of a skewed curve is located farther toward the long tail than is the median. (page 129)

- The **Normal distributions** are described by bell-shaped, symmetric, unimodal density curves. The mean μ and standard deviation σ completely specify the Normal distribution $N(\mu, \sigma)$. The mean is the center of symmetry, and σ is the distance from μ to the change-of-curvature points on either side. (page 130)
- To **standardize** any observation x, subtract the mean of the distribution and then divide by the standard deviation. The resulting ***z*-score** $z = (x - \mu)/\sigma$ says how many standard deviations x lies from the distribution mean. (page 134)
- All Normal distributions are the same when measurements are transformed to the standardized scale. In particular, all Normal distributions satisfy the **68–95–99.7 rule.** (page 132)
- If X has the $N(\mu, \sigma)$ distribution, then the standardized variable $Z = (X - \mu)/\sigma$ has the **standard Normal distribution** $N(0, 1)$. Proportions for any Normal distribution can be calculated by software or from the **standard Normal table** (Table A), which gives the **cumulative proportions** of $Z < z$ for many values of z. (page 135)
- The adequacy of a Normal model for describing a distribution of data is best assessed by a **Normal quantile plot,** which is available in most statistical software packages. A pattern on such a plot that deviates substantially from a straight line indicates that the data are not Normal. (page 141)

Chapter 4: Looking at Data—Relationships

- To study relationships between variables, we must measure the variables on the same cases. (page 154)
- If we think that a variable x may explain or even cause changes in another variable y, we call x an **explanatory variable** and y a **response variable.** (page 156)
- A **scatterplot** displays the relationship between two quantitative variables. Mark values of one variable on the horizontal axis (x axis) and values of the other variable on the vertical axis (y axis). Plot each individual's data as a point on the graph. (page 158)
- Always plot the explanatory variable, if there is one, on the x axis of a scatterplot. Plot the response variable on the y axis. (page 159)
- Plot points with different colors or symbols to see the effect of a categorical variable in a scatterplot. (page 165)
- In examining a scatterplot, look for an overall pattern showing the **form, direction,** and **strength** of the relationship, and then for **outliers** or other deviations from this pattern. (page 160)
- **Form: Linear relationships,** where the points show a straight-line pattern, are an important form of relationship between two variables. Curved relationships and **clusters** are other forms to watch for. (page 160)

- **Direction:** If the relationship has a clear direction, we speak of either **positive association** (high values of the two variables tend to occur together) or **negative association** (high values of one variable tend to occur with low values of the other variable). (page 161)
- **Strength:** The **strength** of a relationship is determined by how close the points in the scatterplot lie to a simple form such as a line. (page 161)
- To display the relationship between a categorical explanatory variable and a quantitative response variable, make a graph that compares the distribution of the response variable for each category of the explanatory variable. (page 166)
- The **correlation** r measures the direction and strength of the linear (straight line) association between two quantitative variables x and y. Although you can calculate a correlation for any scatterplot, r measures only linear relationships. (page 171)
- Correlation indicates the direction of a linear relationship by its sign: $r > 0$ for a positive association and $r < 0$ for a negative association. (page 173)
- Correlation always satisfies $-1 \leq r \leq 1$ and indicates the strength of a relationship by how close it is to -1 or 1. Perfect correlation, $r = \pm 1$, occurs only when the points lie exactly on a straight line. (page 173)
- Correlation ignores the distinction between explanatory and response variables. The value of r is not affected by changes in the unit of measurement of either variable. Correlation is not resistant, so outliers can greatly change the value of r. (page 173)
- A **regression line** is a straight line that describes how a response variable y changes as an explanatory variable x changes. (page 177)
- The most common method of fitting a line to a scatterplot is least squares. The **least-squares regression line** is the straight line $\hat{y} = b_0 + b_1 x$ that minimizes the sum of the squares of the vertical distances of the observed y-values from the line. (page 182)
- You can use a regression line to **predict** the value of y for any value of x by substituting this x into the equation of the line. **Extrapolation** beyond the range of x-values spanned by the data is risky. (page 180)
- The **slope** b_1 of a regression line $\hat{y} = b_0 + b_1 x$ is the rate at which the predicted response $\hat{y}$ changes along the line as the explanatory variable x changes. Specifically, b_1 is the change in $\hat{y}$ when x increases by 1. The numerical value of the slope depends on the units used to measure x and y. (page 183)
- The **intercept** b_0 of a regression line $\hat{y} = b_0 + b_1 x$ is the predicted response $\hat{y}$ when the explanatory variable $x = 0$. This prediction is not particularly useful unless x can actually take values near 0. (page 183)
- The least-squares regression line of y on x is the line with slope $b_1 = rs_y/s_x$ and intercept $b_0 = \bar{y} - b_1\bar{x}$. This line always passes through the point $(\bar{x}, \bar{y})$. (page 184)
- **Correlation** and **regression** are closely connected. The correlation r is the slope of the least-squares regression line when we measure both x and y in standardized units. The square of the correlation r^2 is the fraction of

the variance of one variable that is explained by least-squares regression on the other variable. (page 185)

- You can examine the fit of a regression line by plotting the **residuals,** which are the differences between the observed and predicted values of y. Be on the lookout for points with unusually large residuals and also for nonlinear patterns and uneven variation about the line. (page 191)
- Also look for **influential observations,** individual points that substantially change the regression line. Influential observations are often outliers in the x direction, but they need not have large residuals. (page 194)
- Correlation and regression must be **interpreted with caution.** Plot the data to be sure that the relationship is roughly linear and to detect outliers and influential observations. (page 195)
- **Lurking variables** may explain the relationship between the explanatory and response variables. Correlation and regression can be misleading if you ignore important lurking variables. (page 197)
- We cannot conclude that there is a cause-and-effect relationship between two variables just because they are strongly associated. **High correlation does not imply causation.** (page 198)
- A **two-way table** of counts organizes data about two categorical variables. Values of the **row variable** label the rows that run across the table, and values of the **column variable** label the columns that run down the table. Two-way tables are often used to summarize large amounts of data by grouping outcomes into categories. (page 203)
- The **joint distribution** of the row and column variables is found by dividing the count in each cell by the total number of observations. (page 204)
- The **row totals** and **column totals** in a two-way table give the **marginal distributions** of the two variables separately. It is clearer to present these distributions as percents of the table total. Marginal distributions do not give any information about the relationship between the variables. (page 205)
- To find the **conditional distribution** of the row variable for one specific value of the column variable, look only at that one column in the table. Find each entry in the column as a percent of the column total. (page 207)
- There is a conditional distribution of the row variable for each column in the table. **Comparing conditional distributions is one way to describe the association between the row and the column variables.** It is particularly useful when the column variable is the explanatory variable. When the row variable is explanatory, find the conditional distribution of the column variable for each row and compare these distributions. (page 208)
- **Bar graphs** are a flexible means of presenting categorical data. There is no single best way to describe an association between two categorical variables. (page 208)
- We present data on three categorical variables in a **three-way table,** printed as separate two-way tables for each level of the third variable.

A comparison between two variables that holds for each level of a third variable can be changed or even reversed when the data are **aggregated** by summing over all levels of the third variable. **Simpson's paradox** refers to the reversal of a comparison by aggregation. It is an example of the potential effect of lurking variables on an observed association. (page 210)

PART I EXERCISES

I.1. Compare the smartphone apps. An experiment is designed to compare three different versions of an app for a smartphone. The apps will be given to 30 college-aged women, 30 college-aged men, 30 high-school-aged women, and 30 high-school-aged men. Within each group of 30, 10 will receive each version of the app. After using the app for a week, each subject will rate his or her satisfaction with the app on a scale from 1 to 10.

(a) What are the factors and levels in this experiment?

(b) Who are the experimental units? How many are there?

(c) Draw a sketch that describes the experiment.

I.2. Randomize the smartphone apps. Refer to the previous exercise.

(a) Describe a plan for doing the needed randomization.

(b) Carry out the randomization.

I.3. Evaluate the smartphone apps. Refer to the previous two exercises. You can ask the subjects to respond to five questions in addition to their overall satisfaction with the app. Write five questions that you would use and explain your reason for choosing each.

I.4. Cocurricular activities of college students. Cocurricular activities of college students include participation in clubs, student organizations, and other programs. The National Survey of Student Engagement collects information about cocurricular activities from hundreds of colleges and universities each year.[1] One question asks how many hours per week students spend participating in cocurricular activities. Another asks about satisfaction with the entire educational experience, rated on a scale of 1 to 4.

(a) What are the variables that will be examined?

(b) Are the variables categorical or quantitative? Explain your answer.

(c) What are the possible values for the variables?

(d) What are the cases?

(e) Is there an explanatory variable? Is there a response variable? Give reasons for your answers.

I.5. How will you examine the variables? Refer to the previous exercise. For each of the variables, what graphical and numerical summaries would you use to describe their distributions? Give reasons for your choices.

I.6. How will you examine the relationship? Refer to the previous two exercises.

(a) Explain how you would examine the relationship between hours of cocurricular activity and satisfaction with the entire educational experience. Be specific.

(b) Write a short summary describing how you would interpret a positive association between hours of cocurricular activity and satisfaction with the entire educational experience.

I.7. Does participation in cocurricular activities cause satisfaction? Refer to the previous exercise. Suppose that there is a statistically significant positive association between hours of cocurricular activity and satisfaction with the entire educational experience. Discuss whether or not you can conclude that participation in cocurricular activities causes an increase in satisfaction with the entire educational experience.

I.8. Design an experiment for cocurricular activities and satisfaction. Refer to Exercise I.4. Design an experiment in which you would investigate whether or not participation in cocurricular activity produces higher satisfaction with the entire educational experience.

(a) Describe your experiment in a short paragraph and with a sketch.

(b) Explain how you would perform the randomization needed.

I.9. Design considerations for your experiment. Provide additional details about the experiment you proposed in the previous exercise. Include the following design considerations:

(a) Will you use a control or placebo group? Explain why or why not.

(b) What will you tell the participants about the purpose of the study?

(c) Will you pay the students to participate in the study?

(d) Will you recruit students from a course and give them some course credit for their participation?

I.10. Ethical considerations for your experiment. Refer to the previous two exercises. Discuss your experiment from the viewpoint of ethics.

I.11. Trends in undergraduate student aid. The College Board Advocacy and Policy Center[2] collects data related to sources of funds for college education. One summary gives the amounts of student aid for undergraduate education for 2000–2001 to 2010–2011. The units are dollars per full-time college student in terms of 2010 dollars. Here are the data: STUDENTAIDTOTALS

Year	Dollars
2000–2001	7,987
2001–2002	8,282
2002–2003	8,852
2003–2004	9,623
2004–2005	10,077
2005–2006	10,330
2006–2007	10,542
2007–2008	10,959
2008–2009	11,161
2009–2010	12,648
2010–2011	12,913

(a) Which of the two variables is an explanatory variable? Which is a response variable?

(b) Plot the data and describe the relationship. Is it approximately linear?

(c) Are there any outliers?

(d) Are there any observations that are particularly influential?

I.12. Fit a line to the trend. Refer to the previous exercise. Fit a least-squares line to the data. STUDENTAIDTOTALS

(a) Give the equation of the least-squares line.

(b) Over this period of time, how much is the amount of student aid increasing per year?

I.13. Examine the residuals. Refer to the previous two exercises. STUDENTAIDTOTALS

(a) Make a table of the residuals.

(b) Plot the residuals versus year.

(c) Summarize what you see in the residual plot.

I.14. Sources of funds for college. Refer to Exercise I.11. Another summary gives the amounts of student aid for undergraduate education from various sources for 2011. The units are dollars per full-time college student in terms of 2010 dollars. Here are the data: STUDENTAIDTOTALS

Source	Dollars
Nonfederal student loans	459
Education tax benefits	936
Federal Parent Loans (PLUS) and Grad PLUS Loans	731
Unsubsidized Federal Stafford Loans	2125
Subsidized Federal Stafford Loans	1994
Private and employer grants	464
Institutional grants	2082
Federal Pell Grants	2438
State grants	638
Federal campus-based programs	180
Other federal programs	867

(a) Convert the dollar values to percents.

(b) Choose an appropriate graphical display for the percent data. Give reasons for your choice.

(c) Prepare your display.

(d) Summarize the main features of the data that are evident in the graphical display.

I.15. Baseball scores. The total runs scored in each of the 42 baseball games played in the National League in a recent early-spring week are given in the following table: BASEBALLSCORES

5	8	8	5	6	11	11	5	11	4	3	7	11	5
2	13	2	3	12	7	11	1	9	11	5	3	10	10
10	13	6	9	7	7	13	13	7	5	14	12	13	12

(a) Use a histogram to display these data.

(b) Use a stemplot to display these data.

I.16. Measures of center for the baseball scores. Refer to Exercise I.15. BASEBALLSCORES

(a) Calculate the median.

(b) Calculate the mean.

I.17. Measures of spread for the baseball scores. Refer to Exercise I.15. BASEBALLSCORES

(a) Calculate the interquartile range.

(b) Calculate the standard deviation.

I.18. Summarize the baseball scores. Refer to Exercises I.15 to I.17. DATA FILE BASEBALLSCORES

(a) Based on your work in these exercises, select graphical and numerical summaries that effectively describe the distribution of the baseball scores. Give reasons for your choices.

(b) Write a short paragraph that discusses the distribution of the baseball scores and that includes the summaries you chose in part (a) and other key features such as overall shape and possible outliers.

I.19. Are the baseball scores Normal? Refer to Exercise I.15. DATA FILE BASEBALLSCORES

(a) Generate a Normal quantile plot for these data.

(b) Summarize your findings based on this plot.

I.20. Baseball scores for the American League. Refer to Exercise I.15. Here are data for the same week for the American League: DATA FILE BASEBALLSCORES

10	7	7	5	12	6	9	14	11	5	9	7	7	4	13	15
5	3	11	9	7	3	12	5	6	7	6	10	6	6	6	9
6	9	6	12	3	3	5	11	8	5	27	14	7	5	10	6

Answer the questions in Exercises I.15 to I.19 for these data.

I.21. National League versus American League. Pitchers, who are usually not great batters, do not bat in the American League, whereas in the National League they do. For this reason, some believe that more runs should be scored in American League games. Use the data in Exercises I.15 and I.20 to examine this question. DATA FILE BASEBALLSCORES

I.22. Calcium requirements. The amounts of vitamins and other nutrients that are required by humans are assumed to vary from person to person. Therefore, in setting guidelines for the amounts needed, the Institute of Medicine assumes that requirements follow a distribution. The guidelines, called Dietary Reference Intakes (DRIs), consist of three amounts for each vitamin or nutrient. The Estimated Average Requirement (EAR) is the mean of the requirement distribution, the Recommended Dietary Allowance (RDA) is the mean plus two standard deviations, and the Upper Level Intake (UL) is the amount below which no harm is likely to result. For adults aged 19 to 50, the DRIs for calcium are 800 milligrams per day (mg/day) for the EAR, 1000 mg/day for the RDA, and 2500 for the UL.[3] The Institute of Medicine also assumes that the requirement distribution is Normal.

(a) What are μ and σ for the calcium requirement distribution?

(b) Apply the 68–95–99.7 rule to the distribution. Explain the results.

(c) Find the z-score for the UL. We would like this number to be very large. Why?

(d) What proportion of the population has a requirement at or greater than the UL?

I.23. Your calcium intake is 900 mg/day. Refer to the previous exercise. Suppose that you are an adult aged 19 to 50 and your intake of calcium is 900 mg/day. You don't know what your requirement is other than that you are a member of a population that has the $N(800, 100)$ distribution.

(a) What proportion of adults aged 19 to 50 have a requirement that is greater than your intake of 900 mg/day?

(b) Based on the calculation that you did in part (a), do you think that you should consider eating more foods that are rich in calcium? Explain why or why not.

(c) Suppose that you wanted to consume an amount of calcium that meets the requirement for 97.5% of the population. How many milligrams per day of calcium should you consume?

I.24. Dessert with the meal or after the meal. If you eat dessert after your meal rather than with your meal, are you more likely to eat more dessert? Are you more likely to eat more of your main course? What about the total food consumed? An experiment with 23 children aged 2 to 5 years in a child care center was designed to address this question.[4] Variables describing the subjects included age, gender, and race. Food variables included calories of dessert consumed and calories of main meal consumed, averaged over all days for which data were available. Note that each child was studied under the two conditions: dessert with and dessert after the meal.

(a) Classify each variable described above as categorical or quantitative.

(b) Classify each variable described above as explanatory or response.

(c) What are the cases for this experiment? How many are there?

(d) In this experiment, the main meal was fish and the dessert was a cookie. The study was performed in a midwestern university community. How well do you think that the results would generalize to other main courses, desserts, and communities? Explain your answer.

I.25. Compare the cookie consumption. Refer to Exercise I.24. DATA FILE COOKIES

(a) Describe the distribution of cookie consumption when the cookies are served with the meal. Use numerical and graphical summaries.

(b) Do the same for the distribution of cookie consumption when cookies are served after the meal.

(c) Make a graphical display that allows you to compare the distributions that you described in parts (a) and (b). Do the distributions appear to be similar or different?

I.26. Compare the main-meal consumption. Refer to the previous exercise. Answer the questions in that exercise for main-meal consumption. COOKIES

I.27. Look at the relationships. Let's examine some relationships between the consumptions when the cookie is served with or after the meal. There is no clear way to designate explanatory and response variables here. So let's arbitrarily choose the after-meal condition as the explanatory variable. COOKIES

(a) Plot the cookie consumption when cookies are consumed with the meal versus the cookie consumption when cookies are served after the meal.

(b) Describe the form, direction, and strength of the relationship.

(c) Are there any outliers or influential observations? Explain your answers.

(d) Find the least-squares regression line for these data and plot the line with the data in a graph.

(e) Find the correlation between these two variables.

(f) Answer parts (a) through (e) for main-meal consumption, and compare these results with those that you found for cookie consumption.

I.28. A different way to make the comparisons. In part (c) of Exercise I.25 and in Exercise I.26, you compared the consumption of cookies and the main meal under the two different serving conditions. Here is another way to examine the data to make the comparison. Create a new variable that is the amount of cookie consumed when it is served after the meal minus the amount of cookie consumed when it is served with the meal. Do the same for the amount of the main meal consumed. COOKIES

(a) Use numerical and graphical summaries to describe the distributions of the two new variables.

(b) When you examine the data in this way, does when the cookie is served affect the amount of cookie consumed? Does it affect the amount of the main meal consumed? Write a short paragraph summarizing your conclusions.

(c) Do you prefer the approach to making the comparisons in this exercise to the approach that you used in Exercises I.25 and I.26? Give reasons for your answer.

I.29. Statistical significance. In Exercises I.25, I.26, and I.27, you examined data, made some comparisons, and drew some conclusions. Explain how statistical significance is relevant to the conclusions that you drew in these exercises. In the next part of the text, we will learn the details of how to use statistical significance for exercises such as these.

I.30. Design and ethical issues for the cookie study. Refer to Exercise I.24. Consider designing a similar study. Explain in detail how you would deal with each of the following issues. COOKIES

(a) Randomization.

(b) Double-blinding.

(c) Informed consent.

(d) Confidentiality.

(e) Anonymity.

I.31. Descriptive statistics for the cookie variables. Refer to the four variables in the cookie data set. COOKIES

(a) Use graphical summaries to describe the four distributions.

(b) Use numerical summaries to describe the four distributions.

(c) Which graphical and numerical summaries do you prefer for these data? Give reasons for your choices.

(d) Using the summaries you have chosen, write a short description of the distributions of these four variables.

I.32. A two-way table for the cookie data. Use the cookie data to construct a two-way table. First, define meal consumption for dessert with the meal as low if the number of calories consumed is less than 44, and high if the number of calories consumed is 44 or greater. Do the same for meal consumption for dessert after the meal. Choose appropriate variable names for the two categorical variables that you created. COOKIES

(a) Display the data in a two-way table of counts.

(b) Add the marginal totals to your table.

(c) Find the joint distribution of the two variables (use percents for the entries in your table).

(d) Find the conditional distribution for meal consumption when dessert is after the meal, given meal consumption when dessert is with the meal.

(e) Use bar graphs to display the conditional distributions.

PART TWO

PROBABILITY AND INFERENCE

CHAPTER FIVE

PROBABILITY: THE STUDY OF RANDOMNESS

random sampling
p. 43

When we produce data by random sampling or randomized comparative experiments, the laws of probability answer the question "What would happen if we did this many times?" Games of chance like Texas hold 'em are exciting because the outcomes are determined by the rules of probability. We study these rules here because the reasoning of statistical inference rests on asking, "How often would this method give a correct answer if I used it very many times?"

5.1 Randomness

Toss a coin or choose an SRS. The result can't be known in advance, because the result will vary toss by toss or sample to sample. Nonetheless, there is a regular pattern in the results, a pattern that emerges clearly only after many repetitions. This remarkable fact is the basis for the idea of probability.

EXAMPLE 5.1

Toss a coin 5000 times. When you toss a coin, there are only two possible outcomes, heads or tails. Figure 5.1 shows the results of two trials where a coin is tossed 5000 times. For each number of tosses from 1 to 5000, we have plotted the proportion of those tosses that gave a head. Trial A (red line) begins tail, head, tail, tail. You can see that the proportion of heads for Trial A starts at 0 on the first toss, rises to 0.5 when the second toss gives a head, then falls to 0.33 and 0.25 as we get two more tails. Trial B (blue dotted line), on the other hand, starts with five straight heads, so the proportion of heads is 1 until the sixth toss.

The proportion of tosses that produce heads is quite variable at first. Trial A starts low and Trial B starts high. As we make more and more tosses,

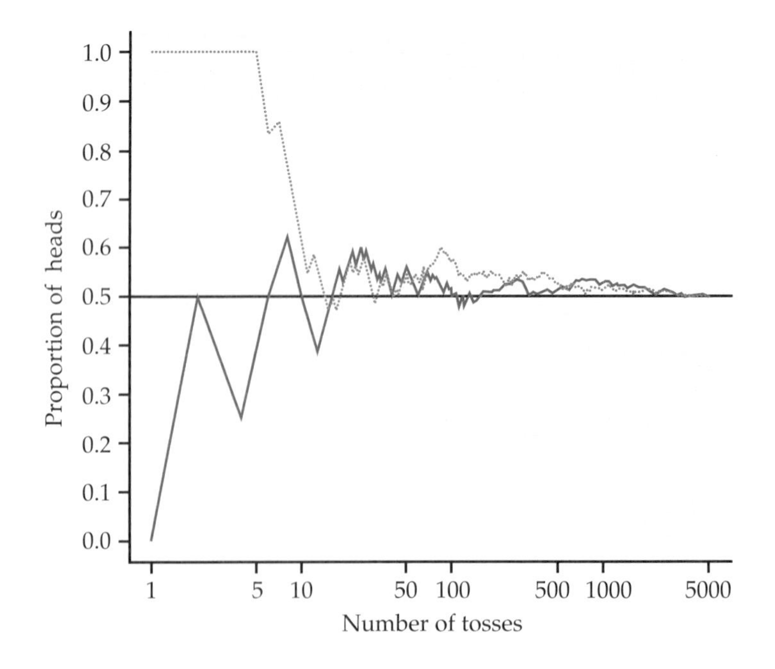

FIGURE 5.1 The proportion of tosses of a coin that give a head varies as we make more tosses. Eventually, however, the proportion approaches 0.5, the probability of a head. This figure shows the results of two trials of 5000 tosses each.

however, the proportions of heads for both trials get close to 0.5 and stay there. If we made yet a third trial at tossing the coin a great many times, the proportion of heads would again settle down to 0.5 in the long run. We say that 0.5 is the *probability* of a head. The probability 0.5 appears as a horizontal line on the graph.

The *Probability* applet on the text Web site animates Figure 5.1. It allows you to choose the probability of a head and simulate any number of tosses of a coin with that probability. Try it. You will see that the proportion of heads gradually settles down close to the chosen probability. Equally important, you will also see that the proportion in a small or moderate number of tosses can be far from the probability.

Probability describes only what happens in the long run. Most people expect chance outcomes to show more short-term regularity than is actually true.

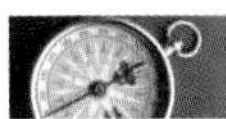

EXAMPLE 5.2

Hypothesis testing and significance levels. In Chapter 6 we will learn about significance levels that are associated with hypothesis testing. The significance level is a probability, typically set to 0.05, that is associated with the performance of the test.

fair coin

In the coin-toss setting, the probability of a head is a characteristic of the coin being tossed. A coin is called **fair** if the probability of a head is 0.5—that is, if it is equally likely to come up heads or tails. If we toss a fair coin once, we do not know if it will come up heads or tails.

The language of probability

"Random" in statistics is not a synonym for "haphazard" but is a description of a kind of order that emerges in the long run. We often encounter the unpredictable side of randomness in our everyday experience, but we rarely see enough repetitions of the same random phenomenon to observe the long-term regularity that probability describes.

You can see that regularity emerging in Figure 5.1. In the very long run, the proportion of tosses that give a head is 0.5. This is the intuitive idea of probability. Probability 0.5 means "occurs half the time in a very large number of trials."

RANDOMNESS AND PROBABILITY

We call a phenomenon **random** if individual outcomes are uncertain but there is nonetheless a regular distribution of outcomes in a large number of repetitions.

The **probability** of any outcome of a random phenomenon is the proportion of times the outcome would occur in a very long series of repetitions.

Not all coins are fair. In fact, most real coins have bumps and imperfections that make the probability of heads a little different from 0.5. The probability might be 0.499999 or 0.500002. For our study of probability in this chapter, we will assume that we know the actual values of probabilities. Thus, we assume things like fair coins, even though we know that real coins are not exactly fair. We do this to learn what kinds of outcomes we are likely to see when we make such assumptions.

When we study statistical inference in later chapters, we look at the situation from the opposite point of view: given that we have observed certain outcomes, what can we say about the probabilities that generated these outcomes?

USE YOUR KNOWLEDGE

5.1 Use Table B. We can use the random digits in Table B in the back of the text to simulate tossing a fair coin. Start at line 119 and read the numbers from left to right. If the number is 0, 1, 2, 3, or 4, you will say that the coin toss resulted in a head; if the number is a 5, 6, 7, 8, or 9, the outcome is tails. Use the first 20 random digits on line 119 to simulate 20 tosses of a fair coin. What is the actual proportion of heads in your simulated sample? Explain why it is not surprising for you to not get exactly 10 heads.

Probability describes what happens in very many trials, and we must actually observe many trials to pin down a probability. In the case of tossing a coin, some diligent people have in fact made thousands of tosses.

EXAMPLE 5.3

Many tosses of a coin. The French naturalist Count Buffon (1707–1788) tossed a coin 4040 times. Result: 2048 heads, or proportion 2048/4040 = 0.5069 for heads.

Around 1900, the English statistician Karl Pearson heroically tossed a coin 24,000 times. Result: 12,012 heads, a proportion of 0.5005.

While imprisoned by the Germans during World War II, the South African statistician John Kerrich tossed a coin 10,000 times. Result: 5067 heads, proportion of heads 0.5067.

Thinking about randomness

That some things are random is an observed fact about the world. The outcome of a coin toss, the time between emissions of particles by a radioactive source, and the sexes of the next litter of lab rats are all random. So is the outcome of a random sample or a randomized experiment.

Probability theory is the branch of mathematics that describes random behavior. Of course, we can never observe a probability exactly. We could always continue tossing the coin, for example. Mathematical probability is an idealization based on imagining what would happen in an indefinitely long series of trials.

The best way to understand randomness is to observe random behavior—not only long-run regularity but unpredictable results of short runs. You can use physical devices such as coins, dice, and cards, but software simulations of random behavior allow faster exploration. As you explore randomness, remember:

independence

- You must have a long series of **independent** trials. That is, the outcome of one trial must not influence the outcome of any other. Imagine a crooked gambling house where the operator of a roulette wheel can stop it where she chooses—she can prevent the proportion of "red" from settling down to a fixed number. These trials are not independent.
- The idea of probability is empirical. Simulations start with given probabilities and imitate random behavior, but we can estimate a real-world probability only by actually observing many trials.
- Nonetheless, simulations are very useful because we need long runs of trials. In situations such as coin tossing, the proportion of an outcome often requires several hundred trials to settle down to the probability of that outcome. The kinds of physical random devices suggested in the exercises are too slow for this. Short runs give only rough estimates of a probability.

The uses of probability

Probability theory originated in the study of games of chance. Tossing dice, dealing shuffled cards, and spinning a roulette wheel are examples of deliberate randomization. In that respect, they are similar to random sampling. Although games of chance are ancient, they were not studied by mathematicians until the sixteenth and seventeenth centuries. It is only a mild simplification to say that probability as a branch of mathematics arose when seventeenth-century French gamblers asked the mathematicians Blaise Pascal and Pierre de Fermat for help.

Gambling is still with us, in casinos and state lotteries. We will make use of games of chance as simple examples that illustrate the principles of probability.

Careful measurements in astronomy and surveying led to further advances in probability in the eighteenth and nineteenth centuries because the results of repeated measurements are random and can be described by distributions much like those arising from random sampling. Similar distributions appear in data on human life span (mortality tables) and in data on lengths or weights in a population of skulls, leaves, or cockroaches.[1]

Now, we employ the mathematics of probability to describe the flow of traffic through a highway system, the Internet, or a computer; the genetic makeup of individuals or populations; the energy states of subatomic particles; the spread of epidemics or rumors; and the rate of return on risky investments. Although we are interested in probability because of its usefulness in statistics, the mathematics of chance is important in many fields of study.

SECTION 5.1 SUMMARY

A **random phenomenon** has outcomes that we cannot predict but that nonetheless have a regular distribution in very many repetitions.

The **probability** of an event is the proportion of times the event occurs in many repeated trials of a random phenomenon.

SECTION 5.1 EXERCISES

For Exercise 5.1, see page 238.

5.2. Graduation rates for bachelor's degree students. In the United States about 56% of students enrolled in college for a bachelor's degree graduate within six years. If you select a U.S. college student at random, the probability is 0.56 that you have selected someone who will graduate within six years. Do you think that this probability would apply to all colleges? Write a short paragraph explaining your answer.

5.3. APPLET **Simulate free throws.** The professional basketball player Diana Taurasi made about 90% of her free throws in a recent season when she was named Most Valuable Player in the league. Use Table B or the *Probability* applet to simulate 100 free throws shot independently by a player who has probability 0.9 of making each shot.

(a) What percent of the 100 shots did she hit?

(b) Examine the sequence of hits and misses. How long was the longest run of shots made? Of shots missed? (Sequences of random outcomes often show runs longer than our intuition thinks likely.)

5.4. Is music playing on the radio? Turn on your favorite music radio station 8 times at least 10 minutes apart. Each time record whether or not music is playing. Calculate the number of times music is playing divided by 8. This number is an estimate of the probability that music is playing when you turn on this station. It is also an estimate of the proportion of time that music is playing on this station.

5.5. Wait 5 seconds between observations. Refer to the previous exercise. Explain why you would not want to wait only 5 seconds between the times you turn on the radio station.

5.6. Winning at craps. The game of craps starts with a "come-out" roll in which the shooter rolls a pair of dice. If the total is 7 or 11, the shooter wins immediately (there are ways that the shooter can win on later rolls if other numbers are rolled on the come-out roll). Roll a pair of dice 25 times and estimate the probability that the shooter wins immediately on the come-out roll. For a pair of perfectly made dice, the probability is 0.2222.

5.7. Side effects of eyedrops. You go to the doctor and she prescribes a medicine for an eye infection that you have. Suppose that the probability of a serious side effect from the medicine is 0.00001. Explain in simple terms what this number means.

5.8. APPLET **Use the *Probability* applet.** The idea of probability is that the *proportion* of heads in many tosses of a balanced coin eventually gets close to 0.5. But does the actual *count* of heads get close to one-half the number of tosses? Let's find out. Set the "Probability of heads" in the *Probability* applet to 0.5 and the number of tosses to 50. You can extend the number of tosses by clicking "Toss" again to get 50 more. Don't click "Reset" during this exercise.

(a) After 50 tosses, what is the proportion of heads? What is the count of heads? What is the difference between the count of heads and 25 (one-half the number of tosses)?

(b) Keep going to 150 tosses. Again record the proportion and count of heads and the difference between the count and 75 (half the number of tosses).

(c) Keep going. Stop at 300 tosses and again at 600 tosses to record the same facts. Although it may take a long time, the laws of probability say that the proportion of heads will always get close to 0.5. On the other hand, the difference between the count of heads and half the number of tosses will vary a lot.

5.2 Probability Models

The idea of probability as a proportion of outcomes in very many repeated trials guides our intuition but is hard to express in mathematical form. A description of a random phenomenon in the language of mathematics is called a

probability model

probability model. To see how to proceed, think first about a very simple random phenomenon, tossing a coin once. When we toss a coin, we cannot know the outcome in advance. What do we know? We are willing to say that the outcome will be either heads or tails. Because the coin appears to be balanced, we believe that each of these outcomes has probability 1/2. This description of coin tossing has two parts:

- A list of possible outcomes
- A probability for each outcome

This two-part description is the starting point for a probability model. We will begin by describing the outcomes of a random phenomenon and then learn how to assign probabilities to the outcomes.

Sample spaces

A probability model first tells us what outcomes are possible.

> **SAMPLE SPACE**
> The **sample space *S*** of a random phenomenon is the set of all possible outcomes.

The name "sample space" is natural in random sampling, where each possible outcome is a sample and the sample space contains all possible samples. To specify S, we must state what constitutes an individual outcome and then state which outcomes can occur. The idea of a sample space, and the freedom we may have in specifying it, are best illustrated by examples.

EXAMPLE 5.4

Sample space for tossing a coin. Toss a coin. There are only two possible outcomes, and the sample space is

$$S = \{\text{heads, tails}\}$$

or, more briefly, $S = \{H, T\}$.

EXAMPLE 5.5

Sample space for random digits. Let your pencil point fall blindly into Table B of random digits. Record the value of the digit it lands on. The possible outcomes are

$$S = \{0, 1, 2, 3, 4, 5, 6, 7, 8, 9\}$$

EXAMPLE 5.6

Sample space for tossing a coin four times. Toss a coin four times and record the results. That's a bit vague. To be exact, record the results of each of the four tosses in order. A typical outcome is then HTTH. Counting shows that there are 16 possible outcomes. The sample space S is the set of all 16 strings of four H's and T's.

Suppose that our only interest is the number of heads in four tosses. Now we can be exact in a simpler fashion. The random phenomenon is to toss a coin four times and count the number of heads. The sample space contains only 5 outcomes:

$$S = \{0, 1, 2, 3, 4\}$$

This example illustrates the importance of carefully specifying what constitutes an individual outcome.

Although these examples seem remote from the practice of statistics, the connection is surprisingly close. Suppose that in conducting an opinion poll you select four people at random from a large population and ask each if he or she favors reducing federal spending on low-interest student loans. The answers are "Yes" or "No." The possible outcomes—the sample space—are exactly as in Example 5.4 if we replace heads by "Yes" and tails by "No."

Similarly, the possible outcomes of an SRS of 1500 people are the same in principle as the possible outcomes of tossing a coin 1500 times. On the other hand, the probabilities of the outcomes might be quite different. One of the great advantages of mathematics is that the essential features of very different phenomena can be described by the same mathematical model.

USE YOUR KNOWLEDGE

5.9 What color is your hair? A student is asked what color hair he or she has. Set up an appropriate sample space for this setting. Note that there is not a single correct answer to this exercise, so give reasons for your choice.

We list all the possible values for the sample spaces described previously. Other sample spaces correspond to variables for which we cannot list all the possible values. Here is an example.

EXAMPLE 5.7

Using software. Most statistical software has a function that will generate a random number between 0 and 1. The sample space is

$$S = \{\text{all numbers between 0 and 1}\}$$

This S is a mathematical idealization. Any specific random number generator produces numbers with some limited number of decimal places so that, strictly speaking, not all numbers between 0 and 1 are possible outcomes. For example, Minitab generates random numbers like 0.736891, with six decimal places. The entire interval from 0 to 1 is easier to think about. It also has the advantage of being a suitable sample space for different software systems that produce random numbers with different numbers of digits.

USE YOUR KNOWLEDGE

5.10 Do you study a lot? You record the time per week that a randomly selected student spends studying. What is the sample space?

A sample space S lists the possible outcomes of a random phenomenon. To complete a mathematical description of the random phenomenon, we must also give the probabilities with which these outcomes occur.

The true long-term proportion of any outcome—say, "exactly 2 heads in four tosses of a coin"—can be found only empirically, and then only approximately. How then can we describe probability mathematically? Rather than immediately attempting to give "correct" probabilities, let's confront the easier task of laying down rules that any assignment of probabilities must satisfy. We need to assign probabilities not only to single outcomes but also to sets of outcomes.

EVENT

An **event** is an outcome or a set of outcomes of a random phenomenon. That is, an event is a subset of the sample space.

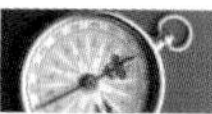

EXAMPLE 5.8

Exactly 2 heads in four tosses. Take the sample space S for four tosses of a coin to be the 16 possible outcomes in the form HTHH. Then "exactly 2 heads" is an event. Call this event A. The event A expressed as a set of outcomes is

$$A = \{\text{HHTT, HTHT, HTTH, THHT, THTH, TTHH}\}$$

In a probability model, events have probabilities. What properties must any assignment of probabilities to events have? Here are some basic facts about any probability model. These facts follow from the idea of probability as "the long-run proportion of repetitions on which an event occurs."

Four facts about probability

1. **Any probability is a number between 0 and 1.** An event with probability 0 never occurs, and an event with probability 1 occurs on every trial. An event with probability 0.5 occurs in half the trials in the long run.
2. **All possible outcomes together must have probability 1.** Because every trial will produce an outcome, the sum of the probabilities for all possible outcomes must be exactly 1.
3. **If two events have no outcomes in common, the probability that one or the other occurs is the sum of their individual probabilities.** If one

event occurs in 40% of all trials, and if a different event occurs in 25% of all trials, and the two can never occur together, then one or the other occurs on 65% of all trials because 40% + 25% = 65%.

4. **The probability that an event does not occur is 1 minus the probability that the event does occur.** If an event occurs in (say) 70% of all trials, it fails to occur in the other 30%. The probability that an event occurs and the probability that it does not occur always add to 100%, or 1.

Probability rules

Formal probability uses mathematical notation to state Facts 1 to 4 more concisely. We use capital letters near the beginning of the alphabet to denote events. If A is any event, we write its probability as $P(A)$. Here are our probability facts in formal language. As you apply these rules, remember that they are just another form of intuitively true facts about long-run proportions.

PROBABILITY RULES

Rule 1. The probability $P(A)$ of any event A satisfies $0 \leq P(A) \leq 1$.

Rule 2. If S is the sample space in a probability model, then $P(S) = 1$.

Rule 3. Two events A and B are **disjoint** if they have no outcomes in common and so can never occur together. If A and B are disjoint,

$$P(A \text{ or } B) = P(A) + P(B)$$

This is the **addition rule for disjoint events.**

Rule 4. The **complement** of any event A is the event that A does not occur, written as A^c. The **complement rule** states that

$$P(A^c) = 1 - P(A)$$

Venn diagram

You may find it helpful to draw a picture to remind yourself of the meaning of complements and disjoint events. A picture like Figure 5.2 that shows the sample space S as a rectangular area and events as areas within S is called a **Venn diagram.** The events A and B in Figure 5.2 are disjoint because they do not overlap. As Figure 5.3 shows, the complement A^c contains exactly the outcomes that are not in A.

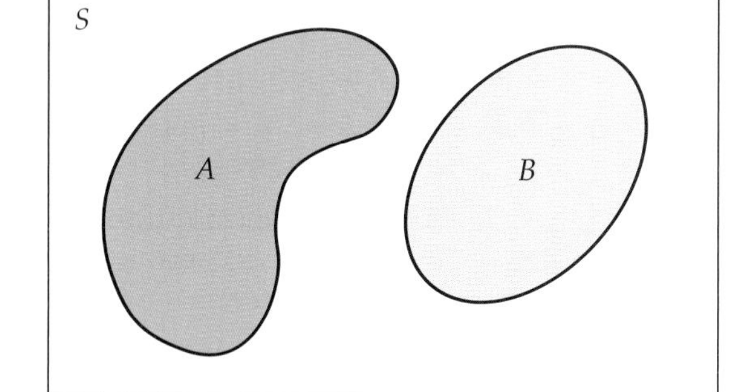

FIGURE 5.2 Venn diagram showing disjoint events A and B. Disjoint events have no common outcomes.

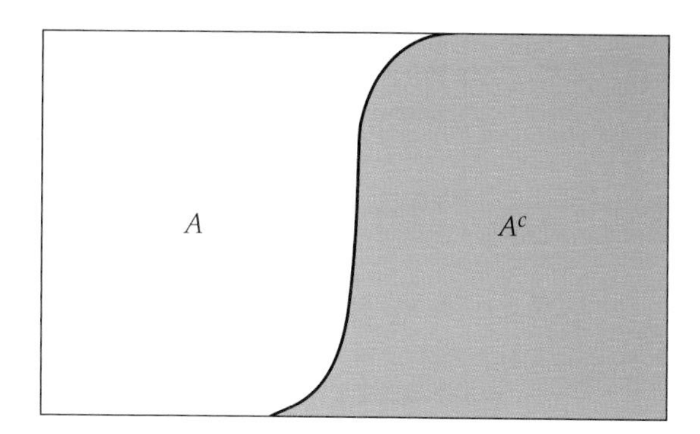

FIGURE 5.3 Venn diagram showing the complement A^c of an event A. The complement consists of all outcomes that are not in A.

EXAMPLE 5.9

Favorite colors. What is your favorite color? Our preferences can be related to our personality, our moods, or particular objects. Here is a probability model for color preferences:[2]

Color	Blue	Green	Purple	Red	Black
Probability	0.42	0.14	0.14	0.08	0.07

Color	Orange	Yellow	Brown	Gray	White
Probability	0.05	0.03	0.03	0.02	0.02

Each probability is between 0 and 1. The probabilities add to 1 because these outcomes together make up the sample space S. Our probability model corresponds to selecting a person at random and asking what is their favorite color.

Let's use the probability Rules 3 and 4 to find some probabilities for favorite colors.

EXAMPLE 5.10

Is it blue or green? What is the probability that a person's favorite color is blue or green? If the favorite is blue, it cannot be green, so these two events are disjoint. Using Rule 3, we find

$$\begin{aligned} P(\text{Blue or Green}) &= P(\text{Blue}) + P(\text{Green}) \\ &= 0.42 + 0.14 = 0.56 \end{aligned}$$

There is a 56% chance that a randomly selected person's favorite color will be blue or green. Suppose that we want to find the probability that the favorite is not purple.

EXAMPLE 5.11

Use the complement rule. To solve this problem, we could use Rule 3 and add the probabilities for Blue, Green, Red, Black, Orange, Yellow, Brown,

Gray, and White. However, it is easier to use the probability that we have for Purple and Rule 4. The event that the favorite is not Purple is the complement of the event that the favorite is Purple. Using our notation for events, we have

$$P(\text{not Purple}) = 1 - P(\text{Purple})$$
$$= 1 - 0.14 = 0.86$$

We see that 86% of people have a favorite color that is not purple.

USE YOUR KNOWLEDGE

5.11 Favorites of Black or White. Find the probability that the favorite color is Black or White.

5.12 Blue, Green, Purple, Red, Black, Orange, or White. Find the probability that the favorite color is Blue, Green, Black, Brown, Gray, or White using Rule 4. Explain why this calculation is easier than finding the answer using Rule 3.

Assigning probabilities: finite number of outcomes

The individual outcomes of a random phenomenon are always disjoint. So the addition rule provides a way to assign probabilities to events with more than one outcome: start with probabilities for individual outcomes and add to get probabilities for events. This idea works well when there are only a finite (fixed and limited) number of outcomes.

PROBABILITIES IN A FINITE SAMPLE SPACE

Assign a probability to each individual outcome. These probabilities must be numbers between 0 and 1 and must have sum 1.

The probability of any event is the sum of the probabilities of the outcomes making up the event.

EXAMPLE 5.12

Benford's law. Faked numbers in tax returns, payment records, invoices, expense account claims, and many other settings often display patterns that aren't present in legitimate records. Some patterns, like too many round numbers, are obvious and easily avoided by a clever crook. Others are more subtle. It is a striking fact that the first digits of numbers in legitimate records often follow a distribution known as **Benford's law.** Here it is (note that a first digit can't be 0):[3]

Benford's law

First digit	1	2	3	4	5	6	7	8	9
Probability	0.301	0.176	0.125	0.097	0.079	0.067	0.058	0.051	0.046

Benford's law usually applies to the first digits of the sizes of similar quantities, such as invoices, expense account claims, and county populations. Investigators can detect fraud by comparing the first digits in records, such as business invoices, with these probabilities.

EXAMPLE 5.13

Find some probabilities for Benford's law. Consider the events

$$A = \{\text{first digit is } 1\}$$
$$B = \{\text{first digit is 6 or greater}\}$$

From the table of probabilities,

$$P(A) = P(1) = 0.301$$
$$P(B) = P(6) + P(7) + P(8) + P(9)$$
$$= 0.067 + 0.058 + 0.051 + 0.046 = 0.222$$

Note that $P(B)$ is not the same as the probability that a first digit is strictly greater than 6. The probability $P(6)$ that a first digit is 6 is included in "6 or greater" but not in "greater than 6."

USE YOUR KNOWLEDGE

5.13 Benford's law. Using the probabilities for Benford's law, find the probability that a first digit is anything other than 1.

5.14 Use the addition rule. Use the addition rule with the probabilities for the events A and B from Example 5.13 to find the probability that a first digit either is 1 or is 6 or greater.

Be careful to apply the addition rule only to disjoint events.

EXAMPLE 5.14

Apply the addition rule to Benford's law. Check that the probability of the event C that a first digit is odd is

$$P(C) = P(1) + P(3) + P(5) + P(7) + P(9) = 0.609$$

The probability

$$P(B \text{ or } C) = P(1) + P(3) + P(5) + P(6) + P(7) + P(8) + P(9) = 0.727$$

is *not* the sum of $P(B)$ and $P(C)$, because events B and C are not disjoint. Outcomes 7 and 9 are common to both events and we want to include their probabilities only once.

Assigning probabilities: equally likely outcomes

Assigning correct probabilities to individual outcomes often requires long observation of the random phenomenon. In some circumstances, however, we are willing to assume that individual outcomes are equally likely because of some balance in the phenomenon. Ordinary coins have a physical balance that should make heads and tails equally likely, for example, and the table of random digits comes from a deliberate randomization.

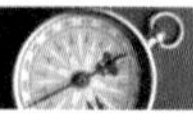

EXAMPLE 5.15

First digits that are equally likely. You might think that first digits are distributed "at random" among the digits 1 to 9 in business records. The 9 possible outcomes would then be equally likely. The sample space for a single digit is

$$S = \{1, 2, 3, 4, 5, 6, 7, 8, 9\}$$

Because the total probability must be 1, the probability of each of the 9 outcomes must be 1/9. That is, the assignment of probabilities to outcomes is

First digit	1	2	3	4	5	6	7	8	9
Probability	1/9	1/9	1/9	1/9	1/9	1/9	1/9	1/9	1/9

The probability of the event B that a randomly chosen first digit is 6 or greater is

$$\begin{aligned} P(B) &= P(6) + P(7) + P(8) + P(9) \\ &= \frac{1}{9} + \frac{1}{9} + \frac{1}{9} + \frac{1}{9} = \frac{4}{9} = 0.444 \end{aligned}$$

Compare this with the Benford's law probability in Example 5.13. A crook who fakes data by using "random" digits will end up with too many first digits 6 or greater and too few 1s and 2s.

In Example 5.15 all outcomes have the same probability. Because there are 9 equally likely outcomes, each must have probability 1/9. Because exactly 4 of the 9 equally likely outcomes are 6 or greater, the probability of this event is 4/9. In the special situation where all outcomes are equally likely, we have a simple rule for assigning probabilities to events.

EQUALLY LIKELY OUTCOMES

If a random phenomenon has k possible outcomes, all equally likely, then each individual outcome has probability $1/k$. The probability of any event A is

$$\begin{aligned} P(A) &= \frac{\text{count of outcomes in } A}{\text{count of outcomes in } S} \\ &= \frac{\text{count of outcomes in } A}{k} \end{aligned}$$

Most random phenomena do not have equally likely outcomes, so the general rule for finite sample spaces is more important than the special rule for equally likely outcomes.

USE YOUR KNOWLEDGE

5.15 Possible outcomes for rolling a die. A die has six sides with 1 to 6 "spots" on the sides. Give the probability distribution for the six possible outcomes that can result when a perfect die is rolled.

Independence and the multiplication rule

Rule 3, the addition rule for disjoint events, describes the probability that *one or the other* of two events A and B will occur in the special situation when A and B cannot occur together because they are disjoint. Our final rule describes the probability that *both* events A and B occur, again only in a special situation. More general rules appear in Section 5.6, but in our study of statistics we will need only the rules that apply to special situations.

Suppose that you toss a fair coin twice. You are counting heads, so two events of interest are

$$A = \{\text{first toss is a head}\}$$
$$B = \{\text{second toss is a head}\}$$

The events A and B are not disjoint. They occur together whenever both tosses give heads. We want to compute the probability of the event $\{A \text{ and } B\}$ that *both* tosses are heads. The Venn diagram in Figure 5.4 illustrates the event $\{A \text{ and } B\}$ as the overlapping area that is common to both A and B.

The coin tossing of Buffon, Pearson, and Kerrich described in Example 5.3 makes us willing to assign probability 1/2 to a head when we toss a coin. So

$$P(A) = 0.5$$
$$P(B) = 0.5$$

What is $P(A \text{ and } B)$? Our common sense says that it is 1/4. The first toss will give a head half the time and then the second will give a head on half of those trials, so both tosses will give heads on $1/2 \times 1/2 = 1/4$ of all trials in the long run. This reasoning assumes that the second toss still has probability 1/2 of a head after the first has given a head. This is true—we can verify it by tossing a coin twice many times and observing the proportion of heads on the second toss after the first toss has produced a head. We say that the events "head on the first toss" and "head on the second toss" are *independent*. Here is our final probability rule.

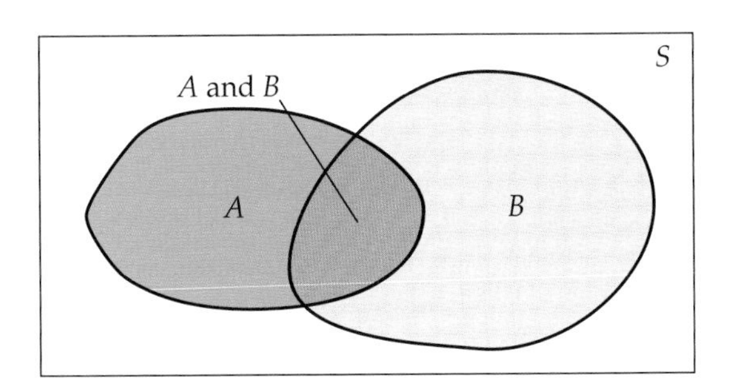

FIGURE 5.4 Venn diagram showing the event $\{A \text{ and } B\}$. This event consists of outcomes common to A and B.

THE MULTIPLICATION RULE FOR INDEPENDENT EVENTS

Rule 5. Two events A and B are **independent** if knowing that one occurs does not change the probability that the other occurs. If A and B are independent,

$$P(A \text{ and } B) = P(A)P(B)$$

This is the **multiplication rule for independent events.**

Our definition of independence is rather informal. We will make this informal idea precise in Section 5.6. In practice, though, we rarely need a precise definition of independence, because independence is usually *assumed* as part of a probability model when we want to describe random phenomena that seem to be physically unrelated to each other. Here is an example of independence.

EXAMPLE 5.16

Coins do not have a memory. Because a coin has no memory and most coin tossers cannot influence the fall of the coin, it is safe to assume that successive coin tosses are independent. For a fair coin, after we see the outcome of the first toss, we still assign probability 1/2 to heads on the second toss.

USE YOUR KNOWLEDGE

5.16 Three tails in three tosses. What is the probability of obtaining three tails on three tosses of a fair coin?

From independence to dependence

Here is an example of a situation where there are dependent events.

EXAMPLE 5.17

Dependent events in cards. The colors of successive cards dealt from the same deck are not independent. A standard 52-card deck contains 26 red and 26 black cards. For the first card dealt from a shuffled deck, the probability of a red card is $26/52 = 0.50$ because the 52 possible cards are equally likely. Once we see that the first card is red, we know that there are only 25 reds among the remaining 51 cards. The probability that the second card is red is therefore only $25/51 = 0.49$. Knowing the outcome of the first deal changes the probabilities for the second.

USE YOUR KNOWLEDGE

5.17 The probability of a second ace. A deck of 52 cards contains 4 aces, so the probability that a card drawn from this deck is an ace is 4/52. If we know that the first card drawn is an ace, what is the probability that the second card drawn is also an ace? Using the idea of independence, explain why this probability is not 4/52.

Here is another example of a situation where events are dependent.

EXAMPLE 5.18

Taking a test twice. If you take an IQ test or other mental test twice in succession, the two test scores are not independent. The learning that occurs on the first attempt influences your second attempt. If you learn a lot, then your second test score might be a lot higher than your first test score.

When independence is part of a probability model, the multiplication rule applies. Here is an example.

EXAMPLE 5.19

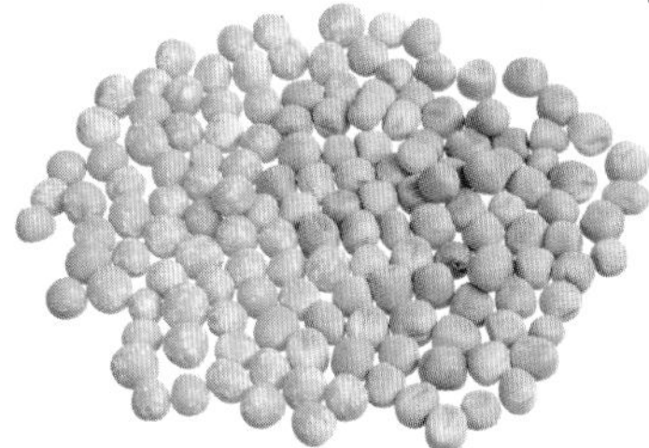

Mendel's peas. Gregor Mendel used garden peas in some of the experiments that revealed that inheritance operates randomly. The seed color of Mendel's peas can be either green or yellow. Two parent plants are "crossed" (one pollinates the other) to produce seeds. Each parent plant carries two genes for seed color, and each of these genes has probability 1/2 of being passed to a seed. The two genes that the seed receives, one from each parent, determine its color. The parents contribute their genes independently of each other.

Suppose that both parents carry the G and the Y genes. The seed will be green if both parents contribute a G gene. Otherwise, it will be yellow. If M is the event that the male contributes a G gene and F is the event that the female contributes a G gene, then the probability of a green seed is

$$\begin{aligned} P(M \text{ and } F) &= P(M)P(F) \\ &= (0.5)(0.5) = 0.25 \end{aligned}$$

In the long run, 1/4 of all seeds produced by crossing these plants will be green.

The multiplication rule applies only to independent events. You cannot use it if events are not independent. Here is a distressing example of misuse of the multiplication rule.

EXAMPLE 5.20

Sudden infant death syndrome. Sudden infant death syndrome (SIDS) causes babies to die suddenly (often in their cribs) with no explanation. Deaths from SIDS have been greatly reduced by placing babies on their backs, but as yet no cause is known.

When more than one SIDS death occurs in a family, the parents are sometimes accused. One "expert witness" popular with prosecutors in England told juries that there is only a 1 in 73 million chance that two children in the same family could have died naturally.

Here's his calculation: the rate of SIDS in a nonsmoking middle-class family is 1 in 8500. So the probability of two deaths is

$$\frac{1}{8500} \times \frac{1}{8500} = \frac{1}{72,250,000}$$

Several women were convicted of murder on this basis, without any direct evidence that they harmed their children.

As the Royal Statistical Society said, this reasoning is nonsense. It assumes that SIDS deaths in the same family are independent events. The cause of SIDS is unknown: "There may well be unknown genetic or environmental factors that predispose families to SIDS, so that a second case within the family becomes much more likely."[4] The British government decided to review the cases of 258 parents convicted of murdering their babies.

The multiplication rule $P(A \text{ and } B) = P(A)P(B)$ holds if A and B are independent but not otherwise. The addition rule $P(A \text{ or } B) = P(A) + P(B)$ holds if A and B are disjoint but not otherwise. Resist the temptation to use these simple formulas when the circumstances that justify them are not present.

You must also be certain not to confuse disjointness and independence. Disjoint events cannot be independent. If A and B are disjoint, then the fact that A occurs tells us that B cannot occur—look again at Figure 5.2. Unlike disjointness or complements, independence cannot be pictured by a Venn diagram, because it involves the probabilities of the events rather than just the outcomes that make up the events.

Applying the probability rules

If two events A and B are independent, then their complements A^c and B^c are also independent and A^c is independent of B. Suppose, for example, that 75% of all registered voters in a suburban district are Republicans. If an opinion poll interviews two voters chosen independently, the probability that the first is a Republican and the second is not a Republican is $(0.75)(0.25) = 0.1875$.

The multiplication rule also extends to collections of more than two events, provided that all are independent. Independence of events A, B, and C means that no information about any one or any two can change the probability of the remaining events.

The formal definition is a bit messy. Fortunately, independence is usually assumed in setting up a probability model. We can then use the multiplication rule freely.

By combining the rules we have learned, we can compute probabilities for rather complex events. Here is an example.

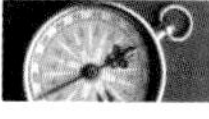

EXAMPLE 5.21

HIV testing. Many people who come to clinics to be tested for HIV, the virus that causes AIDS, don't come back to learn the test results. Clinics now use "rapid HIV tests" that give a result in a few minutes.

The false-positive rate for a diagnostic test is the probability that a person with no disease will have a positive test result. For the rapid HIV tests, the Food and Drug Administration (FDA) has established 2% as the maximum false-positive rate for a rapid HIV test.[5]

If a clinic uses a test that meets the FDA standard and tests 50 people who are free of HIV antibodies, what is the probability that at least 1 false-positive will occur?

It is reasonable to assume as part of the probability model that the test results for different individuals are independent. The probability that the test is positive for a single person is 0.02, so the probability of a negative result is $1 - 0.02 = 0.98$ by the complement rule. The probability of at least 1 false-positive among the 50 people tested is therefore

$$\begin{aligned} P(\text{at least 1 positive}) &= 1 - P(\text{no positives}) \\ &= 1 - P(50 \text{ negatives}) \\ &= 1 - 0.98^{50} \\ &= 1 - 0.3642 = 0.6358 \end{aligned}$$

There is approximately a 64% chance that at least 1 of the 50 people will test positive for HIV, even though no one has the virus.

Concern about excessive numbers of false-positives led the New York City Department of Health and Mental Hygiene to suspend the use of one particular rapid HIV test.[6]

SECTION 5.2 SUMMARY

A **probability model** for a random phenomenon consists of a sample space S and an assignment of probabilities P.

The **sample space** S is the set of all possible outcomes of the random phenomenon. Sets of outcomes are called **events.** P assigns a number $P(A)$ to an event A as its probability.

The **complement** A^c of an event A consists of exactly the outcomes that are not in A. Events A and B are **disjoint** if they have no outcomes in common. Events A and B are **independent** if knowing that one event occurs does not change the probability we would assign to the other event.

Any **assignment of probability** must obey the rules that state the basic properties of probability:

Rule 1. $0 \leq P(A) \leq 1$ for any event A.

Rule 2. $P(S) = 1$.

Rule 3. Addition rule: If events A and B are **disjoint,** then $P(A \text{ or } B) = P(A) + P(B)$.

Rule 4. Complement rule: For any event A, $P(A^c) = 1 - P(A)$.

Rule 5. Multiplication rule: If events A and B are **independent,** then $P(A \text{ and } B) = P(A)P(B)$.

SECTION 5.2 EXERCISES

For Exercise 5.9, see page 242; for Exercise 5.10, see page 243; for Exercises 5.11 and 5.12, see page 246; for Exercises 5.13 and 5.14, see page 247; for Exercise 5.15, see page 249; for Exercise 5.16, see page 250; and for Exercise 5.17, see page 250.

5.18. What's wrong? In each of the following scenarios, there is something wrong. Describe what is wrong and give a reason for your answer.

(a) If the sample space consists of two outcomes, then each outcome has probability 0.5.

(b) If we select a digit at random, then the probability of selecting a 2 is 0.2.

(c) If the probability of A is 0.2, the probability of B is 0.3, and the probability of A and B is 0.5, then A and B are independent.

5.19. Evaluating Web page designs. You are a Web page designer and you set up a page with five different links. A user of the page can click on one of the links or he or she can leave that page. Describe the sample space for the outcome of a visit to your Web page.

5.20. Record the length of time spent on the page. Refer to the previous exercise. You also decide to measure the length of time a visitor spends on your page. Give the sample space for this measure.

5.21. Ringtones. What are the popular ringtones? The Web site `funtonia.com` updates its list of top ringtones often. Here are probabilities for the top 10 ringtones listed by the site recently:[7]

Ringtone	Probability
Empire State of Mind	0.180
Baby by Me	0.136
Forever	0.114
Party in the USA	0.107
Fireflies	0.103
Bad Romance	0.081
I Can Transform Ya	0.075
Down	0.070
I Gotta Feeling	0.068
Money to Blow	0.066

(a) What is the probability that a randomly selected ringtone from this list is either Empire State of Mind or I Gotta Feeling?

(b) What is the probability that a randomly selected ringtone from this list is not Empire State of Mind and not I Gotta Feeling? Be sure to show how you computed this answer.

5.22. More ringtones. Refer to the previous exercise.

(a) If two ringtones are selected independently, what is the probability that both are Party in the USA?

(b) Describe in words the complement of the event described in part (a) of this exercise. Find the probability of this event.

5.23. Distribution of blood types. All human blood can be "ABO-typed" as one of O, A, B, or AB, but the distribution of the types varies a bit among groups of people. Here is the distribution of blood types for a randomly chosen person in the United States:[8]

Blood type	A	B	AB	O
U.S. probability	0.42	0.11	?	0.44

(a) What is the probability of type AB blood in the United States?

(b) Maria has type B blood. She can safely receive blood transfusions from people with blood types O and B. What is the probability that a randomly chosen person from the United States can donate blood to Maria?

5.24. Blood types in Ireland. The distribution of blood types in Ireland differs from the U.S. distribution given in the previous exercise:

Blood type	A	B	AB	O
Ireland probability	0.35	0.10	0.03	0.52

Choose a person from the United States and a person from Ireland at random, independently of each other. What is the probability that both have type O blood? What is the probability that both have the same blood type?

5.25. Are the probabilities legitimate? In each of the following situations, state whether or not the given assignment of probabilities to individual outcomes is legitimate, that is, satisfies the rules of probability. If not, give specific reasons for your answer.

(a) Choose a college student at random and record gender and enrollment status: $P(\text{female full-time}) = 0.44$, $P(\text{female part-time}) = 0.56$, $P(\text{male full-time}) = 0.46$, $P(\text{male part-time}) = 0.54$.

(b) Deal a card from a shuffled deck: $P(\text{clubs}) = 16/52$, $P(\text{diamonds}) = 12/52$, $P(\text{hearts}) = 12/52$, $P(\text{spades}) = 12/52$.

(c) Roll a die and record the count of spots on the up-face: $P(1) = 1/3$, $P(2) = 0$, $P(3) = 1/6$, $P(4) = 1/3$, $P(5) = 1/6$, $P(6) = 0$.

5.26. French and English in Canada. Canada has two official languages, English and French. Choose a Canadian at random and ask, "What is your mother tongue?" Here is the distribution of responses, combining many separate languages from the broad Asian/Pacific region:[9]

Language	English	French	Asian/Pacific	Other
Probability	0.59	?	0.07	0.11

(a) What probability should replace "?" in the distribution?

(b) What is the probability that a Canadian's mother tongue is not English? Explain how you computed your answer.

5.27. Education levels of young adults. Choose a young adult (aged 25 to 34 years) at random. The probability is 0.12 that the person chosen did not complete high school, 0.31 that the person has a high school diploma but no further education, and 0.29 that the person has at least a bachelor's degree.

(a) What must be the probability that a randomly chosen young adult has some education beyond high school but does not have a bachelor's degree?

(b) What is the probability that a randomly chosen young adult has at least a high school education?

5.28. CHALLENGE **Loaded dice.** There are many ways to produce crooked dice. To *load* a die so that 6 comes up too often and 1 (which is opposite 6) comes up too seldom, add a bit of lead to the filling of the spot on the 1 face. Because the spot is solid plastic, this tampering works even with transparent dice. If a die is loaded so that 6 comes up with probability 0.21 and the probabilities of the 2, 3, 4, and 5 faces are not affected, what is the assignment of probabilities to the six faces?

5.29. Rh blood types. Human blood is typed as O, A, B, or AB and also as Rh-positive or Rh-negative. ABO type and Rh-factor type are independent because they are governed by different genes. In the American population, 84% of people are Rh-positive. Use the information about ABO type in Exercise 5.23 to give the probability distribution of blood type (ABO and Rh) for a randomly chosen person.

5.30. Roulette. A roulette wheel has 38 slots, numbered 0, 00, and 1 to 36. The slots 0 and 00 are colored green, 18 of the others are red, and 18 are black. The dealer spins the wheel and at the same time rolls a small ball along the wheel in the opposite direction. The wheel is carefully balanced so that the ball is equally likely to land in any slot when the wheel slows. Gamblers can bet on various combinations of numbers and colors.

(a) What is the probability that the ball will land in any one slot?

(b) If you bet on "red," you win if the ball lands in a red slot. What is the probability of winning?

(c) The slot numbers are laid out on a board on which gamblers place their bets. One column of numbers on the board contains all multiples of 3, that is, 3, 6, 9, . . . , 36. You place a "column bet" that wins if any of these numbers comes up. What is your probability of winning?

5.31. Winning the lottery. A state lottery's Pick 3 game asks players to choose a three-digit number, 000 to 999. The state chooses the winning three-digit number at random, so that each number has probability 1/1000. You win if the winning number contains the digits in your number, in any order.

(a) Your number is 491. What is your probability of winning?

(b) Your number is 222. What is your probability of winning?

5.32. PINs. The personal identification numbers (PINs) for automatic teller machines usually consist of four digits. You notice that most of your PINs have at least one 0, and you wonder if the issuers use lots of 0s to make the numbers easy to remember. Suppose that PINs are assigned at random, so that all four-digit numbers are equally likely.

(a) How many possible PINs are there?

(b) What is the probability that a PIN assigned at random has at least one 0?

5.33. Universal blood donors. People with type O-negative blood are universal donors. That is, any patient can receive a transfusion of O-negative blood. Only 7% of the American population have O-negative blood. If 10 people appear at random to give blood, what is the probability that at least 1 of them is a universal donor?

5.34. CHALLENGE **Axioms of probability.** Show that any assignment of probabilities to events that obeys Rules 2 and 3 on page 244 automatically obeys the complement rule (Rule 4). This implies that a mathematical treatment of probability can start from just Rules 1, 2, and 3. These rules are sometimes called *axioms* of probability.

5.35. CHALLENGE **Independence of complements.** Show that if events A and B obey the multiplication rule, $P(A \text{ and } B) = P(A)P(B)$, then A and the complement B^c of B also obey the multiplication rule, $P(A \text{ and } B^c) = P(A)P(B^c)$. That is, if events A and B are independent, then A and B^c are also independent. (*Hint:* Start by drawing a Venn diagram and noticing that the events "A and B" and "A and B^c" are disjoint.)

Mendelian inheritance. *Some traits of plants and animals depend on inheritance of a single gene. This is called Mendelian inheritance, after Gregor Mendel (1822–1884). Exercises 5.36 to 5.39 are based on the following information about Mendelian inheritance of blood type.*

Each of us has an ABO blood type, which describes whether two characteristics called A and B are present. Every human being has two blood type alleles (gene forms), one inherited from our mother and one from our father. Each of these alleles can be A, B, or O. Which two we inherit determines our blood type. Here is a table that shows what our blood type is for each combination of two alleles:

Alleles inherited	Blood type
A and A	A
A and B	AB
A and O	A
B and B	B
B and O	B
O and O	O

We inherit each of a parent's two alleles with probability 0.5. We inherit independently from our mother and our father.

5.36. Blood types of children. Hannah and Jacob both have alleles A and B.

(a) What blood types can their children have?

(b) What is the probability that their next child has each of these blood types?

5.37. Parents with alleles B and O. Nancy and David both have alleles B and O.

(a) What blood types can their children have?

(b) What is the probability that their next child has each of these blood types?

5.38. Two children. Jennifer has alleles A and O. José has alleles A and B. They have two children. What is the probability that both children have blood type A? What is the probability that both children have the same blood type?

5.39. Three children. Jasmine has alleles A and O. Joshua has alleles B and O.

(a) What is the probability that a child of these parents has blood type O?

(b) If Jasmine and Joshua have three children, what is the probability that all three have blood type O? What is the probability that the first child has blood type O and the next two do not?

5.3 Random Variables

Sample spaces need not consist of numbers. When we toss a coin four times, we can record the outcome as a string of heads and tails, such as HTTH. In statistics, however, we are most often interested in numerical outcomes such as the count of heads in the four tosses.

It is convenient to use a shorthand notation: Let X be the number of heads. If our outcome is HTTH, then $X = 2$. If the next outcome is TTTH, the value of X changes to $X = 1$. The possible values of X are 0, 1, 2, 3, and 4. Tossing a coin four times will give X one of these possible values. Tossing four more times will give X another and probably different value. We call X a *random variable* because its values can vary when the coin tossing is repeated.

RANDOM VARIABLE

A **random variable** is a variable whose value is a numerical outcome of a random phenomenon.

We usually denote random variables by capital letters near the end of the alphabet, such as X or Y. Of course, the random variables of greatest interest to us are outcomes such as the mean $\bar{x}$ of a random sample, for which we will keep the familiar notation.[10]

As we progress from general rules of probability toward statistical inference, we will concentrate on random variables. When a random variable X describes a random phenomenon, the sample space S just lists the possible values of the random variable. We usually do not mention S separately.

There remains the second part of any probability model, the assignment of probabilities to events. There are two main ways of assigning probabilities to the values of a random variable. The two types of probability models that result will dominate our application of probability to statistical inference.

Discrete random variables

We have learned several rules of probability, but only one method of assigning probabilities: state the probabilities of the individual outcomes and assign probabilities to events by summing over the outcomes. The outcome probabilities must be between 0 and 1 and have sum 1. When the outcomes are numerical, they are values of a random variable. We will now attach a name to random variables having probability assigned in this way.[11]

DISCRETE RANDOM VARIABLE

A **discrete random variable** X has a finite number of possible values. The **probability distribution** of X lists the values and their probabilities:

Value of X	x_1	x_2	x_3	$\cdots$	x_k
Probability	p_1	p_2	p_3	$\cdots$	p_k

The probabilities p_i must satisfy two requirements:

1. Every probability p_i is a number between 0 and 1.
2. $p_1 + p_2 + \cdots + p_k = 1$.

Find the probability of any event by adding the probabilities p_i of the particular values x_i that make up the event.

Our definition says that there are k possible values. However, there are situations where k can be infinite. For example, consider tossing a fair coin until you get a head. These situations are beyond the scope of our text.

EXAMPLE 5.22

Grade distributions. A liberal arts college posts the grade distributions for its courses. In a recent semester, students in one section of English 130 received 31% As, 40% Bs, 20% Cs, 4% Ds, and 5% Fs. Choose an English 130 student at random. To "choose at random" means to give every student the same chance to be chosen. The student's grade on a four-point scale (with A = 4) is a random variable X.

The value of X changes when we repeatedly choose students at random, but it is always one of 0, 1, 2, 3, or 4. Here is the distribution of X:

Value of X	0	1	2	3	4
Probability	0.05	0.04	0.20	0.40	0.31

The probability that the student got a B or better is the sum of the probabilities of an A and a B. In the language of random variables,

$$P(X \geq 3) = P(X = 3) + P(X = 4)$$
$$= 0.40 + 0.31 = 0.71$$

USE YOUR KNOWLEDGE

5.40 Will the course satisfy the requirement? Refer to Example 5.22. Suppose that a grade of D or F in English 130 will not count as satisfying a requirement for a major in linguistics. What is the probability that a randomly selected student will not satisfy this requirement?

probability histogram

We can use histograms to show probability distributions as well as distributions of data. Figure 5.5 displays **probability histograms** that compare the probability model for random digits for business records (Example 5.15) with the model given by Benford's law (Example 5.12). The height of each bar shows the probability of the outcome at its base. Because the heights are probabilities, they add to 1. As usual, all the bars in a histogram have the same width. So the areas also display the assignment of probability to outcomes.

Think of these histograms as idealized pictures of the results of very many trials. The histograms make it easy to quickly compare the two distributions.

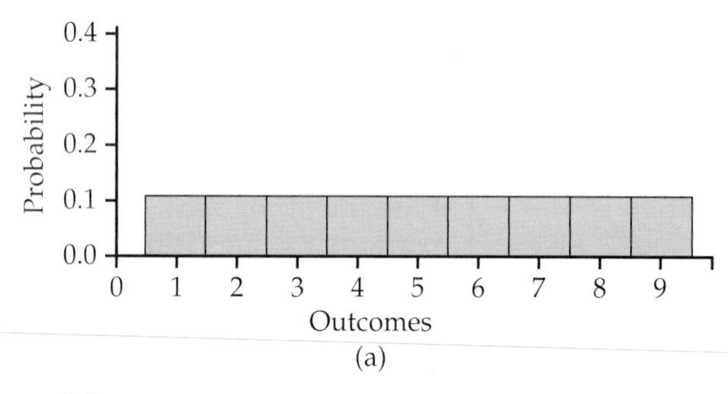

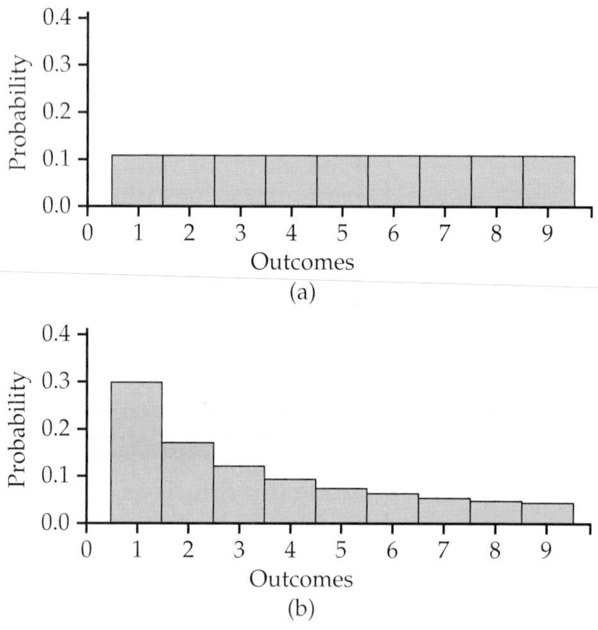

FIGURE 5.5 Probability histograms for (a) random digits 1 to 9 and (b) Benford's law. The height of each bar shows the probability assigned to a single outcome.

EXAMPLE 5.23

Number of heads in four tosses of a coin. What is the probability distribution of the discrete random variable X that counts the number of heads in four tosses of a coin? We can derive this distribution if we make two reasonable assumptions:

- The coin is balanced, so it is fair and each toss is equally likely to give H or T.
- The coin has no memory, so tosses are independent.

The outcome of four tosses is a sequence of heads and tails such as HTTH. There are 16 possible outcomes in all. Figure 5.6 lists these outcomes along

FIGURE 5.6 Possible outcomes in four tosses of a coin, for Example 5.23. The outcomes are arranged by the values of the random variable X, the number of heads.

		HTTH		
		HTHT		
	HTTT	THTH	HHHT	
	THTT	HHTT	HHTH	
	TTHT	THHT	HTHH	
TTTT	TTTH	TTHH	THHH	HHHH
$X = 0$	$X = 1$	$X = 2$	$X = 3$	$X = 4$

with the value of X for each outcome. The multiplication rule for independent events tells us that, for example,

$$P(\text{HTTH}) = \frac{1}{2} \times \frac{1}{2} \times \frac{1}{2} \times \frac{1}{2} = \frac{1}{16}$$

Each of the 16 possible outcomes similarly has probability 1/16. That is, these outcomes are equally likely.

The number of heads X has possible values 0, 1, 2, 3, and 4. These values are *not* equally likely. As Figure 5.6 shows, there is only one way that $X = 0$ can occur: namely, when the outcome is TTTT. So

$$P(X = 0) = \frac{1}{16} = 0.0625$$

The event $\{X = 2\}$ can occur in six different ways, so that

$$P(X = 2) = \frac{\text{count of ways } X = 2 \text{ can occur}}{16}$$

$$= \frac{6}{16} = 0.375$$

We can find the probability of each value of X from Figure 5.6 in the same way. Here is the result:

Value of X	0	1	2	3	4
Probability	0.0625	0.25	0.375	0.25	0.0625

Figure 5.7 is a probability histogram for the distribution in Example 5.23. The probability distribution is exactly symmetric. The probabilities (bar

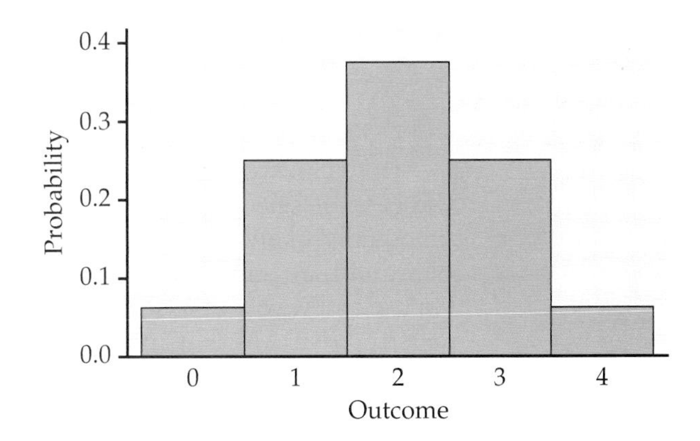

FIGURE 5.7 Probability histogram for the number of heads in four tosses of a coin.

heights) are idealizations of the proportions after very many trials of four tosses of a coin. The actual distribution of proportions observed would be nearly symmetric but is unlikely to be exactly symmetric.

EXAMPLE 5.24

Probability of at least two heads. Any event involving the number of heads observed can be expressed in terms of X, and its probability can be found from the distribution of X. For example, the probability of tossing at least two heads is

$$P(X \geq 2) = 0.375 + 0.25 + 0.0625 = 0.6875$$

The probability of at least one head is most simply found by use of the complement rule:

$$\begin{aligned} P(X \geq 1) &= 1 - P(X = 0) \\ &= 1 - 0.0625 = 0.9375 \end{aligned}$$

Recall that tossing a coin n times is similar to choosing an SRS of size n from a large population and asking a Yes or No question. We will extend the results of Example 5.23 when we return to sampling distributions in the next chapter.

USE YOUR KNOWLEDGE

5.41 Two tosses of a fair coin. Find the probability distribution for the number of heads that appear in two tosses of a fair coin.

Continuous random variables

When we use the table of random digits to select a digit between 0 and 9, the result is a discrete random variable. The probability model assigns probability 1/10 to each of the 10 possible outcomes. Suppose that we want to choose a number at random between 0 and 1, allowing *any* number between 0 and 1 as the outcome. Software random number generators will do this.

You can visualize such a random number by thinking of a spinner (Figure 5.8) that turns freely on its axis and slowly comes to a stop. The pointer can come to rest anywhere on a circle that is marked from 0 to 1. The sample space is now an entire interval of numbers:

$$S = \{\text{all numbers } x \text{ such that } 0 \leq x \leq 1\}$$

How can we assign probabilities to events such as $\{0.3 \leq x \leq 0.7\}$? As in the case of selecting a random digit, we would like all possible outcomes to be equally likely. But we cannot assign probabilities to each individual value of x and then sum, because there are infinitely many possible values. Instead, we use a new way of assigning probabilities directly to events—as *areas under a density curve.* Any density curve has area exactly 1 underneath it, corresponding to total probability 1.

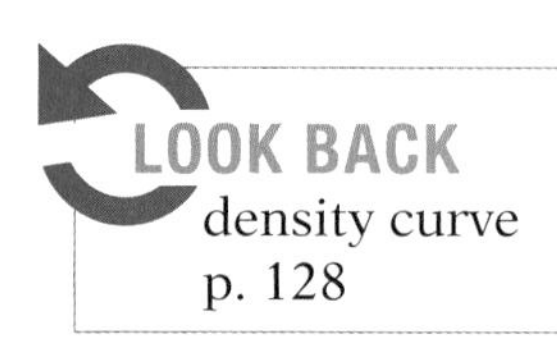

density curve p. 128

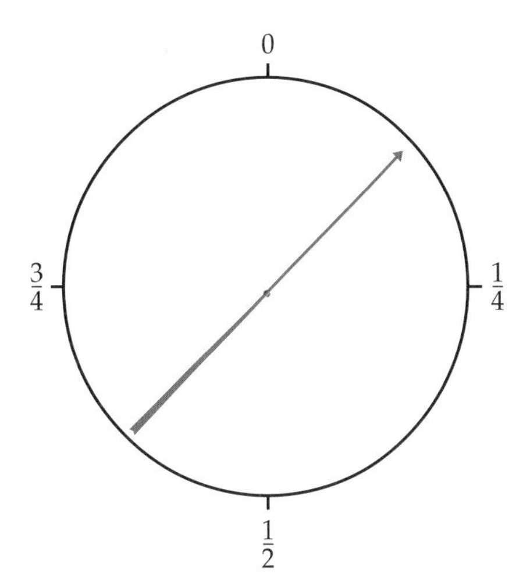

FIGURE 5.8 A spinner that generates a random number between 0 and 1.

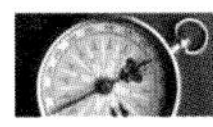

EXAMPLE 5.25

uniform distribution

Uniform random numbers. The random number generator will spread its output uniformly across the entire interval from 0 to 1 as we allow it to generate a long sequence of numbers. The results of many trials are represented by the density curve of a **uniform distribution.** This density curve appears in red in Figure 5.9. It has height 1 over the interval from 0 to 1, and height 0 everywhere else. The area under the density curve is 1: the area of a square with base 1 and height 1. The probability of any event is the area under the density curve and above the event in question.

As Figure 5.9(a) illustrates, the probability that the random number generator produces a number X between 0.3 and 0.7 is

$$P(0.3 \leq X \leq 0.7) = 0.4$$

because the area under the density curve and above the interval from 0.3 to 0.7 is 0.4. The height of the density curve is 1, and the area of a rectangle is the product of height and length, so the probability of any interval of outcomes is just the length of the interval.

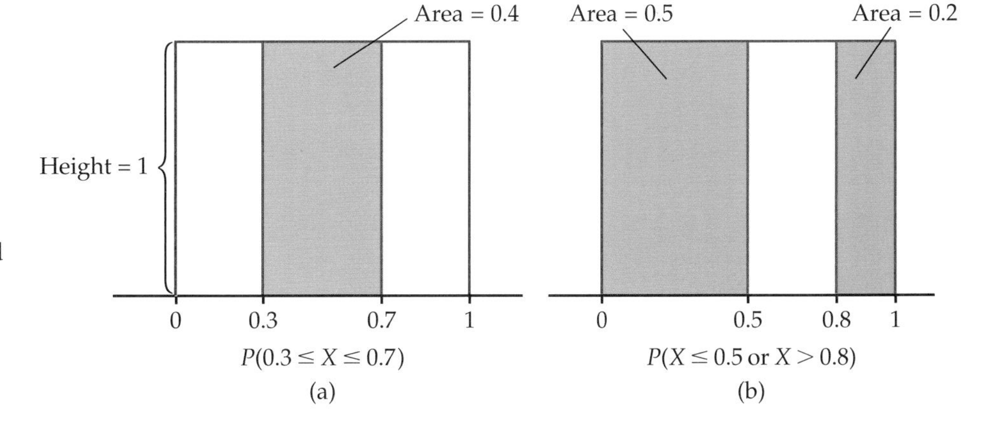

FIGURE 5.9 Assigning probabilities for generating a random number between 0 and 1, for Example 5.25. The probability of any interval of numbers is the area above the interval and under the density curve.

Similarly,

$$P(X \leq 0.5) = 0.5$$
$$P(X > 0.8) = 0.2$$
$$P(X \leq 0.5 \text{ or } X > 0.8) = 0.7$$

Notice that the last event consists of two nonoverlapping intervals, so the total area above the event is found by adding two areas, as illustrated by Figure 5.9(b). This assignment of probabilities obeys all our rules for probability.

USE YOUR KNOWLEDGE

5.42 Find the probability. For the uniform distribution described in Example 5.25, find the probability that X is between 0.1 and 0.4.

Probability as area under a density curve is a second important way of assigning probabilities to events. Figure 5.10 illustrates this idea in general form. We call X in Example 5.25 a *continuous random variable* because its values are not isolated numbers but an entire interval of numbers.

CONTINUOUS RANDOM VARIABLE

A **continuous random variable** X can take all values in an interval of numbers. The **probability distribution** of X is described by a density curve. The probability of any event is the area under the density curve and above the values of X that make up the event.

The probability model for a continuous random variable assigns probabilities to intervals of outcomes rather than to individual outcomes. In fact, all continuous probability distributions assign probability 0 to every individual outcome. Only intervals of values have positive probability. To see that this is true, consider a specific outcome such as $P(X = 0.8)$ in the context of Example 5.25. The probability of any interval is the same as its length. The point 0.8 has no length, so its probability is 0.

Although this fact may seem odd, it makes intuitive, as well as mathematical, sense. The random number generator produces a number between 0.79 and 0.81 with probability 0.02. An outcome between 0.799 and 0.801 has

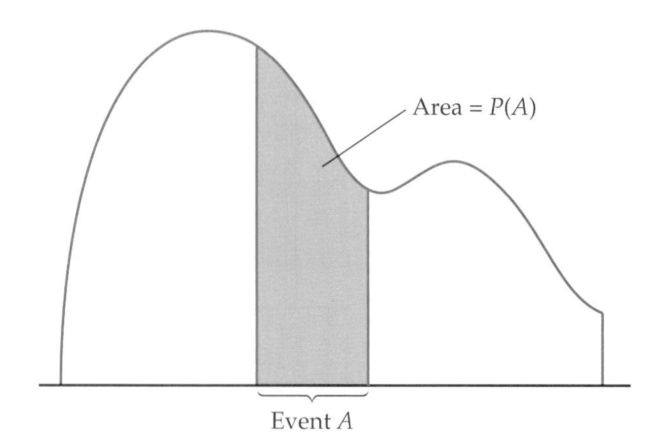

FIGURE 5.10 The probability distribution of a continuous random variable assigns probabilities as areas under a density curve. The total area under any density curve is 1.

probability 0.002. A result between 0.799999 and 0.800001 has probability 0.000002. You see that as we approach 0.8 the probability gets closer to 0. To be consistent, the probability of an outcome *exactly* equal to 0.8 must be 0. Because there is no probability exactly at $X = 0.8$, the two events $\{X > 0.8\}$ and $\{X \geq 0.8\}$ have the same probability. We can ignore the distinction between $>$ and $\geq$ when finding probabilities for continuous (but not discrete) random variables.

Normal distributions as probability distributions

The density curves that are most familiar to us are the Normal curves. Because any density curve describes an assignment of probabilities, *Normal distributions are probability distributions.* Recall that $N(\mu, \sigma)$ is our shorthand for the Normal distribution having mean μ and standard deviation σ. In the language of random variables, if X has the $N(\mu, \sigma)$ distribution, then the standardized variable

$$Z = \frac{X - \mu}{\sigma}$$

is a standard Normal random variable having the distribution $N(0, 1)$.

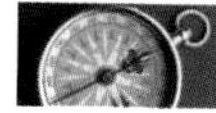

EXAMPLE 5.26

Texting while driving. Texting while driving can be dangerous, but young people want to remain connected. Suppose that 26% of teen drivers text while driving. If we take a sample of 500 teen drivers, what percent would we expect to say that they text while driving?[12]

The proportion $p = 0.26$ is a *parameter* that describes the population of teen drivers. The proportion $\hat{p}$ of the sample who say that they text while driving is a *statistic* used to estimate p. The statistic $\hat{p}$ is a random variable because repeating the SRS would give a different sample of 500 teen drivers and a different value of $\hat{p}$. We will see in the next chapter that $\hat{p}$ has approximately the $N(0.26, 0.0196)$ distribution. The mean 0.26 of this distribution is the same as the population parameter because $\hat{p}$ is an unbiased estimator of p. The standard deviation is controlled mainly by the size of the sample.

What is the probability that the survey result differs from the truth about the population by no more than 3 percentage points? We can use what we learned about Normal distribution calculations to answer this question.

LOOK BACK
Normal distribution calculations
p. 135

Because $p = 0.26$, the survey misses by no more than 3 percentage points if the sample proportion is within 0.23 and 0.29. Figure 5.11 shows this probability as an area under a Normal density curve. You can find it by software or by standardizing and using Table A. From Table A,

$$\begin{aligned} P(0.23 \leq \hat{p} \leq 0.29) &= P\left(\frac{0.23 - 0.26}{0.0196} \leq \frac{\hat{p} - 0.26}{0.0196} \leq \frac{0.29 - 0.26}{0.0196}\right) \\ &= P(-1.53 \leq Z \leq 1.53) \\ &= 0.9370 - 0.0630 = 0.8740 \end{aligned}$$

About 87% of the time, the sample $\hat{p}$ will be within 3 percentage points of the parameter p.

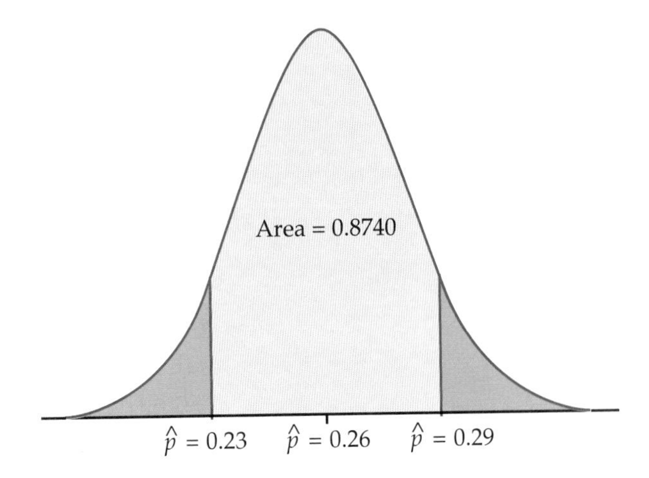

FIGURE 5.11 Probability in Example 5.26 as area under a Normal density curve.

We began this chapter with a general discussion of the idea of probability and the properties of probability models. Two very useful specific types of probability models are distributions of discrete and continuous random variables. In our study of statistics we will employ only these two types of probability models.

SECTION 5.3 SUMMARY

A **random variable** is a variable taking numerical values determined by the outcome of a random phenomenon. The **probability distribution** of a random variable X tells us what the possible values of X are and how probabilities are assigned to those values.

A random variable X and its distribution can be **discrete** or **continuous.**

A **discrete random variable** has finitely many possible values. The probability distribution assigns each of these values a probability between 0 and 1 such that the sum of all the probabilities is exactly 1. The probability of any event is the sum of the probabilities of all the values that make up the event.

A **continuous random variable** takes all values in some interval of numbers. A **density curve** describes the probability distribution of a continuous random variable. The probability of any event is the area under the curve and above the values that make up the event.

Normal distributions are one type of continuous probability distribution.

You can picture a probability distribution by drawing a **probability histogram** in the discrete case or by graphing the density curve in the continuous case.

SECTION 5.3 EXERCISES

For Exercise 5.40, see page 258; for Exercise 5.41, see page 260; and for Exercise 5.42, see page 262.

5.43. What's wrong? In each of the following scenarios, there is something wrong. Describe what is wrong and give a reason for your answer.

(a) The probabilities for a discrete random variable can be negative.

(b) A continuous random variable can take any value between 0 and 1.

(c) Normal distributions are discrete random variables.

5.44. Use of Twitter. Suppose that the population proportion of Internet users who say that they use Twitter or another service to post updates about themselves or to see updates about others is 19%.[13] Think about selecting random samples from a population in which 19% are Twitter users.

(a) Describe the sample space for selecting a single person.

(b) If you select three people, describe the sample space.

(c) Using the results of (b), define the sample space for the random variable that expresses the number of Twitter users in the sample of size 3.

(d) What information is contained in the sample space for part (b) that is not contained in the sample space for part (c)? Do you think this information is important? Explain your answer.

5.45. Use of Twitter. Find the probabilities for parts (a), (b), and (c) of the previous exercise.

5.46. Households and families in government data. In government data, a household consists of all occupants of a dwelling unit, while a family consists of two or more persons who live together and are related by blood or marriage. So all families form households, but some households are not families. Here are the distributions of household size and of family size in the United States:

Number of persons	1	2	3	4	5	6	7
Household probability	0.27	0.33	0.16	0.14	0.06	0.03	0.01
Family probability	0	0.44	0.22	0.20	0.09	0.03	0.02

Make probability histograms for these two discrete distributions, using the same scales. What are the most important differences between the sizes of households and families?

5.47. Discrete or continuous. In each of the following situations decide if the random variable is discrete or continuous and give a reason for your answer.

(a) Your Web page has five different links, and a user can click on one of the links or can leave the page. You record the length of time that a user spends on the Web page before clicking one of the links or leaving the page.

(b) The number of hits on your Web page.

(c) The yearly income of a visitor to your Web page.

5.48. Texas hold 'em. The game of Texas hold 'em starts with each player receiving two cards. Here is the probability distribution for the number of aces in two-card hands:

Number of aces	0	1	2
Probability	0.8507	0.1448	0.0045

(a) Verify that this assignment of probabilities satisfies the requirement that the sum of the probabilities for a discrete distribution must be 1.

(b) Make a probability histogram for this distribution.

(c) What is the probability that a hand contains at least one ace? Show two different ways to calculate this probability.

5.49. Spell-checking software. Spell-checking software catches "nonword errors," which result in a string of letters that is not a word, as when "the" is typed as "teh." When undergraduates are asked to write a 250-word essay (without spell-checking), the number X of nonword errors has the following distribution:

Value of X	0	1	2	3	4
Probability	0.1	0.3	0.3	0.2	0.1

(a) Sketch the probability distribution for this random variable.

(b) Write the event "at least one nonword error" in terms of X. What is the probability of this event?

(c) Describe the event $X \leq 2$ in words. What is its probability? What is the probability that $X < 2$?

5.50. Length of human pregnancies. The length of human pregnancies from conception to birth varies according to a distribution that is approximately Normal with mean 266 days and standard deviation 16 days. Call the length of a randomly chosen pregnancy Y.

(a) Make a sketch of the density curve for this random variable.

(b) What is $P(Y \leq 280)$?

5.51. Tossing two dice. Some games of chance rely on tossing two dice. Each die has six faces, marked with 1, 2, ..., 6 spots. The dice used in casinos are carefully balanced so that each face is equally likely to come up. When two dice are tossed, each of the 36 possible pairs of faces is equally likely to come up. The outcome of interest to a gambler is the sum of the spots on the two up-faces. Call this random variable X.

(a) Write down all 36 possible pairs of faces.

(b) If all pairs have the same probability, what must be the probability of each pair?

(c) Write the value of X next to each pair of faces and use this information with the result of (b) to give the probability distribution of X. Draw a probability histogram to display the distribution.

(d) One bet available in craps wins if a 7 or an 11 comes up on the next roll of two dice. What is the probability of rolling a 7 or an 11 on the next roll?

(e) Several bets in craps lose if a 7 is rolled. If any outcome other than 7 occurs, these bets either win or continue to the next roll. What is the probability that anything other than a 7 is rolled?

5.52. CHALLENGE **Nonstandard dice.** Nonstandard dice can produce interesting distributions of outcomes. You have two balanced, six-sided dice. One is a standard die, with faces having 1, 2, 3, 4, 5, and 6 spots. The other die has three faces with 0 spots and three faces with 6 spots. Find the probability distribution for the total number of spots Y on the up-faces when you roll these two dice.

5.53. Uniform random numbers. Let X be a random number between 0 and 1 produced by the idealized uniform random number generator described in Example 5.25. Find the following probabilities:

(a) $P(X < 0.6)$ **(b)** $P(X \leq 0.6)$

(c) What important fact about continuous random variables does comparing your answers to parts (a) and (b) illustrate?

5.54. Foreign-born residents of California. The Census Bureau reports that 27% of California residents are foreign-born. Suppose that you choose three Californians at random, so that each has probability 0.27 of being foreign-born and the three are independent of each other. Let the random variable W be the number of foreign-born people you chose.

(a) What are the possible values of W?

(b) Look at your three people in order. There are eight possible arrangements of foreign (F) and domestic (D) birth. For example, FFD means the first two are foreign-born and the third is not. All eight arrangements are not equally likely. What is the probability of each one?

(c) What is the value of W for each arrangement in (b)? What is the probability of each possible value of W? (This is the distribution of a Yes/No response for an SRS of size 3. In principle, the same idea works for an SRS of any size.)

5.55. Uniform numbers between 0 and 2. Many random number generators allow users to specify the range of the random numbers to be produced. Suppose that you specify that the range is to be all numbers between 0 and 2. Call the random number generated Y. Then the density curve of the random variable Y has constant height between 0 and 2, and height 0 elsewhere.

(a) What is the height of the density curve between 0 and 2? Draw a graph of the density curve.

(b) Use your graph from (a) and the fact that probability is area under the curve to find $P(Y \leq 1.6)$.

(c) Find $P(0.5 < Y < 1.7)$.

(d) Find $P(Y \geq 0.95)$.

5.56. The sum of two uniform random numbers. Generate *two* random numbers between 0 and 1 and take Y to be their sum. Then Y is a continuous random variable that can take any value between 0 and 2. The density curve of Y is the triangle shown in Figure 5.12.

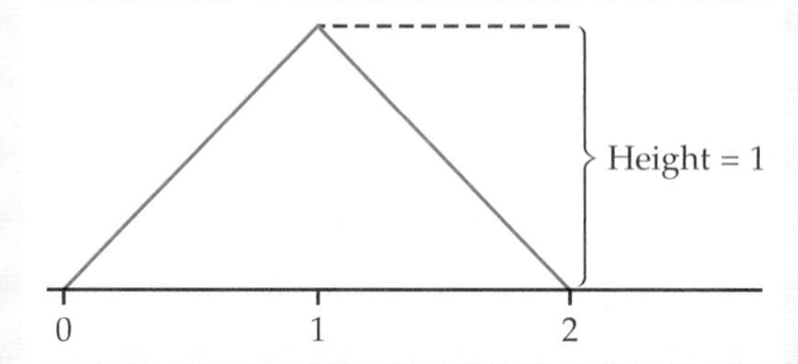

FIGURE 5.12 The density curve for the sum Y of two random numbers, for Exercise 5.56.

(a) Verify by geometry that the area under this curve is 1.

(b) What is the probability that Y is less than 1? Sketch the density curve, shade the area that represents the probability, then find that area. Do this for (c) also.

(c) What is the probability that Y is greater than 0.6?

5.57. How many close friends? How many close friends do you have? Suppose that the number of close friends adults claim to have varies from person to person with mean $\mu = 9$ and standard deviation $\sigma = 2.4$. An opinion poll asks this question of an SRS of 1100 adults. We will see in the next chapter that in this situation the sample mean response $\bar{x}$ has approximately the Normal distribution with mean 9 and standard deviation 0.0724. What is $P(8 \leq \bar{x} \leq 10)$, the probability that the statistic $\bar{x}$ estimates the parameter μ to within ± 1?

5.58. Normal approximation for a sample proportion. A sample survey contacted an SRS of 700 registered voters in Oregon shortly after an election and asked respondents whether they had voted. Voter records show that 56% of registered voters had actually voted. We will see in the next chapter that in this situation the proportion $\hat{p}$ of the sample who voted has approximately the Normal distribution with mean $\mu = 0.56$ and standard deviation $\sigma = 0.019$.

(a) If the respondents answer truthfully, what is $P(0.52 \leq \hat{p} \leq 0.60)$? This is the probability that the statistic $\hat{p}$ estimates the parameter 0.56 within plus or minus 0.04.

(b) In fact, 72% of the respondents said they had voted ($\hat{p} = 0.72$). If respondents answer truthfully, what is $P(\hat{p} \geq 0.72)$? This probability is so small that it is good evidence that some people who did not vote claimed that they did vote.

5.4 Means and Variances of Random Variables

The probability histograms and density curves that picture the probability distributions of random variables resemble our earlier pictures of distributions of data. In describing data, we moved from graphs to numerical measures such as means and standard deviations. Now we will make the same move to expand our descriptions of the distributions of random variables. We can speak of the mean winnings in a game of chance or the standard deviation of the randomly varying number of calls a travel agency receives in an hour. In this section we will learn more about how to compute these descriptive measures and about the laws they obey.

The mean of a random variable

The mean $\bar{x}$ of a set of observations is their ordinary average. The mean of a random variable X is also an average of the possible values of X, but with an important change to take into account the fact that not all outcomes need be equally likely. An example will show what we must do.

EXAMPLE 5.27

The Tri-State Pick 3 lottery. Most states and Canadian provinces have government-sponsored lotteries. Here is a simple lottery wager, from the Tri-State Pick 3 game that New Hampshire shares with Maine and Vermont. You choose a three-digit number, 000 to 999. The state chooses a three-digit winning number at random and pays you \$500 if your number is chosen. Because there are 1000 three-digit numbers, you have probability 1/1000 of winning. Taking X to be the amount your ticket pays you, the probability distribution of X is

Payoff X	\$0	\$500
Probability	0.999	0.001

What is your average payoff from many tickets? The ordinary average of the two possible outcomes \$0 and \$500 is \$250, but that makes no sense as the average because \$500 is much less likely than \$0. In the long run you receive \$500 once in every 1000 tickets and \$0 on the remaining 999 of 1000 tickets. The long-run average payoff is

$$\$500\frac{1}{1000} + \$0\frac{999}{1000} = \$0.50$$

or 50 cents. That number is the mean of the random variable X. (Tickets cost \$1, so in the long run the state keeps half the money you wager.)

If you play Tri-State Pick 3 several times, we would as usual call the mean of the actual amounts you win $\bar{x}$. The mean in Example 5.27 is a different quantity—it is the long-run average winnings you expect if you play a very large number of times.

USE YOUR KNOWLEDGE

5.59 Find the mean of the probability distribution. You toss a fair coin. If the outcome is heads, you win \$1.00. If the outcome is tails, you win nothing. Let X be the amount that you win in a single toss of a coin. Find the probability distribution of this random variable and its mean.

mean μ
p. 130

Just as probabilities are an idealized description of long-run proportions, the mean of a probability distribution describes the long-run average outcome. We can't call this mean $\bar{x}$, so we need a different symbol. The common symbol for the **mean of a probability distribution** is μ, the Greek letter mu. We used μ in Chapter 3 for the mean of a Normal distribution, so this is not a new notation. We will often be interested in several random variables, each having a different probability distribution with a different mean. To remind ourselves that we are talking about the mean of X, we often write μ_X rather than simply μ. In Example 5.27, $\mu_X = \$0.50$.

expected value

Notice that, as often happens, the mean is not a possible value of X. You will often find the mean of a random variable X called the **expected value** of X. This term can be misleading, for we don't necessarily expect one observation on X to be close to the expected value of X.

The mean of any discrete random variable is found just as in Example 5.27. It is an average of the possible outcomes—but it is a weighted average in which each outcome is weighted by its probability. Because the probabilities add to 1, we have total weight 1 to distribute among the outcomes. An outcome that occurs half the time has probability one-half and gets one-half the weight in calculating the mean. Here is the general definition.

MEAN OF A DISCRETE RANDOM VARIABLE

Suppose that X is a discrete random variable whose distribution is

Value of X	x_1	x_2	x_3	$\cdots$	x_k
Probability	p_1	p_2	p_3	$\cdots$	p_k

To find the **mean** of X, multiply each possible value by its probability, then add all the products:

$$\mu_X = x_1p_1 + x_2p_2 + \cdots + x_kp_k$$
$$= \sum x_ip_i$$

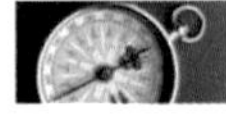

EXAMPLE 5.28

The mean of equally likely first digits. If first digits in a set of data all have the same probability, the probability distribution of the first digit X is then

First digit X	1	2	3	4	5	6	7	8	9
Probability	1/9	1/9	1/9	1/9	1/9	1/9	1/9	1/9	1/9

The mean of this distribution is

$$\mu_X = \left(1 \times \frac{1}{9}\right) + \left(2 \times \frac{1}{9}\right) + \left(3 \times \frac{1}{9}\right) + \left(4 \times \frac{1}{9}\right) + \left(5 \times \frac{1}{9}\right) + \left(6 \times \frac{1}{9}\right)$$
$$+ \left(7 \times \frac{1}{9}\right) + \left(8 \times \frac{1}{9}\right) + \left(9 \times \frac{1}{9}\right)$$
$$= 45 \times \frac{1}{9} = 5$$

Suppose that the random digits in Example 5.28 had a different probability distribution. In Example 5.12 (page 246) we described Benford's law as a probability distribution that describes first digits of numbers in many real situations. Let's calculate the mean for Benford's law.

EXAMPLE 5.29

The mean of the first digits that follow Benford's law. Here is the distribution of the first digit for data that follow Benford's law. We use the letter V for this random variable to distinguish it from the one that we studied in Example 5.28. The distribution of V is

First digit V	1	2	3	4	5	6	7	8	9
Probability	0.301	0.176	0.125	0.097	0.079	0.067	0.058	0.051	0.046

The mean of V is

$$\mu_V = (1)(0.301) + (2)(0.176) + (3)(0.125) + (4)(0.097) + (5)(0.079)$$
$$+ (6)(0.067) + (7)(0.058) + (8)(0.051) + (9)(0.046)$$
$$= 3.441$$

The mean reflects the greater probability of smaller first digits under Benford's law than when first digits 1 to 9 are equally likely.

Figure 5.13 locates the means of X and V on the two probability histograms. Because the discrete uniform distribution of Figure 5.13(a) is symmetric, the mean lies at the center of symmetry. We can't locate the mean of the right-skewed distribution of Figure 5.13(b) by eye—calculation is needed.

What about continuous random variables? The probability distribution of a continuous random variable X is described by a density curve. The mean of the distribution is the point at which the area under the density curve would balance if it were made out of solid material. The mean lies at the center of

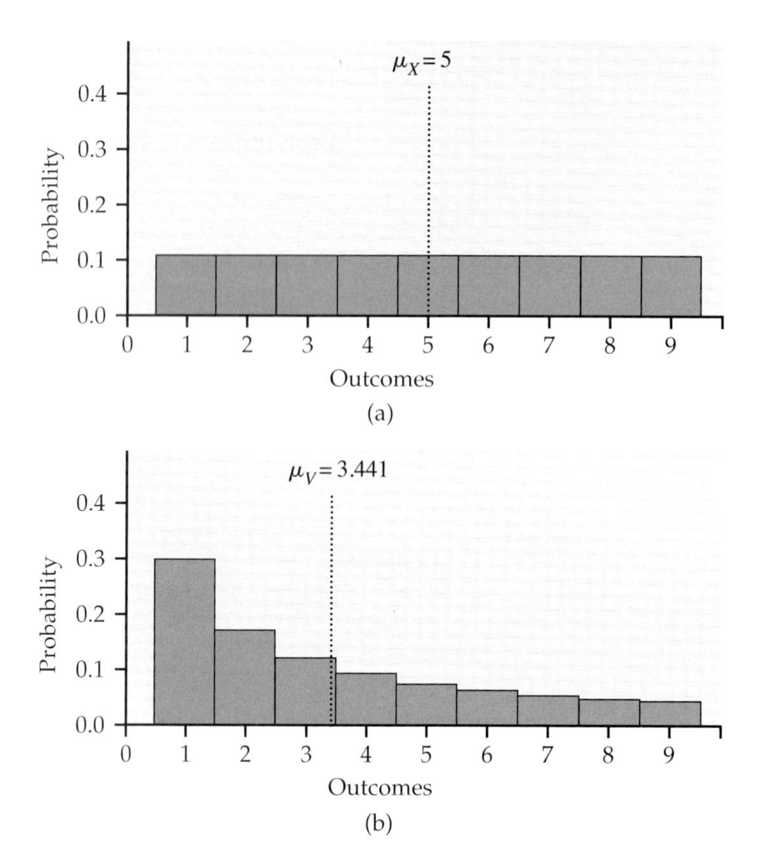

FIGURE 5.13 Locating the mean of a discrete random variable on the probability histogram for (a) digits between 1 and 9 chosen at random; (b) digits between 1 and 9 chosen from records that obey Benford's law.

symmetric density curves such as the Normal curves. Exact calculation of the mean of a distribution with a skewed density curve requires advanced mathematics.[14] The idea that the mean is the balance point of the distribution applies to discrete random variables as well. In the discrete case we have a formula that gives us this point.

Statistical estimation and the law of large numbers

We would like to estimate the mean height μ of the population of all American women between the ages of 18 and 24 years. This μ is the mean μ_X of the random variable X obtained by choosing a young woman at random and measuring her height. To estimate μ, we choose an SRS of young women and use the sample mean $\bar{x}$ to estimate the unknown population mean μ.

Statistics obtained from probability samples are random variables because their values vary in repeated sampling. The distributions of statistics are just the probability distributions of these random variables.

It seems reasonable to use $\bar{x}$ to estimate μ. An SRS should fairly represent the population, so the mean $\bar{x}$ of the sample should be somewhere near the mean μ of the population. Of course, we don't expect $\bar{x}$ to be exactly equal to μ, and we realize that if we choose another SRS, the luck of the draw will probably produce a different $\bar{x}$.

If $\bar{x}$ is rarely exactly right and varies from sample to sample, why is it nonetheless a reasonable estimate of the population mean μ? If we keep on adding observations to our random sample, the statistic $\bar{x}$ is *guaranteed* to get as close as we wish to the parameter μ and then stay that close. We have the comfort of knowing that if we can afford to keep on measuring more women, eventually we will estimate the mean height of all young women very accurately.

This remarkable fact is called the *law of large numbers.* It is remarkable because it holds for *any* population, not just for some special class such as Normal distributions.

> **LAW OF LARGE NUMBERS**
>
> Draw independent observations at random from any population with finite mean μ. Decide how accurately you would like to estimate μ. As the number of observations drawn increases, the mean $\bar{x}$ of the observed values eventually approaches the mean μ of the population as closely as you specified and then stays that close.

The behavior of $\bar{x}$ is similar to the idea of probability. In the long run, the *proportion* of outcomes taking any value gets close to the *probability* of that value, and the *average outcome* gets close to the distribution *mean.* Figure 5.1 (page 236) shows how proportions approach probability in one example. Here is an example of how sample means approach the distribution mean.

EXAMPLE 5.30

Heights of young women. The distribution of the heights of all young women is close to the Normal distribution with mean 64.5 inches and standard deviation 2.5 inches. Suppose that $\mu = 64.5$ were exactly true. Figure 5.14 shows the behavior of the mean height $\bar{x}$ of n women chosen at random from a population whose heights follow the $N(64.5, 2.5)$ distribution. The graph plots the values of $\bar{x}$ as we add women to our sample. The first woman drawn had

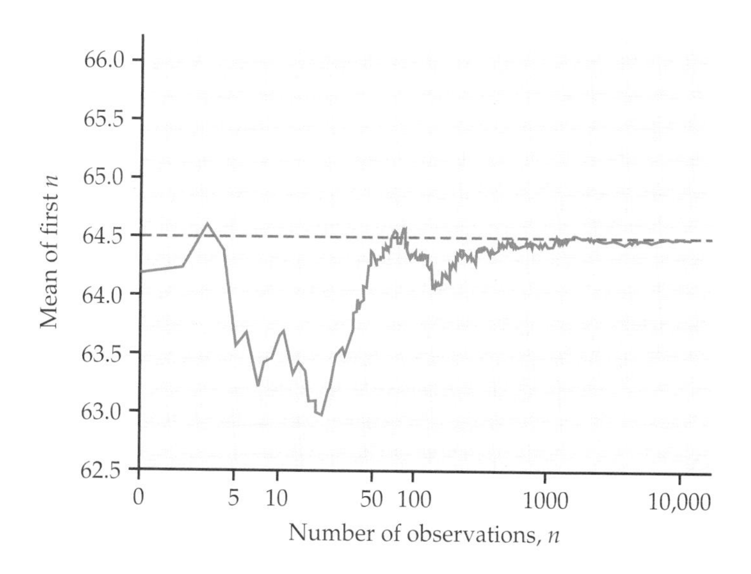

FIGURE 5.14 The law of large numbers in action, for Example 5.30. As we take more observations, the sample mean always approaches the mean of the population.

height 64.21 inches, so the line starts there. The second had height 64.35 inches, so for $n = 2$ the mean is

$$\bar{x} = \frac{64.21 + 64.35}{2} = 64.28$$

This is the second point on the line in the graph.

At first, the graph shows that the mean of the sample changes as we take more observations. Eventually, however, the mean of the observations gets close to the population mean $\mu = 64.5$ and settles down at that value. The law of large numbers says that this *always* happens.

USE YOUR KNOWLEDGE

5.60 Use the *Law of Large Numbers* applet. The *Law of Large Numbers* applet animates a graph like Figure 5.14. Use it to better understand the law of large numbers by making a similar graph.

The mean μ of a random variable is the average value of the variable in two senses. By its definition, μ is the average of the possible values, weighted by their probability of occurring. The law of large numbers says that μ is also the long-run average of many independent observations on the variable.

Thinking about the law of large numbers

The law of large numbers says broadly that the average results of many independent observations are stable and predictable. The gamblers in a casino may win or lose, but the casino will win in the long run because the law of large numbers says what the average outcome of many thousands of bets will be. An insurance company deciding how much to charge for life insurance and a fast-food restaurant deciding how many beef patties to prepare also rely on the fact that averaging over many individuals produces a stable result. It is worth the effort to think a bit more closely about so important a fact.

There is no "law of small numbers"

Both the rules of probability and the law of large numbers describe the regular behavior of chance phenomena *in the long run.* Psychologists have discovered that our intuitive understanding of randomness is quite different from the true laws of chance.[15] For example, most people believe in an incorrect "law of small numbers." That is, we expect even short sequences of random events to show the kind of average behavior that in fact appears only in the long run.

Some teachers of statistics begin a course by asking students to toss a coin 50 times and bring the sequence of heads and tails to the next class. The teacher then announces which students just wrote down a random-looking sequence rather than actually tossing a coin. The faked tosses don't have enough "runs" of consecutive heads or consecutive tails.

Runs of the same outcome don't look random to us but are in fact common. For example, the probability of a run of three or more consecutive heads or tails in just 10 tosses is greater than 0.8.[16] The runs of consecutive heads or consecutive tails that appear in real coin tossing (and that are predicted by the mathematics of probability) seem surprising to us. Because we don't expect to see long runs, we may conclude that the coin tosses are not independent or that some influence is disturbing the random behavior of the coin.

EXAMPLE 5.31

The "hot hand" in basketball. Belief in the law of small numbers influences behavior. If a basketball player makes several consecutive shots, both the fans and her teammates believe that she has a "hot hand" and is more likely to make the next shot. This is doubtful. Careful study suggests that runs of baskets made or missed are no more frequent in basketball than would be expected if each shot were independent of the player's previous shots. Baskets made or missed are just like heads and tails in tossing a coin. (Of course, some players make 30% of their shots in the long run and others make 50%, so a coin-toss model for basketball must allow coins with different probabilities of a head.) Our perception of hot or cold streaks simply shows that we don't perceive random behavior very well.[17]

Our intuition doesn't do a good job of distinguishing random behavior from systematic influences. This is also true when we look at data. We need statistical inference to supplement exploratory analysis of data because probability calculations can help verify that what we see in the data is more than a random pattern.

How large is a large number?

The law of large numbers says that the actual mean outcome of many trials gets close to the distribution mean μ as more trials are made. It doesn't say how many trials are needed to guarantee a mean outcome close to μ. That depends on the *variability* of the random outcomes.

The more variable the outcomes, the more trials are needed to ensure that the mean outcome $\bar{x}$ is close to the distribution mean μ. This is something that casinos understand well. The outcomes of games of chance are variable enough to hold the interest of gamblers. Only the casino plays often enough to rely on the law of large numbers. Gamblers get entertainment. The casino has a business.

Rules for means

You are studying flaws in the painted finish of refrigerators made by your firm. Dimples and paint sags are two kinds of surface flaw. Not all refrigerators have the same number of dimples: many have none, some have one, some two, and so on. You ask for the average number of imperfections on a refrigerator. The inspectors report finding an average of 0.7 dimples and 1.4 sags per refrigerator. How many total imperfections of both kinds (on the average) are there on a refrigerator? That's easy: if the average number of dimples is 0.7 and the average number of sags is 1.4, then counting both gives an average of $0.7 + 1.4 = 2.1$ flaws.

In more formal language, the number of dimples on a refrigerator is a random variable X that varies as we inspect one refrigerator after another. We know only that the mean number of dimples is $\mu_X = 0.7$. The number of paint sags is a second random variable Y, having mean $\mu_Y = 1.4$. (As usual, the subscripts keep straight which variable we are talking about.) The total number of both dimples and sags is another random variable, the sum $X+Y$. Its mean μ_{X+Y} is the average number of dimples and sags together. It is just the sum of the individual means μ_X and μ_Y. That's an important rule for how means of random variables behave.

Here's another rule. The crickets living in a field have mean length 1.2 inches. What is the mean in centimeters? There are 2.54 centimeters in an inch, so the length of a cricket in centimeters is 2.54 times its length in inches. If we multiply every observation by 2.54, we also multiply their average by 2.54. The mean in centimeters must be 2.54×1.2, or about 3.05 centimeters. More formally, the length in inches of a cricket chosen at random from the field is a random variable X with mean μ_X. The length in centimeters is $2.54X$, and this new random variable has mean $2.54\mu_X$.

The point of these examples is that means behave like averages. Here are the rules we need.

RULES FOR MEANS

Rule 1. If X is a random variable and a and b are fixed numbers, then

$$\mu_{a+bX} = a + b\mu_X$$

Rule 2. If X and Y are random variables, then

$$\mu_{X+Y} = \mu_X + \mu_Y$$

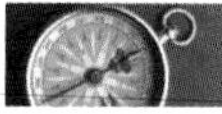

EXAMPLE 5.32

Sales of cars, trucks, and SUVs. Brianna is a sales associate at a large auto dealership. At her commission rate of 25% of gross profit on each vehicle she sells, Brianna expects to earn \$350 for each car sold and \$400 for each truck or SUV sold. Brianna motivates herself by using probability estimates of her sales. For a sunny Saturday in April, she estimates her car sales as follows:

Cars sold	0	1	2	3
Probability	0.3	0.4	0.2	0.1

Brianna's estimate of her truck or SUV sales is

Vehicles sold	0	1	2
Probability	0.4	0.5	0.1

Take X to be the number of cars Brianna sells and Y the number of trucks or SUVs. The means of these random variables are

$$\begin{aligned}\mu_X &= (0)(0.3) + (1)(0.4) + (2)(0.2) + (3)(0.1)\\ &= 1.1 \text{ cars}\\ \mu_Y &= (0)(0.4) + (1)(0.5) + (2)(0.1)\\ &= 0.7 \text{ trucks or SUVs}\end{aligned}$$

Brianna's earnings, at \$350 per car and \$400 per truck or SUV, are

$$Z = 350X + 400Y$$

Combining Rules 1 and 2, her mean earnings are

$$\begin{aligned}\mu_Z &= 350\mu_X + 400\mu_Y \\ &= (350)(1.1) + (400)(0.7) = \$665\end{aligned}$$

This is Brianna's best estimate of her earnings for the day. It's a bit unusual for individuals to use probability estimates, but they are a common tool for business planners.

personal probability

The probabilities in Example 5.32 are **personal probabilities** that describe Brianna's informed opinion about her sales in the coming weekend. Although personal probabilities need not be based on observing many repetitions of a random phenomenon, they must obey the rules of probability if they are to make sense. Personal probability extends the usefulness of probability models to onetime events, but remember that they are subject to the follies of human opinion. Overoptimism is common: 40% of college students think that they will eventually reach the top 1% in income.

USE YOUR KNOWLEDGE

5.61 Find μ_Y. The random variable X has mean $\mu_X = 10$. If $Y = 15 + 8X$, what is μ_Y?

5.62 Find μ_W. The random variable U has mean $\mu_U = 20$, and the random variable V has mean $\mu_V = 20$. If $W = 0.5U + 0.5V$, find μ_W.

The variance of a random variable

The mean is a measure of the center of a distribution. A basic numerical description also requires a measure of the spread, or variability, of the distribution. The variance and the standard deviation are the measures of spread that accompany the choice of the mean to measure center.

Just as for the mean, we need a distinct symbol to distinguish the variance of a random variable from the variance s^2 of a data set. We write the variance of a random variable X as σ_X^2. Once again the subscript reminds us which variable we have in mind.

The definition of the variance σ_X^2 of a random variable is similar to the definition of the sample variance s^2 given in Chapter 3. That is, the variance is an average value of the squared deviation $(X - \mu_X)^2$ of the variable X from its mean μ_X.

Just as for the mean, the average we use is a weighted average in which each outcome is weighted by its probability in order to take account of outcomes that are not equally likely. Calculating this weighted average is straightforward for discrete random variables but requires advanced mathematics in the continuous case. Here is the definition.

VARIANCE OF A DISCRETE RANDOM VARIABLE

Suppose that X is a discrete random variable whose distribution is

Value of X	x_1	x_2	x_3	$\cdots$	x_k
Probability	p_1	p_2	p_3	$\cdots$	p_k

and that μ_X is the mean of X. The **variance** of X is

$$\begin{aligned}\sigma_X^2 &= (x_1 - \mu_X)^2 p_1 + (x_2 - \mu_X)^2 p_2 + \cdots + (x_k - \mu_X)^2 p_k \\ &= \sum (x_i - \mu_X)^2 p_i\end{aligned}$$

The **standard deviation** σ_X of X is the square root of the variance.

EXAMPLE 5.33

Find the mean and the variance. In Example 5.32 we saw that the number X of cars that Brianna hopes to sell has distribution

Cars sold	0	1	2	3
Probability	0.3	0.4	0.2	0.1

We can find the mean and variance of X by arranging the calculation in the form of a table. Both μ_X and σ_X^2 are sums of columns in this table.

x_i	p_i	$x_i p_i$	$(x_i - \mu_X)^2 p_i$		
0	0.3	0.0	$(0 - 1.1)^2(0.3)$	=	0.363
1	0.4	0.4	$(1 - 1.1)^2(0.4)$	=	0.004
2	0.2	0.4	$(2 - 1.1)^2(0.2)$	=	0.162
3	0.1	0.3	$(3 - 1.1)^2(0.1)$	=	0.361
		$\mu_X = 1.1$	σ_X^2	=	0.890

We see that $\sigma_X^2 = 0.89$. The standard deviation of X is $\sigma_X = \sqrt{0.89} = 0.943$. The standard deviation is a measure of the variability of the number of cars Brianna sells. As in the case of distributions for data, the standard deviation of a probability distribution is easiest to understand for Normal distributions.

USE YOUR KNOWLEDGE

5.63 Find the variance and the standard deviation. The random variable X has the following probability distribution:

Value of X	0	2
Probability	0.5	0.5

Find the variance σ_X^2 and the standard deviation σ_X for this random variable.

Rules for variances and standard deviations

What are the facts for variances that parallel Rules 1 and 2 for means? *The mean of a sum of random variables is always the sum of their means, but this addition rule is true for variances only in special situations.*

To understand why, take X to be the percent of a family's after-tax income that is spent and Y the percent that is saved. When X increases, Y decreases by the same amount. Though X and Y may vary widely from year to year, their sum $X + Y$ is always 100% and does not vary at all. Because of this association between the variables X and Y, their variances do not add.

LOOK BACK
independent
p. 249

If random variables are independent, this kind of association between their values is ruled out and their variances do add. Probability models often assume independence when the random variables describe outcomes that appear unrelated to each other. You should ask in each instance whether the assumption of independence seems reasonable.

LOOK BACK
correlation
p. 171

When random variables are not independent, the variance of their sum depends on the correlation between them as well as on their individual variances. In Chapter 4, we met the correlation r between two observed variables measured on the same individuals. We defined the correlation r as an average of the products of the standardized x and y observations.

The correlation between two random variables is defined in the same way, once again using a weighted average with probabilities as weights. We use ρ, the Greek letter rho, for the correlation between two random variables. The correlation ρ is a number between -1 and 1 that measures the direction and strength of the linear relationship between two variables. **The correlation between two independent random variables is zero.**

Returning to family finances, if X is the percent of a family's after-tax income that is spent and Y the percent that is saved, then $Y = 100 - X$. This is a perfect linear relationship with a negative slope, so the correlation between X and Y is $\rho = -1$. With the correlation at hand, we can state the rules for manipulating variances.

RULES FOR VARIANCES AND STANDARD DEVIATIONS

Rule 1. If X is a random variable and a and b are fixed numbers, then

$$\sigma_{a+bX}^2 = b^2\sigma_X^2$$

Rule 2. If X and Y are independent random variables, then

$$\sigma^2_{X+Y} = \sigma^2_X + \sigma^2_Y$$
$$\sigma^2_{X-Y} = \sigma^2_X + \sigma^2_Y$$

This is the **addition rule for variances of independent random variables.**

Rule 3. If X and Y have correlation ρ, then

$$\sigma^2_{X+Y} = \sigma^2_X + \sigma^2_Y + 2\rho\sigma_X\sigma_Y$$
$$\sigma^2_{X-Y} = \sigma^2_X + \sigma^2_Y - 2\rho\sigma_X\sigma_Y$$

This is the **general addition rule for variances of random variables.**

To find the standard deviation, take the square root of the variance.

Because a variance is the average of squared deviations from the mean, multiplying X by a constant b multiplies σ^2_X by the square of the constant. Adding a constant a to a random variable changes its mean but does not change its variability. The variance of $X+a$ is therefore the same as the variance of X. Because the square of -1 is 1, the addition rule says that the variance of a difference of independent random variables is the *sum* of the variances. For independent random variables, the difference $X-Y$ is more variable than either X or Y alone because variations in both X and Y contribute to variation in their difference.

As with data, we prefer the standard deviation to the variance as a measure of the variability of a random variable. *Rule 2 for variances implies that standard deviations of independent random variables do* not *add. To combine standard deviations, use the rules for variances.* For example, the standard deviations of $2X$ and $-2X$ are both equal to $2\sigma_X$ because this is the square root of the variance $4\sigma^2_X$.

EXAMPLE 5.34

Payoff in the Tri-State Pick 3 lottery. The payoff X of a \$1 ticket in the Tri-State Pick 3 game is \$500 with probability 1/1000 and 0 the rest of the time. Here is the combined calculation of mean and variance:

x_i	p_i	$x_i p_i$	$(x_i - \mu_X)^2 p_i$		
0	0.999	0	$(0 - 0.5)^2(0.999)$	=	0.24975
500	0.001	0.5	$(500 - 0.5)^2(0.001)$	=	249.50025
		$\mu_X = 0.5$	σ^2_X	=	249.75

The mean payoff is 50 cents. The standard deviation is $\sigma_X = \sqrt{249.75} =$ \$15.80. It is usual for games of chance to have large standard deviations because large variability makes gambling exciting.

If you buy a Pick 3 ticket, your winnings are $W = X - 1$ because the dollar you paid for the ticket must be subtracted from the payoff. Let's find the mean and variance for this random variable.

EXAMPLE 5.35

Winnings in the Tri-State Pick 3 lottery. By the rules for means, the mean amount you win is

$$\mu_W = \mu_X - 1 = -\$0.50$$

That is, you lose an average of 50 cents on a ticket. The rules for variances remind us that the variance and standard deviation of the winnings $W = X-1$ are the same as those of X. Subtracting a fixed number changes the mean but not the variance.

Suppose now that you buy a \$1 ticket on each of two different days. The payoffs X and Y on the two tickets are independent because separate drawings are held each day. Your total payoff is $X + Y$. Let's find the mean and standard deviation for this payoff.

EXAMPLE 5.36

Two tickets. The mean for the payoff for the two tickets is

$$\mu_{X+Y} = \mu_X + \mu_Y = \$0.50 + \$0.50 = \$1.00$$

Because X and Y are independent, the variance of $X + Y$ is

$$\sigma^2_{X+Y} = \sigma^2_X + \sigma^2_Y = 249.75 + 249.75 = 499.5$$

The standard deviation of the total payoff is

$$\sigma_{X+Y} = \sqrt{499.5} = \$22.35$$

This is not the same as the sum of the individual standard deviations, which is $\$15.80 + \$15.80 = \$31.60$. Variances of independent random variables add. Standard deviations do not.

When we add random variables that are correlated, we need to use the correlation for the calculation of the variance, but not for the calculation of the mean. Here is an example.

EXAMPLE 5.37

Utility bills. Consider a household where the monthly bill for natural gas averages \$125 with a standard deviation of \$75, while the monthly bill for electricity averages \$174 with a standard deviation of \$41. The correlation between the two bills is −0.55.

Let's compute the mean and standard deviation of the sum of the natural-gas bill and the electricity bill. We let X stand for the natural-gas bill and Y stand for the electricity bill. Then the total is $X+Y$. Using the rules for means, we have

$$\mu_{X+Y} = \mu_X + \mu_Y = 125 + 174 = 299$$

To find the standard deviation we first find the variance and then take the square root to determine the standard deviation. From the general addition rule for variances of random variables,

$$\begin{aligned}\sigma^2_{X+Y} &= \sigma^2_X + \sigma^2_Y + 2\rho\sigma_X\sigma_Y \\ &= (75)^2 + (41)^2 + (2)(-0.55)(75)(41) \\ &= 3923\end{aligned}$$

Therefore, the standard deviation is

$$\sigma_{X+Y} = \sqrt{3923} = 63$$

The total of the natural gas bill and the electricity bill has mean \$299 and standard deviation \$63.

The negative correlation in Example 5.37 is due to the fact that, in this household, natural gas is used for heating, and electricity is used for air-conditioning. So, when it is warm, the electricity charges are high and the natural-gas charges are low. When it is cool, the reverse is true. This causes the standard deviation of the sum to be less than it would be if the two bills were uncorrelated (see Exercise 5.71, on page 282).

There are situations where we need to combine several of our rules to find means and standard deviations. Here is an example.

EXAMPLE 5.38

Calcium intake. To get enough calcium for optimal bone health, tablets containing calcium are often recommended to supplement the calcium in the diet. One study designed to evaluate the effectiveness of a supplement followed a group of young people for seven years. Each subject was assigned to take a tablet containing 1000 milligrams of calcium per day (mg/d) or a placebo tablet that was identical except that it had no calcium.[18] A major problem with studies like this one is compliance: subjects do not always take the treatments assigned to them.

In this study, the compliance rate was about 47% toward the end of the seven-year period. The standard deviation was 22%. Calcium from the diet averaged 850 mg/d with a standard deviation of 330 mg/d. The correlation between compliance and diet intake is 0.68.

Let's find the mean and standard deviation for the total calcium intake (calcium from the diet plus calcium from the supplement). We let S stand for the intake from the supplement and D stand for the intake from the diet.

We start with the intake from the supplement. Since the compliance is 47% and the amount in each tablet is 1000 mg, the mean for S is

$$\mu_S = 1000(0.47) = 470$$

Since the standard deviation of the compliance is 22%, the variance of S is

$$\sigma^2_S = 1000^2(0.22)^2 = 48{,}400$$

The standard deviation is

$$\sigma_S = \sqrt{48{,}400} = 220$$

Be sure to verify which rules for means and variances are used in these calculations.

We can now find the mean and standard deviation for the total intake. The mean is

$$\mu_{S+D} = \mu_S + \mu_D = 470 + 850 = 1320$$

and the variance is

$$\sigma^2_{S+D} = \sigma^2_S + \sigma^2_D + 2\rho\sigma_S\sigma_D = 220^2 + 330^2 + 2(0.68)(220)(330) = 256{,}036$$

and the standard deviation is

$$\sigma_{S+D} = \sqrt{256{,}036} = 506$$

The mean of the total calcium intake is 1320 mg/d and the standard deviation is 506 mg/d.

The correlation in this example illustrates an unfortunate fact about compliance and having an adequate diet. Some of the subjects in this study have diets that provide an adequate amount of calcium, while others do not. The positive correlation between compliance and dietary intake tells us that those who have relatively high dietary intakes are more likely to take the assigned supplements. On the other hand, those subjects with relatively low dietary intakes, the ones who need the supplement the most, are less likely to take the assigned supplements.

SECTION 5.4 SUMMARY

The probability distribution of a random variable X, like a distribution of data, has a **mean μ_X** and a **standard deviation σ_X**.

The **law of large numbers** says that the average of the values of X observed in many trials must approach μ.

The **mean μ** is the balance point of the probability histogram or density curve. If X is discrete with possible values x_i having probabilities p_i, the mean is the average of the values of X, each weighted by its probability:

$$\mu_X = x_1p_1 + x_2p_2 + \cdots + x_kp_k$$

The **variance σ^2_X** is the average squared deviation of the values of the variable from their mean. For a discrete random variable,

$$\sigma^2_X = (x_1 - \mu)^2p_1 + (x_2 - \mu)^2p_2 + \cdots + (x_k - \mu)^2p_k$$

The **standard deviation σ_X** is the square root of the variance. The standard deviation measures the variability of the distribution about the mean. It is easiest to interpret for Normal distributions.

The mean and variance of a continuous random variable can be computed from the density curve, but to do so requires more advanced mathematics.

The means and variances of random variables obey the following rules. If a and b are fixed numbers, then

$$\mu_{a+bX} = a + b\mu_X$$
$$\sigma^2_{a+bX} = b^2\sigma^2_X$$

If X and Y are any two random variables having correlation ρ, then

$$\mu_{X+Y} = \mu_X + \mu_Y$$
$$\sigma^2_{X+Y} = \sigma^2_X + \sigma^2_Y + 2\rho\sigma_X\sigma_Y$$
$$\sigma^2_{X-Y} = \sigma^2_X + \sigma^2_Y - 2\rho\sigma_X\sigma_Y$$

If X and Y are **independent,** then $\rho = 0$. In this case,

$$\sigma^2_{X+Y} = \sigma^2_X + \sigma^2_Y$$
$$\sigma^2_{X-Y} = \sigma^2_X + \sigma^2_Y$$

To find the standard deviation, take the square root of the variance.

SECTION 5.4 EXERCISES

For Exercise 5.59, see page 268; for Exercise 5.60, see page 272; for Exercises 5.61 and 5.62, see page 275; and for Exercise 5.63, see page 277.

5.64. What's wrong? In each of the following scenarios, there is something wrong. Describe what is wrong and give a reason for your answer.

(a) If you toss a fair coin three times and get heads all three times, then the probability of getting a tail on the next toss is much greater than a half.

(b) If you multiply a random variable by 10, then the mean is multiplied by 10 and the variance is multiplied by 10.

(c) When finding the mean of the sum of two random variables, you need to know the correlation between them.

5.65. Servings of fruits and vegetables. The following table gives the distribution of the number of servings of fruits and vegetables per day in a population.

Number of servings X	0	1	2	3	4	5
Probability	0.3	0.1	0.1	0.2	0.1	0.2

Find the mean and the standard deviation for this random variable.

5.66. Mean of the distribution for the number of aces. In Exercise 5.48 you examined the probability distribution for the number of aces when you are dealt two cards in the game of Texas hold 'em. Let X represent the number of aces in a randomly selected deal of two cards in this game. Here is the probability distribution for the random variable X:

Value of X	0	1	2
Probability	0.8507	0.1448	0.0045

Find μ_X, the mean of the probability distribution of X.

5.67. Mean of the grade distribution. Example 5.22 gives the distribution of grades (A = 4, B = 3, and so on) in English 130 as

Value of X	0	1	2	3	4
Probability	0.05	0.04	0.20	0.40	0.31

Find the average (that is, the mean) grade in this course.

5.68. Mean of the distributions of errors. Typographical and spelling errors can be either "nonword errors" or "word errors." A nonword error is not a real word, as when "the" is typed as "teh." A word error is a real word, but not the right word, as when "lose" is typed as "loose." When undergraduates are asked to write a 250-word essay (without spell-checking), the number of nonword errors has the following distribution:

Errors	0	1	2	3	4
Probability	0.1	0.3	0.3	0.2	0.1

The number of word errors has this distribution:

Errors	0	1	2	3
Probability	0.4	0.3	0.2	0.1

What are the mean number of nonword errors and the mean number of word errors in an essay?

5.69. Standard deviation of the number of aces. Refer to Exercise 5.66. Find the standard deviation of the number of aces.

5.70. Standard deviation of the grades. Refer to Exercise 5.67. Find the standard deviation of the grade distribution.

5.71. Suppose that the correlation is zero. Refer to Example 5.37 (page 279).

(a) Recompute the standard deviation for the total of the natural-gas bill and the electricity bill assuming that the correlation is zero.

(b) Is this standard deviation larger or smaller than the standard deviation computed in Example 5.37? Explain why.

5.72. Find the mean of the sum. Figure 5.12 (page 266) displays the density curve of the sum $Y = X_1 + X_2$ of two independent random numbers, each uniformly distributed between 0 and 1.

(a) The mean of a continuous random variable is the balance point of its density curve. Use this fact to find the mean of Y from Figure 5.12.

(b) Use the same fact to find the means of X_1 and X_2. (They have the density curve pictured in Figure 5.9, page 261.) Verify that the mean of Y is the sum of the mean of X_1 and the mean of X_2.

5.73. Calcium supplements and calcium in the diet. Refer to Example 5.38 (page 280). Suppose that people who have high intakes of calcium in their diets are more compliant than those who have low intakes. What effect would this have on the calculation of the standard deviation for the total calcium intake? Explain your answer.

5.74. The effect of correlation. Find the mean and standard deviation of the total number of errors (nonword errors plus word errors) in an essay if the error counts have the distributions given in Exercise 5.68 and

(a) the counts of nonword and word errors are independent.

(b) students who make many nonword errors also tend to make many word errors, so that the correlation between the two error counts is 0.5.

5.75. Means and variances of sums. The rules for means and variances allow you to find the mean and variance of a sum of random variables without first finding the distribution of the sum, which is usually much harder to do.

(a) A single toss of a balanced coin has either 0 or 1 head, each with probability 1/2. What are the mean and standard deviation of the number of heads?

(b) Toss a coin four times. Use the rules for means and variances to find the mean and standard deviation of the total number of heads.

(c) Example 5.23 (page 258) finds the distribution of the number of heads in four tosses. Find the mean and standard deviation from this distribution. Your results in parts (b) and (c) should agree.

5.76. CHALLENGE **Toss a four-sided die twice.** Role-playing games like Dungeons & Dragons use many different types of dice. Suppose that a four-sided die has faces marked 1, 2, 3, 4. The intelligence of a character is determined by rolling this die twice and adding 1 to the sum of the spots. The faces are equally likely and the two rolls are independent. What is the average (mean) intelligence for such characters? How spread out are their intelligences, as measured by the standard deviation of the distribution?

5.77. A mechanical assembly. A mechanical assembly (Figure 5.15) consists of a rod with a bearing on each end. The three parts are manufactured independently, and all vary a bit from part to part. The length of the rod has mean 12 centimeters (cm) and standard deviation 0.004 millimeters (mm). The length of a bearing has mean 2 cm and standard deviation 0.001 mm. What are the mean and standard deviation of the total length of the assembly?

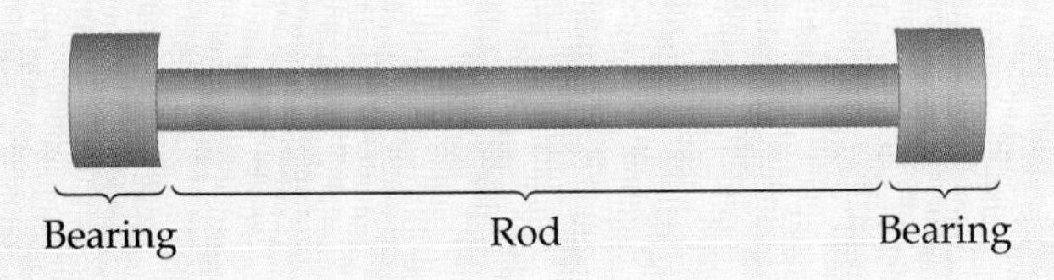

FIGURE 5.15 Sketch of a mechanical assembly, for Exercise 5.77.

5.78. Sums of Normal random variables. Continue your work in the previous exercise. Dimensions of mechanical parts are often roughly Normal. According to the 68–95–99.7 rule, 95% of rods have lengths within $\pm d_1$ of 12 cm and 95% of bearings have lengths within $\pm d_2$ of 2 cm.

(a) What are the values of d_1 and d_2? These are often called the "natural tolerances" of the parts.

(b) Statistical theory says that any sum of independent Normal random variables has a Normal distribution. So the total length of the assembly is roughly Normal. What is the natural tolerance for the total length? It is *not* $d_1 + 2d_2$, because standard deviations don't add.

5.79. Will you assume independence? In which of the following games of chance would you be willing to assume independence of X and Y in making a probability model? Explain your answer in each case.

(a) In blackjack, you are dealt two cards and examine the total points X on the cards (face cards count 10 points). You can choose to be dealt another card and compete based on the total points Y on all three cards.

(b) In craps, the betting is based on successive rolls of two dice. X is the sum of the faces on the first roll, and Y the sum of the faces on the next roll.

5.80. Transform the distribution of heights from centimeters to inches. A report of the National Center for Health Statistics says that the heights of 20-year-old men have mean 176.8 centimeters (cm) and standard deviation 7.2 cm. There are 2.54 centimeters in an inch. What are the mean and standard deviation in inches?

5.81. CHALLENGE **What happens when the correlation is 1?** We know that variances add if the random variables involved are uncorrelated ($\rho = 0$), but not otherwise. The opposite extreme is perfect positive correlation ($\rho = 1$). Show by using the general addition rule for variances that in this case the standard deviations add. That is, $\sigma_{X+Y} = \sigma_X + \sigma_Y$ if $\rho_{XY} = 1$.

5.82. CHALLENGE **A random variable with given mean and standard deviation.** Here is a simple way to create a random variable X that has mean μ and standard deviation σ: X takes only the two values $\mu - \sigma$ and $\mu + \sigma$, each with probability 0.5. Use the definition of the mean and variance for discrete random variables to show that X does have mean μ and standard deviation σ.

5.5 Distributions for Counts

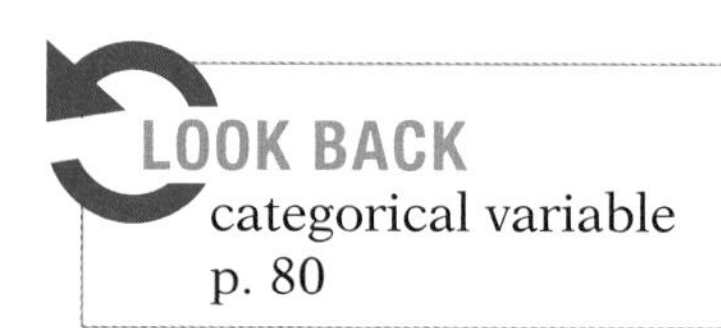

categorical variable p. 80

Counts and proportions are discrete statistics that describe categorical data. We focus our discussion on the simplest case of a random variable with only two possible categories. Here is an example.

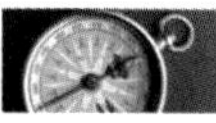

EXAMPLE 5.39

Do parents put too much pressure on their children? A sample survey asks 2000 college students whether they think that parents put too much pressure on their children. We would like to view the responses of these students as representative of a larger population of students who hold similar beliefs. That is, we will view the responses of the sampled students as an SRS from a population.

When there are only two possible outcomes for a random variable, we can summarize the results by giving the count for one of the possible outcomes. We let n represent the sample size, and we use X to represent the random variable that gives the count for the outcome of interest.

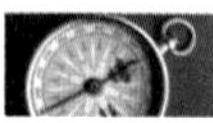

EXAMPLE 5.40

The random variable of interest. In our sample survey of college students, $n = 2000$. We will ask each student in our sample whether he or she feels parents put too much pressure on their children. The variable X is the number of students who think that parents do put too much pressure on their children. Suppose that we observe $X = 840$.

In our example, we chose the random variable X to be the number of students who think that parents put too much pressure on their children. We could have chosen X to be the number of students who do not think that parents put too much pressure on their children. The choice is yours. Often we make the choice based on how we would like to describe the results in a written summary. Which choice do you prefer in this example?

USE YOUR KNOWLEDGE

5.83 Seniors who have taken a statistics course. In a random sample of 250 senior students from your college, 40% reported that they had taken a statistics course. Give n and X for this setting.

5.84 Using the Internet to find a place to live. A poll of 1500 college students asked whether or not they had used the Internet to find a place to live sometime within the past year. There were 825 students who answered "Yes." The other 675 answered "No."

(a) What is n?

(b) Choose one of the two possible outcomes to define the random variable, X. Give a reason for your choice.

(c) What is the value of X?

Just like the sample mean, sample counts are commonly used statistics. Counts, however, are discrete random variables and thus introduce us to a new family of probability distributions.

The binomial distributions for sample counts

The distribution of a count X depends on how the data are produced. Here is a simple but common situation.

THE BINOMIAL SETTING

1. There is a fixed number of observations n.
2. The n observations are all independent.
3. Each observation falls into one of just two categories, which for convenience we call "success" and "failure."
4. The probability of a success, call it p, is the same for each observation.

Think of tossing a coin n times as an example of the binomial setting. Each toss gives either heads or tails and the outcomes of successive tosses are independent. If we call heads a success, then p is the probability of a head and remains the same as long as we toss the same coin. The number of heads we count is a random variable X. The distribution of X, and more generally the distribution of the count of successes in any binomial setting, is completely determined by the number of observations n and the success probability p.

BINOMIAL DISTRIBUTIONS

The distribution of the count X of successes in the binomial setting is called the **binomial distribution** with parameters n and p. The parameter n is the number of observations, and p is the probability of a success on any one observation. The possible values of X are the whole numbers from 0 to n. As an abbreviation, we say that X is $B(n, p)$.

The binomial distributions are an important class of discrete probability distributions. That said, *the most important skill for using binomial distributions is the ability to recognize situations to which they do and don't apply.* This can be done by checking all the facets of the binomial setting. Later in this section we will learn how to assign probabilities to outcomes and how to find the mean and standard deviation of binomial distributions.

EXAMPLE 5.41

Two binomial examples.

(a) Genetics says that children receive genes from their parents independently. Each child of a particular pair of parents has probability 0.25 of having type O blood. If these parents have 3 children, the number who have type O blood is the count X of successes in 3 independent trials with probability 0.25 of a success on each trial. So X has the $B(3, 0.25)$ distribution.

(b) Engineers define reliability as the probability that an item will perform its function under specific conditions for a specific period of time. Replacement heart valves made of animal tissue, for example, have probability 0.77 of performing well for 15 years.[19] The probability of failure is therefore 0.23. It is reasonable to assume that valves in different patients fail or do not fail independently of each other. The number of patients in a group of 500 who will need another valve replacement within 15 years has the $B(500, 0.23)$ distribution.

USE YOUR KNOWLEDGE

5.85 Genetics and blood types. Genetics says that children receive genes from their parents independently. Suppose that each child of a particular pair of parents has probability 0.25 of having type O blood. If these parents have 4 children, what is the distribution of the number who have type O blood? Explain your answer.

5.86 Toss a coin. Toss a fair coin 15 times. Give the distribution of X and the number of heads that you observe.

Binomial distributions in statistical sampling

The binomial distributions are important in statistics when we sample from a population and each observation falls into one of two categories. Here is an example.

EXAMPLE 5.42

Audits of financial records. The financial records of businesses may be audited by state tax authorities to test compliance with tax laws. It is too time-consuming to examine all sales and purchases made by a company during the period covered by the audit.

Suppose that the auditor examines an SRS of 150 sales records out of 10,000 available. One issue is whether each sale was correctly classified as subject to state sales tax or not. Suppose that 800 of the 10,000 sales are incorrectly classified. Is the count X of misclassified records in the sample a binomial random variable?

Choosing an SRS from a population is not quite a binomial setting

After we have selected the first record in Example 5.42, the proportion of bad records in the remaining population is altered. The population proportion of misclassified records is

$$p = \frac{800}{10,000} = 0.08$$

If the first record chosen is bad, the proportion of bad records remaining is $799/9999 = 0.079908$. If the first record is good, the proportion of bad records left is 800/9999 = 0.080008. These proportions are so close to 0.08 that for practical purposes we can act as if removing one record has no effect on the proportion of misclassified records remaining.

The probability that a bad record will be selected as the second record depends upon whether the first record selected was good or bad. Therefore, the selections are not independent and the binomial setting does not apply. However, because the population is large, removing a few items has a very small effect on the composition of the remaining population. Successive inspection results are very nearly independent. We act as if the count X of misclassified sales records in the audit sample has the binomial distribution $B(150, 0.08)$.

stratified sample p. 59

Populations like the one described in Example 5.42 often contain a relatively small number of items with very large values. For this example, these values would be very large sale amounts and likely represent an important group of items to the auditor. An SRS taken from such a population will likely include very few items of this type. Therefore, it is common to use a stratified sample in settings like this. Strata are defined based on dollar value of the sale, and within each stratum, an SRS is taken. The results are then combined to obtain an estimate for the entire population.

SAMPLING DISTRIBUTION OF A COUNT

A population contains proportion p of successes. If the population is much larger than the sample, the count X of successes in an SRS of size n has approximately the binomial distribution $B(n, p)$.

The accuracy of this approximation improves as the size of the population increases relative to the size of the sample. As a rule of thumb, we will use the binomial sampling distribution for counts when the population is at least 10 times as large as the sample.

Finding binomial probabilities

We will later give a formula for the probability that a binomial random variable takes any of its values. In practice, you will rarely have to use this formula for calculations. Some calculators and most statistical software packages calculate binomial probabilities.

EXAMPLE 5.43

The probability of exactly 10 misclassified sales records. In the audit setting of Example 5.42, what is the probability that the audit finds exactly 10 misclassified sales records? What is the probability that the audit finds no more than 10 misclassified records? Figure 5.16 shows the output from one statistical software system. You see that if the count X has the $B(150, 0.08)$ distribution,

$$P(X = 10) = 0.106959$$
$$P(X \leq 10) = 0.338427$$

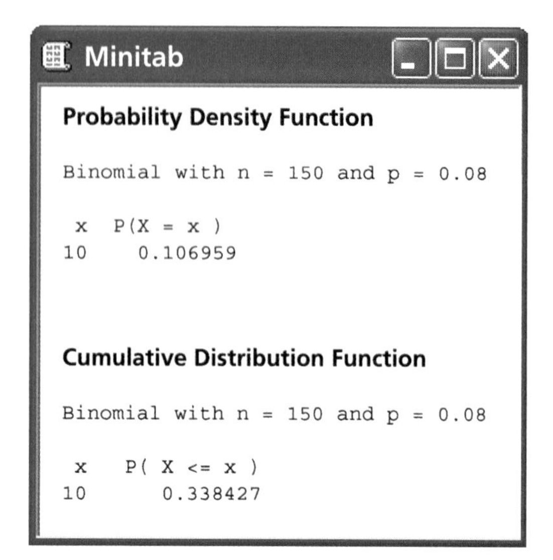

FIGURE 5.16 Binomial probabilities: output from Minitab statistical software, for Example 5.43.

It was easy to request these calculations in the software's menus. Typically, the output supplies more decimal places than we need and sometimes uses labels that may not be helpful. But, as usual with software, we can ignore distractions and find the results we need.

If you do not have suitable computing facilities, you can still shorten the work of calculating binomial probabilities for some values of n and p by looking up probabilities in Table C in the back of this book. The entries in the table are the probabilities $P(X = k)$ of individual outcomes for a binomial random variable X.

EXAMPLE 5.44

The probability histogram. Suppose that the audit in Example 5.42 chose just 15 sales records. What is the probability that no more than 1 of the 15 is misclassified? The count X of misclassified records in the sample has approximately the $B(15, 0.08)$ distribution. Figure 5.17 is a probability histogram for this distribution. The distribution is strongly skewed. Although X can take any whole-number value from 0 to 15, the probabilities of values larger than 5 are so small that they do not appear in the histogram.

We want to calculate

$$P(X \leq 1) = P(X = 0) + P(X = 1)$$

when X has the $B(15, 0.08)$ distribution. To use Table C for this calculation, look opposite $n = 15$ and under $p = 0.08$. This part of the table appears at the left. The entry opposite each k is $P(X = k)$. Blank entries are 0 to four decimal places, so we have omitted most of them here. You see that

$$P(X \leq 1) = P(X = 0) + P(X = 1)$$
$$= 0.2863 + 0.3734 = 0.6597$$

		p
n	***k***	**.08**
15	0	.2863
	1	.3734
	2	.2273
	3	.0857
	4	.0223
	5	.0043
	6	.0006
	7	.0001
	8	
	9	

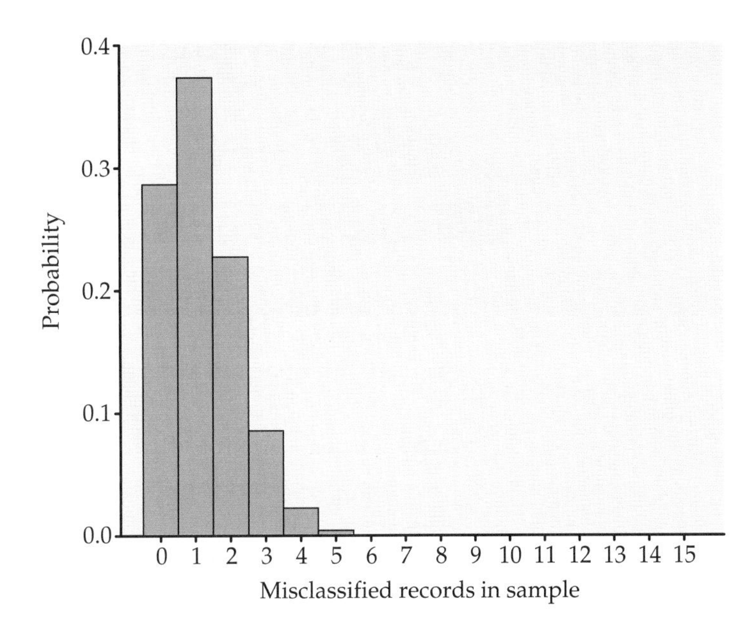

FIGURE 5.17 Probability histogram for the binomial distribution with $n = 15$ and $p = 0.08$, for Example 5.44.

About two-thirds of all samples will contain no more than 1 bad record. In fact, almost 29% of the samples will contain no bad records. The sample of size 15 cannot be trusted to provide adequate evidence about misclassified sales records. A larger number of observations is needed.

The values of p that appear in Table C are all 0.5 or smaller. When the probability of a success is greater than 0.5, restate the problem in terms of the number of failures. The probability of a failure is less than 0.5 when the probability of a success exceeds 0.5. When using the table, always stop to ask whether you must count successes or failures.

EXAMPLE 5.45

Poor sleepers. In a survey of 1125 college students, 65% of the respondents had difficulty sleeping and were classified as poor sleepers. You randomly sample 12 students in your dormitory and classify 10 of them as poor sleepers. Is this an unusually high number of poor sleepers?

To answer this question, assume that the student classifications are independent and that the probability of a "poor" outcome is 0.65. The assumption of independence may not be reasonable if the students study and socialize together. We'll assume, however, that this is not an issue here.

Because the probability of being classified as a poor sleeper is greater than 0.5, we count those students classified as good sleepers to use Table C. Notice that if we have 10 poor sleepers, we then have 2 good sleepers in our sample of size 12.

The probability of being a good sleeper is $1 - 0.65$, or 0.35. The number X of good sleepers out of 12 students has the $B(12, 0.35)$ distribution.

We want the probability of selecting at most 2 good sleepers. This is

$$\begin{aligned} P(X \leq 2) &= P(X = 0) + P(X = 1) + P(X = 2) \\ &= 0.0057 + 0.0368 + 0.1088 = 0.1513 \end{aligned}$$

We would expect to classify 10 or more students as poor sleepers about 15% of the time, or roughly in 3 of every 20 surveys. While this seems like a high number of poor sleepers, this outcome is well within the range of the usual chance variation due to random sampling.

USE YOUR KNOWLEDGE

5.87 Free-throw shooting. Katie is a basketball player who makes 80% of her free throws. In a recent game, she had 10 free throws and missed 6 of them. How unusual is this outcome? Using software, calculator, or Table C, compute $P(X \leq 4)$, where X is the number of free throws made in 10 shots.

5.88 Find the probabilities.

(a) Suppose X has the $B(5, 0.4)$ distribution. Use software, calculator, or Table C to find $P(X = 0)$ and $P(X \geq 3)$.

(b) Suppose X has the $B(5, 0.6)$ distribution. Use software, calculator, or Table C to find $P(X = 5)$ and $P(X \leq 2)$.

(c) Explain the relationship between your answers to parts (a) and (b) of this exercise.

Binomial mean and standard deviation

If a count X has the $B(n, p)$ distribution, what are the mean μ_X and the standard deviation σ_X? We can guess the mean. If we expect 65% of the students to be classified as poor sleepers, the mean number in 12 students should be 65% of 12, or 7.8. That's μ_X when X has the $B(12, 0.65)$ distribution.

Intuition suggests more generally that the mean of the $B(n, p)$ distribution should be np. Can we show that this is correct and also obtain a short formula for the standard deviation? Because binomial distributions are discrete probability distributions, we can find the mean and variance by using the rules for means and variances (pages 274 and 277).

A binomial random variable X is the count of successes in n independent observations that each have the same probability p of success. Let the random variable S_i indicate whether the ith observation is a success or failure by taking the values $S_i = 1$ if a success occurs and $S_i = 0$ if the outcome is a failure. The S_i are independent because the observations are, and each S_i has the same simple distribution:

Outcome	1	0
Probability	p	$1 - p$

From the definition of the mean of a discrete random variable (page 268), we know that the mean of each S_i is

$$\mu_S = (1)(p) + (0)(1 - p) = p$$

Similarly, the definition of the variance shows that $\sigma_S^2 = p(1-p)$. Because each S_i is 1 for a success and 0 for a failure, to find the total number of successes X we add the S_i's:

$$X = S_1 + S_2 + \cdots + S_n$$

Apply the addition rules for means and variances to this sum. To find the mean of X we add the means of the S_i's:

$$\begin{aligned}\mu_X &= \mu_{S_1} + \mu_{S_2} + \cdots + \mu_{S_n} \\ &= n\mu_S = np\end{aligned}$$

Similarly, the variance is n times the variance of a single S, so that $\sigma_X^2 = np(1-p)$. The standard deviation σ_X is the square root of the variance. Here is the result.

BINOMIAL MEAN AND STANDARD DEVIATION

If a count X has the binomial distribution $B(n, p)$, then

$$\begin{aligned}\mu_X &= np \\ \sigma_X &= \sqrt{np(1-p)}\end{aligned}$$

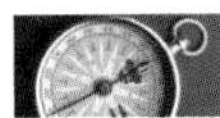

EXAMPLE 5.46

The Helsinki Heart Study. The Helsinki Heart Study asked whether the anticholesterol drug gemfibrozil reduces heart attacks. In planning such an experiment, the researchers must be confident that the sample sizes are large enough to enable them to observe enough heart attacks. The Helsinki study planned to give gemfibrozil to about 2000 men aged 40 to 55 and a placebo to another 2000. The probability of a heart attack during the five-year period of the study for men this age is about 0.04. What are the mean and standard deviation of the number of heart attacks that will be observed in one group if the treatment does not change this probability?

There are 2000 independent observations, each having probability $p = 0.04$ of a heart attack. The count X of heart attacks has the $B(2000, 0.04)$ distribution, so that

$$\begin{aligned}\mu_X &= np = (2000)(0.04) = 80 \\ \sigma_X &= \sqrt{np(1-p)} = \sqrt{(2000)(0.04)(0.96)} = 8.76\end{aligned}$$

The expected number of heart attacks is large enough to permit conclusions about the effectiveness of the drug. In fact, there were 84 heart attacks among the 2035 men actually assigned to the placebo, quite close to the mean. The gemfibrozil group of 2046 men suffered only 56 heart attacks. This is evidence that the drug reduces the chance of a heart attack. In a later chapter we will learn how to determine if this is strong enough evidence to conclude that the drug is effective.

Binomial formula

We can find a formula for the probability that a binomial random variable takes any value by adding probabilities for the different ways of getting exactly that many successes in n observations. Here is the example we will use to show the idea.

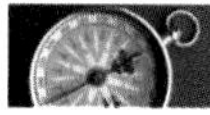

EXAMPLE 5.47

Blood types of children. Each child born to a particular set of parents has probability 0.25 of having blood type O. If these parents have 5 children, what is the probability that exactly 2 of them have type O blood?

The count of children with type O blood is a binomial random variable X with $n = 5$ tries and probability $p = 0.25$ of a success on each try. We want $P(X = 2)$.

Because the method doesn't depend on the specific example, we will use S for success and F for failure. In Example 5.47, S would stand for type O blood. Do the work in two steps.

Step 1: Find the probability that a specific 2 of the 5 tries give successes, say the first and the third. This is the outcome SFSFF. The multiplication rule for independent events tells us that

$$\begin{aligned} P(\text{SFSFF}) &= P(\text{S})P(\text{F})P(\text{S})P(\text{F})P(\text{F}) \\ &= (0.25)(0.75)(0.25)(0.75)(0.75) \\ &= (0.25)^2(0.75)^3 \end{aligned}$$

Step 2: Observe that the probability of *any one* arrangement of 2 S's and 3 F's has this same probability. That's true because we multiply together 0.25 twice and 0.75 three times whenever we have 2 S's and 3 F's. The probability that $X = 2$ is the probability of getting 2 S's and 3 F's in any arrangement whatsoever. Here are all the possible arrangements:

SSFFF	SFSFF	SFFSF	SFFFS	FSSFF
FSFSF	FSFFS	FFSSF	FFSFS	FFFSS

There are 10 of them, all with the same probability. The overall probability of 2 successes is therefore

$$P(X = 2) = 10(0.25)^2(0.75)^3 = 0.2637$$

The pattern of this calculation works for any binomial probability. To use it, we need to count the number of arrangements of k successes in n observations without actually listing them. We use the following fact to do the counting.

> **BINOMIAL COEFFICIENT**
>
> The number of ways of arranging k successes among n observations is given by the **binomial coefficient**
>
> $$\binom{n}{k} = \frac{n!}{k!\,(n-k)!}$$
>
> for $k = 0, 1, 2, \ldots, n$.

factorial

The formula for binomial coefficients uses the **factorial** notation. The factorial $n!$ for any positive whole number n is

$$n! = n \times (n-1) \times (n-2) \times \cdots \times 3 \times 2 \times 1$$

Also, $0! = 1$. Notice that the larger of the two factorials in the denominator of a binomial coefficient will cancel much of the $n!$ in the numerator. For example, the binomial coefficient we need for Example 5.47 is

$$\begin{aligned}\binom{5}{2} &= \frac{5!}{2!\,3!} \\ &= \frac{(5)(4)(3)(2)(1)}{(2)(1) \times (3)(2)(1)} \\ &= \frac{(5)(4)}{(2)(1)} = \frac{20}{2} = 10\end{aligned}$$

This agrees with our previous calculation.

The notation $\binom{n}{k}$ *is not related to the fraction* $\frac{n}{k}$. A helpful way to remember its meaning is to read it as "binomial coefficient n choose k."

Binomial coefficients have many uses in mathematics, but we are interested in them only as an aid to finding binomial probabilities. The binomial coefficient $\binom{n}{k}$ counts the number of ways in which k successes can be distributed among n observations. The binomial probability $P(X = k)$ is this count multiplied by the probability of any specific arrangement of the k successes. Here is the formula we seek.

> **BINOMIAL PROBABILITY**
>
> If X has the binomial distribution $B(n, p)$ with n observations and probability p of success on each observation, the possible values of X are $0, 1, 2, \ldots, n$. If k is any one of these values, the **binomial probability** is
>
> $$P(X = k) = \binom{n}{k} p^k (1-p)^{n-k}$$

Here is an example of the use of the binomial probability formula.

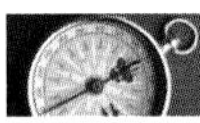

EXAMPLE 5.48

Using the binomial probability formula. The number X of misclassified sales records in the auditor's sample in Example 5.44 has the $B(15, 0.08)$ distribution. The probability of finding no more than 1 misclassified record is

$$\begin{aligned} P(X \leq 1) &= P(X = 0) + P(X = 1) \\ &= \binom{15}{0}(0.08)^0(0.92)^{15} + \binom{15}{1}(0.08)^1(0.92)^{14} \\ &= \frac{15!}{0!\,15!}(1)(0.2863) + \frac{15!}{1!\,14!}(0.08)(0.3112) \\ &= (1)(1)(0.2863) + (15)(0.08)(0.3112) \\ &= 0.2863 + 0.3734 = 0.6597 \end{aligned}$$

The calculation used the facts that $0! = 1$ and that $a^0 = 1$ for any number $a \neq 0$. The result agrees with that obtained from Table C in Example 5.44.

USE YOUR KNOWLEDGE

5.89 Binomial probability. A coin is slightly bent, and as a result the probability of a head is 0.53. Suppose that you toss the coin four times.

(a) Use the binomial formula to find the probability of 3 or more heads.

(b) Compare your answer with the one that you would obtain if the coin were fair.

SECTION 5.5 SUMMARY

A **count** X of successes has the **binomial distribution** $B(n, p)$ in the **binomial setting:** there are n trials, all independent, each resulting in a success or a failure, and each having the same probability p of a success.

Binomial probabilities are most easily found by software. There is an exact formula that is practical for calculations when n is small. Table C contains binomial probabilities for some values of n and p.

The binomial distribution $B(n, p)$ is a good approximation to the **sampling distribution of the count of successes** in an SRS of size n from a large population containing proportion p of successes. We will use this approximation when the population is at least 10 times larger than the sample.

The mean and standard deviation of a **binomial count** X are

$$\mu_X = np$$
$$\sigma_X = \sqrt{np(1-p)}$$

The **binomial probability formula** is

$$P(X = k) = \binom{n}{k} p^k (1-p)^{n-k}$$

where the possible values of X are $k = 0, 1, \ldots, n$. The binomial probability formula uses the **binomial coefficient**

$$\binom{n}{k} = \frac{n!}{k!\,(n-k)!}$$

Here the **factorial** $n!$ is

$$n! = n \times (n-1) \times (n-2) \times \cdots \times 3 \times 2 \times 1$$

for positive whole numbers n, and $0! = 1$. The binomial coefficient counts the number of ways of distributing k successes among n trials.

SECTION 5.5 EXERCISES

For Exercises 5.83 and 5.84, see page 284; for Exercises 5.85 and 5.86, see page 286; for Exercises 5.87 and 5.88, see page 290; and for Exercise 5.89, see page 294.

Most binomial probability calculations required in these exercises can be done by using Table C. Your instructor may request that you use the binomial probability formula or software.

5.90. What's wrong? Explain what is wrong in each of the following scenarios.

(a) In the binomial setting, p is a random variable.

(b) The variance for a binomial count is $\sqrt{p(1-p)/n}$.

(c) The binomial distribution $B(n, p)$ is a continuous distribution.

5.91. Should you use the binomial distribution? In each of the following situations, is it reasonable to use a binomial distribution for the random variable X? Give reasons for your answer in each case. If a binomial distribution applies, give the values of n and p.

(a) A poll of 200 college students asks whether or not you are usually irritable in the morning. X is the number who reply that they are usually irritable in the morning.

(b) You toss a fair coin until a head appears. X is the count of the number of tosses that you make.

(c) Most calls made at random by sample surveys don't succeed in actually talking with a person. Of calls to New York City, only one-twelfth succeed. A survey calls 500 randomly selected numbers in New York City. X is the number of times that a person is reached.

(d) You deal 10 cards from a shuffled deck and count the number X of black cards.

5.92. Should you use the binomial distribution? In each of the following situations, is it reasonable to use a binomial distribution for the random variable X? Give reasons for your answer in each case.

(a) In a random sample of students in a fitness study, X is the mean systolic blood pressure of the sample.

(b) A manufacturer of running shoes picks a random sample of the production of shoes each day for a detailed inspection. Today's sample of 20 pairs of shoes includes 1 pair with a defect.

(c) A nutrition study chooses an SRS of college students. They are asked whether or not they usually eat at least five servings of fruits or vegetables per day. X is the number who say that they do.

5.93. Typographic errors. Typographic errors in a text are either nonword errors (as when "the" is typed as "teh") or word errors that result in a real but incorrect word. Spell-checking software will catch nonword errors but not word errors. Proofreaders who are not professionals catch 70% of word errors. You ask a fellow student to proofread an essay in which you have deliberately made 10 word errors.

(a) If the student matches the usual 70% rate, what is the distribution of the number of errors caught? What is the distribution of the number of errors missed?

(b) Missing 4 or more out of 10 errors seems a poor performance. What is the probability that a proofreader who catches 70% of word errors misses 4 or more out of 10?

5.94. Streaming online music. A survey of United Kingdom music fans revealed that roughly 30% of the teenage music fans are listening to streamed music on their computers every day.[20] You decide to interview a random sample of 20 U.S. teenage music fans. For now, assume that they behave similarly to U.K. teenagers.

(a) What is the distribution of the number who listen to streamed music daily? Explain your answer.

(b) What is the probability that at least 8 of the 20 listen to streamed music daily?

5.95. Typographic errors. Return to the proofreading setting of Exercise 5.93.

(a) What is the mean number of errors caught? What is the mean number of errors missed? You see that these two means must add to 10, the total number of errors.

(b) What is the standard deviation σ of the number of errors caught?

(c) Suppose that a proofreader catches 90% of word errors, so that $p = 0.9$. What is σ in this case? What is σ if $p = 0.99$? What happens to the standard deviation of a binomial distribution as the probability of a success gets close to 1?

5.96. Streaming online music. Recall Exercise 5.94. Suppose that only 25% of the U.S. teenage music fans listen to streamed music daily.

(a) If you interview 20 at random, what is the mean of the count X who listen to streamed music daily?

(b) Repeat the calculations in part (a) for samples of size 200 and 2000. What happens to the mean count of successes as the sample size increases?

5.97. CHALLENGE **Typographic errors.** In the proofreading setting of Exercise 5.93, what is the smallest number of misses m with $P(X \geq m)$ no larger than 0.05? You might consider m or more misses as evidence that a proofreader actually catches fewer than 70% of word errors.

5.98. APPLET **Use the *Probability* applet.** The *Probability* applet simulates tosses of a coin. You can choose the number of tosses n and the probability p of a head. You can therefore use the applet to simulate binomial random variables.

The count of misclassified sales records in Example 5.44 (page 288) has the binomial distribution with $n = 15$ and $p = 0.08$. Set these values for the number of tosses and probability of heads in the applet. Table C shows that the probability of getting a sample with exactly 0 misclassified records is 0.2863. This is the long-run proportion of samples with no bad records. Click "Toss" and "Reset" repeatedly to simulate 25 samples. Record the number of bad records (the count of heads) in each of the 25 samples. What proportion of the 25 samples had exactly 0 bad records? Remember that probability tells us only what happens in the long run.

5.99. Inheritance of blood types. Children inherit their blood type from their parents, with probabilities that reflect the parents' genetic makeup. Children of Juan and Maria each have probability 1/4 of having blood type A and inherit independently of each other. Juan and Maria plan to have 4 children. Let X be the number who have blood type A.

(a) What are n and p in the binomial distribution of X?

(b) Find the probability of each possible value of X, and draw a probability histogram for this distribution.

(c) Find the mean number of children with type A blood, and mark the location of the mean on your probability histogram.

5.100. CHALLENGE **Scuba-diving trips.** The mailing list of an agency that markets scuba-diving trips to the Florida Keys contains 65% males and 35% females. The agency calls 30 people chosen at random from its list.

(a) What is the probability that 20 of the 30 are men? (Use the binomial probability formula.)

(b) What is the probability that the first woman is reached on the fourth call? (That is, the first 4 calls give MMMF.)

5.101. CHALLENGE **Show that these facts are true.** Use the definition of binomial coefficients to show that each of the following facts is true. Then restate each fact in words in terms of the number of ways that k successes can be distributed among n observations.

(a) $\binom{n}{n} = 1$ for any whole number $n \geq 1$.

(b) $\binom{n}{n-1} = n$ for any whole number $n \geq 1$.

(c) $\binom{n}{k} = \binom{n}{n-k}$ for any n and k with $k \leq n$.

5.102. Tossing a die. You are tossing a balanced die that has probability 1/6 of coming up 1 on each toss. Tosses are independent. We are interested in how long we must wait to get the first 1.

(a) The probability of a 1 on the first toss is 1/6. What is the probability that the first toss is not a 1 and the second toss is a 1?

(b) What is the probability that the first two tosses are not 1s and the third toss is a 1? This is the probability that the first 1 occurs on the third toss.

(c) Now you see the pattern. What is the probability that the first 1 occurs on the fourth toss? On the fifth toss?

5.103. CHALLENGE **The geometric distribution.** Generalize your work in the previous exercise. You have independent trials, each resulting in a success or a failure. The probability of a success is p on each trial. The binomial distribution describes the count of successes in a fixed number of trials. Now the number of trials is not fixed. Instead, continue until you get a success. The random variable Y is the number of the trial on which the first success occurs. What are the possible values of Y? What is the probability $P(Y = k)$ for any of these values? (*Comment:* The distribution of the number of trials to the first success is called a **geometric distribution.**)

5.6 General Probability Rules

Our study of probability has concentrated on random variables and their distributions. Now we return to the laws that govern any assignment of probabilities. The purpose of learning more laws of probability is to be able to give probability models for more complex random phenomena. We have already met and used five rules.

> **RULES OF PROBABILITY**
>
> **Rule 1.** $0 \leq P(A) \leq 1$ for any event A
>
> **Rule 2.** $P(S) = 1$
>
> **Rule 3. Addition rule:** If A and B are **disjoint** events, then
>
> $$P(A \text{ or } B) = P(A) + P(B)$$
>
> **Rule 4. Complement rule:** For any event A,
>
> $$P(A^c) = 1 - P(A)$$
>
> **Rule 5. Multiplication rule:** If A and B are **independent** events, then
>
> $$P(A \text{ and } B) = P(A)P(B)$$

General addition rules

Probability has the property that if A and B are disjoint events, then $P(A \text{ or } B) = P(A) + P(B)$. What if there are more than two events, or if the events are not disjoint? These circumstances are covered by more general addition rules for probability.

> **UNION**
>
> The **union** of any collection of events is the event that at least one of the collection occurs.

For two events A and B, the union is the event $\{A \text{ or } B\}$ that A or B or both occur. From the addition rule for two disjoint events we can obtain rules for more general unions. Suppose first that we have several events—say A, B, and C—that are disjoint in pairs. That is, no two can occur at the same time. The Venn diagram in Figure 5.18 illustrates three disjoint events. The addition rule for two disjoint events extends to the following law.

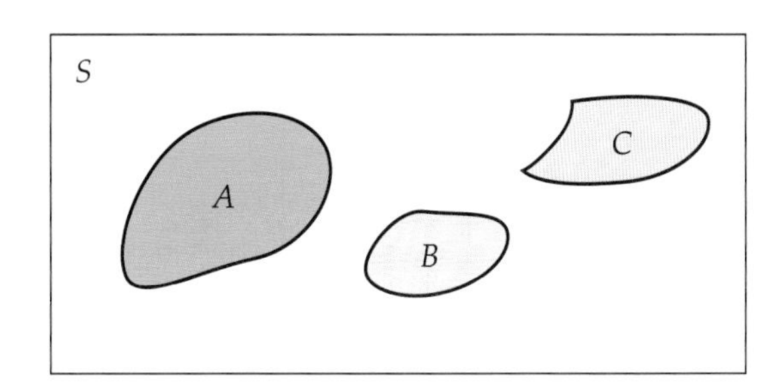

FIGURE 5.18 The addition rule for disjoint events: $P(A \text{ or } B \text{ or } C) = P(A) + P(B) + P(C)$ when events A, B, and C are disjoint.

ADDITION RULE FOR DISJOINT EVENTS

If events A, B, and C are disjoint in the sense that no two have any outcomes in common, then

$$P(\text{one or more of } A, B, C) = P(A) + P(B) + P(C)$$

This rule extends to any number of disjoint events.

EXAMPLE 5.49

Probabilities as areas. Generate a random number X between 0 and 1. What is the probability that the first digit after the decimal point will be odd? The random number X is a continuous random variable whose density curve has constant height 1 between 0 and 1 and is 0 elsewhere. The event that the first digit of X is odd is the union of five disjoint events. These events are

$$0.10 \leq X < 0.20$$
$$0.30 \leq X < 0.40$$
$$0.50 \leq X < 0.60$$
$$0.70 \leq X < 0.80$$
$$0.90 \leq X < 1.00$$

Figure 5.19 illustrates the probabilities of these events as areas under the density curve. Each area is 0.1. The union of the five therefore has probability equal to the sum, or 0.5. As we should expect, a random number is equally likely to begin with an odd or an even digit.

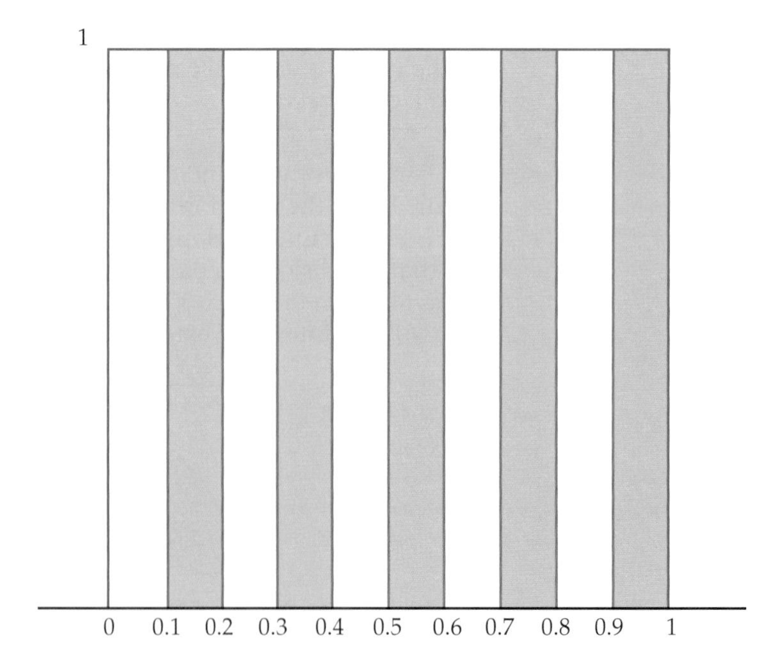

FIGURE 5.19 The probability that the first digit after the decimal point of a random number is odd is the sum of the probabilities of the 5 disjoint events shown. See Example 5.49.

USE YOUR KNOWLEDGE

5.104 Probability that you roll a 3 or a 5. If you roll a die, the probability of each of the six possible outcomes (1, 2, 3, 4, 5, 6) is 1/6. What is the probability that you roll a 3 or a 5?

If events A and B are not disjoint, they can occur simultaneously. The probability of their union is then *less* than the sum of their probabilities. As Figure 5.20 suggests, the outcomes common to both are counted twice when we add probabilities, so we must subtract this probability once. Here is the addition rule for the union of any two events, disjoint or not.

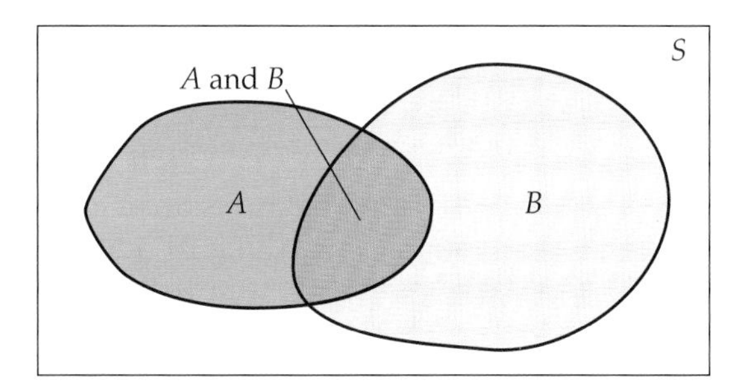

FIGURE 5.20 The union of two events that are not disjoint. The general addition rule says that $P(A \text{ or } B) = P(A) + P(B) - P(A \text{ and } B)$.

GENERAL ADDITION RULE FOR UNIONS OF TWO EVENTS

For any two events A and B,

$$P(A \text{ or } B) = P(A) + P(B) - P(A \text{ and } B)$$

If A and B are disjoint, the event $\{A \text{ and } B\}$ that both occur has no outcomes in it. This *empty event* is the complement of the sample space S and must have probability 0. So the general addition rule includes Rule 3, the addition rule for disjoint events.

EXAMPLE 5.50

Adequate sleep and exercise. Suppose that 40% of adults get enough sleep and 46% exercise regularly. What is the probability that an adult gets enough sleep or exercises regularly? To find this probability, we also need to know the percent who get enough sleep and exercise. Let's assume that 24% do both.

We will use the notation of the general addition rule for unions of two events. Let A be the event that an adult gets enough sleep and let B be the event that a person exercises regularly. We are given that $P(A) = 0.40$, $P(B) = 0.46$, and $P(A \text{ and } B) = 0.24$. Therefore,

$$\begin{aligned} P(A \text{ or } B) &= P(A) + P(B) - P(A \text{ and } B) \\ &= 0.40 + 0.46 - 0.24 \\ &= 0.62 \end{aligned}$$

The probability that an adult gets enough sleep or exercises regularly is 0.62, or 62%.

USE YOUR KNOWLEDGE

5.105 Probability that your roll is even or greater than 4. If you roll a die, the probability of each of the six possible outcomes (1, 2, 3, 4, 5, 6) is 1/6. What is the probability that your roll is even or greater than 4?

Venn diagrams are a great help in finding probabilities for unions because you can just think of adding and subtracting areas. Figure 5.21 shows some events and their probabilities for Example 5.50.

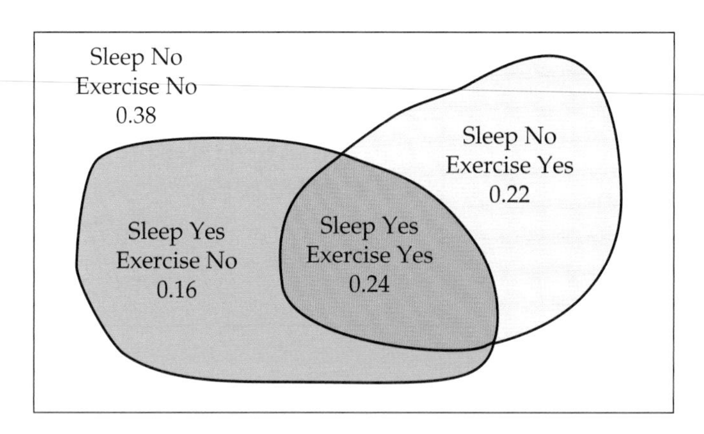

FIGURE 5.21 Venn diagram and probabilities for Example 5.50.

What is the probability that an adult gets adequate sleep but does not exercise? The Venn diagram shows that this is the probability that an individual gets adequate sleep minus the probability that an adult gets adequate sleep and exercises regularly, $0.40 - 0.24 = 0.16$. Similarly, the probability that an adult does not get adequate sleep and exercises regularly is $0.46 - 0.24 = 0.22$. The four probabilities that appear in the figure add to 1 because they refer to four disjoint events whose union is the entire sample space.

Conditional probability

The probability we assign to an event can change if we know that some other event has occurred. This idea is the key to many applications of probability.

EXAMPLE 5.51

Probability of being dealt an ace. Slim is a professional poker player. He stares at the dealer, who prepares to deal. What is the probability that the card dealt to Slim is an ace? There are 52 cards in the deck. Because the deck was carefully shuffled, the next card dealt is equally likely to be any of the cards that Slim has not seen. Four of the 52 cards are aces. So

$$P(\text{ace}) = \frac{4}{52} = \frac{1}{13}$$

This calculation assumes that Slim knows nothing about any cards already dealt. Suppose now that he is looking at 4 cards already in his hand, and that one of them is an ace. He knows nothing about the other 48 cards except that exactly 3 aces are among them. Slim's probability of being dealt an ace *given what he knows* is now

$$P(\text{ace} \mid 1 \text{ ace in 4 visible cards}) = \frac{3}{48} = \frac{1}{16}$$

Knowing that there is 1 ace among the 4 cards that Slim can see changes the probability that the next card dealt is an ace.

conditional probability

The new notation $P(A \mid B)$ is a **conditional probability.** That is, it gives the probability of one event (the next card dealt is an ace) under the condition that we know another event (exactly 1 of the 4 visible cards is an ace). You can read the bar | as "given the information that."

MULTIPLICATION RULE

The probability that both of two events A and B happen together can be found by

$$P(A \text{ and } B) = P(A)P(B \mid A)$$

Here $P(B \mid A)$ is the conditional probability that B occurs given the information that A occurs.

USE YOUR KNOWLEDGE

5.106 The probability of another ace. Refer to Example 5.51. Suppose that 2 of the 4 cards in Slim's hand are aces. What is the probability that the next card dealt to him is an ace?

EXAMPLE 5.52

Downloading music from the Internet. The multiplication rule is just common sense made formal. For example, 29% of Internet users download music files, and 67% of downloaders say that they don't care if the music is copyrighted.[21] So the percent of Internet users who download music (event A)

and don't care about copyright (event B) is 67% of the 29% who download, or

$$(0.67)(0.29) = 0.1943 = 19.43\%$$

The multiplication rule expresses this as

$$P(A \text{ and } B) = P(A) \times P(B \mid A)$$
$$= (0.29)(0.67) = 0.1943$$

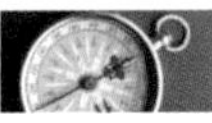

EXAMPLE 5.53

Probability of a favorable draw. Slim is still at the poker table. At the moment, he wants very much to draw two diamonds in a row. As he sits at the table looking at his hand and at the upturned cards on the table, Slim sees 11 cards. Of these, 4 are diamonds. The full deck contains 13 diamonds among its 52 cards, so 9 of the 41 unseen cards are diamonds. To find Slim's probability of drawing 2 diamonds, first calculate

$$P(\text{first card diamond}) = \frac{9}{41}$$

$$P(\text{second card diamond} \mid \text{first card diamond}) = \frac{8}{40}$$

Slim finds both probabilities by counting cards. The probability that the first card drawn is a diamond is 9/41 because 9 of the 41 unseen cards are diamonds. If the first card is a diamond, that leaves 8 diamonds among the 40 remaining cards. So the *conditional* probability of another diamond is 8/40. The multiplication rule now says that

$$P(\text{both cards diamonds}) = \frac{9}{41} \times \frac{8}{40} = 0.044$$

Slim will need luck to draw his diamonds.

USE YOUR KNOWLEDGE

5.107 The probability that the next 2 cards are diamonds. In the setting of Example 5.53, suppose that Slim sees 25 cards and the only diamonds are the 3 in his hand. What is the probability that the next 2 cards dealt to Slim will be diamonds? This outcome would give him 5 cards from the same suit, a very good hand and called a flush.

If $P(A)$ and $P(A \text{ and } B)$ are given, we can rearrange the multiplication rule to produce a *definition* of the conditional probability $P(B \mid A)$ in terms of unconditional probabilities.

DEFINITION OF CONDITIONAL PROBABILITY

When $P(A) > 0$, the **conditional probability** of B given A is

$$P(B \mid A) = \frac{P(A \text{ and } B)}{P(A)}$$

Be sure to keep in mind the distinct roles in $P(B \mid A)$ of the event B whose probability we are computing and the event A that represents the information we are given. The conditional probability $P(B \mid A)$ makes no sense if the event A can never occur, so we require that $P(A) > 0$ whenever we talk about $P(B \mid A)$.

EXAMPLE 5.54

College students. Here is the distribution of U.S. college students classified by age and full-time or part-time status:

Age	Status	
	Full-time	Part-time
15 to 19	0.21	0.02
20 to 24	0.32	0.07
25 to 34	0.10	0.10
35 and over	0.05	0.13

Let's compute the probability that a full-time student is aged 15 to 19. We know that the probability that a student is full-time *and* aged 15 to 19 is 0.21 from the table of probabilities. But what we want here is a conditional probability, given that a student is full-time. Rather than asking about age among all students, we restrict our attention to the subpopulation of students who are full-time. Let

A = the student is a full-time student

B = the student is between 15 and 19 years of age

Our formula is

$$P(B \mid A) = \frac{P(A \text{ and } B)}{P(A)}$$

We read $P(A \text{ and } B) = 0.21$ from the table as we mentioned previously. What about $P(A)$? This is the probability that a student is full-time. Notice that there are four groups of students in our table that fit this description. To find the probability needed, we add the entries:

$$P(A) = 0.21 + 0.32 + 0.10 + 0.05 = 0.68$$

We are now ready to complete the calculation of the conditional probability:

$$\begin{aligned} P(B \mid A) &= \frac{P(A \text{ and } B)}{P(A)} \\ &= \frac{0.21}{0.68} \\ &= 0.31 \end{aligned}$$

The probability that a student is 15 to 19 years of age given that the student is full-time is 0.31.

Here is another way to give the information in the last sentence of this example: 31% of full-time college students are 15 to 19 years old. Which way do you prefer?

USE YOUR KNOWLEDGE

5.108 What rule did we use? In Example 5.54, we calculated $P(B)$. What rule did we use for this calculation? Explain why this rule applies in this setting.

5.109 Find the conditional probability. Refer to Example 5.54. What is the probability that a student is full-time given that the student is 15 to 19 years old? Explain in your own words the difference between this calculation and the one that we did in Example 5.54.

General multiplication rules

The definition of conditional probability reminds us that, in principle, all probabilities, including conditional probabilities, can be found from the assignment of probabilities to events that describe random phenomena. More often, however, conditional probabilities are part of the information given to us in a probability model, and the multiplication rule is used to compute $P(A \text{ and } B)$. This rule extends to more than two events.

The union of a collection of events is the event that *any* of them occur. Here is the corresponding term for the event that *all* of them occur.

INTERSECTION

The **intersection** of any collection of events is the event that *all* the events occur.

To extend the multiplication rule to the probability that all of several events occur, the key is to condition each event on the occurrence of *all* the preceding events. For example, the intersection of three events A, B, and C has probability

$$P(A \text{ and } B \text{ and } C) = P(A)P(B \mid A)P(C \mid A \text{ and } B)$$

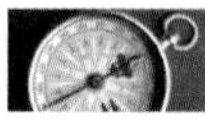

EXAMPLE 5.55

High school athletes and professional careers. Only 5% of male high school basketball, baseball, and football players go on to play at the college level. Of these, only 1.7% enter major league professional sports. About 40% of the athletes who compete in college and then reach the pros have a career of more than three years.[22] Define these events:

$$A = \{\text{competes in college}\}$$
$$B = \{\text{competes professionally}\}$$
$$C = \{\text{pro career longer than three years}\}$$

What is the probability that a high school athlete competes in college and then goes on to have a pro career of more than three years? We know that

$$P(A) = 0.05$$
$$P(B \mid A) = 0.017$$
$$P(C \mid A \text{ and } B) = 0.4$$

The probability we want is therefore

$$P(A \text{ and } B \text{ and } C) = P(A)P(B \mid A)P(C \mid A \text{ and } B)$$
$$= 0.05 \times 0.017 \times 0.4 = 0.00034$$

Only about 3 of every 10,000 high school athletes can expect to compete in college and have a professional career of more than three years. High school students would be wise to concentrate on studies rather than on unrealistic hopes of fortune from pro sports.

Tree diagrams

Probability problems often require us to combine several of the basic rules into a more elaborate calculation. Here is an example that illustrates how to solve problems that have several stages.

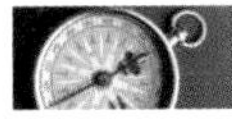

EXAMPLE 5.56

Online chat rooms. Online chat rooms are dominated by the young. Teens are the biggest users. If we look only at adult Internet users (aged 18 and over), 47% of the 18 to 29 age group chat, as do 21% of the 30 to 49 age group and just 7% of those 50 and over. To learn what percent of all Internet users participate in chat, we also need the age breakdown of users. Here it is: 29% of adult Internet users are 18 to 29 years old (event A_1), another 47% are 30 to 49 (event A_2), and the remaining 24% are 50 and over (event A_3).[23]

tree diagram

What is the probability that a randomly chosen user of the Internet participates in chat rooms (event C)? To find out, use the **tree diagram** in Figure 5.22 to organize your thinking. Each segment in the tree is one stage of the problem. Each complete branch shows a path through the two stages. The probability written on each segment is the conditional probability of an Internet user following that segment, given that he or she has reached the node from which it branches.

Starting at the left, an Internet user falls into one of the three age groups. The probabilities of these groups

$$P(A_1) = 0.29 \qquad P(A_2) = 0.47 \qquad P(A_3) = 0.24$$

mark the leftmost branches in the tree. Conditional on being 18 to 29 years old, the probability of participating in chat is $P(C \mid A_1) = 0.47$. So the conditional probability of *not* participating is

$$P(C^c \mid A_1) = 1 - 0.47 = 0.53$$

These conditional probabilities mark the paths branching out from the A_1 node in Figure 5.22. The other two age group nodes similarly lead to two

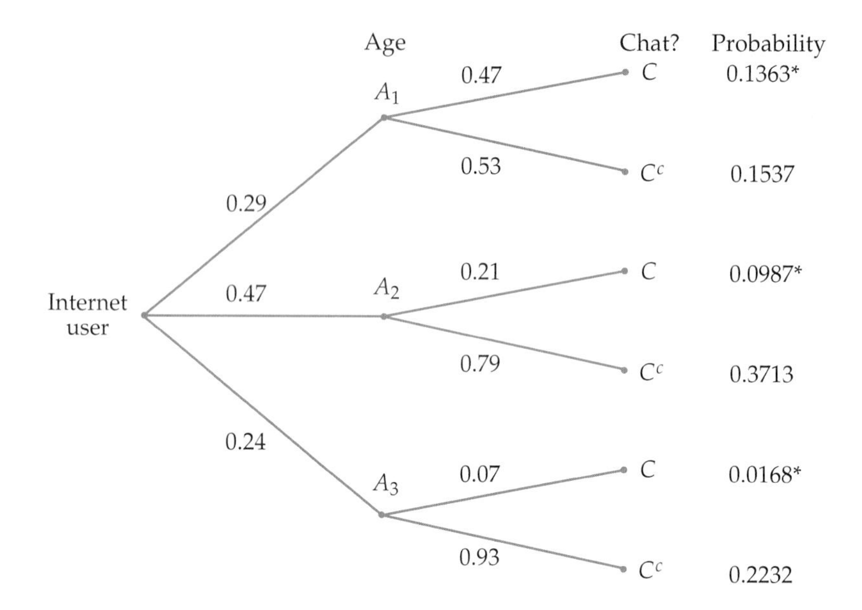

FIGURE 5.22 Tree diagram for Example 5.56. The probability $P(C)$ is the sum of the probabilities of the three branches marked with asterisks (*).

branches marked with the conditional probabilities of chatting or not. The probabilities on the branches from any node add to 1 because they cover all possibilities, given that this node was reached.

There are three disjoint paths to C, one for each age group. By the addition rule, $P(C)$ is the sum of their probabilities. The probability of reaching C through the 18 to 29 age group is

$$P(C \text{ and } A_1) = P(A_1)P(C \mid A_1) = 0.29 \times 0.47 = 0.1363$$

Follow the paths to C through the other two age groups. The probabilities of these paths are

$$P(C \text{ and } A_2) = P(A_2)P(C \mid A_2) = (0.47)(0.21) = 0.0987$$
$$P(C \text{ and } A_3) = P(A_3)P(C \mid A_3) = (0.24)(0.07) = 0.0168$$

The final result is

$$P(C) = 0.1363 + 0.0987 + 0.0168 = 0.2518$$

About 25% of all adult Internet users take part in chat rooms.

It takes longer to explain a tree diagram than it does to use it. Once you have understood a problem well enough to draw the tree, the rest is easy. Tree diagrams combine the addition and multiplication rules. The multiplication rule says that the probability of reaching the end of any complete branch is the product of the probabilities written on its segments. The probability of any outcome, such as the event C that an adult Internet user takes part in chat rooms, is then found by adding the probabilities of all branches that are part of that event.

USE YOUR KNOWLEDGE

5.110 Draw a tree diagram. Refer to Slim's chances of a flush in Exercise 5.107 (page 302). Draw a tree diagram to describe the outcomes for the two cards that he will be dealt. At the first stage, his draw can be a diamond or a nondiamond. At the second stage, he has the same possible outcomes but the probabilities are different.

Bayes's rule

There is another kind of probability question that we might ask in the context of thinking about online chat. What percent of adult chat room participants are aged 18 to 29?

EXAMPLE 5.57

Conditional versus unconditional probabilities. In the notation of Example 5.56 this is the conditional probability $P(A_1 \mid C)$. Start from the definition of conditional probability and then apply the results of Example 5.56:

$$\begin{aligned} P(A_1 \mid C) &= \frac{P(A_1 \text{ and } C)}{P(C)} \\ &= \frac{0.1363}{0.2518} = 0.5413 \end{aligned}$$

Over half of adult chat room participants are between 18 and 29 years old. Compare this conditional probability with the original information (unconditional) that 29% of adult Internet users are between 18 and 29 years old. Knowing that a person chats increases the probability that he or she is young.

We know the probabilities $P(A_1)$, $P(A_2)$, and $P(A_3)$ that give the age distribution of adult Internet users. We also know the conditional probabilities $P(C \mid A_1)$, $P(C \mid A_2)$, and $P(C \mid A_3)$ that a person from each age group chats. Example 5.56 shows how to use this information to calculate $P(C)$. The method can be summarized in a single expression that adds the probabilities of the three paths to C in the tree diagram:

$$P(C) = P(A_1)P(C \mid A_1) + P(A_2)P(C \mid A_2) + P(A_3)P(C \mid A_3)$$

In Example 5.57 we calculated the "reverse" conditional probability $P(A_1 \mid C)$. The denominator 0.2518 in that example came from the previous expression. Put in this general notation, we have another probability law.

BAYES'S RULE

Suppose that $A_1, A_2, \ldots, A_k$ are disjoint events whose probabilities are not 0 and add to exactly 1. That is, any outcome is in exactly one of these events. Then if C is any other event whose probability is not 0 or 1,

$$P(A_i \mid C) = \frac{P(C \mid A_i)P(A_i)}{P(C \mid A_1)P(A_1) + P(C \mid A_2)P(A_2) + \cdots + P(A_k)P(C \mid A_k)}$$

The numerator in Bayes's rule is always one of the terms in the sum that makes up the denominator. The rule is named after Thomas Bayes, who wrestled with arguing from outcomes like C back to the A_i in a book published in 1763. It is far better to think your way through problems like Examples 5.56 and 5.57 than to memorize these formal expressions.

Independence again

The conditional probability $P(B \mid A)$ is generally not equal to the unconditional probability $P(B)$. That is because the occurrence of event A generally gives us some additional information about whether or not event B occurs. If knowing that A occurs gives no additional information about B, then A and B are independent events. The formal definition of independence is expressed in terms of conditional probability.

> **INDEPENDENT EVENTS**
>
> Two events A and B that both have positive probability are **independent** if
>
> $$P(B \mid A) = P(B)$$

This definition makes precise the informal description of independence given in Section 5.2. We now see that the multiplication rule for independent events, $P(A \text{ and } B) = P(A)P(B)$, is a special case of the general multiplication rule, $P(A \text{ and } B) = P(A)P(B \mid A)$, just as the addition rule for disjoint events is a special case of the general addition rule.

SECTION 5.6 SUMMARY

The **complement** A^c of an event A contains all outcomes that are not in A. The **union** $\{A \text{ or } B\}$ of events A and B contains all outcomes in A, in B, or in both A and B. The **intersection** $\{A \text{ and } B\}$ contains all outcomes that are in both A and B, but not outcomes in A alone or B alone.

The **conditional probability** $P(B \mid A)$ of an event B given an event A is defined by

$$P(B \mid A) = \frac{P(A \text{ and } B)}{P(A)}$$

when $P(A) > 0$. In practice, conditional probabilities are most often found from directly available information.

The essential general rules of elementary probability are

Legitimate values: $0 \le P(A) \le 1$ for any event A

Total probability 1: $P(S) = 1$

Complement rule: $P(A^c) = 1 - P(A)$

Addition rule: $P(A \text{ or } B) = P(A) + P(B) - P(A \text{ and } B)$

Multiplication rule: $P(A \text{ and } B) = P(A)P(B \mid A)$

If A and B are **disjoint,** then $P(A \text{ and } B) = 0$. The general addition rule for unions then becomes the special addition rule, $P(A \text{ or } B) = P(A) + P(B)$.

A and B are **independent** when $P(B \mid A) = P(B)$. The multiplication rule for intersections then becomes $P(A \text{ and } B) = P(A)P(B)$.

In problems with several stages, draw a **tree diagram** to organize use of the multiplication and addition rules.

SECTION 5.6 EXERCISES

For Exercise 5.104, see page 299; for Exercise 5.105, see page 300; for Exercise 5.106, see page 301; for Exercise 5.107, see page 302; for Exercises 5.108 and 5.109, see page 304; and for Exercise 5.110, see page 307.

5.111. Dancing and singing. Suppose that 70% of college students like to dance, 40% like to sing, and 30% like to dance and sing. Find the probabilities of the following events:

(a) like to dance and don't like to sing

(b) don't like to dance and like to sing

(c) don't like to dance and don't like to sing

(d) for each of parts (a), (b), and (c), state the rule that you used to find your answer.

5.112. Dancing and singing. Refer to the previous exercise. Draw a Venn diagram, similar to Figure 5.21 (see page 300), showing the probabilities for dancing and singing.

5.113. Lying to a teacher. Suppose that 48% of high school students would admit to lying at least once to a teacher during the past year, and that 25% of students are male and would admit to lying at least once to a teacher during the past year.[24] Assume that 50% of the students are male. What is the probability that a randomly selected student is either male or is a liar? Be sure to show your work and indicate all the rules that you use to find your answer.

5.114. Lying to a teacher. Refer to the previous exercise. Suppose that you select a student from the subpopulation of liars. What is the probability that the student is female? Be sure to show your work and indicate all the rules that you use to find your answer.

5.115. Binge drinking and gender. In a college population, students are classified by gender and whether or not they are frequent binge drinkers. Here are the probabilities:

	Men	Women
Binge drinker	0.11	0.12
Not binge drinker	0.32	0.45

(a) Verify that the sum of the probabilities is 1.

(b) What is the probability that a randomly selected student is not a binge drinker?

(c) What is the probability that a randomly selected male student is not a binge drinker?

(d) Explain why your answers to (b) and (c) are different. Use language that would be understood by someone who has not studied the material in this chapter.

5.116. Find some probabilities. Refer to the previous exercise.

(a) Find the probability that a randomly selected student is a male binge drinker, and find the probability that a randomly selected student is a female binge drinker.

(b) Find the probability that a student is a binge drinker given that the student is male and find the probability that a student is a binge drinker given that the student is female.

(c) Your answer for part (a) gives a higher probability for females, while your answer for part (b) gives a higher probability for males. Interpret your answers in terms of the question of whether there are gender differences in binge-drinking behavior. Decide which comparison you prefer and explain the reasons for your preference.

5.117. Attendance at two-year and four-year colleges. In a large national population of college students, 61% attend four-year institutions and the rest attend two-year institutions. Males make up 44% of the students in the four-year institutions and 41% of the students in the two-year institutions.

(a) Find the four probabilities for each combination of gender and type of institution in the following table. Be sure that your probabilities sum to 1.

	Men	Women
Four-year institution		
Two-year institution		

(b) Consider randomly selecting a female student from this population. What is the probability that she attends a four-year institution?

5.118. Draw a tree diagram. Refer to the previous exercise. Draw a tree diagram to illustrate the probabilities in a situation where you first identify the type of institution attended and then identify the gender of the student.

5.119. Draw a different tree diagram for the same setting. Refer to the previous two exercises. Draw a tree diagram to illustrate the probabilities in a situation where you first identify the gender of the student and then identify the type of institution attended. Explain why the probabilities in this tree diagram are different from those that you used in the previous exercise.

5.120. Education and income. Call a household prosperous if its income exceeds $100,000. Call the household educated if the householder completed college. Select an American household at random, and let A be the event that the selected household is prosperous and B the event that it is educated. Suppose $P(A) = 0.138$, $P(B) = 0.261$, and the probability that a household is both prosperous and educated is $P(A \text{ and } B) = 0.082$. What is the probability $P(A \text{ or } B)$ that the household selected is either prosperous or educated?

5.121. Find a conditional probability. In the setting of the previous exercise, what is the conditional probability that a household is prosperous given that it is educated? Explain why your result shows that events A and B are not independent.

5.122. Draw a Venn diagram. Draw a Venn diagram that shows the relation between the events A and B in Exercise 5.120. Indicate each of the following events on your diagram and use the information in Exercise 5.120 to calculate the probability of each event. Finally, describe in words what each event is.

(a) $\{A \text{ and } B\}$ **(b)** $\{A^c \text{ and } B\}$

(c) $\{A \text{ and } B^c\}$ **(d)** $\{A^c \text{ and } B^c\}$

5.123. Sales of cars and light trucks. Motor vehicles sold to individuals are classified as either cars or light trucks (including SUVs) and as either domestic or imported. In a recent year, 69% of vehicles sold were light trucks, 78% were domestic, and 55% were domestic light trucks. Let A be the event that a vehicle is a car and B the event that it is imported. Write each of the following events in set notation and give its probability.

(a) The vehicle is a light truck.

(b) The vehicle is an imported car.

5.124. Income tax returns. In a recent year, the Internal Revenue Service received 142,978,806 individual tax returns. Of these, 17,993,498 reported an adjusted gross income of at least $100,000, and 392,220 reported at least $1 million.[25] If you know that a randomly chosen return shows an income of $100,000 or more, what is the conditional probability that the income is at least $1 million?

5.125. Conditional probabilities and independence. Using the information in Exercise 5.123, answer these questions.

(a) Given that a vehicle is imported, what is the conditional probability that it is a light truck?

(b) Are the events "vehicle is a light truck" and "vehicle is imported" independent? Justify your answer.

5.126. Job offers. Julie is graduating from college. She has studied biology, chemistry, and computing and hopes to work as a forensic scientist applying her science background to crime investigation. Late one night she thinks about some jobs she has applied for. Let A, B, and C be the events that Julie is offered a job by the

A = Connecticut Office of the Chief Medical Examiner

B = New Jersey Division of Criminal Justice

C = federal Disaster Mortuary Operations Response Team

Julie writes down her personal probabilities for being offered these jobs:

$P(A) = 0.7$ $P(B) = 0.5$ $P(C) = 0.3$
$P(A \text{ and } B) = 0.3$ $P(A \text{ and } C) = 0.1$ $P(B \text{ and } C) = 0.1$
$P(A \text{ and } B \text{ and } C) = 0$

Make a Venn diagram of the events A, B, and C. As in Figure 5.21 (page 300), mark the probabilities of every intersection involving these events and their complements. Use this diagram for Exercises 5.127 to 5.129.

5.127. Find the probability of at least one offer. What is the probability that Julie is offered at least one of the three jobs?

5.128. Find the probability of another event. What is the probability that Julie is offered both the Connecticut and New Jersey jobs but not the federal job?

5.129. Find a conditional probability. If Julie is offered the federal job, what is the conditional probability that she is also offered the New Jersey job? If Julie is offered the New Jersey job, what is the conditional probability that she is also offered the federal job?

5.130. Academic degrees and gender. Here are the projected numbers (in thousands) of earned degrees in the United States in the 2010–2011 academic year, classified by level and by the gender of the degree recipient:[26]

	Bachelor's	Master's	Professional	Doctorate
Female	933	402	51	26
Male	661	260	44	26

(a) Convert this table to a table giving the probabilities for selecting a degree earned and classifying the recipient by gender and degree.

(b) If you choose a degree recipient at random, what is the probability that the person you choose is a woman?

(c) What is the conditional probability that you choose a woman given that the person chosen received a professional degree?

(d) Are the events "choose a woman" and "choose a professional degree recipient" independent? How do you know?

5.131. Find some probabilities. The previous exercise gives the projected number (in thousands) of earned degrees in the United States in the 2010–2011 academic year. Use these data to answer the following questions.

(a) What is the probability that a randomly chosen degree recipient is a man?

(b) What is the conditional probability that the person chosen received a bachelor's degree given that he is a man?

(c) Use the multiplication rule to find the probability of choosing a male bachelor's degree recipient. Check your result by finding this probability directly from the table of counts.

CHAPTER 5 EXERCISES

5.132. Repeat the experiment many times. Here is a probability distribution for a random variable X:

Value of X	-1	2
Probability	0.3	0.7

A single experiment generates a random value from this distribution. If the experiment is repeated many times, what will be the approximate proportion of times that the value is -1? Give a reason for your answer.

5.133. Repeat the experiment many times and take the mean. Here is a probability distribution for a random variable X:

Value of X	-1	2
Probability	0.3	0.7

A single experiment generates a random value from this distribution. If the experiment is repeated many times, what will be the approximate value of the mean of these random variables? Give a reason for your answer.

5.134. Work with a transformation. Here is a probability distribution for a random variable X:

Value of X	1	2
Probability	0.4	0.6

(a) Find the mean and the standard deviation of this distribution.

(b) Let $Y = 3X - 2$. Use the rules for means and variances to find the mean and the standard deviation of the distribution of Y.

(c) For part (b) give the rules that you used to find your answer.

5.135. CHALLENGE **A different transformation.** Refer to the previous exercise. Now let $Y = 3X^2 - 2$.

(a) Find the distribution of Y.

(b) Find the mean and standard deviation for the distribution of Y.

(c) Explain why the rules that you used for part (b) of the previous exercise do not work for this transformation.

5.136. Toss a pair of dice two times. Consider tossing a pair of fair dice two times. For each of the following pairs of events, tell whether they are disjoint, independent, or neither.

(a) $A = 6$ on the first roll, $B = 5$ or less on the first roll.

(b) $A = 6$ on the first roll, $B = 7$ or less on the second roll.

(c) $A = 6$ or less on the second roll, $B = 5$ or less on the first roll.

(d) $A = 6$ or less on the second roll, $B = 5$ or less on the second roll.

5.137. Find the probabilities. Refer to the previous exercise. Find the probabilities for each event.

5.138. Some probability distributions. Here is a probability distribution for a random variable X:

Value of X	2	3	4
Probability	0.3	0.4	0.3

(a) Find the mean and standard deviation for this distribution.

(b) Construct a different probability distribution with the same possible values, the same mean, and a larger standard deviation. Show your work and report the standard deviation of your new distribution.

(c) Construct a different probability distribution with the same possible values, the same mean, and a smaller standard deviation. Show your work and report the standard deviation of your new distribution.

5.139. A fair bet at craps. Almost all bets made at gambling casinos favor the house. In other words, the difference between the amount bet and the mean of the distribution of the payoff is a positive number. An exception is "taking the odds" at the game of craps, a bet that a player can make under certain circumstances. The bet becomes available when a shooter throws a 4, 5, 6, 8, 9, or 10 on the initial roll. This number is called the "point." When a point is rolled, we say that a point has been established. If a 4 is the point, an odds bet can be made that wins if a 4 is rolled before a 7 is rolled. The probability of winning this bet is 1/3 and the payoff for a \$10 bet is \$20. You keep the \$10 you bet and you receive an additional \$20. The same probability of winning and payoff apply for an odds bet on a 10. For an initial roll of 5 or 9, the odds bet has a winning probability of 2/5 and the payoff for a \$10 bet is \$15. Similarly, when the initial roll is 6 or 8, the odds bet has a winning probability of 5/11 and the payoff for a \$10 bet is \$12. Find the mean of the payoff distribution for each of these bets. Then confirm that the bets are fair by showing that the difference between amount bet and the mean of the distribution of the payoff is zero.

5.140. An ancient Korean drinking game. An ancient Korean drinking game involves a 14-sided die. The players

roll the die in turn and must submit to whatever humiliation is written on the up-face: something like "Keep still when tickled on face." Six of the 14 faces are squares. Let's call them A, B, C, D, E, and F for short. The other eight faces are triangles, which we will call 1, 2, 3, 4, 5, 6, 7, and 8. Each of the squares is equally likely. Each of the triangles is also equally likely, but the triangle probability differs from the square probability. The probability of getting a square is 0.72. Give the probability model for the 14 possible outcomes.

5.141. Wine tasters. Two wine tasters rate each wine they taste on a scale of 1 to 5. From data on their ratings of a large number of wines, we obtain the following probabilities for both tasters' ratings of a randomly chosen wine:

	Taster 2				
Taster 1	**1**	**2**	**3**	**4**	**5**
1	0.03	0.02	0.01	0.00	0.00
2	0.02	0.07	0.06	0.02	0.01
3	0.01	0.05	0.25	0.05	0.01
4	0.00	0.02	0.05	0.20	0.02
5	0.00	0.01	0.01	0.02	0.06

(a) Why is this a legitimate assignment of probabilities to outcomes?

(b) What is the probability that the tasters agree when rating a wine?

(c) What is the probability that Taster 1 rates a wine higher than 3? What is the probability that Taster 2 rates a wine higher than 3?

5.142. CHALLENGE **SAT scores.** The College Board finds that the distribution of students' SAT scores depends on the level of education their parents have. Children of parents who did not finish high school have SAT Math scores X with mean 445 and standard deviation 106. Scores Y of children of parents with graduate degrees have mean 566 and standard deviation 109. Perhaps we should standardize to a common scale for equity. Find positive numbers a, b, c, and d such that $a + bX$ and $c + dY$ both have mean 500 and standard deviation 100.

5.143. CHALLENGE **Lottery tickets.** Joe buys a ticket in the Tri-State Pick 3 lottery every day, always betting on 956. He will win something if the winning number contains 9, 5, and 6 in any order. Each day, Joe has probability 0.006 of winning, and he wins (or not) independently of other days because a new drawing is held each day. What is the probability that Joe's first winning ticket comes on the 20th day?

5.144. Genetics of peas. According to genetic theory, the blossom color in the second generation of a certain cross of sweet peas should be red or white in a 3:1 ratio. That is, each plant has probability 3/4 of having red blossoms, and the blossom colors of separate plants are independent.

(a) What is the probability that exactly 9 out of 12 of these plants have red blossoms?

(b) What is the mean number of red-blossomed plants when 120 plants of this type are grown from seeds?

5.145. CHALLENGE **Mathematics degrees and gender.** Of the 21,189 degrees in mathematics given by U.S. colleges and universities in a recent year, 71% were bachelor's degrees, 23% were master's degrees, and the rest were doctorates. Moreover, women earned 44% of the bachelor's degrees, 41% of the master's degrees, and 30% of the doctorates.[27] You choose a mathematics degree at random and find that it was awarded to a woman. What is the probability that it is a bachelor's degree?

5.146. Higher education at two-year and four-year institutions. The following table gives the counts of U.S. institutions of higher education classified as public or private and as two-year or four-year:[28]

	Public	Private
Two-year	639	1894
Four-year	1061	622

Convert the counts to probabilities and summarize the relationship between these two variables using conditional probabilities.

5.147. Odds bets at craps. Refer to the odds bets at craps in Exercise 5.139. Suppose that whenever the shooter has an initial roll of 4, 5, 6, 8, 9, or 10, he or she takes the odds. Here are the probabilities for these initial rolls:

Point	4	5	6	8	9	10
Probability	3/36	4/36	5/36	5/36	4/36	3/36

Draw a tree diagram with the first stage showing the point rolled and the second stage showing whether the point is again rolled before a 7 is rolled. Include a first-stage branch showing the outcome that a point is not established. In this case, the amount bet is zero and the distribution of the winnings is the special random variable that has $P(X = 0) = 1$. For the combined betting system where the player always makes a $10 odds bet when it is available, show that the game is fair.

5.148. Weights and heights of children adjusted for age. The idea of conditional probabilities has many interesting applications, including the idea of a conditional distribution. For example, the National Center for Health Statistics produces distributions for weight and height for children while conditioning on other variables. Visit the Web site `cdc.gov/growthcharts/` and describe the different ways that weight and height distributions are conditioned on other variables.

CHAPTER SIX

INFERENCE FOR PROPORTIONS

6.1 Inference for a Single Proportion
6.2 Comparing Two Proportions

Statistical inference draws conclusions about a population or process based on sample data. It also provides a statement, expressed in terms of probability, of how much confidence we can place in our conclusions. Although there are many specific techniques for inference, there are only a few general types of statistical inference. This chapter introduces the two most common types: *confidence intervals* and *tests of significance*.

Our study of these two types of inference begins with inference about proportions. We frequently collect data on *categorical variables,* such as whether or not a person is employed, the brand name of a cell phone, or the country where a college student studies abroad. In these settings, our data consist of *counts* or of *percents* obtained from counts, and our goal is to say something about the corresponding *population proportions*. We begin in Section 6.1 with inference about a single population proportion. Section 6.2 concerns methods for comparing two proportions.

6.1 Inference for a Single Proportion

What percent of college students favor allowing concealed weapons on campus? What proportion of a company's sales records have an incorrect sales tax classification? What proportion of likely voters approve of the president's conduct in office? Often we want to know about the proportion p of some characteristic in a large population.

Suppose a market research firm interviews a random sample of 2000 adults. The result: 66% think bottled water is cleaner than tap water. That's the truth about the 2000 people in the sample. What is the truth about the almost 235 million American adults who make up the population? Because the sample was chosen at random, it's reasonable to think that these 2000 people represent the entire population fairly well. So the market researchers turn the *fact* that 66% of the *sample* think bottled water is cleaner into an *estimate* that 66% of *all American adults* think this way.

statistical inference

That's a basic move in statistics: **use a fact about a sample to estimate the truth about the whole population.** We call this **statistical inference** because we infer conclusions about the wider population from data on selected individuals.

To think about inference, we must keep straight whether a number describes a sample or a population. Here is the vocabulary we use.

PARAMETERS AND STATISTICS

A **parameter** is a number that describes the **population.** A parameter is a fixed number, but in practice we do not know its value.

A **statistic** is a number that describes a **sample.** The value of a statistic is known when we have taken a sample, but it can change from sample to sample. We often use a statistic to estimate an unknown parameter.

For this market research study, the *parameter* is the proportion (we call it p) of adults who think bottled water is cleaner than tap water. The value of p is unknown, so the firm uses a *statistic* to estimate it, specifically the proportion of adults in the sample who think bottled water is cleaner than tap water.

USE YOUR KNOWLEDGE

6.1 Sexual harassment of college students. A survey of 2036 undergraduate college students aged 18 to 24 reports that 62% of college students say they have encountered some type of sexual harassment while at college.[1] Describe the sample and the population for this setting.

6.2 Web polls. If you connect to the Web site `zdaily.com/polls.htm`, you will be given the opportunity to give your opinion about a different question of human interest each day. Can you apply the ideas about populations and samples that we have just discussed to this poll? Explain why or why not.

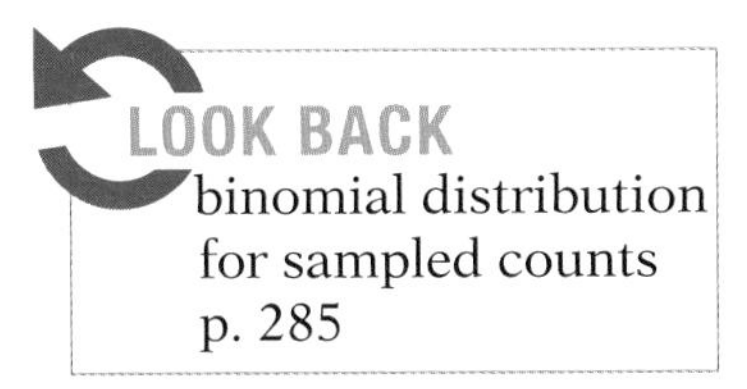

binomial distribution for sampled counts p. 285

Sample proportions

sample proportion

In the previous chapter, we discussed the situation where we draw a random sample of size n from a population and record the count X of "successes." In this chapter, we focus on a closely related statistic that is the estimator of the population parameter p. It is called the **sample proportion**:

$$\hat{p} = \frac{\text{count of successes in sample}}{\text{size of sample}}$$
$$= \frac{X}{n}$$

The market research firm used the sample proportion of 0.66 to estimate the proportion of adults who think bottled water is cleaner. Here is another example.

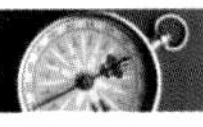

EXAMPLE 6.1

Adults and video games. A DFC Intelligence report estimates that the global video game market will grow from \$66 billion in 2010 to \$81 billion in 2016.[2] What proportion of adults in the United States play these games? A Pew survey, conducted by Princeton Survey Research International, reports that over half of American adults aged 18 and over play video games.[3] The Pew survey used a nationally representative sample of 2054 adults. Of the total, 1063 adults said that they played video games. Here, p is the proportion of adults in the U.S. population who play video games and

$$\hat{p} = \frac{X}{n} = \frac{1063}{2054} = 0.5175$$

is the sample proportion. We estimate that 52% of adults play video games.

USE YOUR KNOWLEDGE

6.3 Bank acquisitions. The American Bankers Association Community Bank Competitiveness Survey had responses from 760 community banks. Of these, 283 reported that they expected to acquire another bank within five years.[4]

(a) What is the sample size n for this survey?

(b) What is the count X of community banks that expect to acquire another bank?

(c) Find the sample proportion $\hat{p}$.

6.4 How often do they play? In the Pew survey described in Example 6.1, those who played video games were asked how often they played. In this subpopulation, 223 adults said that they played every day or almost every day.

(a) What is the sample size n for the subpopulation of U.S. adults who play video games? (*Hint:* Look at Example 6.1.)

(b) What is the count X of those who said that they played every day or almost every day?

(c) Find the sample proportion $\hat{p}$.

While $\hat{p} = 0.518$ in Example 6.1 provides an estimate for the proportion of adults who play video games, we typically want to also know how reliable this estimate is. A second random sample of 2054 adults would have different people in it. It is almost certain that there would not be exactly 1063 positive responses. That is, the value of the statistic $\hat{p}$ varies from sample to sample.

sampling variability

This basic fact is called **sampling variability:** the value of a statistic varies in repeated random sampling. Could it happen that this second random sample of 2054 adults finds that only 38% play video games? If the variation from sample to sample is too great, then we can't trust the results of any one sample.

All of statistical inference is based on one idea: to see how trustworthy a procedure is, ask what would happen if we repeated it many times. In terms of Example 6.1, this means we want to study the distribution of the statistic $\hat{p}$ when multiple SRSs of size $n = 2054$ are drawn. This distribution reveals the sampling variability and allows us to determine how unusual a second sample with $\hat{p} = 0.38$ would be.

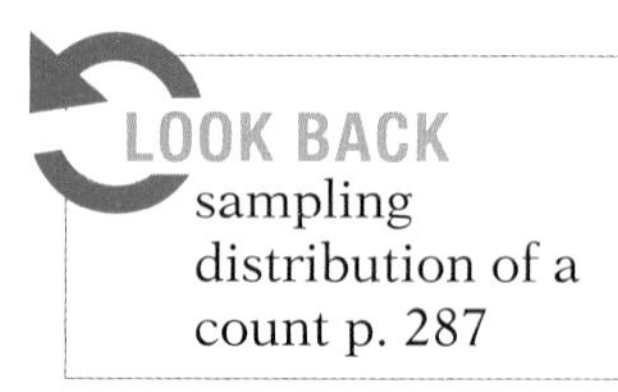

sampling distribution of a count p. 287

Before addressing this specific example, let's apply what we learned in Chapter 5 to describe the distribution of $\hat{p}$ in an SRS of size n. Recall that if the population is much larger than the sample (at least 10 times as large), the distribution of the count X has approximately the binomial distribution $B(n, p)$.[5] Since the sample proportion $\hat{p}$ is just the count X divided by the sample size n, the distribution of $\hat{p}$ is related to that of X. This does *not* mean that the proportion $\hat{p}$ has a binomial distribution. The count X takes whole-number values between 0 and n, but a proportion is always a number between 0 and 1. We can, however, do probability calculations about $\hat{p}$ by restating them in terms of the count X and using binomial methods. Here's an example:

EXAMPLE 6.2

Buying clothes online. A survey by the Consumer Reports National Research Center revealed that 85% of all respondents were very or completely satisfied with their online clothes-shopping experience.[6] It was also reported,

however, that people over the age of 40 were generally more satisfied than younger respondents.

You decide to take a nationwide random sample of 2500 college students and ask if they agree or disagree that "I am very or completely satisfied with my online clothes-shopping experience." Suppose that 60% of all college students would agree if asked this question. In other words, assume that we know that the population parameter $p = 0.60$. For a sample of $n = 2500$ students, what is the probability that the sample proportion who agree is at least 58%?

Since the population of college students is much larger than 10 times the sample, the count X who agree has the binomial distribution $B(2500, 0.60)$. The sample proportion $\hat{p} = X/2500$ does *not* have a binomial distribution, because it is not a count. However, we can translate any question about a sample proportion $\hat{p}$ into a question about the count X. Because 58% of 2500 is 1450,

$$\begin{aligned} P(\hat{p} \geq 0.58) &= P(X \geq 1450) \\ &= P(X = 1450) + P(X = 1451) + \cdots + P(X = 2500) \end{aligned}$$

This is a rather elaborate calculation. We must add more than 1000 binomial probabilities. Software tells us that $P(\hat{p} \geq 0.58) = 0.9802$. Because some software packages cannot handle an n as large as 2500, we need another way to do this calculation.

As a first step, find the mean and standard deviation of a sample proportion. We know how to find the mean and standard deviation of a sample count, so apply the rules for the mean and variance of a constant times a random variable. Here is the result.

LOOK BACK
rules for means
p. 274
rules for variances
p. 277

MEAN AND STANDARD DEVIATION OF A SAMPLE PROPORTION

Let $\hat{p}$ be the sample proportion of successes in an SRS of size n drawn from a large population having population proportion p of successes. The mean and standard deviation of $\hat{p}$ are

$$\mu_{\hat{p}} = p$$

$$\sigma_{\hat{p}} = \sqrt{\frac{p(1-p)}{n}}$$

The formula for $\sigma_{\hat{p}}$ is exactly correct in the binomial setting. It is approximately correct for an SRS from a large population. We will use it when the population is at least 10 times as large as the sample.

EXAMPLE 6.3

The mean and the standard deviation. In reference to Example 6.2 where the population proportion $p = 0.60$, the mean and standard deviation of the proportion of the survey respondents who are very satisfied with their online clothes-shopping experience are

$$\mu_{\hat{p}} = p = 0.60$$

$$\sigma_{\hat{p}} = \sqrt{\frac{p(1-p)}{n}} = \sqrt{\frac{(0.60)(0.40)}{2500}} = 0.0098$$

USE YOUR KNOWLEDGE

6.5 Change the sample size. Suppose that in Example 6.2 a total of only 625 students are sampled. Find the mean and standard deviation of the sample proportion. Compare these values with those of Example 6.3. How does this quartering of n change the mean and standard deviation?

6.6 Find the mean and the standard deviation. If we toss a fair coin 100 times, the number of heads is a random variable that is binomial.

(a) Find the mean and the standard deviation of the sample proportion of heads.

(b) Is your answer to part (a) the same as the mean and the standard deviation of the sample count? Explain your answer.

Sampling distributions

Now that we've found the mean and standard deviation of $\hat{p}$, let's take a closer look at its distribution. Since we want to know what would happen if we took many samples, here's an approach to answer that question:

- Take a large number of samples of size n from the same population.
- Calculate the sample proportion $\hat{p}$ for each sample.
- Make a histogram of the values of $\hat{p}$.
- Examine the distribution displayed in the histogram for shape, center, and spread, as well as outliers or other deviations.

In practice it is too expensive to take many samples from a large population such as all adult U.S. residents. But we can imitate many samples by using random digits. Using random digits from a table or computer software to imitate chance behavior is called **simulation.**

simulation

EXAMPLE 6.4

Simulate a random sample. Recall Example 6.2 (page 318) where $p = 0.60$. We will simulate drawing a random sample of size 100 from the population of college students. Of course, we would not sample in practice if we already knew that $p = 0.60$. We are sampling here to understand how sampling behaves.

We can imitate the population by a table of random digits, with each entry standing for a person. Six of the 10 digits (say 0 to 5) stand for people who are very satisfied with online clothes shopping. The remaining four digits, 6 to 9, stand for those who are not. Because all digits in a random number table are equally likely, this assignment produces a population proportion of students equal to $p = 0.60$. We then imitate an SRS of 100 students from the population by taking 100 consecutive digits from Table B. The statistic $\hat{p}$ is the proportion of 0s to 5s in the sample.

Here are the first 100 entries in Table B, with digits 0 to 5 highlighted:

19**223**	9**5034**	**05**7**5**6	**2**87**13**	96**40**9	**12531**	**42544**	8**2**8**53**
7**3**676	**4**7**150**	99**400**	**01**9**2**7	**2**77**54**	**42**6**4**8	8**2425**	**3**6**2**9**0**
45467	7**1**7**0**9	77**55**8	**000**9**5**				

There are 64 digits between 0 and 5, so $\hat{p} = 64/100 = 0.64$. A second SRS based on the second 100 entries in Table B gives a different result, $\hat{p} = 0.55$. The two sample results are different, and neither is equal to the true population value $p = 0.60$. That's sampling variability.

Simulation is a powerful tool for studying chance. Now that we see how simulation works, it is faster to abandon Table B and to use a computer programmed to generate random numbers.

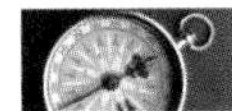

EXAMPLE 6.5

Take many random samples. Figure 6.1 illustrates the process of choosing many samples and finding the sample proportion $\hat{p}$ for each one. Follow the flow of the figure from the population at the left, to choosing an SRS and finding the $\hat{p}$ for this sample, to collecting together the $\hat{p}$'s from many samples. The histogram at the right of the figure shows the distribution of the values of $\hat{p}$ from 1000 separate SRSs of size 100 drawn from a population with $p = 0.60$.

In Example 6.2, we discuss a sample of 2500 students, not 100. Figure 6.2 is parallel to Figure 6.1. It shows the process of choosing 1000 SRSs, each of size 2500, from a population in which the true proportion is $p = 0.60$. The 1000 values of $\hat{p}$ from these samples form the histogram at the right of

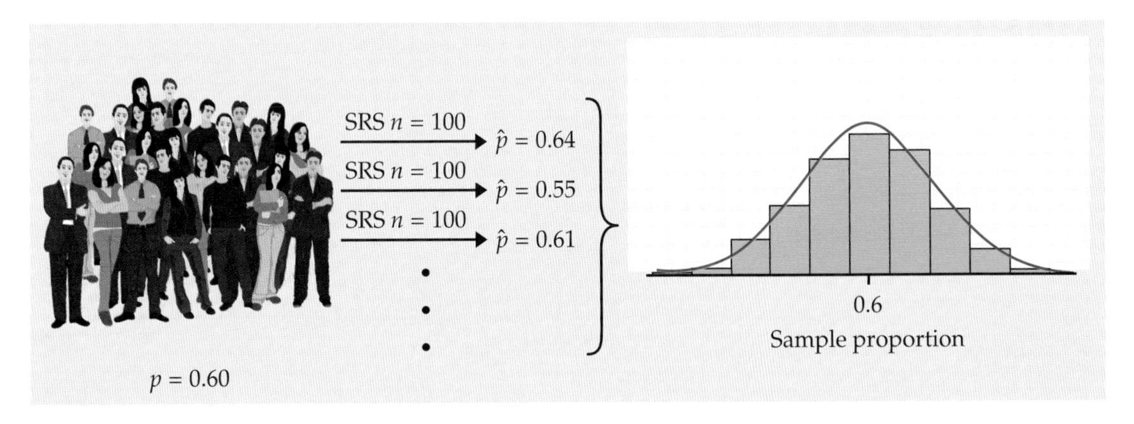

FIGURE 6.1 The results of many SRSs have a regular pattern. Here we draw 1000 SRSs of size 100 from the same population. The population proportion is $p = 0.60$. The histogram shows the distribution of 1000 sample proportions.

the figure. Figures 6.1 and 6.2 are drawn on the same scale. Comparing them shows what happens when we increase the size of our samples from 100 to 2500. These histograms display the *sampling distribution* of the statistic $\hat{p}$ for two sample sizes.

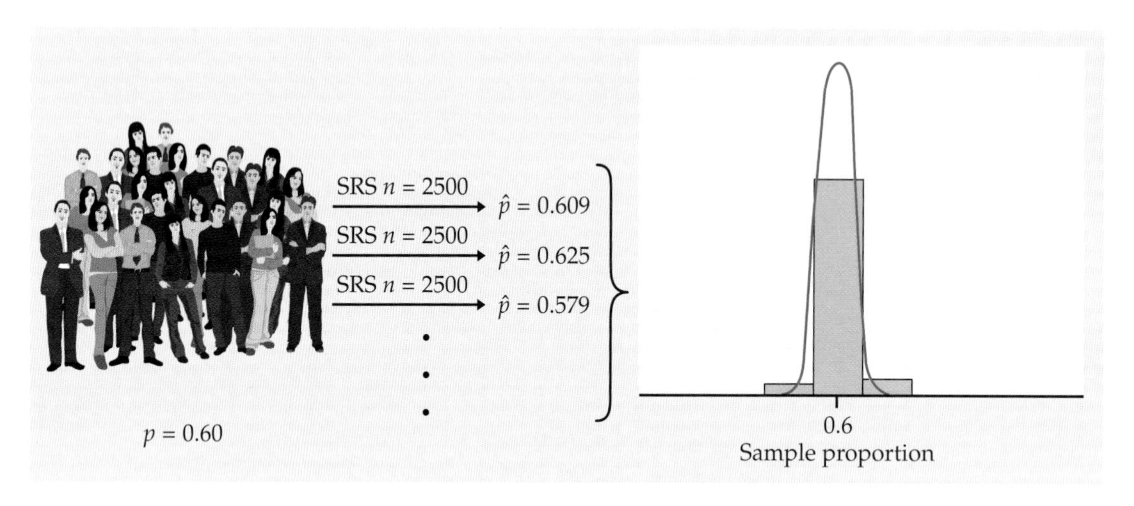

FIGURE 6.2 The distribution of sample proportions for 1000 SRSs of size 2500 drawn from the same population as in Figure 6.1. The two histograms have the same scale. The statistic from the larger sample is less variable.

SAMPLING DISTRIBUTION

The **sampling distribution** of a statistic is the distribution of values taken by the statistic in all possible samples of the same size from the same population.

Strictly speaking, the sampling distribution is the pattern that would emerge if we looked at all possible samples of a particular size from the population. A distribution obtained from a fixed number of samples, like the 1000 SRSs in Figure 6.1, is only an approximation to the sampling distribution. We will see in Chapter 7 that probability theory, the mathematics of chance behavior, can sometimes describe sampling distributions exactly. The interpretation of a sampling distribution is the same, however, whether we obtain it by simulation or by the mathematics of probability.

We can use the tools of data analysis to describe any distribution. Let's apply those tools to Figures 6.1 and 6.2.

- **Shape:** The histograms look Normal. Figure 6.3 is a Normal quantile plot of the values of $\hat{p}$ for our samples of size 100. It confirms that the distribution in Figure 6.1 is close to Normal. The 1000 values for samples of size 2500 in Figure 6.2 are even closer to Normal. The Normal curves drawn through the histograms describe the overall shape quite well.
- **Center:** In both cases, the values of the sample proportion $\hat{p}$ vary from sample to sample, but the values are centered at 0.60. We already knew this to be the case because the mean of $\hat{p}$ is p. Some samples have a $\hat{p}$ less

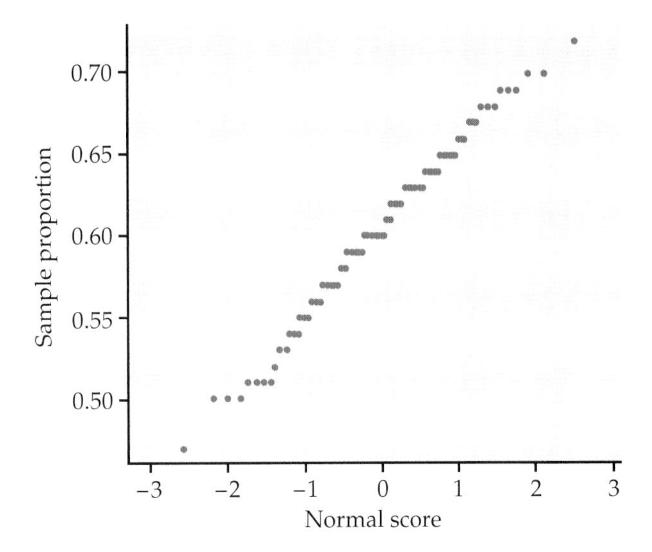

FIGURE 6.3 Normal quantile plot of the sample proportions in Figure 6.1. The distribution is close to Normal except for some granularity due to the fact that sample proportions from a sample size of 100 can take only values that are multiples of 0.01. Because a plot of 1000 points is hard to read, this plot presents only every 10th value.

bias p. 40

than 0.60 and some greater, but there is no tendency to be always low or always high. That is, $\hat{p}$ has no **bias** as an estimator of p. In other words, $\hat{p}$ is an unbiased estimator of p. This is true for both large and small samples.

- **Spread:** The values of $\hat{p}$ from samples of size 2500 are much less spread out than the values from samples of size 100. Less spread means a smaller standard deviation. Earlier we showed that the standard deviation of $\hat{p} = \sqrt{p(1-p)/n}$.

Although these results describe just two sets of simulations, they reflect facts that are true whenever we use random sampling.

USE YOUR KNOWLEDGE

6.7 Effect of sample size on the sampling distribution. You are planning a study and are considering taking an SRS of either 200 or 400 observations. Explain how the sampling distribution would differ for these two scenarios.

Normal approximation for a single proportion

Using simulation, we've shown that the sampling distribution of a sample proportion $\hat{p}$ is close to Normal. We also know that the distribution of $\hat{p}$ is that of a binomial count divided by the sample size n. This seems at first to be a contradiction. To clear up the matter, look at Figure 6.4. This is a probability histogram of the exact distribution of the proportion of contented shoppers $\hat{p}$, based on the binomial distribution $B(2500, 0.60)$. There are hundreds of narrow bars, one for each of the 2501 possible values of $\hat{p}$. Most have probabilities too small to show in a graph. *The probability histogram looks very Normal!*

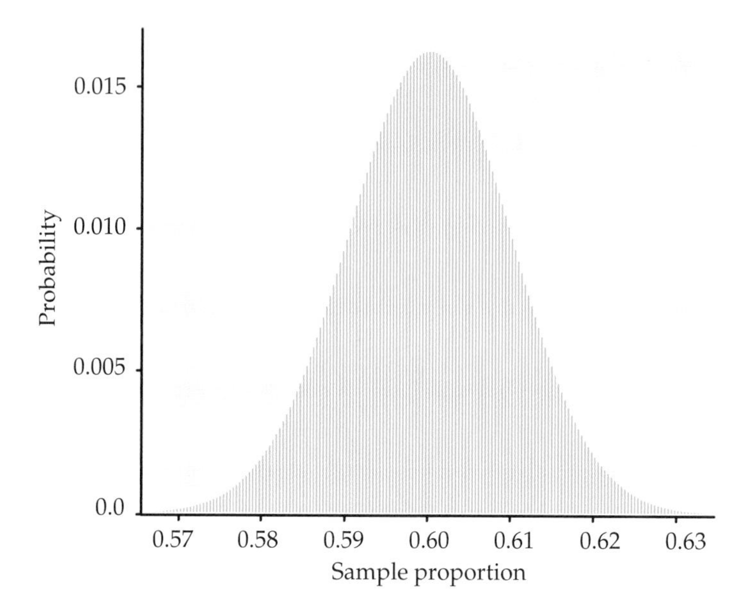

FIGURE 6.4 Probability histogram of the sample proportion $\hat{p}$ based on a binomial count with $n = 2500$ and $p = 0.6$. The distribution is very close to Normal.

> **NORMAL APPROXIMATION FOR A SINGLE PROPORTION**
>
> Draw an SRS of size n from a large population having population proportion p of successes. Let X be the count of successes in the sample and $\hat{p} = X/n$ be the sample proportion of successes. When n is large, the sampling distribution of $\hat{p}$ is approximately Normal:
>
> $$\hat{p} \text{ is approximately } N\left(p, \sqrt{\frac{p(1-p)}{n}}\right)$$
>
> As a rule of thumb, we will use this approximation for values of n and p that satisfy $np \geq 10$ and $n(1 - p) \geq 10$.

This Normal approximation is easy to remember because it says that $\hat{p}$ is Normal, with its usual mean and standard deviation. Whether or not you should use this Normal approximation should depend on how accurate your calculations need to be. For most statistical purposes great accuracy is not required. Our "rule of thumb" for use of this Normal approximation reflects this judgment.

The accuracy of the Normal approximation improves as the sample size n increases. It is most accurate for any fixed n when p is close to 1/2 and least accurate when p is near 0 or 1. You can compare binomial distributions with their Normal approximations by using the *Normal Approximation to Binomial Distributions* applet. This applet allows you to change n or p while watching the effect on the binomial probability histogram and the Normal curve that approximates it.

APPLET

EXAMPLE 6.6

Compare the Normal approximation with the exact calculation. Let's compare the Normal approximation for the calculation of Example 6.2 with the exact calculation from software. We want to calculate $P(\hat{p} \geq 0.58)$ when the sample size is $n = 2500$ and the population proportion is $p = 0.60$. Example 6.3 shows that

$$\mu_{\hat{p}} = p = 0.60$$

$$\sigma_{\hat{p}} = \sqrt{\frac{p(1-p)}{n}} = 0.0098$$

Act as if $\hat{p}$ were Normal with mean 0.60 and standard deviation 0.0098. The approximate probability, as illustrated in Figure 6.5, is

$$P(\hat{p} \geq 0.58) = P\left(\frac{\hat{p} - 0.60}{0.0098} \geq \frac{0.58 - 0.60}{0.0098}\right)$$

$$\doteq P(Z \geq -2.04) = 0.9793$$

That is, about 98% of all samples of size $n = 2500$ from the population of college students have a sample proportion that is at least 0.58. Because the sample was large, this Normal approximation is quite accurate. It misses the software value 0.9802 by only 0.0009.

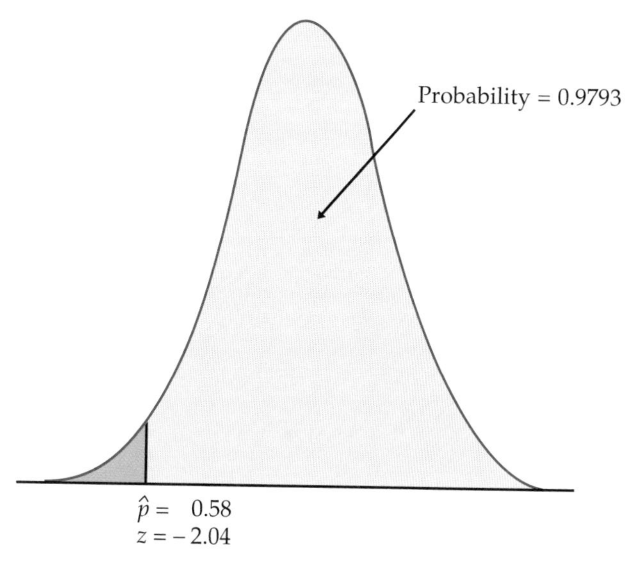

FIGURE 6.5 The Normal probability calculation for Example 6.6.

USE YOUR KNOWLEDGE

6.8 Use the Normal approximation. Suppose we toss a fair coin 100 times. Use the Normal approximation to find the probability that the sample proportion of heads is

(a) between 0.3 and 0.7.

(b) between 0.4 and 0.65.

Statistical confidence

Now that we know the sampling distribution of $\hat{p}$, we can proceed with inference. We will consider only inference procedures based on the Normal

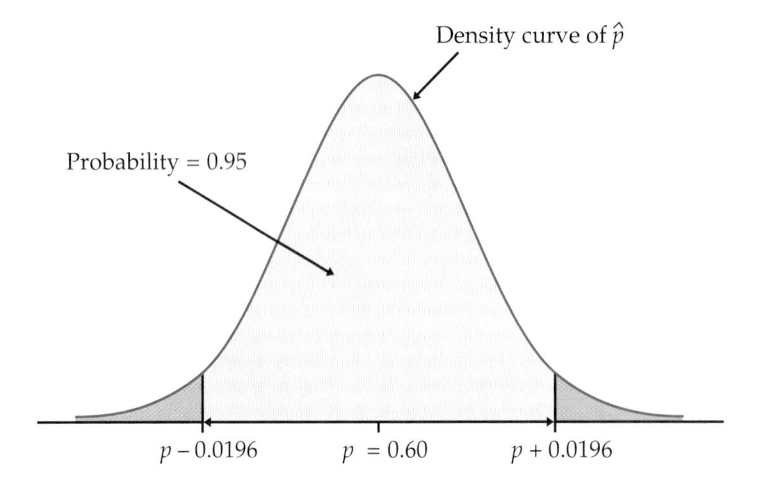

FIGURE 6.6 Distribution of the sample proportion for Example 6.2. The statistic $\hat{p}$ lies within ± 0.0196 points of p in 95% of all samples. This also means that p is within ± 0.0196 points of $\hat{p}$ in those samples.

approximation.[7] In other words, we will assume that the sample size n is sufficiently large so that $\hat{p}$ has approximately the Normal distribution with mean $\mu_{\hat{p}} = p$ and standard deviation $\sigma_{\hat{p}} = \sqrt{p(1-p)/n}$.

Consider the setting of Example 6.2 and this line of thought, which is illustrated by Figure 6.6:

LOOK BACK
68–95–99.7 rule
p. 132

- Given an SRS of $n = 2500$ students and $p = 0.60$, the distribution of $\hat{p}$ is approximately Normal with mean $\mu_{\hat{p}} = 0.60$ and standard deviation $\sigma_{\hat{p}} = \sqrt{0.60(1-0.60)/2500} = 0.0098$ (see Example 6.3).
- The 68–95–99.7 rule says that the probability is about 0.95 that $\hat{p}$ will be within 0.0196 (that is, two standard deviations of $\hat{p}$) of the population proportion p.
- To say that $\hat{p}$ lies within 0.0196 of p is the same as saying that p is within 0.0196 of $\hat{p}$.
- So about 95% of all samples will contain the population proportion p in the interval from $\hat{p} - 0.0196$ to $\hat{p} + 0.0196$.

We have simply restated a fact about the sampling distribution of $\hat{p}$. *The language of statistical inference uses this fact about what would happen in the long run to express our confidence in the results of any one sample.* We cannot know whether our sample is one of the roughly 95% for which the interval catches p or one of the unlucky 5% that does not catch p. The statement that we are 95% confident is shorthand for saying, "We arrived at these numbers by a method that gives correct results 95% of the time."

Confidence intervals

The interval between the values $\hat{p} \pm 0.0196$ is called an approximate *95% confidence interval* for p. Like most confidence intervals we will discuss, this one has the form

$$\text{estimate} \pm \text{margin of error}$$

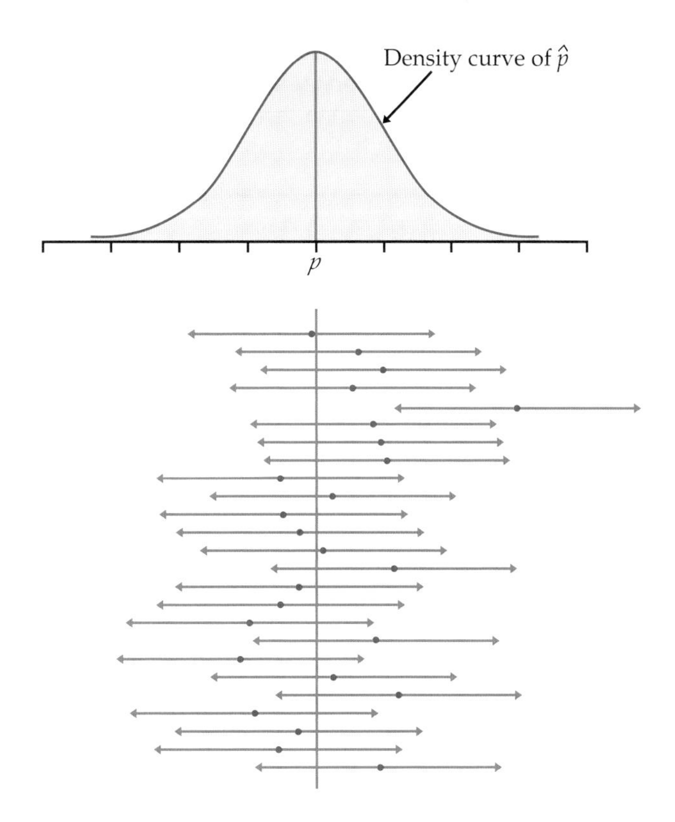

FIGURE 6.7 Twenty-five samples from the same population gave these 95% confidence intervals. In the long run, 95% of all samples give an interval that covers p. The sampling distribution of $\hat{p}$ is shown at the top.

margin of error

The estimate ($\hat{p}$ in this case) is our guess for the value of the unknown parameter based on the sample. The **margin of error** (0.0196 here) reflects how accurate we believe our guess is, based on the variability of the estimate, and how confident we are that the procedure will catch the true population proportion p.

Figure 6.7 illustrates the behavior of 95% confidence intervals in repeated sampling. The center of each interval is at $\hat{p}$ and therefore varies from sample to sample. The approximate sampling distribution of $\hat{p}$ appears at the top of the figure to show the long-term pattern of this variation. The approximate 95% confidence intervals, $\hat{p} \pm 0.0196$, from 25 SRSs of size $n = 2500$ appear below the sampling distribution. The center $\hat{p}$ of each interval is marked by a dot. The arrows on either side of the dot span the confidence interval. All except 1 of the 25 intervals cover the true value of p. In a very large number of samples, approximately 95% of the confidence intervals would contain p.

With the *Confidence Intervals for Proportions* applet, you can construct many diagrams similar to the one displayed in Figure 6.7.

Statisticians have constructed confidence intervals for many different parameters based on a variety of designs for data collection. We will meet a number of these in the following chapters. In all these situations, there are two common and important aspects of a confidence interval:

1. It is an interval of the form (a, b), where a and b are numbers computed from the data.

confidence level

2. It has a property called a **confidence level** that gives the probability of producing an interval that contains the unknown parameter.

Users can choose the confidence level, but 95% is the standard for most situations. Occasionally, 90% or 99% is used. We will use C to stand for the confidence level in decimal form. For example, a 95% confidence level corresponds to $C = 0.95$.

CONFIDENCE INTERVAL

A level C **confidence interval** for a parameter is an interval computed from sample data by a method that has probability C of producing an interval containing the true value of the parameter.

Large-sample confidence interval for a single proportion

We will now construct a level C confidence interval for the proportion p of a population when the data are an SRS of size n and n is sufficiently large so $\hat{p}$ has approximately the Normal distribution.

First, note that the standard deviation $\sigma_{\hat{p}}$ depends upon the unknown parameter p. To estimate this standard deviation using the data, we replace p in the interval formula by the sample proportion $\hat{p}$. We use the term *standard error* for the standard deviation of a statistic that is estimated from data.

STANDARD ERROR

When the standard deviation of a statistic is estimated from the data, the result is called the **standard error** of the statistic. The standard error of the sample proportion is

$$SE_{\hat{p}} = \sqrt{\frac{\hat{p}(1 - \hat{p})}{n}}$$

Second, recall that our construction of an approximate 95% confidence interval for the population proportion began by noting that any Normal distribution has probability about 0.95 within ± 2 standard deviations of its mean. To construct a level C confidence interval we first catch the central C area under a Normal curve. That is, we must find the number z^* such that any Normal distribution has probability C within $\pm z^*$ standard deviations of its mean. Because all Normal distributions have the same standardized form, we can obtain everything we need from the standard Normal curve. Figure 6.8 shows how C and z^* are related. Values of z^* for many choices of C appear in the row labeled z^* at the bottom of Table D. Here are the most important entries from that row:

z^*	1.645	1.960	2.576
C	90%	95%	99%

For values of C not in Table D, use Table A to find z^*.

As Figure 6.8 reminds us, any Normal curve has probability C between the point z^* standard deviations below the mean and the point z^* standard

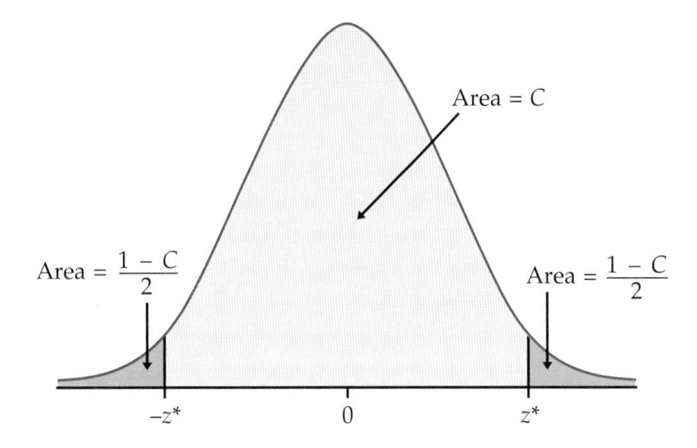

FIGURE 6.8 To construct a level C confidence interval, we must find the number z^*. This is how they are related. The area between $-z^*$ and z^* under the standard Normal curve is C.

deviations above the mean. The sample proportion $\hat{p}$ has the approximate Normal distribution with mean p and standard deviation $\sigma_{\hat{p}}$, so there is probability C that $\hat{p}$ lies between

$$p - z^*\sigma_{\hat{p}} \quad \text{and} \quad p + z^*\sigma_{\hat{p}}$$

This statement about $\hat{p}$'s location relative to the unknown population proportion p is exactly the same as saying that p lies between

$$\hat{p} - z^*\sigma_{\hat{p}} \quad \text{and} \quad \hat{p} + z^*\sigma_{\hat{p}}$$

If we replace the standard deviation of $\hat{p}$ with the standard error, there is approximate probability C that the interval $\hat{p} \pm z^*\text{SE}_{\hat{p}}$ contains p. This interval is our confidence interval. The estimate of the unknown p is $\hat{p}$, and the margin of error is $z^*\text{SE}_{\hat{p}}$.

LARGE-SAMPLE CONFIDENCE INTERVAL FOR A POPULATION PROPORTION

Choose an SRS of size n from a large population with an unknown proportion p of successes. The **sample proportion** is

$$\hat{p} = \frac{X}{n}$$

where X is the number of successes. The **standard error of $\hat{p}$** is

$$\text{SE}_{\hat{p}} = \sqrt{\frac{\hat{p}(1-\hat{p})}{n}}$$

and the **margin of error** for confidence level C is

$$m = z^*\text{SE}_{\hat{p}}$$

where the critical value z^* is the value for the standard Normal density curve with area C between $-z^*$ and z^*. An **approximate level C confidence interval** for p is

$$\hat{p} \pm m$$

Use this interval when the number of successes and failures are both at least 15. For most applications we recommend using 90%, 95%, or 99% confidence.

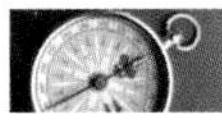

EXAMPLE 6.7

Inference for adults and video games. The sample survey in Example 6.1 found that 1063 of a sample of 2054 adults reported that they played video games. In that example we calculated $\hat{p} = 0.5175$. The standard error is

$$\text{SE}_{\hat{p}} = \sqrt{\frac{\hat{p}(1-\hat{p})}{n}} = \sqrt{\frac{0.5175(1-0.5175)}{2054}} = 0.011026$$

The critical value for 95% confidence is $z^* = 1.96$, so the margin of error is

$$m = 1.96\text{SE}_{\hat{p}} = (1.96)(0.011026) = 0.021610$$

The confidence interval is

$$\hat{p} \pm m = 0.52 \pm 0.02$$

We conclude that we are 95% confident that between 50% and 54% of adults play video games. We do not know if the true parameter p is 51% or 53%, but we are 95% confident that p is somewhere between 50% and 54%. Alternatively, we can say that based on this sample, 52% of adults play video games with a 95% margin of error of 2%.

In performing these calculations we have kept a large number of digits for our intermediate calculations. However, when reporting the results we prefer to use rounded values—for example, "52% with a margin of error of 2%." In this way we focus attention on the important parts that we have found. There is no additional information to be gained by reporting "0.5175 with a margin of error of 0.021610."

Remember that the margin of error in any confidence interval includes only random sampling error. If people do not respond honestly to the questions asked, for example, your estimate is likely to miss by more than the margin of error.

Because the calculations for statistical inference for a single proportion are relatively straightforward, we often do them with a calculator or in a spreadsheet. Figure 6.9 gives output from Minitab and SAS for the data on adults and video games in Example 6.1 (page 317). As usual, the output reports more digits than are useful. *When you use software, be sure to think about how many digits are meaningful for your purposes.* Do not clutter your report with information that is not meaningful. SAS gives the standard error next to the label ASE, which stands for "asymptotic standard error." The SAS output also includes an alternative interval based on an "exact" method.

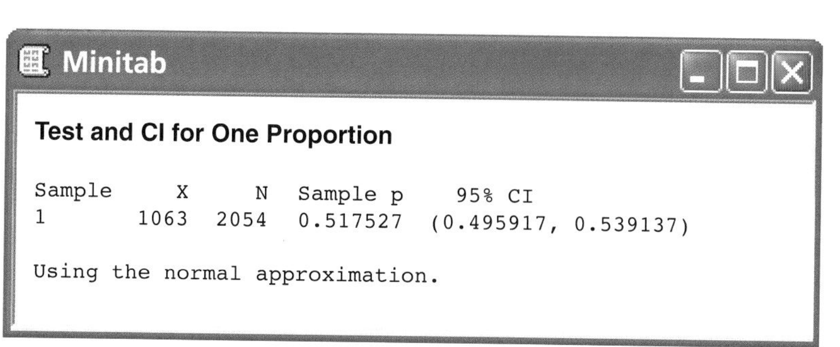

Minitab

Test and CI for One Proportion

```
Sample     X     N  Sample p        95% CI
1       1063  2054  0.517527  (0.495917, 0.539137)

Using the normal approximation.
```

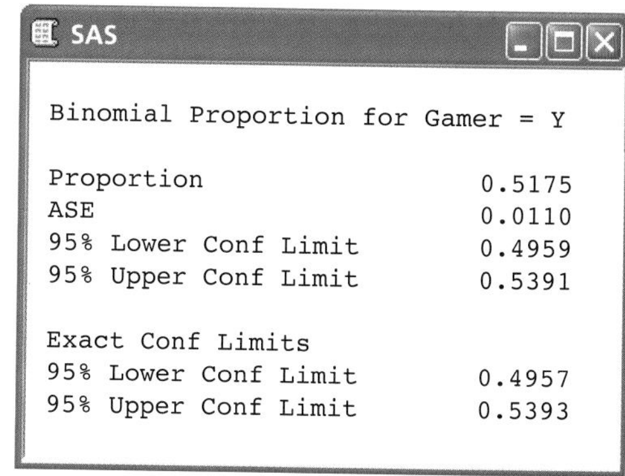

SAS

```
Binomial Proportion for Gamer = Y

Proportion                    0.5175
ASE                           0.0110
95% Lower Conf Limit          0.4959
95% Upper Conf Limit          0.5391

Exact Conf Limits
95% Lower Conf Limit          0.4957
95% Upper Conf Limit          0.5393
```

FIGURE 6.9 Minitab and SAS output for the confidence interval in Example 6.7.

We recommend the large-sample confidence interval for 90%, 95%, and 99% confidence whenever the number of successes and the number of failures are both at least 15. For smaller sample sizes, we recommend exact methods that use the binomial distribution. These are available as the default or as options in many statistical software packages.

USE YOUR KNOWLEDGE

6.9 Bank acquisitions. Refer to Exercise 6.3 (page 317).

(a) Find $SE_{\hat{p}}$, the standard error of $\hat{p}$.

(b) Give the 95% confidence interval for p in the form of estimate plus or minus the margin of error.

(c) Give the confidence interval as an interval of percents.

6.10 How often do they play? Refer to Exercise 6.4 (page 318).

(a) Find $SE_{\hat{p}}$, the standard error of $\hat{p}$.

(b) Give the 95% confidence interval for p in the form of estimate plus or minus the margin of error.

(c) Give the confidence interval as an interval of percents.

Significance tests

The confidence interval is appropriate when our goal is to estimate population parameters. The second common type of inference is directed at a quite different goal: to assess the evidence provided by the data in favor of some claim about the population parameters.

A significance test is a formal procedure for comparing observed data with a hypothesis whose truth we want to assess. The hypothesis is a statement about the population parameters. The results of a test are expressed in terms of a probability that measures how well the data and the hypothesis agree. We use the following example to illustrate these concepts.

EXAMPLE 6.8

College expectations. Each year the Cooperative Institutional Research Program (CIRP) Freshman Survey is administered to first-time incoming students at hundreds of colleges and universities. In 2011, 203,967 freshmen were polled. Of the respondents, 67.5% thought they had a very good chance of having at least a B average in college.[8] You decide to see if the view of incoming freshmen at your large university is more optimistic than this. You draw an SRS of $n = 200$ students and find $X = 147$ of them have this view. Can we conclude from this survey that incoming freshmen at your university have a more optimistic view of their grade point average than the general freshman population?

One way to answer this is to compute the probability of obtaining a sample proportion as large or larger than the observed $\hat{p} = 147/200 = 0.735$ assuming that, in fact, the population proportion at your university is 0.675. Software tells us that $P(X \geq 147) = 0.04$. Because this probability is relatively small, we conclude that observing a sample proportion of 0.735 is surprising when the true proportion is 0.675. The data provide evidence for us to conclude that the incoming freshmen at your university are more optimistic than the overall freshman population.

What are the key steps in this example?

- We started with a question about the population proportion. In this case the population was the incoming freshmen at your university. We wanted to see if this population had a proportion compatible with that of the overall freshman population. In other words, we wanted to compare the university population proportion with 0.675.
- Next we compared the data, $\hat{p} = 0.735$, with the value that comes from the question, $p = 0.675$.
- The result of the comparison is the probability 0.04.

The 0.04 probability is relatively small. Something that happens with probability 0.04 occurs 4 times out of 100. In this case we have two possible explanations:

1. We have observed something that is quite unusual, or
2. The assumption that underlies the calculation, $p = 0.675$, is not true.

Because this probability is small, we prefer the second conclusion: the incoming freshmen at your university are more optimistic than the overall freshman population.

Stating hypotheses

In Example 6.8, we asked whether the observed sample proportion is reasonable if, in fact, the underlying true proportion is 0.675. To answer this, we begin by supposing that the statement following the "if" in the previous sentence is true. In other words, we suppose that the true proportion is 0.675. We then ask whether the data provide evidence against the supposition we have made. If so, we have evidence in favor of an effect (the proportion is larger) that we are seeking. The first step in a test of significance is to state a claim that we will try to find evidence *against*.

> **NULL HYPOTHESIS**
>
> The statement being tested in a test of significance is called the **null hypothesis.** The test of significance is designed to assess the strength of the evidence against the null hypothesis. Usually the null hypothesis is a statement of "no effect" or "no difference."

We abbreviate "null hypothesis" as H_0. A null hypothesis is a statement about the population parameters. For example, the null hypothesis for Example 6.8 is

H_0: There is no difference between the population proportion at your university and that of the overall freshman population, $p = 0.675$.

Note that the null hypothesis refers to the *population* proportion for all freshmen from your university, including those for whom we do not have data.

alternative hypothesis

It is convenient also to give a name to the statement we hope or suspect is true instead of H_0. This is called the **alternative hypothesis** and is abbreviated as H_a. In Example 6.8, the alternative hypothesis states that the population proportion is larger than 0.675 (more optimistic). We write this as

H_a: The population proportion at your university is larger, $p > 0.675$.

Hypotheses always refer to some populations or a model, not to a particular outcome. For this reason, we must state H_0 and H_a in terms of population parameters.

one-sided or two-sided alternatives

Because H_a expresses the effect that we hope to find evidence *for,* we often begin with H_a and then set up H_0 as the statement that the hoped-for effect is not present. Stating H_a is often the more difficult task. It is not always clear, in particular, whether H_a should be **one-sided** or **two-sided,** which refers to whether a parameter differs from its null hypothesis value in a specific direction or in either direction.

The alternative hypothesis should express the hopes or suspicions we bring to the data. *It is cheating to first look at the data and then frame H_a to fit what the data show.* If you do not have a specific direction firmly in mind in advance, as we did in Example 6.8, you must use a two-sided alternative. Moreover, some users of statistics argue that we should always use a two-sided alternative.

USE YOUR KNOWLEDGE

6.11 Food court survey. The food court closest to your dormitory has been redesigned. A survey is planned to determine whether or not students think that the new design is an improvement. Sampled students will respond "Yes" if they think it is an improvement and "No" otherwise. The redesign will be considered a success if at least half the students view the redesign favorably. State the null and alternative hypotheses you would use to examine whether the redesign is a success.

6.12 More on the food court survey. Refer to the previous exercise. Suppose that the food court staff had input into the redesign and expect that 3 of every 4 students will view the redesign favorably. State the null and alternative hypotheses you would use to examine whether or not student opinions are different from staff expectations.

Test statistics

We will learn the form of significance tests in a number of common situations. Here are some principles that apply to most tests and that help in understanding these tests:

- The test is based on a statistic that estimates the parameter that appears in the hypotheses. Usually this is the same estimate we would use in a confidence interval for the parameter. When H_0 is true, we expect the estimate to take a value near the parameter value specified by H_0.
- Values of the estimate far from the parameter value specified by H_0 give evidence against H_0. The alternative hypothesis determines which directions count against H_0.
- To assess how far the estimate is from the parameter, standardize the estimate. In many common situations the test statistic has the form

$$z = \frac{\text{estimate} - \text{hypothesized value}}{\text{standard deviation of the estimate}}$$

test statistic

A **test statistic** measures compatibility between the null hypothesis and the data. We use it for the probability calculation that we need for our test of significance. It is a random variable with a distribution that we know.

In Example 6.8, an SRS of $n = 200$ students resulted in a sample proportion $\hat{p} = 147/200 = 0.735$. We can standardize this estimate based on the Normal approximation (page 324) to get the observed test statistic. For these data it is

$$z = \frac{0.735 - 0.675}{\sqrt{0.675(1 - 0.675)/200}} = 1.812$$

Normal distribution p. 129

This means we have observed a sample estimate that is slightly more than 1.8 standard deviations away from the hypothesized value of the parameter. We will now use facts about the Normal distribution to assess how unusual an observation this far away from the hypothesized value is.

P-values

If all test statistics were Normal, we could base our conclusions on the value of the z test statistic. In fact, the Supreme Court of the United States has said that "two or three standard deviations" ($z = 2$ or 3) is its criterion for rejecting H_0, and this is the criterion used in most applications involving the law. Because not all test statistics are Normal, we translate the value of test statistics into a common language, the language of probability.

A test of significance finds the probability of getting an outcome *as extreme or more extreme than the actually observed outcome.* "Extreme" means "far from what we would expect if H_0 were true." The direction or directions that count as "far from what we would expect" are determined by H_a and H_0.

> ### *P*-VALUE
> The probability, assuming H_0 is true, that the test statistic would take a value as extreme or more extreme than that actually observed is called the ***P*-value** of the test. The smaller the P-value, the stronger the evidence against H_0 provided by the data.

The key to calculating the P-value is the sampling distribution of the test statistic. For the problems we consider in this chapter, we need only the standard Normal distribution for the test statistic z.

In Example 6.8, an SRS of $n = 200$ students resulted in $\hat{p} = 147/200 = 0.735$. This sample proportion corresponds to 1.812 standard deviations away from the hypothesized parameter value of 0.675. Because we are using a one-sided alternative and our expectation is for the proportion to be larger than 0.675, the P-value is the probability that we observed a Z as extreme or more extreme than 1.812. More formally, this probability is

$$P(Z \geq 1.812)$$

where Z has the standard Normal distribution $N(0, 1)$. Using Table A, this probability is equal to 0.035. This is slightly lower than what was reported in Example 6.8, which used the binomial distribution to compute the P-value.

If we had used the two-sided alternative in this example, we would need to consider the probability in both tails of the Normal distribution. In other words, a sample proportion of 0.615, which is just as far away from 0.675 as 0.735, provides just as much evidence against the null hypothesis. The P-value would then be

$$P(|Z| \geq 1.812) = 0.070$$

which is exactly twice as large as the one-sided P-value.

USE YOUR KNOWLEDGE

6.13 The Normal curve and the *P*-value. A test statistic for a two-sided significance test for a population proportion is $z = -1.63$. Sketch a standard Normal curve and mark this value of z on it. Find the P-value and shade the appropriate areas under the curve to illustrate your calculations.

6.14 More on the Normal curve and the *P*-value. A test statistic for a two-sided significance test for a population proportion is $z = 2.42$. Sketch a standard Normal curve and mark this value of z on it. Find the P-value and shade the appropriate areas under the curve to illustrate your calculations.

Statistical significance

We started our discussion of the reasoning of significance tests with the statement of null and alternative hypotheses. We then learned that a test statistic is the tool used to examine the compatibility of the observed data with the null hypothesis. Finally, we translated the test statistic into a P-value to quantify the evidence against H_0. One important final step is needed: to state our conclusion.

We can compare the P-value we calculated with a fixed value that we regard as decisive. This amounts to announcing in advance how much evidence against H_0 we will require to reject H_0. The decisive value of P is called the **significance level.** It is commonly denoted by α. If we choose $\alpha = 0.05$, we are requiring that the data give evidence against H_0 so strong that it would happen no more than 5% of the time (1 time in 20) when H_0 is true. If we choose $\alpha = 0.01$, we are insisting on stronger evidence against H_0, evidence so strong that it would appear only 1% of the time (1 time in 100) if H_0 is in fact true.

significance level

STATISTICAL SIGNIFICANCE

If the P-value is as small or smaller than α, we say that the data are **statistically significant at level α.**

In Example 6.8, the P-value is 0.04 (or 0.035 using the Normal approximation). If we choose $\alpha = 0.05$, we would state that there is enough evidence against H_0 to reject this hypothesis and conclude that the proportion at your university is higher than 0.675.

However, if we required stronger evidence and chose $\alpha = 0.01$, we would say that the data do not provide enough evidence to conclude that the proportion at your university is higher than 0.675. This does not mean we conclude that H_0 is true—that the proportion at your university is equal to 0.675—only that the level of evidence we required to reject H_0 was not met.

Our criminal court system follows a similar procedure in which a defendant is presumed innocent (H_0) until proven guilty. If the level of evidence presented is not strong enough for the jury to find the defendant guilty beyond a reasonable doubt, the defendant is acquitted. Acquittal does not imply innocence, only that the degree of evidence was not strong enough to prove guilt.

We will learn the details of many tests of significance in the following chapters. The proper test statistic is determined by the hypotheses and the data collection design. We use computer software or a calculator to find its numerical value and the P-value. The computer will not formulate your hypotheses for you, however. Nor will it decide if significance testing is appropriate or help you to interpret the P-value that it presents to you. The most difficult and important step is the last one: stating a conclusion.

Significance test for a single proportion

Recall that the sample proportion $\hat{p} = X/n$ is approximately Normal, with mean $\mu_{\hat{p}} = p$ and standard deviation $\sigma_{\hat{p}} = \sqrt{p(1-p)/n}$ (page 324). For confidence intervals, we substitute $\hat{p}$ for p in the last expression to obtain the standard error. When performing a significance test, however, the null hypothesis specifies a value for p, and we assume that this is the true value when calculating the P-value. Therefore, when we test H_0: $p = p_0$, we substitute p_0 into the expression for $\sigma_{\hat{p}}$ and then standardize $\hat{p}$. Here are the details.

LARGE-SAMPLE SIGNIFICANCE TEST FOR A POPULATION PROPORTION

Draw an SRS of size n from a large population with an unknown proportion p of successes. To test the hypothesis H_0: $p = p_0$, compute the **z statistic**

$$z = \frac{\hat{p} - p_0}{\sqrt{\dfrac{p_0(1 - p_0)}{n}}}$$

In terms of a standard Normal random variable Z, the approximate P-value for a test of H_0 against

H_a: $p > p_0$ is $P(Z \geq z)$

H_a: $p < p_0$ is $P(Z \leq z)$

H_a: $p \neq p_0$ is $2P(Z \geq |z|)$

Use the large-sample z significance test as long as the expected number of successes, np_0, and the expected number of failures, $n(1 - p_0)$, are both at least 10.

This large-sample z significance test also relies on the Normal approximation, so if our rule of thumb is not met, or the population is less than 10 times as large as the sample, other procedures should be used. Here is a large-sample example.

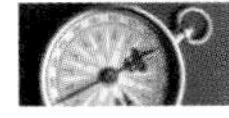

EXAMPLE 6.9

Comparing two sunblock lotions. Your company produces a sunblock lotion designed to protect the skin from exposure to both UVA and UVB radiation from the sun. You hire a testing firm to compare your product with the product sold by your major competitor. The testing firm exposes skin on the backs of a sample of 20 people to UVA and UVB rays and measures the protection provided by each product. For 13 of the subjects, your product provided better protection, while for the other 7 subjects, your competitor's product provided better protection.

Do you have evidence to support a commercial claiming that your product provides superior UVA and UVB protection? To answer this, we first need to state the hypotheses. The parameter of interest is p, the proportion of people who would receive superior UVA and UVB protection from your product. If there is no difference between these two lotions, then we'd expect $p = 0.5$. (In other words, which product works better on someone is like flipping a fair coin.) This is our null hypothesis. As for H_a, we'll use the two-sided alternative even though your company hopes for $p > 0.5$.

So, to answer this claim, we have $n = 20$ subjects and $X = 13$ successes and want to test

$$H_0: p = 0.5$$
$$H_a: p \neq 0.5$$

The expected numbers of successes (your product provides better protection) and failures (your competitor's product provides better protection) are $20 \times 0.5 = 10$ and $20 \times 0.5 = 10$. Both are at least 10, so we can use the z test. The sample proportion is

$$\hat{p} = \frac{X}{n} = \frac{13}{20} = 0.65$$

The test statistic is

$$z = \frac{\hat{p} - p_0}{\sqrt{\dfrac{p_0(1 - p_0)}{n}}} = \frac{0.65 - 0.5}{\sqrt{\dfrac{(0.5)(0.5)}{20}}} = 1.34$$

From Table A we find $P(Z \geq 1.34) = 0.9099$, so the probability in the upper tail is $1 - 0.9099 = 0.0901$. The P-value is the area in both tails, so $P = 2 \times 0.0901 = 0.1802$. Minitab and SAS outputs for this analysis appear in Figure 6.10. We conclude that the sunblock-testing data are compatible with the hypothesis of no difference between your product and your competitor's ($\hat{p} = 0.65$, $z = 1.34$, $P = 0.18$). The data do not provide you with a basis to support your advertising claim.

Note that we used a two-sided hypothesis test when we compared the two sunblock lotions in Example 6.9. In settings like this, we must start with the view that either product could be better if we want to prove a claim of superiority. Thinking or hoping that your product is superior cannot be used to justify a one-sided test.

USE YOUR KNOWLEDGE

6.15 Draw a picture. Draw a picture of a standard Normal curve and shade the tail areas to illustrate the calculation of the P-value for Example 6.9.

6.16 What does the confidence interval tell us? Inspect the outputs in Figure 6.10 and report the confidence interval for the percent of people who would get better sun protection from your product than from your competitor's. Be sure to convert from proportions to percents and to round appropriately. Interpret the confidence interval. In Chapter 8, we'll

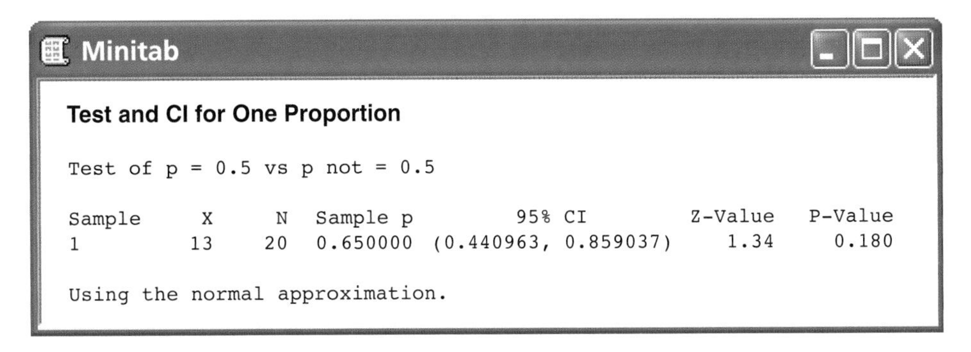

Test and CI for One Proportion

Test of p = 0.5 vs p not = 0.5

Sample	X	N	Sample p	95% CI	Z-Value	P-Value
1	13	20	0.650000	(0.440963, 0.859037)	1.34	0.180

Using the normal approximation.

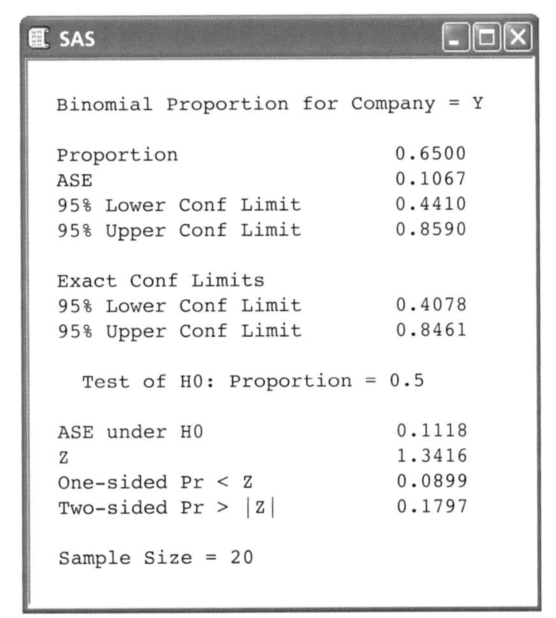

Binomial Proportion for Company = Y

Proportion	0.6500
ASE	0.1067
95% Lower Conf Limit	0.4410
95% Upper Conf Limit	0.8590
Exact Conf Limits	
95% Lower Conf Limit	0.4078
95% Upper Conf Limit	0.8461

Test of H0: Proportion = 0.5

ASE under H0	0.1118
Z	1.3416
One-sided Pr < Z	0.0899
Two-sided Pr > |Z|	0.1797

Sample Size = 20

FIGURE 6.10 Minitab and SAS output for the significance test in Example 6.9.

discuss the relationship between confidence intervals and two-sided tests in more detail.

6.17 The effect of *X*. In Example 6.9, suppose that your product provided better UVA and UVB protection for 15 of the 20 subjects. Perform the significance test using these results and summarize the results.

6.18 The effect of *n*. In Example 6.9, consider what would have happened if you had paid for twice as many subjects to be tested. Assume that the results would be similar to those in your test—that is, 65% of the subjects had better UVA and UVB protection with your product. Perform the significance test and summarize the results.

In Example 6.9, we treated an outcome as a success whenever your product provided better sun protection. Would we get the same results if we defined success as an outcome where your competitor's product was superior? In this setting the null hypothesis is still H_0: $p = 0.5$. You will find that the z test statistic is unchanged except for its sign and that the P-value remains the same.

USE YOUR KNOWLEDGE

6.19 Yes or no? In Example 6.9 we performed a significance test to compare your product with your competitor's. Success was defined as the outcome where your product provided better protection. Now, take the viewpoint of your competitor, where success is defined to be the outcome where your competitor's product provides better protection. In other words, n remains the same (20) but X is now 7.

(a) Perform the two-sided significance test to verify that the P-value remains the same.

(b) Using the outputs in Figure 6.10, construct a 95% confidence interval for the proportion of people who would get better sun protection from your competitor's product. (*Hint:* Use the fact that this proportion is 1 minus the previous proportion.)

We do not often use significance tests for a single proportion, because it is uncommon to have a situation where there is a precise p_0 that we want to test. For physical experiments such as coin tossing or drawing cards from a well-shuffled deck, probability arguments lead to an ideal p_0. Even here, however, it can be argued, for example, that no real coin has a probability of heads *exactly* equal to 0.5. Data from past large samples can sometimes provide a p_0 for the null hypothesis of a significance test. In some types of epidemiology research, for example, "historical controls" from past studies serve as the benchmark for evaluating new treatments. Medical researchers argue about the validity of these approaches, because the past never quite resembles the present. In general, we prefer comparative studies whenever possible.

SECTION 6.1 SUMMARY

A number that describes a population is a **parameter.** A number that can be computed from the data is a **statistic.** The purpose of sampling or experimentation is usually **inference:** using sample statistics to make statements about unknown population parameters.

A statistic from a probability sample or randomized experiment has a **sampling distribution** that describes how the statistic varies in repeated data production. The sampling distribution answers the question "What would happen if we repeated the sample or experiment many times?" Formal statistical inference is based on the sampling distributions of statistics.

Inference about a population proportion p from an SRS of size n is based on the **sample proportion** $\hat{p} = X/n$. When n is large, $\hat{p}$ has approximately the Normal distribution with mean p and standard deviation $\sqrt{p(1-p)/n}$.

The purpose of a **confidence interval** is to estimate an unknown parameter with an indication of how accurate the estimate is and of how confident we are that the result is correct.

Any confidence interval has two parts: an interval computed from the data and a confidence level. The interval often has the form

$$\text{estimate} \pm \text{margin of error}$$

The **confidence level** states the probability that the method will give a correct answer. That is, if you use 95% confidence intervals, in the long run 95% of your intervals will contain the true parameter value. When you apply the method once, you do not know whether your interval gave a correct value (this happens 95% of the time) or not (this happens 5% of the time).

For large samples, the **margin of error for confidence level C** of a proportion is

$$m = z^* \text{SE}_{\hat{p}}$$

where the critical value z^* is the value for the standard Normal density curve with area C between $-z^*$ and z^*, and the

standard error of $\hat{p}$ is

$$SE_{\hat{p}} = \sqrt{\frac{\hat{p}(1-\hat{p})}{n}}$$

The **level C large-sample confidence interval** is

$$\hat{p} \pm m$$

We recommend using this interval for 90%, 95%, and 99% confidence whenever the number of successes and the number of failures are both at least 15.

A **test of significance** is intended to assess the evidence provided by data against a **null hypothesis H_0** in favor of an **alternative hypothesis H_a**.

The hypotheses are stated in terms of population parameters. Usually H_0 is a statement that no effect or no difference is present, and H_a says that there is an effect or difference, in a specific direction (**one-sided alternative**) or in either direction (**two-sided alternative**).

The test is based on a **test statistic.** The ***P*-value** is the probability, computed assuming that H_0 is true, that the test statistic will take a value at least as extreme as that actually observed. Small P-values indicate strong evidence against H_0. Calculating P-values requires knowledge of the sampling distribution of the test statistic when H_0 is true.

If the P-value is as small or smaller than a specified value α, the data are **statistically significant** at significance level α.

Tests of H_0: $p = p_0$ are based on the **z statistic**

$$z = \frac{\hat{p} - p_0}{\sqrt{\frac{p_0(1-p_0)}{n}}}$$

with P-values calculated from the $N(0, 1)$ distribution. Use this procedure when the expected number of successes, np_0, and the expected number of failures, $n(1-p_0)$, are both at least 10.

SECTION 6.1 EXERCISES

For Exercises 6.1 and 6.2, see pages 316–317; for Exercises 6.3 and 6.4, see pages 317–318; for Exercises 6.5 and 6.6, see page 320; for Exercise 6.7, see page 323; for Exercise 6.8, see page 325; for Exercises 6.9 and 6.10, see page 331; for Exercises 6.11 and 6.12, see pages 333–334; for Exercises 6.13 and 6.14, see page 335; for Exercises 6.15 to 6.18, see pages 338–339; and for Exercise 6.19, see page 340.

6.20. What's wrong? Explain what is wrong with each of the following:

(a) You can use a significance test to evaluate the hypothesis H_0: $\hat{p} = 0.6$ versus the two-sided alternative.

(b) The large-sample significance test for a population proportion is based on the binomial distribution.

(c) An approximate 95% confidence interval for an unknown proportion p is $\hat{p}$ plus or minus its standard error.

6.21. What's wrong? Explain what is wrong with each of the following:

(a) The margin of error for a confidence interval used for an opinion poll takes into account the fact that people who did not answer the poll questions may have had different responses from those who did answer the questions.

(b) If the P-value for a significance test is 0.35, we can conclude that the null hypothesis has a 35% chance of being true.

(c) A student project used a confidence interval to describe the results in a final report. The confidence level was 110%.

6.22. Draw some pictures. Consider the binomial setting with $n = 50$ and $p = 0.4$.

(a) The sample proportion $\hat{p}$ will have a distribution that is approximately Normal. Give the mean and the standard deviation of this Normal distribution.

(b) Draw a sketch of this Normal distribution. Mark the location of the mean.

(c) Find a value p^* for which the probability is 95% that $\hat{p}$ will be between $\pm p^*$. Mark these two values on your sketch.

6.23. "Country food" and Inuits. "Country food" for Inuits includes seal, caribou, whale, ducks, fish, and berries and is an important part of the diet of the

aboriginal people called Inuits who inhabit Inuit Nunaat, the northern region of what is now called Canada. A survey of Inuits in Inuit Nunaat reported that 3274 out of 5000 respondents said that at least half of the meat and fish that they eat is country food.[9] Find the sample proportion and a 95% confidence interval for the population proportion of Inuits whose meat and fish consumption consists of at least half country food.

6.24. Most desirable mates. A poll of 5000 residents in Brazil, Canada, China, France, Malaysia, South Africa, and the United States asked about what profession they would prefer a marriage partner to have. The choice receiving the highest percent, 16% of the responses, was doctors, nurses, and other health care professionals.[10]

(a) Find the sample proportion and a 95% confidence interval for the proportion of people who would prefer a doctor, nurse, or other health care professional as a marriage partner.

(b) Convert the estimate and the confidence interval to percents.

6.25. Guitar Hero and Rock Band. An electronic survey of size 7061 reported that 67% of players of Guitar Hero and Rock Band who do not currently play a musical instrument said that they are likely to begin playing a real musical instrument in the next two years.[11] The reports describing the survey do not give the number of respondents who do not currently play a musical instrument.

(a) Explain why it is important to know the number of respondents who do not currently play a musical instrument.

(b) Assume that half of the respondents do not currently play a musical instrument. Find the count of players who said that they are likely to begin playing a real musical instrument in the next two years.

(c) Give a 99% confidence interval for the population proportion who would say that they are likely to begin playing a real musical instrument in the next two years.

(d) The survey collected data from two separate consumer panels. There were 3300 respondents from the LightSpeed consumer panel, and the others were from Guitar Center's proprietary consumer panel. Comment on the sampling procedure used for this survey and how it would influence your interpretation of the findings.

6.26. Guitar Hero and Rock Band. Refer to the previous exercise.

(a) How would the result that you reported in part (c) change if only 25% of the respondents said that they did not currently play a musical instrument?

(b) Do the same calculations if the percent was 75%.

(c) The main conclusion of the survey that appeared in many news stories was that 67% of players of Guitar Hero and Rock Band who do not currently play a musical instrument said that they are likely to begin playing a real musical instrument in the next two years. What can you conclude about the effect of the three scenarios (part (b) in the previous exercise and parts (a) and (b) in this exercise) on the margin of error for the main result?

6.27. Are seniors prepared for class? The National Survey of Student Engagement found that 24% of seniors report that they often or very often went to class without completing readings or assignments.[12] Assume that the sample size of seniors is 276,000.

(a) Find the margin of error for 99% confidence.

(b) Here are some items from the report that summarizes the survey. More than 537,000 students from 751 institutions in the United States and Canada participated. The average response rate was 33% and ranged from 15% to 87%. Institutions pay a participation fee of between $1800 and $7800 based on the size of their undergraduate enrollment. Discuss these as sources of error in this study. How do you think these errors would compare with the error that you calculated in part (a)?

6.28. Confidence level and interval width. Refer to Exercise 6.27. Would a 90% confidence interval be wider or narrower than the one that you found in that exercise? Verify your answer by computing the interval.

6.29. Can we use the *z* test? In each of the following cases state whether or not the Normal approximation to the binomial should be used for a significance test on the population proportion p.

(a) $n = 30$ and H_0: $p = 0.2$.

(b) $n = 30$ and H_0: $p = 0.6$.

(c) $n = 100$ and H_0: $p = 0.5$.

(d) $n = 200$ and H_0: $p = 0.01$.

6.30. Instant versus fresh-brewed coffee. A matched pairs experiment compares the taste of instant versus fresh-brewed coffee. Each subject tastes two unmarked cups of coffee, one of each type, in random order and states which he or she prefers. Of the 40 subjects who participate in the study, 12 prefer the instant coffee. Let p

be the probability that a randomly chosen subject prefers fresh-brewed coffee to instant coffee. (In practical terms, p is the proportion of the population who prefer fresh-brewed coffee.)

(a) Test the claim that a majority of people prefer the taste of fresh-brewed coffee. Report the large-sample z statistic and its P-value.

(b) Draw a sketch of a standard Normal curve and mark the location of your z statistic. Shade the appropriate area that corresponds to the P-value.

(c) Is your result significant at the 5% level? What is your practical conclusion?

6.31. CHALLENGE **Long sermons.** The National Congregations Study collected data in a one-hour interview with a key informant—that is, a minister, priest, rabbi, or other staff person or leader.[13] One question asked concerned the length of the typical sermon. For this question 390 out of 1191 congregations reported that the typical sermon lasted more than 30 minutes.

(a) Use the large-sample inference procedures to construct a 95% confidence interval for the true proportion of congregations in which the typical sermon lasts more than 30 minutes.

(b) The respondents to this question were not asked to use a stopwatch to record the lengths of a random sample of sermons at their congregations. They responded based on their impressions of the sermons. Do you think that ministers, priests, rabbis, or other staff persons or leaders might perceive sermon lengths differently from the people listening to the sermons? Discuss how your ideas would influence your interpretation of the results of this study.

6.32. Do you enjoy driving your car? The Pew Research Center polled $n = 1048$ U.S. drivers and found that 69% enjoyed driving their automobiles.[14]

(a) Construct a 95% confidence interval for the proportion of U.S. drivers who enjoy driving their automobiles.

(b) In 1991, a Gallup Poll reported this percent to be 79%. Using the data from this poll, test the claim that the percent of drivers who enjoy driving their cars has declined since 1991. Report the large-sample z statistic and its P-value.

6.33. Getting angry at other drivers. Refer to Exercise 6.32. The same Pew Poll found that 38% of the respondents "shouted, cursed or made gestures to other drivers" in the last year.

(a) Construct a 95% confidence interval for the true proportion of U.S. drivers who did these actions in the last year.

(b) Does the fact that the respondent is self-reporting these actions affect the way that you interpret the results? Write a short paragraph explaining your answer.

6.34. Cheating during a test. A national survey of high school students conducted by the Josephson Institute of Ethics was sent to 43,000 high school students, and 40,774 were returned. One question asked students if they had cheated during a test in the last school year.[15] Of those who returned the survey, 14,028 responded that they had cheated at least two times in the last year.

(a) What is the sample proportion of respondents who cheated at least twice?

(b) Compute the 95% confidence interval for the true proportion of students who have cheated on at least two tests in the last year.

(c) Compute the nonresponse rate for this study. Does this influence how you interpret these results? Write a short discussion of this issue.

6.35. Pet ownership among older adults. In a study of the relationship between pet ownership and physical activity in older adults,[16] 594 subjects reported that they owned a pet, while 1939 reported that they did not. Give a 95% confidence interval for the proportion of older adults in this population who are pet owners.

6.36. CHALLENGE **Annual income of older adults.** In the study described in the previous exercise, 1434 subjects out of a total of 2533 reported that their annual income was $25,000 or more.

(a) Give a 95% confidence interval for the true proportion of subjects in this population with incomes of at least $25,000.

(b) Do you think that some respondents might not give truthful answers to a question about their income? Discuss the possible effects on your estimate and confidence interval.

(c) In the previous exercise, the question analyzed concerned pet ownership. Compare this question with the income question with respect to the possibility that the respondents were not truthful.

6.37. Dogs sniffing out cancer. Can dogs detect lung cancer by sniffing exhaled breath samples? In one study, researchers performed 125 trials.[17] In each trial, a sniffer dog smelled five breath samples, consisting of four control samples and one cancer sample. A correct response involved the dog lying down next to the cancer sample. Collectively, the dogs correctly identified the cancer sample in 110 of these trials. Construct a 95% confidence interval for the true proportion of times these dogs will correctly identify a lung cancer sample.

6.38. CHALLENGE **Bicycle accidents and alcohol.** In the United States approximately 900 people die in bicycle accidents each year. One study examined the records of 1711 bicyclists aged 15 or older who were fatally injured in bicycle accidents between 1987 and 1991 and were tested for alcohol. Of these, 542 tested positive for alcohol (blood alcohol concentration of 0.01% or higher).[18]

(a) Summarize the data with appropriate descriptive statistics.

(b) To do statistical inference for these data, we think in terms of a model where p is a parameter that represents the probability that a tested bicycle rider is positive for alcohol. Find a 99% confidence interval for p.

(c) Can you conclude from your analysis of this study that alcohol causes fatal bicycle accidents? Explain.

(d) In this study 386 bicyclists had blood alcohol levels above 0.10%, a level defining legally drunk in many states at the time. Give a 99% confidence interval for the proportion who were legally drunk according to this criterion.

6.2 Comparing Two Proportions

Because comparative studies are so common, we often want to compare the proportions of two groups (such as men and women) that have some characteristic. In the previous section, we learned how to estimate a single proportion. Our problem now concerns the comparison of two proportions.

We call the two groups being compared Population 1 and Population 2 and the two population proportions of "successes" p_1 and p_2. The data consist of two independent SRSs, of size n_1 from Population 1 and size n_2 from Population 2. The proportion of successes in each sample estimates the corresponding population proportion. Here is the notation we will use in this section:

Population	Population proportion	Sample size	Count of successes	Sample proportion
1	p_1	n_1	X_1	$\hat{p}_1 = X_1/n_1$
2	p_2	n_2	X_2	$\hat{p}_2 = X_2/n_2$

To compare the two populations, we use the difference between the two sample proportions:

$$D = \hat{p}_1 - \hat{p}_2$$

Normal approximation for the difference between two proportions

LOOK BACK addition rule for means p. 274

Inference procedures for comparing proportions are z procedures based on the Normal approximation and on standardizing the difference D. The first step is to obtain the mean and standard deviation of D. By the addition rule for means, the mean of D is the difference between the means:

$$\mu_D = \mu_{\hat{p}_1} - \mu_{\hat{p}_2} = p_1 - p_2$$

That is, the difference $D = \hat{p}_1 - \hat{p}_2$ between the sample proportions is an unbiased estimator of the population difference $p_1 - p_2$. Similarly, the addition

LOOK BACK addition rule for variances p. 277

rule for variances tells us that the variance of D is the *sum* of the variances:

$$\sigma_D^2 = \sigma_{\hat{p}_1}^2 + \sigma_{\hat{p}_2}^2$$
$$= \frac{p_1(1-p_1)}{n_1} + \frac{p_2(1-p_2)}{n_2}$$

Each of these variances is approximately correct for an SRS from a large population. Throughout this section we will assume that each population is at least 10 times as large as its sample. Therefore, when n_1 and n_2 are large, D is approximately Normal with mean $\mu_D = p_1 - p_2$ and standard deviation

$$\sigma_D = \sqrt{\frac{p_1(1-p_1)}{n_1} + \frac{p_2(1-p_2)}{n_2}}$$

USE YOUR KNOWLEDGE

6.39 Rules for means and variances. Suppose that $p_1 = 0.4$, $n_1 = 25$, $p_2 = 0.5$, $n_2 = 30$. Find the mean and the standard deviation of the sampling distribution of $p_1 - p_2$.

6.40 Effect of the sample sizes. Suppose that $p_1 = 0.4$, $n_1 = 100$, $p_2 = 0.5$, $n_2 = 120$.

(a) Find the mean and the standard deviation of the sampling distribution of $p_1 - p_2$.

(b) The sample sizes here are four times as large as those in the previous exercise, while the population proportions are the same. Compare the results for this exercise with those that you found in the previous exercise. What is the effect of multiplying the sample sizes by 4?

6.41 Rules for means and variances. It is quite easy to verify the formulas for the mean and standard deviation of the difference D.

(a) What are the means and standard deviations of the two sample proportions $\hat{p}_1$ and $\hat{p}_2$?

(b) Use the addition rule for means of random variables: what is the mean of $D = \hat{p}_1 - \hat{p}_2$?

(c) The two samples are independent. Use the addition rule for variances of random variables: what is the variance of D?

Large-sample confidence interval for the difference between two proportions

To obtain a confidence interval for $p_1 - p_2$, we once again replace the unknown parameters in the standard deviation by estimates to obtain an estimated standard deviation, or standard error. Here is the confidence interval we want.

LARGE-SAMPLE CONFIDENCE INTERVAL FOR COMPARING TWO PROPORTIONS

Choose an SRS of size n_1 from a large population having proportion p_1 of successes and an independent SRS of size n_2 from another population having proportion p_2 of successes. The estimate of the difference between the population proportions is

$$D = \hat{p}_1 - \hat{p}_2$$

The **standard error of D** is

$$SE_D = \sqrt{\frac{\hat{p}_1(1-\hat{p}_1)}{n_1} + \frac{\hat{p}_2(1-\hat{p}_2)}{n_2}}$$

and the **margin of error** for confidence level C is

$$m = z^* SE_D$$

where the critical value z^* is the value for the standard Normal density curve with area C between $-z^*$ and z^*. An **approximate level C confidence interval** for $p_1 - p_2$ is

$$D \pm m$$

Use this interval when the number of successes and failures are both at least 10. For most applications we recommend using 90%, 95%, or 99% confidence.

EXAMPLE 6.10

Gender and the proportion of frequent binge drinkers. Many studies have documented binge drinking as a major problem among college students.[19] Here are some data that let us compare men and women:

Population	n	X	$\hat{p} = X/n$
1 (men)	5,348	1,392	0.260
2 (women)	8,471	1,748	0.206
Total	13,819	3,140	0.227

In this table the $\hat{p}$ column gives the sample proportions of frequent binge drinkers.

Let's find a 95% confidence interval for the difference between the proportions of men and of women who are frequent binge drinkers. Output from Minitab and CrunchIt! is given in Figure 6.11. To perform the computations using our formulas, we first find the difference in the proportions:

$$\begin{aligned} D &= \hat{p}_1 - \hat{p}_2 \\ &= 0.260 - 0.206 \\ &= 0.054 \end{aligned}$$

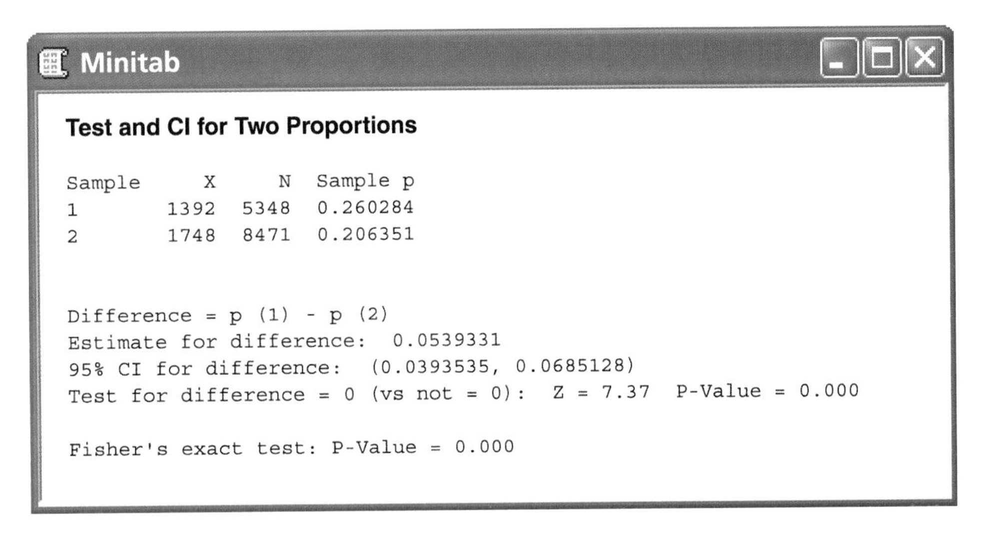

Minitab

Test and CI for Two Proportions

```
Sample      X      N   Sample p
1        1392   5348   0.260284
2        1748   8471   0.206351


Difference = p (1) - p (2)
Estimate for difference:  0.0539331
95% CI for difference:  (0.0393535, 0.0685128)
Test for difference = 0 (vs not = 0):  Z = 7.37  P-Value = 0.000

Fisher's exact test: P-Value = 0.000
```

CrunchIt!

	n	Successes	P-hat
Sample 1	5348	1392	0.2603
Sample 2	8471	1748	0.2064

Difference: 0.05393
StdErr: 0.007439
95% ConfInt: (0.03935, 0.06851)

FIGURE 6.11 Minitab and Crunch It! output for Example 6.10.

Then we calculate the standard error of D:

$$\begin{aligned} SE_D &= \sqrt{\frac{\hat{p}_1(1-\hat{p}_1)}{n_1} + \frac{\hat{p}_2(1-\hat{p}_2)}{n_2}} \\ &= \sqrt{\frac{(0.260)(0.740)}{5348} + \frac{(0.206)(0.794)}{8471}} \\ &= 0.00744 \end{aligned}$$

For 95% confidence, we have $z^* = 1.96$, so the margin of error is

$$\begin{aligned} m &= z^* SE_D \\ &= (1.96)(0.00744) \\ &= 0.015 \end{aligned}$$

The 95% confidence interval is

$$\begin{aligned} D \pm m &= 0.054 \pm 0.015 \\ &= (0.039, 0.069) \end{aligned}$$

With 95% confidence we can say that the difference in the proportions is between 0.039 and 0.069. Alternatively, we can report that the difference in the percent of men who are frequent binge drinkers and the percent of women who are frequent binge drinkers is 5.4%, with a 95% margin of error of 1.5%.

In this example men and women were not sampled separately. The sample sizes are in fact random and reflect the gender distributions of the colleges that were randomly chosen. Two-sample significance tests and confidence intervals are still approximately correct in this situation. The authors of the report note that women are overrepresented partly because 6 of the 140 colleges in the study were women's colleges.

In the example above we chose men to be the first population. Had we chosen women to be the first population, the estimate of the difference would be negative (−0.054). Because it is easier to discuss positive numbers, we generally choose the population with the higher proportion to be the first population.

USE YOUR KNOWLEDGE

6.42 Gender and commercial preference. A study was designed to compare two energy drink commercials. Each participant was shown the commercials in random order and asked to select the better one. Commercial A was selected by 44 out of 100 women and 79 out of 140 men. Give an estimate of the difference in gender proportions that favored Commercial A. Also construct a large-sample 95% confidence interval for this difference.

6.43 Gender and commercial preference, revisited. Refer to Exercise 6.42. Construct a 95% confidence interval for the difference in gender proportions that favor Commercial B. Explain how you could have obtained these results from the calculations you did in Exercise 6.42.

Significance test for a difference in proportions

Although we prefer to compare two proportions by giving a confidence interval for the difference between the two population proportions, it is sometimes useful to test the null hypothesis that the two population proportions are the same.

We standardize $D = \hat{p}_1 - \hat{p}_2$ by subtracting its mean $p_1 - p_2$ and then dividing by its standard deviation

$$\sigma_D = \sqrt{\frac{p_1(1-p_1)}{n_1} + \frac{p_2(1-p_2)}{n_2}}$$

If n_1 and n_2 are large, the standardized difference is approximately $N(0, 1)$. For the large-sample confidence interval, we used sample estimates in place of the unknown population values in the expression for σ_D. Although this approach would lead to a valid significance test, we instead adopt the more common practice of replacing the unknown σ_D with an estimate that takes into account our null hypothesis H_0: $p_1 = p_2$. If these two proportions are equal, then we can view all the data as coming from a single population. Let p denote the common value of p_1 and p_2. Then the standard deviation of $D = \hat{p}_1 - \hat{p}_2$ is

$$\begin{aligned}\sigma_D &= \sqrt{\frac{p(1-p)}{n_1} + \frac{p(1-p)}{n_2}} \\ &= \sqrt{p(1-p)\left(\frac{1}{n_1} + \frac{1}{n_2}\right)}\end{aligned}$$

We estimate the common value of p by the overall proportion of successes in the two samples:

$$\hat{p} = \frac{\text{number of successes in both samples}}{\text{number of observations in both samples}} = \frac{X_1 + X_2}{n_1 + n_2}$$

pooled estimate of p

This estimate of p is called the **pooled estimate** because it combines, or pools, the information from both samples.

To estimate σ_D under the null hypothesis, we substitute $\hat{p}$ for p in the expression for σ_D. The result is a standard error for D that assumes H_0: $p_1 = p_2$:

$$\text{SE}_{Dp} = \sqrt{\hat{p}(1 - \hat{p})\left(\frac{1}{n_1} + \frac{1}{n_2}\right)}$$

The subscript on SE_{Dp} reminds us that we pooled data from the two samples to construct the estimate.

SIGNIFICANCE TEST FOR COMPARING TWO PROPORTIONS

To test the hypothesis

$$H_0: p_1 = p_2$$

compute the z **statistic**

$$z = \frac{\hat{p}_1 - \hat{p}_2}{\text{SE}_{Dp}}$$

where the **pooled standard error** is

$$\text{SE}_{Dp} = \sqrt{\hat{p}(1 - \hat{p})\left(\frac{1}{n_1} + \frac{1}{n_2}\right)}$$

and where

$$\hat{p} = \frac{X_1 + X_2}{n_1 + n_2}$$

In terms of a standard Normal random variable Z, the P-value for a test of H_0 against

H_a: $p_1 > p_2$ is $P(Z \geq z)$

H_a: $p_1 < p_2$ is $P(Z \leq z)$

H_a: $p_1 \neq p_2$ is $2P(Z \geq |z|)$

As a general rule, we will use this test when the number of successes and the number of failures in each of the samples are at least 5.

This z test has the general form given on page 334 and is based on the Normal approximation to the binomial distribution.

EXAMPLE 6.11

Gender and the proportion of frequent binge drinkers: the z test. Are men and women college students equally likely to be frequent binge drinkers? We examine the survey data in Example 6.10 (page 346) to answer this question. Here is the data summary:

Population	n	X	$\hat{p} = X/n$
1 (men)	5,348	1,392	0.260
2 (women)	8,471	1,748	0.206
Total	13,819	3,140	0.227

The sample proportions are certainly quite different, but we will perform a significance test to see if the difference is large enough to lead us to believe that the population proportions are not equal. Formally, we test the hypotheses

$$H_0: p_1 = p_2$$
$$H_a: p_1 \neq p_2$$

The pooled estimate of the common value of p is

$$\hat{p} = \frac{1392 + 1748}{5348 + 8471} = \frac{3140}{13,819} = 0.227$$

Note that this is the estimate on the bottom line of the data summary.

The test statistic is calculated as follows:

$$\text{SE}_{Dp} = \sqrt{(0.227)(0.773)\left(\frac{1}{5348} + \frac{1}{8471}\right)} = 0.007316$$

$$z = \frac{\hat{p}_1 - \hat{p}_2}{\text{SE}_{Dp}} = \frac{0.260 - 0.206}{0.007316}$$
$$= 7.37$$

The P-value is $2P(Z \geq 7.37)$. The largest value of z in Table A is 3.49, so from this table we can conclude $P < 2 \times 0.0002 = 0.0004$. Most software reports this result as 0 or a very small number. Output from Minitab and CrunchIt! is given in Figure 6.12. Minitab reports the P-value as 0.000. This means that the calculated value is less than 0.0005; this is certainly a very small number. CrunchIt! gives < 0.0001.

The exact value is not particularly important. It is clear that we should reject the null hypothesis because the chance of getting these sample results is so small if the null hypothesis is true. For most situations, 0.001 (1 chance in 1000) is sufficiently small. We report: among college students in the study, 26.0% of the men and 20.6% of the women were frequent binge drinkers. The difference is statistically significant ($z = 7.37$, $P < 0.001$).

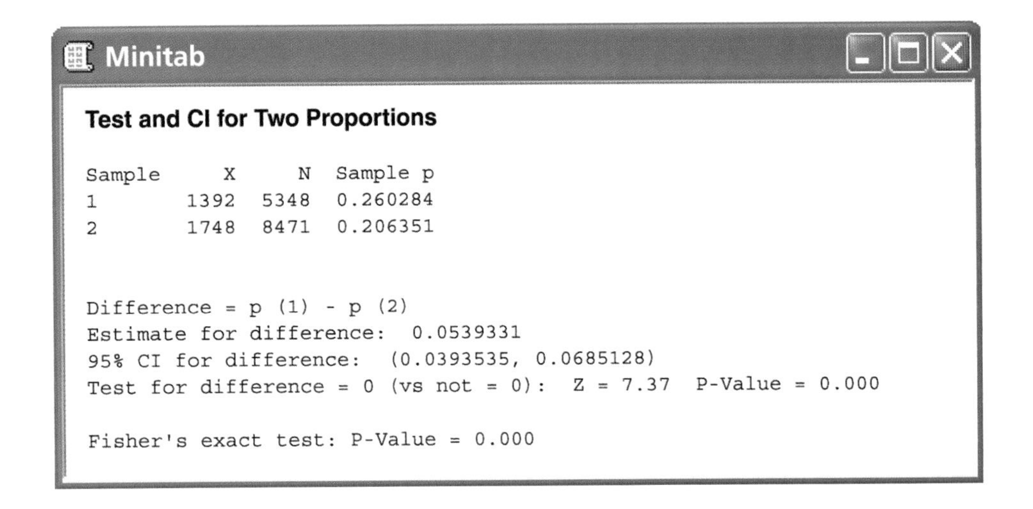

Minitab

Test and CI for Two Proportions

```
Sample      X     N  Sample p
1        1392  5348  0.260284
2        1748  8471  0.206351


Difference = p (1) - p (2)
Estimate for difference:  0.0539331
95% CI for difference:  (0.0393535, 0.0685128)
Test for difference = 0 (vs not = 0):  Z = 7.37  P-Value = 0.000

Fisher's exact test: P-Value = 0.000
```

Null hypothesis: Difference of proportions = 0
Alternative hypothesis: Difference of proportions is not 0

	n	Successes	P-hat
Sample 1	5348	1392	0.2603
Sample 2	8471	1748	0.2064

Difference: 0.05393
PooledStdErr: 0.007319
z statistic: 7.369
P-value: <0.0001

FIGURE 6.12 Minitab and Crunch It! output for Example 6.11.

We could have argued that we expect the proportion to be higher for men than for women in this example. This would justify using the one-sided alternative H_a: $p_1 > p_2$. The P-value would be half of the value obtained for the two-sided test. Because the z statistic is so large, this distinction is of no practical importance.

USE YOUR KNOWLEDGE

6.44 Gender and commercial preference: the z test. Refer to Exercise 6.42 (page 348). Test that the proportions of women and men who liked Commercial A are the same versus the two-sided alternative at the 5% level.

6.45 Changing the alternative hypothesis. Refer to the previous exercise. Does your conclusion change if you test whether the proportion of men who favor Commercial A is larger than the proportion of females? Explain.

SECTION 6.2 SUMMARY

The **large-sample estimate of the difference in two population proportions** is

$$D = \hat{p}_1 - \hat{p}_2$$

where $\hat{p}_1$ and $\hat{p}_2$ are the sample proportions

$$\hat{p}_1 = \frac{X_1}{n_1} \quad \text{and} \quad \hat{p}_2 = \frac{X_2}{n_2}$$

The **standard error of the difference *D*** is

$$SE_D = \sqrt{\frac{\hat{p}_1(1-\hat{p}_1)}{n_1} + \frac{\hat{p}_2(1-\hat{p}_2)}{n_2}}$$

The **margin of error for confidence level *C*** is

$$m = z^* SE_D$$

where z^* is the value for the standard Normal density curve with area C between $-z^*$ and z^*. The **large-sample level *C* confidence interval** is

$$D \pm m$$

We recommend using this interval for 90%, 95%, or 99% confidence when the number of successes and the number of failures in both samples are all at least 10.

Significance tests of H_0: $p_1 = p_2$ use the **z statistic**

$$z = \frac{\hat{p}_1 - \hat{p}_2}{SE_{Dp}}$$

with P-values from the $N(0, 1)$ distribution. In this statistic,

$$SE_{Dp} = \sqrt{\hat{p}(1-\hat{p})\left(\frac{1}{n_1} + \frac{1}{n_2}\right)}$$

and $\hat{p}$ is the **pooled estimate** of the common value of p_1 and p_2:

$$\hat{p} = \frac{X_1 + X_2}{n_1 + n_2}$$

We recommend the use of this test when the number of successes and the number of failures in each of the samples are at least 5.

SECTION 6.2 EXERCISES

For Exercises 6.39 to 6.41, see page 345; for Exercises 6.42 and 6.43, see page 348; and for Exercises 6.44 and 6.45, see page 351.

6.46. Draw a picture. Suppose that there are two binomial populations. For the first, the true proportion of successes is 0.4; for the second, it is 0.5. Consider taking independent samples from these populations, 50 from the first and 60 from the second.

(a) Find the mean and the standard deviation of the distribution of $\hat{p}_1 - \hat{p}_2$.

(b) This distribution is approximately Normal. Sketch this Normal distribution and mark the location of the mean.

(c) Find a value d for which the probability is 0.95 that the difference in sample proportions is within $\pm d$. Mark these values on your sketch.

6.47. What's wrong? For each of the following, explain what is wrong and why.

(a) A z statistic is used to test the null hypothesis that $\hat{p}_1 = \hat{p}_2$.

(b) If two sample proportions are equal, then the sample counts are equal.

(c) A 95% confidence interval for the difference in two proportions includes errors due to nonresponse.

6.48. Podcast downloading. The Podcast Alley Web site recently reported that they have 91,701 podcasts available for downloading, with 6,070,164 episodes.[20] The Pew Research Center performed two surveys about podcast downloading. The first was conducted between February and April 2006 and surveyed 2822 Internet users. It found that 198 of these said that they had downloaded a podcast to listen to it or view it later at least once. In a more recent survey, conducted in May 2008, there were 1553 Internet users. Of this total, 295 said that they had downloaded a podcast to listen to it or view it later.[21]

(a) Refer to the table that appears at the beginning of this section (page 344). Fill in the numerical values of all quantities that are known.

(b) Find the estimate of the difference between the proportion of Internet users who had downloaded podcasts as of February to April 2006 and the proportion as of May 2008.

(c) Is the large-sample confidence interval for the difference in two proportions appropriate to use in this setting? Explain your answer.

(d) Find the 95% confidence interval for the difference.

(e) Convert your estimated difference and confidence interval to percents.

(f) One of the surveys was conducted between February and April, whereas the other was conducted in May. Do you think that this difference should have any effect on the interpretation of the results? Be sure to explain your answer.

6.49. Significance test for podcast downloading. Refer to the previous exercise. Test the null hypothesis that the two proportions are equal versus the two-sided alternative. Report the test statistic with the P-value and summarize your conclusion.

6.50. Are more Internet users downloading podcasts? Refer to the previous two exercises. The ratio of the proportion in the 2008 sample to the proportion in the 2006 sample is about 2.7.

(a) Can you conclude that 2.7 times as many people are downloading podcasts? Explain why or why not.

(b) Can you conclude from the data available that there has been an increase from 2006 to 2008 in the number of people who download podcasts? If your answer is no, explain what additional data you would need or what additional assumptions you would need to be able to draw this conclusion.

6.51. Adult gamers versus teen gamers. A Pew Internet Project Data Memo presented data comparing adult gamers with teen gamers with respect to the devices on which they play. The data are from two surveys. The adult survey had 1063 gamers, while the teen survey had 1064 gamers. The memo reports that 54% of adult gamers played on game consoles (Xbox, PlayStation, Wii, etc.), while 89% of teen gamers played on game consoles.[22]

(a) Refer to the table that appears at the beginning of this section (page 344). Fill in the numerical values of all quantities that are known.

(b) Find the estimate of the difference between the proportion of teen gamers who played on game consoles and the proportion of adults who played on these devices.

(c) Is the large-sample confidence interval for the difference in two proportions appropriate to use in this setting? Explain your answer.

(d) Find the 95% confidence interval for the difference.

(e) Convert your estimated difference and confidence interval to percents.

(f) The adult survey was conducted between October and December 2008, whereas the teen survey was conducted between November 2007 and February 2008. Do you think that this difference should have any effect on the interpretation of the results? Be sure to explain your answer.

6.52. Significance test for gaming on consoles. Refer to the previous exercise. Test the null hypothesis that the two proportions are equal versus the two-sided alternative. Report the test statistic with the P-value and summarize your conclusion.

6.53. Gamers on computers. The report described in Exercise 6.51 also presented data from the same surveys for gaming on computers (desktops or laptops). These devices were used by 73% of adult gamers and by 76% of teen gamers. Answer the questions given in Exercise 6.51 for gaming on computers.

6.54. Significance test for gaming on computers. Refer to the previous exercise. Test the null hypothesis that the two proportions are equal versus the two-sided alternative. Report the test statistic with the P-value and summarize your conclusion.

6.55. Can we compare gaming on consoles with gaming on computers? Refer to the previous four exercises. Do you think that you can use the large-sample confidence intervals for a difference in proportions to compare teens' use of computers with teens' use of consoles? Write a short paragraph giving the reason for your answer. (*Hint:* Look carefully in the box on page 346 giving the assumptions needed for this procedure.)

6.56. CHALLENGE **$\hat{p}_1 - \hat{p}_2$ and the Normal distribution.** Refer to Exercise 6.46. Assume that all the conditions for that exercise remain the same, with the exception that $n_2 = 1000$.

(a) Find the mean and standard deviation of $\hat{p}_1 - \hat{p}_2$.

(b) Find the mean and standard deviation of $\hat{p}_1 - 0.5$.

(c) Because n_2 is very large, we expect $\hat{p}_2$ to be very close to 0.5. How close?

(d) Summarize what you have found in parts (a), (b), and (c) of this exercise. Interpret your results in terms of inference for comparing two proportions when the size of one of the samples is much larger than the size of the other.

6.57. Peer-to-peer music downloading. The NPD Group reported that the percent of Internet users who download music via peer-to-peer (P2P) services was 9% in late 2010, compared with 16% in late 2007.[23] The filing of lawsuits by the recording industry may be a reason why this percent has decreased. Assume that the sample sizes are both 5549, the sample size reported for the 2010 survey. Using a significance test, evaluate whether or not there has been a change in the percent of Internet users who download music via P2P services. Provide all details for the test and summarize your conclusion. Also report a 95% confidence interval for the difference in proportions (2007 versus 2010) and explain what information is provided in the interval that is not given in the significance test results.

6.58. More on downloading music via P2P. Refer to the previous exercise. Because we are not exactly sure about the size of the 2007 sample, redo the calculations for the significance test and the confidence interval under the following assumptions, and summarize the effects of the sample sizes on the results.

(a) The first sample size is 500.

(b) The first sample size is 1500.

(c) The first sample size is 3000.

6.59. Gender bias in textbooks. To what extent do syntax textbooks, which analyze the structure of sentences, illustrate gender bias? A study of this question sampled sentences from 10 texts.[24] One part of the study examined the use of the words "girl," "boy," "man," and "woman." We will call the first two words *juvenile* and the last two *adult*. Is the proportion of female references that are juvenile (girl) equal to the proportion of male references that are juvenile (boy)? Here are data from one of the texts:

Gender	n	X (juvenile)
Female	60	48
Male	132	52

(a) Find the proportion of juvenile references for females and its standard error. Do the same for the males.

(b) Give a 90% confidence interval for the difference and briefly summarize what the data show.

(c) Use a test of significance to examine whether the two proportions are equal.

6.60. Cheating during a test: 2002 versus 2010. In Exercise 6.34 (page 343), you examined the proportion of high school students who cheated on tests at least twice during the past year. Also available are results from other years. A reported 14,028 out of 40,774 students said they cheated at least twice in 2010. A reported 5794 out of 12,121 students said they cheated at least twice in 2002. Give an estimate of the difference between these two proportions with a 90% confidence interval.

6.61. Bicycle accidents, alcohol, and gender. In Exercise 6.38 (page 344) we examined the percent of fatally injured bicyclists tested for alcohol who tested positive. Here we examine the same data with respect to gender.

Gender	n	X (tested positive)
Female	191	27
Male	1520	515

(a) Summarize the data by giving the estimates of the two population proportions and a 95% confidence interval for their difference. Briefly summarize what the data show.

(b) Use a test of significance to examine whether the two proportions are equal.

CHAPTER 6 EXERCISES

6.62. Video game genres. U.S. computer and video game software sales were $25.1 billion in 2010.[25] A survey of 1102 teens collected data about their video game use. According to the survey, the most popular game genres are as summarized in the table at right.[26] Give a 95% confidence interval for the proportion who play games in each of these six genres.

Genre	Examples	Percent who play
Racing	NASCAR, Mario Kart, Burnout	74
Puzzle	Bejeweled, Tetris, Solitaire	72
Sports	Madden, FIFA, Tony Hawk	68
Action	Grand Theft Auto, Devil May Cry, Ratchet and Clank	67
Adventure	Legend of Zelda, Tomb Raider	66
Rhythm	Guitar Hero, Dance Dance Revolution, Lumines	61

6.63. CHALLENGE **Too many errors.** Refer to the previous exercise. The chance that each of the six intervals that you calculated includes the true proportion for that genre is approximately 95%. In other words, the chance that you make an error and your interval misses the true value is approximately 5%.

(a) Explain why the chance that at least one of your intervals does not contain the true value of the parameter is greater than 5%.

(b) One way to deal with this problem is to adjust the confidence level for each interval so that the overall probability of at least one miss is 5%. One simple way to do this is to use a **Bonferroni procedure.** Here is the basic idea: You have an error budget of 5% and you choose to spend it equally on six intervals. Each interval has a budget of $0.05/6 = 0.0083$. So each confidence interval should have a 0.83% chance of missing the true value. In other words, the confidence level for each interval should be $1 - 0.0083 = 0.9917$.

(c) Use Table A to find the value of z for a large-sample confidence interval for a single proportion corresponding to 99.17% confidence.

(d) Calculate the six confidence intervals using the Bonferroni procedure.

6.64. Wireless only. Are customers giving up their landlines and relying on wireless for all of their phone needs? Surveys have collected data to answer this question.[27] In June 2006, 10.5% of households were wireless only. Assume that this survey is based on sampling 15,000 households.

(a) Convert the percent to a proportion. Then use the proportion and the sample size to find the count of households who were wireless only.

(b) Find a 95% confidence interval for the proportion of households that were wireless only in June 2006.

6.65. Change in wireless only. Refer to the previous exercise. The percent increased to 31.6% in June 2011. Assume the same sample size for this sample.

(a) Find the proportion and the count for this sample.

(b) Compute the 95% confidence interval for the proportion.

(c) Convert the estimate and confidence interval in terms of proportions to an estimate and confidence interval in terms of percents.

(d) Find the estimate of the difference between the proportions of households that are wireless only in June 2011 and the households that are wireless only in June 2006.

(e) Give the margin of error for 95% confidence for the difference in proportions.

6.66. Student employment during the school year. A study of 1430 undergraduate students reported that 994 work 10 or more hours a week during the school year. Give a 95% confidence interval for the proportion of all undergraduate students who work 10 or more hours a week during the school year.

6.67. Examine the effect of the sample size. Refer to the previous exercise. Assume a variety of different scenarios where the sample size changes but the proportion in the sample who work 10 or more hours a week during the school year remains the same. Write a short report summarizing your results and conclusions. Be sure to include numerical and graphical summaries of what you have found.

6.68. Using a handheld phone while driving. Refer to Exercise 6.32 (page 343). This same poll found that 58% of the respondents talked on a handheld phone while driving in the last year. Construct a 90% confidence interval for the proportion of U.S. drivers who talked on a handheld phone while driving in the last year.

6.69. Gender and using a handheld phone while driving. Refer to the previous exercise. In this same report, this percent was broken down into 59% for men and 56% for women. Assuming that among the 1048 respondents, there were an equal number of men and women, construct a 95% confidence interval for the difference in these proportions.

6.70. CHALLENGE **Even more on downloading music from the Internet.** The following quotation is from a survey of Internet users. The sample size for the survey was 1371. Since 18% of those surveyed said they download music, the sample size for this subsample is 247.

> Among current music down loaders, 38% say they are downloading less because of the RIAA suits .... About a third of current music down loaders say they use peer-to-peer networks ... 24% of them say they swap files using email and instant messaging; 20% download files from music-related Web sites like those run by music magazines or musician homepages. And while online music services like iTunes are far from trumping the popularity of file-sharing networks, 17% of current music down loaders say they are using these paid services. Overall, 7% of Internet users say they have bought music at these new services at one time or another, including 3% who currently use paid services.[28]

(a) For each percent quoted, give the margin of error. You should express these in percents, as given in the quotation.

(b) Rewrite the paragraph more concisely and include the margins of error.

(c) Pick either side A or side B below and give arguments in favor of the view that you select.

(A) The margins of error should be included because they are necessary for the reader to properly interpret the results.

(B) The margins of error interfere with the flow of the important ideas. It would be better to just report one margin of error and say that all the others are no greater than this number.

If you choose View B, be sure to give the value of the margin of error that you report.

6.71. Parental pressure to succeed in school. A Pew Research Center Poll used telephone interviews to ask American adults if parents are pushing their kids too hard to succeed in school. Of those responding, 64% said parents are placing too little pressure on their children.[29] Assuming that this is an SRS of 1200 U.S. residents over the age of 18, give the 95% margin of error for this estimate.

6.72. Improving the time to repair golf clubs. The Ping Company makes custom-built golf clubs and competes in the $4 billion golf equipment industry. To improve its business processes, Ping decided to seek ISO 9001 certification.[30] As part of this process, a study of the time it took to repair golf clubs that were sent to the company by mail determined that 16% of orders were sent back to the customers in 5 days or less. Ping examined the processing of repair orders and made changes. Following the changes, 90% of orders were completed within 5 days. Assume that each of the estimated percents is based on a random sample of 200 orders.

(a) How many orders were completed in 5 days or less before the changes? Give a 95% confidence interval for the proportion of orders completed in this time.

(b) Do the same for orders after the changes.

(c) Give a 95% confidence interval for the improvement. Express this both for a difference in proportions and for a difference in percents.

6.73. Gallup Poll study. Go to the Gallup Poll Web site at `gallup.com/` and find two Gallup Daily Polls that interest you. Summarize the results of the polls giving margins of error and comparisons of interest. (For this exercise, you may assume that the data come from SRSs.)

6.74. Brand loyalty and the Chicago Cubs. According to literature on brand loyalty, consumers who are loyal to a brand are likely to consistently select the same product. This type of consistency could come from a positive childhood association. To examine brand loyalty among fans of the Chicago Cubs, 371 Cubs fans among patrons of a restaurant located in Lakeview were surveyed prior to a game at Wrigley Field, the Cubs' home field.[31] The respondents were classified as "die-hard fans" or "less loyal fans." Of the 134 die-hard fans, 90.3% reported that they had watched or listened to Cubs games when they were children. Among the 237 less loyal fans, 67.9% said that they had watched or listened as children.

(a) Find the number of die-hard Cubs fans who watched or listened to games when they were children. Do the same for the less loyal fans.

(b) Use a significance test to compare the die-hard fans with the less loyal fans with respect to their childhood experiences relative to the team.

(c) Express the results with a 95% confidence interval for the difference in proportions.

6.75. Brand loyalty in action. The study mentioned in the previous exercise found that two-thirds of the die-hard fans attended Cubs games at least once a month, but only 20% of the less loyal fans attended this often. Analyze these data using a significance test and a confidence interval. Write a short summary of your findings.

6.76. CHALLENGE **More on gender bias in textbooks.** Refer to the study of gender bias and textbooks described in Exercise 6.59 (page 354). Here are the counts of "girl," "woman," "boy," and "man" for all the syntax texts studied. The one we analyzed in Exercise 6.59 was number 6.

	Text Number									
	1	2	3	4	5	6	7	8	9	10
Girl	2	5	25	11	2	48	38	5	48	13
Woman	3	2	31	65	1	12	2	13	24	5
Boy	7	18	14	19	12	52	70	6	128	32
Man	27	45	51	138	31	80	2	27	48	95

For each text perform the significance test to compare the proportions of juvenile references for females and males. Summarize the results of the significance tests for the 10 texts studied. The researchers who conducted the study note that the authors of the last 3 texts are women, while the other 7 texts were written by men. Do you see any pattern that suggests that the gender of the author is associated with the results?

6.77. CHALLENGE **Even more on gender bias in textbooks.** Refer to the previous exercise. Let's now combine the categories "girl" with "woman" and "boy" with "man." For each text calculate the proportion of male references and test the hypothesis that male and female references are equally likely (that is, the proportion of male references is equal to 0.5). Summarize the results of your 10 tests. Is there a pattern that suggests a relation with the gender of the author?

6.78. Parental pressure and gender. The Pew Research Center Poll in Exercise 6.71 also reported that 65% of the men and 62% of the women thought parents are placing too little pressure on their children to succeed in school. Assuming that the respondents were 52% women, compare the proportions with a significance test and give a 95% confidence interval for the difference. Write a summary of your results.

6.79. CHALLENGE **Sample size and the *P*-value.** In this exercise we examine the effect of the sample size on the significance test for comparing two proportions. In each case suppose that $\hat{p}_1 = 0.5$ and $\hat{p}_2 = 0.4$, and take n to be the common value of n_1 and n_2. Use the z statistic to test H_0: $p_1 = p_2$ versus the alternative H_a: $p_1 \neq p_2$. Compute the statistic and the associated P-value for the following values of n: 40, 50, 80, 100, 400, 500, and 1000. Summarize the results in a table. Explain what you observe about the effect of the sample size on statistical significance when the sample proportions $\hat{p}_1$ and $\hat{p}_2$ are unchanged.

6.80. A corporate liability trial. A major court case on the health effects of drinking contaminated water took place in the town of Woburn, Massachusetts. A town well in Woburn was contaminated by industrial chemicals. During the period that residents drank water from this well, there were 16 birth defects among 414 births. In years when the contaminated well was shut off and water was supplied from other wells, there were 3 birth defects among 228 births. The plaintiffs suing the firm responsible for the contamination claimed that these data show that the rate of birth defects was higher when the contaminated well was in use.[32] How statistically significant is the evidence? What assumptions does your analysis require? Do these assumptions seem reasonable in this case?

6.81. CHALLENGE **Attitudes toward student loan debt.** The National Student Loan Survey asked the student loan borrowers in their sample about attitudes toward debt.[33] Here are some of the questions they asked, with the percent who responded in a particular way:

(a) "To what extent do you feel burdened by your student loan payments?" 55.5% said they felt burdened.

(b) "If you could begin again, taking into account your current experience, what would you borrow?" 54.4% said they would borrow less.

(c) "Since leaving school, my education loans have not caused me more financial hardship than I had anticipated at the time I took out the loans." 34.3% disagreed.

(d) "Making loan payments is unpleasant but I know that the benefits of education loans are worth it." 58.9% agreed.

(e) "I am satisfied that the education I invested in with my student loan(s) was worth the investment for career opportunities." 58.9% agreed.

(f) "I am satisfied that the education I invested in with my student loan(s) was worth the investment for personal growth." 71.5% agreed.

Assume that the sample size is 1280 for all of these questions. Compute a 95% confidence interval for each of the questions, and write a short report about what student loan borrowers think about their debt.

CHAPTER SEVEN

INFERENCE FOR MEANS

In addition to categorical variables, we often collect data on *quantitative variables*. In Chapter 3, we learned graphical and numerical tools for describing the distribution of a single quantitative variable and for comparing several variables. In this chapter, we continue our study of the practice of statistical inference with inference about the population mean μ and the difference between two population means $\mu_1 - \mu_2$. Comparing more than two population means requires more elaborate methods, which are presented in Chapters 12 and 13.

As in the previous chapter, we will focus on confidence intervals and significance tests. These procedures for inference are among the most commonly used statistical methods and will allow us to address questions such as

- How many hours per month, on average, does a college student spend watching streaming videos on a cell phone?
- Do male and female college students differ in terms of "social insight," the ability to appraise other people?
- Does the daily number of disruptive behaviors in dementia patients change when there is a full moon?

7.1 Inference for the Mean of a Population

A variety of statistics is used to describe quantitative data. The sample mean, percentiles, and standard deviation are all examples of statistics based on quantitative data. In this chapter we will concentrate on the sample mean. Because sample means are just averages of observations, they are among the most frequently used statistics.

LOOK BACK parameters and statistics p. 316

The sample mean $\bar{x}$ from a sample or an experiment is an estimate of the mean μ of the underlying population, just as a sample proportion $\hat{p}$ is an estimate of a population proportion p. In the previous chapter, we learned that when the data are produced by random sampling or randomized experimentation, a statistic is a random variable and its *sampling distribution* shows how the statistic would vary in repeated data productions. To study inference about a population mean μ, we must first understand the sampling distribution of the sample mean $\bar{x}$.

The sampling distribution of a sample mean

Suppose you plan to survey 1000 college students at your university about their sleeping habits. The sampling distribution of the average hours of sleep per night describes what this average would be if all simple random samples of 1000 students were drawn from the population of college students at your university. In other words, it gives you an idea of what you are likely to see from your survey.

density curves p. 128

Another probability distribution we consider here is the population distribution. Any quantity that can be measured for each member of a population is described by the distribution of its values for all members of the population. This is the context in which we first met distributions as density curves that provide models for the overall pattern of data. Imagine choosing one individual at random from the population. The results of repeated choices have a probability distribution that is the distribution of the population.

EXAMPLE 7.1

Total sleep time of college students. A recent survey describes the distribution of total sleep time among college students as approximately Normal with a mean of 7.02 hours and standard deviation of 1.15 hours.[1] Select a college student at random and obtain his or her sleep time. The result is a random variable X. Prior to the random sampling, we don't know the sleep time of the chosen college student, but we do know that in repeated sampling X will have the same $N(7.02, 1.15)$ distribution that describes the pattern of sleep time in the entire population of college students. We call $N(7.02, 1.15)$ the *population distribution*.

POPULATION DISTRIBUTION

The **population distribution** of a variable is the distribution of its values for all members of the population. The population distribution is also the probability distribution of the variable when we choose one individual at random from the population.

LOOK BACK
SRS p. 56

The population of all college students actually exists, so that we can in principle draw an SRS from it. Sometimes our population of interest does not actually exist. For example, suppose that we are interested in studying final-exam scores in a statistics course and we have the scores of the 34 students who took the course last semester. For the purposes of statistical inference, we might want to consider these 34 students as part of a hypothetical population of similar students who would take this course. In this sense, these students represent not only themselves but also a larger population of similar students.

The key idea is to think of the observations that you have as coming from a population with a probability distribution. This population distribution can be approximately Normal, as in Example 7.1, highly skewed, as we'll see in Example 7.2, or have multiple peaks. In each case, the sampling distribution depends on both the population distribution and the way we collect the data from the population.

USE YOUR KNOWLEDGE

7.1 Number of apps on an iPhone. AppsFire is a service that shares the names of the apps on your iPhone with everyone else using the service. This, in a sense, creates an iPhone app recommendation system. Recently, the service drew a sample of over 1000 AppsFire users and reported a median of 88 downloaded apps per device.[2] State the population of this survey and some likely values from the population distribution. Also explain why you might expect this population distribution to be skewed to the right.

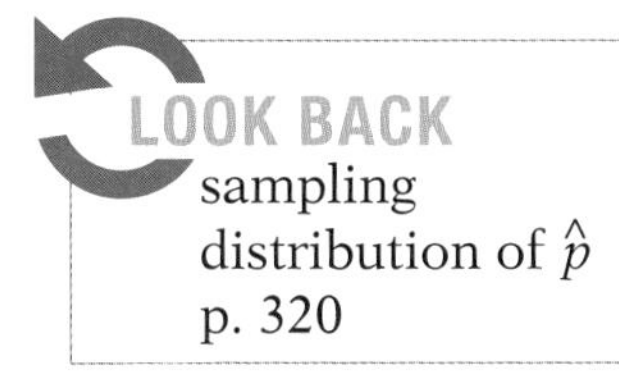
LOOK BACK
sampling distribution of $\hat{p}$ p. 320

We studied the sampling distribution of the sample proportion $\hat{p}$ using simulation. Since the general framework for constructing a sampling distribution is the same for all statistics, let's do the same here to understand the sampling distribution of $\bar{x}$.

EXAMPLE 7.2

DATA FILE CALL_LENGTH

Sample means are approximately Normal. Figure 7.1 illustrates two striking facts about the sampling distribution of a sample mean. Figure 7.1(a) displays the distribution of customer service call lengths for a bank service

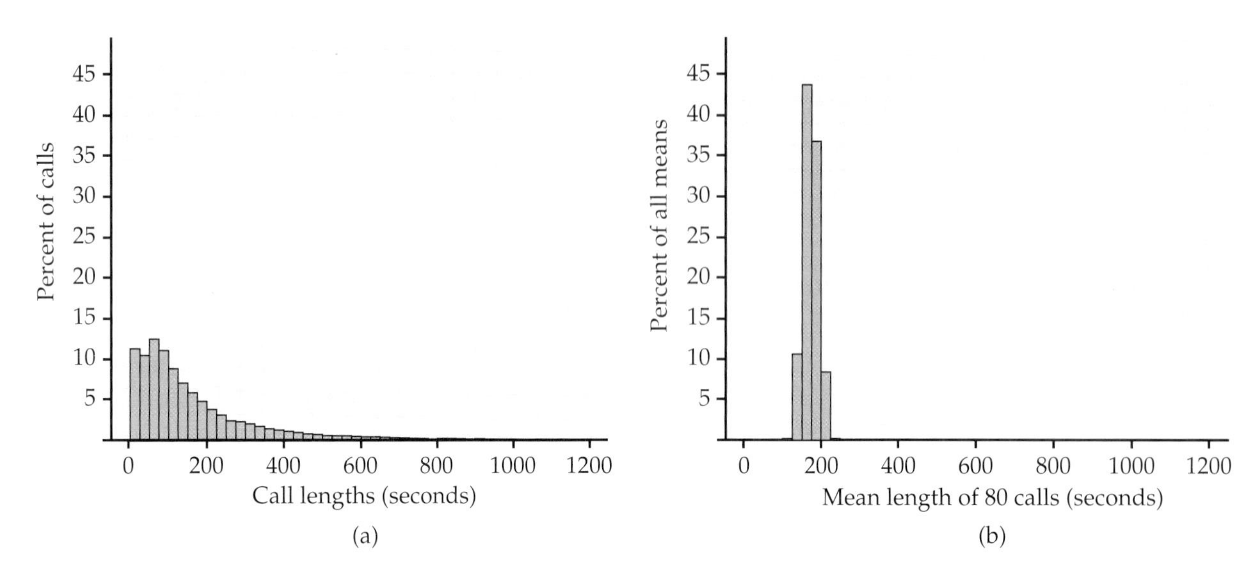

FIGURE 7.1 The distribution of lengths of all customer service calls received by a bank in a month, for Example 7.2. (b) The distribution of the sample means $\bar{x}$ for 500 random samples of size 80 from this population. The scales and histogram classes are exactly the same in both panels.

center for a month. There are more than 30,000 calls in this population.[3] (We omitted a few extreme outliers, calls that lasted more than 20 minutes.) The distribution is extremely skewed to the right. The population mean is μ = 173.95 seconds.

Table 3.2 (page 93) contains the lengths of a sample of 80 calls from this population. The mean of these 80 calls is $\bar{x} = 196.6$ seconds. If we were to take another sample of size 80, we would likely get a different value of $\bar{x}$. This is because this new sample would contain a different set of calls. To find the sampling distribution of $\bar{x}$, we take many SRSs of size 80 and calculate $\bar{x}$ for each sample. Figure 7.1(b) is the distribution of the values of $\bar{x}$ for 500 such random samples. The scales and choice of classes are exactly the same as in Figure 7.1(a), so that we can make a direct comparison.

The sample means are much less spread out than the individual call lengths. What is more, the distribution in Figure 7.1(b) is roughly symmetric rather than skewed. The Normal quantile plot in Figure 7.2 confirms that the distribution is close to Normal.

This example illustrates two important facts about sample means that contribute to the popularity of sample means in statistical inference.

FACTS ABOUT SAMPLE MEANS

1. Sample means are less variable than individual observations.
2. Sample means are more Normal than individual observations.

LOOK BACK
mean and standard deviation of $\hat{p}$ p. 319

From Figure 7.1(b), it also appears that the center of this distribution is near the population mean. As we did with the sample proportion, let's compute the mean and standard deviation of $\bar{x}$.

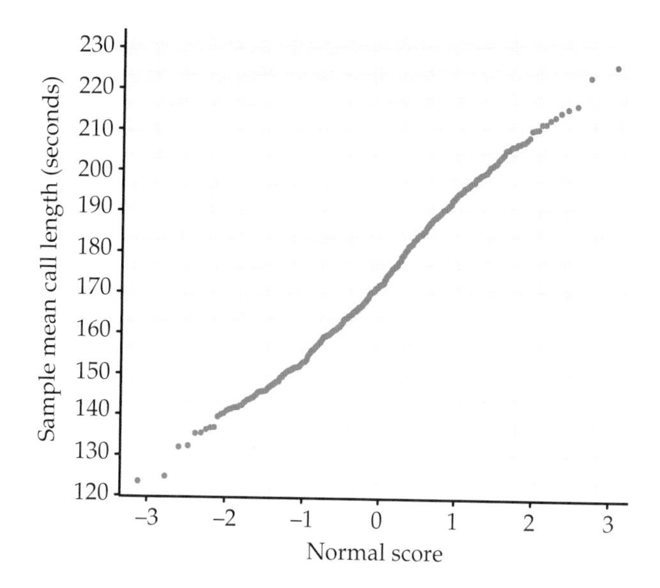

FIGURE 7.2 Normal quantile plot of the 500 sample means in Figure 7.1(b). The distribution is close to Normal.

Consider a random sample of size n from a population, and measure a variable X on each individual in the sample. The n measurements are values of n random variables $X_1, X_2, \ldots, X_n$. A single X_i is a measurement on one individual selected at random from the population and therefore has the distribution of the population. If the population is large relative to the sample, we can consider $X_1, X_2, \ldots, X_n$ to be independent random variables each having the same distribution. This is our probability model for measurements on each individual in an SRS.

The sample mean of an SRS of size n is

LOOK BACK
addition rule for means p. 274

$$\bar{x} = \frac{1}{n}(X_1 + X_2 + \cdots + X_n)$$

If the population has mean μ, then μ is the mean of the distribution of each observation X_i. To get the mean of $\bar{x}$, we use the addition rule for means of random variables. Specifically,

$$\mu_{\bar{x}} = \frac{1}{n}(\mu_{X_1} + \mu_{X_2} + \cdots + \mu_{X_n})$$

$$= \frac{1}{n}(\mu + \mu + \cdots + \mu) = \mu$$

LOOK BACK
unbiased estimator p. 323
addition rule for variances p. 278

That is, *the mean of $\bar{x}$ is the same as the mean of the population.* The sample mean $\bar{x}$ is therefore an unbiased estimator of the unknown population mean μ.

The observations are independent, so the addition rule for variances also applies:

$$\sigma^2_{\bar{x}} = \left(\frac{1}{n}\right)^2 (\sigma^2_{X_1} + \sigma^2_{X_2} + \cdots + \sigma^2_{X_n})$$

$$= \left(\frac{1}{n}\right)^2 (\sigma^2 + \sigma^2 + \cdots + \sigma^2)$$

$$= \frac{\sigma^2}{n}$$

Just as in the case of a sample proportion $\hat{p}$, the variability of $\bar{x}$ about its mean decreases as the sample size grows. Thus, a sample mean from a large sample will usually be very close to the true population mean μ. Here is a summary of these facts.

MEAN AND STANDARD DEVIATION OF A SAMPLE MEAN

Let $\bar{x}$ be the mean of an SRS of size n from a population having mean μ and standard deviation σ. The **mean and standard deviation of $\bar{x}$** are

$$\mu_{\bar{x}} = \mu$$

$$\sigma_{\bar{x}} = \frac{\sigma}{\sqrt{n}}$$

Because the standard deviation of $\bar{x}$ is $\sigma/\sqrt{n}$, it is again true that the standard deviation of the statistic decreases in proportion to the square root of the sample size.

How precisely does a sample mean $\bar{x}$ estimate a population mean μ? Because the values of $\bar{x}$ vary from sample to sample, we must give an answer in terms of the sampling distribution. We know that $\bar{x}$ is an unbiased estimator of μ, so its values in repeated samples are not systematically too high or too low. Most samples will give an $\bar{x}$-value close to μ if the sampling distribution is concentrated close to its mean μ. So the precision of estimation depends on the spread of the sampling distribution.

EXAMPLE 7.3

Standard deviations for sample means of service call lengths. The standard deviation of the population of service call lengths in Figure 7.1(a) is $\sigma = 184.81$ seconds. The length of a single call will often be far from the population mean. If we choose an SRS of 20 calls, the standard deviation of their mean length is

$$\sigma_{\bar{x}} = \frac{184.81}{\sqrt{20}} = 41.32 \text{ seconds}$$

Averaging over more calls reduces the variability and makes it more likely that $\bar{x}$ is close to μ. Our sample size of 80 calls is 4 times 20, so the standard deviation will be half as large:

$$\sigma_{\bar{x}} = \frac{184.81}{\sqrt{80}} = 20.66 \text{ seconds}$$

USE YOUR KNOWLEDGE

7.2 Find the mean and the standard deviation of the sampling distribution. You take an SRS of size 49 from a population with mean 320 and standard deviation 21. Find the mean and standard deviation of the sampling distribution of your sample mean.

7.3 The effect of increasing the sample size. In the setting of the previous exercise, repeat the calculations for a sample size of 196. Explain the effect of the increase in sample size on the sampling distribution's mean and standard deviation.

The central limit theorem

Although we now know the center and spread of the probability distribution of a sample mean $\bar{x}$, we don't know its shape. The shape of the distribution of $\bar{x}$ depends on the shape of the population distribution. Here is one important case: if the population distribution is Normal, then so is the distribution of the sample mean.

SAMPLING DISTRIBUTION OF A SAMPLE MEAN

If a population has the $N(\mu, \sigma)$ distribution, then the sample mean $\bar{x}$ of n independent observations has the $N(\mu, \sigma/\sqrt{n})$ distribution.

This is a somewhat special result. Many population distributions are not Normal. The service call lengths in Figure 7.1(a), for example, are strongly skewed. Yet Figures 7.1(b) and 7.2 show that means of samples of size 80 are close to Normal. One of the most famous facts of probability theory says that, for large sample sizes, the distribution of $\bar{x}$ is close to a Normal distribution. This is true no matter what shape the population distribution has, as long as the population has a finite standard deviation σ. This is the *central limit theorem*. It is much more useful than the fact that the distribution of $\bar{x}$ is exactly Normal if the population is exactly Normal.

CENTRAL LIMIT THEOREM

Draw an SRS of size n from any population with mean μ and finite standard deviation σ. When n is large, the sampling distribution of the sample mean $\bar{x}$ is approximately Normal:

$$\bar{x} \text{ is approximately } N\left(\mu, \frac{\sigma}{\sqrt{n}}\right)$$

EXAMPLE 7.4

How close will the sample mean be to the population mean? With the Normal distribution to work with, we can better describe how precisely a random sample of 80 calls estimates the mean length of all the calls in the population. The population standard deviation for the more than 30,000 calls in the population of Figure 7.1(a) is $\sigma = 184.81$ seconds. From Example 7.3 we know $\sigma_{\bar{x}} = 20.66$ seconds for a sample of size $n = 80$ calls. By the 95 part of the 68–95–99.7 rule, about 95% of all samples of 80 calls will have mean $\bar{x}$ within two standard deviations of μ, that is, within ± 41.32 seconds of μ.

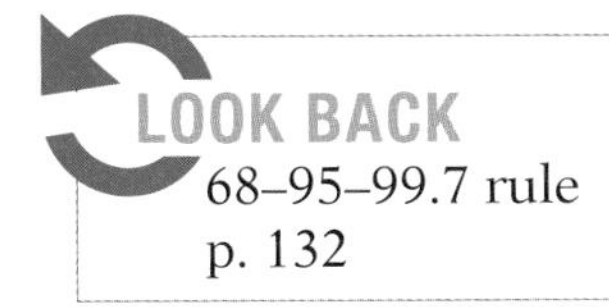

68–95–99.7 rule
p. 132

USE YOUR KNOWLEDGE

7.4 Use the 68–95–99.7 rule. You take an SRS of size 196 from a population with mean 320 and standard deviation 21. According to the central limit theorem, what is the approximate sampling distribution of the sample mean? Use the 95 part of the 68–95–99.7 rule to describe the variability of this sample mean.

For the sample size of $n = 80$ in Example 7.4, the sample mean is not very precise. The population of service call lengths is very spread out, so the sampling distribution of $\bar{x}$ has a large standard deviation.

EXAMPLE 7.5

How can we reduce the standard deviation? In the setting of Example 7.4, if we want to reduce the standard deviation of $\bar{x}$ by a factor of 4, we must take a sample 16 times as large, $n = 16 \times 80$, or 1280. Then

$$\sigma_{\bar{x}} = \frac{184.81}{\sqrt{1280}} = 5.166 \text{ seconds}$$

For samples of size 1280, about 95% of the sample means will be within twice 5.166, or 10.33 seconds, of the population mean μ.

USE YOUR KNOWLEDGE

7.5 The effect of increasing the sample size. In the setting of Exercise 7.4, suppose that we increase the sample size to 1764. Use the 95 part of the 68–95–99.7 rule to describe the variability of this sample mean. Compare your results with those you found in Exercise 7.4.

Example 7.5 reminds us that, if the population is very spread out, the $\sqrt{n}$ in the standard deviation of $\bar{x}$ implies that very large samples are needed to estimate the population mean precisely. The main point of the example, however, is that the central limit theorem allows us to use Normal probability calculations to answer questions about sample means even when the population distribution is not Normal.

How large a sample size n is needed for $\bar{x}$ to be close to Normal depends on the population distribution. More observations are required if the shape of the population distribution is far from Normal. Even for the very skewed call length population, however, samples of size 80 are large enough. Here is a more detailed example.

EXAMPLE 7.6

The central limit theorem in action. Figure 7.3 shows the central limit theorem in action for another very non-Normal population. Figure 7.3(a) displays the density curve of a single observation from the population. The distribution

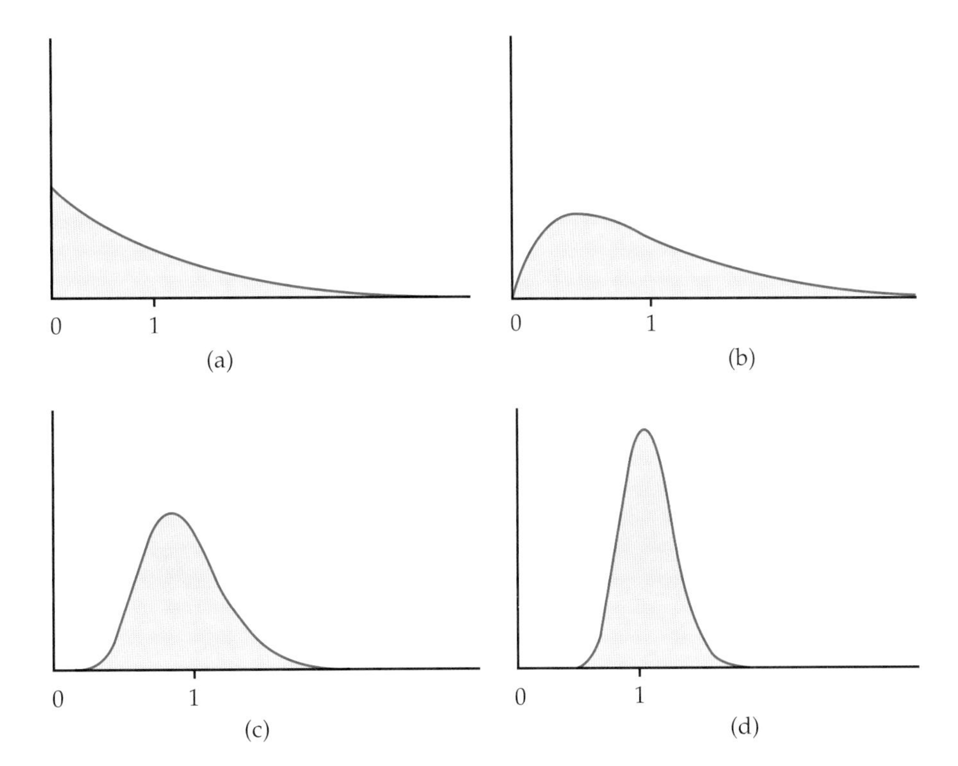

FIGURE 7.3 The central limit theorem in action: the distribution of sample means from a strongly non-Normal population becomes more Normal as the sample size increases. (a) The distribution of 1 observation. (b) The distribution of $\bar{x}$ for 2 observations. (c) The distribution of $\bar{x}$ for 10 observations. (d) The distribution of $\bar{x}$ for 25 observations.

exponential distribution

is strongly right-skewed, and the most probable outcomes are near 0. The mean μ of this distribution is 1, and its standard deviation σ is also 1. This particular continuous distribution is called an **exponential distribution.** Exponential distributions are used as models for variables such as how long an electronic component will last and the time between text messages arriving on your cell phone.

Figures 7.3(b), (c), and (d) are the density curves of the sample means of 2, 10, and 25 observations from this population. As n increases, the shape becomes more Normal. The mean remains at $\mu = 1$, but the standard deviation decreases, taking the value $1/\sqrt{n}$. The density curve for 10 observations is still somewhat skewed to the right but already resembles a Normal curve having $\mu = 1$ and $\sigma = 1/\sqrt{10} = 0.32$. The density curve for $n = 25$ is yet more Normal. The contrast between the shapes of the population distribution and of the distribution of the mean of 10 or 25 observations is striking.

The *Central Limit Theorem* applet animates Figure 7.3. You can slide the sample size n from 1 to 100 and watch both the exact density curve of $\bar{x}$ and the Normal approximation. As you increase n, the two curves move closer together.

EXAMPLE 7.7

Time between text message arrivals. Suppose that the time X between text messages arriving on your cell phone is governed by the exponential distribution with mean $\mu = 2.5$ minutes and standard deviation $\sigma = 2.5$ minutes. You record the times between your next 50 messages. What is the probability that their average exceeds 2.1 minutes?

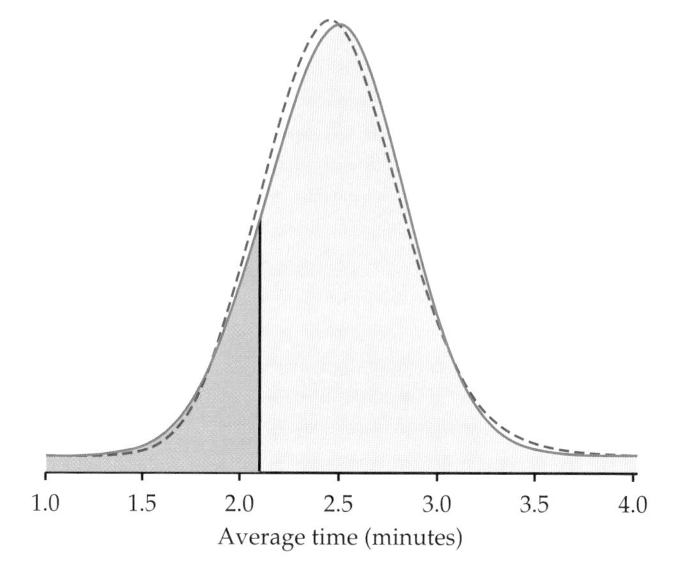

FIGURE 7.4 The exact distribution (dashed) and the Normal approximation from the central limit theorem (solid) for the average time between text messages arriving on your cell phone, for Example 7.7.

The central limit theorem says that the sample mean time $\bar{x}$ (in minutes) between text messages has approximately the Normal distribution with mean equal to the population mean $\mu = 2.5$ minutes and standard deviation

$$\frac{\sigma}{\sqrt{50}} = \frac{2.5}{\sqrt{50}} = 0.354 \text{ minutes}$$

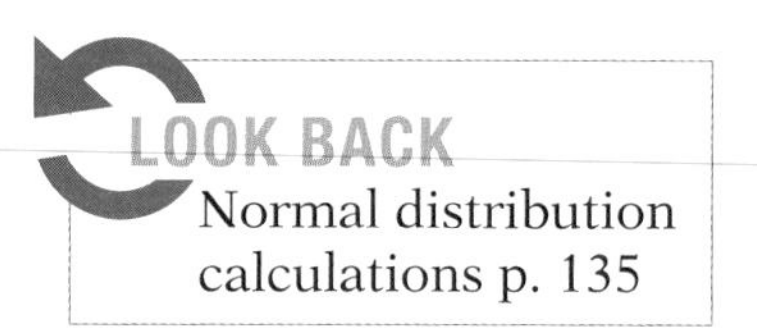
LOOK BACK
Normal distribution calculations p. 135

The distribution of $\bar{x}$ is therefore approximately $N(2.5, 0.354)$. Figure 7.4 shows this Normal curve (solid) and also the actual density curve of $\bar{x}$ (dashed).

The probability we want is $P(\bar{x} > 2.1)$. A Normal distribution calculation (standardizing to z and using Table A) gives this probability as 0.8708. This is the area to the right of 2.1 under the solid Normal curve in Figure 7.4. The exactly correct probability is the area under the dashed density curve in the figure. It is 0.8750. The central limit theorem Normal approximation is off by only about 0.0042.

USE YOUR KNOWLEDGE

7.6 Find a probability. Refer to the example above. Find the probability that the mean time between 50 text messages is less than 2.6 minutes. The exact probability is 0.6279. How does your result compare?

EXAMPLE 7.8

Convert the results to the total time. In reference to Example 7.7, what can we say about the total time between 50 text messages? According to the central limit theorem

$$P(\bar{x} > 2.1) = 0.8708$$

We know that the sample mean is the total time divided by 50, so the event $\{\bar{x} > 2.1\}$ is the same as the event $\{50\bar{x} > 50(2.1)\}$. We can say that the probability is 0.8708 that the total time is $50(2.1) = 105$ minutes (1.75 hours) or greater.

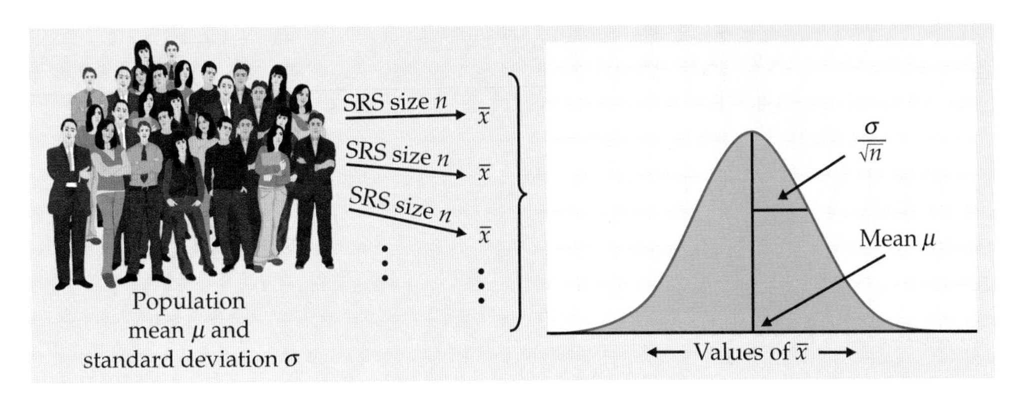

FIGURE 7.5 The sampling distribution of a sample mean $\bar{x}$ has mean μ and standard deviation $\sigma/\sqrt{n}$. The distribution is Normal if the population distribution is Normal; it is approximately Normal for large samples in any case.

Figure 7.5 summarizes the facts about the sampling distribution of $\bar{x}$ in a way that emphasizes the big idea of a sampling distribution.

- Keep taking random samples of size n from a population with mean μ.
- Find the sample mean $\bar{x}$ for each sample.
- Collect all the $\bar{x}$'s and display their distribution.

That's the sampling distribution of $\bar{x}$. A similar approach could be used for a different statistic. Keep this figure in mind as we proceed through statistical inference.

A few more facts

To understand the sampling distribution of $\bar{x}$, we considered both Normal and non-Normal population distributions. Here are three additional facts related to these investigations that we will use or have used in describing methods of statistical inference.

The fact that the sample mean of an SRS from a Normal population has a Normal distribution is a special case of a more general fact: **any linear combination of independent Normal random variables is also Normally distributed.** That is, if X and Y are independent Normal random variables and a and b are any fixed numbers, $aX + bY$ is also Normally distributed, and so it is for any number of Normal variables. In particular, the sum or difference of independent Normal random variables has a Normal distribution. The mean and standard deviation of $aX + bY$ are found as usual from the addition rules for means and variances. These facts are often used in statistical calculations. Here is an example.

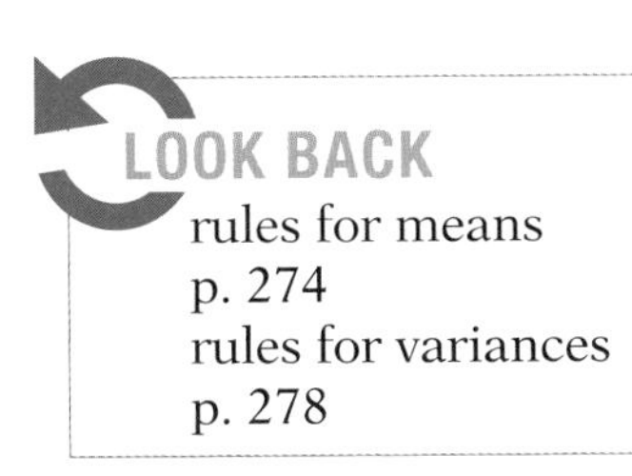

rules for means p. 274
rules for variances p. 278

EXAMPLE 7.9

Getting to and from campus. You live off campus and take the shuttle, provided by your apartment complex, to and from campus. Your time on the shuttle in minutes varies from day to day. The time going to campus X has the $N(20, 4)$ distribution, and the time returning from campus Y varies according

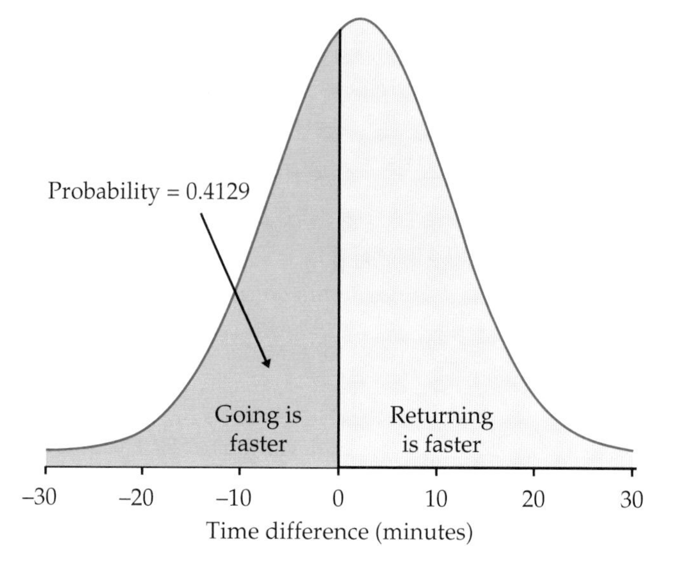

FIGURE 7.6 The Normal probability calculation for Example 7.9. The difference in times going to campus and returning from campus (X − Y) is Normal with mean 2 minutes and standard deviation 8.94 minutes.

to the $N(18, 8)$ distribution. If they vary independently, what is the probability that you will be on the shuttle for less time going to campus?

The difference in times $X - Y$ is Normally distributed, with mean and variance

$$\mu_{X-Y} = \mu_X - \mu_Y = 20 - 18 = 2$$
$$\sigma^2_{X-Y} = \sigma^2_X + \sigma^2_Y = 4^2 + 8^2 = 80$$

Because $\sqrt{80} = 8.94$, $X - Y$ has the $N(2, 8.94)$ distribution. Figure 7.6 illustrates the probability computation:

$$\begin{aligned} P(X < Y) &= P(X - Y < 0) \\ &= P\left[\frac{(X-Y)-2}{8.94} < \frac{0-2}{8.94}\right] \\ &= P(Z < -0.22) = 0.4129 \end{aligned}$$

Although, on average, it takes longer to go to campus than return, the trip to campus will be shorter on roughly two of every five days.

The second useful fact is that **more general versions of the central limit theorem say that the distribution of a sum or average of many small random quantities is close to Normal.** This is true even if the quantities are not independent (as long as they are not too highly correlated). It is also true even if they have different distributions (as long as no single random quantity is so large that it dominates the others). The central limit theorem suggests why the Normal distributions are common models for observed data. Any variable that is a sum of many small influences will have approximately a Normal distribution.

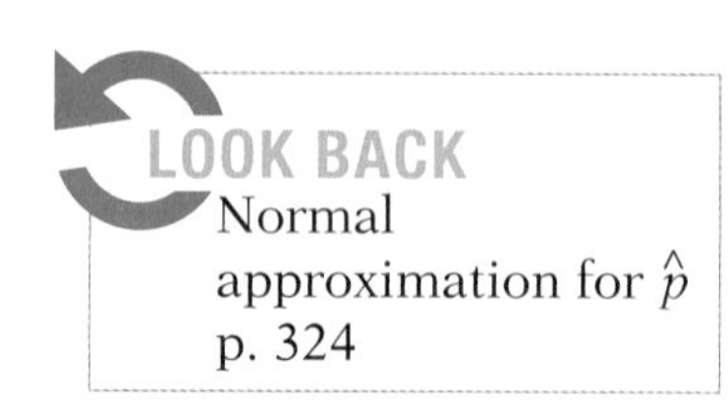
LOOK BACK
Normal approximation for $\hat{p}$
p. 324

Finally, **the central limit theorem also applies to discrete random variables.** An average of discrete random variables will never result in a continuous sampling distribution, but the Normal distribution often serves as a good approximation. The Normal approximation for the sample proportion is one example of this fact. Recall the idea we used to find the mean and variance of a binomial random variable X. We wrote the count X as a sum

$$X = S_1 + S_2 + \cdots + S_n$$

of random variables S_i that take the value 1 if a success occurs on the ith trial and the value 0 otherwise. The variables S_i take only the values 0 and 1 and are far from Normal. The proportion $\hat{p} = X/n$ is the sample mean of the S_i and, like all sample means, is approximately Normal when n is large.

The *t* distributions

Although we now know the approximate sampling distribution of the sample mean, there is one more issue we must address before we proceed with inference about a population mean. This is the fact that the sampling distribution of $\bar{x}$ depends on σ. There is no difficulty when σ is known but rarely will σ be known in practice. Typically, we will estimate σ with the sample standard deviation s. How does the use of s for σ affect inference?

Suppose that we have a simple random sample (SRS) of size n from a Normally distributed population with mean μ and standard deviation σ. The sample mean $\bar{x}$ is then Normally distributed with mean μ and standard deviation $\sigma/\sqrt{n}$. If we follow the same logic as that used for the sample proportion $\hat{p}$, the standardized sample mean, or one-sample z statistic,

$$z = \frac{\bar{x} - \mu}{\sigma/\sqrt{n}}$$

becomes the basis of the procedures for inference about μ. This statistic has the standard Normal distribution $N(0, 1)$.

standard error

When σ is not known, we estimate it with the sample standard deviation s, and then we estimate the standard deviation of $\bar{x}$ by $s/\sqrt{n}$. This quantity is called the **standard error** of the sample mean $\bar{x}$ and we denote it by $SE_{\bar{x}}$. When we substitute the standard error $s/\sqrt{n}$ for the standard deviation $\sigma/\sqrt{n}$ of $\bar{x}$, the standardized sample mean statistic does *not* have a Normal distribution. It has a distribution that is new to us, called a *t distribution*.

THE *t* DISTRIBUTIONS

Suppose that an SRS of size n is drawn from an $N(\mu, \sigma)$ population. Then the **one-sample *t* statistic**

$$t = \frac{\bar{x} - \mu}{s/\sqrt{n}}$$

has the ***t* distribution** with $n - 1$ **degrees of freedom.**

LOOK BACK
degrees of freedom
p. 118

A particular t distribution is specified by giving the *degrees of freedom*. We use $t(k)$ to stand for the t distribution with k degrees of freedom. The degrees of freedom for this t statistic come from the sample standard deviation s in the denominator of t. We showed earlier that s has $n - 1$ degrees of freedom. Thus, there is a different t distribution for each sample size. There are also other t statistics with different degrees of freedom, some of which we will meet later in this chapter.

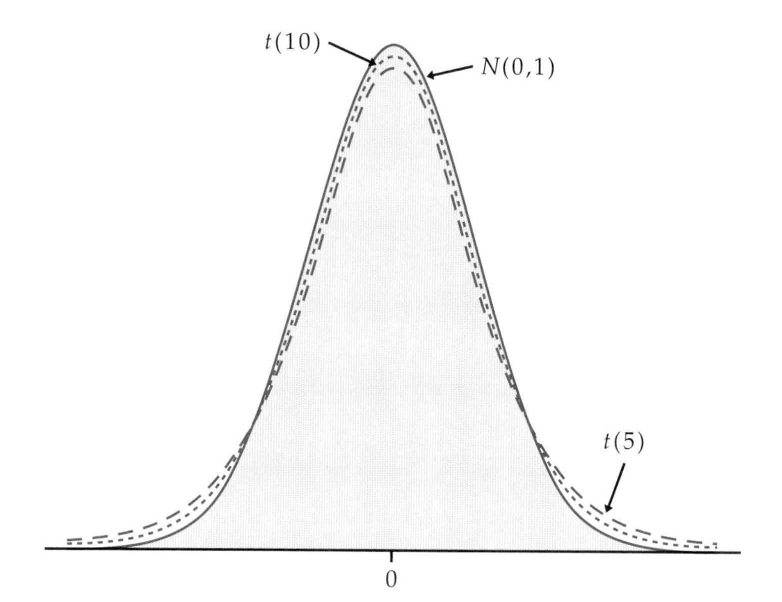

FIGURE 7.7 Density curves for the standard Normal, $t(10)$, and $t(5)$ distributions. All are symmetric with center 0. The t distributions have more probability in the tails than the standard Normal distribution.

The t distributions were discovered in 1908 by William S. Gosset. Gosset was a statistician employed by the Guinness brewing company, which prohibited its employees from publishing their discoveries that were brewing related. In this case, the company let him publish under the pen name "Student" using an example that did not involve brewing. The t distribution is often called "Student's t" in his honor.

The density curves of the $t(k)$ distributions are similar in shape to the standard Normal curve. That is, they are symmetric about 0 and are bell-shaped. Figure 7.7 compares the density curves of the standard Normal distribution and the t distributions with 5 and 10 degrees of freedom. The similarity in shape is apparent, as is the fact that the t distributions have more probability in the tails and less in the center. This greater spread is due to the extra variability caused by substituting the random variable s for the fixed parameter σ. Figure 7.7 also shows that as the degrees of freedom k increase, the $t(k)$ density curve gets closer to the $N(0, 1)$ curve. This reflects the fact that s will be closer to σ as the sample size increases.

Table D in the back of the book gives critical values, labeled t^*, for the t distributions. For convenience, we have labeled the table entries both by the value of p needed for significance tests and by the confidence level C (in percent) required for confidence intervals. The standard Normal critical values are in the bottom row of entries and labeled z^*. As in the case of the Normal table (Table A), computer software often makes Table D unnecessary.

USE YOUR KNOWLEDGE

7.7 Apartment rents. You randomly choose 16 unfurnished one-bedroom apartments from a large number of advertisements in your local newspaper. You calculate that their mean monthly rent is \$658 and their standard deviation is \$111.

(a) What is the standard error of the mean?

(b) What are the degrees of freedom for a one-sample t statistic?

The one-sample t confidence interval

LOOK BACK
large-sample confidence interval for p p. 329

With the t distributions to help us, we can now perform inference for a sample from a population with unknown σ. The *one-sample t confidence interval* is similar in both reasoning and computational detail to the large-sample confidence interval for the population p in Chapter 6. There, the margin of error for the population proportion was $z^*\text{SE}_{\hat{p}}$. Here, we replace $\text{SE}_{\hat{p}}$ by $\text{SE}_{\bar{x}}$ and z^* by t^*. This means that the margin of error for the population mean when we use the data to estimate σ is $t^*s/\sqrt{n}$.

THE ONE-SAMPLE t CONFIDENCE INTERVAL

Suppose that an SRS of size n is drawn from a population having unknown mean μ. A level C **confidence interval for** μ is

$$\bar{x} \pm t^* \frac{s}{\sqrt{n}}$$

where t^* is the value for the $t(n-1)$ density curve with area C between $-t^*$ and t^*. The quantity

$$t^* \frac{s}{\sqrt{n}}$$

is the **margin of error.** This interval is exact when the population distribution is Normal and is approximately correct for large n in other cases.

EXAMPLE 7.10

Watching videos on a mobile phone. The Nielsen Company is a global information and media company and one of the leading suppliers of media information. Recently, they reported that U.S. mobile phone subscribers average 4.3 hours per month watching videos on their phone.[4] Suppose that we want to determine a 95% confidence interval for the average among U.S. college students and draw the following SRS of size 8 from this population:

7 9 2 6 13 10 4 5

The sample mean is

$$\bar{x} = \frac{7+9+\cdots+5}{8} = 7.00$$

and the standard deviation is

$$s = \sqrt{\frac{(7-7.00)^2+(9-7.00)^2+\cdots+(5-7.00)^2}{8-1}} = 3.546$$

with degrees of freedom $n-1=7$. The standard error is

$$\text{SE}_{\bar{x}} = s/\sqrt{n} = 3.546/\sqrt{8} = 1.25$$

From Table D we find $t^* = 2.365$. The 95% confidence interval is

$$\begin{aligned}\bar{x} \pm t^*\frac{s}{\sqrt{n}} &= 7.00 \pm 2.365\frac{3.546}{\sqrt{8}} \\ &= 7.00 \pm (2.365)(1.25) \\ &= 7.00 \pm 2.96 \\ &= (4.0, 10.0)\end{aligned}$$

We are 95% confident that among U.S. college students the average time spent watching videos on a mobile phone is between 4.0 and 10.0 hours per month.

In this example we have given the actual interval (4.0, 10.0) as our answer. Sometimes we prefer to report the mean and margin of error: the mean time is 7.00 hours per month with a margin of error of 2.96 hours.

The use of the t confidence interval in Example 7.10 rests on assumptions that appear reasonable here. First, we assume that our random sample is an SRS from the U.S. population of college student mobile phone users. Second, we assume that the distribution of watching times is Normal. With only 8 observations, this assumption cannot be effectively checked. In fact, because the watching time cannot be negative, we might expect this distribution to be skewed to the right. With these data, however, there are no extreme outliers to suggest a severe departure from Normality.

USE YOUR KNOWLEDGE

7.8 Using Table D. What critical value t^* from Table D should be used to construct

(a) a 95% confidence interval when $n = 15$?

(b) a 99% confidence interval when $n = 25$?

(c) a 90% confidence interval when $n = 29$?

7.9 More on apartment rents. Recall Exercise 7.7 (page 372). Construct a 95% confidence interval for the mean monthly rent of all advertised one-bedroom apartments.

7.10 90% versus 95% confidence interval. For the previous problem, if you were to use 90% confidence, rather than 95% confidence, would the margin of error be larger or smaller? Explain your answer.

The one-sample t test

LOOK BACK
large-sample significance test for p
p. 337

The significance test for the population mean is also very similar to the test that we studied in the last chapter.

THE ONE-SAMPLE t SIGNIFICANCE TEST

Suppose that an SRS of size n is drawn from a population having unknown mean μ. To test the hypothesis H_0: $\mu = \mu_0$ based on an SRS of size n, compute the **one-sample t statistic**

$$t = \frac{\bar{x} - \mu_0}{s/\sqrt{n}}$$

In terms of a random variable T having the $t(n-1)$ distribution, the P-value for a test of H_0 against

$$H_a\colon \mu > \mu_0 \quad \text{is} \quad P(T \geq t)$$

$$H_a\colon \mu < \mu_0 \quad \text{is} \quad P(T \leq t)$$

$$H_a\colon \mu \neq \mu_0 \quad \text{is} \quad 2P(T \geq |t|)$$

These P-values are exact if the population distribution is Normal and are approximately correct for large n in other cases.

EXAMPLE 7.11

Significance test for watching videos on mobile phones. Suppose that we want to test whether the U.S. college student average number of hours watching videos on a mobile phone is different from the reported overall U.S. average at the 0.05 significance level. Specifically, we want to test

$$H_0 : \mu = 4.3$$
$$H_a : \mu \neq 4.3$$

Recall that $n = 8$, $\bar{x} = 7.00$, and $s = 3.546$. The t test statistic is

$$t = \frac{\bar{x} - \mu_0}{s/\sqrt{n}} = \frac{7.00 - 4.3}{3.546/\sqrt{8}} = 2.15$$

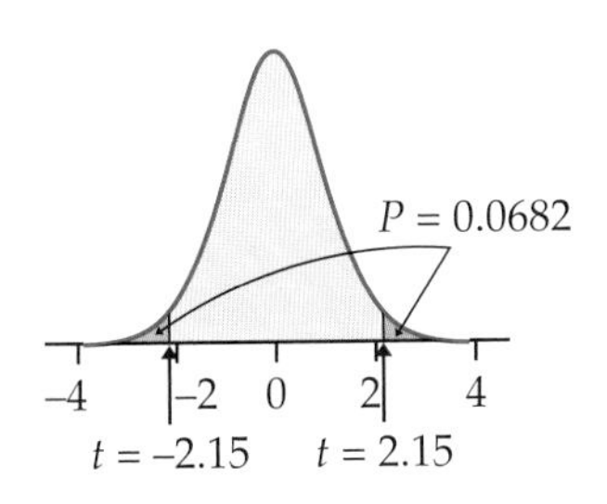

FIGURE 7.8 Sketch of the P-value calculation for Example 7.11.

df = 7		
p	0.05	0.025
t^*	1.895	2.365

This means that the sample mean $\bar{x} = 7.00$ is slightly more than 2.0 standard errors away from the null hypothesized value $\mu = 4.3$. Because the degrees of freedom are $n - 1 = 7$, this t statistic has the $t(7)$ distribution. Figure 7.8 shows that the P-value is $2P(T \geq 2.15)$, where T has the $t(7)$ distribution. From Table D we see that $P(T \geq 1.895) = 0.05$ and $P(T \geq 2.365) = 0.025$.

Therefore, we conclude that the P-value is between $2 \times 0.025 = 0.05$ and $2 \times 0.05 = 0.10$. Software gives the exact value as $P = 0.0682$. These data are compatible with a mean of 4.3 hours per month at the $\alpha = 0.05$ level. In other words, there is not enough evidence to claim that college students spend a different amount of time watching videos on their phones.

In this example we tested the null hypothesis $\mu = 4.3$ hours per month against the two-sided alternative $\mu \neq 4.3$ hours per month because we had no prior suspicion that the average among college students would be larger or smaller. If we were interested in seeing only if it were larger, we would have used a one-sided test.

EXAMPLE 7.12

One-sided test for watching videos on mobile phones. For the mobile phone problem described in the previous example, we want to test whether the U.S. college student average is larger than the overall U.S. population average. Here we test

$$H_0: \mu = 4.3$$

versus

$$H_a: \mu > 4.3$$

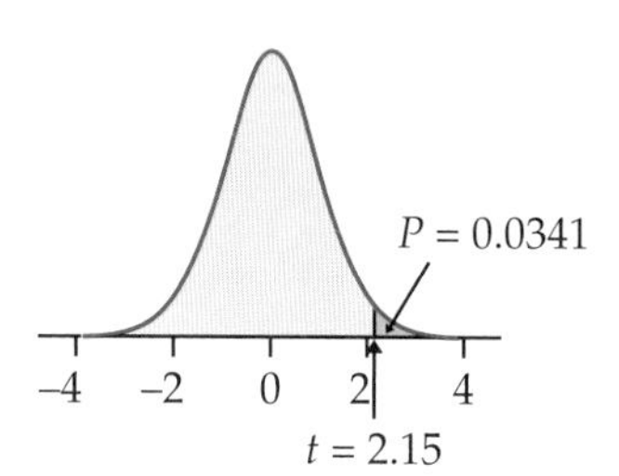

FIGURE 7.9 Sketch of the P-value calculation for Example 7.12.

The t test statistic does not change: $t = 2.15$. As Figure 7.9 illustrates, however, the P-value is now $P(T \geq 2.15)$, half of the value in the previous example. From Table D we can determine that $0.025 < P < 0.05$. Software gives the exact value as $P = 0.0341$. At the $\alpha = 0.05$ level, we reject the null hypothesis and conclude that the U.S. college student average is larger than the U.S. average.

For the mobile phone problem, our conclusion depends on the choice of a one-sided or a two-sided test. This choice needs to be done prior to analysis. If in doubt, always use a two-sided test. *It is wrong to examine the data first and then decide to do a one-sided test in the direction indicated by the data.* In the present circumstance we could use our results from Example 7.11 to justify a one-sided test for *another* sample from the same population.

For small data sets, such as the one in Example 7.11, it is easy to perform the computations for confidence intervals and significance tests with an ordinary calculator. For larger data sets, however, we prefer to use software or a statistical calculator.

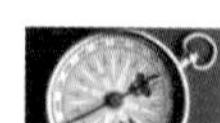

EXAMPLE 7.13

Stock portfolio diversification? An investor with a stock portfolio worth several hundred thousand dollars sued his broker and brokerage firm because lack of diversification in his portfolio led to poor performance.

Table 7.1 gives the rates of return for the 39 months that the account was managed by the broker.[5] Figure 7.10 gives a histogram for these data and Figure 7.11 gives the Normal quantile plot. There are no outliers and the distribution shows no strong skewness. We are reasonably confident that the distribution of $\bar{x}$ is approximately Normal, and we proceed with our inference based on Normal theory.

TABLE 7.1

Monthly rates of return on a portfolio (%)

−8.36	1.63	−2.27	−2.93	−2.70	−2.93	−9.14	−2.64
6.82	−2.35	−3.58	6.13	7.00	−15.25	−8.66	−1.03
−9.16	−1.25	−1.22	−10.27	−5.11	−0.80	−1.44	1.28
−0.65	4.34	12.22	−7.21	−0.09	7.34	5.04	−7.24
−2.14	−1.01	−1.41	12.03	−2.56	4.33	2.35	

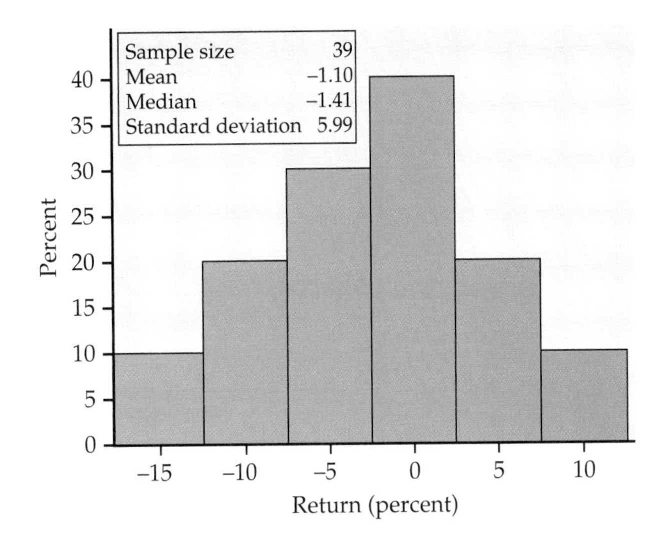

FIGURE 7.10 Histogram for Example 7.13.

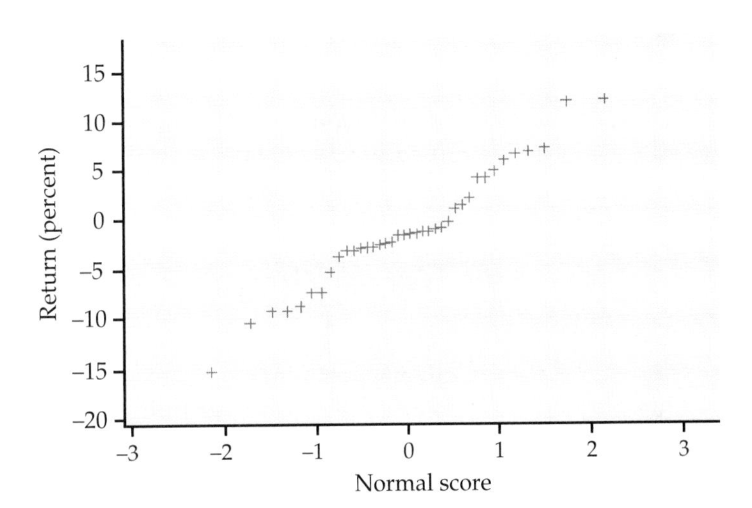

FIGURE 7.11 Normal quantile plot for Example 7.13.

The arbitration panel compared these returns with the average of the Standard and Poor's 500 stock index for the same period. Consider the 39 monthly returns as a random sample from the population of monthly returns the brokerage would generate if it managed the account forever. Are these returns compatible with a population mean of $\mu = 0.95\%$, the S&P 500 average? Our hypotheses are

$$H_0: \mu = 0.95$$
$$H_a: \mu \neq 0.95$$

Minitab and SPSS outputs appear in Figure 7.12. Output from other software will look similar.

Here is one way to report the conclusion: the mean monthly return on investment for this client's account was $\bar{x} = -1.1\%$. This differs significantly from the performance of the S&P 500 stock index for the same period ($t = -2.14$, df = 38, $P = 0.039$).

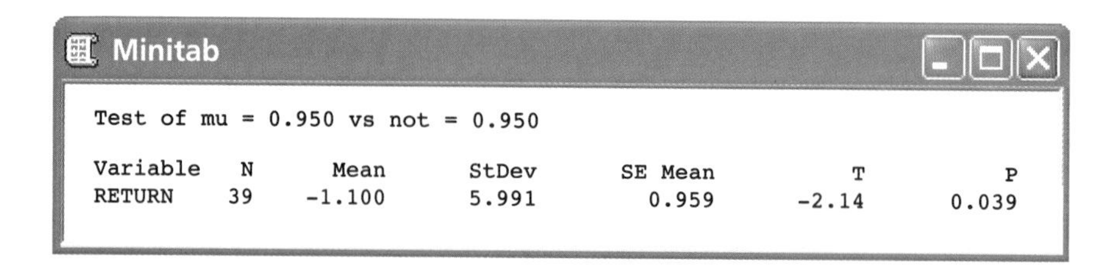

```
Minitab
Test of mu = 0.950 vs not = 0.950

Variable   N     Mean    StDev   SE Mean      T       P
RETURN    39   -1.100    5.991     0.959  -2.14   0.039
```

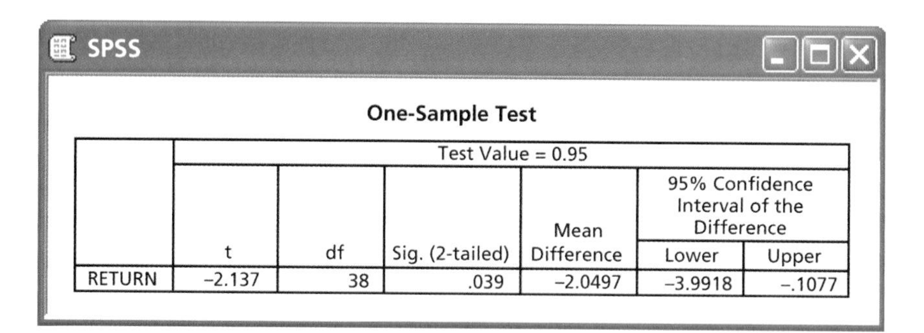

SPSS

One-Sample Test

	Test Value = 0.95					
					95% Confidence Interval of the Difference	
	t	df	Sig. (2-tailed)	Mean Difference	Lower	Upper
RETURN	-2.137	38	.039	-2.0497	-3.9918	-.1077

FIGURE 7.12 Minitab and SPSS output for Example 7.13.

The hypothesis test in Example 7.13 leads us to conclude that the mean return on the client's account differs from that of the S&P 500 stock index. Now let's assess the return on the client's account with a confidence interval.

EXAMPLE 7.14

Estimating the mean monthly return. The mean monthly return on the client's portfolio was $\bar{x} = -1.1\%$ and the standard deviation was $s = 5.99\%$. Figure 7.13 gives the Minitab, SPSS, and Excel outputs for a 95% confidence interval for the population mean μ. Note that Excel gives the margin of error next to the label "Confidence Level(95.0%)" rather than the actual confidence interval. We see that the 95% confidence interval is $(-3.04, 0.84)$, or (from Excel) -1.0997 ± 1.9420.

Because the S&P 500 return, 0.95%, falls outside this interval, we know that μ differs significantly from 0.95% at the $\alpha = 0.05$ level. Example 7.13 gave the actual P-value as $P = 0.039$.

The confidence interval suggests that the broker's management of this account had a long-term mean somewhere between a loss of 3.04% and a gain of 0.84% per month. We are interested not in the actual mean but in the difference between the performance of the client's portfolio and that of the diversified S&P 500 stock index.

EXAMPLE 7.15

Estimating the difference from a standard. Following the analysis accepted by the arbitration panel, we are considering the S&P 500 monthly average return as a constant standard. (It is easy to envision scenarios where

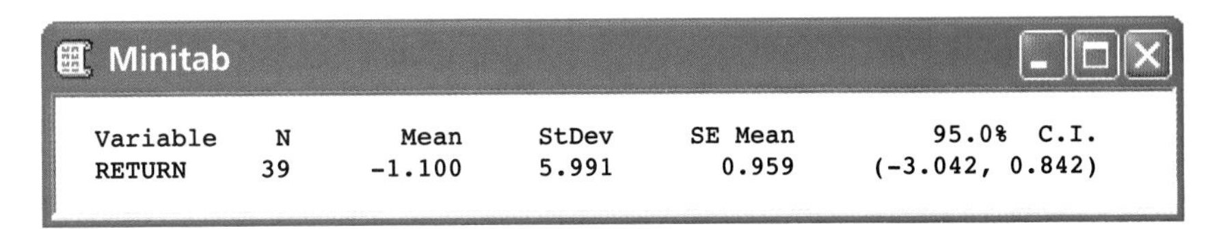

Minitab

Variable	N	Mean	StDev	SE Mean	95.0% C.I.
RETURN	39	-1.100	5.991	0.959	(-3.042, 0.842)

SPSS

Descriptives

			Statistic	Std. Error
RETURN	MEAN		-1.0997	.9593
	95% Confidence Interval for Mean	Lower Bound	-3.0418	
		Upper Bound	.8423	

Excel

	A	B
1	Mean	-1.09974359
2	Standard Error	0.95930991
3	Standard Deviation	5.990888471
4	Count	39
5	Confidence Level(95.0%)	1.942021452

FIGURE 7.13 Minitab, SPSS, and Excel output for Example 7.14.

we would want to treat this type of quantity as random.) The difference between the mean of the investor's account and the S&P 500 is $\bar{x} - \mu = -1.10 - 0.95 = -2.05\%$. In Example 7.14 we found that the 95% confidence interval for the investor's account was $(-3.04,\ 0.84)$. To obtain the corresponding interval for the difference, subtract 0.95 from each of the endpoints. The resulting interval is $(-3.04-0.95,\ 0.84-0.95)$, or $(-3.99,\ -0.11)$. We conclude with 95% confidence that the underperformance was between -3.99% and -0.11%. This interval is presented in the SPSS output of Figure 7.12. This estimate helps to set the compensation owed the investor.

The assumption that these 39 monthly returns represent an SRS from the population of monthly returns is certainly questionable. If the monthly S&P 500 returns were available, an alternative analysis would be to compare the average difference between the monthly returns for this account and for the S&P 500. This method of analysis, where the data are arranged in pairs, is discussed next.

USE YOUR KNOWLEDGE

7.11 Significance test using the *t* distribution. A test of a null hypothesis versus a two-sided alternative gives $t = 2.18$.

(a) The sample size is 18. Is the test result significant at the 5% level? Explain how you obtained your answer.

(b) The sample size is 10. Is the test result significant at the 5% level? Explain how you obtained your answer.
(c) Sketch the two t distributions to illustrate your answers.

7.12 Significance test for apartment rents. Recall Exercise 7.7 (page 372). Does this SRS give good reason to believe that the mean rent of all advertised one-bedroom apartments is greater than $550? State the hypotheses, find the t statistic and its P-value, and state your conclusion.

7.13 Using software. In Example 7.10 (page 373) we calculated the 95% confidence interval for the U.S. college student average number of hours spent per month watching videos on a mobile phone. Use software to compute this interval and verify that you obtain the same interval.

Matched pairs t procedures

The mobile phone problem of Example 7.10 concerns only a single population. We know that comparative studies are usually preferred to single-sample investigations because of the protection they offer against confounding. For that reason, inference about a parameter of a single distribution is less common than comparative inference.

One common comparative design, however, makes use of single-sample procedures. In a matched pairs study, subjects are matched in pairs and their outcomes are compared within each matched pair. For example, an experiment to compare two cell phone packages might use pairs of subjects who are the same age and gender and are at the same income level. The experimenter could toss a coin to assign the two packages to the two subjects in each pair. The idea is that matched subjects are more similar than unmatched subjects, so comparing outcomes within each pair is more efficient. Matched pairs are also common when randomization is not possible. One situation calling for matched pairs is when observations are taken on the same subjects under two different conditions or before and after some intervention. Here is an example.

EXAMPLE 7.16

Does a full moon affect behavior? Many people believe that the moon influences the actions of some individuals. A study of dementia patients in nursing homes recorded various types of disruptive behaviors every day for 12 weeks. Days were classified as moon days if they were in a three-day period centered at the day of the full moon. For each patient the average number of disruptive behaviors was computed for moon days and for all other days. The data for the 15 subjects whose behaviors were classified as aggressive are presented in Table 7.2.[6] The patients in this study are not a random sample of dementia patients. However, we examine their data in the hope that what we find is not unique to this particular group of individuals and applies to other patients who have similar characteristics.

To analyze these paired data, we first subtract the disruptive behaviors for moon days from the disruptive behaviors for other days. These 15 differences form a single sample. They appear in the "Difference" columns in Table 7.2.

TABLE 7.2

Aggressive behaviors of dementia patients

Patient	Moon days	Other days	Difference	Patient	Moon days	Other days	Difference
1	3.33	0.27	3.06	9	6.00	1.59	4.41
2	3.67	0.59	3.08	10	4.33	0.60	3.73
3	2.67	0.32	2.35	11	3.33	0.65	2.68
4	3.33	0.19	3.14	12	0.67	0.69	−0.02
5	3.33	1.26	2.07	13	1.33	1.26	0.07
6	3.67	0.11	3.56	14	0.33	0.23	0.10
7	4.67	0.30	4.37	15	2.00	0.38	1.62
8	2.67	0.40	2.27				

The first patient, for example, averaged 3.33 aggressive behaviors on moon days but only 0.27 aggressive behaviors on other days. The difference 3.33 − 0.27 = 3.06 is what we will use in our analysis.

```
4 | 4 4
3 | 1 1 1 6 7
2 | 1 3 4 7
1 | 6
0 | 0 1 1
```

FIGURE 7.14 Stemplot of differences in aggressive behaviors for Examples 7.16 and 7.17.

Next, we examine the distribution of these differences. Figure 7.14 gives a stemplot of the differences. This plot indicates that there are three patients with very small differences, but there are no indications of extreme outliers or strong skewness. We will proceed with our analysis using the Normality-based methods of this section.

To assess whether there is a difference in aggressive behaviors on moon days versus other days, we test

$$H_0 : \mu = 0$$
$$H_a : \mu \neq 0$$

Here μ is the mean difference in aggressive behaviors, moon versus other days, for patients of this type. The null hypothesis says that aggressive behaviors occur at the same frequency for both types of days, and H_a says that the behaviors on moon days are not the same as on other days.

The 15 differences have

$$\bar{x} = 2.433 \text{ and } s = 1.460$$

The one-sample t statistic is therefore

$$t = \frac{\bar{x} - 0}{s/\sqrt{n}} = \frac{2.433}{1.460/\sqrt{15}} = 6.45$$

The P-value is found from the $t(14)$ distribution (remember that the degrees of freedom are 1 less than the sample size).

df = 14

p	0.001	0.0005
t^*	3.787	4.140

Table D shows that 6.45 lies beyond the upper 0.0005 critical value of the $t(14)$ distribution. Since we are using a two-sided alternative, we know that the P-value is less than two times this value, or 0.0010. Software gives a value that is much smaller, $P = 0.000015$. In practice, there is little difference between these two P-values. The data provide clear evidence in favor of the alternative hypothesis.

A difference this large is very unlikely to occur by chance if there is, in fact, no effect of the moon on aggressive behaviors. In scholarly publications, the details of routine statistical procedures are omitted. Because the sample mean of these differences was greater than 0, our test would be reported in the form: "There was more aggressive behavior on moon days than on other days ($t = 6.45$, df $= 14$, $P < 0.001$)."

Note that we could have justified a one-sided alternative in this example. Based on previous research, we expect more aggressive behaviors on moon days, and the alternative H_a: $\mu > 0$ is reasonable in this setting. The choice of the alternative here, however, has no effect on the conclusion: from Table D we determine that P is less than 0.0005; from software it is 0.000008. These are very small values and we would still report $P < 0.001$. *In most circumstances we cannot be absolutely certain about the direction, and the safest strategy is to use the two-sided alternative.*

The results of the significance test allow us to conclude that dementia patients exhibit more aggressive behaviors in the days around a full moon. What are the implications of the study for the administrators who run the facilities where these patients live? For example, should they increase staff on these days? To make these kinds of decisions, an estimate of the magnitude of the problem, with a margin of error, would be helpful.

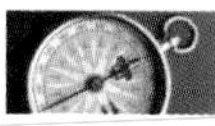

EXAMPLE 7.17

95% confidence interval for the full-moon study. A 95% confidence interval for the mean difference in aggressive behaviors per day requires the critical value $t^* = 2.145$ from Table D. The margin of error is

$$t^* \frac{s}{\sqrt{n}} = 2.145 \frac{1.460}{\sqrt{15}} = 0.81$$

and the confidence interval is

$$\bar{x} \pm t^* \frac{s}{\sqrt{n}} = 2.43 \pm 0.81 = (1.62,\ 3.24)$$

The estimated average difference is 2.43 aggressive behaviors per day, with margin of error 0.81 for 95% confidence. The increase needs to be interpreted in terms of the baseline values. The average number of aggressive behaviors per day on other days is 0.59; on moon days it is 3.02. This is approximately a 400% increase. If aggressive behaviors require a substantial amount of attention by staff, then administrators should be aware of the increased level of these activities during the full-moon period. Additional staff may be needed.

The following are key points to remember concerning matched pairs:

1. A matched pairs analysis is called for when subjects are **matched in pairs** or there are **two measurements or observations on each individual** and we want to examine the difference.

2. For each pair or individual, use the difference between the two measurements as the data for your analysis.
3. Use the one-sample confidence interval and the significance-testing procedures that we learned in this section.

There are several issues to be aware of with respect to our use of the t procedures in Examples 7.16 and 7.17. First, no randomization is possible in a study like this. Our inference procedures assume that there is a process that generates these aggressive behaviors and that the process produces them at possibly different rates during the days near the full moon. Second, many of the patients in these nursing homes did not exhibit any disruptive behaviors. These were not included in our analysis, so our inference is restricted to patients who do exhibit disruptive behaviors.

A final difficulty is that the data show departures from Normality. In a matched pairs analysis, the t procedures are applied to the differences, so we are assuming that the differences are Normally distributed. Figure 7.14 gives a stemplot of the differences. There are 3 patients with very small differences in aggressive behaviors, while the other 12 have a large increase.

We have a dilemma here similar to that in Example 7.10. *The data may not be Normal, and our sample size is very small.* We can try an alternative procedure that does not require the Normality assumption—but there is a price to pay. The alternative procedures have less power to detect differences. Despite these caveats, for Example 7.16 the P-value is so small that we are very confident that we have found an effect of the moon phase on behavior.

USE YOUR KNOWLEDGE

7.14 Comparison of two energy drinks. Consider the following study to compare two popular energy drinks. For each subject, a coin was flipped to determine which drink to rate first. Each drink was rated on a 0 to 100 scale, with 100 being the highest rating.

	Subject				
Drink	**1**	**2**	**3**	**4**	**5**
A	43	83	66	89	78
B	45	78	64	79	71

Is there a difference in preference? State appropriate hypotheses and carry out a matched pairs t test for these data.

7.15 95% confidence interval for the difference in energy drinks. For the companies producing these drinks, the real question is how much difference there is between the two preferences. Use the data above to give a 95% confidence interval for the difference in preference between Drink A and Drink B.

Robustness of the t procedures

The results of one-sample t procedures are exactly correct only when the population is Normal. Real populations are never exactly Normal. The usefulness of the t procedures in practice therefore depends on how strongly they are affected by non-Normality. Procedures that are not strongly affected are called *robust*.

> **ROBUST PROCEDURE**
>
> A statistical inference procedure is called **robust** if the required probability calculations are insensitive to violations of the assumptions made.

resistant measure p. 107

The assumption that the population is Normal rules out outliers, so the presence of outliers shows that this assumption is not valid. The t procedures are not robust against outliers, because $\bar{x}$ and s are not resistant to outliers.

In Example 7.16, three patients have fairly low values of the difference. Whether or not these are outliers is a matter of judgment. If we rerun the analysis without these three patients, the t statistic would increase to 11.89 and the P-value would be much lower. Careful inspection of the records may reveal some characteristic of these patients that distinguishes them from the others in the study. Without such information, it is difficult to justify excluding them from the analysis. *In general, we should be very cautious about discarding suspected outliers, particularly when they make up a substantial proportion of the data, as they do in this example.*

Fortunately, the t procedures are quite robust against non-Normality of the population except in the case of outliers or strong skewness. Larger samples improve the accuracy of P-values and critical values from the t distributions when the population is not Normal. This is true for two reasons:

1. The sampling distribution of the sample mean $\bar{x}$ from a large sample is close to Normal (that's the central limit theorem). Normality of the individual observations is of little concern when the sample is large.
2. As the sample size n grows, the sample standard deviation s will be an accurate estimate of σ whether or not the population has a Normal distribution. This fact is closely related to the law of large numbers.

LOOK BACK
law of large numbers p. 271

Constructing a Normal quantile plot, stemplot, or boxplot to check for skewness and outliers is an important preliminary to the use of t procedures for small samples. For most purposes, the one-sample t procedures can be safely used when $n \geq 15$ unless an outlier or clearly marked skewness is present. *Except in the case of small samples, the assumption that the data are an SRS from the population of interest is more crucial than the assumption that the population distribution is Normal.* Here are practical guidelines for inference on a single mean:[7]

CAUTION

- *Sample size less than 15:* Use t procedures if the data are close to Normal. If the data are clearly non-Normal or if outliers are present, do not use t.
- *Sample size at least 15:* The t procedures can be used except in the presence of outliers or strong skewness.

- *Large samples:* The t procedures can be used even for clearly skewed distributions when the sample is large, roughly $n \geq 40$.

Consider, for example, some of the data we studied in Chapter 3. The service center call lengths in Figure 3.30 (page 142) are strongly skewed to the right. Since there are 80 observations, we could use the t procedures here. On the other hand, many would prefer to use a transformation to make these data more nearly Normal. (See the material on inference for non-Normal populations in Chapter 16.) The data related to time to start a business in Exercise 3.40 (page 107) contain one outlier in a sample of size 25, which makes the use of t procedures more risky. Figure 3.32 (page 143) gives the Normal quantile plot for 105 acidity measurements of rainwater. These data appear to be Normal, and we would apply the t procedures in this case.

SECTION 7.1 SUMMARY

Significance tests and confidence intervals for the mean μ of a Normal population are based on the sample mean $\bar{x}$ of an SRS. Because of the **central limit theorem**, the resulting procedures are approximately correct for other population distributions when the sample is large.

The standardized sample mean, or **one-sample z statistic,**

$$z = \frac{\bar{x} - \mu}{\sigma/\sqrt{n}}$$

has the $N(0, 1)$ distribution. If the standard deviation $\sigma/\sqrt{n}$ of $\bar{x}$ is replaced by the **standard error** $s/\sqrt{n}$, the **one-sample t statistic**

$$t = \frac{\bar{x} - \mu}{s/\sqrt{n}}$$

has the **t distribution** with $n - 1$ degrees of freedom.

There is a t distribution for every positive **degrees of freedom k.** All are symmetric distributions similar in shape to Normal distributions. The $t(k)$ distribution approaches the $N(0, 1)$ distribution as k increases.

A level C **confidence interval for the mean** μ of a Normal population is

$$\bar{x} \pm t^* \frac{s}{\sqrt{n}}$$

where t^* is the value for the $t(n-1)$ density curve with area C between $-t^*$ and t^*. The quantity

$$t^* \frac{s}{\sqrt{n}}$$

is the **margin of error.**

Significance tests for $H_0: \mu = \mu_0$ are based on the t statistic. P-values or fixed significance levels are computed from the $t(n - 1)$ distribution.

A common comparative design is a **matched pairs** study. With this design, subjects are first matched in pairs, and then within each pair, one subject is randomly assigned Treatment 1 and the other is assigned Treatment 2. Alternatively, we can make two measurements or observations on each subject. One-sample procedures are used to analyze matched pairs data by first taking the differences within the matched pairs (or subject) to produce a single sample of differences.

The t procedures are relatively **robust** against non-Normal populations. The t procedures are useful for non-Normal data when $15 \leq n < 40$ unless the data show outliers or strong skewness. When $n \geq 40$, the t procedures can be used even for clearly skewed distributions.

SECTION 7.1 EXERCISES

For Exercise 7.1, see page 361; for Exercises 7.2 and 7.3, see pages 364–365; for Exercise 7.4, see page 366; for Exercise 7.5, see page 366; for Exercise 7.6, see page 368; for Exercises 7.7, see page 372; for Exercises 7.8 to 7.10, see page 374; for Exercises 7.11 to 7.13, see pages 379–380; and for Exercises 7.14 and 7.15, see page 383.

7.16. What is wrong? Explain what is wrong in each of the following statements.

(a) For large n, the distribution of observed values will be approximately Normal.

(b) The 68–95–99.7 rule says that $\bar{x}$ should be within $\mu \pm 2\sigma$ about 95% of the time.

(c) The central limit theorem states that for large n, μ is approximately Normal.

7.17. What is wrong? Explain what is wrong in each of the following scenarios.

(a) If the variance of a population is 10, then the variance of the mean for an SRS of 30 observations from this population will be $10/\sqrt{30}$.

(b) When taking SRSs from a population, larger sample sizes will result in larger standard deviations of the sample mean.

(c) The mean of a sampling distribution of $\bar{x}$ changes when the sample size changes.

7.18. Total sleep time of college students. In Example 7.1, the total sleep time per night among college students was approximately Normally distributed with mean $\mu = 7.02$ hours and standard deviation $\sigma = 1.15$ hours (page 360). Suppose that you plan to take an SRS of size $n = 400$ and compute the average total sleep time.

(a) What is the standard deviation for the average time?

(b) Use the 95 part of the 68–95–99.7 rule to describe the variability of this sample mean.

(c) What is the probability that your average will be below 6.9 hours?

7.19. Determining sample size. Recall the previous exercise. Suppose you want to use a sample size such that about 95% of the averages fall within ±0.05 hours (±3 minutes) of the true mean $\mu = 7.02$.

(a) Based on your answer to part (b) in Exercise 7.18, should the sample size be larger or smaller than 400? Explain.

(b) What standard deviation of the sample mean do you need such that 95% of all samples will have a mean within 0.05 hours of μ?

(c) Using the standard deviation calculated in part (b), determine the number of students you need to sample.

7.20. Number of friends on Facebook. Pew Center's Internet and American Life Project recently reported that the average Facebook user has 245 friends.[8] This distribution only takes integer values, so it is certainly not Normal. We'll also assume that it is skewed to the right with a standard deviation $\sigma = 115$. Consider an SRS of 30 Facebook users.

(a) What are the mean and standard deviation of the total number of friends in this sample?

(b) What are the mean and standard deviation of the mean number of friends per user?

(c) Use the central limit theorem to find the probability that the average number of friends among 30 Facebook users is greater than 275.

7.21. Finding the critical value t^*. What critical value t^* from Table D should be used to calculate the margin of error for a confidence interval for the mean of the population in each of the following situations?

(a) A 95% confidence interval based on $n = 11$ observations.

(b) A 95% confidence interval from an SRS of 22 observations.

(c) A 90% confidence interval from a sample of size 22.

(d) These cases illustrate how the size of the margin of error depends upon the confidence level and the sample size. Summarize these relationships.

7.22. Distribution of the t statistic. Assume that you have a sample of size $n = 18$. Draw a picture of the distribution of the t statistic under the null hypothesis. Use Table D and your picture to illustrate the values of the test statistic that would lead to rejection of the null hypothesis at the 5% level for a two-sided alternative.

7.23. More on the distribution of the t statistic. Repeat the previous exercise for the two situations where the alternative is one-sided.

7.24. One-sided versus two-sided P-values. Computer software reports $\bar{x} = 15.3$ and $P = 0.074$ for a t test of H_0: $\mu = 0$ versus H_a: $\mu \neq 0$. Based on prior knowledge, you can justify testing the alternative H_a: $\mu > 0$. What is the P-value for your significance test?

7.25. More on one-sided versus two-sided P-values. Suppose that $\bar{x} = -15.3$ in the setting of the previous exercise. Would this change your P-value? Use a sketch of the distribution of the test statistic under the null hypothesis to illustrate and explain your answer.

7.26. Number of Facebook friends. Consider the following SRS of $n = 30$ Facebook users from a large university. The number of Facebook friends is recorded. FACEBOOKFRIENDS

437	321	216	452	162	310	284	93	300	99
240	137	532	214	363	194	362	416	395	239
508	426	368	101	279	205	290	252	253	406

(a) Do you think these data are Normally distributed? Use graphical methods to examine the distribution. Write a short summary of your findings.

(b) Is it appropriate to use the t methods of this section to compute a 95% confidence interval for the mean number of Facebook users at this large university? Explain why or why not.

(c) Find the mean, standard deviation, standard error, and margin of error for 95% confidence.

(d) Report the 95% confidence interval for μ, the average number of friends for Facebook users at this large university.

7.27. Rudeness and its effect on onlookers. Many believe that an uncivil environment has a negative effect on people. A pair of researchers performed a series of experiments to test whether witnessing rudeness and disrespect affects task performance.[9] In one study, 34 participants met in small groups and witnessed the group organizer being rude to a "participant" who showed up late for the group meeting. After the exchange, each participant performed an individual brainstorming task in which he or she was asked to produce as many uses for a brick as possible in 5 minutes. The mean number of uses was 7.88 with a standard deviation of 2.35.

(a) Suppose that prior research has shown that the average number of uses a person can produce in 5 minutes under normal conditions is 10. Given that the researchers hypothesize that witnessing this rudeness will decrease performance, state the appropriate null and alternative hypotheses.

(b) Carry out the significance test using a significance level of 0.05. Give the P-value and state your conclusion.

7.28. Fuel efficiency t test. Computers in some vehicles calculate various quantities related to performance. One of these is the fuel efficiency, or gas mileage, usually expressed as miles per gallon (mpg). For one vehicle equipped in this way, the mpg were recorded each time the gas tank was filled, and the computer was then reset.[10] Here are the mpg values for a random sample of 20 of these records: DATA FILE GASMILEAGE

41.5	50.7	36.6	37.3	34.2	45.0	48.0	43.2	47.7	42.2
43.2	44.6	48.4	46.4	46.8	39.2	37.3	43.5	44.3	43.3

(a) Describe the distribution using graphical methods. Is it appropriate to analyze these data using methods based on Normal distributions? Explain why or why not.

(b) Find the mean, standard deviation, standard error, and margin of error for 95% confidence.

(c) Report the 95% confidence interval for μ, the mean mpg for this vehicle based on these data.

7.29. Do you feel lucky? Children in a psychology study were asked to solve some puzzles and were then given feedback on their performance. Then they were asked to rate how luck played a role in determining their scores.[11] This variable was recorded on a 1 to 10 scale with 1 corresponding to very lucky and 10 corresponding to very unlucky. Here are the scores for 60 children: DATA FILE FEELLUCKY

1	10	1	10	1	1	10	5	1	1	8	1	10	2	1
9	5	2	1	8	10	5	9	10	10	9	6	10	1	5
1	9	2	1	7	10	9	5	10	10	10	1	8	1	6
10	1	6	10	10	8	10	3	10	8	1	8	10	4	2

(a) Use graphical methods to display the distribution. Describe any unusual characteristics. Do these characteristics cause you concern about using the t-based methods of this section? Explain your answer.

(b) Give a 95% confidence interval for the mean luck score.

(c) Based on the distribution in part (a), one could argue that there are two distinct groups of children. With that in mind, explain why inference about the mean luck score may not be very useful.

(d) The children in this study were volunteers whose parents agreed to have them participate in the study. To what extent do you think your results would apply to all similar children in this community?

7.30. Nutritional intake in Canadian high-performance male athletes. Since previous studies have reported that elite athletes are often deficient in their nutritional intake (for example, total calories, carbohydrates, protein), a group of researchers decided to evaluate Canadian high-performance athletes.[12] A total of $n = 114$ male athletes from eight Canadian sports centers were surveyed. The average caloric intake was 3077.0 kilocalories/day with a standard deviation of 987.0. The recommended amount is 3421.7. Is there evidence that Canadian high-performance male athletes are deficient in their caloric intake?

(a) State the appropriate H_0 and H_a to test this.

(b) Carry out the test, give the P-value, and state your conclusion.

(c) Construct a 95% confidence interval for the average deficiency in caloric intake.

7.31. Fuel efficiency comparison t test. Refer to Exercise 7.28. In addition to the computer calculating mpg, the driver also recorded the mpg by dividing the miles driven by the number of gallons at fill-up. The driver wants to determine if these calculations are different. MPGCOMPARISON

Fill-up	1	2	3	4	5	6	7	8	9	10
Computer	41.5	50.7	36.6	37.3	34.2	45.0	48.0	43.2	47.7	42.2
Driver	36.5	44.2	37.2	35.6	30.5	40.5	40.0	41.0	42.8	39.2
Fill-up	**11**	**12**	**13**	**14**	**15**	**16**	**17**	**18**	**19**	**20**
Computer	43.2	44.6	48.4	46.4	46.8	39.2	37.3	43.5	44.3	43.3
Driver	38.8	44.5	45.4	45.3	45.7	34.2	35.2	39.8	44.9	47.5

(a) State the appropriate H_0 and H_a.

(b) Carry out the test using a significance level of 0.05. Give the P-value, and then interpret the result.

7.32. CHALLENGE **Food intake and weight gain.** If we increase our food intake, we generally gain weight. Nutrition scientists can calculate the amount of weight gain that would be associated with a given increase in calories. In one study, 16 nonobese adults, aged 25 to 36 years, were fed 1000 calories per day in excess of the calories needed to maintain a stable body weight. The subjects maintained this diet for 8 weeks, so they consumed a total of 56,000 extra calories.[13] According to theory, 3500 extra calories will translate into a weight gain of 1 pound. Therefore, we expect each of these subjects to gain $56{,}000/3500 = 16$ pounds (lb). Here are the weights before and after the 8-week period expressed in kilograms (kg): WEIGHTGAIN

Subject	1	2	3	4	5	6	7	8
Weight before	55.7	54.9	59.6	62.3	74.2	75.6	70.7	53.3
Weight after	61.7	58.8	66.0	66.2	79.0	82.3	74.3	59.3
Subject	**9**	**10**	**11**	**12**	**13**	**14**	**15**	**16**
Weight before	73.3	63.4	68.1	73.7	91.7	55.9	61.7	57.8
Weight after	79.1	66.0	73.4	76.9	93.1	63.0	68.2	60.3

(a) For each subject, subtract the weight before from the weight after to determine the weight change.

(b) Find the mean and the standard deviation for the weight change.

(c) Calculate the standard error and the margin of error for 95% confidence. Report the 95% confidence interval in a sentence that explains the meaning of "95%" in this context.

(d) Convert the mean weight gain in kilograms to mean weight gain in pounds. Because there are 2.2 kg per pound, multiply the value in kilograms by 2.2 to obtain pounds. Do the same for the standard deviation and the confidence interval.

(e) Test the null hypothesis that the mean weight gain is 16 lb. Be sure to specify the null and alternative hypotheses, the test statistic with degrees of freedom, and the P-value. What do you conclude?

(f) Write a short paragraph explaining your results.

7.33. Food intake and NEAT. Nonexercise activity thermogenesis (NEAT) provides a partial explanation for the results you found in the previous exercise. NEAT is energy burned by fidgeting, maintenance of posture, spontaneous muscle contraction, and other activities of daily living. In the study of the previous exercise, the 16 subjects increased their NEAT by 328 calories per day, on average, in response to the additional food intake. The standard deviation was 256.

(a) Test the null hypothesis that there was no change in NEAT versus the two-sided alternative. Summarize the results of the test and give your conclusion.

(b) Find a 95% confidence interval for the change in NEAT. Discuss the additional information provided by the confidence interval that is not evident from the results of the significance test.

7.34. Insurance fraud? Insurance adjusters are concerned about the high estimates they are receiving from Jocko's Garage. To see if the estimates are unreasonably high, each of 10 damaged cars was taken to Jocko's and to another garage and the estimates were recorded. Here are the results (in dollars): JOCKOGARAGE

Car	1	2	3	4	5
Jocko's	1375	1550	1250	1300	900
Other	1250	1300	1250	1200	950
Car	**6**	**7**	**8**	**9**	**10**
Jocko's	1500	1750	3600	2250	2800
Other	1575	1600	3300	2125	2600

(a) For each car, subtract the estimate of the other garage from Jocko's estimate. Find the mean and the standard deviation for these differences.

(b) Test the null hypothesis that there is no difference between the estimates of the two garages. Be sure to specify the null and alternative hypotheses, the test statistic with degrees of freedom, and the P-value. What do you conclude using the 0.05 significance level?

(c) Construct a 95% confidence interval for the difference in estimates.

(d) The insurance company is considering seeking repayment for 1000 claims filed with Jocko's last year. Using your answer to part (c), what repayment would you recommend the insurance company seek? Explain your answer.

7.35. Comparing operators of a DXA machine. Dual-energy X-ray absorptiometry (DXA) is a technique for measuring bone health. One of the most common measures is total body bone mineral content (TBBMC). A highly skilled operator is required to take the measurements. Recently, a new DXA machine was purchased by a research lab, and two operators were trained to take the measurements. TBBMC for eight subjects was measured by both operators.[14] The units are grams (g). A comparison of the means for the two operators provides a check on the training they received and allows us to determine if one of the operators is producing measurements that are consistently higher than the other. Here are the data: TBBMC

	Subject							
Operator	**1**	**2**	**3**	**4**	**5**	**6**	**7**	**8**
1	1.328	1.342	1.075	1.228	0.939	1.004	1.178	1.286
2	1.323	1.322	1.073	1.233	0.934	1.019	1.184	1.304

(a) Take the difference between the TBBMC recorded by Operator 1 and the TBBMC recorded by Operator 2. Describe the distribution of these differences.

(b) Use a significance test to examine the null hypothesis that the two operators have the same mean. Be sure to give the test statistic with its degrees of freedom, the P-value, and your conclusion.

(c) The sample here is rather small, so we may not have much power to detect differences of interest. Use a 95% confidence interval to provide a range of differences that are compatible with these data.

(d) The eight subjects used for this comparison were not a random sample. In fact, they were friends of the researchers whose ages and weights were similar to the types of people who would be measured with this DXA machine. Comment on the appropriateness of this procedure for selecting a sample, and discuss any consequences regarding the interpretation of the significance testing and confidence interval results.

7.36. Another comparison of DXA machine operators. Refer to the previous exercise. TBBMC measures the total amount of mineral in the bones. Another important variable is total body bone mineral density (TBBMD). This variable is calculated by dividing TBBMC by the area corresponding to bone in the DXA scan. The units are grams per squared centimeter (g/cm^2). Here are the TBBMD values for the same subjects: TBBMD

	Subject							
Operator	**1**	**2**	**3**	**4**	**5**	**6**	**7**	**8**
1	4042	3703	2626	2673	1724	2136	2808	3322
2	4041	3697	2613	2628	1755	2140	2836	3287

Analyze these data using the questions in the previous exercise as a guide.

7.37. CHALLENGE **Assessment of a foreign-language institute.** The National Endowment for the Humanities sponsors summer institutes to improve the skills of high school teachers of foreign languages. One such institute hosted 20 French teachers for 4 weeks. At the beginning of the period, the teachers were given the Modern Language Association's listening test of understanding of spoken French. After 4 weeks of immersion in French in and out of class, the listening test was given again. (The actual French spoken in the two tests was different, so that simply taking the first test should not improve the score on the second test.) The maximum possible score on the test is 36.[15] Here are the data: FRENCHTEST

Teacher	Pretest	Posttest	Gain
1	32	34	2
2	31	31	0
3	29	35	6
4	10	16	6
5	30	33	3
6	33	36	3
7	22	24	2
8	25	28	3
9	32	26	−6
10	20	26	6
11	30	36	6
12	20	26	6
13	24	27	3
14	24	24	0
15	31	32	1
16	30	31	1
17	15	15	0
18	32	34	2
19	23	26	3
20	23	26	3

The "Gain" column is the difference between the posttest score and pretest score and represents the improvement for each teacher.

(a) State appropriate null and alternative hypotheses for examining the question of whether or not the course improves French spoken-language skills.

(b) Describe the gain data. Use numerical and graphical summaries.

(c) Perform the significance test. Give the test statistic, the degrees of freedom, and the P-value. Summarize your conclusion.

(d) Give a 95% confidence interval for the mean improvement.

7.38. Length of calls to a customer service center. Refer to the lengths of calls to a customer service center in Table 3.2 (page 93). Give graphical and numerical summaries for these data. Compute a 95% confidence interval for the mean call length. Comment on the validity of your interval. CALLCENTER80

7.39. IQ test scores. Refer to the IQ test scores for fifth-grade students in Table 3.1 (page 91). Give numerical and graphical summaries of the data and compute a 95% confidence interval. Comment on the validity of the interval. IQ

7.2 Comparing Two Means

A psychologist wants to compare male and female college students' impressions of personality based on selected Facebook pages. A nutritionist is interested in the effect of increased calcium on blood pressure. A bank wants to know which of two incentive plans will most increase the use of its debit cards. Two-sample problems such as these are among the most common situations encountered in statistical practice.

TWO-SAMPLE PROBLEMS

- The goal of inference is to compare the responses in two groups.
- Each group is considered to be a sample from a distinct population.
- The responses in each group are independent of those in the other group.

LOOK BACK randomized comparative experiment p. 42

A two-sample problem can arise from a randomized comparative experiment that randomly divides the subjects into two groups and exposes each group to a different treatment. Comparing random samples separately selected from two populations is also a two-sample problem.

Unlike the matched pairs designs studied earlier, there is no matching of the units in the two samples. In fact, the two samples may be of different sizes. As a result, inference procedures for two-sample data differ from those for matched pairs.

We can present two-sample data graphically by a back-to-back stemplot (for small samples) or by side-by-side boxplots (for larger samples). Now we will apply the ideas of formal inference in this setting. When both population distributions are symmetric, and especially when they are at least approximately Normal, a comparison of the mean responses in the two populations is most often the goal of inference.

We have two independent samples, from two distinct populations (such as subjects given a treatment and those given a placebo). The same variable is measured for both samples. We will call the variable x_1 in the first population and x_2 in the second because the variable may have different distributions in the two populations. Here is the notation that we will use to describe the two populations:

Population	Variable	Mean	Standard deviation
1	x_1	μ_1	σ_1
2	x_2	μ_2	σ_2

We want to compare the two population means, either by giving a confidence interval for $\mu_1 - \mu_2$ or by testing the hypothesis of no difference, H_0: $\mu_1 = \mu_2$.

Inference is based on two independent SRSs, one from each population. Here is the notation that describes the samples:

Population	Sample size	Sample mean	Sample standard deviation
1	n_1	$\bar{x}_1$	s_1
2	n_2	$\bar{x}_2$	s_2

Throughout this section, the subscripts 1 and 2 show the population to which a parameter or a sample statistic refers.

The sampling distribution of the difference between two means

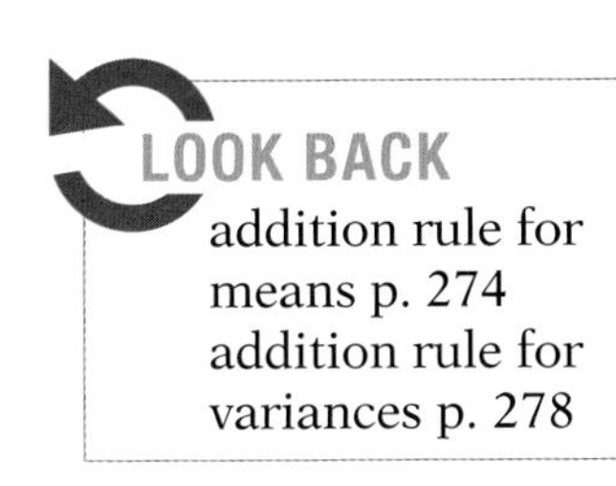

LOOK BACK
addition rule for means p. 274
addition rule for variances p. 278

The natural estimator of the difference $\mu_1 - \mu_2$ is the difference between the sample means, $\bar{x}_1 - \bar{x}_2$. If we are to base inference on this statistic, we must know its sampling distribution. First, the mean of the difference $\bar{x}_1 - \bar{x}_2$ is the difference of the means $\mu_1 - \mu_2$. This follows from the addition rule for means and the fact that the mean of any $\bar{x}$ is the mean of its population. To compute the variance, we use the addition rule for variances. Because the samples are independent, their sample means $\bar{x}_1$ and $\bar{x}_2$ are independent random variables. Thus, the variance of the difference $\bar{x}_1 - \bar{x}_2$ is the sum of their variances, which is

$$\frac{\sigma_1^2}{n_1} + \frac{\sigma_2^2}{n_2}$$

We now know the mean and variance of the distribution of $\bar{x}_1 - \bar{x}_2$ in terms of the parameters of the two populations. If the two population distributions are both Normal, then the distribution of $\bar{x}_1 - \bar{x}_2$ is also Normal. This is true because each sample mean alone is Normally distributed and because a difference between independent Normal random variables is also Normal (page 369).

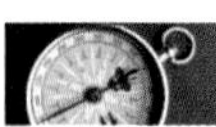

EXAMPLE 7.18

Heights of 10-year-old girls and boys. A fourth-grade class has 12 girls and 8 boys. The children's heights are recorded on their tenth birthdays. What is the chance that the girls are taller than the boys? Of course, it is very unlikely that all the girls are taller than all the boys. We translate the question into the following: what is the probability that the mean height of the 12 girls is greater than the mean height of the 8 boys?

Based on information from the National Health and Nutrition Examination Survey,[16] we assume that the heights (in inches) of 10-year-old girls are $N(56.4, 2.7)$ and the heights of 10-year-old boys are $N(55.7, 3.8)$. The heights of the students in our class are assumed to be random samples from these populations. The two distributions are shown in Figure 7.15(a).

The difference $\bar{x}_1 - \bar{x}_2$ between the female and male mean heights varies in different random samples. The sampling distribution has mean

$$\mu_1 - \mu_2 = 56.4 - 55.7 = 0.7 \text{ inch}$$

and variance

$$\frac{\sigma_1^2}{n_1} + \frac{\sigma_2^2}{n_2} = \frac{2.7^2}{12} + \frac{3.8^2}{8}$$
$$= 2.41$$

The standard deviation of the difference in sample means is therefore $\sqrt{2.41} = 1.55$ inches.

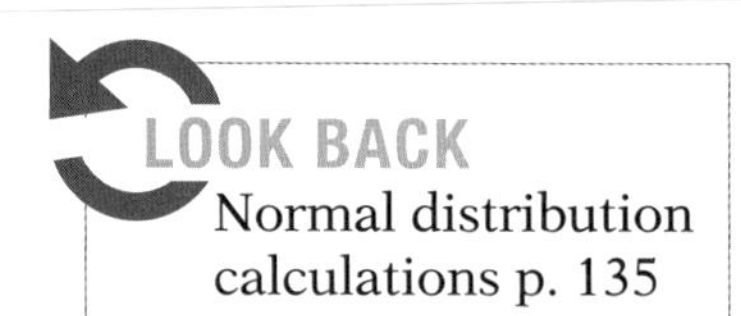

LOOK BACK
Normal distribution calculations p. 135

If the heights vary Normally, the difference in sample means is also Normally distributed. The distribution of the difference in heights is shown in Figure 7.15(b). We standardize $\bar{x}_1 - \bar{x}_2$ by subtracting its mean (0.7) and dividing by its standard deviation (1.55). Therefore, the probability that the girls

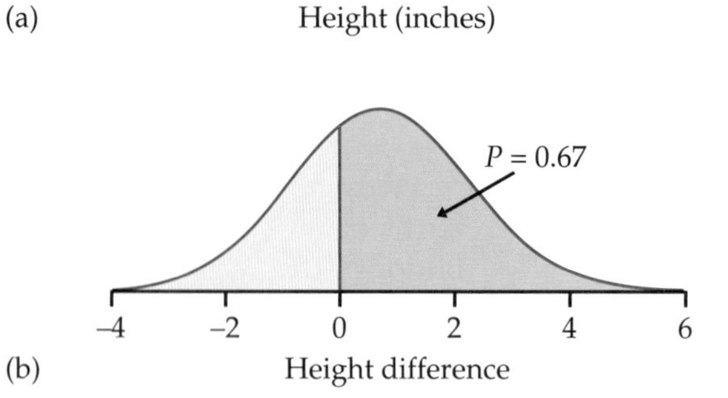

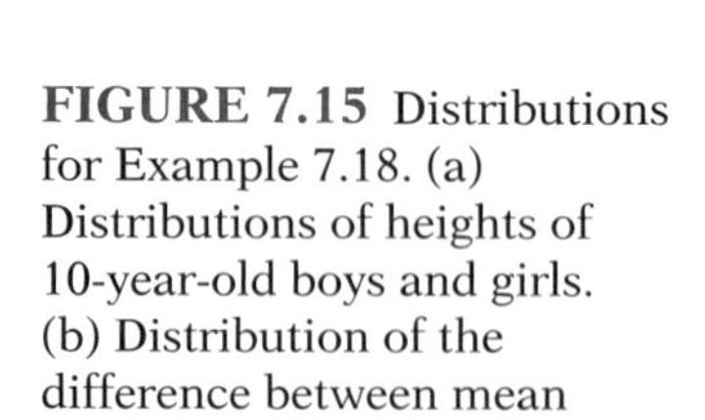

FIGURE 7.15 Distributions for Example 7.18. (a) Distributions of heights of 10-year-old boys and girls. (b) Distribution of the difference between mean heights of 12 girls and 8 boys.

are taller than the boys is

$$P(\bar{x}_1 - \bar{x}_2 > 0) = P\left[\frac{(\bar{x}_1 - \bar{x}_2) - 0.7}{1.55} > \frac{0 - 0.7}{1.55}\right]$$
$$= P(Z > -0.45) = 0.6736$$

Even though the population mean height of 10-year-old girls is greater than the population mean height of 10-year-old boys, the probability that the sample mean of the girls is greater than the sample mean of the boys in our class is only 67%. *Large samples are needed to see the effects of small differences.*

CAUTION

As Example 7.18 reminds us, any Normal random variable has the $N(0, 1)$ distribution when standardized. We have arrived at a new z statistic.

TWO-SAMPLE z STATISTIC

Suppose that $\bar{x}_1$ is the mean of an SRS of size n_1 drawn from an $N(\mu_1, \sigma_1)$ population and that $\bar{x}_2$ is the mean of an independent SRS of size n_2 drawn from an $N(\mu_2, \sigma_2)$ population. Then the **two-sample z statistic**

$$z = \frac{(\bar{x}_1 - \bar{x}_2) - (\mu_1 - \mu_2)}{\sqrt{\frac{\sigma_1^2}{n_1} + \frac{\sigma_2^2}{n_2}}}$$

has the standard Normal $N(0, 1)$ sampling distribution.

In the unlikely event that both population standard deviations are known, the two-sample z statistic is the basis for inference about $\mu_1 - \mu_2$. Exact z procedures are seldom used, however, because σ_1 and σ_2 are rarely known. As in the one-sample situation, we will consider the more useful t procedures.

The two-sample t procedures

Suppose now that the population standard deviations σ_1 and σ_2 are not known. We estimate them by the sample standard deviations s_1 and s_2 from our two samples. Following the pattern of the one-sample case, we substitute the standard errors for the standard deviations used in the two-sample z statistic. The result is the **two-sample t statistic:**

two-sample t statistic

$$t = \frac{(\bar{x}_1 - \bar{x}_2) - (\mu_1 - \mu_2)}{\sqrt{\frac{s_1^2}{n_1} + \frac{s_2^2}{n_2}}}$$

Unfortunately, this statistic does *not* have a t distribution. A t distribution replaces the $N(0, 1)$ distribution only when a single standard deviation (σ) in a z statistic is replaced by its sample standard deviation (s). In this case, we replace two standard deviations (σ_1 and σ_2) by their estimates (s_1 and s_2), which does not produce a statistic having a t distribution.

Nonetheless, we can approximate the distribution of the two-sample t statistic by using the $t(k)$ distribution with an **approximation for the degrees of freedom k.** We use these approximations to find approximate values of t^* for confidence intervals and to find approximate P-values for significance tests.

approximations for the degrees of freedom

Here are two approximations:

1. Use a value of k that is calculated from the data. In general, it will not be a whole number.
2. Use k equal to the smaller of $n_1 - 1$ and $n_2 - 1$.

In practice, the choice of approximation rarely makes a difference in our conclusion. Most statistical software uses the first option to approximate the $t(k)$ distribution for two-sample problems unless the user requests another method. Using this approximation without software is a bit complicated. We provide details of this approximation on the text Web site.

If you are not using software, the second approximation is preferred. This approximation is appealing because it is conservative.[17] That is, margins of error for the level C confidence intervals are a bit larger than they need to be, so the true confidence level is larger than C. Similarly, for significance testing, the true P-values are a bit smaller than those we obtain from this approximation. Thus, for tests at a fixed significance level, we are a little less likely to reject H_0 when it is true.

The two-sample *t* significance test

Except for differences in the t statistic and associated degrees of freedom, the significance test for the difference between two population means is very similar to the one-sample t test.

THE TWO-SAMPLE *t* SIGNIFICANCE TEST

Suppose that an SRS of size n_1 is drawn from a Normal population with unknown mean μ_1 and that an independent SRS of size n_2 is drawn from another Normal population with unknown mean μ_2. To test the hypothesis H_0: $\mu_1 = \mu_2$, compute the **two-sample *t* statistic**

$$t = \frac{\bar{x}_1 - \bar{x}_2}{\sqrt{\dfrac{s_1^2}{n_1} + \dfrac{s_2^2}{n_2}}}$$

and use P-values or critical values for the $t(k)$ distribution, where the value of the degrees of freedom k is approximated by software or by the smaller of $n_1 - 1$ and $n_2 - 1$.

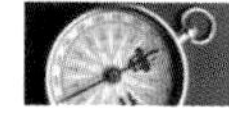

EXAMPLE 7.19

DRP

Directed reading activities assessment. An educator believes that new directed reading activities in the classroom will help elementary school pupils improve some aspects of their reading ability. She arranges for a third-grade class of 21 students to take part in these activities for an eight-week period. A control classroom of 23 third-graders follows the same curriculum without the activities. At the end of the eight weeks, all students are given a Degree of Reading Power (DRP) test, which measures the aspects of reading ability that the treatment is designed to improve. The data appear in Table 7.3.[18]

TABLE 7.3

DRP scores for third-graders

Treatment Group				Control Group			
24	61	59	46	42	33	46	37
43	44	52	43	43	41	10	42
58	67	62	57	55	19	17	55
71	49	54		26	54	60	28
43	53	57		62	20	53	48
49	56	33		37	85	42	

First, examine the data:

Control		Treatment
970	1	
860	2	4
773	3	3
8632221	4	3334699
5543	5	23467789
20	6	127
	7	1
5	8	

The back-to-back stemplot suggests that there is a mild outlier in the control group but no deviation from Normality serious enough to forbid use of t procedures. Separate Normal quantile plots for both groups (Figure 7.16) confirm that both distributions are approximately Normal. The scores of the treatment group appear to be somewhat higher than those of the control group. The summary statistics are

Group	n	$\overline{x}$	s
Treatment	21	51.48	11.01
Control	23	41.52	17.15

Because we hope to show that the treatment (Group 1) is better than the control (Group 2), the hypotheses are

$$H_0: \mu_1 = \mu_2$$
$$H_a: \mu_1 > \mu_2$$

The two-sample t test statistic is

$$t = \frac{\overline{x}_1 - \overline{x}_2}{\sqrt{\dfrac{s_1^2}{n_1} + \dfrac{s_2^2}{n_2}}}$$

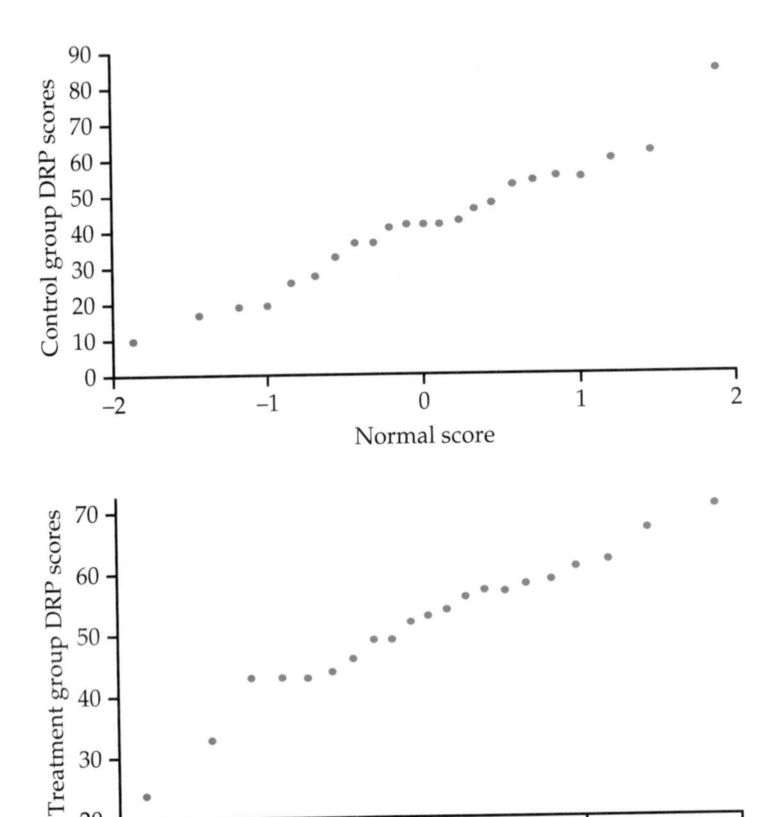

FIGURE 7.16 Normal quantile plots of the DRP scores in Table 7.3, for Example 7.19.

$$= \frac{51.48 - 41.52}{\sqrt{\dfrac{11.01^2}{21} + \dfrac{17.15^2}{23}}}$$

$$= 2.31$$

df = 20

p	0.02	0.01
t^*	2.197	2.528

The P-value for the one-sided test is $P(T \geq 2.31)$. Software gives the approximate P-value as 0.0132 and uses 37.9 as the degrees of freedom. For the second approximation, the degrees of freedom k are equal to the smaller of

$$n_1 - 1 = 21 - 1 = 20 \text{ and } n_2 - 1 = 23 - 1 = 22$$

Comparing 2.31 with the entries in Table D for 20 degrees of freedom, we see that P lies between 0.01 and 0.02. The data strongly suggest that directed reading activity improves the DRP score ($t = 2.31$, df = 20, $0.01 < P < 0.02$).

If your software gives P-values for only the two-sided alternative, $2P(T \geq |t|)$, you need to divide the reported value by 2 *after checking that the means differ in the direction specified by the alternative hypothesis.*

USE YOUR KNOWLEDGE

7.40 Comparison of two Web page designs. You want to compare the daily number of hits for two different MySpace page designs that advertise your new indie rock band. You assign the next 30 days to either Design A or Design B, 15 days to each.

(a) Would you use a one-sided or two-sided significance test for this problem? Explain your choice.

(b) If you use Table D to find the critical value, what are the degrees of freedom using the second approximation?

(c) If you perform the significance test using $\alpha = 0.05$, how large (positive or negative) must the t statistic be to reject the null hypothesis that the two designs result in the same average number of hits?

7.41 More on the comparison of two Web page designs. Consider the previous problem. If the t statistic for comparing the mean hits for Design A versus Design B was 2.45, what P-value would you report? What would you conclude using $\alpha = 0.05$?

The two-sample t confidence interval

The same ideas that we used for the two-sample t significance tests also apply to *two-sample t confidence intervals*. We can use either software or the conservative approach with Table D to approximate the value of t^*.

THE TWO-SAMPLE t CONFIDENCE INTERVAL

Suppose that an SRS of size n_1 is drawn from a Normal population with unknown mean μ_1 and that an independent SRS of size n_2 is drawn from another Normal population with unknown mean μ_2. The **confidence interval for $\mu_1 - \mu_2$** given by

$$(\bar{x}_1 - \bar{x}_2) \pm t^* \sqrt{\frac{s_1^2}{n_1} + \frac{s_2^2}{n_2}}$$

has confidence level at least C no matter what the population standard deviations may be. Here t^* is the value for the $t(k)$ density curve with area C between $-t^*$ and t^*. The value of the degrees of freedom k is approximated by software or by the smaller of $n_1 - 1$ and $n_2 - 1$.

To complete the analysis of the DRP scores we examined in Example 7.19, we describe the size of the treatment effect. We do this with a confidence interval for the difference between the treatment group and the control group means.

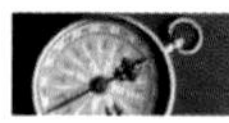

EXAMPLE 7.20

How much improvement? We will find a 95% confidence interval for the mean improvement in the entire population of third-graders. The interval is

$$(\bar{x}_1 - \bar{x}_2) \pm t^* \sqrt{\frac{s_1^2}{n_1} + \frac{s_2^2}{n_2}} = (51.48 - 41.52) \pm t^* \sqrt{\frac{11.01^2}{21} + \frac{17.15^2}{23}}$$

$$= 9.96 \pm 4.31t^*$$

Using software, the degrees of freedom are 37.9 and $t^* = 2.025$. This approximation gives

$$9.96 \pm (4.31 \times 2.025) = 9.96 \pm 8.72 = (1.2,\ 18.7)$$

The conservative approach uses the $t(20)$ distribution. Table D gives $t^* = 2.086$. With this approximation we have

$$9.96 \pm (4.31 \times 2.086) = 9.96 \pm 8.99 = (1.0,\ 18.9)$$

We can see that the conservative approach does, in fact, give a wider interval than the more accurate approximation used by software. However, the difference is pretty small.

We estimate the mean improvement to be about 10 points, but with a margin of error of almost 9 points with either approximation method. Although we have good evidence of some improvement, the data do not allow a very precise estimate of the size of the average improvement.

The design of the study in Example 7.19 is not ideal. Random assignment of students was not possible in a school environment, so existing third-grade classes were used. The effect of the reading programs is therefore confounded with any other differences between the two classes. The classes were chosen to be as similar as possible—for example, in terms of the social and economic status of the students. Extensive pretesting showed that the two classes were, on the average, quite similar in reading ability at the beginning of the experiment. To avoid the effect of two different teachers, the researcher herself taught reading in both classes during the eight-week period of the experiment.

We can therefore be somewhat confident that the two-sample test is detecting the effect of the treatment and not some other difference between the classes. This example is typical of many situations in which an experiment is carried out but randomization is not possible.

USE YOUR KNOWLEDGE

7.42 Two-sample *t* confidence interval. Assume that $\bar{x}_1 = 110$, $\bar{x}_2 = 120$, $s_1 = 8$, $s_2 = 12$, $n_1 = 50$, and $n_2 = 50$. Find a 95% confidence interval for the difference in the corresponding values of μ using the second approximation for degrees of freedom. Does this interval include more or fewer values than a 99% confidence interval? Explain your answer.

7.43 Another two-sample t confidence interval. Assume that $\bar{x}_1 = 110$, $\bar{x}_2 = 120$, $s_1 = 8$, $s_2 = 12$, $n_1 = 10$, and $n_2 = 10$. Find a 95% confidence interval for the difference in the corresponding values of μ using the second approximation for degrees of freedom. Would you reject the null hypothesis that the population means are equal in favor of the two-sided alternative at significance level 0.05? Explain.

Robustness of the two-sample procedures

The two-sample t procedures are more robust than the one-sample t methods. When the sizes of the two samples are equal and the distributions of the two populations being compared have similar shapes, probability values from the t table are quite accurate for a broad range of distributions when the sample sizes are as small as $n_1 = n_2 = 5$.[19] When the two population distributions have different shapes, larger samples are needed.

The guidelines for the use of one-sample t procedures can be adapted to two-sample procedures by replacing "sample size" with the "sum of the sample sizes" $n_1 + n_2$. Specifically,

- *If $n_1 + n_2$ is less than 15:* Use t procedures if the data are close to Normal. If the data in either sample are clearly non-Normal or if outliers are present, do not use t.
- *If $n_1 + n_2$ is at least 15:* The t procedures can be used except in the presence of outliers or strong skewness.
- *Large samples:* The t procedures can be used even for clearly skewed distributions when the sample is large, roughly $n_1 + n_2 \geq 40$.

These guidelines are rather conservative, especially when the two samples are of equal size. *In planning a two-sample study, choose equal sample sizes if you can.* The two-sample t procedures are most robust against non-Normality in this case, and the conservative probability values are most accurate.

Here is an example with reasonably large sample sizes that are not equal. Even if the distributions are not Normal, we are confident that the sample means will be approximately Normal. The two-sample t test is very robust in this case.

EXAMPLE 7.21

Sleep and blood pressure in adolescents. Hypertension is an increasingly common health problem in both adults and adolescents. Childhood hypertension is associated with hypertension in adulthood, a risk factor for cardiovascular disease and death. While several studies have implicated insufficient sleep as a risk factor for hypertension in adults, only a few studies have looked at the relationship between sleep and hypertension in children.

One study examined 238 adolescents between the ages of 13 and 16.[20] Based on in-home monitoring and overnight observance, each child was classified as having either high or low sleep efficiency. Here are the summary statistics of their systolic blood pressures in millimeters of mercury (mmHg):

Sleep efficiency	n	$\overline{x}$	s
Low	61	118.4	9.9
High	177	112.6	7.5

The low-sleep-efficiency children have higher pressures on the average. Can we conclude that the systolic blood pressures of children with low and high sleep efficiency are not the same? Or is this observed difference merely what we could expect to see given the variation among children?

Even though prior evidence suggested that the blood pressure would be elevated in low-sleep-efficiency children, the researchers did not specify a direction for the difference. Thus, the hypotheses are

$$H_0: \mu_1 = \mu_2$$
$$H_a: \mu_1 \neq \mu_2$$

Because the samples are relatively large, we can confidently use the t procedures even though we lack the detailed data and so cannot verify the Normality condition.

The two-sample t statistic is

$$t = \frac{\overline{x}_1 - \overline{x}_2}{\sqrt{\frac{s_1^2}{n_1} + \frac{s_2^2}{n_2}}}$$

$$= \frac{118.4 - 112.6}{\sqrt{\frac{9.9^2}{61} + \frac{7.5^2}{177}}}$$

$$= 4.18$$

df = 60

p	0.0005
t^*	3.460

The conservative approach finds the P-value by comparing 4.18 to critical values for the $t(60)$ distribution because the smaller sample has 61 observations. We must double the table tail area p because the alternative is two-sided.

Our calculated value of t is larger than the $p = 0.0005$ entry in the table. Doubling 0.0005, we conclude that the P-value is less than 0.001. The data give conclusive evidence that the mean systolic blood pressure is higher in low-sleep-efficiency children ($t = 4.18$, df $= 60$, $P < 0.001$).

In this example the exact P-value is very small because $t = 4.18$ says that the observed difference in means is over 4 standard errors above the hypothesized difference of zero ($\mu_1 = \mu_2$). The difference of 5.8 mmHg may not appear that large, but in terms of having an elevated systolic blood pressure, only 6.2% of the high-sleep-efficiency children had elevated pressure, while 26.2% of the low-sleep-efficiency children had elevated pressure.

In this and other examples, we can choose which population to label 1 and which to label 2. After inspecting the data, we chose low-sleep-efficiency children as Population 1 because this choice makes the t statistic a positive number. This avoids any possible confusion from reporting a negative value

for t. *Choosing the population labels is* ***not*** *the same as choosing a one-sided alternative after looking at the data.* Choosing hypotheses after seeing a result in the data is a violation of sound statistical practice.

Inference for small samples

Small samples require special care. We do not have enough observations to examine the distribution shapes, and only extreme outliers stand out. The power of significance tests tends to be low, and the margins of error of confidence intervals tend to be large. Despite these difficulties, we can often draw important conclusions from studies with small sample sizes. If the size of an effect is very large, it should still be evident even if the n's are small.

EXAMPLE 7.22

Sleep efficiency and blood pressure. In the setting of Example 7.21, let's consider a much smaller study that collects systolic blood pressures from only 5 children in each sleep efficiency group. Also, given the results of this past example, we choose the one-sided alternative. The data are

Sleep efficiency	Systolic blood pressure (mmHg)				
Low	110	118	128	126	119
High	113	120	102	108	114

First, examine the distributions with a back-to-back stemplot.

Low		High
	10	2
	10	8
0	11	34
98	11	
	12	0
86	12	

While there is variation among pressures within each group, there is also a noticeable separation. The high-sleep-efficiency group contains 4 of the 5 lowest pressures, and the low-efficiency group contains 4 of the 5 highest blood pressures. A significance test can confirm whether this pattern can arise just by chance or if the low-efficiency group has a higher mean. We test

$$H_0: \mu_1 = \mu_2$$
$$H_a: \mu_1 > \mu_2$$

The blood pressure is higher in the low-efficiency group ($t = 2.00$, df $= 7.98$, $P = 0.0404$). The difference in sample means is 8.8 mmHg.

Figure 7.17 (next page) gives outputs for this analysis from several software packages. Although the formats differ, the basic information is the same. All

FIGURE 7.17 SAS, Excel, JMP, and SPSS output for Example 7.22.

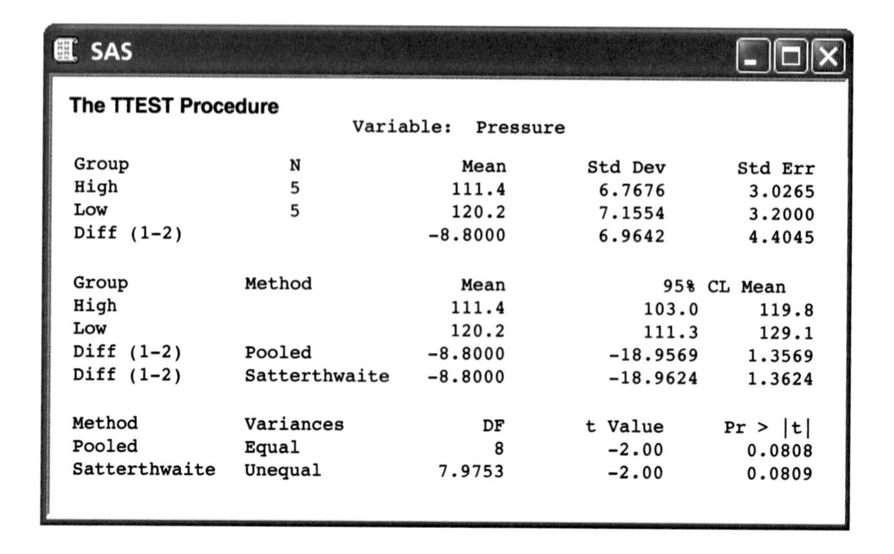

SAS

The TTEST Procedure

Variable: Pressure

Group	N	Mean	Std Dev	Std Err
High	5	111.4	6.7676	3.0265
Low	5	120.2	7.1554	3.2000
Diff (1-2)		-8.8000	6.9642	4.4045

Group	Method	Mean	95% CL Mean	
High		111.4	103.0	119.8
Low		120.2	111.3	129.1
Diff (1-2)	Pooled	-8.8000	-18.9569	1.3569
Diff (1-2)	Satterthwaite	-8.8000	-18.9624	1.3624

Method	Variances	DF	t Value	Pr > |t|
Pooled	Equal	8	-2.00	0.0808
Satterthwaite	Unequal	7.9753	-2.00	0.0809

Excel

	A	B	C
1	**t-Test: Two-Sample Assuming Unequal Variances**		
2			
3		*Low*	*High*
4	Mean	120.2	111.4
5	Variance	51.2	45.8
6	Observations	5	5
7	Hypothesized Mean Difference	0	
8	df	8	
9	t Stat	1.99793708	
10	P(T<=t) one-tail	0.040386978	
11	t Critical one-tail	1.859548033	
12	P(T<=t) two-tail	0.080773957	
13	t Critical two-tail	2.306004133	

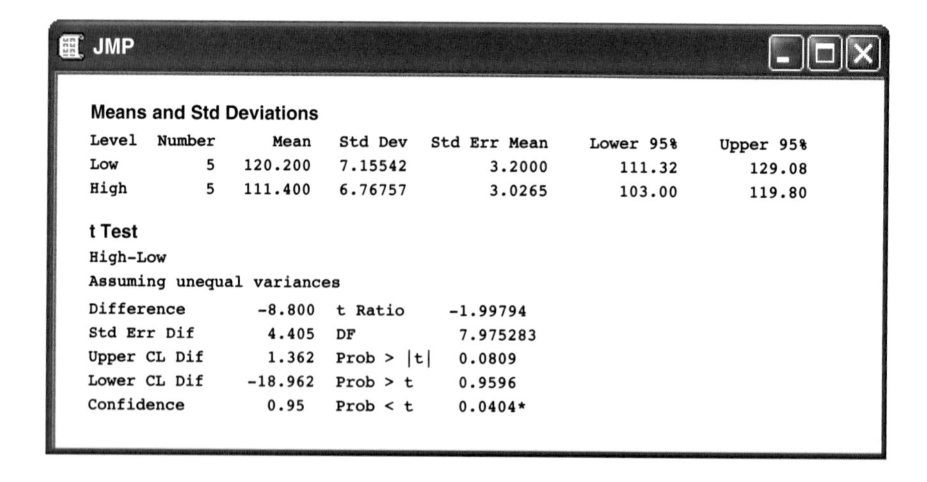

JMP

Means and Std Deviations

Level	Number	Mean	Std Dev	Std Err Mean	Lower 95%	Upper 95%
Low	5	120.200	7.15542	3.2000	111.32	129.08
High	5	111.400	6.76757	3.0265	103.00	119.80

t Test

High-Low

Assuming unequal variances

Difference	-8.800	t Ratio	-1.99794
Std Err Dif	4.405	DF	7.975283
Upper CL Dif	1.362	Prob > |t|	0.0809
Lower CL Dif	-18.962	Prob > t	0.9596
Confidence	0.95	Prob < t	0.0404*

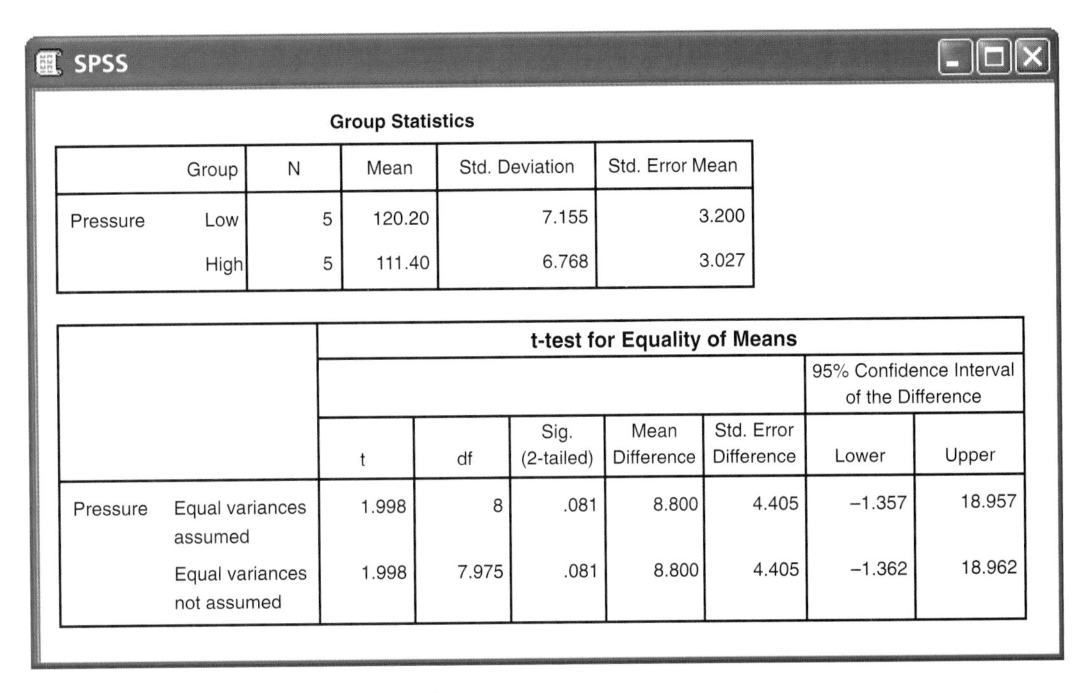
SPSS

Group Statistics

	Group	N	Mean	Std. Deviation	Std. Error Mean
Pressure	Low	5	120.20	7.155	3.200
	High	5	111.40	6.768	3.027

		t-test for Equality of Means						
							95% Confidence Interval of the Difference	
		t	df	Sig. (2-tailed)	Mean Difference	Std. Error Difference	Lower	Upper
Pressure	Equal variances assumed	1.998	8	.081	8.800	4.405	−1.357	18.957
	Equal variances not assumed	1.998	7.975	.081	8.800	4.405	−1.362	18.962

FIGURE 7.17 (*Continued*)

report the sample sizes, the sample means and standard deviations (or variances), the t statistic, and its P-value. All agree that the P-value is small, though some give more detail than others. Software often labels the groups in alphabetical order. In this example, High is then the first population and $t = -2.00$, the negative of our result.

Always check the means first and report the statistic (you may need to change the sign) in an appropriate way. Be sure to also mention the size of the effect you observed, such as "The mean systolic blood pressure for low-sleep-efficiency children was 8.8 mmHg higher than in the high-sleep-efficiency group."

There are two other things to notice in the outputs. First, SAS and SPSS give results only for the two-sided alternative. To get the P-value for the one-sided alternative, we must first check the mean difference to make sure it is in the proper direction. If it is, we divide the given P-value by 2. Also, SAS and SPSS report the results of *two t* procedures: a special procedure that assumes that the two population variances are equal and the general two-sample procedure that we have just studied. We don't recommend these "equal-variances" procedures that statistical software often provides. We will describe them, however, because this information is helpful for those who plan to read Chapters 12 and 13 on the comparison of several means.

The pooled two-sample t procedures

There is one situation in which a t statistic for comparing two means has exactly a t distribution. This is when the two Normal population distributions have the *same* standard deviation. As we've done with other t statistics, we will first develop the z statistic and, from it, the t statistic. In this case, notice that we need to substitute only a single standard error going from the z to the t statistic. This is why the resulting t statistic has a t distribution.

Call the common—and still unknown—standard deviation of both populations σ. Both sample variances s_1^2 and s_2^2 estimate σ^2. The best way to combine these two estimates is to average them with weights equal to their degrees of freedom. This gives more weight to the sample variance from the larger sample, which is reasonable. The resulting estimator of σ^2 is

$$s_p^2 = \frac{(n_1 - 1)s_1^2 + (n_2 - 1)s_2^2}{n_1 + n_2 - 2}$$

pooled estimator of σ^2

This is called the **pooled estimator of σ^2** because it combines the information in both samples.

When both populations have variance σ^2, the addition rule for variances says that $\bar{x}_1 - \bar{x}_2$ has variance equal to the *sum* of the individual variances, which is

$$\frac{\sigma^2}{n_1} + \frac{\sigma^2}{n_2} = \sigma^2\left(\frac{1}{n_1} + \frac{1}{n_2}\right)$$

The standardized difference of means in this equal-variance case is therefore

$$z = \frac{(\bar{x}_1 - \bar{x}_2) - (\mu_1 - \mu_2)}{\sigma\sqrt{\dfrac{1}{n_1} + \dfrac{1}{n_2}}}$$

This is a special two-sample z statistic for the case in which the populations have the same σ. Replacing the unknown σ by the estimate s_p gives a t statistic. The degrees of freedom are $n_1 + n_2 - 2$, the sum of the degrees of freedom of the two sample variances. This statistic is the basis of the pooled two-sample t inference procedures.

THE POOLED TWO-SAMPLE t PROCEDURES

Suppose that an SRS of size n_1 is drawn from a Normal population with unknown mean μ_1 and that an independent SRS of size n_2 is drawn from another Normal population with unknown mean μ_2. Suppose also that the two populations have the same standard deviation. A level C **confidence interval for $\mu_1 - \mu_2$** is

$$(\bar{x}_1 - \bar{x}_2) \pm t^* s_p\sqrt{\frac{1}{n_1} + \frac{1}{n_2}}$$

Here t^* is the value for the $t(n_1 + n_2 - 2)$ density curve with area C between $-t^*$ and t^*, and s_p is the **pooled estimator of the standard deviation,**

$$s_p = \sqrt{\frac{(n_1 - 1)s_1^2 + (n_2 - 1)s_2^2}{n_1 + n_2 - 2}}$$

To test the hypothesis H_0: $\mu_1 = \mu_2$, compute the **pooled two-sample *t* statistic**

$$t = \frac{\bar{x}_1 - \bar{x}_2}{s_p\sqrt{\dfrac{1}{n_1} + \dfrac{1}{n_2}}}$$

In terms of a random variable T having the $t(n_1 + n_2 - 2)$ distribution, the P-value for a test of H_0 against

$$H_a: \mu_1 > \mu_2 \text{ is } P(T \geq t)$$
$$H_a: \mu_1 < \mu_2 \text{ is } P(T \leq t)$$
$$H_a: \mu_1 \neq \mu_2 \text{ is } 2P(T \geq |t|)$$

EXAMPLE 7.23

Calcium and blood pressure. Does increasing the amount of calcium in our diet reduce blood pressure? Examination of a large sample of people revealed a relationship between calcium intake and blood pressure, but such observational studies do not establish causation. Animal experiments, however, showed that calcium supplements do reduce blood pressure in rats, justifying an experiment with human subjects.

A randomized comparative experiment gave one group of 10 black men a calcium supplement for 12 weeks. The control group of 11 black men received a placebo that appeared identical. (In fact, a block design with black and white men as the blocks was used. We will look only at the results for blacks, because the earlier survey suggested that calcium is more effective for blacks.) The experiment was double-blind.

Table 7.4 gives the seated systolic blood pressure for all subjects at the beginning and end of the 12-week period, in millimeters of mercury (mmHg).

TABLE 7.4

Seated systolic blood pressure

Calcium Group			Placebo Group		
Begin	**End**	**Decrease**	**Begin**	**End**	**Decrease**
107	100	7	123	124	−1
110	114	−4	109	97	12
123	105	18	112	113	−1
129	112	17	102	105	−3
112	115	−3	98	95	3
111	116	−5	114	119	−5
107	106	1	119	114	5
112	102	10	114	112	2
136	125	11	110	121	−11
102	104	−2	117	118	−1
			130	133	−3

Because the researchers were interested in decreasing blood pressure, Table 7.4 also shows the decrease for each subject. An increase appears as a negative entry.[21]

As usual, we first examine the data. To compare the effects of the two treatments, take the response variable to be the amount of the decrease in blood pressure. Inspection of the data reveals that there are no outliers. Side-by-side boxplots and Normal quantile plots (Figures 7.18 and 7.19) give a more detailed

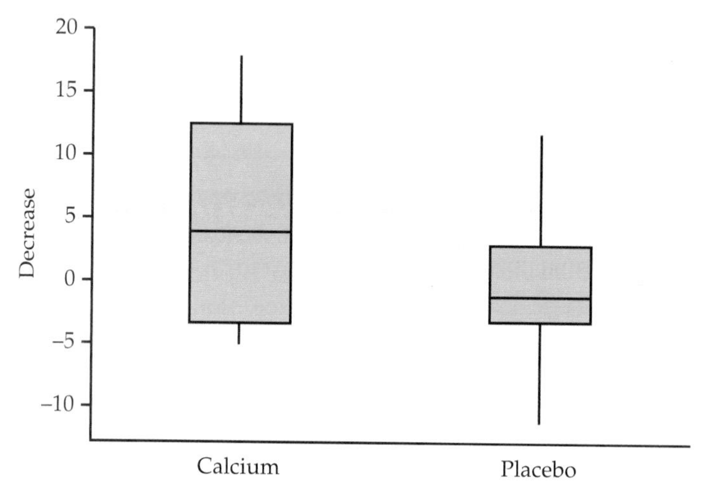

FIGURE 7.18 Side-by-side boxplots of the decrease in blood pressure from Table 7.4, for Example 7.23.

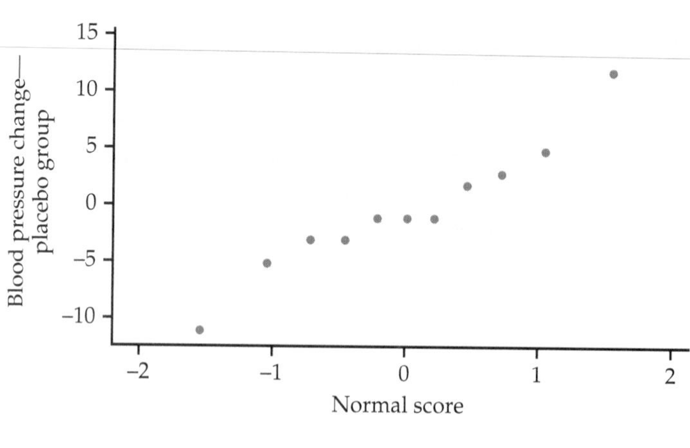

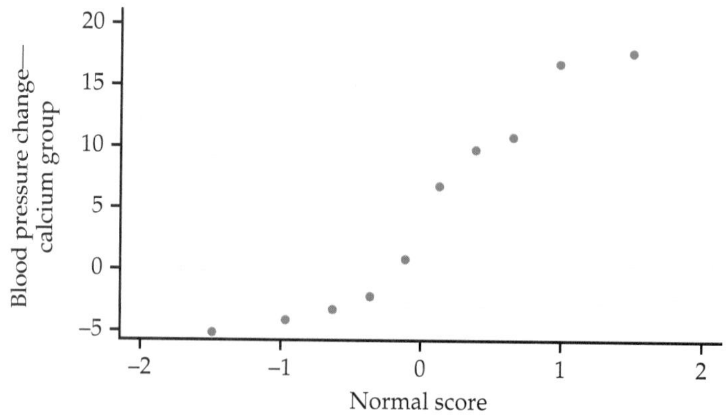

FIGURE 7.19 Normal quantile plots of the change in blood pressure from Table 7.4, for Example 7.23.

picture. The calcium group has a somewhat short left tail, but there are no severe departures from Normality that will prevent use of t procedures. To examine the question of the researchers who collected these data, we perform a significance test.

EXAMPLE 7.24

Does increased calcium reduce blood pressure? Take Group 1 to be the calcium group and Group 2 to be the placebo group. The evidence that calcium lowers blood pressure more than a placebo is assessed by testing

$$H_0: \mu_1 = \mu_2$$
$$H_a: \mu_1 > \mu_2$$

Here are the summary statistics for the decrease in blood pressure:

Group	Treatment	n	$\bar{x}$	s
1	Calcium	10	5.000	8.743
2	Placebo	11	−0.273	5.901

The calcium group shows a drop in blood pressure, and the placebo group has a small increase. The sample standard deviations do not rule out equal population standard deviations. A difference this large will often arise by chance in samples this small. We are willing to assume equal population standard deviations. The pooled sample variance is

$$s_p^2 = \frac{(n_1 - 1)s_1^2 + (n_2 - 1)s_2^2}{n_1 + n_2 - 2}$$

$$= \frac{(10 - 1)8.743^2 + (11 - 1)5.901^2}{10 + 11 - 2} = 54.536$$

so that

$$s_p = \sqrt{54.536} = 7.385$$

The pooled two-sample t statistic is

$$t = \frac{\bar{x}_1 - \bar{x}_2}{s_p\sqrt{\frac{1}{n_1} + \frac{1}{n_2}}}$$

$$= \frac{5.000 - (-0.273)}{7.385\sqrt{\frac{1}{10} + \frac{1}{11}}}$$

$$= \frac{5.273}{3.227} = 1.634$$

The P-value is $P(T \geq 1.634)$, where T has the $t(19)$ distribution.

df = 19		
p	0.10	0.05
t^*	1.328	1.729

From Table D we can see that P falls between the $\alpha = 0.10$ and $\alpha = 0.05$ levels. Statistical software gives the exact value $P = 0.059$. The experiment found evidence that calcium reduces blood pressure, but the evidence falls a bit short of the traditional 5% and 1% levels.

Sample size strongly influences the P-value of a test. An effect that fails to be significant at a specified level α in a small sample can be significant in a larger sample. In the light of the rather small samples in Example 7.24, the evidence for some effect of calcium on blood pressure is rather good. The published account of the study combined these results for blacks with the results for whites and adjusted for pretest differences among the subjects. Using this more detailed analysis, the researchers were able to report a P-value of 0.008.

Of course, a P-value is almost never the last part of a statistical analysis. To make a judgment regarding the size of the effect of calcium on blood pressure, we need a confidence interval.

EXAMPLE 7.25

How different are the calcium and placebo groups? We estimate that the effect of calcium supplementation is the difference between the sample means of the calcium and the placebo groups, $\bar{x}_1 - \bar{x}_2 = 5.273$ mmHg. A 90% confidence interval for $\mu_1 - \mu_2$ uses the critical value $t^* = 1.729$ from the $t(19)$ distribution. The interval is

$$(\bar{x}_1 - \bar{x}_2) \pm t^* s_p \sqrt{\frac{1}{n_1} + \frac{1}{n_2}} = [5.000 - (-0.273)] \pm (1.729)(7.385)\sqrt{\frac{1}{10} + \frac{1}{11}}$$

$$= 5.273 \pm 5.579$$

We are 90% confident that the difference in means is in the interval (−0.306, 10.852). The calcium treatment reduced blood pressure by about 5.3 mmHg more than a placebo on the average, but the margin of error for this estimate is 5.6 mmHg.

The pooled two-sample t procedures are anchored in statistical theory and so have long been the standard version of the two-sample t in textbooks. *But they require the assumption that the two unknown population standard deviations are equal.* This assumption is hard to verify. The pooled t procedures are therefore a bit risky. They are reasonably robust against both non-Normality and unequal standard deviations when the sample sizes are nearly the same. When the samples are quite different in size, the pooled t procedures become sensitive to unequal standard deviations and should be used with caution unless the samples are large. Unequal standard deviations are quite common. In particular, it is not unusual for the spread of data to increase when the center gets larger. Statistical software often calculates both the pooled and the unpooled t statistics, as in Figure 7.17.

USE YOUR KNOWLEDGE

7.44 Sleep efficiency revisited. Figure 7.17 (pages 402–403) gives the outputs from four software packages for comparing the systolic blood pressures across two sleep efficiency groups. Some of the software reports both pooled and unpooled analyses. Which outputs give the pooled results? What are the pooled t and its P-value?

SECTION 7.2 SUMMARY

Significance tests and confidence intervals for the difference between the means μ_1 and μ_2 of two Normal populations are based on the difference $\bar{x}_1 - \bar{x}_2$ of the sample means from two independent SRSs. Because of the **central limit theorem,** the resulting procedures are approximately correct for other population distributions when the sample sizes are large.

When independent SRSs of sizes n_1 and n_2 are drawn from two Normal populations with parameters μ_1, σ_1 and μ_2, σ_2 the **two-sample z statistic**

$$z = \frac{(\bar{x}_1 - \bar{x}_2) - (\mu_1 - \mu_2)}{\sqrt{\dfrac{\sigma_1^2}{n_1} + \dfrac{\sigma_2^2}{n_2}}}$$

has the $N(0, 1)$ distribution.

The **two-sample t statistic**

$$t = \frac{(\bar{x}_1 - \bar{x}_2) - (\mu_1 - \mu_2)}{\sqrt{\dfrac{s_1^2}{n_1} + \dfrac{s_2^2}{n_2}}}$$

does *not* have a t distribution. However, good approximations are available.

Conservative inference procedures for comparing μ_1 and μ_2 are obtained from the two-sample t statistic by using the $t(k)$ distribution with degrees of freedom k equal to the smaller of $n_1 - 1$ and $n_2 - 1$.

More accurate probability values can be obtained by estimating the degrees of freedom from the data. This is the usual procedure for statistical software.

An approximate level C **confidence interval** for $\mu_1 - \mu_2$ is given by

$$(\bar{x}_1 - \bar{x}_2) \pm t^* \sqrt{\frac{s_1^2}{n_1} + \frac{s_2^2}{n_2}}$$

Here, t^* is the value for the $t(k)$ density curve with area C between $-t^*$ and t^*, where k is computed from the data by software or is the smaller of $n_1 - 1$ and $n_2 - 1$. The quantity

$$t^* \sqrt{\frac{s_1^2}{n_1} + \frac{s_2^2}{n_2}}$$

is the **margin of error.**

Significance tests for H_0: $\mu_1 = \mu_2$ use the **two-sample t statistic**

$$t = \frac{\bar{x}_1 - \bar{x}_2}{\sqrt{\dfrac{s_1^2}{n_1} + \dfrac{s_2^2}{n_2}}}$$

The P-value is approximated using the $t(k)$ distribution where k is estimated from the data using software or is the smaller of $n_1 - 1$ and $n_2 - 1$.

The guidelines for practical use of two-sample t procedures are similar to those for one-sample t procedures. Equal sample sizes are recommended.

If we can assume that the two populations have equal variances, **pooled two-sample t procedures** can be used. These are based on the **pooled estimator of the unknown common variance σ^2:**

$$s_p^2 = \frac{(n_1 - 1)s_1^2 + (n_2 - 1)s_2^2}{n_1 + n_2 - 2}$$

and the $t(n_1 + n_2 - 2)$ distribution.

SECTION 7.2 EXERCISES

For Exercises 7.40 and 7.41, see page 397; for Exercises 7.42 and 7.43, see pages 398–399; for Exercise 7.44, see page 409.

In exercises that call for two-sample t procedures, you may use either of the two approximations for the degrees of freedom that we have discussed: the value given by your software or the smaller of n_1-1 and n_2-1. Be sure to state clearly which approximation you have used.

7.45. What is wrong? In each of the following situations explain what is wrong and why.

(a) A researcher wants to test H_0: $\bar{x}_1 = \bar{x}_2$ versus the two-sided alternative H_a: $\bar{x}_1 \neq \bar{x}_2$.

(b) A study recorded the IQ scores of 100 college freshmen. The scores of the 56 males in the study were compared with the scores of all 100 freshmen using the two-sample methods of this section.

(c) A two-sample t statistic gave a P-value of 0.94. From this we can reject the null hypothesis with 90% confidence.

(d) A researcher is interested in testing the one-sided alternative H_a: $\mu_1 < \mu_2$. The significance test gave $t = 2.15$. Since the P-value for the two-sided alternative is 0.036, he concluded that his P-value was 0.018.

7.46. Basic concepts. For each of the following, answer the question and give a short explanation of your reasoning.

(a) A 95% confidence interval for the difference between two means is reported as (0.8, 2.3). What can you conclude about the results of a significance test of the null hypothesis that the population means are equal versus the two-sided alternative?

(b) Will larger samples generally give a larger or a smaller margin of error for the difference between two sample means?

7.47. More basic concepts. For each of the following, answer the question and give a short explanation of your reasoning.

(a) A significance test for comparing two means gave $t = -2.18$ with 10 degrees of freedom. Can you reject the null hypothesis that the μ's are equal versus the two-sided alternative at the 5% significance level?

(b) Answer part (a) for the one-sided alternative that the difference in means is negative.

7.48. Effect of the confidence level. Assume that $\bar{x}_1 = 100$, $\bar{x}_2 = 110$, $s_1 = 18$, $s_2 = 15$, $n_1 = 50$, and $n_2 = 40$. Find a 95% confidence interval for the difference in the corresponding values of μ. Does this interval include more or fewer values than a 99% confidence interval? Explain your answer.

7.49. Sadness and spending. The "misery is not miserly" phenomenon refers to how sadness affects a person's spending judgment. In a recent study, 31 young adults were given \$10 and randomly assigned to either a sad or a neutral group. The participants in the sad group watched a video about the death of a boy's mentor (from the film *The Champ*) and those in the neutral group watched a video on the Great Barrier Reef. After the video, each participant was offered the chance to trade \$0.50 increments of the \$10 for an insulated water bottle.[22] Here are the data as follows: DATA FILE SADNESS

Group	Purchase price (\$)								
Neutral	0.00	2.00	0.00	1.00	0.50	0.00	0.50		
	2.00	1.00	0.00	0.00	0.00	0.00	1.00		
Sad	3.00	4.00	0.50	1.00	2.50	2.00	1.50	0.00	1.00
	1.50	1.50	2.50	4.00	3.00	3.50	1.00	3.50	

(a) Examine each group's prices graphically. Is use of the t procedures appropriate for these data? Carefully explain your answer.

(b) Make a table with the sample size, mean, and standard deviation for each of the two groups.

(c) State appropriate null and alternative hypotheses for comparing these two groups.

(d) Perform the significance test at the $\alpha = 0.05$ level, making sure to report the test statistic, degrees of freedom, and P-value. What is your conclusion?

(e) Construct a 95% confidence interval for the mean difference in purchase price between the two groups.

7.50. Wine labels with animals? Traditional brand research argues that successful logos are ones that are highly relevant to the product they represent. However, a market research firm recently reported that nearly 20% of all table wine brands introduced in the last three years feature an animal on the label. Since animals have little to do with the product, why are marketers using this tactic?

Some researchers have proposed that consumers who are "primed" — in other words, they've thought about the image earlier in an unrelated context — process visual information more easily.[23] To demonstrate this, the researchers randomly assigned participants to either a primed or a nonprimed group. Each participant was asked to indicate his or her attitude toward a product on a seven-point scale (from 1 = dislike very much to 7 = like very much). A bottle of MagicCoat pet shampoo, with a picture of a collie on the label, was the product. Before giving this score, however, participants were asked to do a word find where four of the words were common across groups (pet, grooming, bottle, label) and four were either related to the image (dog, collie, puppy, woof) or image conflicting (cat, feline, kitten, meow). The following table contains the responses listed from smallest to largest: DATA FILE BRANDPREFERENCE

Group	Brand attitude
Primed	2 2 3 3 3 4 4 4 4 4 4 4 4 4 4 4 5 5 5 5 5 5 5
Nonprimed	1 1 2 2 3 3 3 3 3 3 3 3 3 3 3 3 4 4 4 5

(a) Examine the scores of each group graphically. Is it appropriate to use the two-sample t procedures? Explain your answer.

(b) Test whether these two groups show the same preference for this product. Use a two-sided alternative hypothesis and a significance level of 5%.

(c) Construct a 95% confidence interval for the difference in average preference.

(d) Write a short summary of your conclusions.

7.51. Drive-thru speaker clarity. *QSRMagazine.com* surveyed 689 adults on their drive-thru window experiences at quick-service restaurants.[24] One question was "Thinking about your most recent drive-thru experience, please rate how satisfied you were with the clarity of communication through the speaker." Responses ranged from "Very Dissatisfied (1)" to "Very Satisfied (5)." The following table breaks down the responses according to gender: DATA FILE SPEAKERCLARITY

	Rating				
Gender	1	2	3	4	5
Female	5	44	48	183	188
Male	5	29	30	91	66

(a) Report the means and standard deviations of the rating for the male and female participants separately.

(b) Comment on the appropriateness of t procedures for these data.

(c) Test whether males and females are, on average, equally satisfied with speaker clarity. Use a two-sided alternative hypothesis and a significance level of 5%.

(d) Construct a 95% confidence interval for the difference in average satisfaction.

(e) Given the coarseness of the rating, the owner of the Sir Beef-a-lot chain considers only a difference in the means of at least 0.25 units as meaningful. Based on your results in parts (c) and (d), what would you tell this owner?

7.52. Diet and mood. Researchers were interested in comparing the long-term psychological effects on dieters who were on a high-carbohydrate, low-fat (LF) diet with those on a high-fat, low-carbohydrate (LC) diet.[25] A total of 106 overweight and obese participants were randomly assigned to one of these two energy-restricted diets. At 52 weeks a total of 32 LC dieters and 33 LF dieters remained. Mood was assessed using a total mood disturbance score (TMDS), where a lower score is associated with a less negative mood. A summary of these results follows:

Group	n	$\bar{x}$	s
LC	32	47.3	28.3
LF	33	19.3	25.8

(a) Is there a difference in the TMDS at Week 52? Test the null hypothesis that the dieters' average mood in the two groups is the same. Use a significance level of 0.05.

(b) Critics of this study focus on the specific LC diet (that is, the science) and the dropout rate. Explain why the dropout rate is important to consider when drawing conclusions from this study.

7.53. Dust exposure at work. Exposure to dust at work can lead to lung disease later in life. One study measured the workplace exposure of tunnel construction workers.[26] Part of the study compared 115 drill and blast workers with 220 outdoor concrete workers. Total dust exposure was measured in milligram years per cubic meter (mg.y/m^3). The mean exposure for the drill and blast workers was 18.0 mg.y/m^3 with a standard deviation of 7.8 mg.y/m^3. For the outdoor concrete workers, the corresponding values were 6.5 mg.y/m^3 and 3.4 mg.y/m^3.

(a) The sample included all workers for a tunnel construction company who received medical examinations as part of routine health checkups. Discuss the extent to which you think these results apply to other, similar types of workers.

(b) Use a 95% confidence interval to describe the difference in the exposures. Write a sentence that gives the interval and provides the meaning of 95% confidence.

(c) Test the null hypothesis that the exposures for these two types of workers are the same. Justify your choice of a one-sided or two-sided alternative. Report the test statistic, the degrees of freedom, and the P-value. Give a short summary of your conclusion.

(d) The authors of the article describing these results note that the distributions are somewhat skewed. Do you think that this fact makes your analysis invalid? Give reasons for your answer.

7.54. Not all dust is the same. Not all dust particles that are in the air around us cause problems for our lungs. Some particles are too large and stick to other areas of our body before they can get to our lungs. Others are so small that we can breathe them in and out and they will not deposit on our lungs. The researchers in the study described in the previous exercise also measured respirable dust. This is dust that deposits in our lungs when we breathe it. For the drill and blast workers, the mean exposure to respirable dust was 6.3 mg.y/m^3 with a standard deviation of 2.8 mg.y/m^3. The corresponding values for the outdoor concrete workers were 1.4 mg.y/m^3 and 0.7 mg.y/m^3. Analyze these data using the questions in the previous exercise as a guide.

7.55. Change in portion size. A recent study of food portion sizes reported that over a 17-year period, the average size of a soft drink consumed by Americans aged 2 years and older increased from 13.1 ounces (oz) to 19.9 oz. The authors state that the difference is statistically significant with $P < 0.01$.[27] Explain what additional information you would need to compute a confidence interval for the increase, and outline the procedure that you would use for the computations. Do you think that a confidence interval would provide useful additional information? Explain why or why not.

7.56. Beverage consumption. The results in the previous exercise were based on two national surveys with a very large number of individuals. Here is a study that also looked at beverage consumption, but the sample sizes are much smaller. One part of this study compared 20 children who were 7 to 10 years old with 5 children who were 11 to 13.[28] The younger children consumed an average of 8.2 oz of sweetened drinks per day, while the older ones averaged 14.5 oz. The standard deviations were 10.7 oz and 8.2 oz, respectively.

(a) Do you think that it is reasonable to assume that these data are Normally distributed? Explain why or why not. (*Hint:* Think about the 68–95–99.7 rule.)

(b) Using the methods in this section, test the null hypothesis that the two groups of children consume equal amounts of sweetened drinks versus the two-sided alternative. Report all details of the significance-testing procedure with your conclusion.

(c) Give a 95% confidence interval for the difference in means.

(d) Do you think that the analyses performed in parts (b) and (c) are appropriate for these data? Explain why or why not.

(e) The children in this study were all participants in an intervention study at the Cornell Summer Day Camp at Cornell University. To what extent do you think that these results apply to other groups of children?

7.57. Study design is important! Recall Exercise 7.40 (page 397). You are concerned that day of the week may affect online sales. So to compare the two Web page designs, you choose two successive weeks in the middle of a month. You flip a coin to assign one Monday to the first design and the other Monday to the second. You repeat this for each of the seven days of the week. You now have 7 hit amounts for each design. It is *incorrect* to use the two-sample t test to see if the mean hits differ for the two designs. Carefully explain why.

7.58. New computer monitors? The purchasing department has suggested that all new computer monitors for your company should be flat screens. You want data to assure you that employees will like the new screens. The next 20 employees needing a new computer are the subjects for an experiment.

(a) Label the employees 01 to 20. Randomly choose 10 to receive flat screens. The remaining 10 get standard monitors.

(b) After a month of use, employees express their satisfaction with their new monitors by responding to the statement "I like my new monitor" on a scale from 1 to 5, where 1 represents "strongly disagree," 2 is "disagree," 3 is "neutral," 4 is "agree," and 5 stands for "strongly agree." The employees with the flat screens have average satisfaction 4.8 with standard deviation 0.7. The employees with the standard monitors have average 3.0 with standard deviation 1.5. Give a 95% confidence interval for the difference in the mean satisfaction scores for all employees.

(c) Would you reject the null hypothesis that the mean satisfaction for the two types of monitors is the same versus the two-sided alternative at significance level 0.05? Use your confidence interval to answer this question. Explain why you do not need to calculate the test statistic.

7.59. Why randomize? Refer to the previous exercise. A coworker suggested that you give the flat screens to the next 10 employees who need new screens and the standard monitor to the following 10. Explain why your randomized design is better.

7.60. Does ad placement matter? Corporate advertising tries to enhance the image of the corporation. A study compared two ads from two sources, the *Wall Street Journal* and the *National Enquirer.* Subjects were asked to pretend that their company was considering a major investment in Performax, the fictitious sportswear firm in the ads. Each subject was asked to respond to the question "How trustworthy was the source in the sportswear company ad for Performax?" on a 7-point scale. Higher values indicated more trustworthiness.[29] Here is a summary of the results:

Ad source	n	$\bar{x}$	s
Wall Street Journal	66	4.77	1.50
National Enquirer	61	2.43	1.64

(a) Compare the two sources of ads using a t test. Be sure to state your null and alternative hypotheses, the test statistic with degrees of freedom, the P-value, and your conclusion.

(b) Give a 95% confidence interval for the difference.

(c) Write a short paragraph summarizing the results of your analyses.

7.61. Sales of a small appliance across months. A market research firm supplies manufacturers with estimates of the retail sales of their products from samples of retail stores. Marketing managers are prone to look at the estimate and ignore sampling error. Suppose that an SRS of 70 stores this month shows mean sales of 53 units of a small appliance, with standard deviation 12 units. During the same month last year, an SRS of 55 stores gave mean sales of 50 units, with standard deviation 10 units. An increase from 50 to 53 is a rise of 6%. The marketing manager is happy because sales are up 6%.

(a) Use the two-sample t procedure to give a 95% confidence interval for the difference in mean number of units sold at all retail stores.

(b) Explain in language that the manager can understand why he cannot be certain that sales rose by 6%, and that in fact sales may even have dropped.

7.62. An improper significance test. A friend has performed a significance test of the null hypothesis that two means are equal. His report states that the null hypothesis is rejected in favor of the alternative that the first mean is larger than the second. In a presentation on his work, he notes that the first sample mean was larger than the second mean and this is why he chose this particular one-sided alternative.

(a) Explain what is wrong with your friend's procedure and why.

(b) Suppose that he reported $t = 1.70$ with a P-value of 0.06. What is the correct P-value that he should report?

7.63. Breast-feeding versus baby formula. A study of iron deficiency among infants compared samples of infants following different feeding regimens. One group contained breast-fed infants, while the children in another group were fed a standard baby formula without any iron supplements. Here are summary results on blood hemoglobin levels (a measure of iron in the blood) at 12 months of age:[30]

Group	n	$\bar{x}$	s
Breast-fed	23	13.3	1.7
Formula	19	12.4	1.8

(a) Is there significant evidence that the mean hemoglobin level is higher among breast-fed babies? State H_0 and H_a and carry out a t test. Give the P-value. What is your conclusion?

(b) Give a 95% confidence interval for the mean difference in hemoglobin level between the two populations of infants.

(c) State the assumptions that your procedures in parts (a) and (b) require to be valid.

7.64. Revisiting the sadness and spending study. In Exercise 7.49 (page 410), the purchase prices of a water bottle were compared using the two-sample t procedures that do not assume equal standard deviations. Compare the means using a significance test and find the 95% confidence interval for the difference using the pooled methods. How do the results compare with those you obtained in Exercise 7.49?

7.65. Revisiting wine labels with animals. In Exercise 7.50 (page 410), the attitudes toward a product were compared using the two-sample t procedures that do not assume equal standard deviations. Compare the means using a significance test and find the 95% confidence interval for the difference using the pooled methods. How do the results compare with those you obtained in Exercise 7.50? DATA FILE BRANDPREFERENCE

7.66. Revisiting drive-thru speaker clarity. In Exercise 7.51 (page 411), you compared men's satisfaction with drive-thru speaker clarity with that of women using the two-sample t procedures that do not assume equal standard deviations. Compare the means using a significance test and find the 95% confidence interval for the difference using the pooled methods. How do the results compare with those you obtained in Exercise 7.51? DATA FILE SPEAKERCLARITY

7.67. Revisiting the sleep efficiency study. Example 7.21 (page 399) gives summary statistics for systolic blood pressure in high-sleep-efficiency and low-sleep-efficiency children. The two sample standard deviations are somewhat similar, so we may be willing to assume equal population standard deviations. Calculate the pooled t test statistic and its degrees of freedom from the summary statistics. Use Table D to assess significance. How do your results compare with the unpooled analysis in the example?

7.68. CHALLENGE **Revisiting the dust exposure study.** The data on occupational exposure to dust that we analyzed in Exercise 7.53 (page 411) come from two groups of workers that are quite different in size. This complicates pooling because the sample that is larger will dominate the calculations.

(a) Find the pooled estimate of the standard deviation. Write a short summary comparing it with the estimates of the standard deviations that come from each group.

(b) Find the standard error of the difference in sample means that you would use for the method that does not assume equal variances. Do the same for the pooled approach. Compare these two estimates with each other.

(c) Perform the significance test and find the 95% confidence interval using the pooled methods. How do these results compare with those you found in Exercise 7.53?

(d) Exercise 7.54 has data for the same workers but for respirable dust. Here the standard deviations differ more than those in Exercise 7.53 do. Answer parts (a) through (c) for these data. Write a summary of what you have found in this exercise.

CHAPTER 7 EXERCISES

7.69. LSAT scores. The scores of four senior roommates on the Law School Admission Test (LSAT) are

154, 139, 144, 128

Find the mean, the standard deviation, and the standard error of the mean. Is it appropriate to calculate a confidence interval based on these data? Explain why or why not. DATA FILE LSAT

7.70. Converting a two-sided P-value. You use statistical software to perform a significance test of the null hypothesis that two means are equal. The software reports P-values for the two-sided alternative. Your alternative is that the first mean is greater than the second mean.

(a) The software reports $t = 2.08$ with a P-value of 0.07. Would you reject H_0 with $\alpha = 0.05$? Explain your answer.

(b) The software reports $t = -2.08$ with a P-value of 0.07. Would you reject H_0 with $\alpha = 0.05$? Explain your answer.

7.71. Degrees of freedom and confidence interval width. As the degrees of freedom increase, the t distributions get closer and closer to the z ($N(0, 1)$) distribution. One way to see this is to look at how the value of t^* for a 95% confidence interval changes with the degrees of freedom. Make a plot with degrees of freedom from 2 to 100 on the x axis and t^* on the y axis. Draw a horizontal line on the plot corresponding to the value of $z^* = 1.96$. Summarize the main features of the plot.

7.72. Degrees of freedom and t^*. Refer to the previous exercise. Make a similar plot and summarize its features for the value of t^* for a 90% confidence interval.

7.73. Sample size and margin of error. The margin of error for a confidence interval depends on the confidence level, the standard deviation, and the sample size. Fix the confidence level at 95% and the standard deviation at 1 to examine the effect of the sample size. Find the margin of error for sample sizes of 5 to 100 by increments of 5—that is, let $n = 5, 10, 15, \ldots, 100$. Plot the margins of error versus the sample sizes and summarize the relationship.

7.74. More on sample size and margin of error. Refer to the previous exercise. Make a similar plot and summarize its features for a 99% confidence interval.

7.75. Which design? The following situations all require inference about a mean or means. Identify each as (1) a single sample, (2) matched pairs, or (3) two independent samples. Explain your answers.

(a) Your customers are college students. You are interested in comparing the interest in a new product that you are developing between those students who live in the dorms and those who live elsewhere.

(b) Your customers are college students. You are interested in comparing which of two new product labels is more appealing.

(c) Your customers are college students. You are interested in assessing their interest in a new product.

7.76. Which design? The following situations all require inference about a mean or means. Identify each as (1) a single sample, (2) matched pairs, or (3) two independent samples. Explain your answers.

(a) You want to estimate the average age of your store's customers.

(b) You do an SRS survey of your customers every year. One of the questions in the survey asks about customer satisfaction on a seven-point scale with the response 1 indicating "very dissatisfied" and 7 indicating "very satisfied." You want to see if the mean customer satisfaction has improved from last year.

(c) You ask an SRS of customers their opinions on each of two new floor plans for your store.

7.77. Number of critical food violations. The results of a major city's restaurant inspections are available through its online newspaper.[31] Critical food violations are those that put patrons at risk of getting sick and must be corrected immediately by the restaurant. An SRS of $n = 200$ inspections from the more than 14,000 inspections since January 2009 was collected in which $\bar{x} = 0.65$ critical violations and $s = 0.88$ critical violations.

(a) Test the hypothesis that the average number of critical violations is less than 0.75 using a significance level of 0.05. State the two hypotheses, the test statistic, and P-value.

(b) Construct a 95% confidence interval for the average number of critical violations and summarize your result.

(c) Which of the two summaries (significance test versus confidence interval) do you find more helpful in this case? Explain your answer.

(d) These data are integers ranging from 0 to 9. The data are also skewed to the right, with 70% of the values either a 0 or a 1. Given this information, do you feel that use of the t procedures is appropriate? Explain your answer.

7.78. Brain training. The assessment of computerized brain-training programs is a rapidly growing area of research. Researchers are now focusing on whom this training benefits most, what brain functions can be best improved, and which products are most effective. A recent study looked at 487 community-dwelling adults aged 65 and older, each randomly assigned to one of two training groups. In one group, the participants used a computerized program 1 hour per day. In the other, DVD-based educational programs were shown with quizzes following each video. The training period lasted 8 weeks. The response was the improvement in a composite score obtained from an auditory memory/attention survey given before and after the 8 weeks.[32] The results are summarized in the following table:

Group	n	$\bar{x}$	s
Computer program	242	3.9	8.28
DVD program	245	1.8	8.33

(a) Given that other studies show a benefit of computerized brain training, state the null and alternative hypotheses.

(b) Report the test statistic, its degrees of freedom, and the P-value. What is your conclusion using significance level $\alpha = 0.05$?

(c) Can you conclude that this computerized brain training always improves a person's auditory memory better than the DVD program does? If not, explain why.

7.79. CHALLENGE **Alcohol consumption and body composition.** Individuals who consume large amounts of alcohol do not use the calories from this source as efficiently as calories from other sources. One study examined the effects of moderate alcohol consumption on body composition and the intake of other foods. Fourteen subjects participated in a crossover design study in which they either drank wine for the first 6 weeks and then abstained for the next 6 weeks or vice versa.[33] During the period when they drank wine, the subjects, on average, lost 0.4 kilograms (kg) of body weight. When they did not drink wine, they lost an average of 1.1 kg. The standard deviation of the difference between the weight lost under these two conditions is 8.6 kg. During the wine period, they consumed an average of 2589 calories. With no wine, the mean consumption was 2575. The standard deviation of the difference was 210.

(a) Compute the differences in means and the standard errors for comparing body weight and caloric intake under the two experimental conditions.

(b) A report of the study indicated that there were no significant differences in these two outcome measures. Verify this result for each measure, giving the test statistic, degrees of freedom, and the P-value.

(c) One concern with studies such as this, with a small number of subjects, is that there may not be sufficient power to detect differences that are potentially important. Address this question by computing 95% confidence intervals for the two measures and discuss the information provided by the intervals.

(d) Here are some other facts about the study. All subjects were males between the ages of 21 and 50 years who weighed between 68 and 91 kg. They were all from the same city. During the wine period, subjects were told to consume two 135 milliliter (ml) servings of red wine and no other alcohol. The entire 6-week supply was given to each subject at the beginning of the period. During the other period, subjects were instructed to refrain from any use of alcohol. All subjects reported that they complied with these instructions, except for three subjects who said that they drank no more than three to four 12-ounce bottles of beer during the no-alcohol period. Discuss how these factors could influence the interpretation of the results.

7.80. The wine makes the meal? In a recent study, 39 diners were given a free glass of Cabernet Sauvignon to accompany a French meal.[34] Although the wine was identical, half the bottle labels claimed that the wine was from California and the other half claimed that the wine was from North Dakota. The following table summarizes the grams of entrée and wine consumed during the meal.

	Wine label	n	Mean	St. dev.
Entrée	California	24	499.8	87.2
	North Dakota	15	439.0	89.2
Wine	California	24	100.8	23.3
	North Dakota	15	110.4	9.0

Did those patrons who thought that the wine was from California consume more? Analyze the data and write a report summarizing your work. Be sure to include details regarding the statistical methods you used, your assumptions, and your conclusions.

7.81. Study design information. In the previous study, diners were seated alone or in groups of two, three, four, and, in one case, nine (for a total of $n = 16$ tables). Also, each table, not each patron, was randomly assigned a particular wine label. Does this information alter how you might do the analysis in the previous problem? Explain your answer.

7.82. CHALLENGE **Revisiting the small-sample exercise.** Recall Example 7.22 (page 401). This is a case where the sample sizes are quite small. With only 5 observations per group, we have very little information to make a judgment about whether or not the population standard deviations are equal. The potential gain from pooling is large when the sample sizes are small. Assume that we will perform a two-sided test using the 5% significance level. DATA FILE BPANDSLEEP

(a) Find the critical value for the unpooled t test statistic that does not assume equal variances. Use the minimum of $n_1 - 1$ and $n_2 - 1$ for the degrees of freedom.

(b) Find the critical value for the pooled t test statistic.

(c) How does comparing these critical values show an advantage of the pooled test?

7.83. CHALLENGE **Can mockingbirds learn to identify specific humans?** A central question in urban ecology is why some animals adapt well to the presence of humans and others do not. The following results summarize part of a study of the Northern Mockingbird (*Mimus polyglottos*) that took place on a campus of a large university.[35] For four consecutive days, the same human approached a nest and stood 1 meter (m) away for 30 seconds, placing his or her hand on the rim of the nest. On the fifth day, a new person did the same thing. Each day, the distance of the human from the nest when the bird flushed (flew away) was recorded. This was repeated for 24 nests. The human intruder varied his or her appearance (that is, wore different clothes) over the four days. We report results only for Days 1, 4, and 5 here.

	Flush distance (m)	
Day	Mean	s
1	6.1	4.9
4	15.1	7.3
5	4.9	5.3

(a) Explain why this should be treated as a matched pairs design.

(b) Unfortunately, the research article does not provide the standard error of the difference, only the standard error of the mean flush distance for each day. However, we can use the general addition rule for variances (page 278) to approximate it. Assuming that the correlation between the flush distance at Day 1 and Day 4 for each nest is $\rho = 0.40$, what is the standard deviation for the difference in distance?

(c) Using your result in part (b), test the hypothesis that there is no difference in the flush distance between these two days. Use a significance level of 0.05.

(d) Repeat parts (b) and (c), but now compare Day 1 and Day 5.

(e) Write a brief summary of your conclusions.

7.84. A comparison of female high school students. A study was performed to determine the prevalence of the "female athlete triad" (low energy availability, menstrual dysfunction, and low bone mineral density) in high school students.[36] A total of 80 high school athletes and 80 sedentary students were assessed. The following table summarizes several measured characteristics:

	Athletes		Sedentary	
Characteristic	$\bar{x}$	s	$\bar{x}$	s
Body fat (%)	25.61	5.54	32.51	8.05
Body mass index	21.60	2.46	26.41	2.73
Calcium deficit (mg)	297.13	516.63	580.54	372.77
Glasses of milk/day	2.21	1.46	1.82	1.24

(a) For each of the characteristics, test the hypothesis that the means are the same in the two groups. Use a significance level of 0.05 for each test.

(b) Write a short report summarizing your results.

7.85. Competitive prices? A retailer entered into an exclusive agreement with a supplier who guaranteed to provide all products at competitive prices. The retailer eventually began to purchase supplies from other vendors who offered better prices. The original supplier filed a legal action claiming violation of the agreement. In defense, the retailer had an audit performed on a random sample of invoices. For each audited invoice, all purchases made from other suppliers were examined, and the prices were compared with those offered by the original supplier. For each invoice, the percent of purchases for which the alternate supplier offered a lower price than the original supplier was recorded.[37] Here are the data:

0	100	0	100	33	34	100	48	78	100	77	100	38
68	100	79	100	100	100	100	100	100	89	100	100	

Report the average of the percents with a 95% margin of error. Do the sample invoices suggest that the original supplier's prices are not competitive on the average? COMPPRICE

7.86. Weight-loss programs. In a study of the effectiveness of weight-loss programs, 47 subjects who were at least 20% overweight took part in a group support program for 10 weeks. Private weighings determined each subject's weight at the beginning of the program and 6 months after the program's end. The matched pairs t test was used to assess the significance of the average weight loss. The paper reporting the study said, "The subjects lost a significant amount of weight over time, $t(46) = 4.68$, $p < 0.01$." It is common to report the results of statistical tests in this abbreviated style.[38]

(a) Why was the matched pairs statistic appropriate?

(b) Explain to someone who knows no statistics but is interested in weight-loss programs what the practical conclusion is.

(c) The paper follows the tradition of reporting significance only at fixed levels such as $\alpha = 0.01$. In fact, the results are more significant than "$p < 0.01$" suggests. What can you say about the P-value of the t test?

7.87. Behavior of pet owners. On the morning of March 5, 1996, a train with 14 tankers of propane derailed near the center of the small Wisconsin town of Weyauwega. Six of the tankers were ruptured and burning when the 1700 residents were ordered to evacuate the town. Researchers study disasters like this so that effective relief efforts can be designed for future disasters. About half the households with pets did not evacuate all of their pets. A study conducted after the derailment focused on problems associated with retrieval of the pets after the evacuation and characteristics of the pet owners. One of the scales measured "commitment to adult animals," and the people who evacuated all or some of their pets were compared with those who did not evacuate any of their pets. Higher scores indicate that the pet owner is more likely to take actions that benefit the pet.[39] Here are the data summaries:

Group	n	$\bar{x}$	s
Evacuated all or some pets	116	7.95	3.62
Did not evacuate any pets	125	6.26	3.56

Analyze the data and prepare a short report describing the results.

7.88. Occupation and diet. Do the diets of various occupational groups differ? A British study of this question compared 98 drivers and 83 conductors of London double-decker buses.[40] The conductors' jobs require more physical activity. The article reporting the study gives the data as "Mean daily consumption (± se)." Some of the study results follow:

	Drivers	Conductors
Total calories	2821 ± 44	2844 ± 48
Alcohol (grams)	0.24 ± 0.06	0.39 ± 0.11

(a) What does "se" stand for? Give $\bar{x}$ and s for each of the four sets of measurements.

(b) Is there significant evidence at the 5% level that conductors consume more calories per day than do drivers? Use the two-sample t method to give a P-value, and then assess significance.

(c) How significant is the observed difference in mean alcohol consumption? Use two-sample t methods to obtain the P-value.

(d) Give a 95% confidence interval for the mean daily alcohol consumption of London double-decker bus conductors.

(e) Give a 99% confidence interval for the difference in mean daily alcohol consumption between drivers and conductors.

7.89. Occupation and diet, continued. Use of the pooled two-sample t test is justified in part (b) of the previous exercise. Explain why. Find the P-value for the pooled t statistic, and compare it with your result in the previous exercise.

7.90. Conditions for inference. The report cited in Exercise 7.88 says that the distribution of alcohol consumption among the individuals studied is "grossly skew." Do you think that this skewness prevents the use of the two-sample t test for equality of means? Explain your answer.

7.91. More on conditions for inference. Suppose that your state contains 85 school corporations, and each corporation reports its expenditures per pupil. Is it proper to apply the one-sample t method to these data to give a 95% confidence interval for the average expenditure per pupil? Explain your answer.

7.92. CHALLENGE **Male and female computer science students.** Is there a difference between the average SAT scores of males and females? The CSDATA data set gives the Math (SATM) and Verbal (SATV) scores for a group of 224 computer science majors. The variable SEX indicates whether each individual is male or female. DATA FILE CSDATA

(a) Compare the two distributions graphically, and then use the two-sample t test to compare the average SATM scores of males and females. Is it appropriate to use the pooled t test for this comparison? Write a brief summary of your results and conclusions that refers to both versions of the t test. Also give a 95% confidence interval for the difference in the means.

(b) Answer part (a) for the SATV scores.

(c) The students in the CSDATA data set were all computer science majors who began college during a particular year. To what extent do you think that your results would generalize to (i) computer science students entering in different years, (ii) computer science majors at other colleges and universities, and (iii) college students in general?

7.93. CHALLENGE **House prices.** How much more would you expect to pay for a house that has four bedrooms than for a house that has three? Here are some data for West Lafayette, Indiana.[41] These are the asking prices (in dollars) that the owners have set for their houses. DATA FILE HOUSEPRICE

Four-bedroom houses

529,000	227,500	310,000	429,900	300,000	290,000
279,900	392,500	186,500	255,000	249,995	225,000
174,900	137,500				

Three-bedroom houses

132,500	182,000	167,900	145,000	359,900	245,900
230,000	218,000	227,500	173,000	135,000	84,900
79,000	179,900	188,900	182,000	230,000	197,000

(a) Plot the asking prices for the two sets of houses and describe the two distributions.

(b) Test the null hypothesis that the mean asking prices for the two sets of houses are equal versus the two-sided alternative. Give the test statistic with degrees of freedom, the P-value, and your conclusion.

(c) Would you consider using a one-sided alternative for this analysis? Explain why or why not.

(d) Give a 95% confidence interval for the difference in mean asking prices.

(e) These data are not SRSs from a population. Give a justification for use of the two-sample t procedures in this case.

7.94. CHALLENGE **More on house prices.** Go to the Web site `www.realtor.com` and select two geographical areas of interest to you. You will compare the prices of similar types of houses in these two areas. State clearly how you define the areas and the type of houses. For example, you can use city names or zip codes to define the area and you can select single-family houses or condominiums. We view these houses as representative of the asking prices of houses for these areas at the time of your search. If the search gives a large number of houses, select a random sample. Be sure to explain exactly how you do this. Use the methods you have learned in this chapter to compare the asking prices. Be sure to include a graphical summary.

CHAPTER EIGHT

DELVING INTO THE USE OF INFERENCE

8.1 Estimating with Confidence
8.2 Use and Abuse of Significance Tests

In the previous two chapters, we studied the most common methods of statistical inference: *confidence intervals* and *tests of significance*. An important message to take away from those chapters is that regardless of the choice of parameter, the underlying reasoning and construction of these two inference methods remained constant. In this chapter, we discuss some additional properties of these procedures as well as some cautions to consider when using them. While we again focus our discussion on inference for proportions and means, these properties and warnings apply to many other settings and parameters.

8.1 Estimating with Confidence

The confidence interval is useful when our goal is to estimate population parameters. As we saw in the previous two chapters, a confidence interval often takes the form

$$\text{estimate} \pm \text{margin of error}$$

The estimate is our guess for the value of the unknown parameter, and the margin of error is a combination of the accuracy of the estimate and our confidence that the procedure will result in an interval that contains the true parameter value.

How confidence intervals behave

High confidence and a small margin of error are desirable characteristics of a confidence interval. High confidence says that our method almost always gives correct answers. A small margin of error says that we have pinned down the parameter quite precisely.

A wise user of statistics never plans data collection without at the same time planning the inference. With a confidence interval, the user chooses the confidence level C and the sample size n. How do different choices of these values affect the margin of error?

Let's investigate the choice of C using the confidence interval for a population proportion p. Recall that the margin of error for this confidence interval is $z^*\sqrt{\hat{p}(1-\hat{p})/n}$, where n is the sample size, $\hat{p}$ is the sample proportion, and z^* is the critical value for the selected confidence level C (page 329). Because we don't know the value of $\hat{p}$ until we've gathered the data, there are two ways to make this quantity smaller:

- Use a lower level of confidence (smaller C).
- Increase the sample size (larger n).

For most problems you will choose a confidence level of 90%, 95%, or 99%, so z^* will be 1.645, 1.960, or 2.576, respectively. Notice that for these choices of confidence level z^* is smaller for lower confidence (smaller C). The bottom row of Table D also shows this but for more confidence levels. *A smaller z^* leads to a smaller margin of error.*

EXAMPLE 8.1

How the confidence level affects the confidence interval. In Example 6.7 (page 330), we constructed a 95% confidence interval for the proportion of adults who play video games. The sample proportion was $\hat{p} = 0.5175$ with a margin of error of 0.022. Suppose instead that we wanted 99% confidence. Table D tells us that for 99% confidence, $z^* = 2.576$. The margin of error for 99% confidence based on 2054 observations is

$$\begin{aligned} m &= z^* \sqrt{\frac{\hat{p}(1-\hat{p})}{n}} \\ &= 2.576 \sqrt{\frac{0.5175(1-0.5175)}{2054}} \\ &= 0.0284 \end{aligned}$$

and the 99% confidence interval is

$$\begin{aligned} \hat{p} \pm m &= 0.52 \pm 0.03 \\ &= (0.49,\ 0.55) \end{aligned}$$

Requiring 99%, rather than 95%, confidence has increased the margin of error from 0.02 to 0.03. Figure 8.1 compares the two intervals.

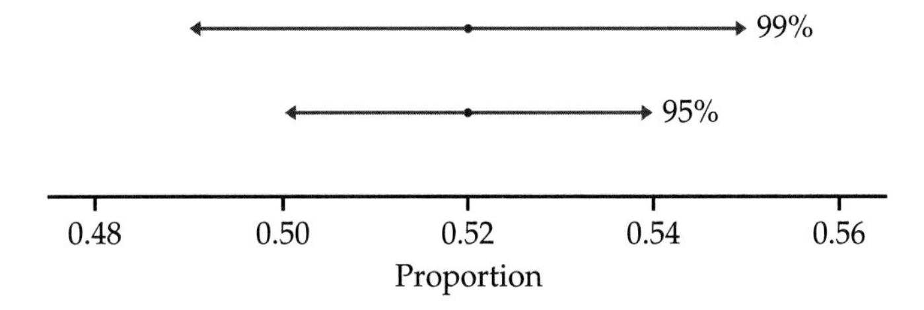

FIGURE 8.1 Confidence intervals for Example 8.1. The larger the value of C, the wider the interval.

Besides the confidence level C, the user also selects the sample size n. Suppose that the researchers who designed this survey of adults had used a different sample size. How would this affect the confidence interval? We can answer this question by changing the sample size in our calculations and assuming that the sample proportion remains the same.

EXAMPLE 8.2

How the sample size affects the confidence interval. Let's assume that the sample proportion of adults who play video games is 0.5175, as in Example 6.7. But suppose that the sample size is 8216. The margin of error for 95% confidence is

$$\begin{aligned} m &= z^* \sqrt{\frac{\hat{p}(1-\hat{p})}{n}} \\ &= 1.960 \sqrt{\frac{0.5175(1-0.5175)}{8216}} \\ &= 0.0108 \end{aligned}$$

and the 95% confidence interval is

$$\begin{aligned} \hat{p} \pm m &= 0.52 \pm 0.01 \\ &= (0.51,\ 0.53) \end{aligned}$$

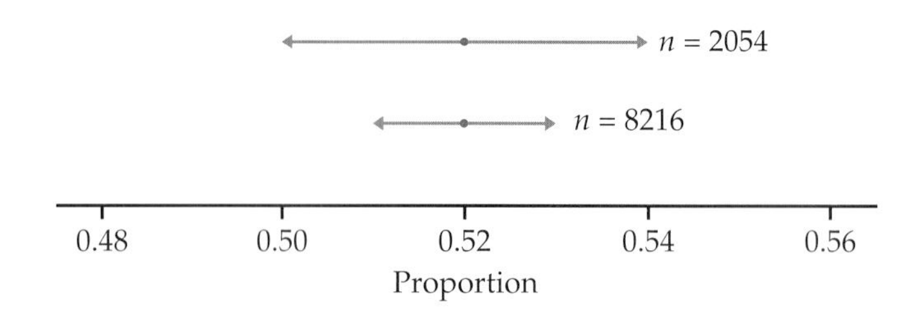

FIGURE 8.2 Confidence intervals for $n = 8216$ and $n = 2054$, for Example 8.2. A sample size four times as large results in a confidence interval that is half as wide.

Notice that the margin of error for this example is half as large as the margin of error that we computed in Example 6.7 (page 330). The only change that we made was to assume that the sample size is 8216 rather than 2054. This sample size is exactly four times the size of the original 2054. Thus, we halve the margin of error when we increase the sample size by a factor of 4. Figure 8.2 illustrates the effect in terms of the intervals.

These two approaches to reduce the length of the interval are shared by all confidence intervals in common use. For example, consider the confidence interval for a population mean. In this setting, the margin of error is $t^*s/\sqrt{n}$. For any row in Table D, t^* is smaller for smaller levels of confidence just as z^* (the last row of the table) is. Also, because $\sqrt{n}$ is in the denominator, increasing the sample size by a factor of four decreases the standard error by half (assuming s remains the same). However, the effect on the margin of error is even greater in this setting because the increase in n also increases the degrees of freedom, thereby reducing t^* too.

one-sample t confidence interval p. 373

Choosing a sample size

You can arrange to have both high confidence and a small margin of error by choosing an appropriate sample size. Again, let's first focus on the margin of error for the large-sample confidence interval for a population proportion:

$$m = z^*\text{SE}_{\hat{p}} = z^*\sqrt{\frac{\hat{p}(1-\hat{p})}{n}}$$

Choosing a confidence level C fixes the critical value z^*. The margin of error also depends on the value of $\hat{p}$ and the sample size n. Because we don't know the value of $\hat{p}$ until we gather the data, we must guess a value to use in the sample size calculations. We will call the guessed value p^*. There are two common ways to get p^*:

1. Use the sample estimate from a pilot study or from similar studies done earlier.
2. Use $p^* = 0.5$. Because the margin of error is largest when $\hat{p} = 0.5$, this choice gives a sample size that is at least as large as what we really need for the confidence level we choose. It is a safe choice no matter what the data later show.

Once we have chosen p^* and the margin of error m that we want, we can find the n we need to achieve this margin of error. Here is the result.

SAMPLE SIZE FOR DESIRED MARGIN OF ERROR FOR A PROPORTION p

The level C confidence interval for a proportion p will have a margin of error approximately equal to a specified value m when the sample size satisfies

$$n = \left(\frac{z^*}{m}\right)^2 p^*(1 - p^*)$$

Here z^* is the critical value for confidence level C, and p^* is a guessed value for the proportion of successes in the future sample.

The margin of error will be less than or equal to m if p^* is chosen to be 0.5. The sample size required when $p^* = 0.5$ is

$$n = \frac{1}{4}\left(\frac{z^*}{m}\right)^2$$

The value of n obtained by this method is not particularly sensitive to the choice of p^* when p^* is fairly close to 0.5. However, if the value of p is likely to be smaller than about 0.3 or larger than about 0.7, use of $p^* = 0.5$ may result in a sample size that is much larger than needed.

Do notice that this method makes no reference to the size of the *population*. It is the size of the *sample* that determines the margin of error. The size of the population does not influence the sample size we need (as long as the population is much larger than the sample).

Lastly, this formula does not account for collection costs. In practice, taking observations costs time and money. The required sample size based on this formula may be impossibly expensive. In those situations, one might consider a larger margin of error and/or a lower confidence level to be acceptable.

EXAMPLE 8.3

Planning a survey of students. A large university is interested in assessing student satisfaction with the overall campus environment. The plan is to distribute a questionnaire to an SRS of students, but before proceeding, the university wants to determine how many students to sample. The questionnaire asks about a student's degree of satisfaction with various student services, each measured on a five-point scale. The university is interested in the proportion p of students who are satisfied (that is, who choose either "satisfied" or "very satisfied," the two highest levels on the five-point scale).

The university wants to estimate p with 95% confidence and a margin of error less than or equal to 3%, or 0.03. For planning purposes, they are willing to use $p^* = 0.5$. To find the sample size required, we calculate

$$n = \frac{1}{4}\left(\frac{z^*}{m}\right)^2 = \frac{1}{4}\left(\frac{1.96}{0.03}\right)^2 = 1067.1$$

Round up to get $n = 1068$. (*Always round up*. Rounding down would give a margin of error slightly greater than 0.03.)

Similarly, for a 2.5% margin of error we have (after rounding up)

$$n = \frac{1}{4}\left(\frac{1.96}{0.025}\right)^2 = 1537$$

and for a 2% margin of error,

$$n = \frac{1}{4}\left(\frac{1.96}{0.02}\right)^2 = 2401$$

News reports frequently describe the results of surveys with sample sizes between 1000 and 1500 and a margin of error of about 3%. These surveys generally use sampling procedures more complicated than simple random sampling, so the calculation of confidence intervals is more involved than what we have studied. The calculations in Example 8.3 nonetheless show in principle how such surveys are planned.

In practice, many factors influence the choice of a sample size. The following example illustrates one set of factors.

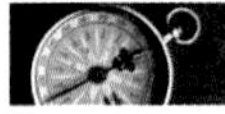

EXAMPLE 8.4

Assessing interest in Pilates classes. The Division of Recreational Sports (Rec Sports) at a major university is responsible for offering comprehensive recreational programs, services, and facilities to the students. Rec Sports is continually examining its programs to determine how well it is meeting the needs of the students. Rec Sports is considering adding some new programs and would like to know how much interest there is in a new exercise program based on the Pilates method.[1] They will take a survey of undergraduate students. In the past, they emailed short surveys to all undergraduate students. The response rate obtained in this way was about 5%. This time they will send emails to a simple random sample of the students and will follow up with additional emails and eventually a phone call to get a higher response rate. Because of limited staff and the work involved with the follow-up, they would like to use a sample size of about 200. One of the questions they will ask is "Have you ever heard about the Pilates method of exercise?"

The main purpose of the survey is to estimate various sample proportions for undergraduate students. We will focus here on the one question regarding the Pilates method. Will the proposed sample size of $n = 200$ be adequate to provide Rec Sports with the needed information? To address this question, we calculate the margins of error of 95% confidence intervals for various values of $\hat{p}$.

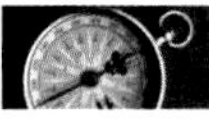

EXAMPLE 8.5

Margins of error. In the Rec Sports survey, the margin of error of a 95% confidence interval for any value of $\hat{p}$ and $n = 200$ is

$$\begin{aligned} m &= z^*\text{SE}_{\hat{p}} \\ &= 1.96\sqrt{\frac{\hat{p}(1-\hat{p})}{200}} \\ &= 0.139\sqrt{\hat{p}(1-\hat{p})} \end{aligned}$$

The results for various values of $\hat{p}$ are

$\hat{p}$	m	$\hat{p}$	m
0.05	0.030	0.60	0.068
0.10	0.042	0.70	0.064
0.20	0.055	0.80	0.055
0.30	0.064	0.90	0.042
0.40	0.068	0.95	0.030
0.50	0.070		

Rec Sports judged these margins of error to be acceptable, and they used a sample size of 200 in their survey.

The table in Example 8.5 illustrates two points. First, the margins of error for $\hat{p} = 0.05$ and $\hat{p} = 0.95$ are the same. The margins of error will always be the same for $\hat{p}$ and $1 - \hat{p}$. This is a direct consequence of the form of the confidence interval. Second, the margin of error varies only between 0.064 and 0.070 as $\hat{p}$ varies from 0.3 to 0.7, and the margin of error is greatest when $\hat{p} = 0.5$, as we claimed earlier. It is true in general that the margin of error will vary relatively little for values of $\hat{p}$ between 0.3 and 0.7. Therefore, when planning a study, it is not necessary to have a very precise guess for p. If $p^* = 0.5$ is used and the observed $\hat{p}$ is between 0.3 and 0.7, the actual interval will be a little shorter than needed but the difference will be small.

It is important to emphasize that these calculations consider only the effects of sampling variability that are quantified in the margin of error. Other sources of error, such as nonresponse and possible misinterpretation of questions, are not included in the table of margins of error for Example 8.5. Rec Sports is trying to minimize these kinds of errors. They did a pilot study using a small group of current users of their facilities to check the wording of the questions, and they devised a careful plan to follow up with the students who did not respond to the initial email.

As with most opinion surveys, Rec Sports is asking more than one question, resulting in more than one sample proportion. *When determining the sample size n in these cases, it is common to take into account the fact that multiple confidence intervals will be constructed.* This will result in a larger n than our calculations specify here. An example of one such approach, which inflates the value of z^*, was described in Exercise 6.63 (page 355).

USE YOUR KNOWLEDGE

8.1 Make a plot. Use the values of $\hat{p}$ and m in Example 8.5 to draw a plot of the sample proportion versus the margin of error. Summarize the major features of your plot.

8.2 Confidence level and sample size. Refer to Example 8.3 (page 423). Suppose that the university was interested in a 90% confidence interval with margin of error 0.03. Would the required sample size be smaller or larger than 1068 students? Verify this by performing the calculation.

Similar calculations can be done to determine n when interested in a population mean. In this setting, recall that the margin of error is $t^*s/\sqrt{n}$. Besides the confidence level C and sample size n, this margin of error depends on the sample standard deviation s, which is known only once the data are collected. We handle this unknown s in the same way we handled $\hat{p}$ in the margin of error for a proportion. We guess at its value using results from a pilot study or from similar studies done earlier.

It is always better to use a value of the standard deviation that is a little larger than what is expected. This may give a sample size that is a little larger than needed but it helps avoid the situation where the margin of error is larger than desired.

Given an estimate for s, which we call s^*, the approach to find n is similar to what we did for a proportion. The added complication is that t^* depends not only on the confidence level C but also on the sample size n. Here are the details.

SAMPLE SIZE FOR DESIRED MARGIN OF ERROR FOR A MEAN μ

The level C confidence interval for a mean μ will have an expected margin of error less than or equal to a specified value m when the sample size is such that

$$m \geq t^*s^*/\sqrt{n}$$

Here t^* is the critical value for confidence level C with $n - 1$ degrees of freedom, and s^* is the guessed value for the population standard deviation.

Finding the smallest sample size n that satisfies this requirement can be done using the following iterative search:

1. Get an initial sample size by replacing t^* with z^* so $n = (z^*s^*/m)^2$.
2. Use this sample size to get t^* and see if $m \geq t^*s^*/\sqrt{n}$.
3. If the requirement is satisfied, then this n is the needed sample size. If the requirement is not satisfied, increase n by 1 and go back to Step 2.

Here is an example.

EXAMPLE 8.6

Planning a survey of college students. In Example 7.10 (page 373), we calculated a 95% confidence interval for the mean hours per month a college student watches videos on his or her phone. The margin of error based on an SRS of $n = 8$ students was 2.96 hours. Suppose that a new study is being planned and the goal is to have a margin of error of 1.5 hours. How many students need to be sampled? The sample standard deviation in Example 7.10 was 3.55. To be a bit conservative, we'll guess that the population standard deviation is 4.00.

1. To compute an initial n, we replace t^* with z^*. This results in

$$n = \left(\frac{z^*s^*}{m}\right)^2 = \left[\frac{1.96(4.00)}{1.5}\right]^2 = 27.32$$

Round up to get $n = 28$.

2. We now check to see if this sample size satisfies the requirement when we switch back to t^*. For $n = 28$, we have $n - 1 = 27$ degrees of freedom and $t^* = 2.052$. Using this value, the expected margin of error is

$$2.052(4.00)/\sqrt{28} = 1.55$$

This is larger than $m = 1.5$, so the requirement is not satisfied.

3. The following table summarizes these calculations for additional choices of n.

n	$t^*s^*/\sqrt{n}$
28	1.55
29	1.52
30	1.49
31	1.47

The requirement is first satisfied when $n = 30$. We need to sample at least $n = 30$ students for the expected margin of error to be no more than 1.5 hours.

Unfortunately, the actual number of usable observations is often less than what we plan at the beginning of a study. This is particularly true of data collected in surveys but is an important consideration in most studies. Careful study designers often assume a nonresponse rate or dropout rate that specifies what proportion of the originally planned sample will fail to provide data. We use this information to calculate the sample size to be used at the start of the study. For example, if in the preceding survey we expect only 25% of those students to respond, we would need to start with a sample size of $4 \times 30 = 120$ to obtain usable information from 30 students.

USE YOUR KNOWLEDGE

8.3 Starting salaries. In a recent survey by the National Association of Colleges and Employers, the average starting salary for computer science majors was reported to be \$63,017.[2] You are planning to do a survey of starting salaries for recent computer science majors from your university. Using an estimated standard deviation of \$13,500, what sample size do you need to have a margin of error equal to \$1100 with 95% confidence?

8.4 Changes in sample size. Suppose that in the setting of the previous exercise you have the resources to contact 1000 recent graduates. If all respond, will your margin of error be larger or smaller than \$1100? What if only 50% respond? Verify your answers by performing the calculations.

Two-sided significance tests and confidence intervals

In the previous two chapters, we described confidence intervals and significance tests separately, with each method directed at a different goal of a research study. There is, however, a relationship between these two types of inference.

Recall the basic idea of a confidence interval. We construct an interval that will include the true value of the parameter with a specified probability C. Suppose that we use a 95% confidence interval ($C = 0.95$). Then the values of the parameter that are not in our interval would seem to be incompatible with the data. This sounds like a significance test with $\alpha = 0.05$ (or 5%) as our standard for drawing a conclusion. The following example demonstrates that this is correct.

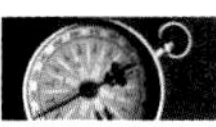

EXAMPLE 8.7

Watching videos on a mobile phone. In Example 7.11 (page 375), we tested whether the U.S. college student average is different from the overall U.S. average of 4.3 hours. Software gave us a P-value of $P = 0.0682$, so we concluded, at the 0.05 significance level, that there was not enough evidence to claim there was a difference.

What if, instead, we wanted to test whether the U.S. college student average is different from $\mu = 3.6$ hours, the U.S. average of those who classify themselves as white.[3] In this case, the t test statistic is

$$t = \frac{\bar{x} - \mu_0}{s/\sqrt{n}} = \frac{7.00 - 3.6}{3.546/\sqrt{8}}$$
$$= 2.71$$

The P-value is $2P(T \geq 2.71)$, where T has the $t(7)$ distribution. From Table D we see that $P(T \geq 2.517) = 0.02$ and $P(T \geq 2.998) = 0.01$. Thus, the P-value is between $2 \times 0.02 = 0.04$ and $2 \times 0.01 = 0.02$. At the 0.05 significance level, there is enough evidence to conclude that the U.S. college student average is larger than the average of 3.6 hours.

In Example 7.10 (page 373), we concluded, based on an SRS of $n = 8$ students, that the mean hours per month a college student watches videos on his or her mobile phone is 7.00 hours with a 95% margin of error of 2.96 hours. In other words, we are 95% confident that the unknown mean μ is in the interval (4.04, 9.96).

Notice the relationship between the conclusions of the significance tests in Example 8.7 and the locations of μ_0 in relation to the confidence interval. We could not reject the null hypothesis that $\mu_0 = 4.3$ hours and $\mu_0 = 4.3$ lies inside the 95% confidence interval. We could reject the null hypothesis that $\mu = 3.6$ hours and $\mu_0 = 3.6$ falls *outside* the 95% confidence interval. Figure 8.3 illustrates both cases.

For the population mean μ (or difference in means), the calculation of the significance test is very similar to the calculation for a confidence interval. In fact, a two-sided test at significance level α can be carried out directly from a confidence interval with confidence level $C = 1 - \alpha$.

> **TWO-SIDED SIGNIFICANCE TESTS AND CONFIDENCE INTERVALS**
>
> A level α two-sided significance test rejects a hypothesis H_0: $\mu = \mu_0$ exactly when the value μ_0 falls outside a level $1 - \alpha$ confidence interval for μ.

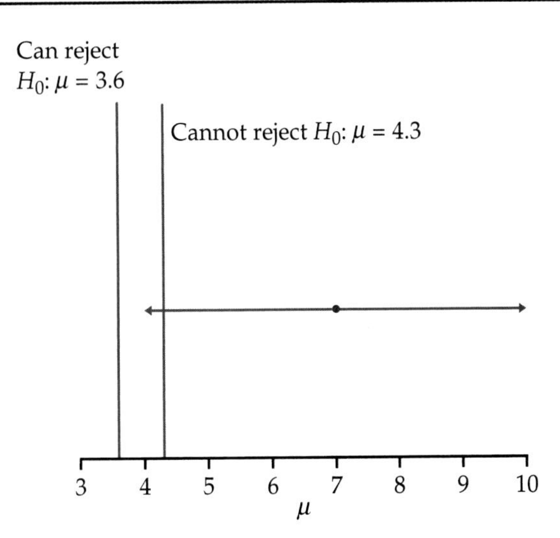

FIGURE 8.3 The link between two-sided significance tests and confidence intervals. For the study described in Example 8.7, values of μ falling outside a 95% confidence interval can be rejected at the 5% significance level; values falling inside the interval cannot be rejected. This holds for any significance level α and $1 - \alpha$ confidence interval.

USE YOUR KNOWLEDGE

8.5 Two-sided significance tests and confidence intervals. The P-value for a two-sided test of the null hypothesis H_0: $\mu = 35$ is 0.027.

(a) Does the 95% confidence interval include the value 35? Explain.

(b) Does the 99% confidence interval include the value 35? Explain.

8.6 More on two-sided tests and confidence intervals. A 95% confidence interval for a population mean is (45, 57).

(a) Can you reject the null hypothesis that $\mu = 58$ against the two-sided alternative at the 5% significance level? Explain.

(b) Can you reject the null hypothesis that $\mu = 55$ against the two-sided alternative at the 5% significance level? Explain.

Some cautions

We have already seen that small margins of error and high confidence can require large numbers of observations. You should also be keenly aware that *any formula for inference is correct only in specific circumstances*. If the government required statistical procedures to carry warning labels like those on drugs, most inference methods would have long labels. Our handy formulas for estimating a population proportion and mean come with the following list of warnings for the user:

- The data should be an SRS from the population. We are completely safe if we actually did a randomization and drew an SRS. We are not in great danger if the data can plausibly be thought of as independent observations from a population. That is the case in Examples 8.4 to 8.6, where we redefine our population to correspond to survey respondents.

- The formulas are not correct for probability sampling designs more complex than an SRS. Correct methods for other designs are available. We will not discuss confidence intervals for the multistage or stratified samples that we discussed in Chapter 2. If you plan such samples, be sure that you (or your statistical consultant) know how to carry out the inference you desire.
- There is no correct method for inference from data haphazardly collected with bias of unknown size. Fancy formulas cannot rescue badly produced data.
- The confidence interval formula for a proportion assumes that the population size is at least 10 times as large as the sample. If this is not the case, the resulting large-sample interval will be too wide. Ask your consultant about appropriate methods to be used in these situations.

resistant measure p. 107

- Because $\bar{x}$ is not a resistant measure, outliers can have a large effect on the confidence interval for the mean. You should search for outliers and try to correct them or justify their removal before computing the interval. If the outliers cannot be removed, ask your statistical consultant about procedures that are not sensitive to outliers.
- If the sample size is small and the response of interest from the population is not Normal, the true confidence level for the population mean will be different from the value C used in computing the interval. Examine your data carefully for skewness and other signs of non-Normality. The interval relies only on the distribution of $\bar{x}$, which even for quite small sample sizes is much closer to Normal than that of the individual observations. When $n \geq 15$, the confidence level is not greatly disturbed by non-Normal populations unless extreme outliers or quite strong skewness are present.

The most important caution concerning confidence intervals is a consequence of the first of these warnings. *The margin of error in a confidence interval covers only random sampling errors.* The margin of error is obtained from the sampling distribution and indicates how much error can be expected because of chance variation in randomized data production. The practical conduct of the survey influences the trustworthiness of its results in ways that are not included in the announced margin of error.

Practical difficulties such as undercoverage and nonresponse in a sample survey can cause errors that are larger than the random sampling error. This often happens when the sample size is large (so that the standard error is small). Remember this unpleasant fact when reading the results of an opinion poll or other sample survey.

Every inference procedure that we will meet has its own list of warnings. Because many of the warnings are similar to those we have mentioned, we will not print the full warning label each time. It is easy to state (from the mathematics of probability) conditions under which a method of inference is exactly correct. These conditions are *never* fully met in practice. For example, no population is exactly Normal.

Deciding when a statistical procedure should be used in practice often requires judgment assisted by exploratory analysis of the data. Mathematical facts are therefore only a part of statistics. The difference between statistics and mathematics can be stated thus: mathematical theorems are true; statistical methods are often effective when used with skill.

USE YOUR KNOWLEDGE

8.7 Self-report survey. As part of Sallie Mae's recent credit card usage study, a self-report survey was distributed to 5800 undergraduates. Of the 292 returned surveys, only 249 were complete and used in the analysis. Based on these responses, the report states, "Ninety-two percent of undergraduate credit cardholders charged textbooks, school supplies, or other direct education expenses, up from 85 percent when the study was last conducted, in 2004."[4] The reported margin of error is 4%. Do you think that this small margin of error is a good measure of the accuracy of the survey's results? Explain your answer.

SECTION 8.1 SUMMARY

The purpose of a **confidence interval** is to estimate an unknown parameter with an indication of how accurate the estimate is and of how confident we are that the result is correct.

Other things being equal, the margin of error of a confidence interval decreases as

- the confidence level C decreases, and
- the sample size n increases.

The **sample size** required to obtain a confidence interval with approximate margin of error m for a proportion is found from

$$n = \left(\frac{z^*}{m}\right)^2 p^*(1 - p^*)$$

where p^* is a guessed value for the proportion, and z^* is the standard Normal critical value for the desired level of confidence. To ensure that the margin of error of the interval is less than or equal to m no matter what $\hat{p}$ may be, use

$$n = \frac{1}{4}\left(\frac{z^*}{m}\right)^2$$

The sample size n required to obtain a confidence interval with an expected margin of error no larger than m for a population mean satisfies the constraint

$$m \geq t^*s^*/\sqrt{n}$$

where t^* is the critical value for the desired level of confidence with $n - 1$ degrees of freedom, and s^* is the guessed value for the population standard deviation.

A specific confidence interval recipe is correct only under specific conditions. The most important conditions concern the method used to produce the data. Other factors such as the form of the population distribution may also be important.

SECTION 8.1 EXERCISES

For Exercises 8.1 and 8.2, see page 425; for Exercises 8.3 and 8.4, see page 427; for Exercises 8.5 and 8.6, see page 429, and for Exercise 8.7, see page 431.

8.8. Confidence interval mistakes and misunderstandings. Suppose that 101 randomly selected members of the Karaoke Channel were asked how much time they typically spend on the site during the week.[5] The sample mean $\bar{x}$ was found to be 4.2 hours, and the sample standard deviation was $s = 2.5$ hours.

(a) Cary Oakey computes the 95% confidence interval for the average time on the site as $4.2 \pm 1.984(2.5/101)$. What is his mistake?

(b) He corrects this mistake and then states that "95% of the members spend between 3.71 and 4.69 hours a week on the site." What is wrong with his interpretation of this interval?

(c) The margin of error is slightly less than half an hour. To reduce this to 15 minutes, Cary says that the sample size needs to be doubled to 202. What is wrong with this statement?

8.9. More confidence interval mistakes and misunderstandings. Suppose that 400 randomly selected alumni of the University of Okoboji were asked to rate the university's counseling services on a 1 to 10 scale. The sample mean $\bar{x}$ was found to be 8.6, and the sample standard deviation was $s = 2.0$. Given the large sample size, use z^* in place of t^* as the critical value.

(a) Ima Bitlost computes the 95% confidence interval for the average satisfaction score as $8.6 \pm 1.96(2.0)$. What is her mistake?

(b) After correcting her mistake in part (a), she states, "I am 95% confident that the sample mean falls between 8.404 and 8.796." What is wrong with this statement?

(c) She quickly realizes her mistake in part (b) and instead states, "The probability that the true mean is between 8.404 and 8.796 is 0.95." What misinterpretation is she making now?

(d) Finally, in her defense for using the Normal distribution to determine the confidence critical value she says, "Because the sample size is quite large, the population of alumni ratings will be approximately Normal." Explain to Ima her misunderstanding and correct this statement.

8.10. Margin of error and the confidence interval. A study based on a sample of size 144 reported a sample proportion of 0.87 and a margin of error of 0.055 for 95% confidence.

(a) Give the 95% confidence interval.

(b) If you wanted 99% confidence for the same study, would your margin of error be greater than, equal to, or less than 0.055? Explain your answer.

8.11. Changing the sample size. Suppose that the sample proportion is 0.50. Make a diagram similar to Figure 8.2 (page 422) that illustrates the effect of sample size on the width of a 95% interval. Use the following sample sizes: 40, 80, 120, and 160. Summarize what the diagram shows.

8.12. Changing the confidence level. A study with 225 respondents had a sample proportion of 0.70. Make a diagram similar to Figure 8.1 (page 421) that illustrates the effect of the confidence level on the width of the interval. Use 80%, 90%, 95%, and 99%. Summarize what the diagram shows.

8.13. Populations sampled and margins of error. Consider the following two scenarios. (A) Take a simple random sample of 100 sophomore college students in your state. (B) Take a simple random sample of 100 sophomore students at your college or university. For each of these samples you will record the amount spent on textbooks used for classes during the fall semester. Which sample should have the smaller margin of error? Explain your answer.

8.14. Planning a mobile gaming study. A recent study by a market research firm found that 38% of tablet gamers and 20% of mobile phone gamers play five or more hours per week.[6] You decide to survey students at your large university and see how they compare.

(a) Using the sample proportion of just the tablet gamers from this previous study, what sample size of tablet gamers would you need for an estimate with a margin of error of 2.5% for 95% confidence?

(b) You expect the proportion of tablet gamers who play more than 5 hours to be higher in your population. What sample size would you need so that the margin of error is less than or equal to 2.5% regardless of the observed $\hat{p}$?

(c) To implement this survey, you will need to take an SRS of students and first screen them to find out if they are tablet gamers. Suppose that only 20% of the students at your university are tablet gamers. What is the size of the initial sample of students that you will have to draw to obtain usable information from the sample size in part (b)?

8.15. More on planning a mobile gaming study. Refer to the previous exercise. Suppose that you will also survey mobile phone gamers.

(a) If you were to repeat part (a) for the mobile phone gamers, would the necessary sample size be smaller, equal to, or larger than the sample size of tablet gamers? Explain your answer without doing any calculations.

(b) If you were to repeat part (b) for the mobile phone gamers, would the necessary sample size be smaller, equal to, or larger than the sample size of tablet gamers? Explain your answer without doing any calculations.

(c) Suppose that 35% of the students at your university are mobile phone gamers and none of the students are both types of gamers. What is the size of the initial sample of students that you will have to draw to obtain the desired usable information for both types of gamers under the conditions of part (b)? Explain your answer.

8.16. What's "in" on campus. The *Student Monitor* surveys 1200 undergraduates from 100 colleges semiannually to understand trends among college students.[7] Recently, the *Student Monitor* reported that 58% of those surveyed agreed that Facebook was "in" on campus. You feel that your campus has shown signs of "Facebook fatigue" and expect the proportion at your school to be lower.

(a) Ignoring the fact that the *Student Monitor* used a more complex sampling design and, instead, treating it as an SRS, what is the margin of error of their estimate for 95% confidence?

(b) You expect the proportion at your school to be lower and could be near 50%. What number of students would you have to sample to make sure the margin of error is no larger than what you calculated in part (a)?

(c) What is the required sample size if you reduced your confidence level to 90%?

8.17. What's "in" on campus, continued. Refer to the previous exercise. To implement this survey, you are going to have the registrar send out an email that contains a link to an online survey to a random selection of students. Past surveys done in this manner have had an initial 40% response rate with an additional 15% response after an email reminder sent a week later. Based on your answer to part (b) in the previous exercise, how many students should the registrar email?

8.18. Apartment rental rates. You hope to rent an unfurnished one-bedroom apartment in Dallas next year. You call a friend who lives there and ask him to give you an estimate of the mean monthly rate. Having taken a statistics course recently, the friend asks about the desired margin of error and confidence level for this estimate. He also tells you that the standard deviation of monthly rents is about $300.

(a) For 95% confidence and a margin of error of $100, how many apartments should the friend randomly sample from the local newspaper?

(b) Suppose that you want the margin of error to be no more than $50. How many apartments should the friend sample?

(c) Why is the sample size in part (b) not just four times larger than the sample size in part (a)?

8.19. More on apartment rental rates. Refer to the previous exercise. Will the 95% confidence interval include approximately 95% of the rents of all unfurnished one-bedroom apartments in this area? Explain why or why not.

8.20. Average hours per week on the Internet. Refer to Exercise 8.16. The *Student Monitor* also reported that the average amount of time spent per week on the Internet was 19.0 hours. You suspect that this amount is far too small for your campus and plan to add this question to your survey.

(a) You feel that a reasonable estimate of the standard deviation is 6.5 hours. What sample size is needed so that the margin of error of your estimate is not larger than 1 hour for 95% confidence?

(b) The distribution of times is likely to be heavily skewed to the right. Do you think that this skewness will invalidate the use of the t confidence interval in this case? Explain your answer.

8.21. Satisfied with your job? Job satisfaction is one of four workplace measures the Gallup-Healthways Well-Being Index tracks among U.S. workers. The question asked is "Are you satisfied or dissatisfied with your job or the work that you do?" In 2011, 87.5% responded that they were satisfied. Material provided with the results of the poll noted:

> *Results are based on telephone interviews conducted as part of the Gallup-Healthways Well-Being Index survey Jan. 1–April 30, 2011, with a random sample of 61,889 adults, aged 18 and older, living in all 50 U.S. states and the District of Columbia, selected using random-digit-dial sampling.*
>
> *For results based on the total sample of national adults, one can say with 95% confidence that the maximum margin of sampling error is 1 percentage point.*[8]

The poll uses a complex multistage sample design, but the sample percent has approximately a Normal sampling distribution.

(a) The announced poll result was 87.5% ± 1%. Can we be certain that the true population percent falls in this interval? Explain your answer.

(b) Explain to someone who knows no statistics what the announced result 87.5% ± 1% means.

(c) What is the standard error of the estimated percent?

(d) Does the announced margin of error include errors due to practical problems such as nonresponse? Explain your answer.

8.22. Accuracy of a laboratory scale. To assess the accuracy of a laboratory scale, a standard weight known to weigh 10 grams is weighed repeatedly. The scale readings are Normally distributed with unknown mean (this mean is 10 grams if the scale has no bias). The standard deviation of the scale readings in the past has been 0.0002 gram.

(a) The weight is measured five times. The mean result is 10.0023 grams. Give a 98% confidence interval for the mean of repeated measurements of the weight.

(b) How many measurements must be averaged to get a margin of error of ±0.0001 with 98% confidence?

8.23. Radio poll. A college radio station invites listeners to enter a dispute about a proposed "pay as you throw" waste collection program. The station asks listeners to call in and state how much each bag of trash should cost. A total of 633 listeners call in. The station calculates the 95% confidence interval for the average fee desired by city residents to be \$0.83 to \$1.28. Is this result trustworthy? Explain your answer.

8.2 Use and Abuse of Significance Tests

Carrying out a test of significance is often quite simple, especially if the P-value is given effortlessly by a computer. Using tests wisely is not so simple. Each test is valid only in certain circumstances, with properly produced data being particularly important. The z and t tests, for example, should bear the same warning label that was attached in Section 8.1 to their corresponding confidence intervals (page 429). Similar warnings accompany the other tests that we will learn.

There are additional caveats that concern tests more than confidence intervals, enough to warrant this separate section. Some hesitation about the unthinking use of significance tests is a sign of statistical maturity.

The reasoning of significance tests has appealed to researchers in many fields, so that tests are widely used to report research results. In this setting H_a is a "research hypothesis" asserting that some effect or difference is present. The null hypothesis H_0 says that there is no effect or no difference. A low P-value represents good evidence that the research hypothesis is true. Here are some comments on the use of significance tests, with emphasis on their use in reporting scientific research.

Choosing a level of significance

The spirit of a test of significance is to give a clear statement of the degree of evidence provided by the sample against the null hypothesis. The P-value does this. It is common practice to report P-values and to describe results as statistically significant whenever $P \leq 0.05$. *However, there is no sharp border between "significant" and "not significant," only increasingly strong evidence as the P-value decreases.* Having both the P-value and the statement that we reject or fail to reject H_0 allows us to draw better conclusions from our data.

EXAMPLE 8.8

Information provided by the P-value. Suppose that the test statistic for a two-sided significance test for a population mean is $t = 2.05$ with 23 degrees of freedom. From Table D we see that this has a P-value between 0.05 and 0.10. The exact value is 0.0519.

We have failed to meet the standard of evidence for $\alpha = 0.05$. However, with the information provided by the P-value, we can also see that the result just barely missed the standard. If the effect in question is interesting and potentially important, we might want to design another study with a larger sample to investigate it further.

Here is another example where the P-value provides useful information beyond that provided by the statement that we reject or fail to reject the null hypothesis.

EXAMPLE 8.9

More on information provided by the *P*-value. We have a test statistic of $t = -4.66$ for a two-sided significance test on a population mean. Software tells us that the P-value is 0.000003. This means that there are 3 chances in 1,000,000 of observing a sample mean this far or farther away from the null hypothesized value of μ. This kind of event is virtually impossible if the null hypothesis is true. There is no ambiguity in the result, so we can clearly reject the null hypothesis.

We frequently report small P-values such as that in the previous example as $P < 0.001$. This corresponds to a chance of 1 in 1000 and is sufficiently small to lead us to a clear rejection of the null hypothesis.

One reason for the common use of $\alpha = 0.05$ is the great influence of Sir R. A. Fisher, the inventor of formal statistical methods for analyzing experimental data. Here is his opinion on choosing a level of significance: "A scientific fact should be regarded as experimentally established only if a properly designed experiment *rarely fails* to give this level of significance."[9]

What statistical significance does not mean

When a null hypothesis ("no effect" or "no difference") can be rejected at the usual level $\alpha = 0.05$, there is good evidence that an effect is present. That effect, however, can be extremely small. *When large samples are available, even tiny deviations from the null hypothesis will be significant.*

EXAMPLE 8.10

It's significant but is it important? Suppose that we are testing the hypothesis of no correlation between two variables. With 400 observations, an observed correlation of only $r = 0.1$ is significant evidence at the $\alpha = 0.05$ level that the correlation in the population is not zero. Figure 8.4 is an example of 400 (x, y) pairs that have an observed correlation of 0.10. The low significance level does *not* mean there is a strong association, only that there is strong evidence of some association. The proportion of the variability in one of the variables that is explained by the other is $r^2 = 0.01$, or 1%.

For practical purposes, we might well decide to ignore this association. *Statistical significance is not the same as practical significance.* Statistical significance rarely tells us about the importance of the experimental results. This depends on the context of the experiment.

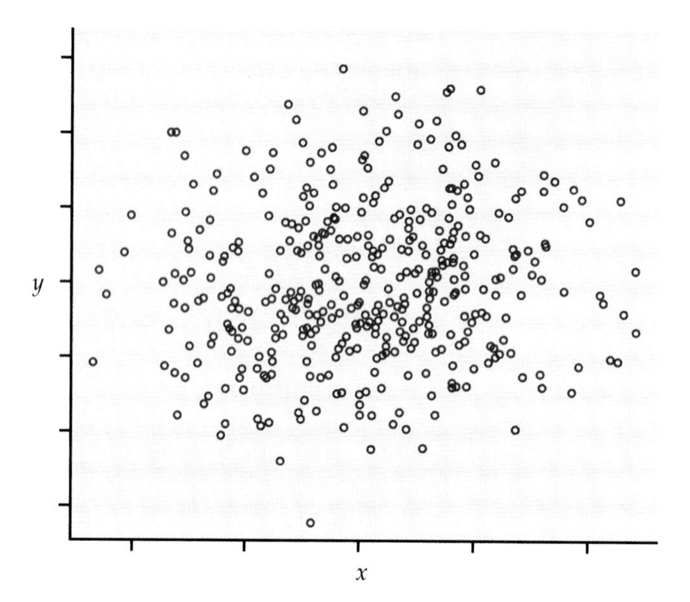

FIGURE 8.4 Scatterplot of $n = 400$ observations with an observed correlation of 0.10. There is not a strong association between the two variables even though there is significant evidence ($P < 0.05$) that the population correlation is not zero.

The remedy for attaching too much importance to statistical significance is to pay attention to the actual experimental results as well as to the P-value. *Always examine your data graphically and numerically before drawing conclusions from statistically significant tests.* Plot your data, as we did in Example 8.10, and examine them carefully. Beware of outliers.

It is usually wise to give a confidence interval for the parameter in which you are interested. Confidence intervals are not used as often as they should be, while tests of significance are perhaps overused.

USE YOUR KNOWLEDGE

8.24 Is it significant? More than 200,000 people worldwide take the Graduate Management Admissions Test (GMAT) each year as they apply for MBA programs. Their scores vary Normally with mean about $\mu = 525$. One hundred students go through a rigorous training program designed to raise their GMAT scores. Test the following hypotheses about the training program

$$H_0 : \mu = 525$$
$$H_a : \mu > 525$$

in each of the following situations:

(a) The students' average score is $\bar{x} = 541.1$ and $s = 98.0$. Is this result significant at the 5% level? Use the 100 degrees of freedom row of Table D in your calculations.

(b) Now suppose that the average score is $\bar{x} = 541.3$ and $s = 98.0$. Is this result significant at the 5% level? Use the 100 degrees of freedom row of Table D in your calculations.

(c) Explain how you would reconcile this difference in significance, especially if any increase greater than 15 points is considered a success.

Don't ignore lack of significance

There is a tendency to conclude that there is no effect whenever a P-value fails to attain the usual 5% standard. A provocative editorial in the *British Medical Journal* entitled "Absence of Evidence Is Not Evidence of Absence" deals with this issue.[10] Here is one of the examples they cite.

EXAMPLE 8.11

Interventions to reduce HIV-1 transmission. A randomized trial of interventions for reducing transmission of HIV-1 reported an incident rate ratio of 1.00, meaning that the intervention group and the control group both had the same rate of HIV-1 infection. The 95% confidence interval was reported as 0.63 to 1.58.[11] The editorial notes that a summary of these results that says the intervention has no effect on HIV-1 infection is misleading. The confidence interval indicates that the intervention may be capable of achieving a 37% decrease in infection. It might also be harmful and produce a 58% increase in infection. Clearly, more data are needed to distinguish between these possibilities.

The situation can be worse. Research in some fields has rarely been published unless significance at the 0.05 level is attained.

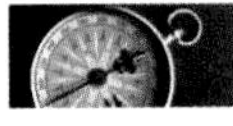

EXAMPLE 8.12

Journal survey of reported significance results. A survey of four journals published by the American Psychological Association showed that of 294 articles using statistical tests, only 8 reported results that did not attain the 5% significance level.[12] It is very unlikely that these were the only 8 studies of scientific merit that did not attain significance at the 0.05 level. Manuscripts describing other studies were likely rejected because of a lack of statistical significance or were never submitted in the first place due to the expectation of rejection. This sort of publication bias is common in many research areas.

In some areas of research, small effects that are detectable only with large sample sizes can be of great practical significance. Data accumulated from a large number of patients taking a new drug may be needed before we can conclude that there are life-threatening consequences for a small number of people.

On the other hand, sometimes a meaningful result is not found significant.

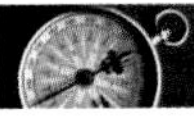

EXAMPLE 8.13

A meaningful but statistically insignificant result. A sample of size 10 gave a correlation of $r = 0.5$ between two variables. The P-value is 0.102 for a two-sided significance test. In many situations, a correlation this large would be interesting and worthy of additional study. When it takes a lot of effort (say, in terms of time or money) to obtain samples, researchers often use small studies like these as pilot projects to gain interest from various funding sources. With financial support, a larger, more powerful study can then be run.

power

Another important aspect of planning a study is to verify that the test you plan to use does have high probability of detecting an effect of the size you hope to find. This probability is the **power** of the test. Power calculations are similar in flavor to determining a sample size that controls the width of a confidence interval and are discussed further on the text Web site.

Statistical inference is not valid for all sets of data

In Chapter 2, we learned that badly designed surveys or experiments often produce invalid results. *Formal statistical inference cannot correct basic flaws in the design.*

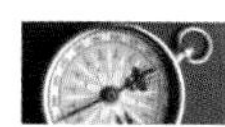

EXAMPLE 8.14

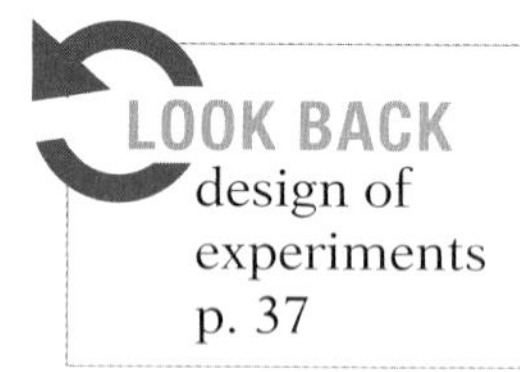
LOOK BACK design of experiments p. 37

English vocabulary and studying a foreign language. There is no doubt that there is a significant difference in English vocabulary scores between high school seniors who have studied a foreign language and those who have not. But because the effect of actually studying a language is confounded with the differences between students who choose language study and those who do not, this statistical significance is hard to interpret. The most plausible explanation is that students who were already good at English chose to study another language. A randomized comparative experiment would isolate the actual effect of language study and so make significance meaningful. However, such an experiment probably could not be done.

Tests of significance and confidence intervals are based on the laws of probability. Randomization in sampling or experimentation ensures that these laws apply. *We must often analyze data that do not arise from randomized samples or experiments. To apply statistical inference to such data, we must have confidence in a probability model for the data.* We can check a probability model by examining the data. If the Normal distribution model appears correct, we can apply the methods of this chapter to do inference about the mean μ.

USE YOUR KNOWLEDGE

8.25 Home security systems. A recent TV advertisement for home security systems said that homes without an alarm system are three times more likely to be broken into. Suppose that this conclusion was obtained by examining an SRS of police records of break-ins and determining whether the percent of homes with alarm systems was significantly smaller than 50%. Explain why the significance of this study is suspect and propose an alternative study that would help clarify the importance of an alarm system.

Beware of searching for significance

Statistical significance is an outcome much desired by researchers. It means (or ought to mean) that you have found an effect that you were looking for. *The reasoning behind statistical significance works well if you decide what effect you are seeking, design an experiment or sample to search for it, and use a test of*

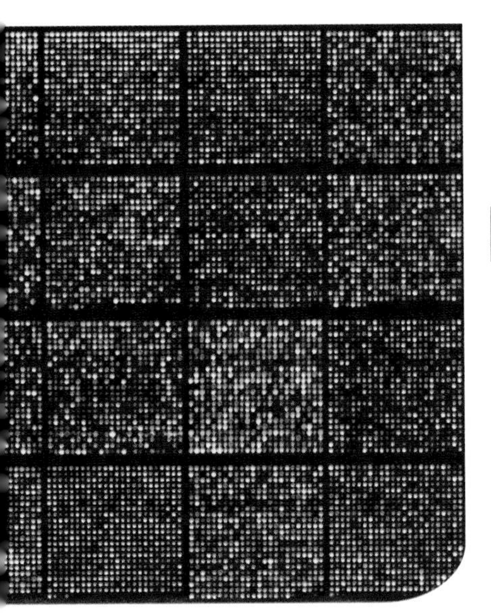

significance to weigh the evidence you get. But because a successful search for a new scientific phenomenon often ends with statistical significance, it is all too tempting to make significance itself the object of the search. There are several ways to do this, none of them acceptable in polite scientific society.

EXAMPLE 8.15

Genomics studies. In genomics experiments, it is common to assess the differences in expression for tens of thousands of genes. If each of these genes was examined separately and statistical significance was declared for all that had P-values that pass the 0.05 standard, we would have quite a mess. In the absence of any real biological effects, we expect that, by chance alone, approximately 5% of these tests will show statistical significance. Much research in genomics is directed toward appropriate ways to deal with this situation.[13]

We do not mean that searching data for suggestive patterns is not proper scientific work. It certainly is. Many important discoveries have been made by accident rather than by design. Exploratory analysis of data is an essential part of statistics.

We do mean that the usual reasoning of statistical inference does not apply when the search for a pattern is successful. *You cannot legitimately test a hypothesis on the same data that first suggested that hypothesis.*

The remedy is clear. Once you have a hypothesis, design a study to search specifically for the effect you now think is there. If the result of this study is statistically significant, you have real evidence.

Type I and Type II errors in a significance test

In a test of significance, we focus on a single hypothesis H_0 and use the sample data to compute the P-value, which measures the strength of evidence against H_0. If the evidence is strong (P-value is small), we reject the null hypothesis and accept the alternative hypothesis H_a. If we cannot reject H_0, we conclude only that there is not sufficient evidence against H_0, not that H_0 is actually true.

With a null hypothesis H_0 that is either actually true or not, our conclusion from the significance test will be either correct or incorrect. There are two types of incorrect conclusions. We could fail to reject H_0 when in fact H_0 is false (H_a is true), or we could reject H_0 when in fact H_0 is true. To help distinguish these two types of error, we give them specific names.

TYPE I AND TYPE II ERRORS

If we reject H_0 (accept H_a) when in fact H_0 is true, this is a **Type I error.**
If we fail to reject H_0 when in fact H_a is true, this is a **Type II error.**

The possibilities are summed up in Figure 8.5. If H_0 is true, our decision either is correct (if we do not reject H_0) or is a Type I error. If H_a is true, our decision either is correct or is a Type II error. Only one error is possible at one time.

		Truth about the population	
		H_0 true	H_a true
Decision based on sample	Reject H_0	Type I error	Correct decision
	Do not reject H_0	Correct decision	Type II error

FIGURE 8.5 The two types of error in a test of significance.

Error probabilities

We can assess the performance of a significance test by determining the probabilities of these two types of error. This is in keeping with the idea that statistical inference is based on probability. Ideally, we'd like both of these probabilities to be zero, but short of sampling the entire population, we cannot guarantee we'll always draw the correct conclusion.

To demonstrate the calculation of these probabilities and their relationships with other test characteristics, we focus on the population proportion p. This significance test uses the standard Normal distribution. Calculations for the population mean μ, which use the t distributions, are discussed on the text Web page.

EXAMPLE 8.16

Outer diameter of a skateboard bearing. The mean outer diameter of a skateboard bearing is supposed to be 22.000 millimeters (mm). If it is too small or too large, the bearing cannot be used. A company that produces these bearings has renovated their production line and wants to make sure the bearings are still being produced properly. Prior to the renovations, 5% of the bearings were unacceptable.

To assess this, they plan to take a random sample of $n = 2000$ bearings off the line and perform a significance test to see if the proportion of unacceptable bearings is different from 0.05. They decide to use a 5% significance level.

This is a test of the hypotheses

$$H_0: p = 0.05$$
$$H_a: p \neq 0.05$$

Because both np_0 and $n(1 - p_0)$ are greater than 10, we can use the large-sample z significance test. The z statistic in this case is

$$z = \frac{\hat{p} - 0.05}{\sqrt{0.05(1 - 0.05)/2000}}$$

and the producer will reject H_0 if

$$z < -1.96 \text{ or } z > 1.96$$

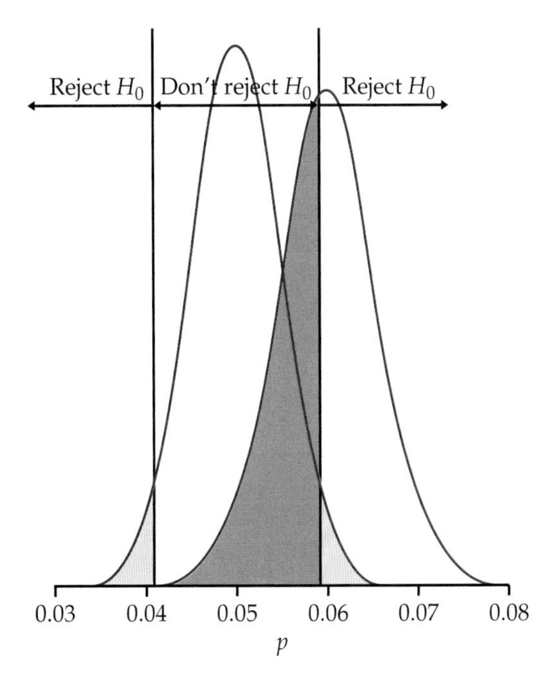

FIGURE 8.6 The two error probabilities for Example 8.16. The probability of a Type I error (yellow area) is the probability of rejecting H_0: $p = 0.05$ when in fact $p = 0.05$. The probability of a Type II error (blue area) is the probability of not rejecting H_0 when in fact $p = 0.06$.

A Type I error is to reject H_0 when in fact $p = 0.05$. What about Type II errors? Because there are many values of p in H_a, we will concentrate on one value. The producer is more concerned about a possible increase in the proportion of unacceptable bearings, so we'll consider $p = 0.06$.

Figure 8.6 shows how the two probabilities of error are obtained from the two sampling distributions of $\hat{p}$, for $p = 0.05$ and for $p = 0.06$. When $p = 0.05$, H_0 is true and to reject H_0 is a Type I error. When $p = 0.06$, not rejecting H_0 is a Type II error. We will now calculate these error probabilities.

The probability of a Type I error is the probability of rejecting H_0 when it is really true. In Example 8.16, this is the probability that $|z| \geq 1.96$ when $p = 0.05$. But this is exactly the significance level of the test. The critical value 1.96 was chosen to make this probability 0.05, so we do not have to compute it again. The definition of "significant at level 0.05" is that sample outcomes this extreme will occur with probability 0.05 when H_0 is true.

SIGNIFICANCE AND TYPE I ERROR

The significance level α of any fixed level test is the probability of a Type I error. That is, α is the probability that the test will reject the null hypothesis H_0 when H_0 is in fact true.

The probability of a Type II error for the particular alternative $p = 0.06$ in Example 8.16 is the probability that the test will fail to reject H_0 when p has this alternative value. The *power* of the test (page 438) against the alternative $p = 0.06$ is just the probability that the test *does* reject H_0.

> **POWER AND TYPE II ERROR**
>
> The power of a fixed level test to detect a particular alternative is 1 minus the probability of a Type II error for that alternative.

It is often easier to consider the probability that the test rejects H_0. Here, the test rejects H_0 when $|z| \geq 1.96$. The test statistic is

$$z = \frac{\hat{p} - 0.05}{\sqrt{0.05(1 - 0.05)/2000}}$$

Some arithmetic shows that the test rejects when either of the following is true:

$$z \geq 1.96 \qquad \text{(in other words, } \hat{p} \geq 0.0596\text{)}$$

$$z \leq -1.96 \qquad \text{(in other words, } \hat{p} \leq 0.0404\text{)}$$

These are disjoint events, so the power is the sum of their probabilities, *computed assuming that the alternative* $p = 0.06$ *is true.* We find that

$$P(\hat{p} \geq 0.0596) = P\left(\frac{\hat{p} - p}{\sqrt{p(1-p)/n}} \geq \frac{0.0596 - 0.06}{\sqrt{0.06(1 - 0.06)/2000}}\right)$$

$$= P(Z \geq -0.08) = 0.5319$$

$$P(\hat{p} \leq 0.0404) = P\left(\frac{\hat{p} - p}{\sqrt{p(1-p)/n}} \leq \frac{0.0404 - 0.06}{\sqrt{0.06(1 - 0.06)/2000}}\right)$$

$$= P(Z \leq -3.68) \doteq 0$$

The power at $p = 0.06$ is about 0.53. The probability of a Type II error is therefore $1 - 0.53$, or 0.47. This means that rejecting the null hypothesis when $p = 0.06$ is comparable to flipping a coin. If it is very important to detect $p = 0.06$, the producer should consider a larger sample size.

SECTION 8.2 SUMMARY

P-values are more informative than the reject-or-not result of a fixed level α test. Beware of placing too much weight on traditional values of α, such as $\alpha = 0.05$.

Very small effects can be highly significant (small P), especially when a test is based on a large sample. A statistically significant effect need not be practically important. Plot the data to display the effect you are seeking, and use confidence intervals to estimate the actual values of parameters.

On the other hand, lack of significance does not imply that H_0 is true, especially when the test has low power.

Significance tests are not always valid. Faulty data collection, outliers in the data, and testing a hypothesis on the same data that suggested the hypothesis can invalidate a test. Many tests run at once will probably produce some significant results by chance alone, even if all the null hypotheses are true.

In a test of significance, the focus is on the hypothesis H_0. A **Type I error** occurs if H_0 is rejected when it is in fact true. A **Type II error** occurs if H_0 is not rejected when in fact H_a is true.

In a fixed level α significance test, the significance level α is the probability of a Type I error, and the power against a specific alternative is 1 minus the probability of a Type II error for that alternative.

SECTION 8.2 EXERCISES

For Exercise 8.24, see page 436; and for Exercise 8.25, see page 438.

8.26. A role as a statistical consultant. You are the statistical expert for a graduate student planning her PhD research. After you carefully present the mechanics of significance testing, she suggests using $\alpha = 0.20$ for the study because she would be more likely to obtain statistically significant results and she *really* needs significant results to graduate. Explain in simple terms why this would not be a good use of statistical methods.

8.27. What do you know? A research report described two results that both achieved statistical significance at the 5% level. The P-value for the first is 0.048; for the second it is 0.0002. Do the P-values add any useful information beyond that conveyed by the statement that both results are statistically significant? Write a short paragraph explaining your views on this question.

8.28. Selective publication based on results. In addition to statistical significance, selective publication can also be due to the observed outcome. A review of 74 FDA-registered studies of antidepressant agents found 38 studies with positive results and 36 studies with negative or questionable results. All but 1 of the 38 positive studies were published. Of the remaining 36, 22 were not published and 11 were published in such a way as to convey a positive outcome.[14] Describe how this selective reporting can have adverse consequences on health care.

8.29. What a test of significance can answer. Explain whether a test of significance can answer each of the following questions.

(a) Is the sample or experiment properly designed?

(b) Is the observed effect compatible with the null hypothesis?

(c) Is the observed effect important?

8.30. Vitamin C and colds. In a study to investigate whether vitamin C will prevent colds, 400 subjects are assigned at random to one of two groups. The experimental group takes a vitamin C tablet daily, while the control group takes a placebo. At the end of the experiment, the researchers calculate the difference between the percents of subjects in the two groups who were free of colds. This difference is statistically significant ($P = 0.03$) in favor of the vitamin C group. Can we conclude that vitamin C has a strong effect in preventing colds? Explain your answer.

8.31. How far do rich parents take us? How much education children get is strongly associated with the wealth and social status of their parents, termed "socioeconomic status," or SES. The SES of parents, however, has little influence on whether children who have graduated from college continue their education. One study looked at whether college graduates took the graduate admission tests for business, law, and other graduate programs. The effects of the parents' SES on taking the LSAT for law school were "both statistically insignificant and small."

(a) What does "statistically insignificant" mean?

(b) Why is it important that the effects were small in size as well as insignificant?

8.32. Do you agree? State whether or not you agree with each of the following statements, and provide a short summary of the reasons for your answers.

(a) If the P-value is larger than 0.05, the null hypothesis is true.

(b) Practical significance is not the same as statistical significance.

(c) We can perform a statistical analysis using any set of data.

(d) If you find an interesting pattern in a set of data, it is appropriate to then use a significance test to determine its significance.

8.33. Practical significance and sample size. Every user of statistics should understand the distinction between statistical significance and practical importance. A sufficiently large sample will declare very small effects statistically significant. Consider a study of elite female Canadian athletes that investigated whether elite athletes are deficient in their nutritional intake.[15] A total of $n = 201$ athletes from eight Canadian sports centers participated in the study. Female athletes were consuming an average of 2403.7 kilocalories per day (kcal/day) with a standard deviation of 880 kcal/day. The recommended amount is 2811.5 kcal/day.

Suppose that a nutritionist is brought in to implement a new health program for these athletes. This program should increase mean caloric intake but not change the standard deviation. Given the standard deviation and how caloric deficient these athletes are, a change in the mean of 50 kcal/day to 2453.7 is of little importance. However, with a large enough sample, this change can be significant. To see this, calculate the P-value for the test of

$$H_0: \mu = 2403.7$$
$$H_a: \mu > 2403.7$$

in each of the following situations:

(a) A sample of 100 athletes; their average caloric intake is $\bar{x} = 2453.7$.

(b) A sample of 500 athletes; their average caloric intake is $\bar{x} = 2453.7$.

(c) A sample of 2500 athletes; their average caloric intake is $\bar{x} = 2453.7$.

8.34. Statistical versus practical significance. A study with 7500 subjects reported a result that was statistically significant at the 5% level. Explain why this result might not be particularly important.

8.35. More on statistical versus practical significance. A study with 14 subjects reported a result that failed to achieve statistical significance at the 5% level. The P-value was 0.051. Write a short summary of how you would interpret these findings.

8.36. CHALLENGE **Find journal articles.** Find two journal articles that report results with statistical analyses. For each article, summarize how the results are reported and write a critique of the presentation. Be sure to include details regarding use of significance testing at a particular level of significance, P-values, and confidence intervals.

8.37. Create an example of your own. For each case, provide an example and explain why it is appropriate.

(a) A set of data or experiment for which statistical inference is not valid.

(b) A set of data or experiment for which statistical inference is valid.

8.38. CHALLENGE **Predicting success of trainees.** What distinguishes managerial trainees who eventually become executives from those who, after expensive training, don't succeed and leave the company? We have abundant data on past trainees—data on their personalities and goals, their college preparation and performance, even their family backgrounds and their hobbies. Statistical software makes it easy to perform dozens of significance tests on these dozens of variables to see which ones best predict later success. We find that future executives are significantly more likely than washouts to have an urban or suburban upbringing and an undergraduate degree in a technical field.

Explain clearly why using these "significant" variables to select future trainees is not wise. Then suggest a follow-up study using this year's trainees as subjects that should clarify the importance of the variables identified by the first study.

8.39. Searching for significance. Give an example of a situation where searching for significance would lead to misleading conclusions.

8.40. More on searching for significance. You perform 1000 significance tests using $\alpha = 0.05$. Assuming that all null hypotheses are true, about how many of the test results would you expect to be statistically significant? Explain how you obtained your answer.

8.41. Interpreting a very small P-value. Assume that you are performing a large number of significance tests. Let n be the number of these tests. How large would n need to be for you to expect about one P-value to be 0.00001 or smaller? Use this information to write an explanation of how to interpret a result that has $P = 0.00001$ in this setting.

8.42. Meaning of "statistically significant." When asked to explain the meaning of "statistically significant at the $\alpha = 0.01$ level," a student says, "This means there is only probability 0.01 that the null hypothesis is true." Is this an essentially correct explanation of statistical significance? Explain your answer.

8.43. More on the meaning of "statistically significant." Another student, when asked why statistical significance appears so often in research reports, says, "Because saying that results are significant tells us that they cannot easily be explained by chance variation alone." Do you think that this statement is essentially correct? Explain your answer.

8.44. Make a recommendation. Your manager has asked you to review a research proposal that includes a section on sample size justification. A careful reading of this section indicates that the power is 25% for detecting an effect that you would consider important. Write a short report for your manager explaining what this means, and make a recommendation on whether or not this study should be run.

8.45. Explain power and sample size. Two studies are identical in all respects except for the sample sizes. Consider the power versus a particular sample size. Will the study with the larger sample size have more power or less power than the one with the smaller sample size? Explain your answer in terms that could be understood by someone with very little knowledge of statistics.

CHAPTER 8 EXERCISES

8.46. Telemarketing wages. An advertisement in the student newspaper asks you to consider working for a telemarketing company. The ad states, "Earn between $500 and $1000 per week." Do you think that the ad is describing a confidence interval? Explain your answer.

8.47. Exercise and statistics exams. A study examined whether exercise affects how students perform on their final exam in statistics. The *P*-value was given as 0.38.

(a) State null and alternative hypotheses that could be used for this study. (*Note:* There is more than one correct answer.)

(b) Do you reject the null hypothesis? State your conclusion in plain language.

(c) What other facts about the study would you like to know for a proper interpretation of the results?

8.48. CHALLENGE **Stress by occupation.** As part of a study on the impact of job stress on smoking, researchers used data from the Health and Retirement Study (HRS) to collect information on 3825 "ever-smoker" individuals who were 50 to 64 years of age.[16] An ever-smoker is someone who was a smoker at some time in his or her life. The HRS survey provided the researchers with 17,043 "person-year" observations. One of the questions on the survey asked a participant how much he or she agrees or disagrees with the statement "My job involves a lot of stress." The answers were coded as a 1 if a participant "strongly agreed" and 0 otherwise. The following table summarizes these responses by occupation.

Occupation	$\hat{p}$	n
Professional	0.23	2447
Managerial	0.22	2552
Administrative	0.17	2309
Sales	0.15	1811
Service	0.13	2592
Mechanical	0.12	1979
Operator	0.12	2782
Farm	0.08	571

(a) For each occupation, construct a 95% confidence interval for the proportion who feel a lot of stress.

(b) Summarize the results. Do there appear to be certain groups of occupations with similar stress levels?

8.49. Blood phosphorus level in dialysis patients. Patients with chronic kidney failure may be treated by dialysis, in which a machine removes toxic wastes from the blood, a function normally performed by the kidneys. Kidney failure and dialysis can cause retention of phosphorus, which must be corrected by changes in diet. A study of the nutrition of dialysis patients measured the level of phosphorus in the blood of several patients on six occasions. Here are the data for one patient (in milligrams of phosphorus per deciliter of blood):[17]

5.4 5.2 4.5 4.9 5.7 6.3

The measurements are separated in time and can be considered an SRS of the patient's blood phosphorus level. Assume that this level varies Normally. DATA FILE PHOSPHORUS

(a) Give a 95% confidence interval for the mean blood phosphorus level.

(b) What sample size would be needed for a margin of error of 0.25 mg/dl for 95% confidence?

8.50. CHALLENGE **Where do you buy?** Consumers can purchase nonprescription medications at food stores, mass merchandise stores such as Kmart and Walmart, or pharmacies. About 45% of consumers make such purchases at pharmacies. What accounts for the popularity of pharmacies, which often charge higher prices?

A study examined consumers' perceptions of overall performance of the three types of store, using a long questionnaire that asked about such things as "neat and attractive store," "knowledgeable staff," and "assistance in choosing among various types of nonprescription medication." A performance score was based on 27 such questions. The subjects were 201 people chosen at random from the Indianapolis telephone directory. Here are the means and standard deviations of the performance scores for the sample:[18]

Store type	$\bar{x}$	s
Food stores	18.67	24.95
Mass merchandisers	32.38	33.37
Pharmacies	48.60	35.62

(a) What population do you think the authors of the study want to draw conclusions about? What population are you certain they can draw conclusions about?

(b) Give a 95% confidence interval for the mean performance for each type of store.

(c) Based on these confidence intervals, are you convinced that consumers think that pharmacies offer higher performance than the other types of store? (In Chapter 12, we will study a statistical method for comparing the means of several groups.)

8.51. CHALLENGE **More than one confidence interval.** As we prepare to take a sample and compute a 95% confidence interval, we know that the probability that the interval we compute will cover the parameter is 0.95. That's the meaning of 95% confidence. If we use several such intervals, however, our confidence that *all* of them give correct results is less than 95%. Suppose that we take independent samples each month for five months and report a 95% confidence interval for each set of data.

(a) What is the probability that all five intervals cover the true means? This probability (expressed as a percent) is our overall confidence level for the five simultaneous statements.

(b) What is the probability that at least four of the five intervals cover the true means?

8.52. Is there interest in a new product? One of your employees has suggested that your company develop a new product. You decide to take a random sample of your customers and ask whether or not there is interest in the new product. The response is on a 1 to 5 scale with 1 indicating "definitely would not purchase"; 2, "probably would not purchase"; 3, "not sure"; 4, "probably would purchase"; and 5, "definitely would purchase." For an initial analysis, you will record the responses 1, 2, and 3 as "No" and 4 and 5 as "Yes." What sample size would you use if you wanted the 95% margin of error to be 0.15 or less?

8.53. More information is needed. Refer to the previous exercise. Suppose that after reviewing the results of the previous survey, you proceeded with preliminary development of the product. Now you are at the stage where you need to decide whether or not to make a major investment to produce and market it. You will use another random sample of your customers, but now you want the margin of error to be smaller. What sample size would you use if you wanted the 95% margin of error to be 0.075 or less?

8.54. Sample size needed for an evaluation. You are planning an evaluation of a semester-long alcohol awareness campaign at your college. Previous evaluations indicate that about 25% of the students surveyed will respond "Yes" to the question "Did the campaign alter your behavior toward alcohol consumption?" How large a sample of students should you take if you want the margin of error for 95% confidence to be about 0.1?

8.55. CHALLENGE **Sample size needed for an evaluation, continued.** The evaluation in the previous exercise will also have questions that have not been asked before, so you do not have previous information about the possible value of p. Repeat the calculation above for the following values of p^*: 0.1, 0.2, 0.3, 0.4, 0.5, 0.6, 0.7, 0.8, and 0.9. Summarize the results in a table and graphically. What sample size will you use?

8.56. Are the customers dissatisfied? An automobile manufacturer would like to know what proportion of its customers are dissatisfied with the service received from their local dealer. The customer relations department will survey a random sample of customers and compute a 95% confidence interval for the proportion that are dissatisfied. From past studies, they believe that this proportion will be about 0.15. Find the sample size needed if the margin of error of the confidence interval is to be no more than 0.02.

8.57. Required sample size for specified margin of error. A new bone study is being planned that will measure the biomarker tartrate-resistant acid phosphatase, or TRAP. In a study of bone turnover in young women, serum TRAP was measured in 31 subjects.[19] The units are units per liter (U/l). The mean was 13.2 U/l with a standard deviation of 6.5 U/l. Based on these values, find the sample size required to provide an estimate of the mean TRAP with a margin of error of 1.5 U/l for 95% confidence.

8.58. CHALLENGE **Adjusting required sample size for dropout.** Refer to the previous exercise. In similar previous studies, about 20% of the subjects drop out before the study is completed. Adjust your sample size requirement to have enough subjects at the end of the study to meet the margin of error criterion.

8.59. CHALLENGE **Calculating sample sizes for the two-sample problem.** For a single proportion the margin of error of a confidence interval is largest for any given sample size n and confidence level C when $\hat{p} = 0.5$. This led us to use $p^* = 0.5$ for planning purposes. The same kind of result is true for the two-sample problem. The margin of error of the confidence interval for the difference between two proportions is largest when $\hat{p}_1 = \hat{p}_2 = 0.5$. You are planning a survey and will calculate a 95% confidence interval for the difference in two proportions when the data are collected. You would like the margin of error of the interval to be less than or equal to 0.075. You will use the same sample size n for both populations.

(a) How large a value of n is needed?

(b) Give a general formula for n in terms of the desired margin of error m and the critical value z^*.

8.60. CHALLENGE **Sample size and margin of error.** In Section 8.1, we studied the effect of the sample size on the margin of error of the confidence interval for a single proportion. In this exercise we perform some calculations to observe this effect for the two-sample problem. Suppose that $\hat{p}_1 = 0.7$ and $\hat{p}_2 = 0.6$, and n represents the common value of n_1 and n_2. Compute the 95% margins of error for the difference in the two proportions for n = 40, 50, 80, 100, 400, 500, and 1000. Present the results in a table and with a graph. Write a short summary of your findings.

8.61. CHALLENGE **Planning a study to compare two means.** Refer to Exercise 7.60 (page 413), where we compared trustworthiness ratings for ads from two different publications. Suppose that you are planning a similar study using two different publications that are not expected to show the differences seen when comparing the *Wall Street Journal* with the *National Enquirer*. If we assume the same standard deviation and sample size for these two publications, the margin of error can be expressed as

$$t^* s\sqrt{2/n}$$

with the degrees of freedom of $t^* = n - 1$. Suppose that you want the margin of error for 95% confidence to be no more than 0.50 points. What sample size is required?

8.62. Power for a different alternative. The power for a two-sided test of the null hypothesis $p = 0.30$ versus the alternative $p = 0.40$ is 0.82. What is the power versus the alternative $p = 0.20$? Explain your answer.

8.63. More on the power for a different alternative. A one-sided test of the null hypothesis $p = 0.40$ versus the alternative $p = 0.55$ has power equal to 0.6. Will the power for the alternative $p = 0.50$ be higher or lower than 0.6? Draw a picture and use this to explain your answer.

PART TWO REVIEW

PROBABILITY AND INFERENCE

In Part II, we were introduced to statistical inference, the process of inferring conclusions about a population or process from sample data. Because statistical inference uses the language of probability to describe how reliable the conclusions are, we began this study (Chapter 5) with a discussion of randomness and the rules of probability that describe the random behavior of a process. These rules of probability were then used in the next two chapters to perform the two most common types of statistical inference, confidence intervals and tests of significance.

In Chapter 6, the focus was on inference about a single proportion and the comparison of two proportions. In Chapter 7, the focus was on inference about a single mean and the comparison of two means. In all settings, a sampling distribution was developed to describe the pattern of the statistic under multiple samples of the same size. This distribution, or an approximation of it based on the central limit theorem, was then used to construct a confidence interval or perform a test of significance. Due to this similarity in approach across settings, we concluded this part with a general discussion of the properties of these inference types and some cautions to consider when using them (Chapter 8).

Chapter 5 Probability: The Study of Randomness

- A **random phenomenon** has outcomes that we cannot predict but that nonetheless have a regular distribution in very many repetitions. (page 237)
- The **probability** of an event is the proportion of times the event occurs in many repeated trials of a random phenomenon. (page 237)
- A **probability model** for a random phenomenon consists of a sample space S and an assignment of probabilities P. (page 241)
- The **sample space S** is the set of all possible outcomes of the random phenomenon. Sets of outcomes are called **events.** P assigns a number $P(A)$ to an event A as its probability. (pages 241 and 243)
- The **complement A^c** of an event A consists of exactly the outcomes that are not in A. Events A and B are **disjoint** if they have no outcomes in common. Events A and B are **independent** if knowing that one event occurs does not change the probability we would assign to the other event. (pages 244–245 and 249)
- Any **assignment of probability** must obey the rules that state the basic properties of probability: (pages 244 and 250)
  - **Rule 1.** $0 \leq P(A) \leq 1$ for any event A.
  - **Rule 2.** $P(S) = 1$.
  - **Rule 3. Addition rule:** If events A and B are **disjoint,** then $P(A \text{ or } B) = P(A) + P(B)$.
  - **Rule 4. Complement rule:** For any event A, $P(A^c) = 1 - P(A)$.
  - **Rule 5. Multiplication rule:** If events A and B are **independent,** then $P(A \text{ and } B) = P(A)P(B)$.
- A **random variable** is a variable taking numerical values determined by the outcome of a random phenomenon. The **probability distribution** of a random variable X tells us what the possible values of X are and how probabilities are assigned to those values. (page 254)
- A random variable X and its distribution can be **discrete** or **continuous.** (pages 257 and 260)
- A **discrete random variable** has finitely many possible values. The probability distribution assigns each of these values a probability between 0 and 1 such that the sum of all the probabilities is exactly 1. The probability of any event is the sum of the probabilities of all the values that make up the event. (page 257)
- A **continuous random variable** takes all values in some interval of numbers. A **density curve** describes the probability distribution of a continuous random variable. The probability of any event is the area under the curve and above the values that make up the event. (page 262) **Normal distributions** are one type of continuous probability distribution. (page 263)
- You can picture a probability distribution by drawing a **probability histogram** in the discrete case or by graphing the density curve in the continuous case. (page 258)
- The probability distribution of a random variable X, like a distribution of data, has a **mean μ_X** and a **standard deviation σ_X.** (pages 267 and 275)

- The **law of large numbers** says that the average of the values of X observed in many trials must approach μ. (page 271)
- The **mean** μ is the balance point of the probability histogram or density curve (page 268). If X is discrete with possible values x_i having probabilities p_i, the mean is the average of the values of X, each weighted by its probability:

$$\mu_X = x_1 p_1 + x_2 p_2 + \cdots + x_k p_k$$

The **variance** σ_X^2 is the average squared deviation of the values of the variable from their mean (page 275). For a discrete random variable,

$$\sigma_X^2 = (x_1 - \mu)^2 p_1 + (x_2 - \mu)^2 p_2 + \cdots + (x_k - \mu)^2 p_k$$

- The **standard deviation** σ_X is the square root of the variance. The standard deviation measures the variability of the distribution about the mean. It is easiest to interpret for Normal distributions. (page 276)
- The mean and variance of a continuous random variable can be computed from the density curve, but to do so requires more advanced mathematics. (page 270)
- The means and variances of random variables obey the following rules. If a and b are fixed numbers, then

$$\mu_{a+bX} = a + b\mu_X$$
$$\sigma_{a+bX}^2 = b^2 \sigma_X^2$$

If X and Y are any two random variables having correlation ρ, then

$$\mu_{X+Y} = \mu_X + \mu_Y$$
$$\sigma_{X+Y}^2 = \sigma_X^2 + \sigma_Y^2 + 2\rho\sigma_X\sigma_Y$$
$$\sigma_{X-Y}^2 = \sigma_X^2 + \sigma_Y^2 - 2\rho\sigma_X\sigma_Y$$

If X and Y are **independent,** then $\rho = 0$. In this case,

$$\sigma_{X+Y}^2 = \sigma_X^2 + \sigma_Y^2$$
$$\sigma_{X-Y}^2 = \sigma_X^2 + \sigma_Y^2$$

(page 277) To find the standard deviation, take the square root of the variance. (page 278)

- A **count** X of successes has the **binomial distribution** $B(n, p)$ in the **binomial setting:** there are n trials, all independent, each resulting in a success or a failure, and each having the same probability p of a success. (page 285) **Binomial probabilities** are most easily found by software. There is an exact formula that is practical for calculations when n is small. Table C contains binomial probabilities for some values of n and p. (page 287)
- The binomial distribution $B(n, p)$ is a good approximation to the **sampling distribution of the count of successes** in an SRS of size n from a large population containing proportion p of successes. We will use this approximation when the population is at least 10 times larger than the sample. (page 287)
- The mean and standard deviation of a **binomial count** X are

$$\mu_X = np$$
$$\sigma_X = \sqrt{np(1-p)}$$

The **binomial probability formula** is

$$P(X = k) = \binom{n}{k} p^k (1 - p)^{n-k}$$

where the possible values of X are $k = 0, 1, \ldots, n$. The binomial probability formula uses the **binomial coefficient**

$$\binom{n}{k} = \frac{n!}{k!\,(n-k)!}$$

Here the **factorial** $n!$ is

$$n! = n \times (n-1) \times (n-2) \times \cdots \times 3 \times 2 \times 1$$

for positive whole numbers n, and $0! = 1$. The binomial coefficient counts the number of ways of distributing k successes among n trials. (page 293)

- The **complement** A^c of an event A contains all outcomes that are not in A. The **union** $\{A \text{ or } B\}$ of events A and B contains all outcomes in A, in B, or in both A and B. The **intersection** $\{A \text{ and } B\}$ contains all outcomes that are in both A and B, but not outcomes in A alone or B alone. (page 297)
- The **conditional probability** $P(B \mid A)$ of an event B given an event A is defined by

$$P(B \mid A) = \frac{P(A \text{ and } B)}{P(A)}$$

when $P(A) > 0$. In practice, conditional probabilities are most often found from directly available information. (page 303)

- The essential general rules of elementary probability are (page 297)
  - Legitimate values: $0 \leq P(A) \leq 1$ for any event A
  - Total probability 1: $P(S) = 1$
  - Complement rule: $P(A^c) = 1 - P(A)$
  - Addition rule: $P(A \text{ or } B) = P(A) + P(B) - P(A \text{ and } B)$
  - Multiplication rule: $P(A \text{ and } B) = P(A)P(B \mid A)$
- If A and B are **disjoint,** then $P(A \text{ and } B) = 0$. The general addition rule for unions then becomes the special addition rule, $P(A \text{ or } B) = P(A) + P(B)$. (page 298)
- A and B are **independent** when $P(B \mid A) = P(B)$. The multiplication rule for intersections then becomes $P(A \text{ and } B) = P(A)P(B)$. (page 308)
- In problems with several stages, draw a **tree diagram** to organize use of the multiplication and addition rules. (page 305)

Chapter 6 Inference for Proportions

- A number that describes a population is a **parameter.** A number that can be computed from the data is a **statistic.** The purpose of sampling or experimentation is usually **inference:** using sample statistics to make statements about unknown population parameters. (page 316)
- A statistic from a probability sample or randomized experiment has a **sampling distribution** that describes how the statistic varies in repeated data production. The sampling distribution answers the question, "What would happen if we repeated the sample or experiment many times?"

Formal statistical inference is based on the sampling distributions of statistics. (page 318)

- Inference about a population proportion p from an SRS of size n is based on the **sample proportion** $\hat{p} = X/n$. When n is large, $\hat{p}$ has approximately the Normal distribution with mean p and standard deviation $\sqrt{p(1-p)/n}$. (page 324)
- The purpose of a **confidence interval** is to estimate an unknown parameter with an indication of how accurate the estimate is and of how confident we are that the result is correct. (page 326)
- Any confidence interval has two parts: an interval computed from the data and a confidence level (page 326). The interval often has the form

$$\text{estimate} \pm \text{margin of error}$$

- The **confidence level** states the probability that the method will give a correct answer. That is, if you use 95% confidence intervals, in the long run 95% of your intervals will contain the true parameter value. When you apply the method once, you do not know whether your interval gave a correct value (this happens 95% of the time) or not (this happens 5% of the time). (page 328)
- For large samples, the **margin of error for confidence level *C*** of a proportion is

$$m = z^* \text{SE}_{\hat{p}}$$

where the critical value z^* is the value for the standard Normal density curve with area C between $-z^*$ and z^*, and the **standard error of $\hat{p}$** is

$$\text{SE}_{\hat{p}} = \sqrt{\frac{\hat{p}(1-\hat{p})}{n}}$$

(page 329)

- The **level *C* large-sample confidence interval** is

$$\hat{p} \pm m$$

where m is the margin of error defined in the previous bullet. We recommend using this interval for 90%, 95%, and 99% confidence whenever the number of successes and the number of failures are both at least 15. (page 329)

- A **test of significance** is intended to assess the evidence provided by data against a **null hypothesis H_0** in favor of an **alternative hypothesis H_a.** (page 333)
- The hypotheses are stated in terms of population parameters. Usually H_0 is a statement that no effect or no difference is present, and H_a says that there is an effect or difference, in a specific direction (**one-sided alternative**) or in either direction (**two-sided alternative**). (page 333)
- The test is based on a **test statistic.** The ***P*-value** is the probability, computed assuming that H_0 is true, that the test statistic will take a value at least as extreme as that actually observed. Small P-values indicate strong evidence against H_0. Calculating P-values requires

knowledge of the sampling distribution of the test statistic when H_0 is true. (page 335)

- If the P-value is as small or smaller than a specified value α, the data are **statistically significant** at significance level α. (page 336)
- Tests of H_0: $p = p_0$ are based on the **z statistic**

$$z = \frac{\hat{p} - p_0}{\sqrt{\dfrac{p_0(1 - p_0)}{n}}}$$

with P-values calculated from the $N(0, 1)$ distribution. Use this procedure when the expected number of successes, np_0, and the expected number of failures, $n(1 - p_0)$, are both greater than 10. (page 337)

- The **large-sample estimate of the difference in two population proportions** is

$$D = \hat{p}_1 - \hat{p}_2$$

where $\hat{p}_1$ and $\hat{p}_2$ are the sample proportions

$$\hat{p}_1 = \frac{X_1}{n_1} \quad \text{and} \quad \hat{p}_2 = \frac{X_2}{n_2}$$

(page 344)

- The **standard error of the difference D** is

$$\text{SE}_D = \sqrt{\frac{\hat{p}_1(1 - \hat{p}_1)}{n_1} + \frac{\hat{p}_2(1 - \hat{p}_2)}{n_2}}$$

(page 345)

- The **margin of error for confidence level C** is

$$m = z^* \text{SE}_D$$

where z^* is the value for the standard Normal density curve with area C between $-z^*$ and z^*. (page 346)

- The **large-sample level C confidence interval** is

$$D \pm m$$

with D and m as defined in the previous bullets. We recommend using this interval for 90%, 95%, or 99% confidence when the number of successes and the number of failures in both samples are all at least 10. (page 346)

- Significance tests of H_0: $p_1 = p_2$ use the **z statistic**

$$z = \frac{\hat{p}_1 - \hat{p}_2}{\text{SE}_{Dp}}$$

with P-values from the $N(0, 1)$ distribution (page 349). In this statistic,

$$\text{SE}_{Dp} = \sqrt{\hat{p}(1 - \hat{p})\left(\frac{1}{n_1} + \frac{1}{n_2}\right)}$$

and $\hat{p}$ is the **pooled estimate** of the common value of p_1 and p_2:

$$\hat{p} = \frac{X_1 + X_2}{n_1 + n_2}$$

Chapter 7 Inference for Means

- Significance tests and confidence intervals for the mean μ of a Normal population are based on the sample mean $\bar{x}$ of an SRS. Because of the **central limit theorem,** the resulting procedures are approximately correct for other population distributions when the sample is large. (page 365)
- The standardized sample mean, or **one-sample z statistic,**

$$z = \frac{\bar{x} - \mu}{\sigma/\sqrt{n}}$$

has the $N(0, 1)$ distribution. If the standard deviation $\sigma/\sqrt{n}$ of $\bar{x}$ is replaced by the **standard error** $s/\sqrt{n}$, the **one-sample t statistic**

$$t = \frac{\bar{x} - \mu}{s/\sqrt{n}}$$

has the ***t* distribution** with $n - 1$ degrees of freedom. (page 371)

- There is a t distribution for every positive **degrees of freedom k.** All are symmetric distributions similar in shape to Normal distributions. The $t(k)$ distribution approaches the $N(0, 1)$ distribution as k increases. (page 372)
- A level C **confidence interval for the mean** μ of a Normal population is

$$\bar{x} \pm t^* \frac{s}{\sqrt{n}}$$

where t^* is the value for the $t(n-1)$ density curve with area C between $-t^*$ and t^*. The quantity

$$t^* \frac{s}{\sqrt{n}}$$

is the **margin of error.** (page 373)

- Significance tests for H_0: $\mu = \mu_0$ are based on the t statistic. P-values or fixed significance levels are computed from the $t(n-1)$ distribution. (page 374)
- A common comparative design is a **matched pairs** study. With this design, subjects are first matched in pairs, and then within each pair, one subject is randomly assigned Treatment 1 and the other is assigned Treatment 2. Alternatively, we can make two measurements or observations on each subject. One-sample procedures are used to analyze matched pairs data by first taking the differences within the matched pairs to produce a single sample of differences. (page 380)
- The t procedures are relatively **robust** against non-Normal populations. The t procedures are useful for non-Normal data when $15 \leq n < 40$ unless the data show outliers or strong skewness. When $n \geq 40$, the t procedures can be used even for clearly skewed distributions. (page 384)

- Significance tests and confidence intervals for the difference between the means μ_1 and μ_2 of two Normal populations are based on the difference $\bar{x}_1 - \bar{x}_2$ of the sample means from two independent SRSs. Because of the **central limit theorem,** the resulting procedures are approximately correct for other population distributions when the sample sizes are large. (page 393)
- When independent SRSs of sizes n_1 and n_2 are drawn from two Normal populations with parameters μ_1, σ_1 and μ_2, σ_2 the **two-sample z statistic**

$$z = \frac{(\bar{x}_1 - \bar{x}_2) - (\mu_1 - \mu_2)}{\sqrt{\frac{\sigma_1^2}{n_1} + \frac{\sigma_2^2}{n_2}}}$$

has the $N(0, 1)$ distribution. (page 393)
- The **two-sample t statistic**

$$t = \frac{(\bar{x}_1 - \bar{x}_2) - (\mu_1 - \mu_2)}{\sqrt{\frac{s_1^2}{n_1} + \frac{s_2^2}{n_2}}}$$

does *not* have a t distribution. (page 393)
- **Conservative inference procedures** for comparing μ_1 and μ_2 are obtained from the two-sample t statistic by using the $t(k)$ distribution with degrees of freedom k equal to the smaller of $n_1 - 1$ and $n_2 - 1$. (page 394)
- **More accurate probability values** can be obtained by estimating the degrees of freedom from the data. This is the usual procedure for statistical software. (page 394)
- An approximate level C **confidence interval** for $\mu_1 - \mu_2$ is given by

$$(\bar{x}_1 - \bar{x}_2) \pm t^* \sqrt{\frac{s_1^2}{n_1} + \frac{s_2^2}{n_2}}$$

Here, t^* is the value for the $t(k)$ density curve with area C between $-t^*$ and t^*, where k is computed from the data by software or is the smaller of $n_1 - 1$ and $n_2 - 1$. The quantity

$$t^* \sqrt{\frac{s_1^2}{n_1} + \frac{s_2^2}{n_2}}$$

is the **margin of error.** (page 397)
- Significance tests for H_0: $\mu_1 = \mu_2$ use the **two-sample t statistic**

$$t = \frac{\bar{x}_1 - \bar{x}_2}{\sqrt{\frac{s_1^2}{n_1} + \frac{s_2^2}{n_2}}}$$

The P-value is approximated using the $t(k)$ distribution where k is estimated from the data using software or is the smaller of $n_1 - 1$ and $n_2 - 1$. (page 394)

- The guidelines for practical use of two-sample t procedures are similar to those for one-sample t procedures. Equal sample sizes are recommended. (page 397)
- If we can assume that the two populations have equal variances, **pooled two-sample t procedures** can be used. These are based on the **pooled estimator of the unknown common variance σ^2:**

$$s_p^2 = \frac{(n_1 - 1)s_1^2 + (n_2 - 1)s_2^2}{n_1 + n_2 - 2}$$

and the $t(n_1 + n_2 - 2)$ distribution. (page 404)

Chapter 8 Delving into the Use of Inference

- The purpose of a **confidence interval** is to estimate an unknown parameter with an indication of how accurate the estimate is and of how confident we are that the result is correct. (page 420)
- Other things being equal, the margin of error of a confidence interval decreases as
  - the confidence level C decreases (page 420), and
  - the sample size n increases (page 421).
- The **sample size** required to obtain a confidence interval with approximate margin of error m for a proportion is found from

$$n = \left(\frac{z^*}{m}\right)^2 p^*(1 - p^*)$$

where p^* is a guessed value for the proportion, and z^* is the standard Normal critical value for the desired level of confidence. (page 423)
- To ensure that the margin of error of the interval is less than or equal to m no matter what $\hat{p}$ may be, use

$$n = \frac{1}{4}\left(\frac{z^*}{m}\right)^2$$

(page 423)
- The sample size n required to obtain a confidence interval with an expected margin of error no larger than m for a population mean satisfies the constraint

$$m \geq t^* s^* / \sqrt{n}$$

where t^* is the critical value for the desired level of confidence with $n - 1$ degrees of freedom, and s^* is the guessed value for the population standard deviation. (page 426)
- A specific confidence interval recipe is correct only under specific conditions. The most important conditions concern the method used to produce the data. Other factors such as the form of the population distribution may also be important. (page 429)
- P-values are more informative than the reject-or-not result of a fixed level α test. Beware of placing too much weight on traditional values of α, such as $\alpha = 0.05$. (page 434)

- Very small effects can be highly significant (small P), especially when a test is based on a large sample. A statistically significant effect need not be practically important. Plot the data to display the effect you are seeking, and use confidence intervals to estimate the actual values of parameters. (page 435)
- On the other hand, lack of significance does not imply that H_0 is true, especially when the test has low power. (page 437)
- Significance tests are not always valid. Faulty data collection, outliers in the data, and testing a hypothesis on the same data that suggested the hypothesis can invalidate a test. Many tests run at once will probably produce some significant results by chance alone, even if all the null hypotheses are true. (page 438)
- In a test of significance, the focus is on the hypothesis H_0. A **Type I error** occurs if H_0 is rejected when it is in fact true. A **Type II error** occurs if H_0 is not rejected when in fact H_a is true. (page 439)
- In a fixed level α significance test, the significance level α is the probability of a Type I error, and the power against a specific alternative is 1 minus the probability of a Type II error for that alternative. (page 442)

PART II EXERCISES

II.1. Canadian provinces and territories. Here is a list of the 13 provinces and territories in Canada with their populations according to the 2011 census:[1]

Province or territory	Population (thousands)
Ontario	13,373.0
Quebec	7,979.7
British Columbia	4,573.3
Alberta	3,779.4
Manitoba	1,250.6
Saskatchewan	1,057.9
Nova Scotia	945.4
New Brunswick	755.5
Newfoundland and Labrador	510.6
Prince Edward Island	145.9
Northwest Territories	43.7
Yukon	34.7
Nunavut	33.3

Consider selecting a Canadian province or territory at random, where each province or territory is equally likely to be picked.

(a) Describe the sample space.

(b) Give the selection probabilities.

(c) What is the probability that Ontario is not selected?

(d) The three territories in Canada are Northwest Territories, Nunavut, and Yukon. What is the probability that a territory is selected?

(e) What events did you need to consider when you answered parts (c) and (d) of this exercise?

II.2. A random variable for Canadian provinces and territories. Suppose that you select a Canadian province or territory at random and record X, the population (in thousands).

(a) Is X a random variable? Explain your answer.

(b) Give the probability distribution of X.

(c) Explain how you would calculate the mean of X, then calculate it.

(d) Explain how you would calculate the variance and the standard deviation of X. You do not need to do the calculations, but describe the needed calculations in detail.

II.3. A random variable. Suppose that the random variable X has mean 10 and variance 25.

(a) Find the mean and standard deviation of $2X$.

(b) Find the mean and standard deviation of $-2X$.

II.4. Two random variables. Refer to the previous exercise. Suppose that the random variable Y has mean 20 and variance 36.

(a) Find the mean and standard deviation of $2X + Y$, assuming that the correlation between X and Y is 0.

(b) Repeat part (b), assuming that the correlation between X and Y is 0.5.

II.5. Facebook friends. You can become a Facebook friend with someone in two different ways. Your friend can request the friendship, or you can request the friendship. Explain how the way that you became friends with someone can be viewed in terms of the binomial setting. Be sure to explain the meaning of X, n, and p.

II.6. Compute some probabilities for Facebook friends. Suppose that 60% of your Facebook friends became your friends because you asked them to be. You select a simple random sample of 20 of your friends.

(a) What is the probability that 12 of them became your friend because you asked them?

(b) What is the probability that between 9 and 15 of them became your friends because you asked them? Include the possibilities of 9 and 15 in your calculations.

(c) What are the mean and the standard deviation of the number who became your friends because you asked them?

II.7. Find some probabilities. Refer to Exercise II.1.

(a) Find the probability that Nunavut is selected given that a territory is selected.

(b) Are the events "province is selected" and "a territory is selected" independent? Explain why or why not.

(c) Are the events "a province is selected" and "a territory is selected" disjoint? Explain why or why not.

II.8. Facebook use in college. Because of Facebook's rapid rise in popularity among college students, there is a great deal of interest in the relationship between Facebook use and academic performance. One study collected information on $n = 1839$ undergraduate students to look at the relationships among frequency of Facebook use, participation in Facebook activities, time spent preparing for class, and overall GPA.[2] In this study, 8% of the students reported spending no time on Facebook.

(a) Find the sample proportion and the 95% margin of error for the proportion of undergraduates who spend no time on Facebook.

(b) Convert your estimate and margin of error to percents and interpret your findings.

(c) The authors state:

> *All students surveyed were US residents admitted through the regular admissions process at a 4-year, public, primarily residential institution in the northeastern United States (N = 3866). Students were sent a link to a survey hosted on SurveyMonkey.com, a survey-hosting website, through their university-sponsored email accounts. For the students who did not participate immediately, two additional reminders were sent, 1 week apart. Participants were offered a chance to enter a drawing to win one of 90 $10 Amazon.com gift cards as incentive. A total of 1839 surveys were completed for an overall response rate of 48%.*

Discuss how these factors influence your interpretation of the results of this survey.

II.9. Studying for class. Refer to the previous exercise. Overall, students reported preparing for class an average of 706 minutes per week with a standard deviation of 526 minutes. Students also reported spending an average of 106 minutes per day on Facebook with a standard deviation of 93 minutes.

(a) Construct a 95% confidence interval for the average number of minutes per week a student prepares for class.

(b) Construct a 95% confidence interval for the average number of minutes per week a student spends on Facebook. (*Hint:* Be sure to convert from minutes per day to minutes per week.)

(c) Explain why you might expect the population distributions of these two variables to be highly skewed to the right. Do you think this fact makes your confidence intervals invalid? Explain your answer.

II.10. Comparing means. Refer to the previous exercise. Suppose that you wanted to compare the average minutes per week spent on Facebook with the average minutes per week spent preparing for class.

(a) Provide an estimate of this difference.

(b) Explain why it is incorrect to use the two-sample t test to see if the means differ.

II.11. Dietary supplementation in Canadian athletes. Dietary supplements are often used by athletes to enhance performance, training, and recovery. There are, however, potential risks of taking supplements, especially for athletes who are subject to doping control. In Canada, a recent study was performed to evaluate the dietary supplementation practices of high-performance Canadian athletes.[3] Surveys were administered to $n = 440$ athletes associated with the country's eight Canadian Sport Centres.

(a) Overall, 383 athletes reported having taken a dietary supplement within the previous six months. Construct a 95% confidence interval for the proportion of high-performance Canadian athletes who have taken a supplement in the previous six months.

(b) Test the hypothesis that no more than 85% of the Canadian athletes use supplements. State the two hypotheses, the test statistic, and P-value.

II.12. Dietary supplementation by centre. Refer to the previous exercise. A majority of the athletes came from the Sport Centres in Ontario and in Calgary, Alberta. Below is a table summarizing the number of athletes and the number who used supplements at each of the centres.

Centre	Athletes	Supplements
Calgary	109	100
Ontario	145	121

Test the hypothesis that the proportion of athletes using supplements is the same at the two centres. State the two hypotheses, the test statistic, P-value, and your conclusion.

II.13. Potential bias? Refer to the previous two exercises. Registered dietitians were trained to administer the survey. Elite athletes were then recruited from participants of group workshops or one-on-one counseling sessions conducted by one of these trained dietitians. The surveys were distributed and completed prior to the beginning of the workshop or session. It was assumed that an athlete's completion of the survey signified his or her consent to be in the study. Write a brief paragraph that discusses possible selection and response biases that could arise based on this data collection process.

II.14. Roulette. You've been hired by the Nevada Gaming Commission to look into a suspicious roulette wheel. This roulette wheel has 2 green slots among its 38 slots, and patrons have complained that green occurs too often. You plan to observe many spins and record the number of times that green occurs. Your goal is to form a 95% confidence interval for the proportion of times green occurs with a margin of error no bigger than 0.005. How many spins do you plan to observe? Write a brief summary explaining your calculations.

II.15. Roulette, continued. Refer to the previous exercise. Suppose that you decided instead to record the number of times a colored slot other than green occurs. Would this change the number of spins you plan to observe? Explain your answer.

II.16. Planning a new study. Refer to Exercise II.8 (page 459). Suppose that you are planning a similar study of undergraduates at your large university. You want to sample enough students so that the 95% margin of error for each of these two variables is no more than 20 minutes. What sample size do you need for your study? Make sure to show your calculations.

II.17. Interracial friendships in college. A recent study utilized the random roommate assignment process of a small college to investigate the interracial mix of friends among students in college.[4] As part of this study, the researchers looked at 238 white students who were randomly assigned a roommate in their first year and recorded the proportion of friends (not including the first-year roommate) who were black. The following table summarizes the results, broken down by roommate race, for the middle of the first and third years of college.

Middle of First Year			
Randomly assigned	n	$\bar{x}$	s
Black roommate	41	0.085	0.134
White roommate	197	0.063	0.112

Middle of Third Year			
Randomly assigned	n	$\bar{x}$	s
Black roommate	41	0.146	0.243
White roommate	197	0.062	0.154

(a) Proportions are not Normally distributed. Explain why it may still be appropriate to use the t procedures for these data.

(b) For each year, state the null and alternative hypotheses for comparing these two groups.

(c) For each year, perform the significance test at the $\alpha = 0.05$ level, making sure to report the test statistic, degrees of freedom, and P-value.

(d) Write a one-paragraph summary of your conclusions from these two tests.

II.18. Interracial friendships in college, continued. Refer to the previous exercise. For each year, construct a 95% confidence interval for the difference in means $\mu_1 - \mu_2$ and describe how these intervals can be used to test the null hypotheses in part (b).

II.19. Do you agree? For each of the following statements, comment on whether you agree or disagree. If you disagree, explain why.

(a) An important reason to strive for a large sample size is that the bigger the sample, the smaller the bias.

(b) If the sample size is large, the shape of a histogram of the response variable is approximately Normal.

(c) For the same data set, a 95% confidence interval for the mean is wider than a 90% confidence interval.

II.20. Do you disagree? For each of the following statements, comment on whether you agree or disagree. If you disagree, explain why.

(a) In terms of resistance to extreme/unusual observations, the sample mean and variance are very resistant.

(b) When comparing the means from two populations, a test equivalent to the two-sample t test is to construct a 95% confidence interval for each population mean and reject the null hypothesis that the means are the same if the confidence intervals do not overlap.

(c) Given a 95% confidence interval for $\mu_1 - \mu_2$, one can determine whether to reject or fail to reject the hypothesis that the means are the same at the $\alpha = 0.05$ significance level.

II.21. True or false? Terry Aki performs a matched pairs t test and the P-value is 0.033. For each of the following statements, indicate whether it is true or false. If it is false, explain why.

(a) If $\alpha = 0.05$, we reject the null hypothesis.

(b) There is a 3.3% chance that the null hypothesis is true.

(c) If $\alpha = 0.05$, a Type I error occurred.

(d) If $\alpha = 0.01$, a Type II error may have occurred.

II.22. True or false? Wes Consin performs a two-sample t test and the P-value is 0.082. For each of the following statements, indicate whether it is true or false. If it is false, explain why.

(a) If $\alpha = 0.05$, we accept the null hypothesis.

(b) If $\alpha = 0.05$, a Type II error may have occurred.

(c) If H_a is one-sided, the P-value for the two-sided alternative is 0.041.

(d) A 95% confidence interval for the difference in means $\mu_1 - \mu_2$ will not contain 0.

II.23. Risk behaviors by gender. A recent study investigated the association of well-being with various health risk behaviors in college students.[5] The sample consisted of 9515 undergraduate students, 6945 females and 2570 males. The following table presents the sample proportions, by gender, of various risky behaviors performed in the last 30 days.

Behavior	Female	Male
Marijuana use	0.179	0.274
Unprotected sex	0.343	0.291
Sex while drunk/high	0.266	0.279
Driving while drunk/high	0.194	0.264

(a) For each of these behaviors, test whether there is a gender difference in the proportions, using the $\alpha = 0.05$ significance level.

(b) Since, for each behavior, the probability of a Type I error is 0.05, the probability of at least one Type I error is greater than 0.05. Let's adjust for this using the Bonferroni procedure. Suppose that there are a total of 10 behaviors so that the significance level for each test should be $0.05/10 = 0.005$. Does this change in significance level alter any of your conclusions?

(c) Suppose that differences in proportions greater than 0.06 are considered of practical significance. For which risk factors do you find both practical and statistical significance?

II.24. Two-sample t test versus matched pairs t test. Consider the following data set. The data were actually collected in pairs, and each row represents a pair. DATA FILE PAIRED

Group 1	Group 2
48.86	48.88
50.60	52.63
51.02	52.55
47.99	50.94
54.20	53.02
50.66	50.66
45.91	47.78
48.79	48.44
47.76	48.92
51.13	51.63

(a) Suppose that we ignore the fact that the data were collected in pairs and mistakenly treat this as a two-sample problem. Compute the sample mean and variance for each group. Then compute the two-sample t statistic, degrees of freedom, and P-value.

(b) Now analyze the data in the proper way. Compute the sample mean and variance of the differences. Then compute the t statistic, degrees of freedom, and P-value.

(c) Describe the differences in the two test results.

II.25. Two-sample *t* test versus matched pairs *t* test, continued. Refer to the previous exercise. Perhaps an easier way to see the major difference in the two analysis approaches for these data is by computing 95% confidence intervals for the mean difference.

(a) Compute the 95% confidence interval using the two-sample t confidence interval.

(b) Compute the 95% confidence interval using the matched pairs t confidence interval.

(c) Compare the estimates (that is, the centers of the intervals) and margins of error. What is the major difference between the two approaches for these data?

II.26. Insomnia and pain. A recent study looked at the relationship between insomnia and pain among parents and adolescents.[6] A total of 259 adolescents (69 with current insomnia) and their parents (256 middle-aged adults, 78 with insomnia) were recruited for the study. For each participant, the perceived level of overall pain was determined using an average score measured on 7 pain-related items extracted from the Somatic Symptom Inventory (SSI-28). The following table summarizes the results.

Adolescents			
Group	n	$\bar{x}$	s
Insomnia	69	12.0	4.4
Non-insomnia	190	10.1	3.7
Middle-aged adults			
Group	n	$\bar{x}$	s
Insomnia	78	14.7	4.4
Non-insomnia	178	12.1	4.1

(a) For each age category, state the null and alternative hypotheses for comparing the two groups.

(b) For each age category, perform the significance test at the $\alpha = 0.05$ level, making sure to report the test statistic, degrees of freedom, and P-value.

(c) Write a one-paragraph summary of your conclusions.

II.27. Insomnia and pain, continued. Refer to the previous exercise.

(a) For each age category, compute a 95% confidence interval for the difference in mean scores.

(b) Does it appear that one age category has a larger difference in group means than the other? Explain your answer.

(c) Suppose that you decide to test whether these differences are statistically significant. Explain why it might not be appropriate to use two-sample t procedures here. (*Hint:* Think back to how the data were collected.)

II.28. Texting and teens. A 2011 teen-driving study by Liberty Mutual Insurance and Students against Destructive Decisions (SADD) reports that 53% of the 2294 high school students surveyed say that they text while they drive at least sometimes, and 28% admit doing so often or very often.[7]

(a) Construct a 90% confidence interval for the proportion of high school students who say that they text while driving at least sometimes.

(b) Construct a 90% confidence interval for the proportion of high school students who say that they text while driving often or very often.

II.29. Texting and teens, continued. Refer to the previous exercise. In this same survey, 59% of the teens believe that texting while driving is very/extremely distracting.

(a) Construct a 95% confidence interval for the proportion of teens who find texting while driving very distracting.

(b) In 2009, it was reported that 48% of teens found texting while driving very distracting. Test the hypothesis that there has been a rise in this proportion since 2009 using the $\alpha = 0.05$ significance level.

PART THREE

TOPICS IN INFERENCE

CHAPTER NINE

ANALYSIS OF TWO-WAY TABLES

We return to our study of methods for analyzing categorical data in this chapter. Inference about proportions in one-sample and two-sample settings was the focus of Chapter 6. We now study how to compare two or more populations when the response variable has two or more categories and how to test whether two categorical variables are independent. A single statistical test handles both of these cases.

The first section of this chapter gives the basics of statistical inference that are appropriate in these settings. A second section describes a goodness-of-fit test. The methods in this chapter answer questions such as

- Are men and women equally likely to suffer lingering fear symptoms after watching scary movies at a young age?
- Is there an association between texting while driving and automobile accidents?
- You take a random sample of students from your school. You know the true proportions of men and women enrolled. Does your sample come close to reproducing the proportions?

9.1 Inference for Two-Way Tables

When we studied inference for two proportions in Chapter 6, we started summarizing the raw data by giving the number of observations in each population (n) and how many of these were classified as "successes" (X).

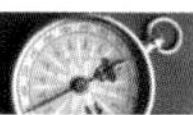

EXAMPLE 9.1

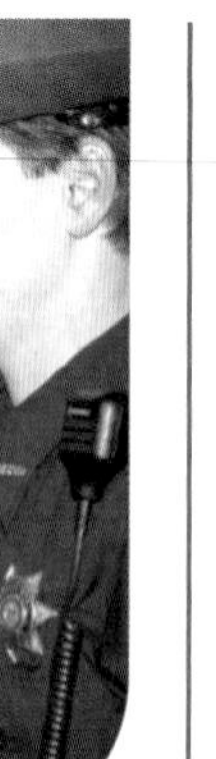

Gender and the proportion of frequent binge drinkers. In Example 6.10 (page 346), we compared the proportions of male and female college students who engage in frequent binge drinking. The following table summarizes the data used in this comparison:

Population	n	X	$\hat{p} = X/n$
1 (men)	5,348	1,392	0.260
2 (women)	8,471	1,748	0.206
Total	13,819	3,140	0.227

We see that 26.0% of the men are frequent binge drinkers. On the other hand, only 20.6% of the women are frequent binge drinkers. Therefore, these data suggest that the men are 5.4% more likely to be frequent binge drinkers, with a 95% margin of error of 1.5%.

LOOK BACK
two-way table p. 203

In this chapter, we consider a different summary of the data. Rather than recording just the count of binge drinkers, we record counts of all the outcomes in a two-way table.

EXAMPLE 9.2

Two-way table of frequent binge drinking and gender. Here is the two-way table classifying students by gender and whether or not they are frequent binge drinkers:

Two-way table for frequent binge drinking and gender

	Gender		
Frequent binge drinker	**Men**	**Women**	**Total**
Yes	1,392	1,748	3,140
No	3,956	6,723	10,679
Total	5,348	8,471	13,819

$r \times c$ table

We use the term **$r \times c$ table** to describe a two-way table of counts with r rows and c columns. The two categorical variables in the 2×2 table of Example 9.2 are "Frequent binge drinker" and "Gender." "Frequent binge drinker" is the row variable, with values "Yes" and "No," and "Gender" is the column variable, with values "Men" and "Women." Since the objective in this example is to compare the genders, we view "Gender" as an explanatory variable, and therefore, we make it the column variable. The row variable is a categorical response variable, "Frequent binge drinker." The next example presents another two-way table.

EXAMPLE 9.3

Lingering symptoms from frightening movies. There is a growing body of literature demonstrating that early exposure to frightening movies is associated with lingering fright symptoms. As part of a class on media effects, college students were asked to write narrative accounts of their exposure to frightening movies before the age of 13. More than one-fourth of the respondents said that some of the fright symptoms were still present in waking life.[1] The following table breaks down these results by gender:

Observed numbers of students

	Gender		
Ongoing fright symptoms	**Male**	**Female**	**Total**
Yes	7	29	36
No	31	50	81
Total	38	79	117

The two categorical variables in this example are "Ongoing fright symptoms," with values "Yes" and "No," and "Gender," with values "Male" and "Female." Again we view "Gender" as an explanatory variable and "Ongoing fright symptoms" as a categorical response variable.

CrunchIt!

	Male	Female	All
Yes	7 19.44 18.42 5.983	29 80.56 36.71 24.79	36 100 30.77 30.77
No	31 38.27 81.58 26.50	50 61.73 63.29 42.74	81 100 69.23 69.23
All	38 32.48 100 32.48	79 67.52 100 67.52	117 100 100 100

Count
% of Row
% of Col
% of Total

Chi-squared statistic:	4.028
df:	1
P-value:	0.04474

FIGURE 9.1 Computer output for the data in Example 9.3.

conditional distributions p. 207

In Section 4.5 we discussed two-way tables and the basics about joint, marginal, and conditional distributions. There we learned that the key to examining the relationship between two categorical variables is to look at conditional distributions. Figure 9.1 shows the output from CrunchIt! for the data of Example 9.3. Check this figure carefully. Be sure that you can identify the joint distribution, the marginal distributions, and the conditional distributions.

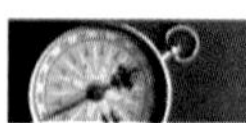

EXAMPLE 9.4

Two-way table of ongoing fright symptoms and gender. To compare the frequency of lingering fright symptoms across genders, we examine column percents. Here they are, rounded from the output for clarity:

Column percents for gender

	Gender	
Ongoing fright symptoms	**Male**	**Female**
Yes	18%	37%
No	82%	63%
Total	100%	100%

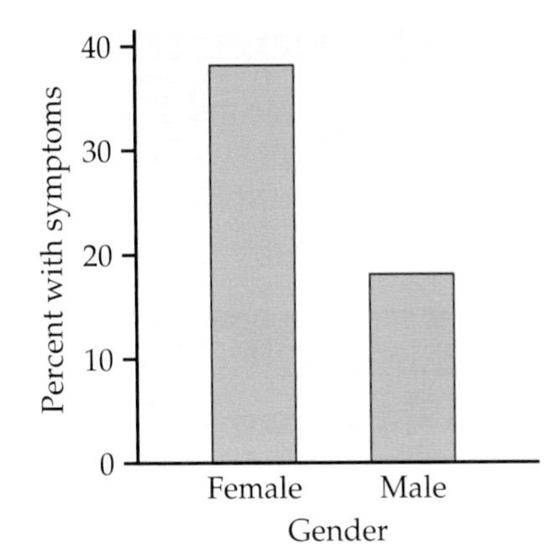

FIGURE 9.2 Bar graph of the percents of male and female students with lingering fright symptoms, for Example 9.4.

The "Total" row reminds us that 100% of the male and female students have been classified as having ongoing fright symptoms or not. (The sums sometimes differ slightly from 100% because of roundoff error.) The bar graph in Figure 9.2 compares the percents. The data reveal a clear relationship: 37% of the women have ongoing fright symptoms, as opposed to only 18% of the men.

The difference between the percents of students with lingering fears is reasonably large. A statistical test will tell us whether or not this difference can be plausibly attributed to chance. Specifically, if there is no association between gender and having ongoing fright symptoms, how likely is it that a sample would show a difference as large or larger than that displayed in Figure 9.2? In the remainder of this section we discuss the significance test to examine this question.

The hypothesis: no association

The null hypothesis H_0 of interest in a two-way table is that there is *no association* between the row variable and the column variable. In Example 9.3, this null hypothesis says that gender and having ongoing fright symptoms are not related. The alternative hypothesis H_a is that there is an association between these two variables. The alternative H_a does not specify any particular direction for the association. For two-way tables in general, the alternative includes many different possibilities. Because it includes all sorts of possible associations, we cannot describe H_a as either one-sided or two-sided.

In our example, the hypothesis H_0 that there is no association between gender and having ongoing fright symptoms is equivalent to the statement that the distributions of the ongoing fright symptoms variable are the same across the genders. For other two-way tables, where the columns correspond to independent samples from c distinct populations, there are c distributions for the row variable, one for each population. The null hypothesis then says that the c distributions of the row variable are identical. This hypothesis is sometimes called the **homogeneity hypothesis.** The alternative hypothesis is that the distributions are not all the same.

homogeneity hypothesis

Expected cell counts

expected cell counts

To test the null hypothesis in $r \times c$ tables, we compare the observed cell counts with **expected cell counts** calculated under the assumption that the null hypothesis is true. A numerical summary of the comparison will be our test statistic.

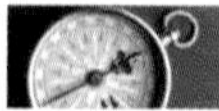

EXAMPLE 9.5

Expected counts from software. The observed and expected counts for the ongoing fright symptoms example appear in the Minitab computer output shown in Figure 9.3. The expected counts are given as the second entry in each cell. For example, in the first cell the observed count is 7 and the expected count is 11.69.

How is this expected count obtained? Look at the percents in the right margin of the table in Figure 9.1. We see that 30.77% of all students had ongoing fright symptoms. If the null hypothesis of no relation between gender and ongoing fright is true, we expect this overall percent to apply to both men and women. In particular, we expect 30.77% of the men to have lingering fright symptoms. Since there are 38 men, the expected count is 30.77% of 38, or 11.69. The other expected counts are calculated in the same way.

Minitab

```
Rows: Symptom   Columns: Gender

          1_Male  2_Female     All

1_Yes          7        29      36
           11.69     24.31   36.00

2_No          31        50      81
           26.31     54.69   81.00

All           38        79     117
           38.00     79.00  117.00

Cell Contents:      Count
                    Expected count


Pearson Chi-Square = 4.028, DF = 1, P-Value = 0.045
```

FIGURE 9.3 Minitab computer output for Example 9.5.

The reasoning of Example 9.5 leads to a simple formula for calculating expected cell counts. To compute the expected count of men with ongoing fright symptoms, we multiplied the proportion of students with fright symptoms (36/117) by the number of men (38). From Figures 9.1 and 9.3 we see that the numbers 36 and 38 are the row and column totals for the cell of interest and that 117 is n, the total number of observations in the table. The expected cell count is therefore the product of the row and column totals divided by the table total.

$$\text{expected cell count} = \frac{\text{row total} \times \text{column total}}{n}$$

The chi-square test

To test H_0 that there is no association between the row and column classifications, we use a statistic that compares the entire set of observed counts with the set of expected counts. To compute this statistic,

- First, take the difference between each observed count and its corresponding expected count, and square these values so that they are all 0 or positive.
- Since a large difference means less if it comes from a cell that is expected to have a large count, divide each squared difference by the expected count. This is a type of standardization.
- Finally, sum over all cells.

LOOK BACK
standardizing
p. 134

The result is called the *chi-square statistic* X^2. The chi-square statistic was proposed by the English statistician Karl Pearson (1857–1936) in 1900. It is the oldest inference procedure still used in its original form.

CHI-SQUARE STATISTIC

The **chi-square statistic** is a measure of how much the observed cell counts in a two-way table diverge from the expected cell counts. The formula for the statistic is

$$X^2 = \sum \frac{(\text{observed count} - \text{expected count})^2}{\text{expected count}}$$

where "observed" represents an observed cell count, "expected" represents the expected count for the same cell, and the sum is over all $r \times c$ cells in the table.

LOOK BACK
Normal approximation for a single proportion
p. 324

If the expected counts and the observed counts are very different, a large value of X^2 will result. Large values of X^2 provide evidence against the null hypothesis. To obtain a P-value for the test, we need the sampling distribution of X^2 under the assumption that H_0 (no association between the row and column variables) is true. We once again use an approximation, related to the Normal approximation for proportions. The result is a new distribution, the **chi-square distribution,** which we denote by χ^2 (χ is the lowercase Greek letter chi).

chi-square distribution χ^2

Like the t distributions, the χ^2 distributions form a family described by a single parameter, the degrees of freedom. We use χ^2(df) to indicate a particular member of this family. Figure 9.4 (next page) displays the density curves of the $\chi^2(2)$ and $\chi^2(4)$ distributions. As you can see in the figure, χ^2 distributions take only positive values and are skewed to the right. Table F in the back of the book gives upper critical values for the χ^2 distributions.

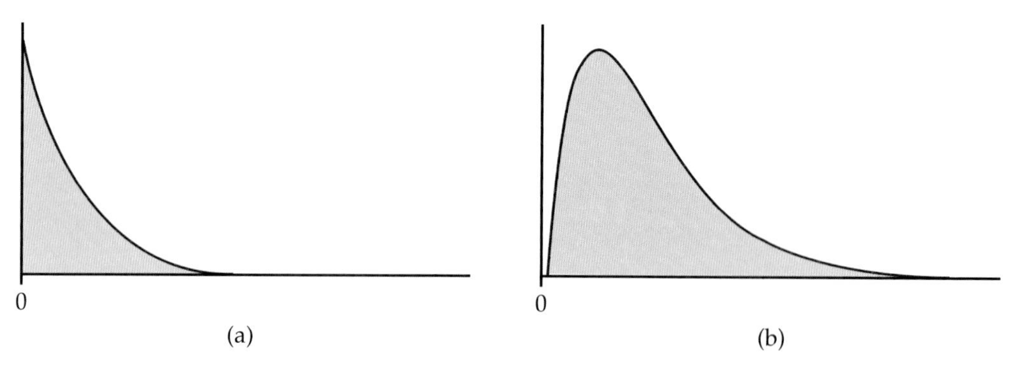

FIGURE 9.4 (a) The $\chi^2(2)$ density curve. (b) The $\chi^2(4)$ density curve.

CHI-SQUARE TEST FOR TWO-WAY TABLES

The null hypothesis H_0 is that there is no association between the row and column variables in a two-way table. The alternative is that these variables are related.

If H_0 is true, the chi-square statistic X^2 has approximately a χ^2 distribution with $(r - 1)(c - 1)$ degrees of freedom.

The P-value for the **chi-square test** is

$$P(\chi^2 \geq X^2)$$

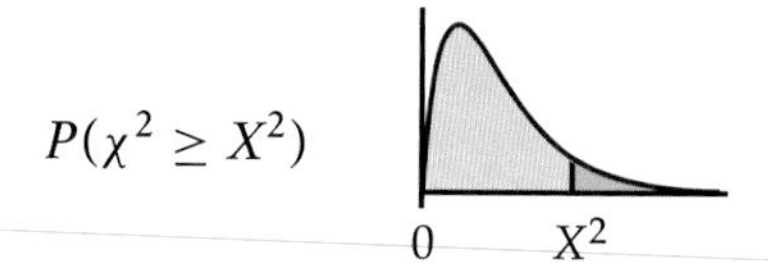

where χ^2 is a random variable having the χ^2(df) distribution with df $= (r - 1)(c - 1)$.

The chi-square test always uses the upper tail of the χ^2 distribution, because any deviation from the null hypothesis makes the statistic larger. The approximation of the distribution of X^2 by χ^2 becomes more accurate as the cell counts increase. Moreover, it is more accurate for tables larger than 2×2 tables. For tables larger than 2×2, we will use this approximation whenever the average of the expected counts is 5 or more and the smallest expected count is 1 or more. For 2×2 tables, we require that all four expected cell counts be 5 or more.[2]

EXAMPLE 9.6

Chi-square significance test from software. The results of the chi-square significance test for the ongoing fright symptoms example appear in the computer outputs in Figures 9.1 and 9.3, labeled "Chi-Square" and "Pearson Chi-Square," respectively. Because all the expected cell counts are moderately large, the χ^2 distribution provides an accurate P-value. We see that $X^2 = 4.03$, df $= 1$, and $P = 0.045$. As a check we verify that the degrees of freedom are correct for a 2×2 table:

$$\text{df} = (r - 1)(c - 1) = (2 - 1)(2 - 1) = 1$$

The chi-square test confirms that the data contain evidence against the null hypothesis that there is no relationship between gender and ongoing fright symptoms. Under H_0, the chance of obtaining a value of X^2 greater than or equal to the calculated value of 4.03 is small—fewer than 5 times in 100.

Here we expressed the null hypothesis in terms of an association between gender and ongoing fright symptoms. We can also state it in terms of a homogeneity hypothesis. Specifically, the null hypothesis is that the distribution of fright symptoms is the same for both genders. The significance test is sometimes called the **homogeneity test** in this setting.

homogeneity test

The test does not provide insight into the nature of the relationship between the variables. It is up to us to see that the data show that women are more likely to have lingering fright symptoms. You should always accompany a chi-square test by percents such as those in Example 9.4 and Figure 9.2 and by a description of the nature of the relationship.

confounding p. 36

The observational study of Example 9.3 cannot tell us whether gender is a *cause* of lingering fright symptoms. The association may be explained by confounding with other variables. For example, other research has shown that there are gender differences in the social desirability of admitting fear.[3] *Our data don't allow us to investigate possible confounding variables.* Often a randomized comparative experiment can settle the issue of causation, but we cannot randomly assign gender to each student. The researcher who published the data of our example states merely that women are more likely to have lingering fright symptoms and that this conclusion is consistent with other studies.

The chi-square test and the z test

A comparison of the proportions of "successes" in two populations leads to a 2×2 table. We can compare two population proportions either by the chi-square test or by the two-sample z test from Section 6.2. In fact, *these tests always give exactly the same result,* because the X^2 statistic is equal to the square of the z statistic, and $\chi^2(1)$ critical values are equal to the squares of the corresponding $N(0, 1)$ critical values. The advantage of the z test is that we can test either one-sided or two-sided alternatives. The chi-square test always tests the two-sided alternative. Of course, the chi-square test can compare more than two populations, whereas the z test compares only two.

USE YOUR KNOWLEDGE

9.1 Comparison of conditional distributions. Consider the following 2×2 table:

Observed counts

	Explanatory variable		
Response variable	**1**	**2**	**Total**
Yes	70	90	160
No	130	110	240
Total	200	200	400

(a) Compute the conditional distribution of the response variable for each of the two explanatory-variable categories.

(b) Display the distributions graphically.

(c) Write a short paragraph describing the two distributions and how they differ.

9.2 Expected cell counts and the chi-square test. Refer to Exercise 9.1. You decide to use the chi-square test to compare these two conditional distributions.

(a) What is the expected count for the first cell (observed count is 70)?

(b) Computer software gives you $X^2 = 4.17$. What are the degrees of freedom for this statistic?

(c) Using Table F, give an appropriate bound on the P-value.

SECTION 9.1 SUMMARY

The **null hypothesis** for $r \times c$ tables of count data is that there is no relationship between the row variable and the column variable.

Expected cell counts under the null hypothesis are computed using the formula

$$\text{expected count} = \frac{\text{row total} \times \text{column total}}{n}$$

The null hypothesis is tested by the **chi-square statistic,** which compares the observed counts with the expected counts:

$$X^2 = \sum \frac{(\text{observed} - \text{expected})^2}{\text{expected}}$$

Under the null hypothesis, X^2 has approximately the χ^2 distribution with $(r-1)(c-1)$ degrees of freedom. The P-value for the test is

$$P(\chi^2 \geq X^2)$$

where χ^2 is a random variable having the $\chi^2(\text{df})$ distribution with $\text{df} = (r-1)(c-1)$.

The chi-square approximation is adequate for practical use when the average expected cell count is 5 or greater and all individual expected counts are 1 or greater, except in the case of 2×2 tables. All four expected counts in a 2×2 table should be 5 or greater.

9.2 Goodness of Fit

In the previous section, we discussed the use of the chi-square test to compare categorical-variable distributions of c populations. We now consider a slight variation on this scenario where we compare a sample from one population with a hypothesized distribution. Here is an example that illustrates the basic ideas.

EXAMPLE 9.7

Sampling in the Adequate Calcium Today (ACT) study. The ACT study was designed to examine relationships among bone growth patterns, bone development, and calcium intake. Participants were over 14,000 adolescents from six states: Arizona (AZ), California (CA), Hawaii (HI), Indiana, (IN),

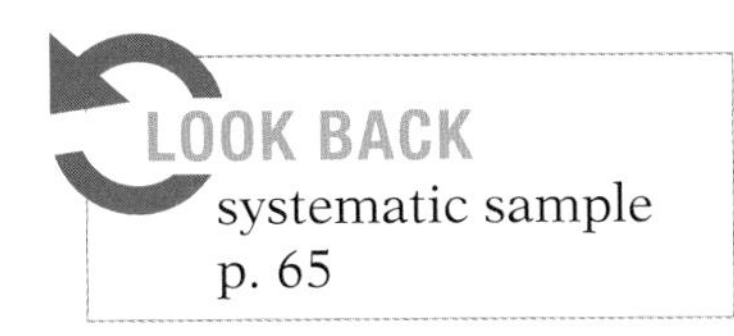

Nevada (NV), and Ohio (OH). After the major goals of the study were completed, the investigators decided to do an additional analysis of the written comments made by the participants during the study. The comments were stored in containers that were filled as the comments were received. Therefore, each container had a mixture of comments from different states.

Because the number of participants was so large, a sampling plan was devised to select sheets containing the written comments of approximately 10% of the participants. A systematic sample of every 10th comment sheet was retrieved from each storage container for analysis.[4] Here are the counts for each of the six states:

Number of study participants in the samples						
AZ	**CA**	**HI**	**IN**	**NV**	**OH**	**Total**
167	257	257	297	107	482	1567

There were 1567 study participants in the sample. We will use the proportions of students from each of the states in the original sample as the population values.[5] Here are the proportions:

Population proportions						
AZ	**CA**	**HI**	**IN**	**NV**	**OH**	**Total**
0.105	0.172	0.164	0.188	0.070	0.301	100.000

Does the sample have bias with respect to the selection of comments from different states? Let's see how well our sample reflects the state population proportions. We start by computing expected counts. Since 10.5% of the population is from Arizona, we expect the sample to have about 10.5% from Arizona. Therefore, since the sample has 1567 subjects, our expected count for Arizona is

$$\text{expected count for Arizona} = 0.105(1567) = 164.535$$

Here are the expected counts for all six states:

Expected counts						
AZ	**CA**	**HI**	**IN**	**NV**	**OH**	**Total**
164.54	269.52	256.99	294.60	109.69	471.67	1567.01

USE YOUR KNOWLEDGE

9.3 Why is the sum 1567.01? Refer to the table of expected counts in Example 9.7. Explain why the sum of the expected counts is 1567.01 and not 1567.

9.4 Calculate the expected counts. Refer to Example 9.7. Find the expected counts for the other five states. Report your results with three places after the decimal as we did for Arizona.

As we saw with the expected counts in the analysis of two-way tables in the previous section of this chapter, we do not really expect the observed counts to be *exactly* equal to the expected counts. Different samples under the same conditions will give different counts. We do, however, expect the *average* of these counts to be equal to the expected counts when the null hypothesis is true. How close should the observed counts from a single sample and the expected counts be if the null hypothesis is true?

We can think of our table of observed counts in Example 9.7 as a one-way table with six cells, each with a count of the number of subjects sampled from a particular state. Our question of interest is translated into a null hypothesis that the observed proportions of students in the six states can be viewed as random samples from the subjects in the ACT study. The alternative is that the process generating the observed counts (systematic sampling in this case) does not provide samples that are compatible with this hypothesis. In other words, the alternative means that there is some bias in the way that we selected the subjects.

Our analysis of these data is very similar to the analyses of two-way tables that we studied in Section 9.1. We have already computed the expected counts. We now construct a chi-square statistic that measures how far the observed counts are from the expected counts. Here is a summary of the procedure.

THE CHI-SQUARE GOODNESS-OF-FIT TEST

Data for n observations of a categorical variable with k possible outcomes are summarized as observed counts, $n_1, n_2, \ldots, n_k$ in k cells. A null hypothesis specifies probabilities $p_1, p_2, \ldots, p_k$ for the possible outcomes.

For each cell, multiply the total number of observations n by the specified probability to determine the expected counts:

$$\text{expected count} = np_i$$

The **chi-square statistic** measures how much the observed cell counts differ from the expected cell counts. The formula for the statistic is

$$X^2 = \sum \frac{(\text{observed count} - \text{expected count})^2}{\text{expected count}}$$

The degrees of freedom are $k - 1$, and P-values are computed from the chi-square distribution.

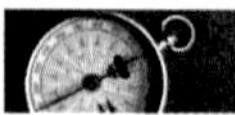

EXAMPLE 9.8

The goodness-of-fit test for the ACT study. For Arizona, the observed count is 167. In Example 9.7, we calculated the expected count, 164.535. The contribution to the chi-square statistic for Arizona is

$$\frac{(\text{observed count} - \text{expected count})^2}{\text{expected count}} = \frac{(167 - 164.535)^2}{164.535} = 0.0369$$

We use the same approach to find the contributions to the chi-square statistic for the other five states. The sum of these six values is the chi-square statistic,

$$X^2 = 0.93$$

The degrees of freedom are the number of cells minus 1, df $= 6 - 1 = 5$. We calculate the P-value using Table F or software. From Table F, we can determine $P > 0.25$. We conclude that the observed counts are compatible with the hypothesized proportions. The data do not provide any evidence that our systematic sample was biased with respect to selection of subjects from different states.

Software output from Minitab and SPSS for this problem is given in Figure 9.5. Both report the P-value as 0.968. Note that the SPSS output includes a column titled "Residual." For tables of counts, a residual for a cell is defined as

$$\text{residual} = \frac{\text{observed count} - \text{expected count}}{\sqrt{\text{expected count}}}$$

Note that the chi-square statistic is the sum of the squares of these residuals.

Some software packages do not provide routines for computing the chi-square goodness-of-fit test. However, there is a very simple trick that can be used to produce the results from software that can analyze two-way tables. Make a two-way table where the first column contains k cells with the observed counts. Make a second column with counts that correspond *exactly* to the probabilities specified by the null hypothesis, with a very large number of observations. For example, make the count in the smallest cell at least 10,000. Then compute χ^2 and perform the significance test as we did in Section 9.1.

USE YOUR KNOWLEDGE

9.5 Compute the chi-square statistic. For each of the other five states, compute the contribution to the chi-square statistic using the method illustrated for Arizona in Example 9.8. Use the expected counts that you calculated in Exercise 9.4 for these calculations. Show that the sum of these values is the chi-square statistic.

9.6 Distribution of M&M colors. Mars, Inc. has varied the mix of colors for M&M's Milk Chocolate Candies over the years. These changes in color blends are the result of consumer preference tests. Most recently, the color distribution is reported to be 13% brown, 14% yellow, 13% red, 20% orange, 24% blue, and 16% green.[6] You open up a 14-ounce bag of M&M's and find 61 brown, 59 yellow, 49 red, 77 orange, 141 blue, and 88 green. Use a goodness of fit test to examine how well this bag fits the percents stated by Mars, Inc.

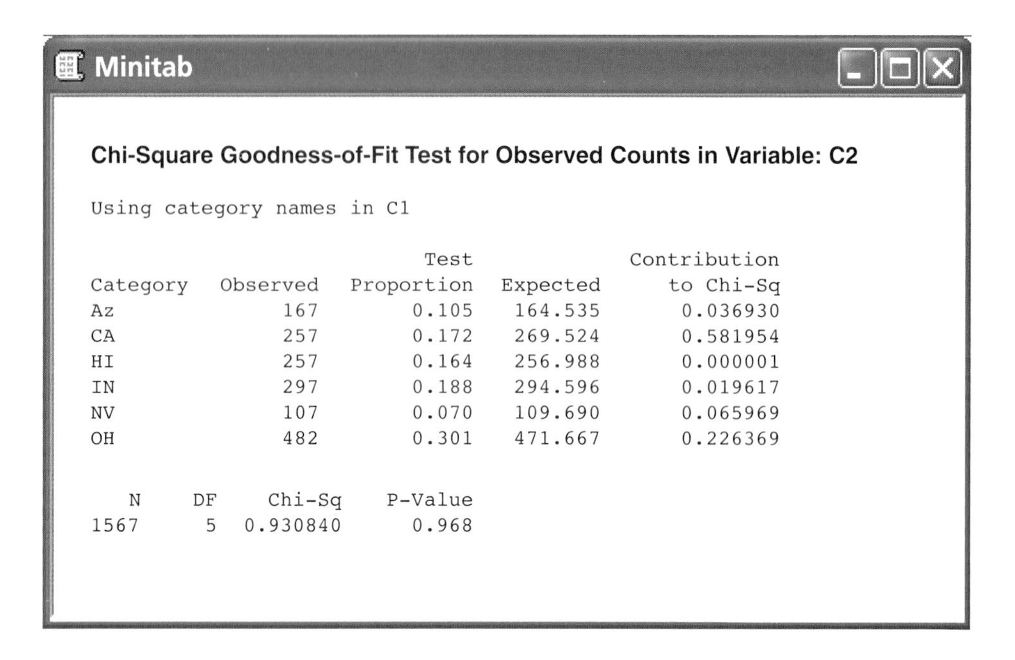

Minitab

Chi-Square Goodness-of-Fit Test for Observed Counts in Variable: C2

```
Using category names in C1

                              Test                 Contribution
Category  Observed  Proportion  Expected    to Chi-Sq
Az             167       0.105   164.535     0.036930
CA             257       0.172   269.524     0.581954
HI             257       0.164   256.988     0.000001
IN             297       0.188   294.596     0.019617
NV             107       0.070   109.690     0.065969
OH             482       0.301   471.667     0.226369

   N   DF    Chi-Sq  P-Value
1567    5  0.930840    0.968
```

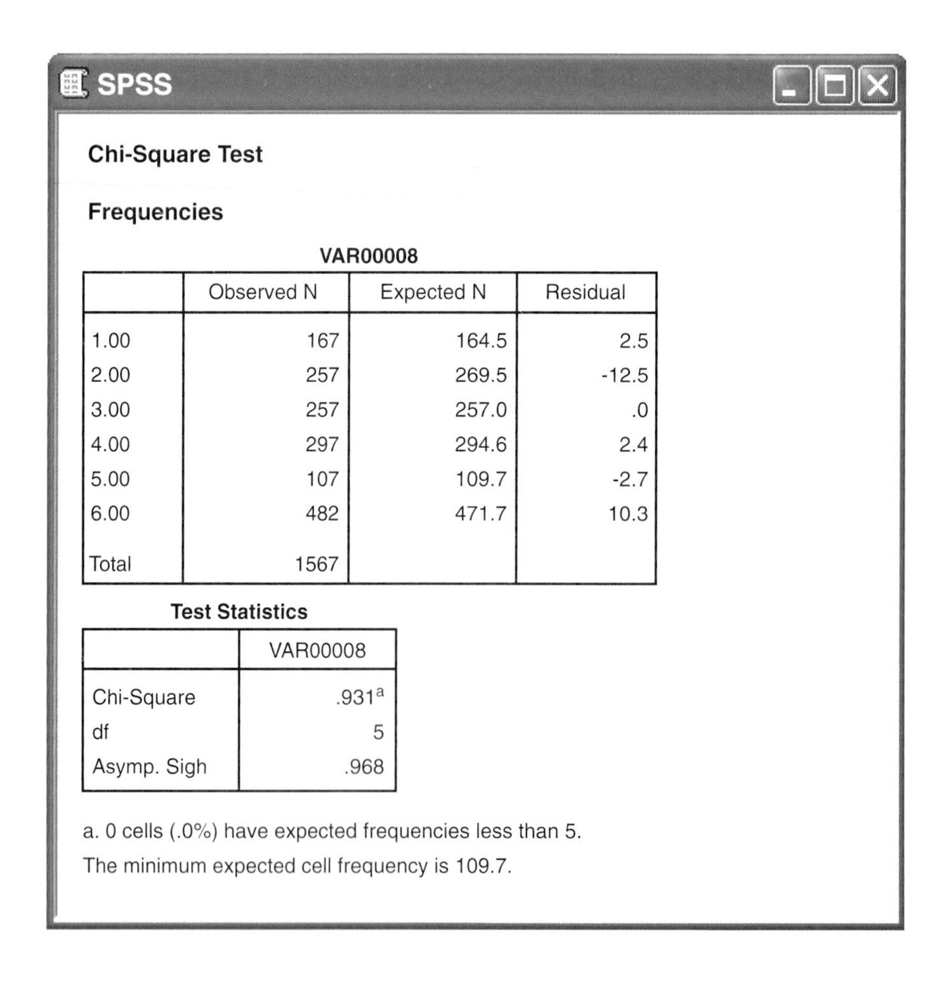

SPSS

Chi-Square Test

Frequencies

VAR00008

	Observed N	Expected N	Residual
1.00	167	164.5	2.5
2.00	257	269.5	-12.5
3.00	257	257.0	.0
4.00	297	294.6	2.4
5.00	107	109.7	-2.7
6.00	482	471.7	10.3
Total	1567		

Test Statistics

	VAR00008
Chi-Square	.931[a]
df	5
Asymp. Sigh	.968

a. 0 cells (.0%) have expected frequencies less than 5.
The minimum expected cell frequency is 109.7.

FIGURE 9.5 (a) Minitab and (b) SPSS output for Example 9.8.

SECTION 9.2 SUMMARY

The **chi-square goodness-of-fit test** is used to compare the sample distribution of a categorical variable from a population with a hypothesized distribution. The data for n observations with k possible outcomes are summarized as observed counts, $n_1, n_2, \ldots, n_k$ in k cells. The **null hypothesis** specifies probabilities $p_1, p_2, \ldots, p_k$ for the possible outcomes.

The analysis of these data is similar to the analyses of two-way tables discussed in Section 9.1. For each cell, the **expected count** is determined by multiplying the total number of observations n by the specified probability p_i. The null hypothesis is tested by the usual **chi-square statistic,** which compares the observed counts, n_i, with the expected counts. Under the null hypothesis, X^2 has approximately the χ^2 distribution with df $= k - 1$.

CHAPTER 9 EXERCISES

For Exercises 9.1 and 9.2, see pages 473–474; for Exercises 9.3 and 9.4, see pages 475–476; and for Exercises 9.5 and 9.6, see page 477.

9.7. Health care fraud. Most errors in billing insurance providers for health care services involve honest mistakes by patients, physicians, or others involved in the health care system. However, fraud is a serious problem. The National Health Care Anti-fraud Association estimates that approximately $68 billion is lost to health care fraud each year.[7] When fraud is suspected, an audit of randomly selected billings is often conducted. The selected claims are then reviewed by experts, and each claim is classified as allowed or not allowed. The distributions of the amounts of claims are frequently highly skewed, with a large number of small claims and a small number of large claims. Since simple random sampling would likely be overwhelmed by small claims and would tend to miss the large claims, stratification is often used. See the section on stratified sampling in Chapter 2 (page 59). Here are data from an audit that used three strata based on the sizes of the claims (small, medium, and large):[8] DATA FILE BILLINGERRORS

Stratum	Sampled claims	Number not allowed
Small	57	12
Medium	17	10
Large	5	2

(a) Construct the 3 × 2 table of counts for these data and include the marginal totals.

(b) Find the percents of claims that were not allowed in each of the three strata.

(c) To perform a significance test, combine the medium and large strata. Explain why we do this.

(d) State an appropriate null hypothesis to be tested for these data.

(e) Perform the significance test and report your test statistic with its degrees of freedom, and the P-value. State your conclusion.

9.8. Population estimates. Refer to the previous exercise. One reason to do an audit such as this is to estimate the number of claims that would not be allowed if all claims in a population were examined by experts. We have an estimate of the proportion of claims from each stratum based on our sample. With our simple random sampling of claims from each stratum, we have an unbiased estimate of the corresponding population proportion for each stratum. Therefore, if we take the sample proportions and multiply by the population sizes, we have the estimates that we need. Here are the population sizes for the three strata:

Stratum	Claims in stratum
Small	3342
Medium	246
Large	58

(a) For each stratum, estimate the total number of claims that would not be allowed if all claims in the stratum had been audited.
(b) (*Optional*) Give margins of error for your estimates. (*Hint:* You first need to find standard errors for your sample estimates, as discussed in Chapter 6 [page 328]. Then you need to use the rules for variances in Chapter 5 [page 277] to find the standard errors for the population estimates. Finally, you need to multiply by z^* to determine the margins of error.)

9.9. DFW rates. One measure of student success for colleges and universities is the percent of admitted students who graduate. Studies indicate that a key issue in retaining students is their performance in so-called gateway courses.[9] These are courses that serve as prerequisites for other key courses that are essential for student success. One measure of student performance in these courses is the DFW rate, the percent of students who receive grades of D, F, or W (withdraw).

A major project was undertaken to reduce the DFW rate in a gateway course at a large midwestern university. The course curriculum was revised to make it more relevant to the majors of the students taking the course, a small group of excellent teachers taught the course, technology (including clickers and online homework) was introduced, and student support outside the classroom was increased. The following table gives data on the DFW rates for the course for three years.[10] In Year 1, the traditional course was given; in Year 2, a few changes were introduced; and in Year 3, the course was substantially revised.

Year	DFW rate	Number of students taking course
Year 1	42.3%	2408
Year 2	24.9%	2325
Year 3	19.9%	2126

Do you think that the changes in this gateway course had an impact on the DFW rate? Write a report giving your answer to this question. Support your answer with an analysis of the data.

9.10. Class attendance and DFW rates. One study that looked at DFW rates surveyed 719 students who were enrolled in one or more of seven gateway courses in business, mathematics, and science.[11] If a student was enrolled in more than one course, then a single course was randomly selected for analysis. In the survey, students were asked how often they attended the gateway class. Here are the data:

Class attendance	ABC rate	DFW rate
Less than 50%	2%	5%
51% to 74%	8%	14%
75% to 94%	25%	30%
95% or more	66%	51%
Total	100%	100%

In this table students are classified as earning an A, B, or C in the gateway course or as earning a D or an F or withdrawing from the course. Notice that the data are given in terms of the marginal distributions of class attendance for each (ABC or DFW) group. In the survey, there were 539 students in the ABC group and 180 students in the DFW group.

(a) Use the data given to construct the 4×2 table of counts, and add the marginal totals to the table. Sum the row totals and the marginal totals separately to verify that you have the correct total sample size, 719.

(b) Test the null hypothesis that there is no association between class attendance and a DFW grade. Give the test statistic, its degrees of freedom, and the P-value.

(c) Summarize your conclusions in a short paragraph.

(d) Can you conclude that the data indicate that not going to class causes a student to get a bad grade? Can you conclude that the data are consistent with this scenario?

9.11. When do Canadian students enter private career colleges? A survey of over 13,000 Canadian students who enrolled in private career colleges was conducted to understand student participation in the private postsecondary educational system.[12] In one part of the survey, students were asked about their field of study and about when they entered college. Here are the results:

Field of study	Number of students	Time of Entry: Right after high school	Later
Trades	942	34%	66%
Design	584	47%	53%
Health	5085	40%	60%
Media/IT	3148	31%	69%
Service	1350	36%	64%
Other	2255	52%	48%

In this table, the second column gives the number of students in each field of study. The next two columns give the marginal distribution of time of entry for each field of study.

(a) Use the data provided to make the 6×2 table of counts for this problem.

(b) Analyze the data.

(c) Write a summary of your conclusions. Be sure to include the results of your significance testing as well as a graphical summary.

9.12. Government loans for Canadian students in private career colleges. Refer to the previous exercise. The survey also asked about how these college students paid for their education. A major source of funding was government loans. Here are the survey percents of Canadian private students who use government loans to finance their education, by field of study:

Field of study	Number of students	Percent using government loans
Trades	942	45
Design	599	53
Health	5234	55
Media/IT	3238	55
Service	1378	60
Other	2300	47

(a) Construct the 6 × 2 table of counts for this exercise.

(b) Test the null hypothesis that the percent of students using government loans to finance their education does not vary with field of study. Be sure to provide all the details for your significance test.

(c) Summarize your analysis and conclusions. Be sure to include a graphical summary.

9.13. Other funding for Canadian students in private career colleges. Refer to the previous exercise. Another major source of funding was parents, family, or spouse. The following table gives the survey percents of Canadian private students who rely on these sources to finance their education, by field of study:

Field of study	Number of students	Percent using parents/family/spouse
Trades	942	20
Design	599	37
Health	5234	26
Media/IT	3238	16
Service	1378	18
Other	2300	41

Answer the questions in the previous exercise for these data.

9.14. Remote deposit capture. The Federal Reserve has called remote deposit capture (RDC) "the most important development the (U.S) banking industry has seen in years." This service allows users to scan checks and to transmit the scanned images to a bank for posting.[13] In its annual survey of community banks, the American Bankers Association asked banks whether or not they offered this service.[14] Here are the results classified by the asset size (in millions of dollars) of the bank: DATA FILE RDCASSET

	Offer RDC	
Asset size	Yes	No
Under $100	63	309
$101–$200	59	132
$201 or more	112	85

(a) Summarize the results of this survey question numerically and graphically.

(b) Test the null hypothesis that there is no association between the size of a bank, measured by assets, and whether or not it offers RDC. Report the test statistic, the P-value, and your conclusion.

9.15. Exercise and adequate sleep. A survey of 656 boys and girls who were 13 to 18 years old asked about adequate sleep and other health-related behaviors. The recommended amount of sleep is six to eight hours per night.[15] In the survey 54% of the respondents reported that they got less than this amount of sleep on school nights. An exercise scale was developed and was used to classify the students as above or below the median in this domain. Here is the 2 × 2 table of counts with students classified as getting or not getting adequate sleep and by the exercise variable: DATA FILE SLEEP

	Exercise	
Enough sleep	High	Low
Yes	151	115
No	148	242

Note that you answered parts (a) through (c) of this exercise if you completed Exercise 4.103 (page 212).

(a) Find the distribution of adequate sleep for the high exercisers.

(b) Do the same for the low exercisers.

(c) Summarize the relationship between adequate sleep and exercise using the results of parts (a) and (b).

(d) Perform the significance test for examining the relationship between exercise and getting enough sleep. Give the test statistic and the P-value (with a sketch similar to the one on page 472) and summarize your conclusion. Be sure to include numerical and graphical summaries.

9.16. Lying to a teacher. One of the questions in a survey of high school students asked about lying to teachers.[16] The following table gives the numbers of students who said that they lied to a teacher at least once during the past year, classified by gender: LYING

	Gender	
Lied at least once	**Male**	**Female**
Yes	3,228	10,295
No	9,659	4,620

(a) Add the marginal totals to the table.

(b) Calculate appropriate percents to describe the results of this survey.

(c) Summarize your findings in a short paragraph.

(d) Test the null hypothesis that there is no association between gender and lying to teachers. Give the test statistic and the P-value (with a sketch similar to the one on page 472) and summarize your conclusion. Be sure to include numerical and graphical summaries.

9.17. Trust and honesty in the workplace. The students surveyed in the study described in the previous exercise were also asked whether they thought trust and honesty were essential in business and the workplace. Here are the counts classified by gender: TRUST

	Gender	
Trust and honesty are essential	**Male**	**Female**
Agree	11,724	14,169
Disagree	1,163	746

Answer the questions in the previous exercise for these data.

9.18. Why not use a chi-square test? As part of the study on ongoing fright symptoms due to exposure to horror films at a young age, the following table was created based on the written responses from 119 students.[17] Explain why a chi-square test is not appropriate for the data in this table.

Percent of students who reported each problem

	Type of Problem (%)			
	Bedtime		Waking	
Movie or video	**Short term**	**Enduring**	**Short term**	**Enduring**
Poltergeist ($n = 29$)	68	7	64	32
Jaws ($n = 23$)	39	4	83	43
Nightmare on Elm Street ($n = 16$)	69	13	37	31
Thriller (music video) ($n = 16$)	40	0	27	7
It ($n = 24$)	64	0	64	50
The Wizard of Oz ($n = 12$)	75	17	50	8
E.T. ($n = 11$)	55	0	64	27

9.19. Age and time status of U.S. college students. The Census Bureau provides estimates of numbers of people in the United States classified in various ways.[18] Let's look at college students. The following table gives us data to examine the relation between age and full-time or part-time status. The numbers in the table are expressed as thousands of U.S. college students. USCOLLEGE1

U.S. college students by age and status: October 2004

	Status	
Age (years)	**Full-time**	**Part-time**
15–19	3553	329
20–24	5710	1215
25–34	1825	1864
35 and over	901	1983

(a) Give the joint distribution of age and status for this table.

(b) What is the marginal distribution of age? Display the results graphically.

(c) What is the marginal distribution of status? Display the results graphically.

(d) Compute the conditional distribution of age for each of the two status categories. Display the results graphically.

(e) Write a short paragraph describing the distributions and how they differ.

9.20. Time status versus gender for the 20 to 24 age category. Refer to Exercise 9.19. The table below breaks down the 20 to 24 age category by gender. USCOLLEGE2

	Gender		
Status	**Male**	**Female**	**Total**
Full-time	2719	2991	5710
Part-time	535	680	1215
Total	3254	3671	6925

(a) Compute the marginal distribution for gender. Display the results graphically.

(b) Compute the conditional distribution of status for males and for females. Display the results graphically and comment on how these distributions differ.

(c) If you wanted to test the null hypothesis that there is no difference between these two conditional distributions, what would the expected cell counts be for the full-time status row of the table?

(d) Computer software gives $X^2 = 5.17$. Using Table F, give an appropriate bound for the P-value and state your conclusions at the 5% level.

9.21. Waking versus bedtime symptoms. As part of the study on ongoing fright symptoms due to exposure to horror films at a young age, the following table was presented to describe the lasting impact these movies have had during bedtime and waking life: FRIGHTSYMPTOMS

	Waking symptoms	
Bedtime symptoms	**Yes**	**No**
Yes	36	33
No	33	17

(a) What percent of the students have lasting waking-life symptoms?

(b) What percent of the students have both waking-life and bedtime symptoms?

(c) Test whether there is an association between waking-life and bedtime symptoms. State the null and alternative hypotheses, the X^2 statistic, and the P-value.

9.22. Find the degrees of freedom and P-value. For each of the following situations give the degrees of freedom and an appropriate bound on the P-value (give the exact value if you have software available) for the X^2 statistic for testing the null hypothesis of no association between the row and column variables.

(a) A 2×2 table with $X^2 = 1.25$.

(b) A 4×4 table with $X^2 = 18.34$.

(c) A 2×8 table with $X^2 = 24.21$.

(d) A 5×3 table with $X^2 = 12.17$.

9.23. Can you construct the joint distribution from the marginal distributions? Here are the row and column totals for a two-way table with two rows and two columns:

a	b	50
c	d	150
100	100	200

Find *two different* sets of counts a, b, c, and d for the body of the table. This demonstrates that the relationship between two variables cannot be obtained solely from the two marginal distributions of the variables.

9.24. Construct a table with no association. Construct a 3×2 table of counts where there is no apparent association between the row and column variables.

9.25. CHALLENGE **Gender versus motivation for volunteer service.** A study examined patterns and characteristics of the volunteer service performed by young people from high school through early adulthood.[19] Here are some data that can be used to compare males and females on participation in unpaid volunteer service or community service and motivation for participation:

	Motivation			
Gender	**Strictly voluntary**	**Court-ordered**	**Other**	**Non-participants**
Men	31.9%	2.1%	6.3%	59.7%
Women	43.7%	1.1%	6.5%	48.7%

Note that the percents in each row sum to 100%. Graphically compare the volunteer-service profiles of men and women. Describe any striking differences.

9.26. CHALLENGE Gender versus motivation for volunteer service, continued. Refer to the previous exercise. Recompute the table for participants only. To do this, take the entries for each motivation and divide by the percent of participants. Do this separately for each gender. Verify that the percents sum to 100% for each gender. Give a graphical summary to compare the motivation of men and women who are participants. Compare this with the summary that you wrote for the previous exercise, and write a short paragraph describing similarities and differences in these two views of the data.

9.27. Dieting trends among male and female undergraduates. A study of undergraduates looked at gender differences in dieting trends.[20] There were 181 women and 105 men who participated in the survey. The following table summarizes whether a student tried a low-fat diet or not by gender:

	Gender	
Tried low-fat diet	**Women**	**Men**
Yes	35	8
No		

(a) Fill in the missing cells of the table.

(b) Summarize the data numerically and graphically.

(c) Test that there is no association between gender and the likelihood of trying a low-fat diet. Summarize the results.

9.28. Sexual imagery in magazine ads. In what ways do advertisers in magazines use sexual imagery to appeal to youth? One study classified each of 1509 full-page or larger ads as "not sexual" or "sexual," according to the amount and style of the dress of the male or female model in the ad. The ads were also classified according to the target readership of the magazine.[21] Here is the two-way table of counts: DATA FILE MAGAZINEADS

	Magazine readership			
Model dress	**Women**	**Men**	**General interest**	**Total**
Not sexual	351	514	248	1113
Sexual	225	105	66	396
Total	576	619	314	1509

(a) Summarize the data numerically and graphically.

(b) Perform the significance test that compares the model dress for the three categories of magazine readership. Summarize the results of your test and give your conclusion.

(c) All the ads were taken from the March, July, and November issues of six magazines in one year. Discuss this fact from the viewpoint of the validity of the significance test and the interpretation of the results.

9.29. Intended readership of ads with sexual imagery. The ads in the study described in the previous exercise were also classified according to the age group of the intended readership. Here is a summary of the data:

	Magazine readership age group	
Model dress	**Young adult**	**Mature adult**
Not sexual	72.3%	76.1%
Sexual	27.7%	23.9%
Number of ads	1006	503

Using parts (a) and (b) in the previous exercise as a guide, analyze these data and write a report summarizing your work.

9.30. Identity theft. A study of identity theft looked at how well consumers protect themselves from this increasingly prevalent crime. The behaviors of 61 college students were compared with the behaviors of 59 nonstudents.[22] One of the questions was "When asked to create a password, I have used either my mother's maiden name, or my pet's name, or my birth date, or the last four digits of my social security number, or a series of consecutive numbers." For the students, 22 agreed with this statement while 30 of the nonstudents disagreed.

(a) Display the data in a two-way table and perform the chi-square test. Summarize the results.

(b) Reanalyze the data using the methods for comparing two proportions that we studied in Chapter 6. Compare the results and verify that the chi-square statistic is the square of the z statistic.

(c) The students in this study were junior and senior college students from two sections of a course in Internet marketing at a large northeastern university. The nonstudents were a group of individuals who were recruited to attend commercial focus groups on the West

Coast conducted by a lifestyle marketing organization. Discuss how the method of selecting the subjects in this study relates to the conclusions that can be drawn from it.

9.31. CHALLENGE **Student-athletes and gambling.** A survey of student-athletes that asked questions about gambling behavior classified students according to the National Collegiate Athletic Association (NCAA) division.[23] The percents of male student-athletes who reported wagering on collegiate sports are given here along with the numbers of respondents in each division:

Division	I	II	III
Percent	17.2%	21.0%	24.4%
Number	5619	2957	4089

(a) Use a significance test to compare the percents for the three NCAA divisions. Give details and a short summary of your conclusion.

(b) The percents in the preceding table are given in the NCAA report, but the numbers of male student-athletes in each division who responded to the survey question are estimated based on other information in the report. To what extent do you think this has an effect on the results? (*Hint:* Rerun your analysis a few times, with slightly different numbers of students but the same percents.)

(c) Some student-athletes may be reluctant to provide this kind of information, even in a survey where there is no possibility that they can be identified. Discuss how this fact may affect your conclusions.

(d) The chi-square test for this set of data assumes that the responses of the student-athletes are independent. However, some of the students are at the same school and even on the same team. Discuss how you think this might affect the results.

9.32. Hummingbirds of Saint Lucia. *Eulampis jugularis* is a type of hummingbird that lives in the forest preserves of the Caribbean island of Saint Lucia. The males and the females of this species have bills that are shaped somewhat differently. Researchers who study these birds thought that the bill shape might be related to the shape of the flowers that they visit for food. The researchers observed 49 females and 21 males. Of the females, 20 visited the flowers of *Heliconia bihai*, while none of the males visited these flowers.[24] Display the data in a two-way table and perform the chi-square test. Summarize the results and give a brief statement of your conclusion. Your two-way table has a count of zero in one cell. Does this invalidate your significance test? Explain why or why not.

9.33. Changing majors. A task force set up to examine retention of students in the majors that they chose when starting college examined data on transfers to other majors.[25] Here are some data giving counts of students classified by initial major and the area that they transferred to:

	Area transferred to				
Initial major	Engineering	Management	Liberal arts	Other	Total
Biology	13	25	158		398
Chemistry	16	15	19		114
Mathematics	3	11	20		72
Physics	9	5	14		61

Complete the table by computing the values for the "Other" column. Write a short paragraph explaining what conclusions you can draw about the relationship between initial major and area transferred to. Be sure to include numerical and graphical summaries as well as the details of your significance test.

9.34. CHALLENGE **Cracks in veneer.** Many furniture pieces are built with veneer, a thin layer of fine wood that is fastened to less expensive wood products underneath. Face checks are cracks that sometimes develop in the veneer. When face checks appear, the furniture needs to be reconstructed. Because this is a fairly expensive process, researchers seek ways to minimize the occurrence of face checks by controlling the manufacturing process. In one study, the type of adhesive used was one of the factors examined.[26] Because of the way that the veneer is cut, it has two different sides, called loose and tight, either of which can face out. Here is a table giving the numbers of veneer panels with and without face checks for two different adhesives, PVA and UF. Separate columns are given for the loose side and the tight side. DATA FILE VENEER

	Loose Side		Tight Side	
	Face checks		Face checks	
Adhesive	**No**	**Yes**	**No**	**Yes**
PVA	10	54	44	20
UF	21	43	37	27

Analyze the data. Write a summary of your results concerning the relationship between the adhesive and the occurrence of face checks. Be sure to include numerical and graphical summaries as well as the details of your significance tests.

9.35. Are Mexican Americans less likely to be selected as jurors? *Castaneda v. Partida* is a case where the Supreme Court review used the phrase "two or three standard deviations" as a criterion for statistical significance. There were 181,535 persons eligible for jury duty, of whom 143,611 were Mexican Americans. Of the 870 people selected for jury duty, 339 were Mexican Americans. We are interested in finding out if there is an association between being a Mexican American and being selected as a juror. Formulate this problem using a two-way table of counts. Construct the 2×2 table using the variables Mexican American or not and juror or not. Find the X^2 statistic and its P-value. Take the square root to obtain the z statistic for comparing two proportions. This is the statistic that the Supreme Court used in its criterion for statistical significance.

9.36. CHALLENGE **Evaluation of a herbal remedy.** A study designed to evaluate the effects of the herbal remedy *Echinacea purpurea* randomly assigned healthy children who were 2 to 11 years old to receive either echinacea or a placebo.[27] Each time a child had an upper respiratory infection (URI), treatment with echinacea or the placebo was given for the duration of the URI. The dose for the echinacea was based on the age of the child according to the recommendation of the manufacturer. The echinacea children had 329 URIs, while the placebo children had 367 URIs. For each URI many variables were measured. One of these was the parental assessment of the illness severity. Here are the data: HERBALASSESS

	Group	
Parental assessment	**Echinacea**	**Placebo**
Mild	153	170
Moderate	128	157
Severe	48	40

They also recorded the presence or absence of various types of *adverse events*. Here is a summary:

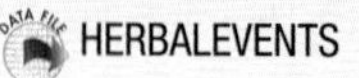

	Group	
Adverse event	**Echinacea**	**Placebo**
Itchiness	13	7
Rash	24	10
"Hyper" behavior	30	23
Diarrhea	38	34
Vomiting	22	21
Headache	33	24
Stomachache	52	41
Drowsiness	63	48
Other	63	48
Any adverse event	152	146

(a) Analyze the parental assessment data. Write a summary of your analysis and conclusion. Be sure to include graphical and numerical summaries.

(b) Analyze each adverse event. Display the results graphically in a single graph. Make a table of the relevant descriptive statistics.

(c) Use a statistical significance test to compare the echinacea URIs with those of the placebo URIs for each type of adverse event. Summarize the results in a table and write a short report giving your conclusions about the effect of echinacea on URIs in healthy children who are 2 to 11 years old. Explain why you need to analyze each type of adverse event separately rather than performing a chi-square test on the preceding 10×2 table.

(d) One concern about analyzing several outcome variables in situations like this is that we may be able to find statistical significance by chance if we look at a sufficiently large number of outcome variables. Explain why this is a concern in general but is not a concern that is important for the interpretation of the results of your analysis here.

(e) The authors of the paper describing these results note that the unit of analysis for their computations is the URI and not the child. They state that similar results were found using more sophisticated statistical methods. Based on the descriptive statistics that you have computed, are you inclined to agree or to disagree with this statement of the authors? Explain your answer.

(f) This study was published in the *Journal of the American Medical Association* (*JAMA*) and was criticized in an article that appeared in *Alternative and Complementary Therapies*.[28] Three herbalists gave responses to the original article. Among their criticisms were (i) the dose of echinacea was too low; (ii) the treatment should have been given before the URI, not at the onset of symptoms; (iii) we should be skeptical of any positive trials on pharmaceuticals or negative trials on natural remedies that are published in *JAMA*; and (iv) *Echinacea angustifolia* (not *E. purpurea*) should have been used in combination with other herbs. Discuss these criticisms and write a summary of your opinions regarding echinacea.

9.37. Use of academic assistance services. The 2005 National Survey of Student Engagement reported on the use of campus services during the first year of college.[29] In terms of academic assistance (for example, tutoring, writing lab), 43% never used the services, 35% sometimes used the services, 15% often used the services, and 7% very often used the services. You decide to see if your large university has this same distribution. You survey first-year students and obtain the counts 79, 83, 36, and 12, respectively. Use a goodness-of-fit test to examine how well your university reflects the national average.

9.38. Goodness of fit to a standard Normal distribution. Computer software generated 500 random numbers that should look as if they are from the standard Normal distribution. They are categorized into five groups: (1) less than or equal to −0.6; (2) greater than −0.6 and less than or equal to −0.1; (3) greater than −0.1 and less than or equal to 0.1; (4) greater than 0.1 and less than or equal to 0.6; and (5) greater than 0.6. The counts in the five groups are 139, 102, 41, 78, and 140, respectively. Find the probabilities for these five intervals using Table A. Then compute the expected number for each interval for a sample of 500. Finally, perform the goodness-of-fit test and summarize your results.

9.39. More on the goodness of fit to a standard Normal distribution. Refer to the previous exercise. Use software to generate your own sample of 500 standard Normal random variables, and perform the goodness-of-fit test. Choose a different set of intervals from the ones used in the previous exercise.

9.40. Goodness of fit to the uniform distribution. Computer software generated 500 random numbers that should look as if they are from the uniform distribution on the interval 0 to 1 (see page 261). They are categorized into five groups: (1) less than or equal to 0.2; (2) greater than 0.2 and less than or equal to 0.4; (3) greater than 0.4 and less than or equal to 0.6; (4) greater than 0.6 and less than or equal to 0.8; and (5) greater than 0.8. The counts in the five groups are 114, 92, 108, 101, and 85, respectively. The probabilities for these five intervals are all the same. What is this probability? Compute the expected number for each interval for a sample of 500. Finally, perform the goodness-of-fit test and summarize your results.

9.41. More on goodness-of-fit to the uniform distribution. Refer to the previous exercise. Use software to generate your own sample of 500 uniform random variables on the interval from 0 to 1, and perform the goodness-of-fit test. Choose a different set of intervals from the ones used in the previous exercise.

9.42. CHALLENGE **Suspicious results?** An instructor who assigned an exercise similar to the one described in the previous exercise received homework from a student who reported a P-value of 0.999. The instructor suspected that the student did not use the computer for the assignment but just made up some numbers for the homework. Why was the instructor suspicious? How would this scenario change if there were 1000 students in the class?

9.43. Is there a random distribution of trees? The Wade Tract in Thomas County, Georgia, is an old-growth forest of longleaf pine trees (*Pinus palustris*) that has survived in a relatively undisturbed state since before the settlement of the area by Europeans.[30]

Foresters who study these trees are interested in how the trees are distributed in the forest. Is there some sort of clustering, resulting in regions of the forest with more trees than others? Or are the tree locations random, resulting in no particular patterns? DATA FILE LONGLEAF1

Here is one way to examine this question. First, we divide the tract into four equal parts, or quadrants, in the east–west direction. Call the four parts, Q_1 to Q_4. Then we take a random sample of 100 trees and count the number of trees in each quadrant. Here are the data:

Quadrant	Q_1	Q_2	Q_3	Q_4
Count	18	22	39	21

(a) If the trees are randomly distributed, we expect to find 25 trees in each quadrant. Why? Explain your answer.

(b) We do not really expect to get *exactly* 25 trees in each quadrant. Why? Explain your answer.

(c) Perform the goodness-of-fit test for these data to determine if these trees are randomly scattered. Write a short report giving the details of your analysis and your conclusion.

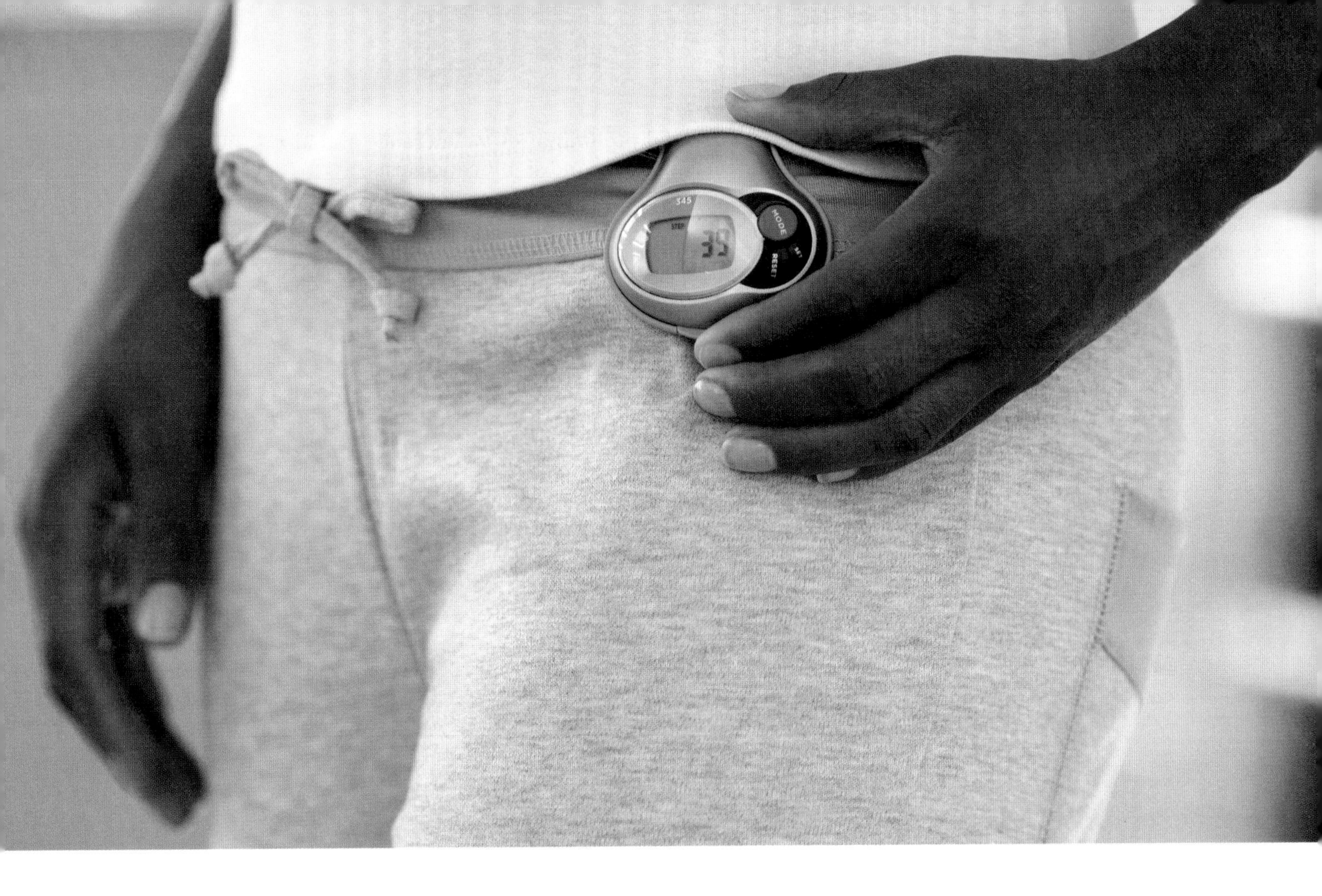

CHAPTER TEN

INFERENCE FOR REGRESSION

In this chapter we return to the study of relationships between variables and describe methods for inference when there is a single quantitative response variable and a single quantitative explanatory variable. The descriptive tools we learned in Chapter 4—scatterplots, least-squares regression, and correlation—are essential preliminaries to inference and also provide a foundation for confidence intervals and significance tests.

We first met the sample mean $\overline{x}$ in Chapter 3 as a measure of the center of a collection of observations. Later we learned that when the data are a random sample from a population, the sample mean is an estimate of the population mean μ. In Chapter 7, we used $\overline{x}$ as the basis for confidence intervals and significance tests for inference about μ.

LOOK BACK
least-squares regression p. 181

Now we will follow the same approach for the problem of fitting straight lines to data. In Chapter 4 we met the least-squares regression line $\hat{y} = b_0 + b_1 x$ as a description of a straight-line relationship between a response variable y and an explanatory variable x. At that point we did not distinguish between sample and population. Now we will think of the least-squares line computed from a sample as an estimate of a *true* regression line for the population.

Following the common practice of using Greek letters for population parameters, we will write the population line as $\beta_0 + \beta_1 x$. This notation reminds us that the intercept b_0 of the fitted line estimates the intercept β_0 of the population line, and the slope b_1 estimates the slope β_1.

The methods detailed in this chapter will help us answer questions such as

- Is the trend in the annual number of tornadoes reported in the United States linear? If so, what is the average yearly increase in the number of tornadoes? How many are predicted for next year?
- Among college students, is there a linear relationship between body mass index and physical activity level measured by a pedometer?

10.1 Simple Linear Regression

Statistical model for linear regression

Simple linear regression studies the relationship between a response variable y and a single explanatory variable x. We expect that different values of x will produce different mean responses. We encountered a similar but simpler situation in Chapter 7 when we discussed methods for comparing two population means. Figure 10.1 illustrates the statistical model for a comparison of blood

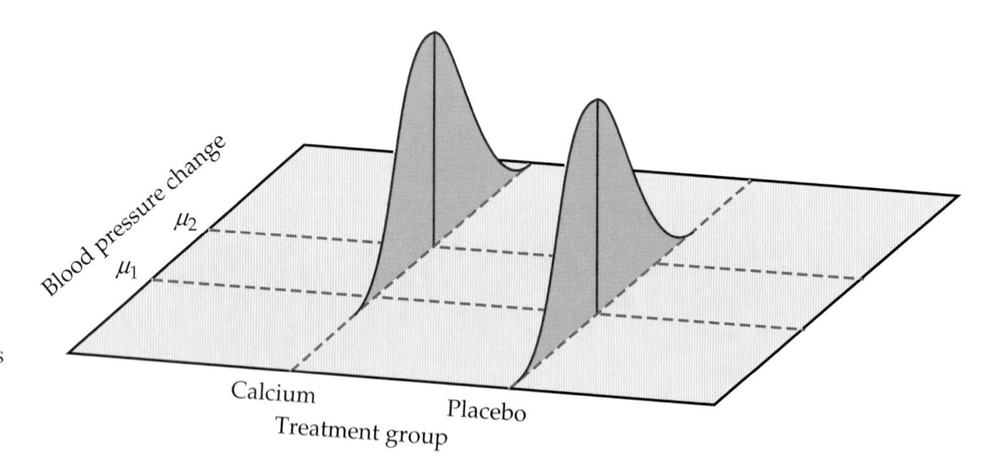

FIGURE 10.1 The statistical model for comparing responses to two treatments. The mean response varies with the treatment.

pressure change in two groups of experimental subjects, one group taking a calcium supplement and the other a placebo. We can think of the treatment (placebo or calcium) as the explanatory variable in this example. This model has two important parts:

- The mean change in blood pressure may be different in the two populations. These means are labeled μ_1 and μ_2 in Figure 10.1.
- Individual changes vary within each population according to a Normal distribution. The two Normal curves in Figure 10.1 describe these responses. These Normal distributions have the same spread, indicating that the population standard deviations are assumed to be equal.

subpopulation

In linear regression, the explanatory variable x is quantitative and can have many different values. Imagine, for example, giving different amounts x of calcium to different groups of subjects. We can think of the values of x as defining different **subpopulations,** one for each possible value of x. Each subpopulation consists of all individuals in the population having the same value of x. If we conducted an experiment with five different amounts of calcium, we could view these values as defining five different subpopulations.

simple linear regression

The statistical model for simple linear regression also assumes that for each value of x the observed values of the response variable y are Normally distributed with a mean that depends on x. We use μ_y to represent these means. In general, the means μ_y can change as x changes according to any sort of pattern. In **simple linear regression,** we assume that the means all lie on a line when plotted against x. To summarize, this model also has two important parts:

- The mean of the response variable y changes as x changes. The means all lie on a straight line. That is, $\mu_y = \beta_0 + \beta_1 x$.
- Individual responses of y with the same x vary according to a Normal distribution. This variation, measured by the standard deviation σ, is the same for all values of x.

population regression line

This statistical model is pictured in Figure 10.2. The line describes how the mean response μ_y changes with x. This is the **population regression line.** The three Normal curves show how the response y will vary for three different values of the explanatory variable x.

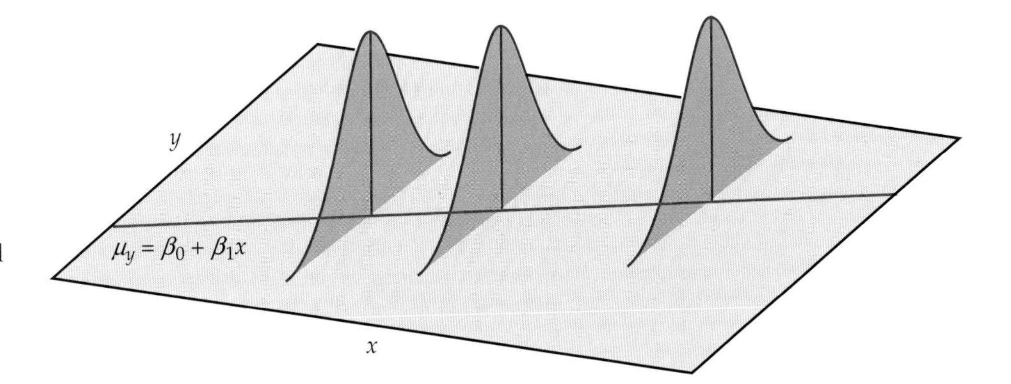

FIGURE 10.2 The statistical model for linear regression. The mean response is a straight-line function of the explanatory variable.

Data for simple linear regression

The data for a linear regression are observed values of y and x. The model takes each x to be a known quantity. In practice, x may not be exactly known. *If the error in measuring x is large, more advanced inference methods are needed.* The response y for a given x is a random variable. The linear regression model describes the mean and standard deviation of this random variable y. These unknown parameters must be estimated from the data.

We will use the following example to explain the fundamentals of simple linear regression. Because regression calculations in practice are always done by statistical software, we will rely on computer output for the arithmetic. We give an example on the text Web site that illustrates how to do the work with a calculator if software is unavailable.

EXAMPLE 10.1

DATA FILE PA_BMI

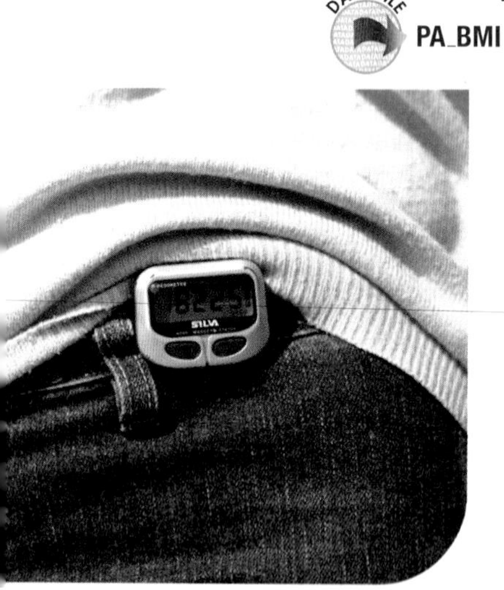

Relationship between physical activity and BMI. Decrease in physical activity is considered to be a major contributor to the increase in the prevalence of being overweight or obese in the general adult population. Because the prevalence of physical inactivity among college students is similar to that among the adult population, many researchers believe that a clearer understanding of college students' physical activity behaviors is needed to develop early interventions. As part of a study, researchers looked at the relationship between physical activity (PA), measured with a pedometer, and body mass index (BMI) among college students.[1] Each participant wore a pedometer for a week and the average number of steps per day (in thousands) was recorded. Various body composition variables, including BMI (in kilograms per square meter), were also measured. For this example, we focus on a sample of 100 female undergraduates.

Before starting our analysis, it is appropriate to consider the extent to which the results can reasonably be generalized. In the original study, undergraduate volunteers were obtained at a large public southeastern university through classroom announcements and campus flyers.

The potential for bias should always be considered when obtaining volunteers. In this case, the participants were screened, and those with severe health issues as well as varsity athletes were excluded. As a result, the researchers considered these volunteers as an SRS from the population of female undergraduates at *this* university. The researchers also limited the extent to which they were willing to generalize their results, stating that similar investigations at universities of different sizes and in other climates of the United States are needed.

In the statistical model for predicting BMI from physical activity, subpopulations are defined by the explanatory variable, physical activity. We could think about sampling women from this university, each averaging the same number of steps per day—say, 9000. Differences in genetic makeup, lifestyle, and diet would be sources of variation that would result in different values of BMI for this subpopulation.

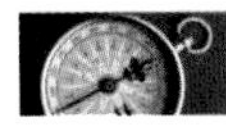

EXAMPLE 10.2

scatterplot p. 158

Graphical display of the BMI and physical activity. We start our analysis with a graphical display of the data. Figure 10.3 is a plot of BMI versus physical activity for our sample of 100 observations. We use the variable names BMI and PA. The least-squares regression line is also shown in the plot. There is a negative association between BMI and PA that appears to be approximately linear. There is also a considerable amount of scatter about this least-squares regression line.

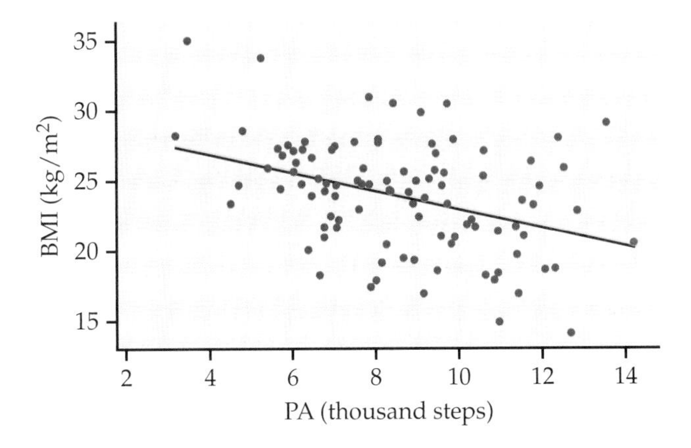

FIGURE 10.3 Scatterplot of BMI versus physical activity (PA), with the least-squares line, for Example 10.2.

Always start with a graphical display of the data. There is no point in trying to do statistical inference if the data do not, at least approximately, meet the assumptions that are the foundation for our inference. Now that we have confirmed that the relationship is approximately linear, we return to predicting BMI for different subpopulations, defined by the explanatory variable physical activity.

Our statistical model assumes that the BMI values are Normally distributed with a mean μ_y that depends upon x in a linear way. Specifically,

$$\mu_y = \beta_0 + \beta_1 x$$

This population regression line gives the average BMI for all values of x. We cannot observe this line, because the observed responses y vary about their means.

The statistical model for linear regression consists of the population regression line and a description of the variation of y about the line. This was displayed in Figure 10.2 with the line and the three Normal curves. The following equation expresses this idea:

$$\text{DATA} = \text{FIT} + \text{RESIDUAL}$$

The FIT part of the model consists of the subpopulation means, given by the expression $\beta_0 + \beta_1 x$. The RESIDUAL part represents deviations of the data from the line of population means. We assume that these deviations are Normally distributed with standard deviation σ. We use ϵ (the lowercase Greek letter epsilon) to stand for the RESIDUAL part of the statistical model. A response y is the sum of its mean and a chance deviation ϵ from the mean. The deviations

ϵ represent "noise," that is, variation in y due to other causes that prevent the observed (x, y)-values from forming a perfectly straight line on the scatterplot.

> **SIMPLE LINEAR REGRESSION MODEL**
>
> Given n observations of the explanatory variable x and the response variable y,
>
> $$(x_1, y_1), \quad (x_2, y_2), \quad \ldots, \quad (x_n, y_n)$$
>
> the **statistical model for simple linear regression** states that the observed response y_i when the explanatory variable takes the value x_i is
>
> $$y_i = \beta_0 + \beta_1 x_i + \epsilon_i$$
>
> Here $\beta_0 + \beta_1 x_i$ is the mean response when $x = x_i$. The deviations ϵ_i are assumed to be independent and Normally distributed with mean 0 and standard deviation σ.
>
> The **parameters** of the model are β_0, β_1, and σ.

Because the means μ_y lie on the line $\mu_y = \beta_0 + \beta_1 x$, they are all determined by β_0 and β_1. Once we have estimates of β_0 and β_1, the linear relationship determines the estimates of μ_y for all values of x. Linear regression allows us to do inference not only for subpopulations for which we have data but also for those corresponding to x's not present in the data.

Estimating the regression parameters

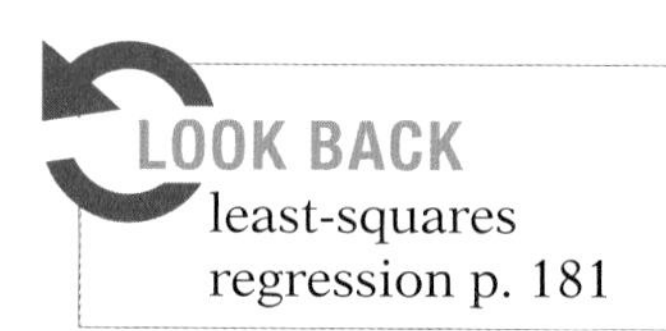
LOOK BACK
least-squares regression p. 181

The method of least squares presented in Chapter 4 fits a line to summarize a relationship between the observed values of an explanatory variable and a response variable. Now we want to use the least-squares line as a basis for inference about a population from which our observations are a sample. We can do this only when the statistical model just presented holds. In that setting, the slope b_1 and intercept b_0 of the least-squares line

$$\hat{y} = b_0 + b_1 x$$

estimate the slope β_1 and the intercept β_0 of the population regression line.

Using the formulas from Chapter 4 (page 182), the slope of the least-squares line is

$$b_1 = r\frac{s_y}{s_x}$$

and the intercept is

$$b_0 = \bar{y} - b_1\bar{x}$$

LOOK BACK
correlation p. 171

Here, r is the correlation between y and x, s_y is the standard deviation of y, and s_x is the standard deviation of x. Notice that if the slope is 0, so is the correlation, and vice versa.

As a consequence of the model assumptions on the deviations ϵ_i, the estimates b_0 and b_1 are Normally distributed with means β_0 and β_1 and standard deviations that can be estimated from the data. In fact, even if the ϵ_i are not

Normally distributed, a general form of the central limit theorem tells us that the distributions of b_0 and b_1 will be approximately Normal.

Thus, our procedures for inference about the population regression line will be similar in flavor to those described in Chapter 7 for the population mean and difference in means. Moreover, we need to watch out for outliers and influential observations, as they can invalidate the results of inference for regression.

The predicted value of y for a given value x^* of x is the point on the least-squares line $\hat{y} = b_0 + b_1 x^*$. This predicted value is an unbiased estimator of the mean response μ_y when $x = x^*$. The **residual** is

residual

$$\begin{aligned} e_i &= \text{observed response} - \text{predicted response} \\ &= y_i - \hat{y}_i \\ &= y_i - b_0 - b_1 x_i \end{aligned}$$

The residuals e_i correspond to the model deviations ϵ_i. The e_i sum to 0, and the ϵ_i come from a population with mean 0.

The remaining parameter to be estimated is σ, which measures the variation of y about the population regression line. Because this parameter is the standard deviation of the model deviations, it should come as no surprise that we use the residuals to estimate it. As usual, we work first with the variance and take the square root to obtain the standard deviation.

For simple linear regression, the estimate of σ^2 is the average squared residual

$$\begin{aligned} s^2 &= \frac{\sum e_i^2}{n-2} \\ &= \frac{\sum (y_i - \hat{y}_i)^2}{n-2} \end{aligned}$$

LOOK BACK
sample variance
p. 116

We average by dividing the sum by $n - 2$ to make s^2 an unbiased estimate of σ^2. The sample variance of n observations uses the divisor $n - 1$ for this same reason. The quantity $n - 2$ is called the **degrees of freedom** for s^2. The estimate of σ is given by

degrees of freedom

$$s = \sqrt{s^2}$$

We will use statistical software to calculate the regression for predicting BMI from physical activity for Example 10.1. In entering the data, we chose the names PA for the explanatory variable and BMI for the response. *It is good practice to use names, rather than just x and y, to remind yourself which data the output describes.*

EXAMPLE 10.3

Statistical software output for BMI and physical activity. Figure 10.4 gives the outputs for three commonly used statistical software packages and Excel. Other software will give similar information. The SPSS output reports estimates of our three parameters as $b_0 = 29.578$, $b_1 = -0.655$, and $s = 3.6549$. Be sure that you can find these entries in this output and the corresponding values in the other outputs.

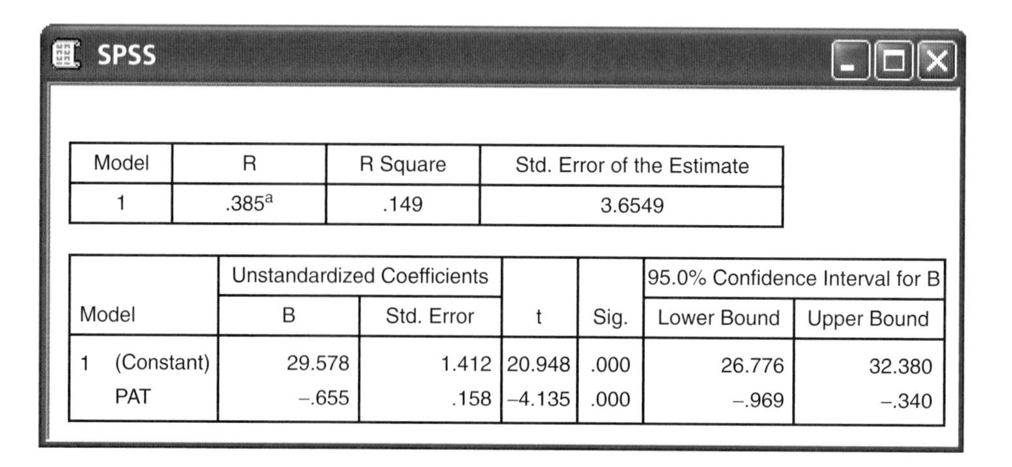

SPSS

Model	R	R Square	Std. Error of the Estimate
1	.385[a]	.149	3.6549

Model		Unstandardized Coefficients		t	Sig.	95.0% Confidence Interval for B	
		B	Std. Error			Lower Bound	Upper Bound
1	(Constant)	29.578	1.412	20.948	.000	26.776	32.380
	PAT	−.655	.158	−4.135	.000	−.969	−.340

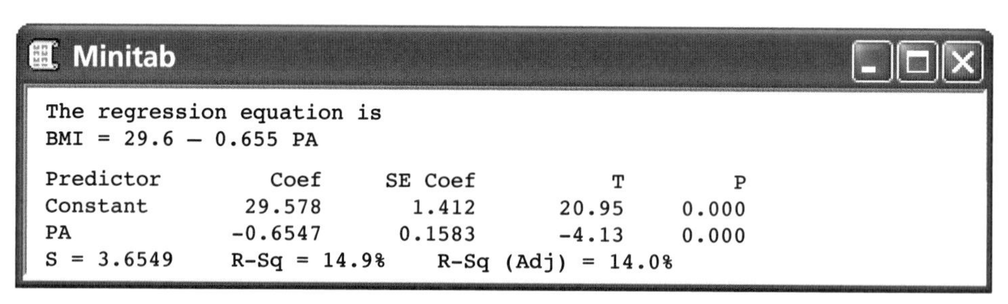

Minitab

```
The regression equation is
BMI = 29.6 – 0.655 PA

Predictor        Coef    SE Coef        T        P
Constant       29.578      1.412    20.95    0.000
PA            -0.6547     0.1583    -4.13    0.000
S = 3.6549    R-Sq = 14.9%    R-Sq (Adj) = 14.0%
```

Excel

	A	B	C	D	E	F	G
1	SUMMARY OUTPUT						
2							
3	*Regression Statistics*						
4	Multiple R	0.385409059					
5	R Square	0.148540143					
6	Adjusted R Square	0.139851777					
7	Standard Error	3.65488311					
8	Observations	100					
9							
10	ANOVA						
11		*df*	*SS*	*MS*	*F*	*Significance F*	
12	Regression	1	228.3771867	228.377	17.09644	7.50303E-05	
13	Residual	98	1309.100713	13.3582			
14	Total	99	1537.4779				
15							
16		*Coefficients*	*Standard Error*	*t Stat*	*P-value*	*Lower 95%*	*Upper 95%*
17	Intercept	29.57824714	1.411978287	20.9481	5.71E-38	28.77622225	32.38027203
18	X Variable 1	−0.65468576	0.158336132	−4.1347	7.5E-05	−0.96889865	−0.34047287

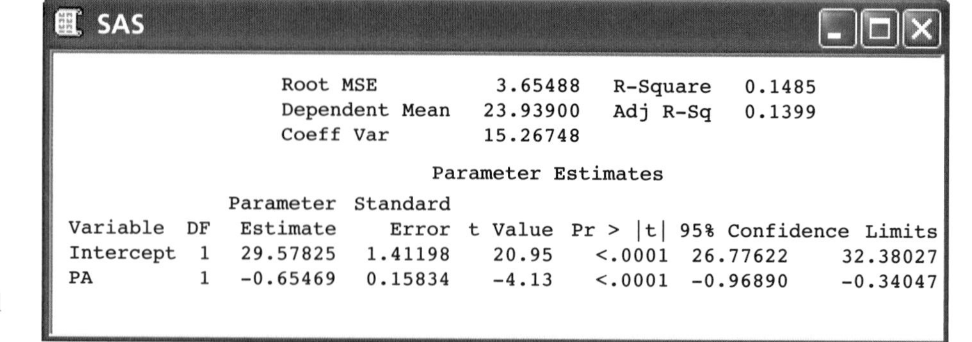

SAS

```
          Root MSE          3.65488   R-Square   0.1485
          Dependent Mean   23.93900   Adj R-Sq   0.1399
          Coeff Var        15.26748

                      Parameter Estimates

              Parameter  Standard
Variable  DF   Estimate     Error  t Value  Pr > |t|  95% Confidence Limits
Intercept  1   29.57825   1.41198    20.95    <.0001  26.77622     32.38027
PA         1   -0.65469   0.15834    -4.13    <.0001  -0.96890     -0.34047
```

FIGURE 10.4 Regression output from SPSS, Minitab, Excel, and SAS for the physical activity example.

The least-squares regression line is the straight line that is plotted in Figure 10.3. We would report it as

$$\widehat{\text{BMI}} = 29.578 - 0.655\text{PA}$$

with a model standard deviation of $s = 3.65$. Note that the number of digits provided varies with the software used, and we have rounded off the values to three significant digits. *It is important to avoid cluttering up your report of the results of a statistical analysis with many digits that are not relevant.* Software often reports many more digits than are meaningful or useful.

The outputs contain other information that we will ignore for now. Computer outputs often give more information than we want or need. This is done to reduce user frustration when a software package does not print out the particular statistics wanted for an analysis. *The experienced user of statistical software learns to ignore the parts of the output that are not needed for the current problem.*

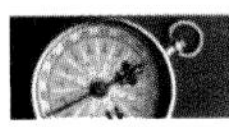

EXAMPLE 10.4

Predicted values and residuals for BMI. We can now use the least-squares regression equation to find the predicted BMI corresponding to any value of PA. Suppose that a female college student averages 8000 steps per day. We predict that this person will have a BMI of

$$29.578 - 0.655(8) = 24.338$$

If her actual BMI is 25.655, then the residual would be

$$y - \hat{y} = 25.655 - 24.338 = 1.317$$

Now that we have fitted a line, we should check the conditions that the simple linear regression model imposes on this fit. *There is no point in trying to do statistical inference if the data do not, at least approximately, meet the conditions that are the foundation for the inference.*

This check is done through a visual examination of the residuals for Normality, outliers and influential observations, and any remaining patterns in the data. We usually plot the residuals both against the case number (especially if this reflects the order in which the observations were collected) and against the explanatory variable. For this example, we will just look at the explanatory variable.

Figure 10.5 (next page) gives a plot of the residuals versus physical activity, with a smooth function fitted to the data. Some software packages provide smoothing algorithms, which can be used to help investigate patterns or trends in the data. The smooth-function fit here suggests that the residuals increase slightly at both low and high physical activity levels. This pattern in the residuals could mean that a curved relationship between BMI and physical activity would better fit the data. It also could just be chance variation.

Notice that there is a large positive residual at each end of physical activity range. Since the effect does not appear to be particularly large, we will ignore this possible curved relationship for the present analysis and investigate it further in Exercise 11.15 (page 533).

In Figure 10.5 the spread of the residuals is roughly uniform across the range of PA, suggesting that the assumption of a common standard deviation

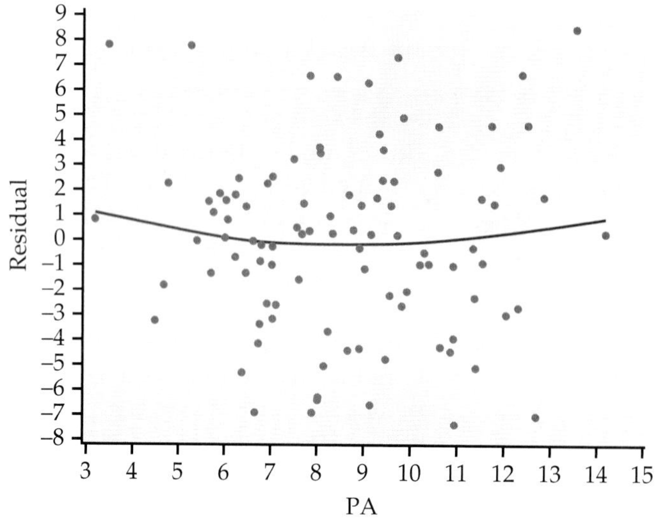

FIGURE 10.5 Plot of residuals versus physical activity (PA), with a smooth function, for the physical activity example.

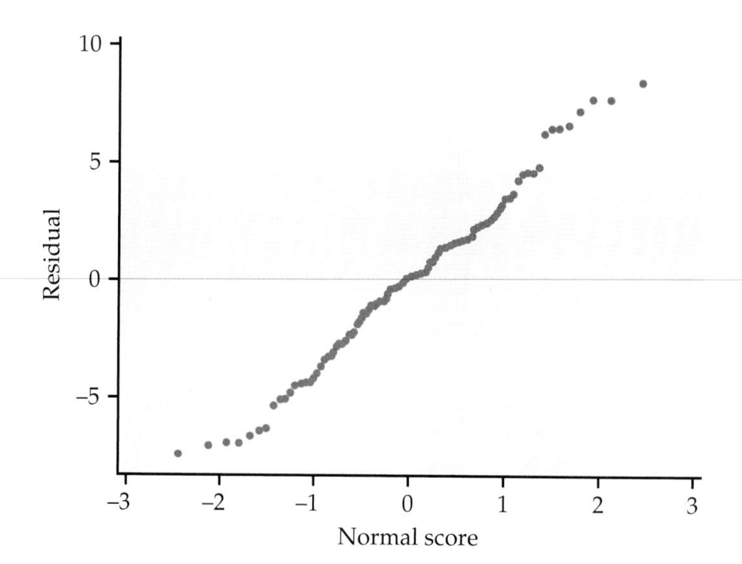

FIGURE 10.6 Normal quantile plot of the residuals for the physical activity example.

is reasonable. There also do not appear to be any outliers or influential observations. Finally, Figure 10.6 is a Normal quantile plot of the residuals. Because the plot looks fairly straight, we are confident that we do not have a serious violation of our assumption that the model deviations are Normally distributed.

USE YOUR KNOWLEDGE

10.1 Understanding a linear regression model. Consider a linear regression model with $\mu_y = 50.3 + 1.7x$ and standard deviation $\sigma = 4.2$.

(a) What is the slope of the population regression line?

(b) Explain clearly what this slope says about the change in the mean of y for a change in x.

(c) What is the subpopulation mean when $x = 7$?

(d) Between what two values would approximately 95% of the observed responses y fall when $x = 7$?

10.2 More on PA and BMI. Refer to Examples 10.3 (page 495) and 10.4 (page 497).

(a) What is the average BMI for a woman who averages 9500 steps per day?

(b) If an observed BMI at $x = 9.5$ were 22.8, what is the residual?

(c) Suppose that you want to use the estimated population regression line to examine the average BMI for a woman who averages 2000, 7000, 12,000, and 17,000 steps per day. Discuss the appropriateness of using the equation to predict BMI for each of these activity levels.

Transforming variables

We started our analysis of Example 10.1 with a scatterplot to check whether the relationship between BMI and physical activity could be summarized with a straight line. We followed that with a residual plot (Figure 10.5) and a Normal quantile plot (Figure 10.6) to check Normality and any remaining patterns in the data. *A check of model assumptions should always be done prior to inference.*

When the relationship between y and x is not linear, sometimes we can make it linear by a transformation of one or both of the variables. Here is an example.

EXAMPLE 10.5

DATA FILE MPH_MPG

Relationship between speed driven and fuel efficiency. Computers in some vehicles calculate various quantities related to the vehicle's performance. One of these is the fuel efficiency, or gas mileage, expressed as miles per gallon (mpg). Another is the average speed in miles per hour (mph). For one vehicle equipped in this way, average mpg and mph were recorded each time the gas tank was filled, and the computer was then reset.[2] How does the speed at which the vehicle is driven affect the fuel efficiency? We will work with a simple random sample of 60 observations.

Our statistical modeling for this data set is concerned with the process by which speed affects the fuel efficiency. Except possibly for the owner, no one cares about the particular vehicle. The results are interesting only if they can be applied to other, similar vehicles that are driven under similar conditions. Although we would not expect the parameters that describe the relationship between speed and fuel efficiency to be *exactly* the same for similar vehicles, we would expect to find qualitatively similar results.

EXAMPLE 10.6

Graphical display of the fuel efficiency relationship. Figure 10.7 is a plot of fuel efficiency versus speed for our sample of 60 observations. We use the variable names MPG and MPH. The least-squares regression line and a

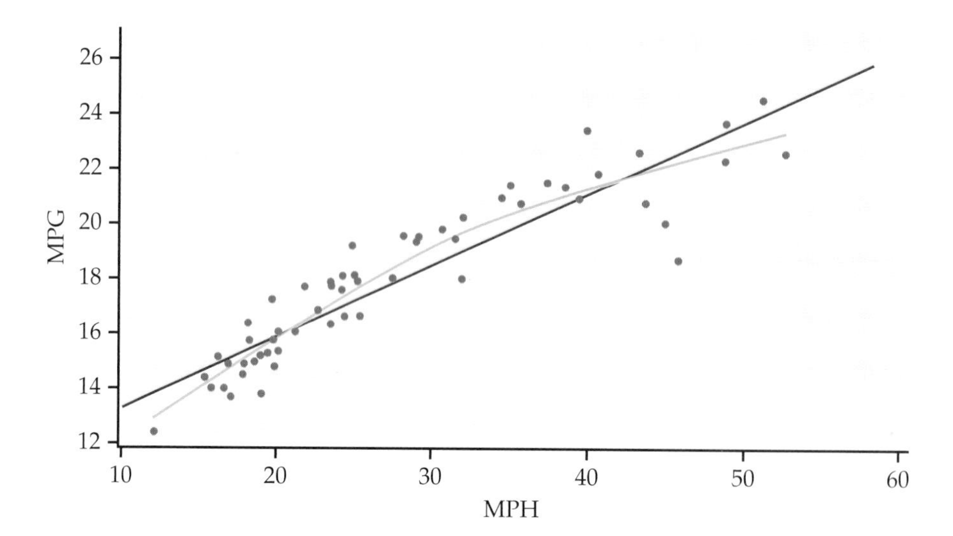

FIGURE 10.7 Scatterplot of MPG versus MPH, with a smooth function (blue) and the least-squares line (red), for Example 10.6. The relationship between MPG and MPH does not appear to be linear.

smooth function are also shown in the plot. Although there is a positive association between MPG and MPH, the fit is not linear. The smooth function shows us that the relationship levels off somewhat with increasing speed.

Given this nonlinearity, we have to make a choice about how to proceed. One approach would be to confine our interest to speeds that are 30 mph or less, a region where it appears that a line would be a good fit to the data. Another possibility is to consider a transformation that will make the relationship approximately linear for the entire set of data.

EXAMPLE 10.7

LOOK BACK
log transformation
p. 164

Is this relationship linear? One type of function that resembles the smooth-function fit in Figure 10.7 is a logarithm. Therefore, we will examine the effect of transforming speed by taking the natural logarithm. The result is shown in Figure 10.8. In this plot, the smooth function and the line are quite close. We are satisfied that the relationship between the log of speed and fuel efficiency is approximately linear for this set of data.

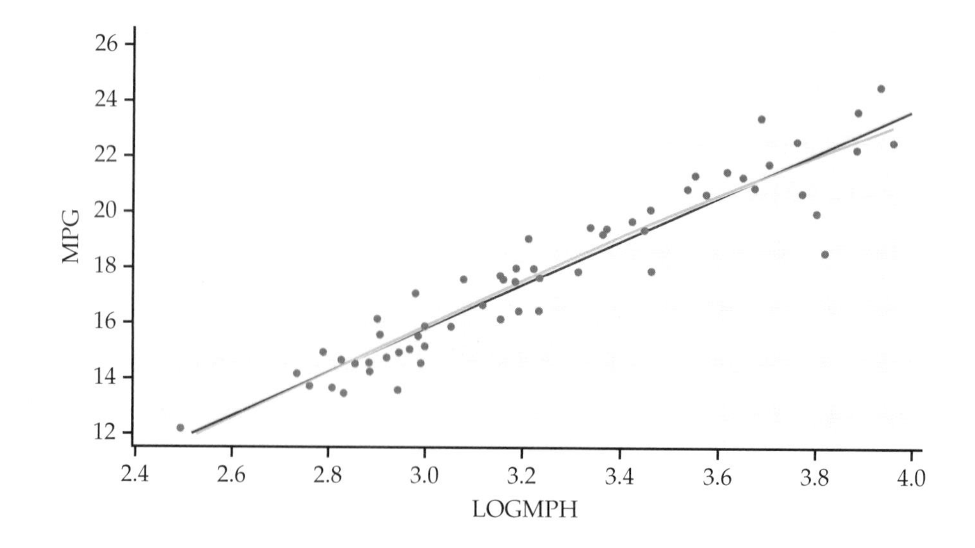

FIGURE 10.8 Scatterplot of MPG versus the logarithm of MPH, with a smooth function (blue) and the least-squares line (red), for Example 10.7. Here, the line and smooth function are very close.

Although this transformation has resulted in an approximately linear relationship, there are still other assumptions of the simple linear model that must be met. In this example, we can show that these assumptions are also satisfied, so statistical inference can be performed. *In other cases, transforming a variable may help linearity but harm the Normality and constant-variance assumptions.* In those cases a more sophisticated model is needed.

SECTION 10.1 SUMMARY

The statistical model for **simple linear regression** assumes that the means of the response variable y fall on a line when plotted against x, with the observed y's varying Normally about these means. For n observations, this model can be written

$$y_i = \beta_0 + \beta_1 x_i + \epsilon_i$$

where $i = 1, 2, \ldots, n$, and the ϵ_i are assumed to be independent and Normally distributed with mean 0 and standard deviation σ. Here $\beta_0 + \beta_1 x_i$ is the mean response when $x = x_i$. The **parameters** of the model are β_0, β_1, and σ.

The **population regression line** intercept and slope, β_0 and β_1, are estimated by the intercept and slope of the **least-squares regression line,** b_0 and b_1. The parameter σ is estimated by

$$s = \sqrt{\frac{\sum e_i^2}{n-2}}$$

where the e_i are the **residuals**

$$e_i = y_i - \hat{y}_i$$

Prior to inference, always examine the residuals for Normality, constant variance, and any other remaining patterns in the data. **Plots of the residuals** both against the case number and against the explanatory variable are usually done as part of this examination.

Sometimes a **transformation** of one or both of the variables can make their relationship linear. However, these transformations can harm the assumptions of Normality and constant variance.

10.2 Inference in Simple Linear Regression

Given the simple linear regression model, we will now learn how to do inference about

- the slope β_1 and the intercept β_0 of the population regression line,
- the mean response μ_y for a given value of x, and
- an individual future response y for a given value of x.

Chapter 7 presented confidence intervals and significance tests for means and differences in means. In each case, inference rested on the standard errors of estimates and on t distributions. Inference in a linear regression is similar in principle. For example, the confidence intervals have the form

$$\text{estimate} \pm t^* \text{SE}_{\text{estimate}}$$

where t^* is a critical point of a t distribution. The formulas for the estimate and standard error, however, are more complicated.

Inference about the slope β_1 and intercept β_0

Confidence intervals and tests for the population slope and intercept are based on the Normal sampling distributions of the estimates b_1 and b_0. The standard deviations of these estimates are multiples of σ, the model parameter that describes the variability about the true regression line. Because we do not know σ, we estimate it by s, the variability of the data about the least-squares line. When we do this, we move from the Normal distribution to t distributions with degrees of freedom $n-2$, the degrees of freedom of s. We give formulas for the standard errors SE_{b_1} and SE_{b_0} on the text Web site. Here we will concentrate on the basic ideas and let the computer do the computations.

INFERENCE FOR REGRESSION SLOPE

A **level C confidence interval for the slope** β_1 is

$$b_1 \pm t^* SE_{b_1}$$

In this expression t^* is the value for the $t(n-2)$ density curve with area C between $-t^*$ and t^*.

To test the hypothesis H_0: $\beta_1 = 0$, compute the **test statistic**

$$t = \frac{b_1}{SE_{b_1}}$$

The **degrees of freedom** are $n-2$. In terms of a random variable T having the $t(n-2)$ distribution, the P-value for a test of H_0 against

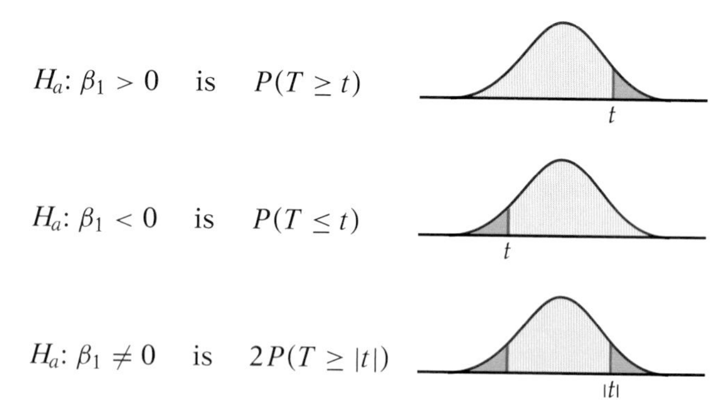

H_a: $\beta_1 > 0$ is $P(T \geq t)$
t
H_a: $\beta_1 < 0$ is $P(T \leq t)$
t
H_a: $\beta_1 \neq 0$ is $2P(T \geq |t|)$
|t|

Confidence intervals and significance tests for the intercept β_0 are exactly the same, replacing b_1 and SE_{b_1} by b_0 and its standard error SE_{b_0}. Although computer outputs often include a test of $H_0: \beta_0 = 0$, this information usually has little practical value. From the equation for the population regression line, $\mu_y = \beta_0 + \beta_1 x$, we see that β_0 is the mean response corresponding to $x = 0$. In many practical situations, this subpopulation does not exist or is not interesting.

On the other hand, the test of H_0: $\beta_1 = 0$ is quite useful. When we substitute $\beta_1 = 0$ in the model, the x term drops out and we are left with

$$\mu_y = \beta_0$$

This model says that the mean of y does *not* vary with x. In other words, all the y's come from a single population with mean β_0, which we would estimate by $\bar{y}$. The hypothesis H_0: $\beta_1 = 0$ therefore says that there is *no straight-line relationship between y and x* and that linear regression of y on x is of no value for predicting y.

EXAMPLE 10.8

Statistical software output, continued. The computer outputs in Figure 10.4 (page 496) for the BMI problem contain the information needed for inference about the regression slope and intercept. Let's look at the SPSS output. The column labeled Std. Error gives the standard errors of the estimates. The value of SE_{b_1} appears on the line labeled with the variable name for the explanatory variable, PA. It is given as 0.158. In a summary we would report that the regression coefficient for the average steps per day is −0.655 with a standard error of 0.158.

The t statistic and P-value for the test of H_0: $\beta_1 = 0$ against the two-sided alternative H_a: $\beta_1 \neq 0$ appear in the columns labeled t and Sig. We can verify the t calculation from the formula for the standardized estimate:

$$t = \frac{b_1}{SE_{b_1}} = \frac{-0.655}{0.158} = -4.14$$

The P-value is given as 0.000. This is a rounded number and from that information we can conclude that $P < 0.0005$. The other outputs in Figure 10.4 also indicate that the P-value is very small. We will report the result as $P < 0.001$ because 1 chance in 1000 is sufficiently small for us to decisively reject the null hypothesis.

We have found a statistically significant linear relationship between physical activity and BMI. The estimated slope is more than 4 standard errors away from zero. Because this is highly unlikely to happen if the true slope is zero, we have strong evidence for our claim that the slope is different from zero.

Note, however, that this is not the same as concluding that we have found a strong relationship between the response and explanatory variables in this example. We saw in Figure 10.3 (page 493) that there was a lot of scatter about the regression line. *A very small P-value for the significance test for a zero slope does not necessarily imply that we have found a strong relationship.*

A confidence interval provides additional information about this relationship.

EXAMPLE 10.9

Confidence interval for the slope. A confidence interval for β_1 requires a critical value t^* from the $t(n-2) = t(98)$ distribution. In Table D there are entries for 80 and 100 degrees of freedom. The values for these rows are very similar. To be conservative, we will use the larger critical value, for 80 degrees of freedom. Find the confidence level values at the bottom of the table. In the 95% confidence column, the entry for 80 degrees of freedom is $t^* = 1.990$.

To compute the 95% confidence interval for β_1 we combine the estimate of the slope with the margin of error:

$$b_1 \pm t^* \mathrm{SE}_{b_1} = -0.655 \pm (1.990)(0.158)$$
$$= -0.655 \pm 0.314$$

The interval is $(-0.969, -0.341)$. This agrees with the intervals given by the software outputs that provide this information in Figure 10.4. We estimate that an increase of 1000 steps per day is associated with a decrease in BMI of between 0.341 and 0.969 $\mathrm{kg/m^2}$.

Note that the intercept in this example is not of practical interest. It estimates average BMI when the activity level is 0, a value that isn't realistic. For this reason, we do not compute a confidence interval for β_0.

USE YOUR KNOWLEDGE

10.3 Significance test for the slope. Test the null hypothesis that the slope is zero versus the two-sided alternative in each of the following settings:

(a) $n = 25$, $\hat{y} = 5.3 + 1.10x$, and $\mathrm{SE}_{b_1} = 0.58$

(b) $n = 25$, $\hat{y} = 5.3 + 2.10x$, and $\mathrm{SE}_{b_1} = 0.58$

(c) $n = 100$, $\hat{y} = 5.3 + 1.10x$, and $\mathrm{SE}_{b_1} = 0.58$

10.4 95% confidence intervals for the slope. For each of the settings in the previous exercise, find a 95% confidence interval for the slope.

Confidence intervals for mean response

Besides drawing inference about the slope (and sometimes the intercept) in a linear regression, we may want to use the estimated regression line to make predictions about the response y at certain values of x. We may be interested in the mean response for different subpopulations or in the response of future observations at different values of x. In either case, we'd want an estimate and associated margin of error.

For any specific value of x, say x^*, the mean of the response y in this subpopulation is given by

$$\mu_y = \beta_0 + \beta_1 x^*$$

To estimate this mean from the sample, we substitute the estimates b_0 and b_1 for β_0 and β_1:

$$\hat{\mu}_y = b_0 + b_1 x^*$$

A confidence interval for μ_y adds to this estimate a margin of error based on the standard error $\mathrm{SE}_{\hat{\mu}}$. (The formula for the standard error is given on the text Web site.)

CONFIDENCE INTERVAL FOR A MEAN RESPONSE

A **level C confidence interval for the mean response** μ_y when x takes the value x^* is

$$\hat{\mu}_y \pm t^* \mathrm{SE}_{\hat{\mu}}$$

where t^* is the value for the $t(n-2)$ density curve with area C between $-t^*$ and t^*.

Many computer programs calculate confidence intervals for the mean response corresponding to each of the x-values in the data. Some can calculate an interval for any value x^* of the explanatory variable. We will use a plot to illustrate these intervals.

EXAMPLE 10.10

Confidence intervals for the mean response. Figure 10.9 shows the upper and lower confidence limits on a graph with the data and the least-squares line. The 95% confidence limits appear as dashed curves. For any x^*, the confidence interval for the mean response extends from the lower dashed curve to the upper dashed curve. The intervals are narrowest for values of x^* near the mean of the observed x's and widen as x^* moves away from $\bar{x}$.

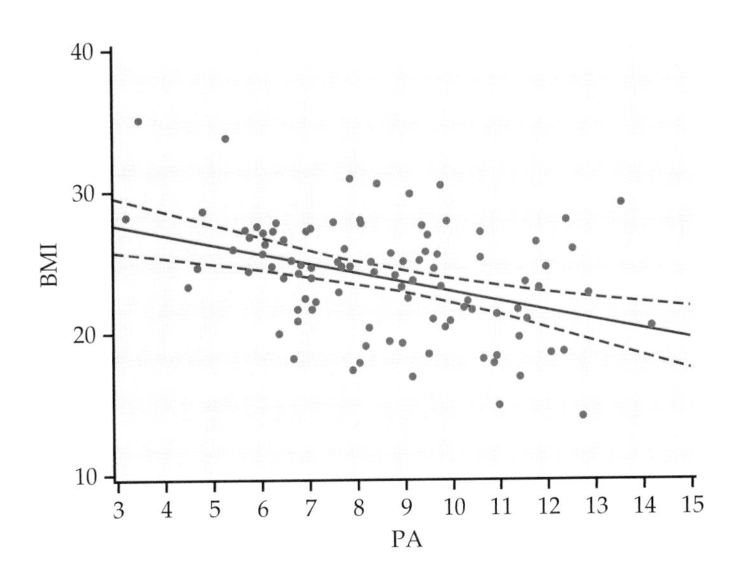

FIGURE 10.9 The 95% confidence limits (dashed curves) for the mean response for the physical activity example.

Some software will do these calculations directly if you input a value for the explanatory variable. Others will calculate the intervals for each value of x in the data set. Creating a new data set with an additional observation with x equal to the value of interest and y missing will often work.

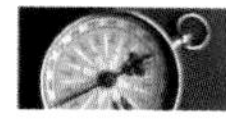

EXAMPLE 10.11

Confidence interval for an average of 9000 steps per day. Let's find the confidence interval for the average BMI at $x = 9.0$. Our predicted BMI is

$$\begin{aligned}\widehat{\text{BMI}} &= 29.578 - 0.655\text{PA} \\ &= 29.578 - 0.655(9.0) \\ &= 23.7\end{aligned}$$

Software tells us that the 95% confidence interval for the mean response is 23.0 to 24.4 kg/m^2.

If we sampled many women who averaged 9000 steps per day, we would expect the average BMI to be between 23.0 and 24.4 kg/m^2. Note that many of the observations in Figure 10.9 lie outside the confidence bands. *These confidence intervals do not tell us what BMI to expect for a single observation at a particular average number of steps per day.* We need a different kind of interval for this purpose.

Prediction intervals

In the last example, we predicted the average BMI for an average of 9000 steps per day. Suppose that we now want to predict a future observation of BMI for a woman averaging 9000 steps per day. Our best guess of BMI is what we obtained using the regression equation, that is, 23.7 kg/m^2. The margin of error, on the other hand, is larger because it is harder to predict an individual value than to predict the mean.

The predicted response y for an individual case with a specific value x^* of the explanatory variable x is

$$\hat{y} = b_0 + b_1 x^*$$

This is the same as the expression for $\hat{\mu}_y$. That is, the fitted line is used both to estimate the mean response when $x = x^*$ and to predict a single future response. We use the two notations $\hat{\mu}_y$ and $\hat{y}$ to remind ourselves of these two distinct uses.

prediction interval

A useful prediction should include a margin of error to indicate its accuracy. The interval used to predict a future observation is called a **prediction interval.** Although the response y that is being predicted is a random variable, the interpretation of a prediction interval is similar to that for a confidence interval. Consider doing the following many times:

- Draw a sample of n observations (x_i, y_i) and then one additional observation (x^*, y).
- Calculate the 95% prediction interval for y when $x = x^*$ using the sample of size n.

Then 95% of the prediction intervals will contain the value of y for the additional observation. In other words, the probability that this method produces an interval that contains the value of a future observation is 0.95.

The form of the prediction interval is very similar to that of the confidence interval for the mean response. The difference is that the standard error $SE_{\hat{y}}$ used in the prediction interval includes both the variability due to the fact that the least-squares line is not exactly equal to the true regression line *and* the variability of the future response variable y around the subpopulation mean. (The formula for $SE_{\hat{y}}$ appears on the text Web site.)

PREDICTION INTERVAL FOR A FUTURE OBSERVATION

A **level C prediction interval for a future observation** on the response variable y from the subpopulation corresponding to x^* is

$$\hat{y} \pm t^* SE_{\hat{y}}$$

where t^* is the value for the $t(n-2)$ density curve with area C between $-t^*$ and t^*.

Again, we use a graph to illustrate the results.

EXAMPLE 10.12

Prediction intervals for BMI. Figure 10.10 shows the upper and lower prediction limits, along with the data and the least-squares line. The 95% prediction limits are indicated by the dashed curves. Compare this figure with Figure 10.9 (page 505), which shows the 95% confidence limits drawn to the same scale. The upper and lower limits of the prediction intervals are farther from the least-squares line than are the confidence limits. This results in most, but not all, of the observations in Figure 10.10 lying within the prediction bands.

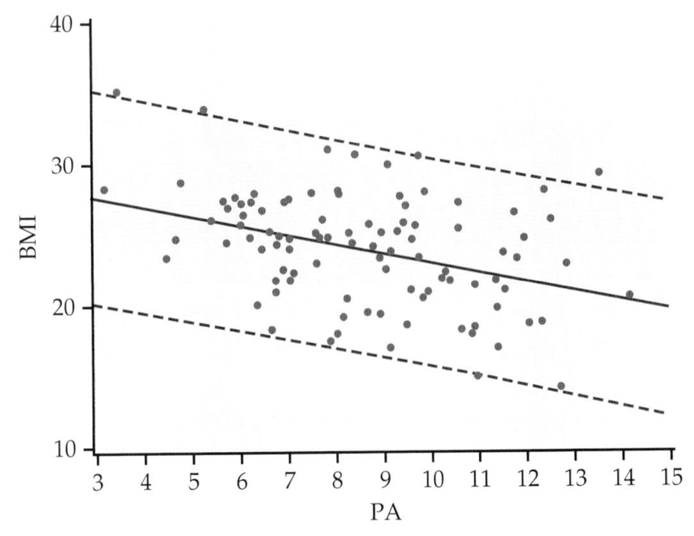

FIGURE 10.10 The 95% prediction limits (dashed curves) for individual responses for the physical activity example. Compare with Figure 10.9. The limits are wider because the margins of error incorporate the variability about the subpopulation means.

The comparison of Figures 10.9 and 10.10 reminds us that the interval for a single future observation must be larger than an interval for the mean of its subpopulation.

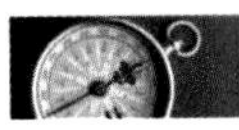

EXAMPLE 10.13

Prediction interval for an average of 9000 steps per day. Let's find the prediction interval for a future observation of BMI when a college-aged woman averages 9000 steps per day. The predicted value is the same as the estimate of the average BMI that we calculated in Example 10.11, that is, 23.7 kg/m^2. Software tells us that the 95% prediction interval is 16.4 to 31.0 kg/m^2. This interval is extremely wide, covering BMI values that are classified as underweight and obese. Because of the large amount of scatter about the regression line, prediction intervals here are relatively useless.

While a larger sample would better estimate the population regression line, it would not reduce the degree of scatter about the line. In other words, prediction intervals for BMI given activity level will always be wide. This example clearly demonstrates that a very small P-value for the significance test for a zero slope does not necessarily imply that we have found a strong predictive relationship.

USE YOUR KNOWLEDGE

10.5 Margin of error for the predicted mean. Refer to Example 10.11. What is the 95% margin of error of $\hat{\mu}_y$ when $x = 9.0$? Would you expect this margin of error to be larger, smaller, or the same for 90% confidence? Explain.

10.6 Margin of error for the predicted response. Refer to Example 10.13. What is the 95% margin of error of $\hat{y}$ when $x = 9.0$? Would you expect the 95% margin of error to be larger, smaller, or the same for $x = 10.0$? Explain.

SECTION 10.2 SUMMARY

A **level C confidence interval for β_1** is

$$b_1 \pm t^* \mathrm{SE}_{b_1}$$

where t^* is the value for the $t(n-2)$ density curve with area C between $-t^*$ and t^*.

The **test of the hypothesis** H_0: $\beta_1 = 0$ is based on the ***t* statistic**

$$t = \frac{b_1}{\mathrm{SE}_{b_1}}$$

and the $t(n-2)$ distribution. This tests whether there is a straight-line relationship between y and x. There are similar formulas for confidence intervals and tests for β_0, but these are meaningful only in special cases.

The **estimated mean response** for the subpopulation corresponding to the value x^* of the explanatory variable is

$$\hat{\mu}_y = b_0 + b_1 x^*$$

A **level C confidence interval for the mean response** is

$$\hat{\mu}_y \pm t^* \mathrm{SE}_{\hat{\mu}}$$

where t^* is the value for the $t(n-2)$ density curve with area C between $-t^*$ and t^*.

The **estimated value of the response variable** y for a future observation from the subpopulation corresponding to the value x^* of the explanatory variable is

$$\hat{y} = b_0 + b_1 x^*$$

A **level C prediction interval** for the estimated response is

$$\hat{y} \pm t^* \mathrm{SE}_{\hat{y}}$$

where t^* is the value for the $t(n-2)$ density curve with area C between $-t^*$ and t^*. The standard error for the prediction interval is larger than that for the confidence interval because it also includes the variability of the future observation around its subpopulation mean.

CHAPTER 10 EXERCISES

For Exercises 10.1 and 10.2, see pages 498–499; for Exercises 10.3 and 10.4, see page 504; and for Exercises 10.5 and 10.6, see page 508.

10.7. Constructing confidence intervals for the change in BMI. Refer to Example 10.9 (page 503). For the following three changes in average steps per day, construct a 95% confidence interval for the change in BMI.

(a) PA increases from 9.2 to 10.2.

(b) PA decreases from 10.5 to 9.5.

(c) PA increases from 11.8 to 12.3.

10.8. What's wrong? For each of the following, explain what is wrong and why.

(a) The slope describes the change in x for a change in y.

(b) The population regression line is $y = b_0 + b_1x$.

(c) A 95% confidence interval for the mean response is the same width regardless of x.

10.9. What's wrong? For each of the following, explain what is wrong and why.

(a) The parameters of the simple linear regression model are b_0, b_1, and s.

(b) To test H_0: $b_1 = 0$, use a t test.

(c) For a particular value of the explanatory variable x, the confidence interval for the mean response will be wider than the prediction interval for a future observation.

10.10. Public university tuition: 2000 versus 2008. Table 10.1 shows the in-state undergraduate tuition and required fees for 33 public universities in 2000 and 2008.[3] DATA FILE TUITION

(a) Plot the data with the 2000 tuition on the x axis and describe the relationship. Are there any outliers or unusual values? Does a linear relationship between the tuition in 2000 and 2008 seem reasonable?

(b) Run the simple linear regression and give the equation of the least-squares regression line.

(c) Obtain the residuals and plot them versus the 2000 tuition amount. Is there anything unusual in the plot?

(d) Do the residuals appear to be approximately Normal? Explain.

(e) Give the null and alternative hypotheses for examining the relationship between 2000 and 2008 tuition amounts.

(f) Write down the test statistic and P-value for the hypotheses stated in part (e). State your conclusions.

10.11. More on public university tuition. Refer to the previous exercise.

(a) Construct a 95% confidence interval for the slope. What does this interval tell you about the annual percent increase in tuition between 2000 and 2008?

(b) What percent of the variability in 2008 tuition is explained by a linear regression model using the 2000 tuition?

TABLE 10.1

In-state tuition and fees (in dollars) for 33 public universities

School	2000	2008	School	2000	2008	School	2000	2008
Penn State	7,018	13,706	Virginia	4,335	9,300	Iowa State	3,132	5,524
Pittsburgh	7,002	13,642	Indiana	4,405	8,231	Oregon	3,819	6,435
Michigan	6,926	11,738	Texas A&M	3,374	7,844	Iowa	3,204	6,544
Rutgers	6,333	11,540	Texas	3,575	8,532	Washington	3,761	6,802
Illinois	4,994	12,106	Cal.–Irvine	3,970	8,046	Nebraska	3,450	6,584
Minnesota	4,877	10,634	Cal.–San Diego	3,848	8,062	Kansas	2,725	7,042
Mich. State	5,432	10,690	Cal.–Berkeley	4,047	7,656	Colorado	3,188	7,278
Ohio State	4,383	8,679	UCLA	3,698	8,310	North Carolina	2,768	5,397
Maryland	5,136	8,005	Purdue	3,872	7,750	Arizona	2,348	5,542
Cal.–Davis	4,072	9,497	Wisconsin	3,791	7,569	Florida	2,256	3,256
Missouri	4,726	7,386	Buffalo	4,715	6,385	Georgia Tech.	3,308	6,040

(c) The tuition at Stat U was $5100 in 2000. What is the predicted tuition in 2008?

(d) The tuition at Moneypit U was $8700 in 2000. What is the predicted tuition in 2008?

(e) Discuss the appropriateness of using the fitted equation to predict tuition for each of these universities.

10.12. U.S. versus overseas stock returns. Returns on common stocks in the United States and overseas appear to be growing more closely correlated as economies become more interdependent. Suppose that the following population regression line connects the total annual returns (in percent) on two indexes of stock prices:

MEAN OVERSEAS RETURN = $-0.4+1.06\times$U.S. RETURN

(a) What is β_0 in this line? What does this number say about overseas returns when the U.S. market is flat (0% return)?

(b) What is β_1 in this line? What does this number say about the relationship between U.S. and overseas returns?

(c) We know that overseas returns will vary in years having the same return on U.S. common stocks. Write the regression model based on the population regression line given above. What part of this model allows overseas returns to vary when U.S. returns remain the same?

10.13. Beer and blood alcohol. How well does the number of beers a student drinks predict his or her blood alcohol content (BAC)? Sixteen student volunteers at Ohio State University drank a randomly assigned number of 12-ounce cans of beer. Thirty minutes later, a police officer measured their BAC. Here are the data:[4]

Student	1	2	3	4	5	6	7	8
Beers	5	2	9	8	3	7	3	5
BAC	0.10	0.03	0.19	0.12	0.04	0.095	0.07	0.06

Student	9	10	11	12	13	14	15	16
Beers	3	5	4	6	5	7	1	4
BAC	0.02	0.05	0.07	0.10	0.085	0.09	0.01	0.05

The students were equally divided between men and women and differed in weight and usual drinking habits. Because of this variation, many students don't believe that number of drinks predicts BAC well. DATA FILE BAC

(a) Make a scatterplot of the data. Find the equation of the least-squares regression line for predicting BAC from number of beers and add this line to your plot. Briefly summarize what your data analysis shows.

(b) Is there significant evidence that drinking more beers increases BAC on the average in the population of all students? State hypotheses, give a test statistic and P-value, and state your conclusion.

(c) Steve thinks he can drive legally 30 minutes after he drinks 5 beers. The legal limit is BAC = 0.08. Give a 90% prediction interval for Steve's BAC. Can he be confident he won't be arrested if he drives and is stopped?

10.14. School budget and number of students. Suppose that there is a linear relationship between the number of students x in an elementary school and the annual budget y. Write a population regression model to describe this relationship.

(a) Which parameter in your model is the fixed cost in the budget (for example, the salary of the principal and some administrative costs) that does not change as x increases?

(b) Which parameter in your model shows how total cost changes when there are more students in the school? Do you expect this number to be greater than 0 or less than 0?

(c) Actual data from schools will not fit a straight line exactly. What term in your model allows variation among schools of the same size x?

10.15. Performance bonuses. In the National Football League (NFL), performance bonuses now account for roughly 25% of player compensation.[5] Does tying a player's salary to performance bonuses result in better individual or team success on the field? The Web site usatoday.com has payroll data for most current NFL players. Cbssports.com has a player rating system that uses game statistics.[6] Focusing on linebackers, let's look at the relationship between a player's overall 2008 rating and the percent of his 2008 salary devoted to incentive payments. DATA FILE LINEBACKERS

(a) Use numerical and graphical methods to describe the two variables and summarize your results.

(b) Both variable distributions are non-Normal. Does this necessarily pose a problem for performing linear regression? Explain.

(c) Construct a scatterplot of the data and describe the relationship. Are there any outliers or unusual values? Does a linear relationship between the percent of salary and the player rating seem reasonable? Is it a very strong relationship? Explain.

(d) Run the simple linear regression and give the equation of the least-squares regression line.

(e) Obtain the residuals and assess whether the assumptions for the linear regression analysis are reasonable. Include all plots and numerical summaries used in doing this assessment.

10.16. CHALLENGE **Performance bonuses, continued.** Refer to the previous exercise. DATA FILE LINEBACKERS

(a) Now run the simple linear regression for the variables sqrt(rating) and percent of salary from incentive payments.

(b) Obtain the residuals and assess whether the assumptions for the linear regression analysis are reasonable. Include all plots and numerical summaries used in doing this assessment.

(c) Construct a 95% confidence interval for the square root increase in rating given a 1% increase in the percent of salary from incentive payments.

(d) Consider the values 0%, 20%, 40%, 60%, and 80% of salary from incentives. Compute the predicted rating for this model and the one in the previous exercise. For the model in this problem, you will need to square the predicted value to get back to the original units.

(e) Plot the predicted values versus the percent and connect those values from the same model. For which regions of percent do the predicted values from the two models vary the most?

(f) Based on the comparison of regression models (both predicted values and residuals), which model do you prefer? Explain.

10.17. Assessment value versus sales price. Real estate is typically reassessed each year to calculate property taxes. This assessed value, however, is not necessarily the same as the fair market value of the property. Table 10.2 summarizes an SRS of 30 houses recently sold in a midwestern city.[7] Both variables are measured in thousands of dollars. DATA FILE HOMESALES

(a) Inspect the data. How many have a selling price greater than the assessed value? Do you think this trend would be true for the larger population of all houses recently sold? Explain your answer.

(b) Make a scatterplot with assessed value on the horizontal axis. Briefly describe the relationship between assessed value and selling price.

(c) Report the least-squares regression line for predicting selling price from assessed value.

(d) Obtain the residuals and plot them versus assessed value. Is there anything unusual to report? If so, explain.

(e) Do the residuals appear to be approximately Normal? Explain your answer.

(f) Based on your answers to parts (b), (d), and (e), do the assumptions for the linear regression analysis appear reasonable? Explain your answer.

10.18. Assessment value versus sales price, continued. Refer to the previous exercise. DATA FILE HOMESALES

(a) Calculate the predicted selling prices for houses currently assessed at $155,000, $220,000, and $285,000.

(b) Suppose that these houses sold for $142,900, $224,000, and $286,000, respectively. Calculate the residual for each of these sales.

(c) Construct a 95% confidence interval for both the slope and the intercept.

(d) Using the result from part (c), compare the estimated regression line with $y = x$, which says that, on average, the selling price is equal to the assessed value. Is there evidence that this model is not reasonable? In other

TABLE 10.2

Assessed value and sales price (in $ thousands) of 30 houses in a midwestern city

Property	Sales price	Assessed value	Property	Sales price	Assessed value	Property	Sales price	Assessed value
1	167.9	152.7	11	230.0	225.4	21	283.0	303.9
2	168.0	163.8	12	230.0	170.4	22	269.0	233.7
3	155.0	167.6	13	222.5	200.4	23	255.0	233.6
4	158.5	127.3	14	225.5	209.6	24	285.0	234.2
5	159.9	155.7	15	220.0	205.2	25	146.0	145.1
6	162.0	169.0	16	216.0	220.9	26	128.0	108.3
7	165.0	187.1	17	215.0	194.9	27	126.5	136.2
8	174.5	153.6	18	228.0	231.4	28	129.9	113.3
9	175.0	167.1	19	209.0	224.2	29	150.0	121.4
10	159.0	148.9	20	267.0	235.1	30	195.0	184.0

words, is the selling price typically larger or smaller than the assessed value? Explain your answer.

10.19. Is the number of tornadoes increasing? The Storm Prediction Center of the National Oceanic and Atmospheric Administration maintains listings of tornadoes, floods, and other weather phenomena. Table 10.3 summarizes the annual number of tornadoes in the United States between 1953 and 2009.[8] DATA FILE TORNADOES

(a) Make a plot of the total number of tornadoes by year. Does a linear trend over years appear reasonable? Are there any outliers or unusual patterns? Explain your answer.

(b) Run the simple linear regression and summarize the results, making sure to construct a 95% confidence interval for the average annual increase in the number of tornadoes.

(c) Obtain the residuals and plot them versus year. Is there anything unusual in the plot?

(d) Are the residuals Normal? Justify your answer.

(e) The number of tornadoes in 2004 is much larger than expected under this linear model. Remove this observation and rerun the simple linear regression. Compare these results with the results in part (b).

10.20. Are the two fuel efficiency measurements similar? Refer to Exercise 7.28 (page 387). In addition to the computer calculating mpg, the driver also recorded the mpg by dividing the miles driven by the amount of gallons at fill-up. The driver wants to determine if these calculations are different.

Fill-up	1	2	3	4	5	6	7	8	9	10
Computer	41.5	50.7	36.6	37.3	34.2	45.0	48.0	43.2	47.7	42.2
Driver	36.5	44.2	37.2	35.6	30.5	40.5	40.0	41.0	42.8	39.2

Fill-up	11	12	13	14	15	16	17	18	19	20
Computer	43.2	44.6	48.4	46.4	46.8	39.2	37.3	43.5	44.3	43.3
Driver	38.8	44.5	45.4	45.3	45.7	34.2	35.2	39.8	44.9	47.5

(a) Consider the driver's mpg calculations as the explanatory variable. Plot the data and describe the

TABLE 10.3

Annual number of tornadoes in the United States between 1953 and 2009

Year	Number of tornadoes	Year	Number of tornadoes	Year	Number of tornadoes	Year	Number of tornadoes
1953	421	1968	660	1983	931	1998	1449
1954	550	1969	608	1984	907	1999	1340
1955	593	1970	653	1985	684	2000	1076
1956	504	1971	888	1986	764	2001	1213
1957	856	1972	741	1987	656	2002	934
1958	564	1973	1102	1988	702	2003	1372
1959	604	1974	947	1989	856	2004	1819
1960	616	1975	920	1990	1133	2005	1194
1961	697	1976	835	1991	1132	2006	1103
1962	657	1977	852	1992	1298	2007	1098
1963	464	1978	788	1993	1176	2008	1691
1964	704	1979	852	1994	1082	2009	1156
1965	906	1980	866	1995	1235		
1966	585	1981	783	1996	1173		
1967	926	1982	1046	1997	1148		

relationship. Are there any outliers or unusual values? Does a linear relationship seem reasonable?

(b) Run the simple linear regression and give the equation of the least-squares regression line.

(c) Summarize the results. Does it appear that the computer and driver calculations are the same? Explain.

10.21. CHALLENGE **Breaking strength of wood.** Exercise 4.126 (page 218) gives the modulus of elasticity (MOE) and the modulus of rupture (MOR) for 32 plywood specimens. Because measuring MOR involves breaking the wood but measuring MOE does not, we would like to predict the destructive test result, MOR, using the nondestructive test result, MOE. DATA FILE MOEMOR

(a) Describe the distribution of MOR using graphical and numerical summaries. Do the same for MOE.

(b) Make a plot of the two variables. Which should be plotted on the x axis? Give a reason for your answer.

(c) Give the statistical model for this analysis, run the analysis, summarize the results, and write a short summary of your conclusions.

(d) Examine the assumptions needed for the analysis. Are you satisfied that there are no serious violations that would cause you to question the validity of your conclusions?

10.22. CHALLENGE **Predicting water quality.** The index of biotic integrity (IBI) is a measure of the water quality of streams. IBI and land use measures for a collection of streams in the Ozark Highland ecoregion of Arkansas were collected as part of a study.[9] Table 10.4 gives the data for IBI, the percent of the watershed area that was forest, and the area of the watershed in square kilometers for streams in the original sample with area less than or equal to 70 square kilometers. DATA FILE IBI

(a) Use numerical and graphical methods to describe the variable IBI. Do the same for area. Summarize your results.

(b) Plot the data for IBI and area and describe the relationship. Are there any outliers or unusual patterns?

(c) Give the statistical model for simple linear regression for this problem.

(d) State the null and alternative hypotheses for examining the relationship between IBI and area.

(e) Run the simple linear regression and summarize the results.

(f) Obtain the residuals and plot them versus area. Is there anything unusual in the plot?

(g) Do the residuals appear to be approximately Normal? Give reasons for your answer.

(h) Do the assumptions for the analysis of these data using the model you gave in part (c) appear to be reasonable? Explain your answer.

10.23. CHALLENGE **More on predicting water quality.** The researchers who conducted the study described in the previous exercise also recorded the percent of the watershed area that was forest for each of the streams. The data are also given in Table 10.4. Analyze these data using the questions in the previous exercise as a guide. DATA FILE IBI

10.24. Comparing the analyses. In Exercises 10.22 and 10.23, you used two different explanatory variables to predict IBI. Summarize the two analyses and compare the

TABLE 10.4

Watershed area (in square kilometers), percent forest, and index of biotic integrity (IBI)

Area	Forest	IBI	Area	Forest	IBI	Area	Forest	IBI	Area	Forest	IBI	Area	Forest	IBI
21	0	47	29	0	61	31	0	39	32	0	59	34	0	72
34	0	76	49	3	85	52	3	89	2	7	74	70	8	89
6	9	33	28	10	46	21	10	32	59	11	80	69	14	80
47	17	78	8	17	53	8	18	43	58	21	88	54	22	84
10	25	62	57	31	55	18	32	29	19	33	29	39	33	54
49	33	78	9	39	71	5	41	55	14	43	58	9	43	71
23	47	33	31	49	59	18	49	81	16	52	71	21	52	75
32	59	64	10	63	41	26	68	82	9	75	60	54	79	84
12	79	83	21	80	82	27	86	82	23	89	86	26	90	79
16	95	67	26	95	56	26	100	85	28	100	91			

results. If you had to choose between the two explanatory variables for predicting IBI, which one would you prefer? Give reasons for your answer. IBI

10.25. How an outlier can affect statistical significance. Consider the data in Table 10.4 and the relationship between IBI and the percent of watershed area that was forest. The relationship between these two variables is almost significant at the 0.05 level. In this exercise you will demonstrate the potential effect of an outlier on statistical significance. Investigate what happens when you decrease the IBI to 0.0 for (1) an observation with 0% forest and (2) an observation with 100% forest. Write a short summary of what you learn from this exercise. IBI

10.26. Predicting water quality for an area of 40 km^2. Refer to Exercise 10.22. IBI

(a) Find a 95% confidence interval for mean response corresponding to an area of 40 km^2.

(b) Find a 95% prediction interval for a future response corresponding to an area of 40 km^2.

(c) Write a short paragraph interpreting the meaning of the intervals in terms of Ozark Highland streams.

(d) Do you think that these results can be applied to other streams in Arkansas or in other states? Explain why or why not.

10.27. Compare the predictions. Consider Case 37 in Table 10.4 (8th row, 2nd section). For this case the area is 10 km^2 and the percent forest is 63%. A predicted index of biotic integrity based on area was computed in Exercise 10.22, while one based on percent forest was computed in Exercise 10.23. Compare these two estimates and explain why they differ. Use the idea of a prediction interval to interpret these results. IBI

10.28. Math pretest predicts success? Can a pretest on mathematics skills predict success in a statistics course? The 82 students in an introductory statistics class took a pretest at the beginning of the semester. The least-squares regression line for predicting the score y on the final exam from the pretest score x was $\hat{y} = 16.23 + 0.82x$. The standard error of b_1 was 0.38.

(a) Test the null hypothesis that there is no linear relationship between the pretest score and the score on the final exam against the two-sided alternative.

(b) Would you reject this null hypothesis versus the one-sided alternative that the slope is positive? Explain your answer.

10.29. Stocks and bonds. How is the flow of investors' money into stock mutual funds related to the flow of money into bond mutual funds? Table 10.5 shows the net new money flowing into stock and bond mutual funds in the years 1985 to 2008, in billions of dollars.[10] "Net" means that funds flowing out are subtracted from those flowing in. If more money leaves than arrives, the net flow will be negative. To eliminate the effect of inflation, all dollar amounts are in "real dollars" with constant buying power equal to that of a dollar in the year 2000. MONEYFLOW

(a) Make a scatterplot with cash flow into stock funds as the explanatory variable. Find the least-squares line for predicting net bond investments from net stock investments. What do the data suggest?

(b) Is there statistically significant evidence that there is some straight-line relationship between the flows of cash into bond funds and stock funds? State hypotheses, give a test statistic and its P-value, and state your conclusion.

TABLE 10.5

Net new money (in $ billions) flowing into stock and bond mutual funds, 1985 to 2008

Year	Stocks	Bonds	Year	Stocks	Bonds	Year	Stocks	Bonds
1985	12.8	100.8	1993	151.3	84.6	2001	31.1	85.0
1986	34.6	161.8	1994	133.6	−72.0	2002	−25.8	134.4
1987	28.8	10.6	1995	140.1	−6.8	2003	143.0	29.4
1988	−23.3	−5.8	1996	238.2	3.3	2004	161.8	−9.4
1989	8.3	−1.4	1997	243.5	30.0	2005	119.7	27.7
1990	17.1	9.2	1998	165.9	79.2	2006	134.9	51.9
1991	50.6	74.6	1999	194.3	−6.2	2007	76.8	90.1
1992	97.0	87.1	2000	309.0	−48.0	2008	−187.4	24.9

(c) Remove the data for 2008 and refit the remaining years. Is there now statistically significant evidence of a straight-line relationship?

(d) How would you report these results in a manuscript? In other words, how would you handle the change in statistical significance caused by this one observation?

10.30. SAT versus ACT. The SAT and the ACT are the two major standardized tests that colleges use to evaluate candidates. Most students take just one of these tests. However, some students take both. Consider the scores of 60 students who did this. How can we relate the two tests? DATA FILE SAT_ACT

(a) Plot the data with SAT on the x axis and ACT on the y axis. Describe the overall pattern and any unusual observations.

(b) Find the least-squares regression line and draw it on your plot. Give the results of the significance test for the slope.

(c) What is the correlation between the SAT and ACT tests?

10.31. CHALLENGE **SAT versus ACT, continued.** Refer to the previous exercise. Find the predicted value of ACT for each observation in the data set. DATA FILE SAT_ACT

(a) What is the mean of these predicted values? Compare it with the actual mean of the ACT scores.

(b) Compare the standard deviation of the predicted values with the standard deviation of the actual ACT scores. If least-squares regression is used to predict ACT scores for a large number of students such as these, the average predicted value will be accurate but the variability of the predicted scores will be too small.

(c) Find the SAT score for a student who is one standard deviation above the mean ($z = (x - \bar{x})/s = 1$). Find the predicted ACT score and standardize this score. (Use the means and standard deviations from this set of data for these calculations.)

(d) Repeat part (c) for a student whose SAT score is one standard deviation below the mean ($z = -1$).

(e) What do you conclude from parts (c) and (d)? Perform additional calculations for different z's if needed.

10.32. CHALLENGE **Matching standardized scores.** Refer to the previous two exercises. An alternative to the least-squares method is based on matching standardized scores. Specifically, we set

$$\frac{y - \bar{y}}{s_y} = \frac{x - \bar{x}}{s_x}$$

and solve for y. Let's use the notation $y = a_0 + a_1 x$ for this line. The slope is $a_1 = s_y/s_x$ and the intercept is $a_0 = \bar{y} - a_1\bar{x}$. Compare these expressions with the formulas for the least-squares slope and intercept (page 494). DATA FILE SAT_ACT

(a) Using the data in the previous exercise, find the values of a_0 and a_1.

(b) Plot the data with the least-squares line and the new prediction line.

(c) Use the new line to find predicted ACT scores. Find the mean and the standard deviation of these scores. How do they compare with the mean and standard deviation of the actual ACT scores?

10.33. Leaning Tower of Pisa. The Leaning Tower of Pisa is an architectural wonder. Engineers concerned about the tower's stability have done extensive studies of its increasing tilt. Measurements of the lean of the tower over time provide much useful information. The following table gives measurements for the years 1975 to 1987. The variable "lean" represents the difference between where a point on the tower would be if the tower were straight and where it actually is. The data are coded as tenths of a millimeter in excess of 2.9 meters, so that the 1975 lean, which was 2.9642 meters, appears in the table as 642. Only the last two digits of the year were entered into the computer.[11] DATA FILE PISA

Year	75	76	77	78	79	80	81	82	83	84	85	86	87
Lean	642	644	656	667	673	688	696	698	713	717	725	742	757

(a) Plot the data. Does the trend in lean over time appear to be linear?

(b) What is the equation of the least-squares line? What percent of the variation in lean is explained by this line?

(c) Give a 99% confidence interval for the average rate of change (tenths of a millimeter per year) of the lean.

10.34. More on the Leaning Tower of Pisa. Refer to the previous exercise.

(a) In 1918 the lean was 2.9071 meters. (The coded value is 71.) Using the least-squares equation for the years 1975 to 1987, calculate a predicted value for the lean in 1918. (Note that you must use the coded value 18 for year.)

(b) Although the least-squares line gives an excellent fit to the data for 1975 to 1987, this pattern did not extend back to 1918. Write a short statement explaining why this

conclusion follows from the information available. Use numerical and graphical summaries to support your explanation.

10.35. Predicting the lean in 2012. Refer to the previous two exercises.

(a) How would you code the explanatory variable for the year 2012?

(b) The engineers working on the Leaning Tower of Pisa were most interested in how much the tower would lean if no corrective action was taken. Use the least-squares equation to predict the tower's lean in the year 2012. (Note: The tower was renovated in 2001 to make sure it does not fall down.)

(c) To give a margin of error for the lean in 2012, would you use a confidence interval for a mean response or a prediction interval? Explain your choice.

10.36. Length, width, and weight of perch. Here are data for 12 perch caught in a lake in Finland:[12] DATA FILE PERCH

Weight (grams)	Length (cm)	Width (cm)	Weight (grams)	Length (cm)	Width (cm)
5.9	8.8	1.4	300.0	28.7	5.1
100.0	19.2	3.3	300.0	30.1	4.6
110.0	22.5	3.6	685.0	39.0	6.9
120.0	23.5	3.5	650.0	41.4	6.0
150.0	24.0	3.6	820.0	42.5	6.6
145.0	25.5	3.8	1000.0	46.6	7.6

In this exercise we will examine different models for predicting weight.

(a) Run the regression using length to predict weight. Do the same using width as the explanatory variable. Summarize the results. Be sure to include the value of r^2.

(b) Plot weight versus length and weight versus width. Include the least-squares lines in these plots. Do these relationships appear to be linear? Explain your answer.

10.37. CHALLENGE **Transforming the perch data.** Refer to the previous exercise.

(a) Try to find a better model using a transformation of length. One possibility is to use the square. Make a plot and perform the regression analysis. Summarize the results.

(b) Do the same for width.

10.38. Resting metabolic rate and exercise. Metabolic rate, the rate at which the body consumes energy, is important in studies of weight gain, dieting, and exercise. The following table gives data on the lean body mass and resting metabolic rate for 12 women and 7 men who are subjects in a study of dieting. Lean body mass, given in kilograms, is a person's weight leaving out all fat. Metabolic rate is measured in calories burned per 24 hours, the same calories used to describe the energy content of foods. The researchers believe that lean body mass is an important influence on metabolic rate. DATA FILE METRATE

Subject	Sex	Mass	Rate	Subject	Sex	Mass	Rate
1	M	62.0	1792	11	F	40.3	1189
2	M	62.9	1666	12	F	33.1	913
3	F	36.1	995	13	M	51.9	1460
4	F	54.6	1425	14	F	42.4	1124
5	F	48.5	1396	15	F	34.5	1052
6	F	42.0	1418	16	F	51.1	1347
7	M	47.4	1362	17	F	41.2	1204
8	F	50.6	1502	18	M	51.9	1867
9	F	42.0	1256	19	M	46.9	1439
10	M	48.7	1614				

(a) Make a scatterplot of the data, using different symbols or colors for men and women. Summarize what you see in the plot.

(b) Run the regression to predict metabolic rate from lean body mass for the women in the sample and summarize the results. Do the same for the men.

10.39. CHALLENGE **Resting metabolic rate and exercise, continued.** Refer to the previous exercise. It is tempting to conclude that there is a strong linear relationship for the women but no relationship for the men. Let's look at this issue a little more carefully.

(a) Find the confidence interval for the slope in the regression equation that you ran for the females. Do the same for the males. What do these suggest about the possibility that these two slopes are the same? (The formal method for making this comparison is a bit complicated and is beyond the scope of this chapter.)

(b) The formula for the standard error of the regression slope has the term $\sqrt{\Sigma(x_i - \bar{x})^2}$ in the denominator. Find this quantity for the females; do the same for the males. How do these calculations help to explain the results of the significance tests?

(c) Suppose that you were able to collect additional data for males. How would you use lean body mass in deciding which subjects to choose?

10.40. CHALLENGE **Inference over different ranges of *X*.** Think about what would happen if you analyzed a subset of a set of data by analyzing only data for a restricted range of values of the explanatory variable. What results would you expect to change? Examine your ideas by analyzing the fuel efficiency data described in Example 10.5 (page 499). First, run a regression of MPG versus MPH using all cases. This least-squares regression line is shown in Figure 10.7. Next, run a regression of MPG versus MPH for only those cases with speed less than or equal to 30 mph. Note that this corresponds to 3.4 in the log scale. Finally, do the same analysis with a restriction on the response variable. Run the analysis with only those cases with fuel efficiency less than or equal to 20 mpg. Write a summary comparing the effects of these two restrictions with each other and with the complete data set results. DATA FILE MPH_MPG

CHAPTER ELEVEN

MULTIPLE REGRESSION

In Chapter 10 we presented methods for inference in the setting of a linear relationship between a response variable y and a *single* explanatory variable x. In this chapter, we use *more than one* explanatory variable to explain or predict a single response variable.

Many of the ideas that we encountered in our study of simple linear regression carry over to the multiple linear regression setting. For example, the descriptive tools we learned in Chapter 4—scatterplots, least-squares regression, and correlation—are still essential preliminaries to inference and also provide a foundation for confidence intervals and significance tests.

The introduction of several explanatory variables leads to many additional considerations. In this short chapter we cannot explore all these issues. Rather, we will outline some basic facts about inference in the multiple regression setting. We then illustrate the analysis with a case study whose purpose was to predict success in college based on several high school achievement scores.

11.1 Inference for Multiple Regression

Population multiple regression equation

The simple linear regression model assumes that the mean of the response variable y depends on the explanatory variable x according to a linear equation

$$\mu_y = \beta_0 + \beta_1 x$$

For any fixed value of x, the response y varies Normally around this mean and has a standard deviation σ that is the same for all values of x.

In the multiple regression setting, the response variable y depends on p explanatory variables, which we will denote by $x_1, x_2, \ldots, x_p$. The mean response depends on these explanatory variables according to a linear function

$$\mu_y = \beta_0 + \beta_1 x_1 + \beta_2 x_2 + \cdots + \beta_p x_p$$

population regression equation

Similar to simple linear regression, this expression is the **population regression equation** and the observed values of y vary about their means given by this equation.

Just as we did in simple linear regression, we can also think of this model in terms of subpopulations of responses. Here, each subpopulation corresponds to a particular set of values for *all* of the explanatory variables $x_1, x_2, \ldots, x_p$. In each subpopulation, y varies Normally with a mean given by the population regression equation. The regression model assumes that the standard deviation σ of the responses is the same in all subpopulations.

EXAMPLE 11.1

Predicting early success in college. Our case study uses data collected at a large university on all first-year computer science majors in a particular year.[1] The purpose of the study was to attempt to predict success in the early university years. One measure of success was the cumulative grade point average (GPA) after three semesters. Among the explanatory variables recorded at the time the students enrolled in the university were average high school grades in mathematics (HSM), science (HSS), and English (HSE).

We will use high school grades to predict the response variable GPA. There are $p = 3$ explanatory variables: x_1 = HSM, x_2 = HSS, and x_3 = HSE. The high school grades are coded on a scale from 1 to 10, with 10 corresponding to A, 9 to A−, 8 to B+, and so on. These grades define the subpopulations. For example, the straight-C students are the subpopulation defined by HSM = 4, HSS = 4, and HSE = 4.

One possible multiple regression model for the subpopulation mean GPAs is

$$\mu_{\text{GPA}} = \beta_0 + \beta_1\text{HSM} + \beta_2\text{HSS} + \beta_3\text{HSE}$$

For the straight-C subpopulation of students, the model gives the subpopulation mean as

$$\mu_{\text{GPA}} = \beta_0 + \beta_1 4 + \beta_2 4 + \beta_3 4$$

Data for multiple regression

The data for a simple linear regression problem consist of observations (x_i, y_i) of the two variables. Because there are several explanatory variables in multiple regression, the notation needed to describe the data is more elaborate. Each observation or case consists of a value for the response variable and for each of the explanatory variables. Call x_{ij} the value of the jth explanatory variable for the ith case. The data are then

$$\begin{array}{l}\text{Case 1: } (x_{11}, x_{12}, \ldots, x_{1p}, y_1)\\ \text{Case 2: } (x_{21}, x_{22}, \ldots, x_{2p}, y_2)\\ \quad\vdots\\ \text{Case } n\text{: } (x_{n1}, x_{n2}, \ldots, x_{np}, y_n)\end{array}$$

Here, n is the number of cases and p is the number of explanatory variables. Data are often entered into computer regression programs in this format. Each row is a case and each column corresponds to a different variable.

The data for Example 11.1, with several additional explanatory variables, appear in this format in the CSDATA data file. Figure 11.1 shows the first 5 rows. Grade point average (GPA) is the response variable, followed by $p = 6$ explanatory variables. There are a total of $n = 224$ students in this data set.

Excel

	A	B	C	D	E	F	G	H
1	obs	gpa	hsm	hss	hse	satm	satv	sex
2	1	3.32	10	10	10	670	600	1
3	2	2.26	6	8	5	700	640	1
4	3	2.35	8	6	8	640	530	1
5	4	2.08	9	10	7	670	600	1
6	5	3.38	8	9	8	540	580	1

FIGURE 11.1 Format of data set for Example 11.1 in an Excel spreadsheet.

USE YOUR KNOWLEDGE

11.1 Describing a multiple regression. As part of a study entitled "Predicting Success for Actuarial Students in Undergraduate Mathematics Courses," data from 106 Bryant University actuarial graduates were obtained.[2] The researchers were interested in describing how students' overall math grade point averages are explained by SAT Math and Verbal scores, class rank, and Bryant University's mathematics placement score.

(a) What is the response variable?

(b) What is n, the number of cases?

(c) What is p, the number of explanatory variables?

(d) What are the explanatory variables?

Multiple linear regression model

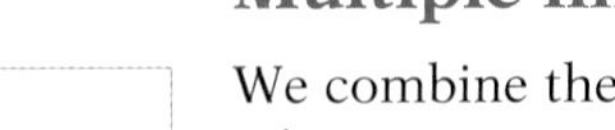

LOOK BACK
DATA = FIT + RESIDUAL
p. 493

We combine the population regression equation and assumptions about variation to construct the multiple linear regression model. The subpopulation means describe the FIT part of our statistical model. The RESIDUAL part represents the variation of observations about the means.

We will use the same notation for the residual that we used in the simple linear regression model. The symbol ϵ represents the deviation of an individual observation from its subpopulation mean.

We assume that these deviations are Normally distributed with mean 0 and an unknown standard deviation σ that does not depend on the values of the x variables. *We can check these assumptions by examining the residuals in the same way that we did for simple linear regression.*

MULTIPLE LINEAR REGRESSION MODEL

The **statistical model for multiple linear regression** is

$$y_i = \beta_0 + \beta_1 x_{i1} + \beta_2 x_{i2} + \cdots + \beta_p x_{ip} + \epsilon_i$$

for $i = 1, 2, \ldots, n$.

The **mean response** μ_y is a linear function of the explanatory variables:

$$\mu_y = \beta_0 + \beta_1 x_1 + \beta_2 x_2 + \cdots + \beta_p x_p$$

The **deviations** ϵ_i are independent and Normally distributed with mean 0 and standard deviation σ. In other words, the deviations are an SRS from the $N(0, \sigma)$ distribution.

The **parameters** of the model are $\beta_0, \beta_1, \beta_2, \ldots, \beta_p$, and σ.

The assumption that the subpopulation means are related to the regression coefficients β by the equation

$$\mu_y = \beta_0 + \beta_1 x_1 + \beta_2 x_2 + \cdots + \beta_p x_p$$

implies that we can estimate all subpopulation means from estimates of the β's. To the extent that this equation is accurate, we have a useful tool for describing how the mean of y varies with the collection of x's.

USE YOUR KNOWLEDGE

11.2 Understanding the fitted regression line. Researchers study the rate of water loss (% body weight per minute) of a tropical bedbug exposed to different combinations of temperature x_1 and relative humidity x_2. The fitted regression equation for their multiple regression is

$$\hat{y} = -2.7 + 0.05x_1 + 0.04x_2$$

(a) If $x_1 = 30°\text{C}$ and $x_2 = 80\%$, what is the predicted value of y?

(b) For the answer to part (a) to be valid, is it necessary that the values $x_1 = 30$ and $x_2 = 80$ correspond to a case in the data set? Explain why or why not.

(c) If you hold x_2 at a fixed value, what is the effect of an increase of 5°C on the predicted value of y?

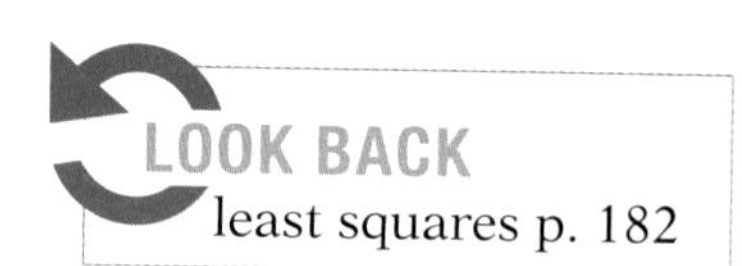

Estimation of the multiple regression parameters

Similar to simple linear regression, we use the method of least squares to obtain estimators of the regression coefficients β. The details, however, are more complicated. Let

$$b_0, b_1, b_2, \ldots, b_p$$

denote the estimators of the parameters

$$\beta_0, \beta_1, \beta_2, \ldots, \beta_p$$

For the ith observation, the predicted response is

$$\hat{y}_i = b_0 + b_1x_{i1} + b_2x_{i2} + \cdots + b_px_{ip}$$

LOOK BACK
residual p. 495

The ith residual, the difference between the observed and the predicted response, is therefore

$$\begin{aligned} e_i &= \text{observed response} - \text{predicted response} \\ &= y_i - \hat{y}_i \\ &= y_i - b_0 - b_1x_{i1} - b_2x_{i2} - \cdots - b_px_{ip} \end{aligned}$$

The method of least squares chooses the values of the b's that make the sum of the squared residuals as small as possible. In other words, the parameter estimates $b_0, b_1, b_2, \ldots, b_p$ minimize the quantity

$$\sum(y_i - b_0 - b_1x_{i1} - b_2x_{i2} - \cdots - b_px_{ip})^2$$

The formula for the least-squares estimates is complicated. We will focus on understanding the principle on which it is based and let software do the computations.

The parameter σ^2 measures the variability of the responses about the population regression equation. As in the case of simple linear regression, we estimate σ^2 by an average of the squared residuals. The estimator is

$$s^2 = \frac{\sum e_i^2}{n-p-1} = \frac{\sum (y_i - \hat{y}_i)^2}{n-p-1}$$

LOOK BACK
degrees of freedom
p. 495

The quantity $n-p-1$ is the degrees of freedom associated with s^2. The degrees of freedom equal the sample size, n, minus $p+1$, the number of β's we must estimate to fit the model. In the simple linear regression case, there is just one explanatory variable, so $p = 1$ and the degrees of freedom are $n-2$. To estimate σ, we use

$$s = \sqrt{s^2}$$

Confidence intervals and significance tests for regression coefficients

We can obtain confidence intervals and perform significance tests for each of the regression coefficients β_j as we did in simple linear regression. The standard errors of the b's have more complicated formulas, but all are multiples of s. We again rely on statistical software to do the calculations.

CONFIDENCE INTERVALS AND SIGNIFICANCE TESTS FOR β_j

A **level *C* confidence interval** for β_j is

$$b_j \pm t^* \text{SE}_{b_j}$$

where SE_{b_j} is the standard error of b_j and t^* is the value for the $t(n-p-1)$ density curve with area C between $-t^*$ and t^*.

To test the hypothesis H_0: $\beta_j = 0$, compute the ***t* statistic**

$$t = \frac{b_j}{\text{SE}_{b_j}}$$

In terms of a random variable T having the $t(n-p-1)$ distribution, the P-value for a test of H_0 against

H_a: $\beta_j > 0$ is $P(T \geq t)$

H_a: $\beta_j < 0$ is $P(T \leq t)$

H_a: $\beta_j \neq 0$ is $2P(T \geq |t|)$

LOOK BACK
confidence intervals for mean response
p. 505
prediction intervals
p. 507

Because regression is often used for prediction, we may wish to use multiple regression models to construct confidence intervals for a mean response and prediction intervals for a future observation. The basic ideas are the same as in the simple linear regression case.

In most software systems, the same commands that give confidence and prediction intervals for simple linear regression work for multiple regression. The only difference is that we specify a list of explanatory variables rather than a single variable. Modern software allows us to perform these rather complex calculations without an intimate knowledge of all the computational details. This frees us to concentrate on the meaning and appropriate use of the results.

ANOVA table for multiple regression

analysis of variance

The usual computer output for regression includes additional calculations called **analysis of variance.** Analysis of variance, often abbreviated ANOVA, is essential for multiple regression and for comparing several means (Chapters 12 and 13). Analysis of variance summarizes information about the sources of variation in the data. It is based on the framework already mentioned:

$$\text{DATA} = \text{FIT} + \text{RESIDUAL}$$

The total variation in the response y is expressed by the deviations $y_i - \bar{y}$. If these deviations were all 0, all observations would be equal and there would be no variation in the response. There are two reasons why the individual observations y_i are not all equal to their mean $\bar{y}$.

1. The responses y_i correspond to different values of the explanatory variables and will differ because of that. The fitted value $\hat{y}_i$ estimates the mean response for different subpopulations. The differences $\hat{y}_i - \bar{y}$ reflect the variation in mean response due to differences in the explanatory variables. This variation is accounted for by the regression equation, because the $\hat{y}$'s lie exactly on the regression line.
2. Individual observations will vary about their mean because of variation within the subpopulation of responses. This variation is represented by the residuals $y_i - \hat{y}_i$ that record the scatter of the actual observations about the fitted regression equation.

The overall deviation of any y observation from the mean of the y's is the sum of these two deviations:

$$y_i - \bar{y} = (\hat{y}_i - \bar{y}) + (y_i - \hat{y}_i)$$

In terms of deviations, this equation expresses the idea that DATA = FIT + RESIDUAL.

Several times, we have measured variation by an average of squared deviations. If we square each of the preceding three deviations and then sum over all n observations, it is an algebraic fact that the sums of squares add:

$$\sum(y_i - \bar{y})^2 = \sum(\hat{y}_i - \bar{y})^2 + \sum(y_i - \hat{y}_i)^2$$

We rewrite this equation as

$$\text{SST} = \text{SSM} + \text{SSE}$$

where

$$\text{SST} = \sum(y_i - \bar{y})^2$$
$$\text{SSM} = \sum(\hat{y}_i - \bar{y})^2$$
$$\text{SSE} = \sum(y_i - \hat{y}_i)^2$$

sum of squares

The SS in each abbreviation stands for **sum of squares,** and the T, M, and E stand for total, model, and error, respectively. ("Error" here stands for deviations from the line, which might better be called "residual" or "unexplained variation.")

The total variation, as expressed by SST, is the sum of the variation due to the straight-line model (SSM) and the variation due to deviations from this model (SSE). This partition of the variation in the data between two sources is the heart of analysis of variance.

If there were no subpopulations, all the y's could be viewed as coming from a single population with mean μ_y. The variation of the y's would then be described by the sample variance

$$s_y^2 = \frac{\sum(y_i - \bar{y})^2}{n - 1}$$

The numerator in this expression is SST. The denominator is the total degrees of freedom, or simply DFT.

Just as the total sum of squares SST is the sum of SSM and SSE, the total degrees of freedom DFT is the sum of DFM and DFE, the degrees of freedom for the model and for the error:

$$\text{DFT} = \text{DFM} + \text{DFE}$$

The model has p explanatory variables, so the degrees of freedom for this source are $\text{DFM} = p$. Because $\text{DFT} = n - 1$, this leaves $\text{DFE} = n - p - 1$ as the degrees of freedom for error. *It is always a good idea to calculate the degrees of freedom by hand and then check that your software agrees with your calculations. In this way you can verify that your software is using the number of cases and number of explanatory variables that you intended.*

mean square

For each source, the ratio of the sum of squares to the degrees of freedom is called the **mean square,** or simply MS. The general formula for a mean square is

$$\text{MS} = \frac{\text{sum of squares}}{\text{degrees of freedom}}$$

Each mean square is an average squared deviation. MST is just s_y^2, the sample variance that we would calculate if all the data came from a single population. MSE is also familiar to us:

$$\text{MSE} = s^2 = \frac{\sum(y_i - \hat{y}_i)^2}{n - p - 1}$$

It is our estimate of σ^2, the variance about the estimated regression equation.

SUMS OF SQUARES, DEGREES OF FREEDOM, AND MEAN SQUARES

Sums of squares represent variation present in the responses. They are calculated by summing squared deviations. **Analysis of variance** partitions the total variation between two sources.

The sums of squares are related by the formula

$$\text{SST} = \text{SSM} + \text{SSE}$$

That is, the total variation is partitioned into two parts, one due to the model and one due to deviations from the model.

Degrees of freedom are associated with each sum of squares. They are related in the same way:

$$\text{DFT} = \text{DFM} + \text{DFE}$$

To calculate **mean squares,** use the formula

$$\text{MS} = \frac{\text{sum of squares}}{\text{degrees of freedom}}$$

ANOVA table

The ANOVA calculations are displayed in an *analysis of variance table,* often abbreviated **ANOVA table.** Here is the format of the table for multiple regression:

Source	Degrees of freedom	Sum of squares	Mean square	F
Model	p	$\sum(\hat{y}_i - \bar{y})^2$	SSM/DFM	MSM/MSE
Error	$n - p - 1$	$\sum(y_i - \hat{y}_i)^2$	SSE/DFE	
Total	$n - 1$	$\sum(y_i - \bar{y})^2$	SST/DFT	

The column labeled F denotes a ratio of mean squares that can be used to jointly assess the ability of the collection of predictors to explain y.

The ANOVA F test

The ratio MSM/MSE is an F statistic for testing the null hypothesis

$$H_0\colon \beta_1 = \beta_2 = \cdots = \beta_p = 0$$

against the alternative hypothesis

$$H_a\colon \text{at least one of the } \beta_j \text{ is not } 0$$

The null hypothesis says that none of the explanatory variables are predictors of the response variable when used in the form expressed by the multiple regression equation. The alternative states that at least one of them is a predictor of the response variable.

F distributions

We call it an F statistic because when H_0 is true, F has the $F(p, n-p-1)$ distribution. The ***F* distributions** are a family of distributions with two parameters: the degrees of freedom in the numerator and denominator of the F statistic. The degrees of freedom for this F distribution are those associated with the model and error in the ANOVA table.

The F distributions are another of R. A. Fisher's contributions to statistics and are called F in his honor. The numerator degrees of freedom are always mentioned first. Interchanging the degrees of freedom changes the distribution, so the order is important. Our brief notation will be $F(j, k)$ for the F distribution with j degrees of freedom in the numerator and k degrees of freedom in the denominator. The F distributions are not symmetric but are right-skewed. The density curve in Figure 11.2 illustrates the shape.

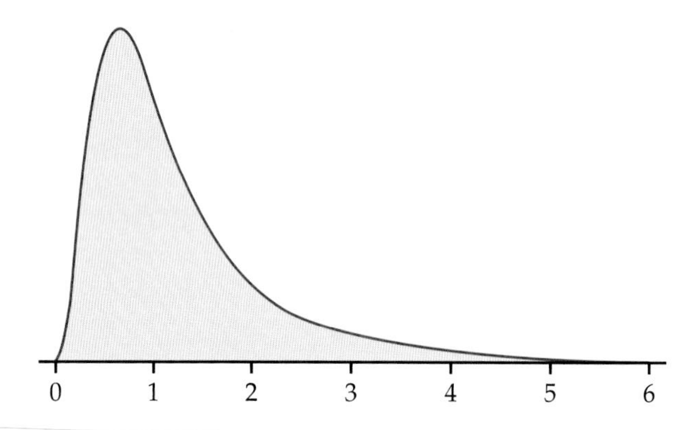

FIGURE 11.2 The density curve for the $F(9, 10)$ distribution. The F distributions are skewed to the right.

Because mean squares cannot be negative, the F statistic takes only positive values and the F distribution has no probability below 0. The peak of the F density curve is near 1. Large values of F provide evidence against H_0. Almost all software packages contain another column in the ANOVA table that gives the P-value of this F test. Table E in the back of the book gives upper p critical values of numerous F distributions for $p = 0.10$, 0.05, 0.025, 0.01, and 0.001.

ANALYSIS OF VARIANCE F TEST

In the multiple regression model, the hypothesis

$$H_0\colon \beta_1 = \beta_2 = \cdots = \beta_p = 0$$

is tested against the alternative hypothesis

$$H_a\colon \text{at least one of the } \beta_j \text{ is not } 0$$

by the **analysis of variance *F* statistic**

$$F = \frac{\text{MSM}}{\text{MSE}}$$

The ***P*-value** is the probability that a random variable having the $F(p, n-p-1)$ distribution is greater than or equal to the calculated value of the F statistic.

A common error in the use of multiple regression is to assume that all the regression coefficients are statistically different from zero whenever the F statistic has a small P-value. Be sure that you understand the difference between the null and alternative hypotheses of the ANOVA F test and the t tests for individual coefficients discussed earlier.

Squared multiple correlation R^2

Another useful statistic from the ANOVA table is the ratio of SSM to SST. Because SST = SSM + SSE, this ratio can be interpreted as the proportion of variation in y explained by the explanatory variables $x_1, x_2, \ldots, x_p$.

THE SQUARED MULTIPLE CORRELATION R^2

The statistic

$$R^2 = \frac{\text{SSM}}{\text{SST}} = \frac{\sum(\hat{y}_i - \bar{y})^2}{\sum(y_i - \bar{y})^2}$$

is the proportion of the variation of the response variable y that is explained by the explanatory variables $x_1, x_2, \ldots, x_p$ in a multiple linear regression.

multiple correlation coefficient

Often, R^2 is multiplied by 100 and expressed as a percent. The square root of R^2, called the **multiple correlation coefficient,** is the correlation between the observations y_i and the predicted values $\hat{y}_i$.

USE YOUR KNOWLEDGE

11.3 Significance tests for regression coefficients. Recall Exercise 11.1 (page 522). Due to missing values for some students, only 86 students were used in the multiple regression analysis. The following table contains the estimated coefficients and standard errors:

Variable	Estimate	SE
Intercept	−0.764	0.651
SAT Math	0.00156	0.00074
SAT Verbal	0.00164	0.00076
High school rank	1.470	0.430
Bryant University placement	0.889	0.402

(a) All the estimated coefficients for the explanatory variables are positive. Is this what you would expect? Explain.

(b) What are the degrees of freedom for the model and error?

(c) Test the significance of each coefficient and state your conclusions.

11.4 ANOVA table for multiple regression. Use the following information to perform the ANOVA F test and compute R^2.

Source	Degrees of freedom	Sum of squares
Model	50	75
Error		
Total	53	515

11.2 Exploring Multiple Regression through a Case Study

Preliminary analysis: examining the variables

In this section we illustrate multiple regression by analyzing the data from the study described in Example 11.1. The response variable is the cumulative GPA, on a 4-point scale, after three semesters of college. The explanatory variables previously mentioned are average high school grades, represented by HSM, HSS, and HSE. We also examine the SAT Mathematics and SAT Verbal scores as explanatory variables. We have data for $n = 224$ students in the study. We use SAS, Excel, and Minitab to illustrate the outputs that are given by most software.

The first step in the analysis is to carefully examine each of the variables. Means, standard deviations, and minimum and maximum values appear in Figure 11.3. The minimum value for the SAT Mathematics (SATM) variable appears to be rather extreme; it is $(595 - 300)/86 = 3.43$ standard deviations below the mean. We do not discard this case at this time but will take care in our subsequent analyses to see if it has an excessive influence on our results. The mean for the SATM score is higher than the mean for the Verbal score (SATV), as we might expect for a group of computer science majors. The two standard deviations are about the same.

The means of the three high school grade variables are similar, with the mathematics grades being a bit higher. The standard deviations for the high school grade variables are very close to each other. The mean GPA is 2.635 on a 4-point scale, with standard deviation 0.779.

SAS

Variable	N	Mean	Std Dev	Minimum	Maximum
GPA	224	2.6352232	0.7793949	0.1200000	4.0000000
SATM	224	595.2857143	86.4014437	300.0000000	800.0000000
SATV	224	504.5491071	92.6104591	285.0000000	760.0000000
HSM	224	8.3214286	1.6387367	2.0000000	10.0000000
HSS	224	8.0892857	1.6996627	3.0000000	10.0000000
HSE	224	8.0937500	1.5078736	3.0000000	10.0000000

FIGURE 11.3 Descriptive statistics for the computer science student case study.

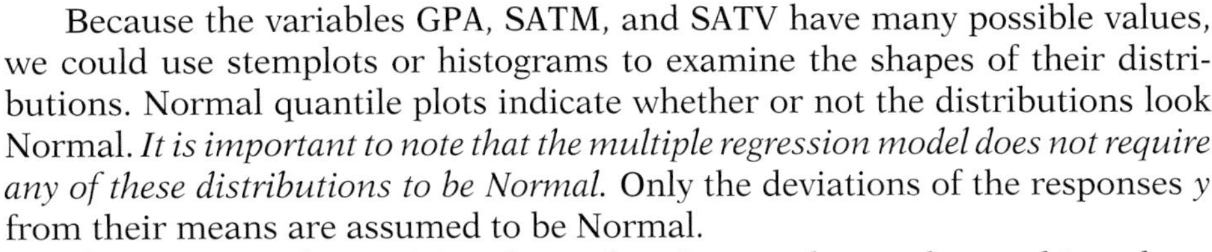

Because the variables GPA, SATM, and SATV have many possible values, we could use stemplots or histograms to examine the shapes of their distributions. Normal quantile plots indicate whether or not the distributions look Normal. *It is important to note that the multiple regression model does not require any of these distributions to be Normal.* Only the deviations of the responses y from their means are assumed to be Normal.

The purpose of examining these plots is to understand something about each variable alone before attempting to use it in a complicated model. *Extreme values of any variable should be noted and checked for accuracy.* If found to be correct, the cases with these values should be carefully examined to see if they are truly exceptional and perhaps do not belong in the same analysis with the other cases. When our data on computer science majors are examined in this way, no obvious problems are evident.

The high school grade variables HSM, HSS, and HSE have relatively few values and are best summarized by giving the relative frequencies for each possible value. The output in Figure 11.4 provides these summaries. The

SAS

hsm	Frequency	Percent	Cumulative Frequency	Cumulative Percent
2	1	0.45	1	0.45
3	1	0.45	2	0.89
4	4	1.79	6	2.68
5	6	2.68	12	5.36
6	23	10.27	35	15.63
7	28	12.50	63	28.13
8	36	16.07	99	44.20
9	59	26.34	158	70.54
10	66	29.46	224	100.00

hss	Frequency	Percent	Cumulative Frequency	Cumulative Percent
3	1	0.45	1	0.45
4	7	3.13	8	3.57
5	9	4.02	17	7.59
6	24	10.71	41	18.30
7	42	18.75	83	37.05
8	31	13.84	114	50.89
9	50	22.32	164	73.21
10	60	26.79	224	100.00

hse	Frequency	Percent	Cumulative Frequency	Cumulative Percent
3	1	0.45	1	0.45
4	4	1.79	5	2.23
5	5	2.23	10	4.46
6	23	10.27	33	14.73
7	43	19.20	76	33.93
8	49	21.88	125	55.80
9	52	23.21	177	79.02
10	47	20.98	224	100.00

FIGURE 11.4 The distributions of the high school grade variables.

distributions are all skewed, with a large proportion of high grades (10 = A and 9 = A$^-$). Again we emphasize that these distributions need not be Normal.

Relationships between pairs of variables

correlation p. 171

The second step in our analysis is to examine the relationships between all pairs of variables. Scatterplots and correlations are our tools for studying two-variable relationships. The correlations appear in Figure 11.5. The output includes the P-value for the test of the null hypothesis that the population correlation is 0 versus the two-sided alternative for each pair. Thus, we see that the correlation between GPA and HSM is 0.44, with a P-value of 0.000 (i.e., $P < 0.0005$), whereas the correlation between GPA and SATV is 0.11, with a P-value of 0.087. The first is statistically significant by any reasonable standard, and the second is slightly larger than the 0.05 significance level.

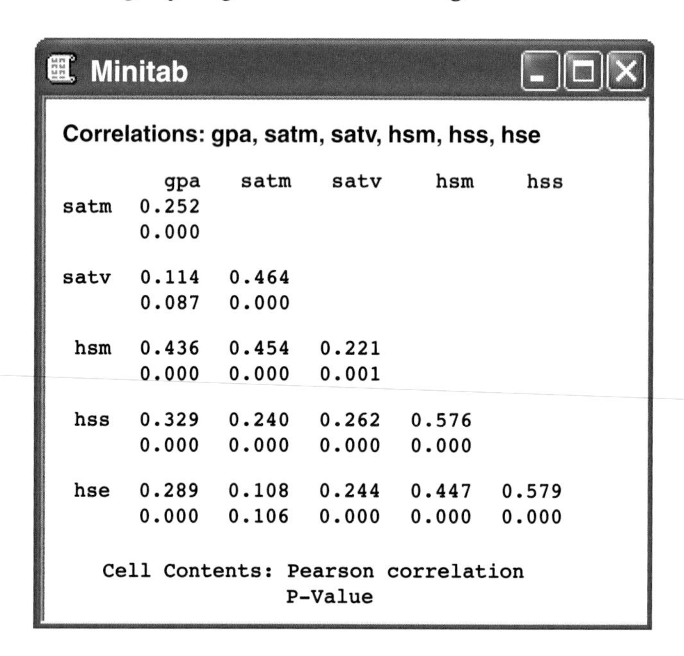

Minitab

Correlations: gpa, satm, satv, hsm, hss, hse

	gpa	satm	satv	hsm	hss
satm	0.252				
	0.000				
satv	0.114	0.464			
	0.087	0.000			
hsm	0.436	0.454	0.221		
	0.000	0.000	0.001		
hss	0.329	0.240	0.262	0.576	
	0.000	0.000	0.000	0.000	
hse	0.289	0.108	0.244	0.447	0.579
	0.000	0.106	0.000	0.000	0.000

Cell Contents: Pearson correlation
P-Value

FIGURE 11.5 Correlations between pairs of case study variables.

The high school grades all have higher correlations with GPA than do the SAT scores. As we might expect, math grades have the highest correlation (r = 0.44), followed by science grades (0.33) and then English grades (0.29). The two SAT scores have a rather high correlation with each other (0.46), and the high school grades also correlate well with each other (0.45 to 0.58). SATM correlates well with HSM (0.45), less well with HSS (0.24), and rather poorly with HSE (0.11). The correlations of SATV with the three high school grades are about equal, ranging from 0.22 to 0.26.

It is important to keep in mind that by examining pairs of variables we are seeking a better understanding of the data. *Even though the correlation of a particular explanatory variable with the response variable does not achieve statistical significance, it may still be a useful (and statistically significant) predictor in a multiple regression.*

Numerical summaries such as correlations are useful, but plots are generally more informative when seeking to understand data. Plots tell us whether the numerical summary gives a fair representation of the data.

For a multiple regression, each pair of variables should be plotted. For the six variables in our case study, this means that we should examine 15 plots. In general, there are $p + 1$ variables in a multiple regression analysis with p explanatory variables, so that $p(p + 1)/2$ plots are required.

Multiple regression is a complicated procedure. If we do not do the necessary preliminary work, we are in serious danger of producing useless or misleading results. We leave the task of making these plots as an exercise.

USE YOUR KNOWLEDGE

11.5 Pairwise relationships among variables in the CSDATA data set. Using a statistical package, generate the pairwise correlations and scatterplots discussed previously. Comment on any unusual patterns or observations. DATA FILE CSDATA

Regression on high school grades

To explore the relationship between the explanatory variables and our response variable GPA, we run several multiple regressions. The explanatory variables fall into two classes. High school grades are represented by HSM, HSS, and HSE, and standardized tests are represented by the two SAT scores.

We begin our analysis by using the high school grades to predict GPA. Figure 11.6 gives the multiple regression output.

The output contains an ANOVA table, some additional descriptive statistics, and information about the parameter estimates. When examining any ANOVA table, it is a good idea to first verify the degrees of freedom. This

SAS

Dependent Variable: GPA

Analysis of Variance

Source	DF	Sum of Squares	Mean Square	F Value	Prob>F
Model	3	27.71233	9.23744	18.861	<.0001
Error	220	107.75046	0.48977		
C Total	223	135.46279			

Root MSE	0.69984	R-Square	0.2046
Dep Mean	2.63522	Adj R-sq	0.1937
C.V.	26.55711		

Parameter Estimates

Variable	DF	Parameter Estimate	Standard Error	T for H0: Parameter=0	Prob > |T|
INTERCEP	1	0.589877	0.29424324	2.005	0.0462
HSM	1	0.168567	0.03549214	4.749	0.0001
HSS	1	0.034316	0.03755888	0.914	0.3619
HSE	1	0.045102	0.03869585	1.166	0.2451

FIGURE 11.6 Multiple regression output for regression using high school grades to predict GPA.

ensures that we have not made some serious error in specifying the model for the software or in entering the data. Because there are $n = 224$ cases, we have DFT $= n - 1 = 223$. The three explanatory variables give DFM $= p = 3$ and DFE $= n - p - 1 = 223 - 3 = 220$.

The ANOVA F statistic is 18.86, with a P-value of 0.0001. Under the null hypothesis

$$H_0: \beta_1 = \beta_2 = \beta_3 = 0$$

the F statistic has an $F(3, 220)$ distribution. According to this distribution, the chance of obtaining an F statistic of 18.86 or larger is less than 0.0001. We therefore conclude that at least one of the three regression coefficients for the high school grades is different from 0 in the population regression equation.

In the descriptive statistics that follow the ANOVA table, we find that Root MSE is 0.6998. This value is the square root of the MSE given in the ANOVA table and is s, the estimate of the parameter σ of our model. The value of R^2 is 0.20. That is, 20% of the observed variation in the GPA scores is explained by linear regression on high school grades.

Although the P-value is very small, the model does not explain very much of the variation in GPA. Remember, a small P-value does not necessarily tell us that we have a large effect, particularly when the sample size is large.

From the Parameter Estimates section of the computer output, we obtain the fitted regression equation

$$\widehat{\text{GPA}} = 0.590 + 0.169\text{HSM} + 0.034\text{HSS} + 0.045\text{HSE}$$

Let's find the predicted GPA for a student with an A– average in HSM, B+ in HSS, and B in HSE. The explanatory variables are HSM $= 9$, HSS $= 8$, and HSE $= 7$. The predicted GPA is

$$\begin{aligned}\widehat{\text{GPA}} &= 0.590 + 0.169(9) + 0.034(8) + 0.045(7) \\ &= 2.7\end{aligned}$$

Recall that the t statistics for testing the regression coefficients are obtained by dividing the estimates by their standard errors. Thus, for the coefficient of HSM we obtain the t-value given in the output by calculating

$$t = \frac{b}{\text{SE}_b} = \frac{0.168567}{0.03549214} = 4.749$$

The P-values appear in the last column. Note that these P-values are for the two-sided alternatives. HSM has a P-value of 0.0001, and we conclude that the regression coefficient for this explanatory variable is significantly different from 0. The P-values for the other explanatory variables (0.36 for HSS and 0.25 for HSE) do not achieve statistical significance.

Interpreting these results

The significance tests for the individual regression coefficients seem to contradict the impression obtained by examining the correlations in Figure 11.5. In that display we see that the correlation between GPA and HSS is 0.33 and the correlation between GPA and HSE is 0.29. The P-values for both of these correlations are < 0.0005. In other words, if we used HSS alone in a regression to predict GPA, or if we used HSE alone, we would obtain statistically significant regression coefficients.

This phenomenon is not unusual in multiple regression analysis. Part of the explanation lies in the correlations between HSM and the other two explanatory variables. These are rather high (at least compared with the other correlations in Figure 11.5). The correlation between HSM and HSS is 0.58, and that between HSM and HSE is 0.45.

Thus, when we have a regression model that contains all three high school grades as explanatory variables, there is considerable overlap of the predictive information contained in these variables. *The significance tests for individual regression coefficients assess the significance of each predictor variable assuming that all other predictors are included in the regression equation.*

Given that we use a model with HSM and HSS as predictors, the coefficient of HSE is not statistically significant. Similarly, given that we have HSM and HSE in the model, HSS does not have a significant regression coefficient. HSM, however, adds significantly to our ability to predict GPA even after HSS and HSE are already in the model.

Unfortunately, we cannot conclude from this analysis that the *pair* of explanatory variables HSS and HSE contributes nothing significant to our model for predicting GPA once HSM is in the model. The impact of relations among the several explanatory variables on fitting models for the response is the most important new phenomenon encountered in moving from simple linear regression to multiple regression. We can only hint at the many complicated problems that arise.

Examining the residuals

As in simple linear regression, we should always examine the residuals as an aid to determining whether the multiple regression model is appropriate for the data. Because there are several explanatory variables, we must examine several residual plots. It is usual to plot the residuals versus the predicted values $\hat{y}$ and also versus each of the explanatory variables. Look for outliers, influential observations, evidence of a curved (rather than linear) relation, and anything else unusual. Again, we leave the task of making these plots as an exercise. The plots all appear to show more or less random noise around the center value of 0.

If the deviations ϵ in the model are Normally distributed, the residuals should be Normally distributed. Figure 11.7 presents a Normal quantile plot of the residuals. The distribution appears to be approximately Normal. There

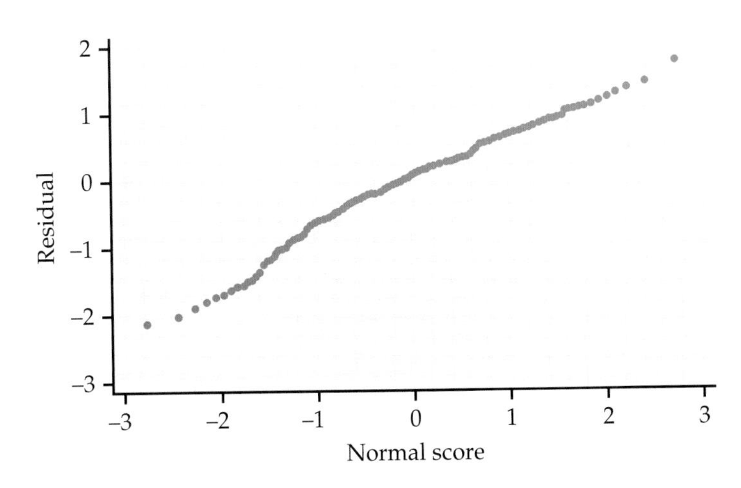

FIGURE 11.7 Normal quantile plot of the residuals from the high school grades model. There are no important deviations from Normality.

are many other specialized plots that help detect departures from the multiple regression model. Discussion of these, however, is more than we can undertake in this chapter.

USE YOUR KNOWLEDGE

11.6 Residual plots for the CSDATA analysis. Using a statistical package, fit the linear model with HSM and HSE as predictors and obtain the residuals and predicted values. Plot the residuals versus the predicted values, HSM, and HSE. Are the residuals more or less randomly dispersed around zero? Comment on any unusual patterns. CSDATA

Refining the model

Because the variable HSS has the largest P-value of the three explanatory variables (see Figure 11.6) and therefore appears to contribute the least to our explanation of GPA, we rerun the regression using only HSM and HSE as explanatory variables. The SAS output appears in Figure 11.8. The F statistic indicates that we can reject the null hypothesis that the regression coefficients for the two explanatory variables are both 0. The P-value is still less than 0.0001.

The value of R^2 has dropped very slightly compared with our previous run, from 0.2046 to 0.2016. Thus, dropping HSS from the model resulted in the loss of very little explanatory power. The measure s of variation about the fitted equation (Root MSE in the printout) is nearly identical for the two regressions, another indication that we lose very little dropping HSS. The t statistics for the

SAS

Dependent Variable: GPA

Analysis of Variance

Source	DF	Sum of Squares	Mean Square	F Value	Prob>F
Model	2	27.30349	13.65175	27.894	<.0001
Error	221	108.15930	0.48941		
C Total	223	135.46279			

Root MSE	0.69958	R-Square	0.2016
Dep Mean	2.63522	Adj R-sq	0.1943
C.V.	26.54718		

Parameter Estimates

Variable	DF	Parameter Estimate	Standard Error	T for H0: Parameter=0	Prob > |T|
INTERCEP	1	0.624228	0.29172204	2.140	0.0335
HSM	1	0.182654	0.03195581	5.716	0.0001
HSE	1	0.060670	0.03472914	1.747	0.0820

FIGURE 11.8 Multiple regression output for regression using HSM and HSE to predict GPA.

individual regression coefficients indicate that HSM is still clearly significant ($P < 0.0001$), while the statistic for HSE is larger than before (1.747 versus 1.166) and approaches the traditional 0.05 level of significance ($P = 0.082$).

Comparison of the fitted equations for the two multiple regression analyses tells us something more about the intricacies of this procedure. For the first run we have

$$\widehat{\text{GPA}} = 0.590 + 0.169\text{HSM} + 0.034\text{HSS} + 0.045\text{HSE}$$

whereas the second gives us

$$\widehat{\text{GPA}} = 0.624 + 0.183\text{HSM} + 0.061\text{HSE}$$

Eliminating HSS from the model changes the regression coefficients for all the remaining variables and the intercept. This phenomenon occurs quite generally in multiple regression. *Individual regression coefficients, their standard errors, and significance tests are meaningful only when interpreted in the context of the other explanatory variables in the model.*

Regression on SAT scores

We now turn to the problem of predicting GPA using the two SAT scores. Figure 11.9 gives the output. The fitted model is

$$\widehat{\text{GPA}} = 1.289 + 0.002283\text{SATM} - 0.000025\text{SATV}$$

The degrees of freedom are as expected: 2, 221, and 223. The F statistic is 7.476, with a P-value of 0.0007.

SAS

Dependent Variable: GPA

Analysis of Variance

Source	DF	Sum of Squares	Mean Square	F Value	Prob>F
Model	2	8.58384	4.29192	7.476	0.0007
Error	221	126.87895	0.57411		
C Total	223	135.46279			

Root MSE	0.75770	R-Square	0.0634
Dep Mean	2.63522	Adj R-sq	0.0549
C.V.	28.75287		

Parameter Estimates

Variable	DF	Parameter Estimate	Standard Error	T for H0: Parameter=0	Prob > |T|
INTERCEP	1	1.288677	0.37603684	3.427	0.0007
SATM	1	0.002283	0.00066291	3.444	0.0007
SATV	1	-0.000024562	0.00061847	-0.040	0.9684

FIGURE 11.9 Multiple regression output for regression using SAT scores to predict GPA.

We conclude that the regression coefficients for SATM and SATV are not both 0. Recall that we obtained a P-value less than 0.0001 when we used high school grades to predict GPA. Both multiple regression equations are highly significant, but this obscures the fact that the two models have quite different explanatory power. For the SAT regression, $R^2 = 0.0634$, whereas for the high school grades model, even with only HSM and HSE (Figure 11.8), we have $R^2 = 0.2016$, a value more than three times as large. *Stating that we have a statistically significant result is quite different from saying that an effect is large or important.*

Further examination of the output in Figure 11.9 reveals that the coefficient of SATM is significant ($t = 3.44$, $P = 0.0007$) and that for SATV is not ($t = -0.04$, $P = 0.9684$). For a complete analysis we should carefully examine the residuals. Also, we might want to run the analysis with SATM as the only explanatory variable.

Regression using all variables

We have seen that either the high school grades or the SAT scores give a highly significant regression equation. The mathematics component of each of these groups of explanatory variables appears to be the key predictor. Comparing the values of R^2 for the two models indicates that high school grades are better predictors than SAT scores. Can we get a better prediction equation using all the explanatory variables together in one multiple regression?

To address this question we run the regression with all five explanatory variables. The output appears in Figure 11.10 (next page). The F statistic is 11.69, with a P-value of 0.0001, so at least one of our explanatory variables has a nonzero regression coefficient. This result is not surprising, given that we have already seen that HSM and SATM are strong predictors of GPA. The value of R^2 is 0.2115, not much higher than the value of 0.2046 that we found for the high school grades regression.

Examination of the t statistics and the associated P-values for the individual regression coefficients reveals that HSM is the only one that is significant ($P = 0.0003$). That is, only HSM makes a significant contribution when it is added to a model that already has the other four explanatory variables. Once again it is important to understand that this result does not necessarily mean that the regression coefficients for the four other explanatory variables are *all* 0.

Figure 11.11 (page 540) gives the Excel and Minitab multiple regression outputs for this problem. Although the format and organization of outputs differ among software packages, the basic results are easy to find.

Many statistical software packages provide the capability for testing whether a collection of regression coefficients in a multiple regression model are *all* 0. We will now use this approach to address two interesting questions about this set of data. We did not discuss such tests in Section 11.1, but the basic idea is quite simple and discussed in Exercise 11.20 (page 544).

Test for a collection of regression coefficients

In the context of the multiple regression model with all five predictors, we ask first whether or not the coefficients for the two SAT scores are both 0. In other words, do the SAT scores add any significant predictive information to that already contained in the high school grades? To be fair, we also ask the

```
SAS

Dependent Variable: GPA

          Analysis of Variance

                           Sum of          Mean
Source          DF        Squares        Square       F Value      Prob>F

Model            5       28.64364       5.72873        11.691      <.0001
Error          218      106.81914       0.49000
C Total        223      135.46279

    Root MSE         0.70000     R-Square         0.2115
    Dep Mean         2.63522     Adj R-sq         0.1934
    C.V.            26.56311

                            Parameter Estimates

                        Parameter       Standard       T for H0:
Variable     DF          Estimate          Error     Parameter=0    Prob > |T|

INTERCEP      1          0.326719     0.39999643           0.817        0.4149
SATM          1          0.000944     0.00068566           1.376        0.1702
SATV          1         -0.000408     0.00059189          -0.689        0.4915
HSM           1          0.145961     0.03926097           3.718        0.0003
HSS           1          0.035905     0.03779841           0.950        0.3432
HSE           1          0.055293     0.03956869           1.397        0.1637

Test: SAT     Numerator:         0.4657  DF:    2   F value:    0.9503
              Denominator:     0.489996  DF:  218   Prob>F:     0.3882

Test: HS      Numerator:         6.6866  DF:    3   F value:   13.6462
              Denominator:     0.489996  DF:  218   Prob>F:     0.0001
```

FIGURE 11.10 Multiple regression output for regression using all variables to predict GPA.

complementary question: do the high school grades add any significant predictive information to that already contained in the SAT scores?

The answers are given in the last two parts of the output in Figure 11.10. For the first test we see that $F = 0.9503$. Under the null hypothesis that the two SAT coefficients are 0, this statistic has an $F(2, 218)$ distribution and the P-value is 0.39. We conclude that the SAT scores are not significant predictors of GPA in a regression that already contains the high school scores as predictor variables. Recall that the model with just SAT scores has a highly significant F statistic. We now see that whatever predictive information is in the SAT scores can also be found in the high school grades. In this sense, the SAT scores are unnecessary.

The test statistic for the three high school grade variables is $F = 13.6462$. Under the null hypothesis that these three regression coefficients are 0, the statistic has an $F(3, 218)$ distribution and the P-value is 0.0001. We conclude that high school grades contain useful information for predicting GPA that is not contained in SAT scores.

Excel

	A	B	C	D	E	F	G
1	SUMMARY OUTPUT						
2							
3	*Regression Statistics*						
4	Multiple R	0.459837234					
5	R Square	0.211450282					
6	Adjusted R Square	0.193364279					
7	Standard Error	0.699997195					
8	Observations	224					
9							
10	ANOVA						
11		*df*	*SS*	*MS*	*F*	*Significance F*	
12	Regression	5	28.64364489	5.728729	11.69138	5.06E-10	
13	Residual	218	106.8191439	0.489996			
14	Total	223	135.4627888				
15							
16		*Coefficients*	*Standard Error*	*tStet*	*P-value*	*Lower 95%*	*Upper 95%*
17	intercept	0.326718739	0.399996431	0.816804	0.414932	–0.461636967	1.115074446
18	HSM	0.14596108	0.039260974	3.717714	0.000256	0.068581358	0.223340801
19	HSS	0.03590532	0.037798412	0.949916	0.343207	–0.03859183	0.11040247
20	HSE	0.055292581	0.039568691	1.397382	0.163719	–0.022693622	0.133278785
21	SATM	0.000843593	0.000685657	1.376187	0.170176	–0.000407774	0.002294959
22	SATV	0.00040785	0.000591893	–0.68906	0.491518	–0.001574415	0.00075816

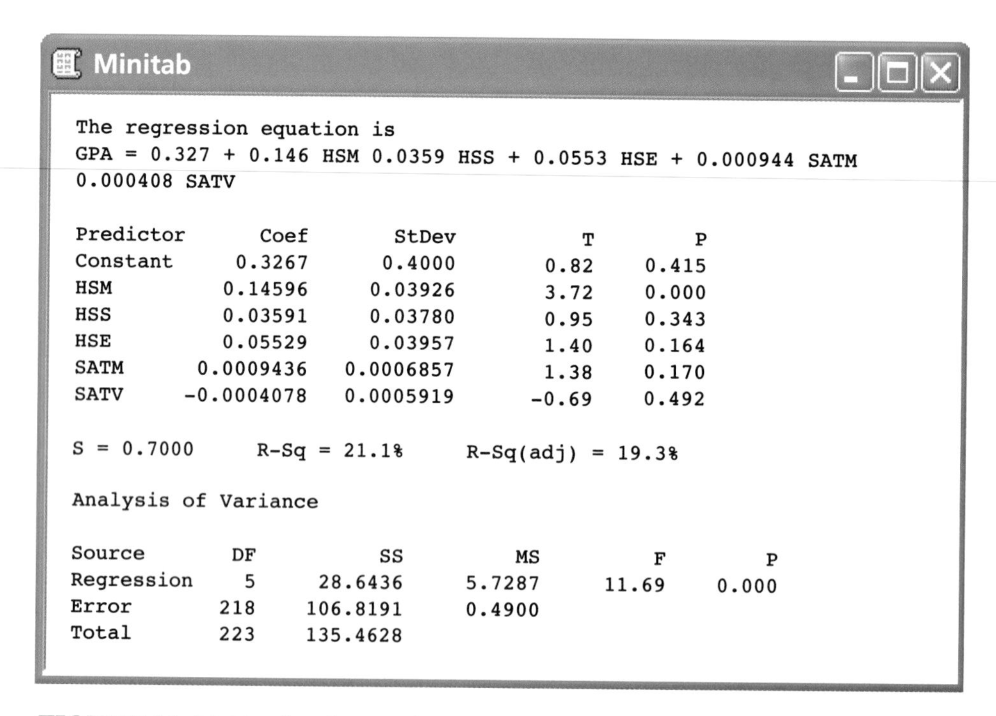

Minitab

```
The regression equation is
GPA = 0.327 + 0.146 HSM 0.0359 HSS + 0.0553 HSE + 0.000944 SATM
0.000408 SATV

Predictor        Coef       StDev        T       P
Constant       0.3267      0.4000     0.82   0.415
HSM           0.14596     0.03926     3.72   0.000
HSS           0.03591     0.03780     0.95   0.343
HSE           0.05529     0.03957     1.40   0.164
SATM        0.0009436   0.0006857     1.38   0.170
SATV       -0.0004078   0.0005919    -0.69   0.492

S = 0.7000     R-Sq = 21.1%     R-Sq(adj) = 19.3%

Analysis of Variance

Source       DF        SS        MS       F       P
Regression    5   28.6436    5.7287   11.69   0.000
Error       218  106.8191    0.4900
Total       223  135.4628
```

FIGURE 11.11 Excel and Minitab multiple regression outputs for regression using all variables to predict GPA.

Of course, our statistical analysis of these data does not imply that SAT scores are less useful than high school grades for predicting college grades for all groups of students. We have studied a select group of students—computer science majors—from a specific university. Generalizations to other situations are beyond the scope of inference based on these data alone.

CHAPTER 11 SUMMARY

Data for multiple linear regression consist of the values of a response variable y and p explanatory variables $x_1, x_2, \ldots, x_p$ for n cases. We write the data and enter them into software in the form

	Variables				
Individual	x_1	x_2	...	x_p	y
1	x_{11}	x_{12}	...	x_{1p}	y_1
2	x_{21}	x_{22}	...	x_{2p}	y_2
⋮					
n	x_{n1}	x_{n2}	...	x_{np}	y_n

The statistical model for **multiple linear regression** with response variable y and p explanatory variables $x_1, x_2, \ldots, x_p$ is

$$y_i = \beta_0 + \beta_1 x_{i1} + \beta_2 x_{i2} + \cdots + \beta_p x_{ip} + \epsilon_i$$

where $i = 1, 2, \ldots, n$. The ϵ_i are assumed to be independent and Normally distributed with mean 0 and standard deviation σ. The **parameters** of the model are $\beta_0, \beta_1, \beta_2, \ldots, \beta_p$, and σ.

The **multiple regression equation** predicts the response variable by a linear relationship with all the explanatory variables:

$$\hat{y} = b_0 + b_1 x_1 + b_2 x_2 + \cdots + b_p x_p$$

The β's are estimated by $b_0, b_1, b_2, \ldots, b_p$, which are obtained by the **method of least squares.** The parameter σ is estimated by

$$s = \sqrt{\text{MSE}} = \sqrt{\frac{\sum e_i^2}{n - p - 1}}$$

where the e_i are the **residuals,**

$$e_i = y_i - \hat{y}_i$$

Always examine the **distribution of the residuals** and plot them against the explanatory variables prior to inference.

A **level *C* confidence interval for** β_j is

$$b_j \pm t^* \text{SE}_{b_j}$$

where t^* is the value for the $t(n - p - 1)$ density curve with area C between $-t^*$ and t^*.

The test of the hypothesis H_0: $\beta_j = 0$ is based on the ***t* statistic**

$$t = \frac{b_j}{\text{SE}_{b_j}}$$

and the $t(n - p - 1)$ distribution.

The estimate b_j of β_j and the test and confidence interval for β_j are all based on a specific multiple linear regression model. The results of all these procedures change if other explanatory variables are added to or deleted from the model.

The **ANOVA table** for a multiple linear regression gives the degrees of freedom, sums of squares, and mean squares for the model, error, and total sources of variation. The **ANOVA *F* statistic** is the ratio MSM/MSE and is used to test the null hypothesis

$$H_0: \beta_1 = \beta_2 = \cdots = \beta_p = 0$$

If H_0 is true, this statistic has an $F(p, n-p-1)$ distribution.

The **squared multiple correlation** is given by the expression

$$R^2 = \frac{\text{SSM}}{\text{SST}}$$

and is interpreted as the proportion of the variability in the response variable y that is explained by the explanatory variables $x_1, x_2, \ldots, x_p$ in the multiple linear regression.

CHAPTER 11 EXERCISES

For Exercise 11.1, see page 522; for Exercise 11.2, see page 523; for Exercises 11.3 and 11.4, see pages 529–530; for Exercise 11.5, see page 533; and for Exercise 11.6, see page 536.

11.7. What's wrong? In each of the following situations, explain what is wrong and why.

(a) In a multiple regression with a sample size of 39 and 3 explanatory variables, the test statistic for the null hypothesis H_0: $b_2 = 0$ is a t statistic that follows the $t(35)$ distribution when the null hypothesis is true.

(b) The multiple correlation coefficient gives the proportion of the variation in the response variable that is explained by the explanatory variables.

(c) A small P-value for the ANOVA F test implies that all explanatory variables are significantly different from zero.

11.8. What's wrong? In each of the following situations, explain what is wrong and why.

(a) One of the assumptions for multiple regression is that the distribution of each explanatory variable is Normal.

(b) The smaller the P-value for the ANOVA F test, the greater the explanatory power of the model.

(c) All explanatory variables that are significantly correlated with the response variable will have a statistically significant regression coefficient in the multiple regression model.

(d) The multiple correlation coefficient gives the average correlation between the response variable and each explanatory variable in the model.

11.9. 95% confidence intervals for regression coefficients. In each of the following settings, give a 95% confidence interval for the coefficient of x_1.

(a) $n = 27$, $\hat{y} = 10.6 + 6.7x_1 + 7.9x_2$, $\text{SE}_{b_1} = 3.3$.

(b) $n = 53$, $\hat{y} = 10.6 + 6.7x_1 + 7.9x_2$, $\text{SE}_{b_1} = 3.3$.

(c) $n = 27$, $\hat{y} = 10.6 + 6.7x_1 + 7.9x_2 + 5.2x_3$, $\text{SE}_{b_1} = 3.3$.

(d) $n = 124$, $\hat{y} = 10.6 + 6.7x_1 + 7.9x_2 + 5.2x_3$, $\text{SE}_{b_1} = 3.3$.

11.10. More on significance tests for regression coefficients. For each of the settings in the previous exercise, test the null hypothesis that the coefficient of x_1 is zero versus the two-sided alternative.

11.11. Constructing the ANOVA table. Seven explanatory variables are used to predict a response variable using a multiple regression. There are 135 observations.

(a) Write the statistical model that is the foundation for this analysis. Also include a description of all assumptions.

(b) Outline the analysis of variance table giving the sources of variation and numerical values for the degrees of freedom.

11.12. More on constructing the ANOVA table. A multiple regression analysis of 67 cases was performed with 6 explanatory variables. Suppose that SSM = 25.6 and SSE = 110.5.

(a) Find the value of the F statistic for testing the null hypothesis that the coefficients of all the explanatory variables are zero.

(b) What are the degrees of freedom for this statistic?

(c) Find bounds on the P-value using Table E. Show your work.

(d) What proportion of the variation in the response variable is explained by the explanatory variables?

11.13. Predicting energy-drink consumption. Energy-drink advertising consistently emphasizes a physically active lifestyle and often features extreme sports and risk taking. Are these typical characteristics of an energy-drink consumer? A researcher decided to examine the links between energy-drink consumption, sport-related (jock) identity, and risk taking.[3] She invited over 1500 undergraduate students enrolled in large introductory-level courses at a public university to participate. Each participant had to complete a 45-minute anonymous questionnaire. From this questionnaire jock identity and risk-taking scores were obtained where the higher the score, the stronger the trait. She ended up with 795 respondents. The following table summarizes the results of a multiple regression analysis using the frequency of energy-drink consumption in the past 30 days as the response variable:

Explanatory variable	b
Age	−0.02
Sex (1 = female, 0 = male)	−0.11**
Race (1 = nonwhite, 0 = white)	−0.02
Ethnicity (1 = Hispanic, 0 = non-Hispanic)	0.10**
Parental education	0.02
College GPA	−0.01
Jock identity	0.05
Risk taking	0.19***

A superscript of ** means that the individual coefficient t test had a P-value less than 0.01, and a superscript of *** means that the test had a P-value less than 0.001. All other P-values were greater than 0.05.

(a) The overall F statistic is reported to be 8.11. What are the degrees of freedom associated with this statistic?

(b) R is reported to be 0.28. What percent of the variation in energy-drink consumption is explained by the model? Is this a highly predictive model? Explain.

(c) Interpret each of the regression coefficients that are significant.

(d) The researcher states, "Controlling for gender, age, race, ethnicity, parental educational achievement, and college GPA, each of the predictors (risk taking and jock identity) was positively associated with energy-drink consumption frequency." Explain what is meant by "controlling for" these variables and how this helps strengthen her assertion that jock identity and risk taking are positively associated with energy-drink consumption.

11.14. Is the number of tornadoes increasing? In Exercise 10.19 (page 512), data on the number of tornadoes in the United States between 1953 and 2009 were analyzed to see if there was a linear trend over time. Many argue that the probability of sighting a tornado has increased over time because there are more people living in the United States. Let's investigate this by including the U.S. census count as an additional explanatory variable. DATA FILE TORNADOES

(a) Using numerical and graphical summaries, describe the relationship between each pair of variables.

(b) Perform a multiple regression using both year and population count as explanatory variables. Write down the fitted model.

(c) Obtain the residuals from part (b). Plot them versus the two explanatory variables and generate a Normal quantile plot. What do you conclude?

(d) Test the hypothesis that there is a linear increase over time. State the null and alternative hypotheses, test statistic, and P-value. What is your conclusion?

11.15. CHALLENGE **Checking for a polynomial relationship.** When looking at the residuals from the simple linear model of BMI versus physical activity (PA), Figure 10.5 (page 498) suggested a possible curvilinear relationship. Let's investigate this further. Multiple regression can be used to fit the polynomial curve of degree q, $y = \beta_0 + \beta_1 x + \beta_2 x^2 + \cdots + \beta_q x^q$, through the creation of additional explanatory variables x^2, x^3, etc. Let's investigate a quadratic fit ($q = 2$) for the physical activity problem. DATA FILE PA_BMI

(a) It is often best to subtract the sample mean $\overline{x}$ before creating the necessary explanatory variables. In this case, the average thousand steps per day is 8.614. Create new explanatory variables $x_1 = (\text{PA} - 8.614)$ and $x_2 = (\text{PA} - 8.614)^2$ and run a multiple regression for BMI using the explanatory variables x_1 and x_2. Write down the fitted regression line.

(b) The regression model that just included PA had $R^2 = 14.9$. What is the R^2 with the inclusion of this quadratic term?

(c) Obtain the residuals from part (a) and check the multiple regression assumptions. Are there any remaining patterns in the data? Are the residuals approximately Normal? Explain.

(d) Test the hypothesis that the coefficient of the variable $(\text{PA} - 8.614)^2$ is equal to 0. Report the t statistic, degrees of freedom, and P-value. Does the quadratic term contribute significantly to the fit? Explain your answer.

11.16. Architectural firm billings. A summary of firms engaged in commercial architecture in the Indianapolis, Indiana, area provides characteristics including total annual billing in billions of dollars and the number of architects, engineers, and staff employed.[4] Consider developing a model to predict total billing. DATA FILE ARCHITECT

(a) Using numerical and graphical summaries, describe the distribution of total billing and the number of architects, engineers, and staff.

(b) For each of the 6 pairs of variables, use graphical and numerical summaries to describe the relationship.

(c) Carry out a multiple regression. Report the fitted regression equation and the value of the regression standard error s.

(d) Analyze the residuals from the multiple regression. Are there any concerns?

(e) The firm HCO did not report its total billing but employs 3 architects, 1 engineer, and 17 staff members. What is the predicted total billing for this firm?

The following six exercises use the MOVIES data set. This data set contains an SRS of 35 movies released in 2008. This sample was collected from the Internet Movie Database (IMDb) to see if information available soon after a movie's theatrical release can successfully predict total revenue.[5] All dollar amounts are measured in millions of U.S. dollars. DATA FILE MOVIES

11.17. Predicting movie revenue: preliminary analysis. The response variable is a movie's total U.S. revenue (USRevenue). Let's consider as explanatory variables the movie's budget (Budget); opening weekend revenue (Opening); the number of theaters (Theaters) in which the movie was shown during the opening weekend; and the movie's IMDb rating (Opinion), which is on a 1 to 10 scale (10 being best). Although this rating is updated continuously, we'll assume that the current rating is the rating at the end of the first week.

(a) Using numerical and graphical summaries, describe the distribution of each explanatory variable. Are there any unusual observations that should be monitored?

(b) Using numerical and graphical summaries, describe the relationship between each pair of explanatory variables.

11.18. Predicting movie revenue: simple linear regressions. Now let's look at the response variable and its relationship with each explanatory variable.

(a) Using numerical and graphical summaries, describe the distribution of the response variable, USRevenue.

(b) This variable is not Normally distributed. Does this violate one of our key model assumptions? Explain.

(c) Generate scatterplots of each explanatory variable and USRevenue. Do all these relationships look linear? Explain what you see.

11.19. Predicting movie revenue: multiple linear regression. Now consider fitting a model using all the explanatory variables.

(a) Write out the statistical model for this analysis, making sure to specify all assumptions.

(b) Run the multiple regression model and specify the fitted regression equation.

(c) Obtain the residuals from part (b) and check assumptions. Comment on any unusual residuals or patterns in the residuals.

(d) What percent of the variability in USRevenue is explained by this model?

11.20. CHALLENGE **A simpler model.** Refer to the previous exercise. In the multiple regression analysis using all four variables, Theaters and Budget appear to be the least helpful (given that the other two explanatory variables are in the model).

(a) Perform a new analysis using only the movie's opening-weekend revenue and IMDb rating. Give the estimated regression equation for this analysis.

(b) What percent of the variability in USRevenue is explained by this model?

(c) In this chapter we discussed the F test for a collection of regression coefficients. In most cases, this capability is provided by the software. When it is not, the test can be performed using the R^2-values from the full and reduced models. The test statistic is

$$F = \left(\frac{n-p-1}{q}\right)\left(\frac{R_1^2 - R_2^2}{1 - R_1^2}\right)$$

with q and $n-p-1$ degrees of freedom. R_1^2 is the value for the full model and R_2^2 is the value for the reduced model. Here $n = 35$ movies, $p = 4$ variables in the full model, and $q = 2$ variables that were removed to form the reduced model. Plug in the values of R^2 from part (b) of this exercise and part (d) of the previous exercise and compute the test statistic and P-value. Do Theaters and Budget combined add any significant predictive information beyond what is already contained in Opening and Opinion?

11.21. Predicting U.S. movie revenue. Refer to the previous two exercises. *Get Smart* was released in 2008, had a budget of \$80.0 million, was shown in 3911 theaters grossing \$38.7 million during the first weekend, and had an IMDb rating of 6.8. Use your software to construct

(a) A 95% prediction interval based on the model with all four predictors.

(b) A 95% prediction interval based on the model using only opening-weekend revenue and IMDb rating.

(c) Compare the two intervals. Do the models give similar predictions?

11.22. Effect of potential outliers. Consider the simpler model of Exercise 11.20 for this analysis.

(a) Two movies have much larger U.S. revenues than those predicted. Which ones are they, and how much more revenue did they obtain compared with that predicted?

(b) Remove these two movies and redo the multiple regression. Make a table giving the regression coefficients and their standard errors, t statistics, and P-values.

(c) Compare these results with those from Exercise 11.20. How does the removal of these outlying movies impact the estimated model?

(d) Obtain the residuals from this reduced data set and graphically examine their distribution. Do the residuals appear approximately Normal? Explain your answer.

The following three exercises use the RANKINGS data set. Since 2004, the Times Higher Education Supplement has provided a worldwide yearly ranking of universities. A total score for each university is calculated based on 13 performance indicators brought together into five main categories, which are Teaching (30%), Research (30%), Citations (30%), Industry Income (2.5%), and International Outlook (7.5%). The percents represent the contributions of each score to the total. For our purposes here, we will assume that these weights are unknown and will focus on the development of a model for the total score based on the first three explanatory variables. The report includes a table for the top 200 universities.[6] *The RANKING data set is a random sample of 75 of these universities. Although this is not a random sample of all universities, we will consider it to be one.* DATA FILE RANKINGS

11.23. Annual ranking of world universities. Let's consider developing a model to predict total score based on the teaching score (TEACH), research score (RESEARCH), and citations score (CITATIONS).

(a) Using numerical and graphical summaries, describe the distribution of each explanatory variable.

(b) Using numerical and graphical summaries, describe the relationship between each pair of explanatory variables.

11.24. Looking at the simple linear regressions. Now let's look at the relationship between each explanatory variable and the total score.

(a) Generate scatterplots for each explanatory variable and the total score. Do these relationships all look linear?

(b) Compute the correlation between each explanatory variable and the total score. Are certain explanatory variables more strongly associated with the total score?

11.25. Multiple linear regression model. Now consider a regression model using all three explanatory variables.

(a) Write out the statistical model for this analysis, making sure to specify all assumptions.

(b) Run the multiple regression model and specify the fitted regression equation.

(c) Generate a 95% confidence interval for each coefficient. Should any of these intervals contain 0? Explain.

(d) What percent of the variation in total score is explained by this model? What is the estimate for σ?

11.26. Understanding the tests of significance. Using a new software package, you ran a multiple regression. The output reported an F statistic with $P < 0.05$, but none of the t tests for the individual coefficients were significant ($P > 0.05$). Does this mean that there is something wrong with the software? Explain your answer.

The following three exercises use the HAPPINESS data set. The World Database of Happiness is an online registry of scientific research on the subjective appreciation of life. It is available at `worlddatabaseofhappiness.eur.nl` *and is directed by Dr. Ruut Veenhoven, Erasmus University, Rotterdam. One inventory presents the "average happiness" score for various nations between 2007 and 2008. This average is based on individual responses from numerous general population surveys to a general life satisfaction (well-being) question. Scores ranged between 0 (dissatisfied) to 10 (satisfied). The NationMaster Web site,* `www.nationmaster.com`*, contains a collection of statistics associated with various nations. For this data set, the factors considered are the Gini Index: the Gini coefficient measures the degree of inequality in the distribution of income (higher score = greater inequality); the degree of corruption in government (higher score = less corruption); average life expectancy; and the degree of democracy (higher score = more political liberties).* DATA FILE HAPPINESS

11.27. Predicting a nation's "average happiness" score. Consider the five statistics for each nation: LSI, the average life satisfaction score; GINI, the Gini coefficient of income inequality; CORRUPT, the degree of corruption in government; LIFE, the average life expectancy; and DEMOCRACY, a measure of civil and political liberties.

(a) Using numerical and graphical summaries, describe the distribution of each variable.

(b) Using numerical and graphical summaries, describe the relationship between each pair of variables.

11.28. Building a multiple linear regression model. Let's now build a model to predict the life satisfaction score, LSI.

(a) Consider a simple linear regression using GINI as the explanatory variable. Run the regression and summarize the results. Be sure to check assumptions.

(b) Now consider a model using GINI and LIFE. Run the multiple regression and summarize the results. Again be sure to check assumptions.

(c) Now consider a model using GINI, LIFE, and DEMOCRACY. Run the multiple regression and summarize the results. Again be sure to check assumptions.

(d) Now consider a model using all four explanatory variables. Again summarize the results and check assumptions.

11.29. Selecting from among several models. Refer to the results from the previous exercise.

(a) Make a table giving the estimated regression coefficients, standard errors, t statistics, and P-values.

(b) Describe how the coefficients and P-values change for the four models.

(c) Based on the table of coefficients, suggest another model. Run that model, summarize the results, and compare it with the other ones. Which model would you choose to explain LSI? Explain.

The following 11 exercises use the PCB data set. Polychlorinated biphenyls (PCBs) are a collection of synthetic compounds, called congeners, that are particularly toxic to fetuses and young children. Although PCBs are no longer produced in the United States, they are still found in the environment. Since human exposure to these PCBs is mainly through the consumption of fish, the Environmental Protection Agency (EPA) monitors PCB levels in fish. Unfortunately, there are 209 different congeners. Measuring all of them in a fish specimen is expensive and time-consuming. You've been asked to see if the total amount of PCBs in a specimen can be estimated with only a few, easily quantifiable congeners.[7] *If this can be done, costs can be greatly reduced.* DATA FILE PCB

11.30. Relationship among PCB congeners. Consider the following variables: PCB (the total amount of PCB)

and four congeners: PCB52, PCB118, PCB138, and PCB180.

(a) Using numerical and graphical summaries, describe the distribution of each of these variables.

(b) Using numerical and graphical summaries, describe the relationship between each pair of variables in this set.

11.31. Predicting the total amount of PCB. Use the four congeners, PCB52, PCB118, PCB138, and PCB180, in a multiple regression to predict PCB.

(a) Write the statistical model for this analysis. Include all assumptions.

(b) Run the regression and summarize the results.

(c) Examine the residuals. Do they appear to be approximately Normal? When you plot them versus each of the explanatory variables, are any patterns evident?

11.32. Adjusting for potential outliers. The examination of the residuals in part (c) of the previous exercise suggests that there may be two outliers, one with a high residual and one with a low residual.

(a) Because of safety issues, we are more concerned about underestimating PCB in a specimen than about overestimating. Give the specimen number for each of the two suspected outliers. Which one corresponds to an overestimate of PCB?

(b) Rerun the analysis with the two suspected outliers deleted, summarize these results, and compare them with those you obtained in the previous exercise.

11.33. More on predicting the total amount of PCB. Run a regression to predict PCB using the variables PCB52, PCB118, and PCB138. Note that this is similar to the analysis that you did in Exercise 11.31 except that PCB180 is not included as an explanatory variable.

(a) Summarize the results.

(b) In this analysis, the regression coefficient for PCB118 is not statistically significant. Give the estimate of the coefficient and the associated P-value.

(c) Find the estimate of the coefficient for PCB118 and the associated P-value for the model analyzed in Exercise 11.31.

(d) Using the results in parts (b) and (c), write a short paragraph explaining how the inclusion of other variables in a multiple regression can have an effect on the estimate of a particular coefficient and the results of the associated significance test.

11.34. Multiple regression model for total TEQ. Dioxins and furans are other classes of chemicals that can cause undesirable health effects similar to those caused by PCB. The three types of chemicals are combined using toxic equivalent scores (TEQs), which attempt to measure the health effects on a common scale. The PCB data set contains TEQs for PCB, dioxins, and furans. The variables are called TEQPCB, TEQDIOXIN, and TEQFURAN. The data set also includes the total TEQ, defined to be the sum of these three variables.

(a) Consider using a multiple regression to predict TEQ using the three components TEQPCB, TEQDIOXIN, and TEQFURAN as explanatory variables. Write the multiple regression model in the form

$$\text{TEQ} = \beta_0 + \beta_1 \text{TEQPCB} + \beta_2 \text{TEQDIOXIN} + \beta_3 \text{TEQFURAN} + \epsilon$$

Give numerical values for the parameters β_0, β_1, β_2, and β_3.

(b) The multiple regression model assumes that the ϵ's are Normal with mean zero and standard deviation σ. What is the numerical value of σ?

(c) Use software to run this regression and summarize the results.

11.35. CHALLENGE **Multiple regression model for total TEQ, continued.** The information summarized in TEQ is used to assess and manage risks from these chemicals. For example, the World Health Organization (WHO) has established the tolerable daily intake (TDI) as 1 to 4 TEQs per kilogram of body weight per day. Therefore, it would be very useful to have a procedure for estimating TEQ using just a few variables that can be measured cheaply. Use the four PCB congeners, PCB52, PCB118, PCB138, and PCB180, in a multiple regression to predict TEQ. Give a description of the model and assumptions, summarize the results, examine the residuals, and write a summary of what you have found.

11.36. CHALLENGE **Predicting total amount of PCB using transformed variables.** Because distributions of variables such as PCB, the PCB congeners, and TEQ tend to be skewed, researchers frequently analyze the logarithms of the measured variables. Create a data set that has the logs of each of the variables in the PCB data set. Note that zero is a possible value for PCB126; most software packages will eliminate these cases when you request a log transformation.

(a) If you do not do anything about the 16 zero values of PCB126, what does your software do with these cases? Is there an error message of some kind?

(b) If you attempt to run a regression to predict the log of PCB using the log of PCB126 and the log of PCB52, are the cases with the zero values of PCB126 eliminated? Do you think that is a good way to handle this situation?

(c) The smallest nonzero value of PCB126 is 0.0052. One common practice when taking logarithms of measured values is to replace the zeros by one-half of the smallest observed value. Create a logarithm data set using this procedure; that is, replace the 16 zero values of PCB126 by 0.0026 before taking logarithms. Use numerical and graphical summaries to describe the distributions of the log variables.

11.37. CHALLENGE **Predicting total amount of PCB using transformed variables, continued.** Refer to the previous exercise.

(a) Use numerical and graphical summaries to describe the relationship between each pair of log variables.

(b) Compare these summaries with the summaries that you produced in Exercise 11.30 for the measured variables.

11.38. CHALLENGE **Even more on predicting total amount of PCB using transformed variables.** Use the log data set that you created in Exercise 11.36 to find a good multiple regression model for predicting the log of PCB. Use only log PCB variables for this analysis. Write a report summarizing your results.

11.39. CHALLENGE **Predicting total TEQ using transformed variables.** Use the log data set that you created in Exercise 11.36 to find a good multiple regression model for predicting the log of TEQ. Use only log PCB variables for this analysis. Write a report summarizing your results and comparing them with the results that you obtained in the previous exercise.

11.40. Interpretation of coefficients in log PCB regressions. Use the results of your analysis of the log PCB data in Exercise 11.38 to write an explanation of how regression coefficients, standard errors of regression coefficients, and tests of significance for explanatory variables can change depending on what other explanatory variables are included in the multiple regression analysis.

CHAPTER TWELVE

ONE-WAY ANALYSIS OF VARIANCE

Many of the most effective statistical studies are comparative. For example, we may wish to compare customer satisfaction of men and women who use an online fantasy football site or compare the responses to various treatments in a clinical trial. We display these comparisons with back-to-back stemplots or side-by-side boxplots, and we measure them with five-number summaries or with means and standard deviations.

When only two groups are compared, Chapter 7 provides the tools we need to answer the question "Is the difference between groups statistically significant?" Two-sample t procedures compare the means of two Normal populations, and we saw that these procedures are sufficiently robust to be widely useful.

In this chapter, we will compare any number of means by techniques that generalize the two-sample t and share its robustness and usefulness. These methods will allow us to address comparisons such as

- Does a user's number of Facebook friends affect his or her social attractiveness?
- On average, which of 5 brands of automobile tires wears longest?
- Among three therapies for lung cancer, is there a difference in the mean time a patient remains tumor free?

12.1 Inference for One-Way Analysis of Variance

When comparing different populations or treatments, the data are subject to sampling variability. For example, we would not expect to observe exactly the same sales data if we mailed an advertising offer to different random samples of households. We also wouldn't expect a new group of cancer patients to provide the same set of relapse-free survival times. We therefore pose the question for inference in terms of the *mean* response.

In Chapter 7 we met procedures for comparing the means of two populations. We now extend those methods to problems involving more than two populations. The statistical methodology for comparing several means is called analysis of variance, or simply **ANOVA.** In this and the following section, we will examine the basic ideas and assumptions that are needed for ANOVA. Although the details differ, many of the concepts are similar to those discussed in the two-sample case.

ANOVA

We will consider two ANOVA techniques. When there is only one way to classify the populations, we use **one-way ANOVA** to analyze the data. We call this categorical explanatory variable a **factor.** For example, to compare the mean tread lifetimes of 5 specific brands of tires we use one-way ANOVA with tire brand as our factor. This chapter presents the details for one-way ANOVA.

one-way ANOVA

factor

In many other comparison studies, there is more than one way to classify the populations. For the tire study, the researcher may also want to consider temperature. Are there tire brands that do relatively better in the summer heat? Analyzing the effect of two factors, tire brand and temperature, requires **two-way ANOVA.** This technique will be discussed in Chapter 13.

two-way ANOVA

Data for one-way ANOVA

One-way analysis of variance is a statistical method for comparing several population means. We draw a simple random sample (SRS) from each population and use the data to test the null hypothesis that the population means are all equal. Consider the following two examples.

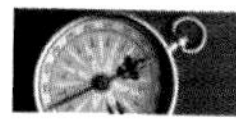

EXAMPLE 12.1

Choosing the best video game box art. An upstart video game company wants to compare three different box art styles for a series of video games that will be offered for sale at electronics stores. The company is interested in whether there is a style that better captures shoppers' attention and results in more sales. To investigate this, they randomly assign each of 60 stores to one of the three box art styles and record the number of games that are sold in a one-week period.

EXAMPLE 12.2

Average age of coffeehouse customers. How do five coffeehouses around campus differ in the demographics of their customers? Are certain coffeehouses more popular among graduate students? Do professors tend to favor one coffeehouse? A market researcher asks 50 customers of each coffeehouse to respond to a questionnaire. One variable of interest is the customer's age.

These two examples are similar in that

- There is a single quantitative response variable measured on many units; the units are stores in the first example and customers in the second.
- The goal is to compare several populations: stores displaying different box art styles in the first example and customers of five coffeehouses in the second.

LOOK BACK
observation versus experiment p. 35

There is, however, an important difference. Example 12.1 describes an experiment in which each store is randomly assigned to a box art style. Example 12.2 is an observational study in which customers are selected during a particular time period and not all are likely to agree to provide data. These samples of customers are not random samples, but we will treat them as such because we believe the selective sampling and nonresponse are ignorable sources of bias. This will not always be the case. *Always consider the various sources of bias in an observational study.*

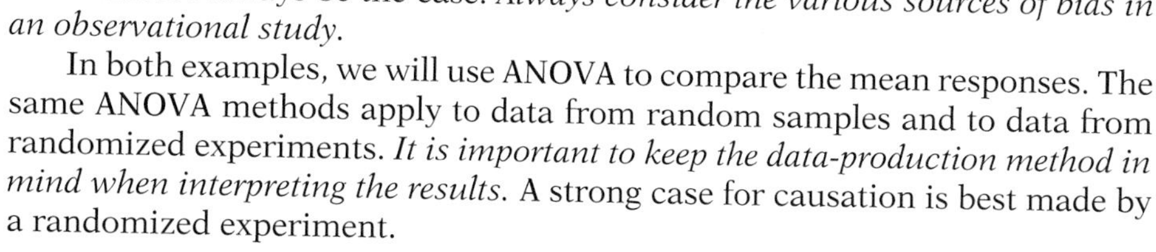

In both examples, we will use ANOVA to compare the mean responses. The same ANOVA methods apply to data from random samples and to data from randomized experiments. *It is important to keep the data-production method in mind when interpreting the results.* A strong case for causation is best made by a randomized experiment.

Comparing means

The question we ask in ANOVA is "Do all groups have the same population mean?" We will often use the term *groups* for the populations to be compared in a one-way ANOVA. To answer this question, we compare the sample means. Figure 12.1 (next page) displays the sample means for Example 12.1. It appears that box art style 2 has the highest average sales. But is the observed difference

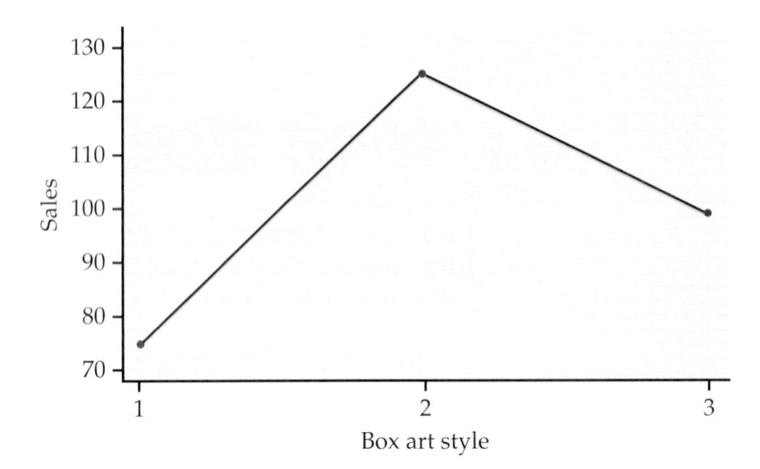

FIGURE 12.1 Mean sales of video games for three different box art styles.

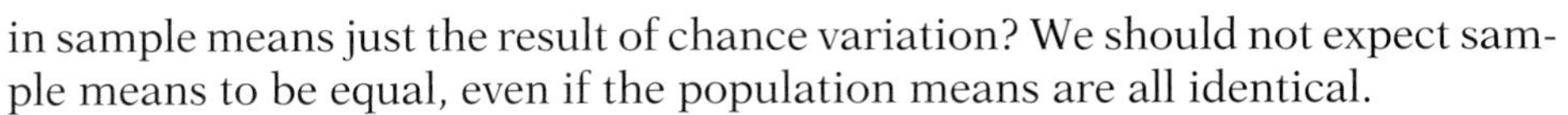

in sample means just the result of chance variation? We should not expect sample means to be equal, even if the population means are all identical.

The purpose of ANOVA is to assess whether the observed differences among sample means are *statistically significant.* Could a variation among the three sample means this large be plausibly due to chance, or is it good evidence for a difference among the population means? This question can't be answered from the sample means alone. Because the standard deviation of a sample mean $\bar{x}$ is the population standard deviation σ divided by $\sqrt{n}$, the answer also depends upon both the variation within the groups of observations and the sizes of the samples.

LOOK BACK
standard deviation of $\bar{x}$ p. 364

Side-by-side boxplots help us see the within-group variation. Compare Figures 12.2(a) and 12.2(b). The sample medians are the same in both figures, but the large variation within the groups in Figure 12.2(a) suggests that the differences among the sample medians could be due simply to chance variation. The data in Figure 12.2(b) are much more convincing evidence that the populations differ.

Even the boxplots omit essential information, however. To assess the observed differences, we must also know how large the samples are. Nonetheless, boxplots are a good preliminary display of the data.

Although ANOVA compares means and boxplots display medians, these two measures of center will be close together for distributions that are nearly

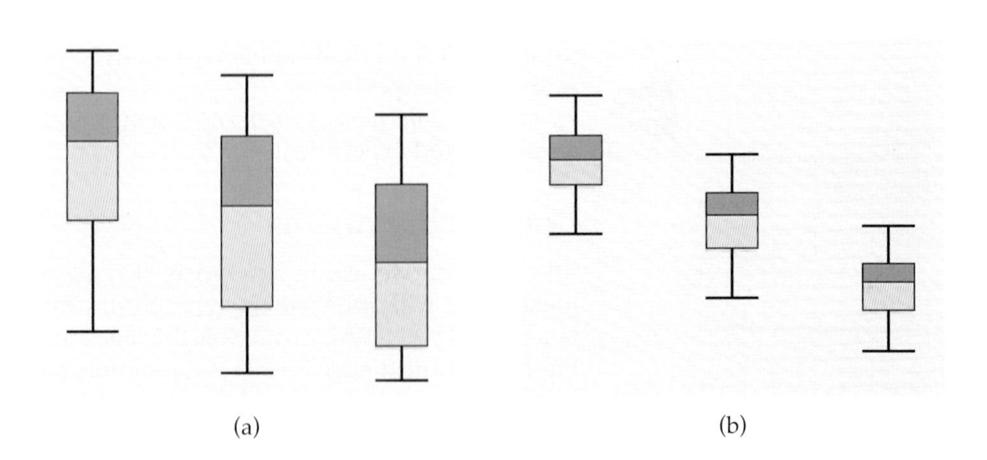

FIGURE 12.2 (a) Side-by-side boxplots for three groups with large within-group variation. The differences among centers may be just chance variation. (b) Side-by-side boxplots for three groups with the same centers as in panel (a) but with small within-group variation. The differences among centers are more likely to be significant.

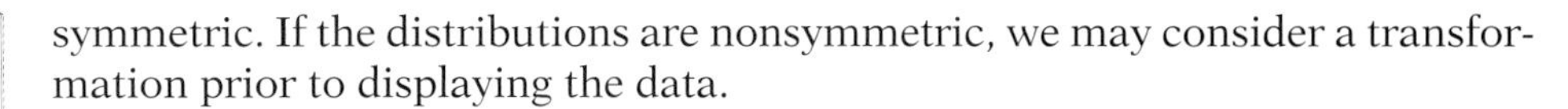

symmetric. If the distributions are nonsymmetric, we may consider a transformation prior to displaying the data.

The two-sample *t* statistic

LOOK BACK
pooled two-sample
t statistic p. 404

Two-sample t statistics compare the means of two populations. If the two populations are assumed to have equal but unknown standard deviations and the sample sizes are both equal to n, the t statistic is

$$t = \frac{\bar{x}_1 - \bar{x}_2}{s_p\sqrt{\frac{1}{n} + \frac{1}{n}}} = \frac{\sqrt{\frac{n}{2}}(\bar{x}_1 - \bar{x}_2)}{s_p}$$

The square of this t statistic is

$$t^2 = \frac{\frac{n}{2}(\bar{x}_1 - \bar{x}_2)^2}{s_p^2}$$

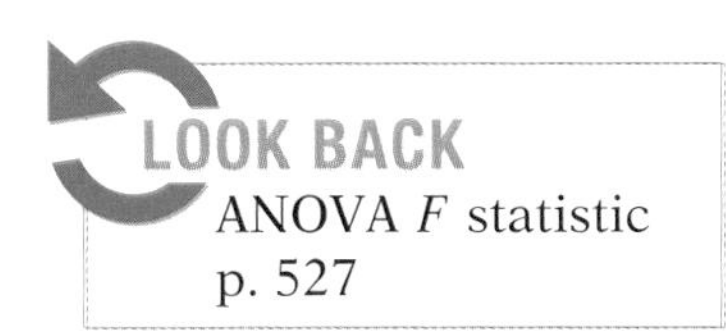

If we use ANOVA to compare two populations, the ANOVA F statistic is exactly equal to this t^2. We can therefore learn something about how ANOVA works by looking carefully at the statistic in this form.

between-group variation

The numerator in the t^2 statistic measures the variation **between** the groups in terms of the difference between their sample means $\bar{x}_1$ and $\bar{x}_2$ and the common sample size n. The numerator can be large because of a large difference between the sample means or because the common sample size is large.

within-group variation

The denominator measures the variation **within** groups by s_p^2, the pooled estimator of the common variance. If the within-group variation is small, the same variation between the groups produces a larger statistic and thus a more significant result.

Although the general form of the F statistic is more complicated, the idea is the same. To assess whether several populations all have the same mean, we compare the variation *among* the means of several groups with the variation *within* groups. Because we are comparing variation, the method is called *analysis of variance.*

An overview of ANOVA

ANOVA tests the null hypothesis that the population means are *all* equal. The alternative is that they are not all equal. This alternative could be true because all the means are different or simply because one of them differs from the rest. This situation is more complex than comparing just two populations. If we reject the null hypothesis, we need to perform some further analysis to draw conclusions about which population means differ from which others and by how much.

The computations needed for an ANOVA are more lengthy than those for the t test. For this reason we generally use computer programs to perform the calculations. Automating the calculations frees us from the burden of arithmetic and allows us to concentrate on interpretation.

Complicated computations performed by a computer, however, do not guarantee a valid statistical analysis. *We should always start our ANOVA with a careful examination of the data using graphical and numerical summaries.* Just as in linear regression, outliers and extreme deviations from Normality can invalidate the computed results.

EXAMPLE 12.3

Number of Facebook friends. A feature of each Facebook user's profile is the number of Facebook "friends," an indicator of the user's social network connectedness. Among college students on Facebook, the average number of Facebook friends has been estimated to be around 281.[1]

Offline, having more friends is associated with higher ratings of positive attributes such as likability and trustworthiness. Is this also the case with Facebook friends?

An experiment was run to examine the relationship between the number of Facebook friends and the user's perceived social attractiveness.[2] A total of 134 undergraduate participants were randomly assigned to observe one of five Facebook profiles. Everything about the profile was the same except the number of friends, which appeared on the profile as 102, 302, 502, 702, or 902.

After viewing the profile, each participant was asked to fill out a questionnaire on the physical and social attractiveness of the profile user. Each attractiveness score is an average of several seven-point questionnaire items, ranging from 1 (strongly disagree) to 7 (strongly agree). Here is a summary of the data for the social attractiveness score:

Number of friends	n	$\bar{x}$	s
102	24	3.82	1.00
302	33	4.88	0.85
502	26	4.56	1.07
702	30	4.41	1.43
902	21	3.99	1.02

Histograms for the five groups are given in Figure 12.3. Note that the heights of the bars in the histograms are percents rather than counts. This is commonly done when the group sample sizes vary. Figure 12.4 gives side-by-side boxplots for these data. We see that the scores covered the entire range of possible values, from 1.0 to 7.0. We also see a lot of overlap in scores across groups. The histograms are relatively symmetric, and with the group sample sizes all more than 15, we can feel confident that the sample means are approximately Normal.

LOOK BACK guidelines for two-sample t procedures p. 399

The five sample means are plotted in Figure 12.5 (page 556). They rise and then fall as the number of friends increases. This suggests that having too many Facebook friends can harm a user's social attractiveness. However, given the variability in the data, this pattern could also just be the result of chance variation. We will use ANOVA to make this determination.

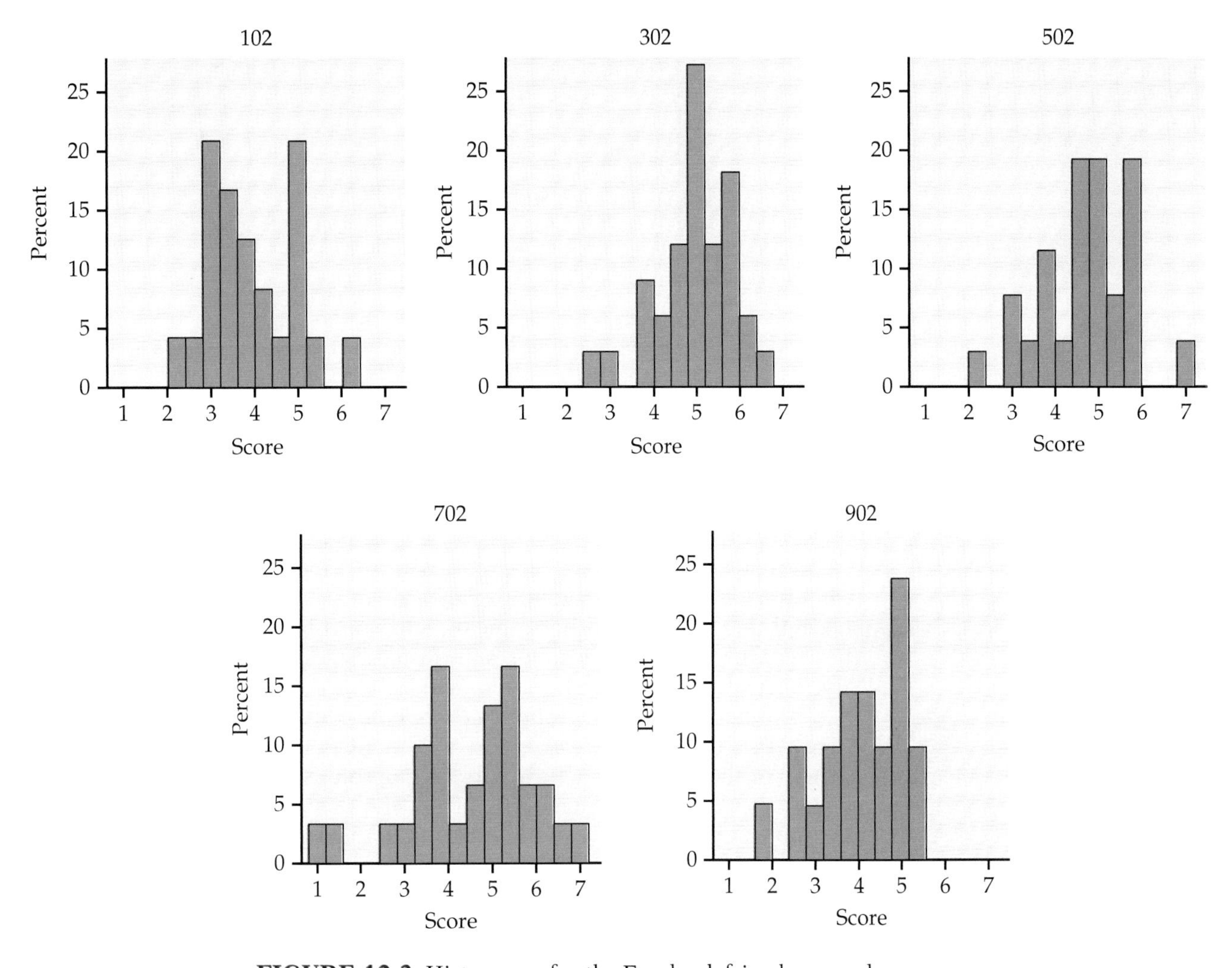

FIGURE 12.3 Histograms for the Facebook friends example.

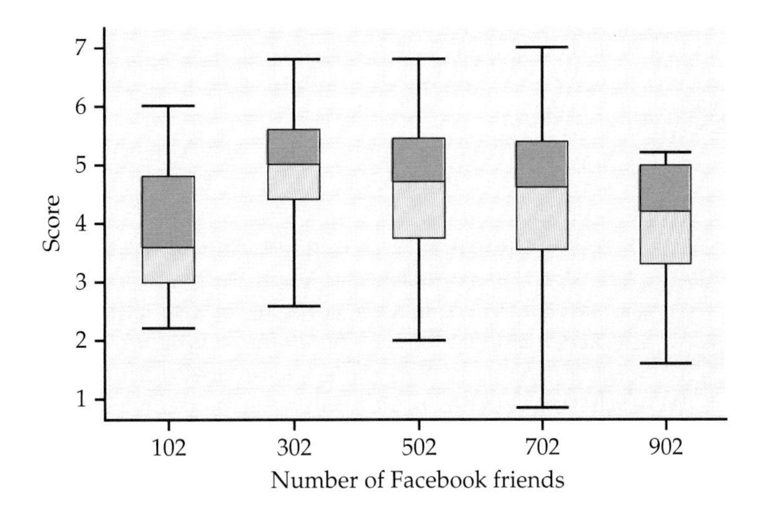

FIGURE 12.4 Side-by-side boxplots for the Facebook friends example.

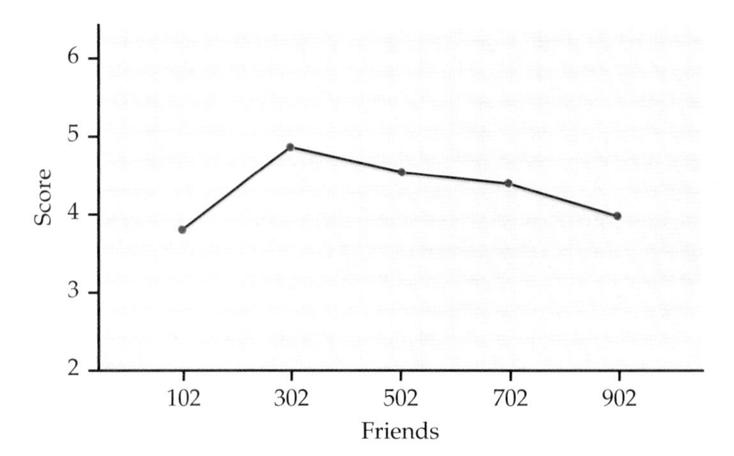

FIGURE 12.5 Social attractiveness score means for the Facebook friends example.

In this setting, we have an experiment in which undergraduate Facebook users were randomly assigned to view one of five Facebook profiles. Each of these profile populations has a mean, and our inference asks questions about these means. The undergraduates in this study were all from the same university. They also volunteered in exchange for course credit.

Formulating a clear definition of the populations being compared with ANOVA can be difficult. Often some expert judgment is required, and different consumers of the results may have differing opinions. Whether one can consider the sample in this study as an SRS from the population of undergraduates at the university or from the population of all undergraduates is open for debate. Regardless, we are more confident in generalizing our conclusions to similar populations when the results are clearly significant than when the level of evidence just barely passes the standard of $\alpha = 0.05$.

We first ask whether or not there is sufficient evidence in the data to conclude that the corresponding population means are not all equal. Our null hypothesis here states that the population mean score is the same for all five Facebook profiles. The alternative is that they are not all the same.

Our inspection of the data for our example suggests that the means may follow a curvilinear relationship. *Rejecting the null hypothesis that the means are all the same using ANOVA is not the same as concluding that all the means are different from one another.* The ANOVA null hypothesis can be false in many different ways. Additional analysis is required to distinguish among these possibilities.

contrasts

When there are particular versions of the alternative hypothesis that are of interest, we use **contrasts** to examine them. In our example, we might want to test whether there is a curvilinear relationship between the number of friends and acceptability score. *Note that, to use contrasts, it is necessary that the questions of interest be formulated before examining the data.* It is cheating to make up these questions after analyzing the data.

multiple comparisons

If we have no specific relations among the means in mind before looking at the data, we instead use a **multiple-comparisons** procedure to determine which pairs of population means differ significantly. In the second section we will explore both contrasts and multiple comparisons in detail.

USE YOUR KNOWLEDGE

12.1 What's wrong? For each of the following, explain what is wrong and why.

(a) ANOVA tests the null hypothesis that the sample means are all equal.

(b) A strong case for causation is best made in an observational study.

(c) You use ANOVA to compare the variances of the populations.

(d) A multiple-comparisons procedure is used to compare a relation among means that was specified prior to looking at the data.

12.2 What's wrong? For each of the following, explain what is wrong and why.

(a) In rejecting the null hypothesis, one can conclude that all the means are different from one another.

(b) A two-way ANOVA can be used only when there are two means to be compared.

(c) The ANOVA F statistic will be large when the within-group variation is much larger than the between-group variation.

The ANOVA model

When analyzing data, the following equation reminds us that we look for an overall pattern and deviations from it:

LOOK BACK
DATA = FIT + RESIDUAL p. 493

$$\text{DATA} = \text{FIT} + \text{RESIDUAL}$$

In the regression model of Chapter 10, the FIT was the population regression line, and the RESIDUAL represented the deviations of the data from this line. We now apply this framework to describe the statistical models used in ANOVA. These models provide a convenient way to summarize the assumptions that are the foundation for our analysis. They also give us the necessary notation to describe the calculations needed.

First, recall the statistical model for a random sample of observations from a single Normal population with mean μ and standard deviation σ. If the observations are

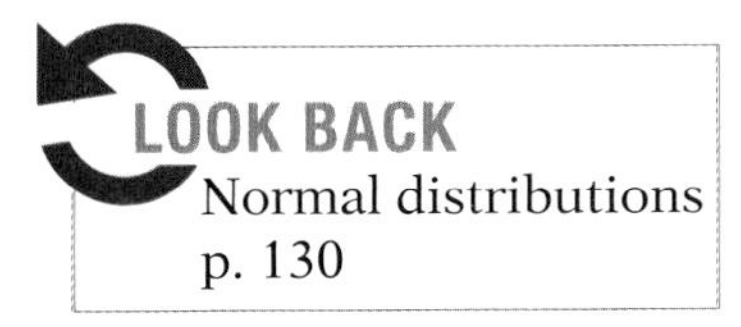

$$x_1, x_2, \ldots, x_n$$

we can describe this model by saying that the x_j are an SRS from the $N(\mu, \sigma)$ distribution. Another way to describe the same model is to think of the x's varying about their population mean. To do this, write each observation x_j as

$$x_j = \mu + \epsilon_j$$

The ϵ_j are then an SRS from the $N(0, \sigma)$ distribution. Because μ is unknown, the ϵ's cannot actually be observed. This second form more closely corresponds to our

$$\text{DATA} = \text{FIT} + \text{RESIDUAL}$$

way of thinking. The FIT part of the model is represented by μ. It is the systematic part of the model, like the line in a regression. The RESIDUAL part is represented by ϵ_j. It represents the deviations of the data from the fit and is due to random, or chance, variation.

There are two unknown parameters in this statistical model: μ and σ. We estimate μ by $\overline{x}$, the sample mean, and σ by s, the sample standard deviation. The differences $e_j = x_j - \overline{x}$ are the sample residuals and correspond to the ϵ_j in the statistical model.

The model for one-way ANOVA is very similar. We take random samples from each of I different populations. The sample size is n_i for the ith population. Let x_{ij} represent the jth observation from the ith population. The I population means are the FIT part of the model and are represented by μ_i. The random variation, or RESIDUAL, part of the model is represented by the deviations ϵ_{ij} of the observations from the means.

THE ONE-WAY ANOVA MODEL

The **one-way ANOVA model** is

$$x_{ij} = \mu_i + \epsilon_{ij}$$

for $i = 1, \ldots, I$ and $j = 1, \ldots, n_i$. The ϵ_{ij} are assumed to be from an $N(0, \sigma)$ distribution. The **parameters of the model** are the population means $\mu_1, \mu_2, \ldots, \mu_I$ and the common standard deviation σ.

Note that the sample sizes n_i may differ, but the standard deviation σ is assumed to be the same in all the populations. Figure 12.6 pictures this model for $I = 3$. The three population means μ_i are different, but the shapes of the three Normal distributions are the same, reflecting the assumption that all three populations have the same standard deviation.

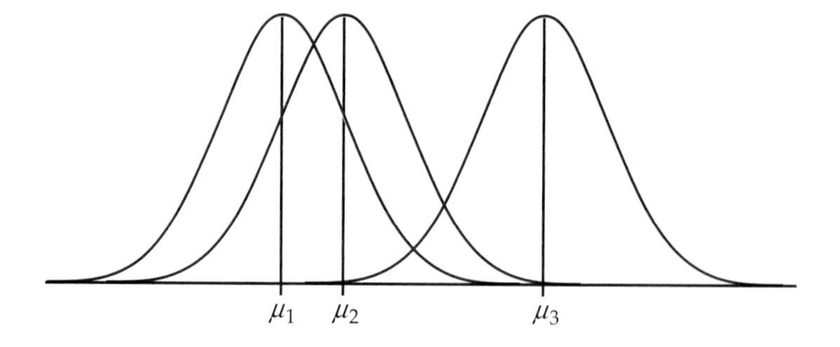

FIGURE 12.6 Model for one-way ANOVA with three groups. The three populations have Normal distributions with the same standard deviation.

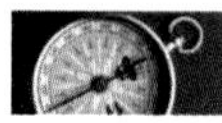

EXAMPLE 12.4

ANOVA model for the Facebook friends study. In the Facebook friends example, there are five profiles that we want to compare, so $I = 5$. The population means $\mu_1, \mu_2, \ldots, \mu_5$ are the mean social attractiveness scores for the profiles with 102, 302, 502, 702, and 902 friends, respectively. The sample sizes n_i are 24, 33, 26, 30, and 21. It is common to use numerical subscripts to distinguish the different means, and some software requires that levels of factors in ANOVA be specified as numerical values. In this situation, it is very important to keep track of what each numerical value represents when drawing conclusions. In our example, we could use numerical values to suggest the actual groups by replacing μ_1 with μ_{102}, μ_2 with μ_{302}, and so on.

The observation $x_{1,1}$ is the social attractiveness score for the first participant who observed the profile with 102 friends. The data for the other participants assigned to this profile are denoted by $x_{1,2}, x_{1,3}, \ldots, x_{1,24}$. Similarly, the data for the other four groups have a first subscript indicating the profile and a second subscript indicating the participants assigned to that profile.

According to the ANOVA model, the score for the first participant is $x_{1,1} = \mu_1 + \epsilon_{1,1}$, where μ_1 is the average score for *all* undergraduates after viewing Profile 1, and $\epsilon_{1,1}$ is the chance variation due to this particular participant. Similarly, the score for the last participant who observed the profile with 902 friends is $x_{5,21} = \mu_5 + \epsilon_{5,21}$, where μ_5 is the average score for *all* undergraduates after viewing Profile 5, and $\epsilon_{5,21}$ is the chance variation due to this participant.

The ANOVA model assumes that these chance variations ϵ_{ij} are independent and Normally distributed with mean 0 and standard deviation σ. For our example, this means that the distribution of scores for Profile i is Normal with mean μ_i and standard deviation σ. The observed scores, however, are numbers ranging from 1.0 to 7.0 by 0.2. These are clearly non-Normal. We are not overly concerned about this violation of model assumptions because our inference is based on the sample means and they will be approximately Normally distributed.

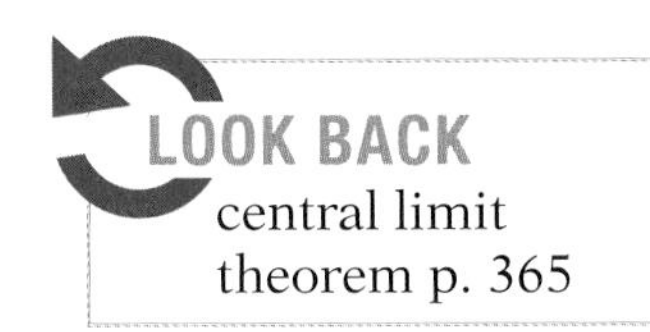

central limit theorem p. 365

Estimates of population parameters

The unknown parameters in the statistical model for ANOVA are the I population means μ_i and the common population standard deviation σ. To estimate μ_i we use the sample mean for the ith group:

$$\bar{x}_i = \frac{1}{n_i} \sum_{j=1}^{n_i} x_{ij}$$

The residuals $e_{ij} = x_{ij} - \bar{x}_i$ reflect the variation about the sample means that we see in the data and are used in the calculations of the sample standard deviations

$$s_i = \sqrt{\frac{\sum_{j=1}^{n_i} (x_{ij} - \bar{x}_i)^2}{n_i - 1}}$$

The ANOVA model assumes that the population standard deviations are all equal. Before estimating σ, it is important check this equality assumption using the sample standard deviations. Unfortunately, formal tests for the equality of standard deviations in several groups lack robustness against non-Normality. (See the text Web site for discussion of a formal test to compare two population variances.)

ANOVA procedures, however, are not extremely sensitive to unequal standard deviations. Thus, we do *not* recommend a formal test of equality of standard deviations as a preliminary to the ANOVA. Instead, we will use the following rule as a guideline.

RULE FOR EXAMINING STANDARD DEVIATIONS IN ANOVA

If the largest standard deviation is less than twice the smallest standard deviation, we can use methods based on the assumption of equal standard deviations, and our results will still be approximately correct.[3]

If we can assume that the population standard deviations are equal, each sample standard deviation s_i is an estimate of σ. To combine these into a single estimate, we use a generalization of the pooling method introduced in Chapter 7 (page 404).

POOLED ESTIMATOR OF σ

Suppose we have sample variances $s_1^2, s_2^2, \ldots, s_I^2$ from I independent SRSs of sizes $n_1, n_2, \ldots, n_I$ from populations with common variance σ^2. The **pooled sample variance**

$$s_p^2 = \frac{(n_1 - 1)s_1^2 + (n_2 - 1)s_2^2 + \cdots + (n_I - 1)s_I^2}{(n_1 - 1) + (n_2 - 1) + \cdots + (n_I - 1)}$$

is an unbiased estimator of σ^2. The **pooled standard deviation**

$$s_p = \sqrt{s_p^2}$$

is the estimate of σ.

Pooling gives more weight to groups with larger sample sizes. If the sample sizes are equal, s_p^2 is just the average of the I sample variances. *Note that s_p is not the average of the I sample standard deviations.*

If it appears that we have unequal standard deviations, we generally try to transform the data so that they are approximately equal. We might, for example, work with $\sqrt{x_{ij}}$ or $\log x_{ij}$. Fortunately, we can often find a transformation that *both* makes the group standard deviations more nearly equal and also makes the distributions of observations in each group more nearly Normal. If the standard deviations are markedly different and cannot be made similar by a transformation, inference requires different methods that are beyond the scope of this book.

EXAMPLE 12.5

Population estimates for the Facebook friends study. In the Facebook friends study, there are $I = 5$ groups and the sample sizes are $n_1 = 24$, $n_2 = 33$, $n_3 = 26$, $n_4 = 30$, and $n_5 = 21$. The sample standard deviations are $s_1 = 1.00$, $s_2 = 0.85$, $s_3 = 1.07$, $s_4 = 1.43$, and $s_5 = 1.02$.

Because the largest standard deviation (1.43) is less than twice the smallest ($2 \times 0.85 = 1.70$), our rule indicates that we can use the assumption of equal population standard deviations.

The pooled variance estimate is

$$s_p^2 = \frac{(n_1 - 1)s_1^2 + (n_2 - 1)s_2^2 + (n_3 - 1)s_3^2 + (n_4 - 1)s_4^2 + (n_5 - 1)s_5^2}{(n_1 - 1) + (n_2 - 1) + (n_3 - 1) + (n_4 - 1) + (n_5 - 1)}$$

$$= \frac{(23)(1.00)^2 + (32)(0.85)^2 + (25)(1.07)^2 + (29)(1.43)^2 + (20)(1.02)^2}{23 + 32 + 25 + 29 + 20}$$

$$= \frac{154.85}{129} = 1.20$$

The pooled standard deviation is

$$s_p = \sqrt{1.20} = 1.10$$

This is our estimate of the common standard deviation σ of the social attractiveness scores for the five profiles.

USE YOUR KNOWLEDGE

12.3 Computing the pooled standard deviation. An experiment was run to compare three timed-release fertilizers in terms of plant growth. The sample sizes were 25, 18, and 28, and the corresponding estimated standard deviations were 4, 5, and 7 centimeters.

(a) Is it reasonable to use the assumption of equal standard deviations when we analyze these data? Give a reason for your answer.

(b) Give the values of the variances for the three groups.

(c) Find the pooled variance.

(d) What is the value of the pooled standard deviation?

12.4 Visualizing the ANOVA model. For each of the following situations, draw a picture of the ANOVA model similar to Figure 12.6 (page 558). Use the numerical values for the μ_i. To sketch the Normal curves, you may want to review the 68–95–99.7 rule on page 132.

(a) $\mu_1 = 18$, $\mu_2 = 12$, $\mu_3 = 13$, and $\sigma = 4$

(b) $\mu_1 = 18$, $\mu_2 = 16$, $\mu_3 = 20$, $\mu_4 = 24$, and $\sigma = 6$

(c) $\mu_1 = 18$, $\mu_2 = 12$, $\mu_3 = 13$, and $\sigma = 2$

Testing hypotheses in one-way ANOVA

LOOK BACK
ANOVA table p. 527

Comparison of several means is accomplished by using an F statistic to compare the variation among groups with the variation within groups. We now show how the F statistic expresses this comparison. Calculations are organized in an ANOVA table, which contains numerical measures of the variation among groups and within groups.

First, we must specify our hypotheses for one-way ANOVA. As before, I represents the number of populations to be compared.

HYPOTHESES FOR ONE-WAY ANOVA

The **null and alternative hypotheses** for one-way ANOVA are

$$H_0: \mu_1 = \mu_2 = \cdots = \mu_I$$

$$H_a: \text{not all of the } \mu_i \text{ are equal}$$

We will now use the Facebook friends example to illustrate how to do a one-way ANOVA. Because the calculations are generally performed using statistical software, we focus on interpretation of the output.

EXAMPLE 12.6

Reading software output. Figure 12.7 gives descriptive statistics generated by SPSS for the ANOVA of the Facebook friends example. Summaries for each profile are given on the first five lines. Besides sample size, the mean, and the standard deviation, this output also gives the minimum and maximum observed values, the standard error of the mean, and the 95% confidence interval for the mean of each profile. The five sample means $\bar{x}_i$ given in the output are estimates of the five unknown population means μ_i.

SPSS

Score — **Descriptives**

	N	Mean	Std. Deviation	Std. Error	95% Confidence Interval for Mean: Lower Bound	95% Confidence Interval for Mean: Upper Bound	Minimum	Maximum
102	24	3.817	.9990	.2039	3.395	4.239	2.2	6.0
302	33	4.879	.8514	.1482	4.577	5.181	2.6	6.4
502	26	4.562	1.0704	.2099	4.129	4.994	2.0	6.8
702	30	4.407	1.4283	.2608	3.873	4.940	1.0	7.0
902	21	3.990	1.0227	.2232	3.525	4.456	1.6	5.2
Total	134	4.382	1.1463	.0990	4.186	4.578	1.0	7.0

FIGURE 12.7 Software output with descriptive statistics for the Facebook friends example.

The output gives the estimates of the standard deviations, s_i, for each group but does not provide s_p, the pooled estimate of the model standard deviation, σ. We could perform the calculation using a calculator, as we did in Example 12.5. We will see an easier way to obtain this quantity from the ANOVA table in Figure 12.8.

Some software packages report s_p as part of the standard ANOVA output. *Sometimes you are not sure whether or not a quantity given by software is what you think it is.* A good way to resolve this dilemma is to do a sample calculation with a simple example to check the numerical results.

Note that s_p is not the standard deviation given in the Total row of Figure 12.7. This quantity is the standard deviation that we would obtain if we viewed the data as a single sample of 134 participants and ignored the possibility that the profile means could be different. As we have mentioned many times before, it is important to use care when reading and interpreting software output.

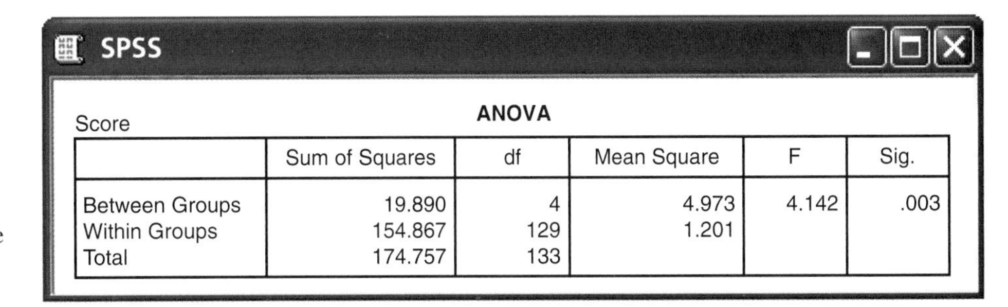

SPSS

Score

ANOVA

	Sum of Squares	df	Mean Square	F	Sig.
Between Groups	19.890	4	4.973	4.142	.003
Within Groups	154.867	129	1.201		
Total	174.757	133			

FIGURE 12.8 Software output giving the ANOVA table for the Facebook friends example.

EXAMPLE 12.7

Reading software output, continued. Additional output generated by SPSS for the ANOVA of the Facebook friends example is given in Figure 12.8. We will discuss some details in the next section. For now, we observe that the results of our significance test are given in the last two columns of the output. The null hypothesis that the five population means are the same is tested by the statistic $F = 4.142$, and the associated P-value is reported as $P = 0.003$. The data provide clear evidence to support the claim that there are some differences among the five profile population means.

The ANOVA table

The information in an analysis of variance is organized in an ANOVA table. To understand the table, it is helpful to think in terms of our

$$\text{DATA} = \text{FIT} + \text{RESIDUAL}$$

view of statistical models. For one-way ANOVA, this corresponds to

$$x_{ij} = \mu_i + \epsilon_{ij}$$

We can think of these three terms as sources of variation. The ANOVA table separates the variation in the data into two parts: the part due to the fit and the remainder, which we call residual.

EXAMPLE 12.8

ANOVA table for the Facebook friends study. The SPSS output in Figure 12.8 gives the sources of variation in the first column. Here, FIT is called Between Groups, RESIDUAL is called Within Groups, and DATA is the last entry, Total. Different software packages use different terms for these sources of variation but the basic concept is common to all. In place of FIT, some software packages use Between Groups, Model, or the name of the factor. Similarly, terms like Within Groups or Error are frequently used in place of RESIDUAL.

variation among groups

The Between Groups row in the table gives information related to the variation **among** group means. In writing ANOVA tables, for this row we will use the generic label "groups" or some other term that describes the factor being studied.

variation within groups

The Within Groups row in the table gives information related to the variation **within** groups. We noted that the term "error" is frequently used for this source of variation, particularly for more general statistical models. This label is most appropriate for experiments in the physical sciences where the observations within a group differ because of measurement error. In business and the biological and social sciences, on the other hand, the within-group variation is often due to the fact that not all firms or plants or people are the same. This sort of variation is not due to errors and is better described as "residual" or "within-group" variation. Nevertheless, we will use the generic label "error" for this source of variation in writing ANOVA tables.

Finally, the Total row in the ANOVA table corresponds to the DATA term in our DATA = FIT + RESIDUAL framework. So, for analysis of variance,

$$\text{DATA} = \text{FIT} + \text{RESIDUAL}$$

translates into

$$\text{Total} = \text{Between Groups} + \text{Within Groups}$$

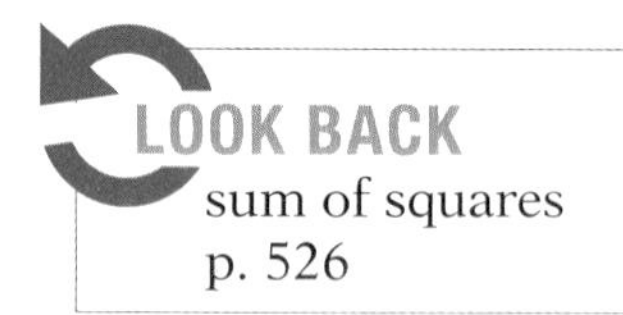
LOOK BACK
sum of squares
p. 526

The second column in the software output given in Figure 12.8 is labeled Sum of Squares. As you might expect, each sum of squares is a sum of squared deviations. We use SSG, SSE, and SST for the entries in this column, corresponding to groups, error, and total. Each sum of squares measures a different type of variation. SST measures variation of the data around the overall mean, $x_{ij} - \overline{x}$. Variation of the group means around the overall mean, $\overline{x}_i - \overline{x}$, is measured by SSG. Finally, SSE measures variation of each observation around its group mean, $x_{ij} - \overline{x}_i$.

EXAMPLE 12.9

ANOVA table for the Facebook friends study, continued. The Sum of Squares column in Figure 12.8 gives the values for the three sums of squares. They are

$$\text{SST} = 174.757$$

$$\text{SSG} = 19.890$$

$$\text{SSE} = 154.867$$

Verify that SST = SSG + SSE for this example.

This fact is true in general: the total variation is always equal to the among-group variation plus the within-group variation. Note that software output frequently gives many more digits than we need, as in this case.

In this example it appears that most of the variation is coming from within groups. However, to assess whether the observed differences in sample means are statistically significant, some additional calculations are needed.

LOOK BACK
degrees of freedom
p. 118

Associated with each sum of squares is a quantity called the degrees of freedom. Because SST measures the variation of all N observations around the overall mean, its degrees of freedom are DFT = $N - 1$. This is the same as the degrees of freedom for the ordinary sample variance with sample size N. Similarly, because SSG measures the variation of the I sample means around the overall mean, its degrees of freedom are DFG = $I - 1$. Finally, SSE is the sum of squares of the deviations $x_{ij} - \overline{x}_i$. Here we have N observations being compared with I sample means, so DFE = $N - I$.

EXAMPLE 12.10

Degrees of freedom for the Facebook friends study. We have $I = 5$ and $N = 134$. Therefore,

$$\text{DFT} = N - 1 = 134 - 1 = 133$$
$$\text{DFG} = I - 1 = 5 - 1 = 4$$
$$\text{DFE} = N - I = 134 - 5 = 129$$

These are the entries in the df column of Figure 12.8.

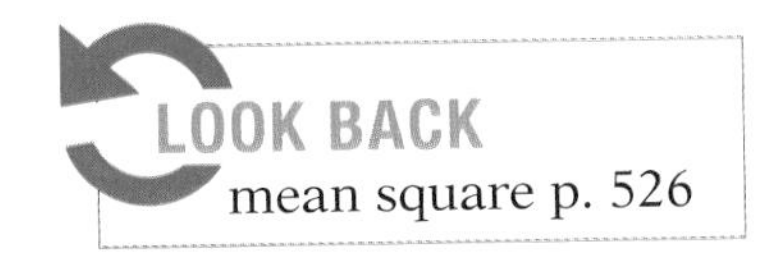

mean square p. 526

Note that the degrees of freedom add in the same way that the sums of squares add. That is, DFT = DFG + DFE.

For each source of variation, we obtain an average squared deviation by computing its mean square, the sum of squares divided by the degrees of freedom. You can verify this by doing the divisions for the values given on the output in Figure 12.8. We compare these mean squares to test whether the population means are all the same.

SUMS OF SQUARES, DEGREES OF FREEDOM, AND MEAN SQUARES

Sums of squares represent variation present in the data. They are calculated by summing squared deviations. In one-way ANOVA, there are three **sources of variation:** groups, error, and total. The sums of squares are related by the formula

$$\text{SST} = \text{SSG} + \text{SSE}$$

Thus, the total variation is composed of two parts, one due to groups and one due to error.

Degrees of freedom are related to the deviations that are used in the sums of squares. The degrees of freedom are related in the same way as the sums of squares are:

$$\text{DFT} = \text{DFG} + \text{DFE}$$

To calculate each **mean square,** divide the corresponding sum of squares by its degrees of freedom.

We can use the mean square for error to find s_p, the pooled estimate of the parameter σ of our model. It is true in general that

$$s_p^2 = \text{MSE} = \frac{\text{SSE}}{\text{DFE}}$$

In other words, the mean square for error is an estimate of the within-group variance, σ^2. The estimate of σ is therefore the square root of this quantity. So,

$$s_p = \sqrt{\text{MSE}}$$

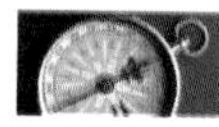

EXAMPLE 12.11

MSE for the Facebook friends study. From the SPSS output in Figure 12.8 we see that the MSE is reported as 1.201. The pooled estimate of σ is therefore

$$s_p = \sqrt{\text{MSE}}$$
$$= \sqrt{1.201} = 1.10$$

This estimate is equal to our calculations of s_p in Example 12.5.

The F test

If H_0 is true, there are no differences among the group means. The ratio MSG/MSE is a statistic that is approximately 1 if H_0 is true and tends to be larger if H_a is true. This is the ANOVA F statistic. In our example, MSG = 4.973 and MSE = 1.201, so the ANOVA F statistic is

$$F = \frac{\text{MSG}}{\text{MSE}} = \frac{4.973}{1.201} = 4.142$$

When H_0 is true, the F statistic has an F distribution that depends upon two numbers: the *degrees of freedom for the numerator* and the *degrees of freedom for the denominator.* These degrees of freedom are those associated with the mean squares in the numerator and denominator of the F statistic. For one-way ANOVA, the degrees of freedom for the numerator are DFG $= I - 1$, and the degrees of freedom for the denominator are DFE $= N - I$. We use the notation $F(I - 1, N - I)$ for this distribution.

The *One-Way ANOVA* applet is an excellent way to see how the value of the F statistic and the P-value depend upon the variability of the data within the groups and the differences between the means. See Exercises 12.20 and 12.21 for use of this applet.

THE ANOVA F TEST

To test the null hypothesis in a one-way ANOVA, calculate the ***F* statistic**

$$F = \frac{\text{MSG}}{\text{MSE}}$$

When H_0 is true, the F statistic has the $F(I - 1, N - I)$ distribution. When H_a is true, the F statistic tends to be large. We reject H_0 in favor of H_a if the F statistic is sufficiently large.

The ***P*-value** of the F test is the probability that a random variable having the $F(I - 1, N - I)$ distribution is greater than or equal to the calculated value of the F statistic.

Tables of F critical values are available for use when software does not give the P-value. Table E in the back of the book contains the F critical values for probabilities $p = 0.100$, 0.050, 0.025, 0.010, and 0.001. For one-way ANOVA we use critical values from the table corresponding to $I - 1$ degrees of freedom in the numerator and $N - I$ degrees of freedom in the denominator. *When determining the P-value, remember that the F test is always one-sided because any differences among the group means tend to make F large.*

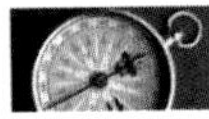

EXAMPLE 12.12

The ANOVA F test for the Facebook friends study. In the Facebook friends study, we found $F = 4.14$. Note that it is standard practice to round F statistics to two places after the decimal point. There are five populations, so the degrees of freedom in the numerator are $\text{DFG} = I - 1 = 4$. For this example, the degrees of freedom in the denominator are $\text{DFE} = N - I = 134 - 5 = 129$. Software provided a P-value of 0.003, so at the 0.05 significance level, we reject H_0 and conclude that the population means are not all the same.

Suppose that $P = 0.003$ was not provided. We'll now run through the process of using the F tables to approximate the P-value. Although we'll rarely need to do this in practice, the process can help in understanding the P-value calculation.

In Table E, we first find the column corresponding to 4 degrees of freedom in the numerator. For the degrees of freedom in the denominator, we see that there are entries for 100 and 200. These entries are very close. To be conservative we use critical values corresponding to 100 degrees of freedom in the denominator since these are slightly larger.

p	Critical value
0.100	2.00
0.050	2.46
0.025	2.92
0.010	3.51
0.001	5.02

We have $F = 4.14$. This is in between the critical value for $P = 0.010$ and $P = 0.001$. Using the table, we can conclude only that $0.001 < P < 0.010$.

The following display shows the general form of a one-way ANOVA table with the F statistic. The formulas in the sum of squares column can be used for calculations in small problems. There are other formulas that are more efficient for hand or calculator use, but ANOVA calculations are usually done by computer software.

Source	Degrees of freedom	Sum of squares	Mean square	F
Groups	$I - 1$	$\sum_{\text{groups}} n_i(\bar{x}_i - \bar{x})^2$	SSG/DFG	MSG/MSE
Error	$N - I$	$\sum_{\text{groups}}(n_i - 1)s_i^2$	SSE/DFE	
Total	$N - 1$	$\sum_{\text{obs}}(x_{ij} - \bar{x})^2$		

coefficient of determination

One other item given by some software for ANOVA is worth noting. For an analysis of variance, we define the **coefficient of determination** as

$$R^2 = \frac{\text{SSG}}{\text{SST}}$$

multiple correlation coefficient p. 529

The coefficient of determination plays the same role as the squared multiple correlation R^2 in a multiple regression. We can easily calculate the value from the ANOVA table entries.

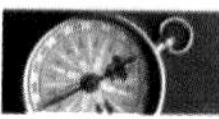

EXAMPLE 12.13

Coefficient of determination for the Facebook friends study. The software-generated ANOVA table for the Facebook friends study is given in Figure 12.8. From that display, we see that SSG = 19.890 and SST = 174.757. The coefficient of determination is

$$R^2 = \frac{\text{SSG}}{\text{SST}} = \frac{19.890}{174.757} = 0.11$$

About 11% of the variation in social attractiveness scores is explained by the different profiles. The other 89% of the variation is due to participant-to-participant variation within each of the profile groups. We can see this breakdown of variation in the histograms of Figure 12.3. Each of the groups has a large amount of variation, and there is a substantial amount of overlap in the distributions.

The fact that we have strong evidence ($P < 0.003$) against the null hypothesis that the five population means are not all the same does not tell us that the distributions of values are far apart.

USE YOUR KNOWLEDGE

12.5 What's wrong? For each of the following, explain what is wrong and why.

(a) Within-group variation is the variation in the data due to the differences in the sample means.

(b) The mean squares in an ANOVA table will add, that is, MST = MSG + MSE.

(c) The pooled estimate s_p is a parameter of the ANOVA model.

(d) A very small P-value implies that the group distributions of responses are far apart.

12.6 Determining the critical value of F. For each of the following situations, state how large the F statistic needs to be for rejection of the null hypothesis at the 0.05 level.

(a) Compare 3 groups with 5 observations per group.

(b) Compare 4 groups with 4 observations per group.

(c) Compare 4 groups with 8 observations per group.

(d) Summarize what you have learned about F distributions from this exercise.

12.2 Comparing the Means

Contrasts

The ANOVA F test gives a general answer to a general question: are the differences among observed group means statistically significant? Unfortunately, a small P-value simply tells us that the group means are not all the same. It does not tell us specifically which means differ from each other. Plotting and inspecting the means give us some indication of where the differences lie, but we would like to supplement inspection with formal inference.

In the ideal situation, specific questions regarding comparisons among the means are posed before the data are collected. We can answer specific questions of this kind and attach a level of confidence to the answers we give. We now explore these ideas through the Facebook friends example.

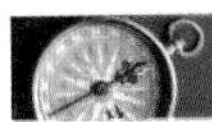

EXAMPLE 12.14

Reporting the results. In the Facebook friends study, we compared the social attractiveness scores for five profiles that varied only in the number of friends. Let's use $\bar{x}_{102}$, $\bar{x}_{302}$, $\bar{x}_{502}$, $\bar{x}_{702}$, and $\bar{x}_{902}$ to represent the five sample means and a similar notation for the population means.

From Figure 12.7 we see that the five sample means are

$$\bar{x}_{102} = 3.82, \bar{x}_{302} = 4.88, \bar{x}_{502} = 4.56, \bar{x}_{702} = 4.41, \text{ and } \bar{x}_{902} = 3.99$$

The null hypothesis we tested was

$$H_0\colon \mu_{102} = \mu_{302} = \mu_{502} = \mu_{702} = \mu_{902}$$

versus the alternative that the five population means are not all the same. We would report these results as $F(4, 129) = 4.14$ with $P = 0.003$. Note that we have given the degrees of freedom for the F statistic in parentheses. Because the P-value is very small, we conclude that the data provide clear evidence that the five population means are not all the same.

However, having evidence that the five population means are not the same does not tell us all we'd like to know. We would like our analysis to provide us with more specific information. For example, the alternative hypothesis is true if

$$\mu_{102} < \mu_{302} = \mu_{502} = \mu_{702} = \mu_{902}$$

or if

$$\mu_{102} = \mu_{302} = \mu_{502} > \mu_{702} = \mu_{902}$$

or if

$$\mu_{102} \neq \mu_{302} \neq \mu_{502} \neq \mu_{702} \neq \mu_{902}$$

When you reject the ANOVA null hypothesis, additional analyses are required to clarify the nature of the differences between the means.

In terms of offline social networks, previous research has shown that the bigger one's social network, the higher one's social attractiveness. In fact, this relationship between the number of friends and social attractiveness appears linear. Therefore, a reasonable question to ask is whether or not this same sort of pattern exists within an online social network. We can take this question and translate it into a testable hypothesis.

EXAMPLE 12.15

An additional comparison of interest. The researchers hypothesize that, unlike an offline social network, the relationship between the number of friends and social attractiveness levels off as the number of friends increases. This can be assessed by comparing changes in the mean scores across various factor levels. These comparisons are simplified because the levels are equally spaced. To compare the change in mean between the lower friend levels with the change in mean between the upper friend levels we construct the following null hypothesis:

$$H_{01}\colon (\mu_{502} - \mu_{102}) - (\mu_{902} - \mu_{502}) = 0$$

We could use the two-sided alternative

$$H_{a1}\colon (\mu_{502} - \mu_{102}) - (\mu_{902} - \mu_{502}) \neq 0$$

but we could also argue that the one-sided alternative

$$H_{a1}\colon (\mu_{502} - \mu_{102}) - (\mu_{902} - \mu_{502}) > 0$$

is appropriate for this problem because we expect there to be a leveling off.

In the example above, we used H_{01} and H_{a1} to designate the null and alternative hypotheses. The reason for this is that there is an additional set of hypotheses to assesses linearity. We use H_{02} and H_{a2} for this set.

EXAMPLE 12.16

Another comparison of interest. This comparison tests the linearity across the factor levels. Here are the null and alternative hypotheses:

$$H_{02}\colon -2\mu_{102} - \mu_{302} + \mu_{702} + 2\mu_{902} = 0$$
$$H_{a2}\colon -2\mu_{102} - \mu_{302} + \mu_{702} + 2\mu_{902} \neq 0$$

Each of H_{01} and H_{02} says that a combination of population means is 0. These combinations of means are called contrasts because the coefficients sum to zero. We use ψ, the Greek letter psi, for contrasts among population means. For our first comparison, we have

$$\begin{aligned}\psi_1 &= (\mu_{502} - \mu_{102}) - (\mu_{902} - \mu_{502})\\ &= -\mu_{102} + 2\mu_{502} - \mu_{902}\\ &= (-1)\mu_{102} + (2)\mu_{502} + (-1)\mu_{902}\end{aligned}$$

and for the second comparison

$$\psi_2 = (-2)\mu_{102} + (-1)\mu_{302} + (1)\mu_{702} + (2)\mu_{902}$$

In each case, the value of the contrast is 0 when H_0 is true. *Note that we have chosen to define the contrasts so that they will be positive when the alternative of*

interest (what we expect) is true. Whenever possible, this is a good idea because it makes some computations easier.

sample contrast

A contrast expresses an effect in the population as a combination of population means. To estimate the contrast, form the corresponding **sample contrast** by using sample means in place of population means. Under the ANOVA assumptions, a sample contrast is a linear combination of independent Normal variables and therefore has a Normal distribution. We can obtain the standard error of a contrast by using the rules for variances. Inference is based on t statistics. Here are the details.

LOOK BACK rules for variances p. 277

CONTRASTS

A **contrast** is a combination of population means of the form

$$\psi = \sum a_i \mu_i$$

where the coefficients a_i sum to 0. The corresponding **sample contrast** is

$$c = \sum a_i \overline{x}_i$$

The **standard error of *c*** is

$$\text{SE}_c = s_p \sqrt{\sum \frac{a_i^2}{n_i}}$$

To test the null hypothesis

$$H_0: \psi = 0$$

use the ***t* statistic**

$$t = \frac{c}{\text{SE}_c}$$

with degrees of freedom DFE that are associated with s_p. The alternative hypothesis can be one-sided or two-sided.

A **level *C* confidence interval for ψ** is

$$c \pm t^* \text{SE}_c$$

where t^* is the value for the t(DFE) density curve with area C between $-t^*$ and t^*.

Because each $\overline{x}_i$ estimates the corresponding μ_i, the addition rule for means tells us that the mean μ_c of the sample contrast c is ψ. In other words, c is an unbiased estimator of ψ. Testing the hypothesis that a contrast is 0 assesses the significance of the effect measured by the contrast. It is often more informative to estimate the size of the effect using a confidence interval for the population contrast.

LOOK BACK addition rule for means p. 274

EXAMPLE 12.17

The contrast coefficients. In our example the coefficients in the contrasts are

$$a_1 = -1, a_2 = 0, a_3 = 2, a_4 = 0, a_5 = -1 \text{ for } \psi_1$$

and

$$a_1 = -2, a_2 = -1, a_3 = 0, a_4 = 1, a_5 = 2 \text{ for } \psi_2$$

where the subscripts 1, 2, 3, 4, and 5 correspond to the profiles with 102, 302, 502, 702, and 902 friends, respectively. In each case the sum of the a_i is 0. We look at inference for each of these contrasts in turn.

EXAMPLE 12.18

Testing the first contrast of interest. The sample contrast that estimates ψ_1 is

$$\begin{aligned} c_1 &= (-1)\overline{x}_{102} + (2)\overline{x}_{502} + (-1)\overline{x}_{902} \\ &= -3.82 + (2)4.56 - 3.99 = 1.31 \end{aligned}$$

with standard error

$$\begin{aligned} \text{SE}_{c_1} &= 1.10\sqrt{\frac{(-1)^2}{24} + \frac{(2)^2}{26} + \frac{(-1)^2}{21}} \\ &= 0.54 \end{aligned}$$

The t statistic for testing $H_{01}\colon \psi_1 = 0$ versus $H_{a1}\colon \psi_1 > 0$ is

$$\begin{aligned} t &= \frac{c_1}{\text{SE}_{c_1}} \\ &= \frac{1.31}{0.54} = 2.43 \end{aligned}$$

Because s_p has 129 degrees of freedom, software using the $t(129)$ distribution gives the one-sided P-value as $P = 0.008$. If we used Table D, we would conclude that $0.005 < P < 0.01$. The P-value is small, so there is strong evidence against H_{01}.

We have evidence to conclude that the rate of change in the attractiveness score at the lower levels is larger than the rate of change at the upper levels. This suggests either a leveling off or a decrease in the attractiveness score as the number of friends increases. The size of the difference can be described with a confidence interval.

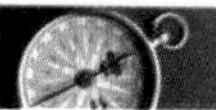

EXAMPLE 12.19

Confidence interval for the first contrast. To find the 95% confidence interval for ψ_1, we combine the estimate with its margin of error:

$$\begin{aligned} c_1 \pm t^*\text{SE}_{c_1} &= 1.31 \pm (1.984)(0.54) \\ &= 1.31 \pm 1.07 \end{aligned}$$

The 1.984 is a conservative estimate of t^* using 100 degrees of freedom. The interval is (0.24, 2.38). We are 95% confident that the change in average score between Profile 3 and Profile 1 is between 0.24 and 2.38 points larger than the change in average score between Profile 5 and Profile 3.

We use the same method for the second contrast.

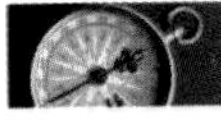

EXAMPLE 12.20

Testing the second contrast of interest. The second sample contrast assesses the linear trend across the levels.

$$\begin{aligned} c_2 &= (-2)\overline{x}_{102} + (-1)\overline{x}_{302} + (1)\overline{x}_{702} + (2)\overline{x}_{902} \\ &= (-2)3.82 + (-1)4.88 + (1)4.41 + (2)3.99 \\ &= -7.64 - 4.88 + 4.41 + 7.98 \\ &= -0.13 \end{aligned}$$

with standard error

$$\begin{aligned} SE_{c_2} &= 1.10\sqrt{\frac{(-2)^2}{24} + \frac{(-1)^2}{33} + \frac{(1)^2}{30} + \frac{(2)^2}{21}} \\ &= 0.71 \end{aligned}$$

The t statistic for assessing the significance of this contrast is

$$t = \frac{-0.13}{0.71} = -0.18$$

The P-value for the two-sided alternative is 0.861. The data do not provide much evidence in favor of a linear trend.

This contrast can be combined with others to assess the various polynomial contributions (for example, linear, quadratic, cubic) to the relationship between attractiveness score and the number of friends. As we saw in Figure 12.5, a quadratic trend appears most prominent. Further discussion of this contrast can be found in Exercise 12.42.

SPSS output for the contrasts is given in Figure 12.9. The results agree with the calculations that we performed in Examples 12.18 and 12.20 except for

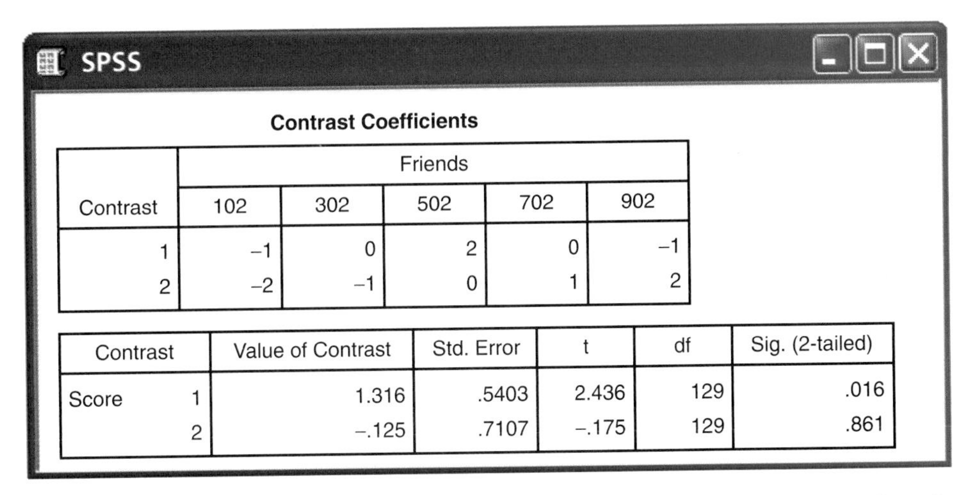

SPSS

Contrast Coefficients

Contrast	Friends 102	302	502	702	902
1	–1	0	2	0	–1
2	–2	–1	0	1	2

Contrast		Value of Contrast	Std. Error	t	df	Sig. (2-tailed)
Score	1	1.316	.5403	2.436	129	.016
	2	–.125	.7107	–.175	129	.861

FIGURE 12.9 Software output giving the contrast analysis for the Facebook friends example.

minor differences due to roundoff error in our calculations. Note that the output does not give the confidence interval that we calculated in Example 12.19. This is easily computed, however, from the contrast estimate and standard error provided in the output.

Some statistical software packages report the test statistics associated with contrasts as F statistics rather than t statistics. These F statistics are the squares of the t statistics described previously. As with much statistical software output, P-values for significance tests are reported for the two-sided alternative.

If the software you are using gives P-values for the two-sided alternative, and you are using the appropriate one-sided alternative, divide the reported P-value by 2. In our example, we argued that a one-sided alternative was appropriate for the first contrast. The software reported the P-value as 0.016, so we can conclude $P = 0.008$. Dividing this value by 2 has no effect on the conclusion.

Questions about population means are expressed as hypotheses about contrasts. A contrast should express a specific question that we have in mind when designing the study. Because the F test answers a very general question, it is less powerful than tests for contrasts designed to answer specific questions.

When contrasts are formulated before seeing the data, inference about contrasts is valid whether or not the ANOVA H_0 of equality of means is rejected. Specifying the important questions before the analysis is undertaken enables us to use this powerful statistical technique.

Multiple comparisons

In many studies, specific questions cannot be formulated in advance of the analysis. If H_0 is not rejected, we conclude that the population means are indistinguishable on the basis of the data given. On the other hand, if H_0 is rejected, we would like to know which pairs of means differ. **Multiple-comparisons methods** address this issue. It is important to keep in mind that multiple-comparisons methods are used only *after rejecting* the ANOVA H_0.

multiple-comparisons methods

EXAMPLE 12.21

Comparing each pair of groups. Return once more to the Facebook friends data with five groups. We can make 10 comparisons between pairs of means. We can write a t statistic for each of these pairs. For example, the statistic

$$t_{12} = \frac{\bar{x}_1 - \bar{x}_2}{s_p\sqrt{\dfrac{1}{n_1} + \dfrac{1}{n_2}}} = \frac{3.82 - 4.88}{1.10\sqrt{\dfrac{1}{24} + \dfrac{1}{33}}} = -3.59$$

compares profiles with 102 and 302 friends. The subscripts on t specify which groups are compared.

The t statistics for two other pairs are

$$t_{23} = \frac{\overline{x}_2 - \overline{x}_3}{s_p\sqrt{\dfrac{1}{n_2} + \dfrac{1}{n_3}}} = \frac{4.88 - 4.56}{1.10\sqrt{\dfrac{1}{33} + \dfrac{1}{26}}} = 1.11$$

and

$$t_{25} = \frac{\overline{x}_2 - \overline{x}_5}{s_p\sqrt{\dfrac{1}{n_2} + \dfrac{1}{n_5}}} = \frac{4.88 - 3.99}{1.10\sqrt{\dfrac{1}{33} + \dfrac{1}{21}}} = 2.90$$

two-sample t procedures p. 390

These t statistics are very similar to the pooled two-sample t statistic for comparing two population means. The difference is that we now have more than two populations, so each statistic uses the pooled estimator s_p from all groups rather than the pooled estimator from just the two groups being compared. This additional information about the common σ increases the power of the tests. The degrees of freedom for all these statistics are DFE = 129, those associated with s_p.

Because we do not have any specific ordering of the means in mind as an alternative to equality, we must use a two-sided approach to the problem of deciding which pairs of means are significantly different.

MULTIPLE COMPARISONS

To perform a **multiple-comparisons procedure,** compute ***t* statistics** for all pairs of means using the formula

$$t_{ij} = \frac{\overline{x}_i - \overline{x}_j}{s_p\sqrt{\dfrac{1}{n_i} + \dfrac{1}{n_j}}}$$

If

$$|t_{ij}| \geq t^{**}$$

we declare that the population means μ_i and μ_j are different. Otherwise, we conclude that the data do not distinguish between them. The value of t^{**} depends upon which multiple-comparisons procedure we choose.

One obvious choice for t^{**} is the upper $\alpha/2$ critical value for the $t(\text{DFE})$ distribution. This choice simply carries out as many separate significance tests of fixed level α as there are pairs of means to be compared. The procedure based on this choice is called the **least-significant differences method,** or simply LSD.

LSD method

LSD has some undesirable properties, particularly if the number of means being compared is large. Suppose, for example, that there are $I = 20$ groups and we use LSD with $\alpha = 0.05$. There are 190 different pairs of means. If we perform 190 t tests, each with an error rate of 5%, our overall error rate will be unacceptably large. We expect about 5% of the 190 to be significant even if the corresponding population means are the same. Since 5% of 190 is 9.5, we expect 9 or 10 false rejections.

The LSD procedure fixes the probability of a false rejection for each single pair of means being compared. It does not control the overall probability of *some* false rejection among all pairs. Other choices of t^{**} control possible errors in other ways. The choice of t^{**} is therefore a complex problem, and a detailed discussion of it is beyond the scope of this text. Many choices for t^{**} are used in practice. One major statistical package lets you select from a list with more than a dozen choices.

Bonferroni method

We will discuss only one of these, called the **Bonferroni method.** Use of this procedure with $\alpha = 0.05$, for example, guarantees that the probability of *any* false rejection among all comparisons made is no greater than 0.05. This is much stronger protection than controlling the probability of a false rejection at 0.05 for *each separate* comparison.

EXAMPLE 12.22

Applying the Bonferroni method. We apply the Bonferroni multiple-comparisons procedure with $\alpha = 0.05$ to the data from the Facebook friends study. The value of t^{**} for this procedure uses $\alpha = 0.05/10 = 0.005$ for each test. From Table D, this value is 2.63. Of the statistics $t_{12} = -3.59$, $t_{23} = 1.11$, and $t_{25} = 2.90$ calculated in the beginning of this section, only t_{12} and t_{25} are significant. These two statistics compare the profile of 302 friends with the two extreme levels.

Of course, we prefer to use software for the calculations.

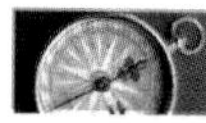

EXAMPLE 12.23

Interpreting software output. The output generated by SPSS for Bonferroni comparisons appears in Figure 12.10. The software uses an asterisk to indicate that the difference in a pair of means is statistically significant. Here, all 10 comparisons are reported. These results agree with the calculations that we performed in Examples 12.21 and 12.22. There are no significant differences except those already mentioned. Note that each comparison is given twice in the output.

The data in the Facebook friends study provide a clear result: up to a certain point, the social attractiveness score increases as the number of friends

SPSS

Multiple Comparisons

Score
Bonferroni

(I) Friends	(J) Friends	Mean Difference (I – J)	Std. Error	Sig.	95% Confidence Interval	
					Lower Bound	Upper Bound
102	302	–1.0621*	.2939	.004	–1.902	–.223
	502	–.7449	.3102	.177	–1.631	.141
	702	–.5900	.3001	.514	–1.447	.267
	902	–.1738	.3274	1.000	–1.109	.761
302	102	1.0621*	.2939	.004	.223	1.902
	502	.3172	.2873	1.000	–.503	1.138
	702	.4721	.2764	.900	–.317	1.262
	902	.8883*	.3059	.043	.015	1.762
502	102	.7449	.3102	.177	–.141	1.631
	302	–.3172	.2873	1.000	–1.138	.503
	702	.1549	.2936	1.000	–.684	.993
	902	.5711	.3215	.780	–.347	1.489
702	102	.5900	.3001	.514	–.267	1.447
	302	–.4721	.2764	.900	–1.262	.317
	502	–.1549	.2936	1.000	–.993	.684
	902	.4162	.3117	1.000	–.474	1.307
902	102	.1738	.3274	1.000	–.761	1.109
	302	–.8883*	.3059	.043	–1.762	–.015
	502	–.5711	.3215	.780	–1.489	.347
	702	–.4162	.3117	1.000	–1.307	.474

*. The mean difference is significant at the 0.05 level.

FIGURE 12.10 Software output giving the multiple-comparisons analysis for the Facebook friends example.

increases, and then it decreases. Unfortunately, with these data, we cannot accurately describe this relationship in more detail. This lack of clarity is not unusual when performing a multiple-comparisons analysis.

Here, the mean associated with 302 friends differs from those for the 102- and 902-friend profiles, but it is not found to be significantly different from the means for the profiles with 502 and 702 friends. To complicate things, the means for profiles with 502 and 702 friends are not found to be significantly different from the means for the 102- and 902-friend profiles.

This kind of apparent contradiction dramatically points out the nature of the conclusions of statistical tests of significance. The conclusion appears to be illogical. If μ_1 is the same as μ_3 and μ_3 is the same as μ_2, doesn't it follow that μ_1 is the same as μ_2? Logically, the answer must be "Yes."

Some of the difficulty can be resolved by noting the choice of words used. In describing the inferences, we talk about failing to detect a difference or concluding that two groups are different. In making logical statements, we say things such as "is the same as." There is a big difference between the two modes of thought. Statistical tests ask, "Do we have adequate evidence to distinguish two means?" It is not illogical to conclude that we have sufficient evidence to distinguish μ_1 from μ_2, but not μ_1 from μ_3 or μ_2 from μ_3.

One way to deal with these difficulties of interpretation is to give confidence intervals for the differences. The intervals remind us that the differences are not known exactly. We want to give *simultaneous confidence intervals,* that is, intervals for *all* differences among the population means at once. Again, we must face the problem that there are many competing procedures—in this case, many methods of obtaining simultaneous intervals.

SIMULTANEOUS CONFIDENCE INTERVALS FOR DIFFERENCES BETWEEN MEANS

Simultaneous confidence intervals for all differences $\mu_i - \mu_j$ between population means have the form

$$(\bar{x}_i - \bar{x}_j) \pm t^{**} s_p \sqrt{\frac{1}{n_i} + \frac{1}{n_j}}$$

The critical values t^{**} are the same as those used for the multiple-comparisons procedure chosen.

The confidence intervals generated by a particular choice of t^{**} are closely related to the multiple-comparisons results for that same method. If one of the confidence intervals includes the value 0, then that pair of means will not be declared significantly different, and vice versa.

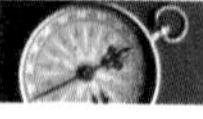

EXAMPLE 12.24

Interpreting software output, continued. The SPSS output for the Bonferroni multiple-comparisons procedure given in Figure 12.10 includes the simultaneous 95% confidence intervals. We can see, for example, that the interval for $\mu_1 - \mu_5$ is -1.63 to 0.14. The fact that the interval includes 0 is consistent with the fact that we failed to detect a difference between these two means using this procedure. Note that the interval for $\mu_5 - \mu_1$ is also provided. This is not really a new piece of information, because it can be obtained from the other interval by reversing the signs and reversing the order, that is, -0.14 to 1.63. So, in fact, we really have only 10 intervals. Use of the Bonferroni procedure provides us with 95% confidence that *all* 10 intervals simultaneously contain the true values of the population mean differences.

Software

We have used SPSS to illustrate the analysis of the Facebook friends data. Other statistical software gives similar output, and you should be able to read it without any difficulty.

EXAMPLE 12.25

Do eyes affect ad response? Research from a variety of fields has found significant effects of eye gaze and eye color on emotions and perceptions such as arousal, attractiveness, and honesty. These findings suggest that a model's eyes may play a role in a viewer's response to an ad.

In one study, students in marketing and management classes of a southern, predominantly Hispanic, university were each presented with one of four portfolios.[4] Each portfolio contained a target ad for a fictional product, Sparkle Toothpaste. Students were asked to view the ad and then respond to questions concerning their attitudes and emotions about the ad and product. All questions were taken from questionnaires previously used in the literature. Each response was on a seven-point scale.

Although the researchers investigated nine attitudes and emotions, we will focus on the viewer's "attitudes toward the brand." This response was obtained by averaging 10 survey questions.

The target ads were created using two digital photographs of a model. In one picture the model is looking directly at the camera so the eyes can be seen. This picture was used in three target ads. The only difference was the model's eyes, which were made to be either brown, blue, or green. In the second picture, the model is in the same pose but looking downward so the eyes are not visible. A total of 222 surveys were used for analysis. The following table summarizes the responses for the four portfolios. Outputs from Excel, SAS, and Minitab are given in Figure 12.11.

Group	n	Mean	Std. Dev.
Blue	67	3.19	1.75
Brown	37	3.72	1.72
Down	41	3.11	1.53
Green	77	3.86	1.67

Excel

	A	B	C	D	E	F	G
1	Anova: Single Factor						
2							
3	SUMMARY						
4	*Groups*	*Count*	*Sum*	*Average*	*Variance*		
5	Blue	67	214	3.19403	3.079055		
6	Brown	37	137.8	3.724324	2.942447		
7	Down	41	127.4	3.107317	2.326695		
8	Green	77	297.2	3.85974	2.775332		
9							
10	ANOVA						
11	*Source of Variation*	*SS*	*df*	*MS*	*F*	*P-value*	*F crit*
12	Between Groups	24.41966	3	8.139886	2.894117	0.036184	2.646014
13	Within Groups	613.1387	218	2.812563			
14							
15	Total	637.5584	221				

FIGURE 12.11 Excel, SAS, and Minitab output for the advertising study in Example 12.25. (*Continued on next page*)

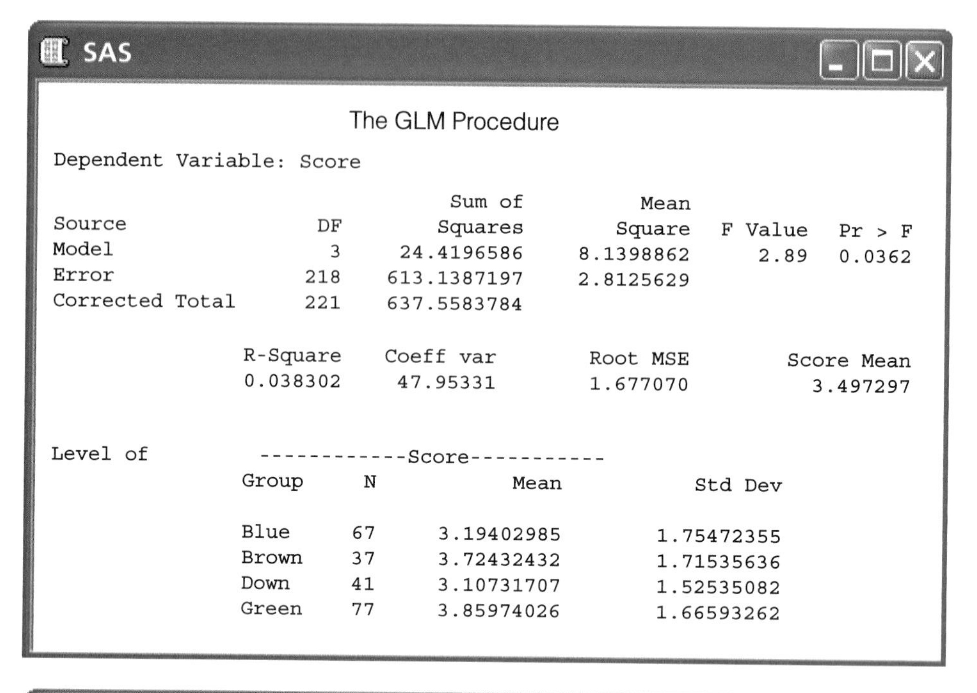

SAS

The GLM Procedure

Dependent Variable: Score

Source	DF	Sum of Squares	Mean Square	F Value	Pr > F
Model	3	24.4196586	8.1398862	2.89	0.0362
Error	218	613.1387197	2.8125629		
Corrected Total	221	637.5583784			

R-Square	Coeff var	Root MSE	Score Mean
0.038302	47.95331	1.677070	3.497297

Level of Group	N	Score Mean	Score Std Dev
Blue	67	3.19402985	1.75472355
Brown	37	3.72432432	1.71535636
Down	41	3.10731707	1.52535082
Green	77	3.85974026	1.66593262

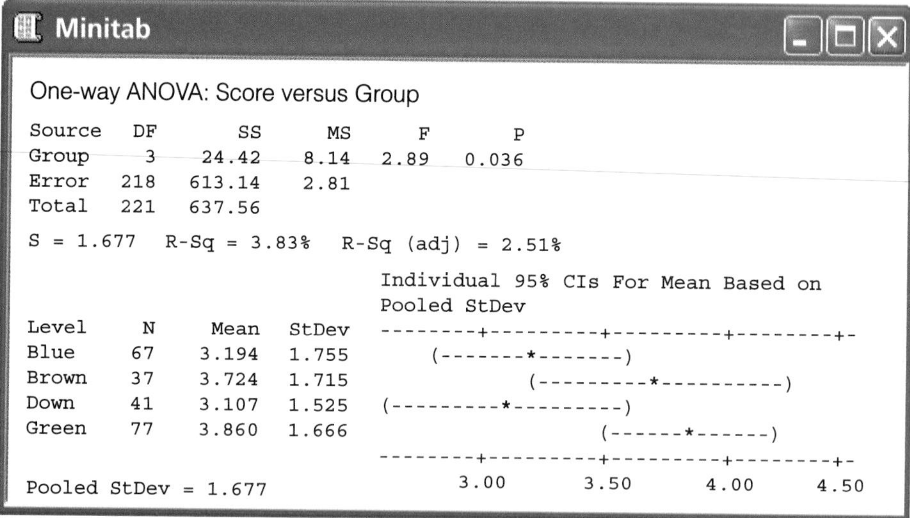

Minitab

One-way ANOVA: Score versus Group

Source	DF	SS	MS	F	P
Group	3	24.42	8.14	2.89	0.036
Error	218	613.14	2.81		
Total	221	637.56			

S = 1.677 R-Sq = 3.83% R-Sq (adj) = 2.51%

```
                                 Individual 95% CIs For Mean Based on
                                 Pooled StDev
Level    N    Mean   StDev  --------+---------+---------+--------+-
Blue    67   3.194   1.755      (-------*-------)
Brown   37   3.724   1.715              (---------*----------)
Down    41   3.107   1.525  (---------*---------)
Green   77   3.860   1.666                    (------*------)
                            --------+---------+---------+--------+-
                                  3.00      3.50      4.00     4.50
Pooled StDev = 1.677
```

FIGURE 12.11 (*Continued*)

There is evidence at the 5% significance level to reject the null hypothesis that the four groups have equal means ($P = 0.036$). In Exercises 12.33 and 12.34, you are asked to perform further inference using contrasts.

USE YOUR KNOWLEDGE

12.7 Why no multiple comparisons? Any pooled two-sample t problem can be run as a one-way ANOVA with $I = 2$. Explain why it is inappropriate to analyze the data using contrasts or multiple-comparisons procedures in this setting.

12.8 Growth of Douglas fir seedlings. An experiment was conducted to compare the growth of Douglas fir seedlings under three different levels of vegetation control (0%, 50%, and 100%). Ten seedlings were randomized to each level of control. The resulting sample means for stem volume were 58, 74, and 120 cubic centimeters (cm^3) with $s_p = 25$ cm^3. The researcher hypothesized that the average growth at 50% control would be less than the averages of the 0% and 100% levels.

(a) What are the coefficients for testing this contrast?

(b) Perform the test. Do the data provide evidence to support this hypothesis?

CHAPTER 12 SUMMARY

One-way analysis of variance (ANOVA) is used to compare several population means based on independent SRSs from each population. The populations are assumed to be Normal with possibly different means and the same standard deviation.

The **null hypothesis** is that the population means are *all* equal. The **alternative hypothesis** is true if there are *any* differences among the population means.

ANOVA is based on separating the total variation observed in the data into two parts: variation **among group means** and variation **within groups.** If the variation among groups is large relative to the variation within groups, we have evidence against the null hypothesis.

An **analysis of variance table** organizes the ANOVA calculations. **Degrees of freedom, sums of squares,** and **mean squares** appear in the table. The ***F* statistic** and its ***P*-value** are used to test the null hypothesis.

To do an analysis of variance, first compute sample means and standard deviations for all groups. Side-by-side boxplots and a plot of the means give an overview of the data. Examine histograms or Normal quantile plots (either for each group separately or for the residuals) to detect outliers or extreme deviations from Normality. Compute the ratio of the largest to the smallest sample standard deviation. If this ratio is less than 2 and the data inspection does not find outliers or severe non-Normality, ANOVA can be performed.

The ANOVA F test shares the **robustness** of the two-sample t test. It is relatively insensitive to moderate non-Normality and unequal variances, especially when the sample sizes are similar.

Specific questions formulated before examination of the data can be expressed as **contrasts.** Tests and confidence intervals for contrasts provide answers to these questions.

If no specific questions are formulated before examination of the data and the null hypothesis of equality of population means is rejected, **multiple-comparisons** methods are used to assess the statistical significance of the differences between pairs of means.

CHAPTER 12 EXERCISES

For Exercises 12.1 and 12.2, see page 557; for Exercises 12.3 and 12.4, see page 561; for Exercises 12.5 and 12.6, see page 568; and for Exercises 12.7 and 12.8, see pages 580–581.

12.9. A one-way ANOVA example. A study compared 5 groups, with 7 observations per group. An F statistic of 2.83 was reported.

(a) Give the degrees of freedom for this statistic and the entries from Table E that correspond to this distribution.

(b) Sketch a picture of this F distribution with the information from the table included.

(c) Based on the table information, how would you report the P-value?

(d) Can you conclude that all pairs of means are different? Explain your answer.

12.10. Calculating the ANOVA *F* test *P*-value. For each of the following situations, find the degrees of freedom for the F statistic and then use Table E to approximate the P-value.

(a) Seven groups are being compared, with 5 observations per group. The value of the F statistic is 2.69.

(b) Five groups are being compared, with 11 observations per group. The value of the F statistic is 2.43.

(c) Six groups are being compared, using 34 total observations. The value of the F statistic is 3.06.

12.11. Calculating the ANOVA F test P-value, continued. For each of the following situations, find the F statistic and the degrees of freedom. Then draw a sketch of the distribution under the null hypothesis and shade in the portion corresponding to the P-value. State how you would report the P-value.

(a) Compare 3 groups, with 11 observations per group, MSE = 50, and MSG = 127.

(b) Compare 4 groups, with 8 observations per group, SSG = 58, and SSE = 182.

12.12. Calculating the pooled standard deviation. An experiment was run to compare three groups. The sample sizes were 29, 32, and 121, and the corresponding estimated standard deviations were 37, 28, and 42.

(a) Is it reasonable to use the assumption of equal standard deviations when we analyze these data? Give a reason for your answer.

(b) Give the values of the variances for the three groups.

(c) Find the pooled variance.

(d) What is the value of the pooled standard deviation?

(e) Explain why your answer in part (d) is much closer to the standard deviation for the third group than to any of the other two standard deviations.

12.13. Describing the ANOVA model. For each of the following situations, identify the response variable and the populations to be compared, and give I, the n_i, and N.

(a) A poultry farmer is interested in reducing the cholesterol level in his marketable eggs. He wants to compare two different cholesterol-lowering drugs added to the hens' standard diet as well as an all-vegetarian diet. He assigns 25 of his hens to each of the three treatments.

(b) A researcher is interested in students' opinions regarding an additional annual fee to support non-income-producing varsity sports. Students were asked to rate their acceptance of this fee on a five-point scale. She received 94 responses, of which 31 were from students who attend varsity football or basketball games only, 18 were from students who also attend other varsity competitions, and 45 were from students who do not attend any varsity games.

(c) A professor wants to evaluate the effectiveness of his teaching assistants. In one class period, the 42 students were randomly divided into three equal-sized groups, and each group was taught power calculations from one of the assistants. At the beginning of the next class, each student took a quiz on power calculations, and these scores were compared.

12.14. Describing the ANOVA model, continued. For each of the following situations, identify the response variable and the populations to be compared, and give I, the n_i, and N.

(a) A developer of a virtual-reality (VR) teaching tool for the deaf wants to compare the effectiveness of different navigation methods. A total of 40 children were available for the experiment, of which equal numbers were randomly assigned to use a joystick, wand, dance mat, or gesture-based pinch gloves. The time (in seconds) to complete a designed VR path is recorded for each child.

(b) To study the effects of pesticides on birds, an experimenter randomly (and equally) allocated 65 chicks to five diets (a control and four with a different pesticide included). After a month, the calcium content (milligrams) in a 1-centimeter length of bone from each chick was measured.

(c) A university sandwich shop wants to compare the effects of providing free food with a sandwich order on sales. The experiment will be conducted from 11:00 A.M. to 2:00 P.M. for the next 20 weekdays. On each day, customers will be offered one of the following: a free drink, free chips, a free cookie, or nothing. Each option will be offered five times.

12.15. Determining the degrees of freedom. Refer to Exercise 12.13. For each situation, give the following:

(a) Degrees of freedom for the model, for error, and for the total

(b) Null and alternative hypotheses

(c) Numerator and denominator degrees of freedom for the F statistic

12.16. Determining the degrees of freedom, continued. Refer to Exercise 12.14. For each situation, give the following:

(a) Degrees of freedom for the model, for error, and for the total

(b) Null and alternative hypotheses

(c) Numerator and denominator degrees of freedom for the F statistic

12.17. Data collection and the interpretation of results. Refer to Exercise 12.13. For each situation, discuss the method of obtaining the data and how this will affect the extent to which the results can be generalized.

12.18. Data collection, continued. Refer to Exercise 12.14. For each situation, discuss the method of obtaining the data and how this will affect the extent to which the results can be generalized.

12.19. Shopping and bargaining in Mexico. Price haggling and other bargaining behaviors among consumers have been observed for a long time. However, research addressing these behaviors, especially in a real-life setting, remains relatively sparse. A group of researchers performed a small study to determine whether gender or nationality of the bargainer has an effect in the final price obtained.[5] The study took place in Mexico because of the prevalence of price haggling in informal markets. Salespersons working at various informal shops were approached by one of three bargainers looking for a specific product. After an initial price was stated by the vendor, bargaining took place. The response was the difference between the initial and the final price of the product. The bargainers were a Spanish-speaking Hispanic male, a Spanish-speaking Hispanic female, and an Anglo non-Spanish-speaking male. The following table summarizes the results:

Bargainer	n	Average reduction
Hispanic male	40	1.055
Anglo male	40	1.050
Hispanic female	40	2.310

(a) To compare the mean reductions in price, what are the degrees of freedom for the ANOVA F statistic?

(b) The reported test statistic is $F = 8.708$. Give an approximate (from a table) or exact (from software) P-value. What do you conclude?

(c) To what extent do you think the results of this study can be generalized? Give reasons for your answer.

12.20. APPLET **The effect of increased variation within groups.** The *One-Way ANOVA* applet lets you see how the F statistic and the P-value depend on the variability of the data within groups and the differences among the means.

(a) The black dots are at the means of the three groups. Move these up and down until you get a configuration that gives a P-value of about 0.01. What is the value of the F statistic?

(b) Now increase the variation within the groups by dragging the mark on the standard deviation scale to the right. Describe what happens to the F statistic and the P-value. Explain why this happens.

12.21. APPLET **The effect of increased variation between groups.** Set the standard deviation for the *One-Way ANOVA* applet at a middle value. Drag the black dots so that they are approximately equal.

(a) What is the F statistic? Give its P-value.

(b) Drag the mean of the second group up and the mean of the third group down. Describe the effect on the F statistic and its P-value. Explain why they change in this way.

12.22. Animals on product labels? Recall Exercise 7.50 (page 410). This experiment actually involved comparing product preference for a group of consumers that were "primed" and two groups of consumers that served as controls.[6] A bottle of MagicCoat pet shampoo was the product, and participants indicated their attitude toward this product on a seven-point scale (from 1 = dislike very much to 7 = like very much). The bottle of shampoo had either a picture of a collie on the label or just the wording. Also, before giving a score, participants were asked to do a word-find puzzle where four of the words were common across groups (pet, grooming, bottle, label) and four were either related to the image (dog, collie, puppy, woof) or image conflicting (cat, feline, kitten, meow). A summary of the groups follows: DATA FILE BRANDPREF1

Group	Label with dog	Dog "primed"	n
1	Y	Y	22
2	Y	N	20
3	N	Y	10

(a) Use graphical and numerical methods to describe the data.

(b) Run the ANOVA and report the results.

(c) Examine the assumptions necessary for inference using your results in part (a) and an examination of the residuals. Summarize your findings.

(d) Use a multiple-comparisons method to compare the three groups. State your conclusions.

12.23. The multiple-play strategy. Multiple play is a bundling strategy in which multiple services are provided over a single network. A common triple-play service these days is Internet, television, and telephone. The market for this service has become a key battleground among telecommunication, cable, and broadband service providers. A study compared the pricing (average monthly cost in U.S. dollars) among triple-play providers using DSL, cable, or fiber platforms.[7] The following table summarizes the results of 47 providers.

Group	n	$\bar{x}$	s
DSL	19	104.49	26.09
Cable	20	119.98	40.39
Fiber	8	83.87	31.78

(a) Plot the means versus the platform type. Does there appear to be a difference in pricing?

(b) Is it reasonable to assume that the variances are equal? Explain.

(c) The F statistic is 3.39. Give the degrees of freedom and either an approximate (from a table) or an exact (from software) P-value. What do you conclude?

12.24. A contrast. Refer to the previous exercise. Use a contrast to compare the fiber platform with the average of the other two. The hypothesis prior to collecting the data is that the fiber platform price would be smaller. Summarize your conclusion.

12.25. Sleep deprivation and reaction times. Sleep deprivation experienced by physicians during residency training and the possible negative consequences are of concern to many in health care. One study of 33 resident anesthesiologists compared their changes from baseline in reaction times on four tasks.[8] Under baseline conditions, the physicians reported getting an average of 7.04 hours of sleep. While on duty, however, the average was 1.66 hours. For each of the tasks, the researchers reported a statistically significant increase in the reaction time when the residents were working in a state of sleep deprivation.

(a) If each task is analyzed separately as the researchers did in their report, what is the appropriate statistical method to use? Explain your answer.

(b) Is it appropriate to use a one-way ANOVA with $I = 4$ to analyze these data? Explain why or why not.

12.26. Restaurant ambiance and consumer behavior. Many studies have investigated the effects of restaurant ambiance on consumer behavior. One study investigated the effects of musical genre on consumer spending.[9] At a single high-end restaurant in England over a three-week period, there was a total of 141 participants; 49 of them were subjected to background pop music (for example, Britney Spears, Culture Club, and Ricky Martin) while dining, 44 to background classical music (for example, Vivaldi, Handel, and Strauss), and 48 to no background music. For each participant, the total food bill, adjusted for time spent dining, was recorded. The following table summarizes the means and standard deviations (in British pounds):

Background music	Mean bill	n	s
Classical	24.130	44	2.243
Pop	21.912	49	2.627
None	21.697	48	3.332
Total	22.531	141	2.969

(a) Plot the means versus the type of background music. Does there appear to be a difference in spending?

(b) Is it reasonable to assume that the variances are equal? Explain.

(c) The F statistic is 10.62. Give the degrees of freedom and either an approximate (from a table) or an exact (from software) P-value. What do you conclude?

(d) Refer back to part (a). Without doing any formal analysis, describe the pattern in the means that is likely responsible for your conclusion in part (c).

(e) To what extent do you think the results of this study can be generalized to other settings? Give reasons for your answer.

12.27. The effects of two stimulant drugs. An experimenter was interested in investigating the effects of two stimulant drugs (labeled A and B). She divided 20 rats equally into 5 groups (placebo, Drug A low, Drug A high, Drug B low, and Drug B high) and, 20 minutes after injection of the drug, recorded each rat's activity level (higher score is more active). The following table summarizes the results:

Treatment	$\bar{x}$	s^2
Placebo	14.00	9.00
Low A	15.25	14.00
High A	18.25	12.25
Low B	15.75	6.75
High B	22.50	11.00

(a) Plot the means versus the type of treatment. Does there appear to be a difference in the activity levels? Explain.

(b) Is it reasonable to assume that the variances are equal? Explain your answer, and if reasonable, compute s_p.

(c) Give the degrees of freedom for the F statistic.

(d) The F statistic is 4.28. Find the associated P-value and state your conclusions.

12.28. CHALLENGE **Exam accommodations and end-of-term grades.** The Americans with Disabilities Act (ADA) requires that students with learning disabilities (LD) and/or attention deficit disorder (ADD) be given certain accommodations when taking examinations. One study designed to assess the effects of these accommodations examined the relationship between end-of-term grades and the number of accommodations given.[10] The researchers reported the mean grades with sample sizes and standard deviations versus the number of accommodations in a table similar to this:

Accommodations	Mean grade	n	s
0	2.7894	160	0.85035
1	2.8605	38	0.83068
2	2.5757	37	0.82745
3	2.6286	7	1.03072
4	2.4667	3	1.66233
Total	2.7596	245	0.85701

(a) Plot the means versus the number of accommodations. Is there a pattern evident?

(b) A large number of digits are reported for the means and the standard deviations. Do you think that all these digits are necessary? Give reasons for your answer and describe how you would report these results.

(c) Should we pool to obtain an estimate of an assumed standard deviation for these data? Explain your answer and give the pooled estimate if your answer is "Yes."

(d) The small numbers of observations with 3 or 4 accommodations lead to estimates that are highly variable in these groups compared with the other groups. Inclusion of groups with relatively few observations in an ANOVA can also lead to low power. We could eliminate these two levels from the analysis or we could combine them with the 37 observations in the group above to form a new group with 2 or more accommodations. Which of these options do you prefer? Give reasons for your answer.

12.29. Exam accommodations study, continued. Refer to the previous exercise.

(a) The 245 grades reported in the table were from a sample of 61 students who completed three, four, or five courses during a spring term at one college and who were qualified to receive accommodations. Students in the sample were self-identified, in the sense that they had to request qualification. Even when qualified, some students chose not to request accommodations for some or all of their courses. Based on these facts, would you advise that ANOVA methods be used for these data? Explain your answer. (The authors did not present the results of an ANOVA in their publication.)

(b) To what extent do you think the results of this study can be generalized to other settings? Give reasons for your answer.

(c) Most reasonable approaches to the analysis of these data would conclude that the data fail to provide evidence that the number of accommodations is related to the mean grades. Does this imply that the accommodations are not needed or does it suggest that they are effective? Discuss your answer.

12.30. Overall standard deviation versus the pooled standard deviation. The last line of the summary table given in Exercise 12.28 gives the mean and the standard deviation for all the data combined. Compare this standard deviation with the pooled standard deviation that you would use as an estimate of the model standard deviation. Explain why you would expect this standard deviation to be larger than the pooled standard deviation.

12.31. Do we experience emotions differently? Do people from different cultures experience emotions differently? One study designed to examine this question collected data from 416 college students from five different cultures.[11] The participants were asked to record, on a 1 (never) to 7 (always) scale, how much of the time they typically felt eight specific emotions. These were averaged to produce the global emotion score for each participant. Here is a summary of this measure:

Culture	n	Mean (s)
European American	46	4.39 (1.03)
Asian American	33	4.35 (1.18)
Japanese	91	4.72 (1.13)
Indian	160	4.34 (1.26)
Hispanic American	80	5.04 (1.16)

Note that the convention of giving the standard deviations in parentheses after the means saves a great deal of space in a table such as this.

(a) From the information given, do you think that we need to be concerned that a possible lack of Normality in the data will invalidate the conclusions that we might draw using ANOVA to analyze the data? Give reasons for your answer.

(b) Is it reasonable to use a pooled standard deviation for these data? Why or why not?

(c) The ANOVA F statistic was reported as 5.69. Give the degrees of freedom and either an approximate (from a table) or an exact (from software) P-value. Sketch a picture of the F distribution that illustrates the P-value. What do you conclude?

(d) Without doing any additional formal analysis, describe the pattern in the means that appears to be responsible for your conclusion in part (c). Are there pairs of means that are quite similar?

12.32. CHALLENGE **The emotion study, continued.** Refer to the previous exercise. The experimenters also measured emotions in some different ways. For a period of a week, each participant carried a device that sounded an alarm at random times during a 3-hour interval 5 times a day. When the alarm sounded, participants recorded several mood ratings indicating their emotions for the time immediately preceding the alarm. These responses were combined to form two variables: frequency, the number of emotions recorded, expressed as a percent; and intensity, an average of the intensity scores measured on a scale of 0 to 6. At the end of the one-week experimental period, the subjects were asked to recall the percent of time that they experienced different emotions. This variable was called "recall." Here is a summary of the results:

Culture	n	Frequency mean (s)	Intensity mean (s)	Recall mean (s)
European American	46	82.87 (18.26)	2.79 (0.72)	49.12 (22.33)
Asian American	33	72.68 (25.15)	2.37 (0.60)	39.77 (23.24)
Japanese	91	73.36 (22.78)	2.53 (0.64)	43.98 (22.02)
Indian	160	82.71 (17.97)	2.87 (0.74)	49.86 (21.60)
Hispanic American	80	92.25 (8.85)	3.21 (0.64)	59.99 (24.64)
F statistic		11.89	13.10	7.06

(a) For each response variable state whether or not it is reasonable to use a pooled standard deviation to analyze these data. Give reasons for your answer.

(b) Give the degrees of freedom for the F statistics and find the associated P-values. Summarize what you can conclude from these ANOVA analyses.

(c) Summarize the means, paying particular attention to similarities and differences across cultures and across variables. Include the means from the previous exercise in your summary.

(d) The European American and Asian American subjects were from the University of Illinois, the Japanese subjects were from two universities in Tokyo, the Indian subjects were from eight universities in or near Calcutta, and the Hispanic American subjects were from California State University at Fresno. Participants were paid \$25 or an equivalent monetary incentive for the Japanese and Indians. Ads were posted on or near the campuses to recruit volunteers for the study. Discuss how these facts influence your conclusions and the extent to which you would generalize the results.

(e) The percents of female students in the samples were as follows: European American, 83%; Asian American, 67%; Japanese, 63%; Indian, 64%; and Hispanic American, 79%. Use a chi-square test to compare these proportions (see Section 9.1, page 466), and discuss how this information influences your interpretation of the results that you have found in this exercise.

12.33. Writing contrasts. Return to the eye study described in Example 12.25 (page 579). Let μ_1, μ_2, μ_3, and μ_4 represent the mean scores for blue, brown, gaze down, and green eyes.

(a) Because a majority of the population sampled is Hispanic (eye color predominantly brown), we want to compare the average score of the brown eyes with the average of the other two eye colors. Write a contrast that expresses this comparison.

(b) Write a contrast to compare the average score when the model is looking at you versus the score when looking down.

12.34. Analyzing contrasts. Answer the following questions for the two contrasts that you defined in Exercise 12.33. DATA FILE EYES

(a) For each contrast give H_0 and an appropriate H_a. In choosing the alternatives you should use information given in the description of the problem, but you may not consider any impressions obtained by inspection of the sample means.

(b) Find the values of the corresponding sample contrasts c_1 and c_2.

(c) Calculate the standard errors SE_{c_1} and SE_{c_2}.

(d) Give the test statistics and approximate P-values for the two significance tests. What do you conclude?

(e) Compute 95% confidence intervals for the two contrasts.

12.35. CHALLENGE **A comparison of tropical flower varieties.** Different varieties of the tropical flower *Heliconia* are fertilized by different species of hummingbirds. Over time, the lengths of the flowers and the forms of the hummingbirds' beaks have evolved to match each other. Here are data on the lengths in millimeters of three varieties of these flowers on the island of Dominica:[12] DATA FILE HUMMINGBIRD

H. bihai							
47.12	46.75	46.81	47.12	46.67	47.43	46.44	46.64
48.07	48.34	48.15	50.26	50.12	46.34	46.94	48.36

H. caribaea red							
41.90	42.01	41.93	43.09	41.47	41.69	39.78	40.57
39.63	42.18	40.66	37.87	39.16	37.40	38.20	38.07
38.10	37.97	38.79	38.23	38.87	37.78	38.01	

H. caribaea yellow							
36.78	37.02	36.52	36.11	36.03	35.45	38.13	37.1
35.17	36.82	36.66	35.68	36.03	34.57	34.63	

Do a complete analysis that includes description of the data and a significance test to compare the mean lengths of the flowers for the three species.

12.36. CHALLENGE **The two-sample t test and one-way ANOVA.** Refer to Exercise 7.67 (page 414). In that exercise, you were asked to calculate the pooled t statistic for comparing high-sleep- and low-sleep-efficiency children. Formulate this problem as an ANOVA and report the results of this analysis. Verify that $F = t^2$.

12.37. CHALLENGE **Taking the log of the response variable.** The distributions of the flower lengths in Exercise 12.35 are somewhat skewed. Take natural logs of the lengths and reanalyze the data. Write a summary of your results and include a comparison with the results you found in Exercise 12.35. DATA FILE HUMMINGBIRD

12.38. CHALLENGE **Do poets die young?** According to William Butler Yeats, "She is the Gaelic muse, for she gives inspiration to those she persecutes. The Gaelic poets die young, for she is restless, and will not let them remain long on earth." One study designed to investigate this issue examined the age at death for writers from different cultures and genders.[13] Three categories of writers examined were novelists, poets, and nonfiction writers. The ages at death for female writers in these categories from North America are given in Table 12.1. Most of the writers are from the United States, but Canadian and Mexican writers are also included. DATA FILE DEADPOETS

(a) Use graphical and numerical methods to describe the data.

(b) Examine the assumptions necessary for ANOVA. Summarize your findings.

(c) Run the ANOVA and report the results.

(d) Use a contrast to compare the poets with the two other types of writers. Do you think that the quotation from Yeats justifies the use of a one-sided alternative for examining this contrast? Explain your answer.

(e) Use another contrast to compare the novelists with the nonfiction writers. Explain your choice for an alternative hypothesis for this contrast.

(f) Use a multiple-comparisons procedure to compare the three means. How do the conclusions from this approach compare with those using the contrasts?

12.39. College dining. University and college food service operations have been trying to keep up with the growing expectations of consumers with regard to the overall campus dining experience. Since customer satisfaction has been shown to be associated with repeat patronage and new customers through word-of-mouth, a public university in the Midwest took a sample of patrons from their eating establishments and asked them about

TABLE 12.1

Age at death for women writers

Type	Age at death (years)														
Novels	57	90	67	56	90	72	56	90	80	74	73	86	53	72	86
($n = 67$)	82	74	60	79	80	79	77	64	72	88	75	79	74	85	71
	78	57	54	50	59	72	60	77	50	49	73	39	73	61	90
	77	57	72	82	54	62	74	65	83	86	73	79	63	72	85
	91	77	66	75	90	35	86								
Poems	88	69	78	68	72	60	50	47	74	36	87	55	68	75	78
($n = 32$)	85	69	38	58	51	72	58	84	30	79	90	66	45	70	48
	31	43													
Nonfiction	74	86	87	68	76	73	63	78	83	86	40	75	90	47	91
($n = 24$)	94	61	83	75	89	77	86	66	97						

their overall dining satisfaction.[14] The following table summarizes the results for three groups of patrons:

Category	$\bar{x}$	n	s
Student with meal plan	3.44	489	0.804
Faculty meal plan	4.04	69	0.824
Student with no meal plan	3.47	212	0.657

(a) Is it reasonable to use a pooled standard deviation for these data? Why or why not? If yes, compute it.

(b) The ANOVA F statistic was reported as 17.66. Give the degrees of freedom and either an approximate (from a table) or an exact (from software) P-value. Sketch a picture of the F distribution that illustrates the P-value. What do you conclude?

(c) Prior to performing this survey, food service operations thought that satisfaction among faculty would be higher than satisfaction among students. Use the results in the table to test this contrast. Make sure to specify the null and alternative hypotheses, test statistic, and P-value.

12.40. Developing marketing strategies for travel to Hawaii. In 1997 approximately one-third of all tourists to Hawaii were from Japan. Since that time the percent has steadily decreased and is now around 20%.[15] To better understand the reasons for travel to Hawaii, a group of researchers surveyed 315 Japanese tourists who planned to visit Hawaii. The tourists were divided into groups based on their purpose for travel. They were (1) honeymoon, (2) fraternal association, (3) sports, (4) leisure, and (5) business. Their responses to various survey questions were compared across these groups. The responses were on a scale ranging from 1 (strongly disagree) to 7 (strongly agree). The following table summarizes the mean response and F test statistic for several questions.

Question	Group 1 $n = 34$	Group 2 $n = 56$	Group 3 $n = 105$	Group 4 $n = 26$	Group 5 $n = 94$	F
I'd like to experience native Hawaiian culture	3.97	4.26	4.25	5.33	4.23	2.46
I'd prefer a group tour to an individual one	3.18	3.38	2.39	2.58	2.98	3.97
I'd like to experience ocean sports	4.71	4.59	4.58	5.33	4.02	2.46
I respect Hawaiian residents' customs	4.88	5.39	5.14	5.83	5.46	1.62

(a) What are the numerator and denominator degrees of freedom for these F tests?

(b) The response variable is not Normally distributed. Explain why this should not cause difficulties in using ANOVA.

(c) Using a significance level of $\alpha = 0.05$ for each question, assess whether there are differences in the group means.

(d) For those questions with a significant F statistic, plot the means and describe their pattern.

12.41. Multiple comparisons. Refer to the previous exercise.

(a) Explain why it is inappropriate to perform a multiple-comparisons analysis for the last question.

(b) For each of the other questions, use the Bonferroni or another multiple-comparisons procedure to determine which group means differ significantly. The following table gives the MSE for each question.

Question	MSE
I'd like to experience native Hawaiian culture	3.261
I'd prefer a group tour to an individual one	2.841
I'd like to experience ocean sports	4.285
I respect Hawaiian residents' customs	2.905

Summarize your results in a short report.

12.42. CHALLENGE **Polynomial contrasts.** Recall the Facebook friends study. In Example 12.16 (page 570) we used a contrast to assess the linear trend between the social attractiveness score and number of Facebook friends. With polynomial contrasts, we can assess the contributions of different polynomial trends to the overall pattern. Because the derivation of the coefficients is beyond the scope of this book, we will just investigate the trends here. The coefficients for the linear, quadratic, and cubic trends follow: DATA FILE FRIENDS

Trend	a_1	a_2	a_3	a_4	a_5
Linear	−2	−1	0	1	2
Quadratic	2	−1	−2	−1	2
Cubic	−1	2	0	−2	1

(a) Plot the a_i versus i for the linear trend. Describe the pattern. Suppose that all the μ_i were constant. What would the value of ψ equal?

(b) Plot the a_i versus i for the quadratic trend. Describe the pattern. Suppose that all the μ_i were constant. What would the value of ψ equal? Suppose that $\mu_i = 5i$ (that is, a linear trend). What would the value of ψ equal?

(c) Construct the sample contrasts for the quadratic and cubic trends using the Facebook data.

(d) Test the hypotheses that there is a quadratic and cubic trend. Combine this with the earlier linear trend results. What do you conclude?

12.43. Exercise and healthy bones. Many studies have suggested that there is a link between exercise and healthy bones. Exercise stresses the bones and this causes them to get stronger. One study examined the effect of jumping on the bone density of growing rats.[16] There were three treatments: a control with no jumping, a low-jump condition (the jump height was 30 centimeters), and a high-jump condition (60 centimeters). After 8 weeks of 10 jumps per day, 5 days per week, the bone density of the rats (expressed in milligrams per cubic centimeter) was measured. Here are the data: DATA FILE BONEDENSITY

Group	Bone density (mg/cm^3)									
Control	611	621	614	593	593	653	600	554	603	569
Low jump	635	605	638	594	599	632	631	588	607	596
High jump	650	622	626	626	631	622	643	674	643	650

(a) Make a table giving the sample size, mean, and standard deviation for each group of rats. Is it reasonable to pool the variances?

(b) Run the analysis of variance. Report the F statistic with its degrees of freedom and P-value. What do you conclude?

12.44. Exercise and healthy bones, continued. Refer to the previous exercise. DATA FILE BONEDENSITY

(a) Examine the residuals. Is the Normality assumption reasonable for these data?

(b) Use the Bonferroni or another multiple-comparisons procedure to determine which pairs of means differ significantly. Summarize your results in a short report. Be sure to include a graph.

12.45. Does the type of cooking pot affect iron content? Iron-deficiency anemia is the most common form of malnutrition in developing countries, affecting about 50% of children and women and 25% of men. Iron pots for cooking foods had traditionally been used in many of these countries, but they have been largely replaced by aluminum pots, which are cheaper and lighter. Some research has suggested that food cooked in iron pots will contain more iron than food cooked in other types of pots. One study designed to investigate this issue compared the iron content of some Ethiopian foods cooked in aluminum, clay, and iron pots.[17] One of the foods was *yesiga wet'*, beef cut into small pieces and prepared with several Ethiopian spices. The iron content of four samples of *yesiga wet'* cooked in each of the three types of pots follows. The units are milligrams of iron per 100 grams of cooked food. DATA FILE COOKINGPOT

Type of pot	Iron (mg/100 g food)			
Aluminum	1.77	2.36	1.96	2.14
Clay	2.27	1.28	2.48	2.68
Iron	5.27	5.17	4.06	4.22

(a) Make a table giving the sample size, mean, and standard deviation for each type of pot. Is it reasonable to pool the variances? Note that with the small sample sizes in this experiment, we expect a large amount of variability in the sample standard deviations.

(b) Run the analysis of variance. Report the F statistic with its degrees of freedom and P-value. What do you conclude?

12.46. The cooking pot study, continued. Refer to the previous exercise.

(a) Examine the residuals. Is the Normality assumption reasonable for these data?

(b) Use the Bonferroni or another multiple-comparisons procedure to determine which pairs of means differ significantly. Summarize your results in a short report. Be sure to include a graph.

12.47. A comparison of different types of scaffold material. One way to repair serious wounds is to insert some material as a scaffold for the body's repair cells to use as a template for new tissue. Scaffolds made from extracellular material (ECM) are particularly promising for this purpose. Because they are made from biological material, they serve as an effective scaffold and are then resorbed. Unlike biological material that includes cells, however, they do not trigger tissue rejection reactions in the body. One study compared six types of scaffold material.[18] Three of these were ECMs and the other three were made of inert materials (MAT). There were three mice used per scaffold type. The response measure was the percent of glucose phosphated isomerase (Gpi) cells in the region of the wound. A large value is good, indicating that there are many bone marrow cells sent by the body to repair the tissue. DATA FILE ECM

Material	Gpi (%)		
ECM1	55	70	70
ECM2	60	65	65
ECM3	75	70	75
MAT1	20	25	25
MAT2	5	10	5
MAT3	10	15	10

(a) Make a table giving the sample size, mean, and standard deviation for each of the six types of material. Is it reasonable to pool the variances? Note that the sample sizes are small and the data are rounded.

(b) Run the analysis of variance. Report the F statistic with its degrees of freedom and P-value. What do you conclude?

12.48. A comparison of different types of scaffold material, continued. Refer to the previous exercise. DATA FILE ECM

(a) Examine the residuals. Is the Normality assumption reasonable for these data?

(b) Use the Bonferroni or another multiple-comparisons procedure to determine which pairs of means differ significantly. Summarize your results in a short report. Be sure to include a graph.

(c) Use a contrast to compare the three ECM materials with the three other materials. Summarize your conclusions. How do these results compare with those that you obtained from the multiple-comparisons procedure in part (b)?

12.49. Two contrasts of interest for the stimulant study. Refer to Exercise 12.27 (page 584). There are two comparisons of interest to the experimenter. They are (1) placebo versus the average of the two low-dose treatments and (2) the difference between High A and Low A versus the difference between High B and Low B.

(a) Express each contrast in terms of the means (μ's) of the treatments.

(b) Give an estimate with the standard error for each of the contrasts.

(c) Perform the significance tests for the contrasts. Summarize the results of your tests and your conclusions.

12.50. Changing the response variable. Refer to Exercise 12.47 (page 590), where we compared six types of scaffold material to repair wounds. The data are given as percents ranging from 5 to 75. DATA FILE ECM

(a) Convert these percents into their decimal forms by dividing by 100. Calculate the transformed means, standard deviations, and standard errors and summarize them, along with the sample sizes, in a table.

(b) Explain how you could have calculated the table entries directly from the table you gave in part (a) of Exercise 12.47.

(c) Analyze the percents using analysis of variance. Compare the test statistic, degrees of freedom, P-value, and conclusion you obtain here with the corresponding values that you found in Exercise 12.47.

12.51. More on changing the response variable. Refer to the previous exercise and Exercise 12.47 (page 590). A calibration error was found with the device that measured Gpi, which resulted in a shifted response. Add 5% to each response and redo the calculations. Summarize the effects of transforming the data by adding a constant to all responses. DATA FILE ECM

12.52. CHALLENGE **Linear transformation of the response variable.** Refer to the previous two exercises. Can you suggest a general conclusion regarding what happens to the test statistic, degrees of freedom, P-value, and conclusion when you perform analysis of variance on data that have been transformed by multiplying the raw data by a constant and then adding another constant? (That is, if y are the original data, we analyze y^*, where $y^* = a + by$ and a and $b \neq 0$ are constants.)

12.53. CHALLENGE **Comparing three levels of reading comprehension instruction.** A study of reading comprehension in children compared three methods of instruction.[19] The three methods of instruction are called Basal, DRTA, and Strategies. As is common in such studies, several pretest variables were measured before any instruction was given. One purpose of the pretest was to see if the three groups of children were similar in their comprehension skills. The READING data set gives two pretest measures that were used in this study. Use one-way ANOVA to analyze these data and write a summary of your results.

12.54. CHALLENGE **More on the reading comprehension study.** In the study described in the previous exercise, Basal is the traditional method of teaching, while DRTA and Strategies are two innovative methods based on similar theoretical considerations. The READING data set includes three response variables that the new methods were designed to improve. Analyze these variables using ANOVA methods. Be sure to include multiple comparisons or contrasts as needed. Write a report summarizing your findings.

12.55. CHALLENGE **More on the Facebook friends study.** Refer to the Facebook friends study that we examined in Example 12.3 (page 554). The explanatory variable in this study is the number of Facebook friends, with possible values of 102, 302, 502, 702, and 902. When using analysis of variance we treat the explanatory variable as categorical. An alternative analysis is to use simple linear regression. Perform this analysis and summarize the results. Plot the residuals from the regression model versus the number of Facebook friends. What do you conclude? DATA FILE FRIENDS

12.56. CHALLENGE **Search the Internet.** Search the Internet or your library to find a study that is interesting to you and that used one-way ANOVA to analyze the data. First describe the question or questions of interest and then give the details of how ANOVA was used to provide answers. Be sure to include how the study authors examined the assumptions for the analysis. Evaluate how well the authors used ANOVA in this study. If your evaluation finds the analysis deficient, make suggestions for how it could be improved.

CHAPTER THIRTEEN

TWO-WAY ANALYSIS OF VARIANCE

In this chapter, we move from one-way ANOVA, which compares the means of several populations, to two-way ANOVA. Two-way ANOVA compares the means of populations that can be classified in two ways or the mean responses in two-factor experiments.

Many of the key concepts are similar to those of one-way ANOVA, but the presence of more than one classification factor also introduces some new ideas. We once more assume that the data are approximately Normal and that groups may have different means but the same standard deviation; we again pool to estimate the variance; and we again use F statistics for significance tests.

The major difference between one-way and two-way ANOVA is in the FIT part of the model. We will carefully study this term, and we will find much that is both new and useful. This will allow us to address comparisons such as the following:

- Can zinc supplementation reduce the occurrence and severity of malaria in both nutrient-sufficient and nutrient-deficient African children?
- What effects do the shapes of the flowers of male and female jack-in-the-pulpit plants have on the degree to which insects eat them?
- Do calcium supplements prevent bone loss in elderly people? Does this depend on whether the person is receiving adequate vitamin D?

13.1 The Two-Way ANOVA Model

We begin with a discussion of the advantages of the two-way ANOVA design and illustrate these with some examples. Then we discuss the model and the assumptions.

Advantages of two-way ANOVA

In one-way ANOVA, we classify populations according to one categorical variable, or factor. In the two-way ANOVA model, there are two factors, each with its own number of levels. When we are interested in the effects of two factors, a two-way design offers great advantages over several single-factor studies. Several examples will illustrate these advantages.

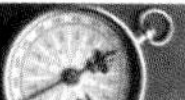

EXAMPLE 13.1

Design 1: Choosing the best box art style and game description. In Example 12.1, a video game company wants to compare three different box art styles. To do this, they plan to randomly assign the three styles equally among 60 electronics stores. The number of games sold during a one-week period is the outcome variable.

Now suppose that a second experiment is planned for the following week to compare four different game descriptions on the back of the box. A similar experimental design will be used, with the four descriptions randomly assigned among the same 60 stores.

Here is an outline of the design of the first experiment:

Style	n
1	20
2	20
3	20
Total	60

And this represents the second experiment:

Description	n
1	15
2	15
3	15
4	15
Total	60

In the first experiment, 20 stores were assigned to each level of the factor, for a total of 60 stores. In the second experiment, 15 stores were assigned to each level of the factor, for a total of 60 stores. The total amount of time for the two experiments is two weeks. Each experiment will be analyzed using one-way ANOVA. The factor in the first experiment is box art style with three levels, and the factor in the second experiment is game description with four levels. Let's now consider combining the two experiments into one.

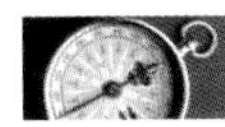

EXAMPLE 13.2

Design 2: Choosing the best box art style and game description. Suppose that we adopt a two-way approach for the video game problem. There are two factors: style and description. Since style has three levels and description has four levels, this is a 3 × 4 design. This gives a total of 12 possible combinations of style and description. With a total of 60 stores, we could assign each combination of style and description to 5 stores. The number of video games sold during a one-week period is the outcome variable.

Here is an outline of the two-way design with the sample sizes:

	Description				
Style	**1**	**2**	**3**	**4**	**Total**
1	5	5	5	5	20
2	5	5	5	5	20
3	5	5	5	5	20
Total	15	15	15	15	60

cell

Each combination of the factors in a two-way design corresponds to a **cell.** The 3 × 4 ANOVA for the video game experiment has 12 cells, each corresponding to a particular combination of box art style and game description.

With the two-way design for style, notice that we have 20 stores assigned to each level, the same as what we had for the one-way design for style alone. Similarly, there are 15 stores assigned to each level of description. Thus, the two-way design gives us the *same amount* of information for estimating the sales for each level of each factor as we had with the two one-way designs. The difference is that we can collect all this information in only one week (and 60 stores). By combining the two factors into one experiment, we have increased our efficiency by reducing the amount of data to be collected by half.

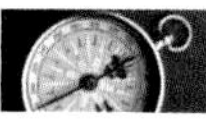

EXAMPLE 13.3

Can dietary supplementation with zinc prevent malaria? Malaria is a serious health problem causing an estimated one million deaths per year, mostly among African children.[1] Several studies, run in Asia, Latin America, and developed countries, have shown zinc supplementation to be an effective control of common infections in children. Can this supplementation program also be effective in Africa, where the primary threat to a child's health is malaria? A group of researchers set out to answer this question.[2]

To design a study to answer this question, the researchers first need to determine an appropriate target group. Since malaria is a serious problem for young children, they will concentrate on children who are 6 months to 5 years of age. A supplement will be prepared that contains either no zinc or 10 milligrams (mg) of zinc. Because the response to zinc may be different in children who lack other important nutrients, they decide to also take this factor into account. Specifically, either their supplement will contain a daily dosage of essential vitamins and minerals or it will not.

EXAMPLE 13.4

Implementing the two-way ANOVA. The factors for the two-way ANOVA are zinc supplementation with two levels and vitamin supplementation with two levels. There are $2 \times 2 = 4$ cells in the study. The researchers plan to enroll 600 children and randomly assign 150 to each of the cells. One outcome variable will be a measure of the child's T cell immune response.

Here is a table that summarizes the design:

	Vitamins		
Zinc	**No**	**Yes**	**Total**
No	150	150	300
Yes	150	150	300
Total	300	300	600

This example illustrates another advantage of two-way designs. Although the researchers are primarily interested in the possible benefit of zinc supplementation, they also included vitamin supplementation in the design because they thought the zinc effect may be different in children who are nutritionally deficient.

Consider an alternative, one-way design where we assign 300 children to the two levels of zinc and ignore nutritional status. With this design, we will have the same number of children at each of the zinc levels, so in this way, it is similar to our two-way design.

However, suppose that there are, in fact, differences due to nutritional status. In this case, the one-way ANOVA would assign this variation to the RESIDUAL (within groups) part of the model. In the two-way ANOVA, vitamin supplementation is included as a factor, and therefore this variation is included in the FIT part of the model. Whenever we can move variation from RESIDUAL to FIT, we reduce the σ of our model and increase the power of our tests.

EXAMPLE 13.5

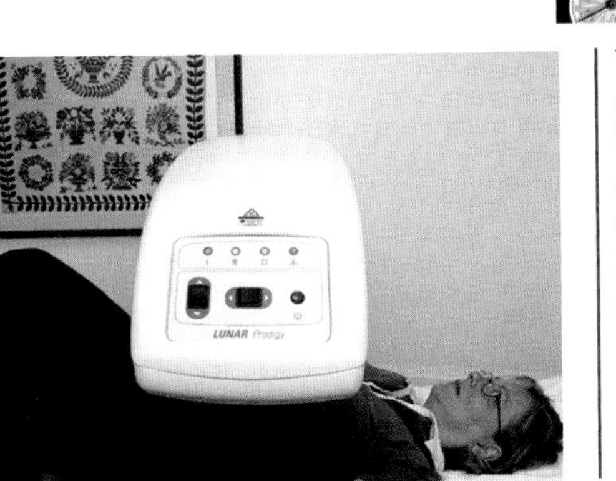

Vitamin D and osteoporosis. Osteoporosis is a disease primarily of the elderly. People with osteoporosis have low bone mass and an increased risk of bone fractures. Over 10 million people in the United States, 1.4 million Canadians, and many millions throughout the world have this disease. Adequate calcium in the diet is necessary for strong bones, but vitamin D is also needed for the body to efficiently use calcium. High doses of calcium in the diet will not prevent osteoporosis unless there is adequate vitamin D. Exposure of the skin to the ultraviolet rays in sunlight enables our bodies to make vitamin D. However, elderly people often avoid sunlight, and in northern areas such as Canada, there is not sufficient ultraviolet light to make vitamin D, particularly in the winter months.

Suppose that we want to see if calcium supplements will increase bone mass (or prevent a decrease in bone mass) in an elderly Canadian population. Because of the vitamin D complication, we will make this a factor in our design.

EXAMPLE 13.6

Designing the osteoporosis study. We will use a 2×2 design for our osteoporosis study. The two factors are calcium and vitamin D. The levels of each factor will be zero (placebo) and an amount that is expected to be adequate, 800 milligrams per day (mg/d) for calcium and 300 international units per day (IU/d) for vitamin D. Women between the ages of 70 and 80 will be recruited as subjects. Bone mineral density (BMD) will be measured at the beginning of the study, and supplements will be taken for one year. The change in BMD over the one-year period is the outcome variable. We expect a dropout rate of 20%, and we would like to have about 20 subjects providing data in each group at the end of the study. We will therefore recruit 100 subjects and randomly assign 25 to each treatment combination.

Here is a table that summarizes the design with the sample sizes at baseline:

	Vitamin D		
Calcium	**Placebo**	**300 IU/d**	**Total**
Placebo	25	25	50
800 mg/d	25	25	50
Total	50	50	100

This example illustrates a third reason for using two-way designs. The effectiveness of the calcium supplement on BMD depends on having adequate vitamin D. We call this an **interaction.** In contrast, the average values for the calcium effect and the vitamin D effect are represented as **main effects.** The two-way model represents FIT as the sum of a main effect for each of the two factors *and* an interaction. One-way designs that vary a single factor and hold other factors fixed cannot discover interactions. We will discuss interactions more fully in a later section.

interaction

main effects

These examples illustrate several reasons why two-way designs are preferable to one-way designs.

ADVANTAGES OF TWO-WAY ANOVA

1. It is more efficient to study two factors simultaneously rather than separately.
2. We can reduce the residual variation in a model by including a second factor thought to influence the response.
3. We can investigate interactions between factors.

These considerations also apply to study designs with more than two factors. We will be content to explore only the two-way case. The choice of sampling design or experimental design is fundamental to any statistical study. *Factors and levels must be carefully selected by an individual or team who understands both the statistical models and the issues that the study will address.*

The two-way ANOVA model

When discussing two-way models in general, we will use the labels A and B for the two factors. For particular examples and when using statistical software, it is better to use meaningful names for these categorical variables. Thus, in Example 13.2 we would say that the factors are *box art style* and *description*, and in Example 13.4 we would say that the factors are *zinc* and *vitamin* supplementation.

The numbers of levels of the factors are often used to describe the model. Again using our earlier examples, we would say that Example 13.2 represents a 3×4 ANOVA and Example 13.4 illustrates a 2×2 ANOVA. In general, Factor A will have I levels and Factor B will have J levels. Therefore, we call the general two-way problem an $I \times J$ ANOVA.

ASSUMPTIONS FOR TWO-WAY ANOVA

We have independent SRSs of size n_{ij} from each of $I \times J$ Normal populations. The population means μ_{ij} may differ, but all populations have the same standard deviation σ. The μ_{ij} and σ are unknown parameters.

Let x_{ijk} represent the kth observation from the population having Factor A at level i and Factor B at level j. The statistical model is

$$x_{ijk} = \mu_{ij} + \epsilon_{ijk}$$

for $i = 1, \ldots, I$ and $j = 1, \ldots, J$ and $k = 1, \ldots, n_{ij}$, where the deviations ϵ_{ijk} are from an $N(0, \sigma)$ distribution.

In a two-way design, every level of A appears in combination with every level of B, so that $I \times J$ groups are compared. The sample size for level i of Factor A and level j of Factor B is n_{ij}.[3] The total number of observations is

$$N = \sum n_{ij}$$

LOOK BACK
one-way model
p. 558

Similar to the one-way model, the FIT part is the group means μ_{ij}, and the RESIDUAL part is the deviations ϵ_{ijk} of the individual observations from their group means. To estimate a group mean μ_{ij} we use the sample mean of the observations in the samples from this group:

$$\overline{x}_{ij} = \frac{1}{n_{ij}} \sum_{k} x_{ijk}$$

The k below the $\sum$ means that we sum the n_{ij} observations that belong to the (i, j)th sample.

The RESIDUAL part of the model contains the unknown σ. We calculate the sample variances for each SRS and pool these to estimate σ^2:

$$s_p^2 = \frac{\sum(n_{ij} - 1)s_{ij}^2}{\sum(n_{ij} - 1)}$$

Just as in one-way ANOVA, the numerator in this fraction is SSE and the denominator is DFE. Also, DFE is the total number of observations minus the number of groups. That is, DFE $= N - IJ$. The estimator of σ is s_p.

Main effects and interactions

In this section we will further explore the FIT part of the two-way ANOVA, which is represented in the model by the population means μ_{ij}. The two-way design gives some structure to the set of means μ_{ij}.

So far, because we have independent samples from each of $I \times J$ groups, we have presented the problem as a one-way ANOVA with IJ groups. Each population mean μ_{ij} is estimated by the corresponding sample mean $\overline{x}_{ij}$, and we can calculate sums of squares and degrees of freedom as in one-way ANOVA. In accordance with the conventions used by many computer software packages, we use the term *model* when discussing the sums of squares and degrees of freedom calculated as in one-way ANOVA with IJ groups. Thus, SSM is a model sum of squares constructed from deviations of the form $\overline{x}_{ij} - \overline{x}$, where $\overline{x}$ is the average of all the observations and $\overline{x}_{ij}$ is the mean of the (i, j)th group. Similarly, DFM is simply $IJ - 1$.

In two-way ANOVA, the terms SSM and DFM can be further broken down into terms corresponding to a main effect for A, a main effect for B, and an AB interaction. Each of SSM and DFM is then a sum of terms:

$$\text{SSM} = \text{SSA} + \text{SSB} + \text{SSAB}$$

and

$$\text{DFM} = \text{DFA} + \text{DFB} + \text{DFAB}$$

The term SSA represents variation among the means for the different levels of Factor A. Because there are I such means, DFA = $I - 1$ degrees of freedom. Similarly, SSB represents variation among the means for the different levels of Factor B, with DFB = $J - 1$.

Interactions are a bit more involved. We can see that SSAB, which is SSM − SSA − SSB, represents the variation in the model that is not accounted for by the main effects. By subtraction we see that its degrees of freedom are

$$\text{DFAB} = (IJ - 1) - (I - 1) - (J - 1) = (I - 1)(J - 1)$$

There are many kinds of interactions. The easiest way to study them is through examples.

EXAMPLE 13.7

Investigating differences in sugar-sweetened beverage consumption. Consumption of sugar-sweetened beverages (SSBs) has been linked to type 2 diabetes and obesity. One study used data from the National Health and Nutrition Examination Survey (NHANES) to estimate SSB consumption among adults. More than 13,000 individuals provided data for this study. Adults were divided into three age categories: 20 to 44, 45 to 64, and 65 years and older.[4] Here are the means for the number of calories in SSBs consumed per day during 1988 to 1994 and 1999 to 2004:

	Year		
Age (years)	**1994**	**2004**	**Mean**
20–44	231	289	260
45–64	124	160	142
$\geq$ 65	68	83	76
Mean	141	177	159

The table includes averages of the means in the rows and columns. For example, in 1988 to 1994 the mean of calories in SSBs consumed among adults is

$$\frac{231 + 124 + 68}{3} = 141$$

Similarly, the corresponding value for 1999 to 2004 is

$$\frac{289 + 160 + 83}{3} = 177.3$$

marginal means

which is rounded to 177 in the table. These averages are called **marginal means** (because of their location at the *margins* of such tabulations). The grand mean can be obtained by averaging either set of marginal means.

Figure 13.1 is a plot of the group means. From the plot we see that the calories in SSBs consumed by each group in 1994 are less than those consumed in 2004. In statistical language, there is a main effect for year. We also see that

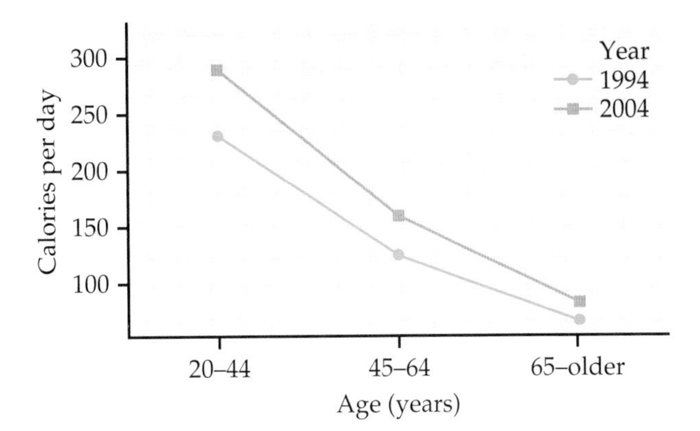

FIGURE 13.1 Plot of the mean calories of sugar-sweetened beverages consumed per day in 1988 to 1994 and 1999 to 2004, for Example 13.7.

the means are different across age categories. This means that there is a main effect for age. These main effects can be described by differences between the marginal means. For example, the mean for 1988 to 1994 is 141 calories and then increases 36 calories to 177 calories in 1999 to 2004. Similarly, the mean for adults 20 to 44 years old is 260; it drops by 118 calories to 142 for adults 45 to 64 years old, and then drops an additional 66 calories to 76 for adults older than 64.

To examine two-way ANOVA data for a possible interaction, always construct a plot similar to Figure 13.1. In this case, it is debatable whether the two profiles should be considered parallel. Profiles that are roughly parallel indicate that there is *no* clear interaction between the two factors. When no interaction is present, the marginal means (and main effects) provide a reasonable description of the two-way table of means.

When there is an interaction, the marginal means do not tell the whole story. For example, with these data, the marginal mean difference between years (main effect for year) is 36 calories. This equals the difference in calories for the 45 to 64 age class, so it adequately describes the change for this age group. However, the mean difference between years for the 20 to 44 age group is 58 calories, so this main effect underestimates the increase in calories by 22 calories per day. Likewise, this main effect overestimates the difference for the oldest age class by 21 calories per day. If these differences of roughly 20 calories per day are scientifically meaningful, then we would say that there is evidence for an interaction. In other words, the main effect for year does not adequately describe the change for *all* age groups.

Interactions come in many shapes and forms. When we find an interaction, a careful examination of the means is needed to properly interpret the data. Simply stating that interactions are significant tells us very little. Plots of the group means are essential. Here is another example.

EXAMPLE 13.8

Eating in groups. Some research has shown that people eat more when they eat in groups. One possible mechanism for this phenomenon is that they may spend more time eating when in a larger group. A study designed to examine this idea measured the length of time spent (in minutes) eating lunch in different settings.[5] Here are some data from this study:

Lunch setting	Number of People Eating 1	2	3	4	5 or more	Mean
Workplace	12.6	23.0	33.0	41.1	44.0	30.7
Fast-food restaurant	10.7	18.2	18.4	19.7	21.9	17.8
Mean	11.6	20.6	25.7	30.4	32.9	24.2

Figure 13.2 gives the plot of the means for this example. The patterns are not parallel, so it appears that we have an interaction. Meals take longer when there are more people present, but this phenomenon is much greater for the meals consumed at work. For fast-food eating, the meal durations are fairly similar when there is more than one person present.

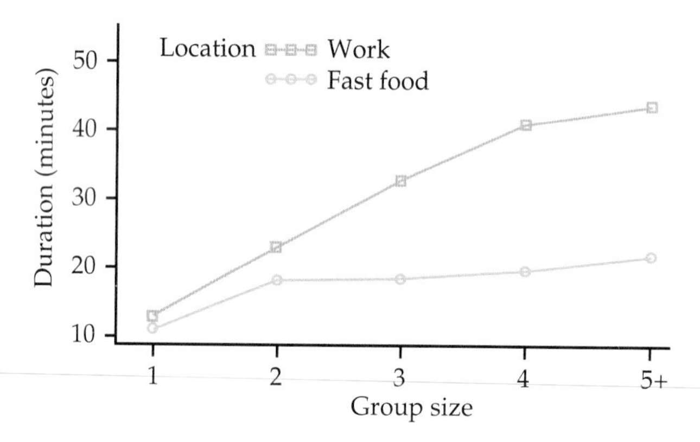

FIGURE 13.2 Plot of mean meal duration versus lunch setting and group size, for Example 13.8.

A different kind of interaction is present in the next example. Here, we must be very cautious in our interpretation of the main effects since one of them can lead to a distorted conclusion.

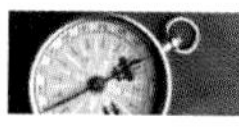

EXAMPLE 13.9

We got the beat? When we hear music that is familiar to us, we can quickly pick up the beat and our mind synchronizes with the music. However, if the music is unfamiliar, it takes us longer to synchronize. In a study that investigated the theoretical framework for this phenomenon, French and Tunisian nationals listened to French and Tunisian music.[6] Each subject was asked to tap in time with the music being played. A synchronization score, recorded in milliseconds, measured how well the subjects synchronized with the music. A higher score indicates better synchronization. Six songs of each music type were used. Here are the means:

Nationality	Music French	Tunisian	Mean
French	950	750	850
Tunisian	760	1090	925
Mean	855	920	887

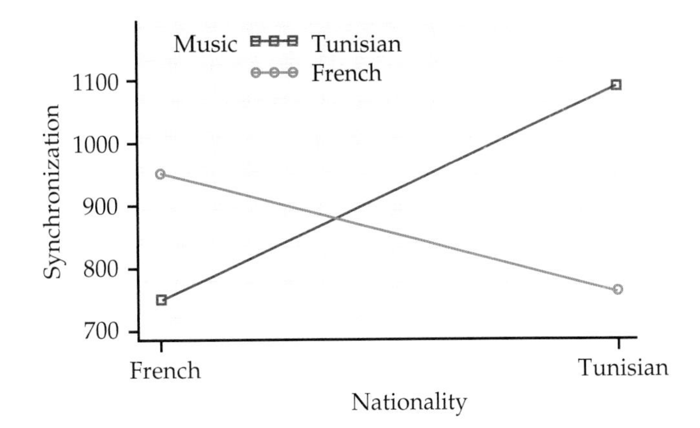

FIGURE 13.3 Plot of mean synchronization score versus type of music for French and Tunisian nationals, for Example 13.9.

The means are plotted in Figure 13.3. In the study the researchers were not interested in main effects. Their theory predicted the interaction that we see in the figure. Subjects synchronize better with music from their own culture.

The interaction in Figure 13.3 is very different from those that we saw in Figures 13.1 and 13.2. These examples illustrate the point that it is necessary to plot the means and carefully describe the patterns when interpreting an interaction.

The design of the study in Example 13.9 allows us to examine two main effects and an interaction. However, this setting does not meet all the assumptions needed for statistical inference using the two-way ANOVA framework of this chapter. *As with one-way ANOVA, we require that observations be independent.*

In this study, we have a design that has each subject contributing data for two types of music, so these two scores will be dependent. The framework is similar to the matched pairs setting. The design is called a **repeated-measures design.** More advanced texts on statistical methods cover this important design.

repeated-measures design

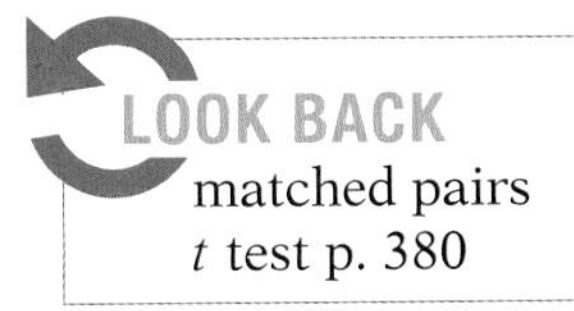

matched pairs t test p. 380

USE YOUR KNOWLEDGE

13.1 What's wrong? For each of the following, explain what is wrong and why.

(a) A two-way ANOVA is used when the outcome variable can take only two possible values.

(b) In a 2 × 3 ANOVA each level of Factor A appears with two levels of Factor B.

(c) The FIT part of the model in a two-way ANOVA represents the variation that is sometimes called error or residual.

(d) In an $I \times J$ ANOVA, DFAB $= IJ - 1$.

13.2 What's wrong? For each of the following, explain what is wrong and why.

(a) Parallel profiles of cell means imply that a strong interaction is present.

(b) You can perform a two-way ANOVA only when the sample sizes are the same in all cells.

(c) The estimate s_p^2 is obtained by pooling the marginal sample variances.

(d) When interaction is present, the marginal means are always uninformative.

13.2 Inference for Two-Way ANOVA

Inference for two-way ANOVA involves F statistics for each of the two main effects and an additional F statistic for the interaction. As with one-way ANOVA, the calculations are organized in an ANOVA table.

The ANOVA table for two-way ANOVA

Two-way ANOVA is the statistical analysis for a two-way design with a quantitative response variable. The results of a two-way ANOVA are summarized in an ANOVA table based on splitting the total variation SST and the total degrees of freedom DFT among the two main effects and the interaction. Both the sums of squares (which measure variation) and the degrees of freedom add:

$$\text{SST} = \text{SSA} + \text{SSB} + \text{SSAB} + \text{SSE}$$

$$\text{DFT} = \text{DFA} + \text{DFB} + \text{DFAB} + \text{DFE}$$

The sums of squares are always calculated in practice by statistical software. *When the n_{ij} are not all equal, some methods of analysis can give sums of squares that do not add.*

From each sum of squares and its degrees of freedom we find the mean square in the usual way:

$$\text{mean square} = \frac{\text{sum of squares}}{\text{degrees of freedom}}$$

The significance of each of the main effects and the interaction is assessed by an F statistic that compares the variation due to the effect of interest with the within-group variation. Each F statistic is the mean square for the source of interest divided by MSE. Here is the general form of the two-way ANOVA table:

Source	Degrees of freedom	Sum of squares	Mean square	F
A	$I-1$	SSA	SSA/DFA	MSA/MSE
B	$J-1$	SSB	SSB/DFB	MSB/MSE
AB	$(I-1)(J-1)$	SSAB	SSAB/DFAB	MSAB/MSE
Error	$N-IJ$	SSE	SSE/DFE	
Total	$N-1$	SST		

There are three null hypotheses in two-way ANOVA, with an F test for each. If the effect being tested is zero, the calculated F statistic has an F distribution with numerator degrees of freedom corresponding to the effect and denominator degrees of freedom equal to DFE. Large values of the F statistic lead to rejection of the null hypothesis.

We can test for significance of the main effect of A, the main effect of B, and the AB interaction. *It is generally good practice to examine the test for interaction first, since the presence of a strong interaction may influence the interpretation of the main effects.* Be sure to plot the means as an aid to interpreting the results of the significance tests.

SIGNIFICANCE TESTS IN TWO-WAY ANOVA

To test the main effect of A, use the F statistic

$$F_{\mathrm{A}} = \frac{\mathrm{MSA}}{\mathrm{MSE}}$$

To test the main effect of B, use the F statistic

$$F_{\mathrm{B}} = \frac{\mathrm{MSB}}{\mathrm{MSE}}$$

To test the interaction of A and B, use the F statistic

$$F_{\mathrm{AB}} = \frac{\mathrm{MSAB}}{\mathrm{MSE}}$$

The P-value is the probability that a random variable having an F distribution with numerator degrees of freedom corresponding to the effect and denominator degrees of freedom equal to DFE is greater than or equal to the calculated F statistic.

The following example illustrates how to do a two-way ANOVA. As with the one-way ANOVA, we focus our attention on interpretation of the computer output.

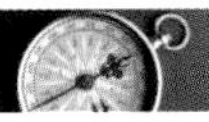

EXAMPLE 13.10

A study of cardiovascular risk factors. A study of cardiovascular risk factors compared runners who averaged at least 15 miles per week with a control group described as "generally sedentary." Both men and women were included in the study.[7] The design is a 2×2 ANOVA with the factors group and gender. There were 200 subjects in each of the four combinations. One of the variables measured was the heart rate after six minutes of exercise on a treadmill. SAS computer analysis produced the outputs in Figure 13.4 and Figure 13.5.

We begin with the usual preliminary examination. From Figure 13.4 we see that the ratio of the largest to the smallest standard deviation is less than 2. Therefore, we are not concerned about violating the assumption of equal population standard deviations. Normal quantile plots (not shown) do not reveal any outliers, and the data appear to be reasonably Normal.

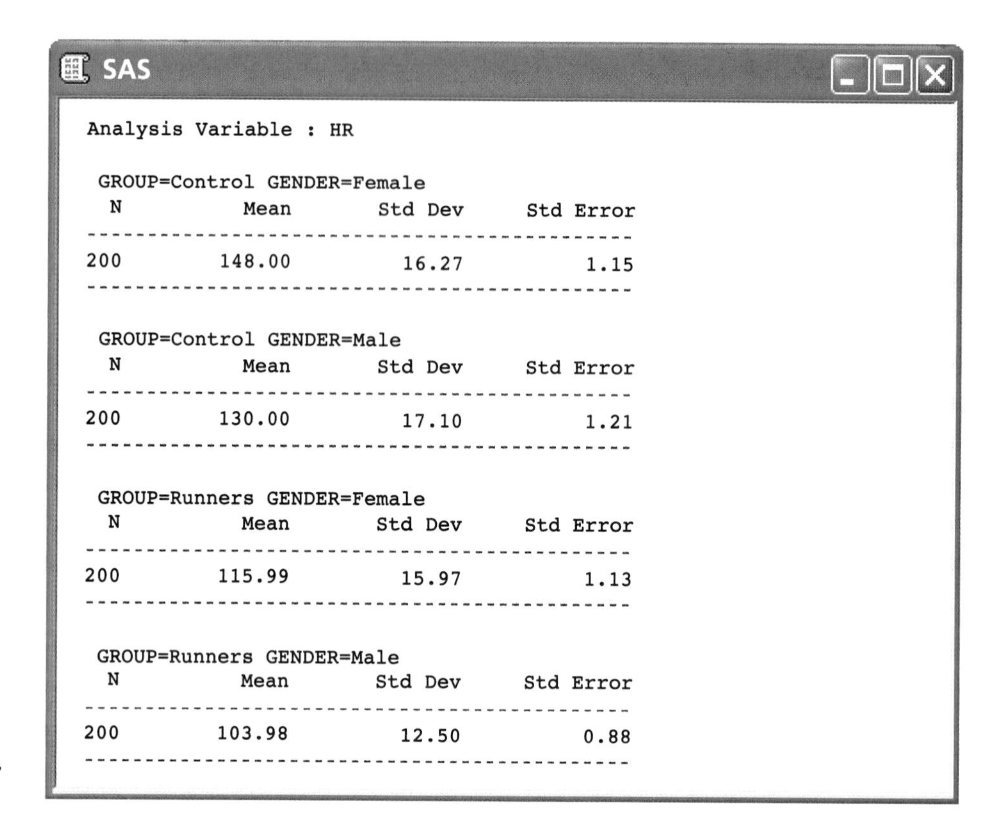

```
SAS

Analysis Variable : HR

 GROUP=Control GENDER=Female
  N          Mean        Std Dev      Std Error
---------------------------------------------
200         148.00         16.27           1.15
---------------------------------------------

 GROUP=Control GENDER=Male
  N          Mean        Std Dev      Std Error
---------------------------------------------
200         130.00         17.10           1.21
---------------------------------------------

 GROUP=Runners GENDER=Female
  N          Mean        Std Dev      Std Error
---------------------------------------------
200         115.99         15.97           1.13
---------------------------------------------

 GROUP=Runners GENDER=Male
  N          Mean        Std Dev      Std Error
---------------------------------------------
200         103.98         12.50           0.88
---------------------------------------------
```

FIGURE 13.4 Summary statistics for heart rates in the four groups of a 2 × 2 ANOVA, for Example 13.10.

```
SAS

General Linear Models Procedure
Dependent Variable: HR

                               Sum of           Mean
Source               DF       Squares         Square   F Value      Pr > F
Model                 3     215256.09       71752.03    296.35      0.0001
Error               796     192729.83         242.12
Corrected Total     799     407985.92

              R-square          C.V.       Root MSE                HR Mean
              0.527607      12.49924         15.560                 124.49

Source               DF      Type I SS   Mean Square   F Value      Pr > F
GROUP                 1      168432.08     168432.08    695.65      0.0001
GENDER                1       45030.00      45030.00    185.98      0.0001
GROUP*GENDER          1        1794.01       1794.01      7.41      0.0066
```

FIGURE 13.5 Two-way ANOVA output for heart rates, for Example 13.10.

The ANOVA table at the top of the output in Figure 13.5 is in effect a one-way ANOVA with four groups: female control, female runner, male control, and male runner. In this analysis, Model has 3 degrees of freedom, and Error has 796 degrees of freedom. The F test ($F = 296.35$) and its associated P-value refer to the hypothesis that all four groups have the same population mean. We are interested in the main effects and interaction, so we ignore this test.

The sums of squares for the group and gender main effects and the group-by-gender interaction appear at the bottom of Figure 13.5 under the heading Type I SS. These sum to the sum of squares for Model. Similarly, the degrees of freedom for these sums of squares sum to the degrees of freedom for Model. Two-way ANOVA splits the variation among the means (expressed by the Model sum of squares) into three parts that reflect the two-way layout.

Because the degrees of freedom are all 1 for the main effects and the interaction, the mean squares are the same as the sums of squares. The F statistics for the three effects appear in the column labeled F Value, and the P-values are under the heading Pr > F. For the group main effect, we verify the calculation of F as follows:

$$F = \frac{\text{MSG}}{\text{MSE}} = \frac{168{,}432}{242.12} = 695.65$$

All three effects are statistically significant. The group effect has the largest F, followed by the gender effect and then the group-by-gender interaction. To interpret these results, we examine the plot of means with bars indicating one standard error in Figure 13.6. Note that the standard errors are quite small due to the large sample sizes. The significance of the main effect for group is due to the fact that the controls have higher average heart rates than the runners for both genders. This is the largest effect evident in the plot.

The significance of the main effect for gender is due to the fact that the females have higher heart rates than the men in both groups. The differences are not as large as those for the group effect, and this is reflected in the smaller value of the F statistic.

The analysis indicates that a complete description of the average heart rates requires consideration of the interaction in addition to the main effects. The two lines in the plot are not quite parallel. This interaction can be described in two ways. The female-male difference in average heart rates is greater for the controls than for the runners. Alternatively, the difference in average heart rates between controls and runners is greater for women than for men. As the plot suggests, the interaction is not large. It is statistically significant because there were 800 subjects in the study.

Two-way ANOVA output for other software is similar to that given by SAS. Figure 13.7 gives the analysis of the heart rate data using Excel and Minitab.

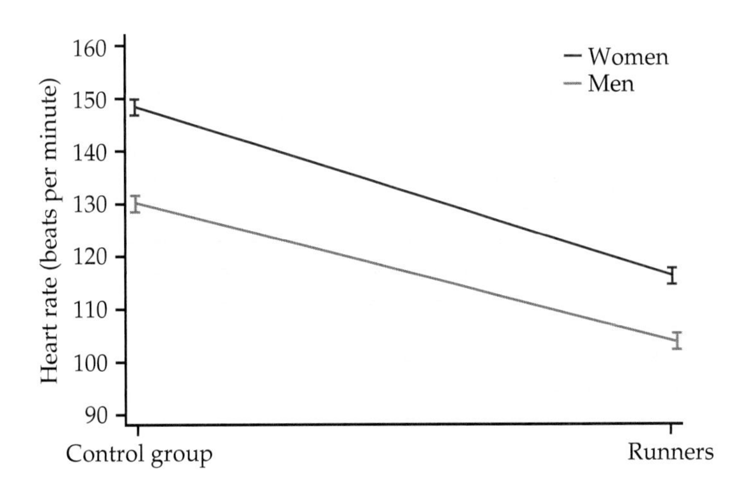

FIGURE 13.6 Plot of the group means with standard errors for heart rates in the 2×2 ANOVA, for Example 13.10.

Excel

	F	G	H	I	J	K	L	M
1	Anova Two Factor With Replication							
2								
3	SUMMARY	Control	Runners	Total				
4	*Female*							
5	Count	200	200	400				
6	Sum	29600	23197	52797				
7	Average	148	115.985	131.9925				
8	Variance	264.7437	255.0902	516.1478				
9								
10	*Male*							
11	Count	200	200	400				
12	Sum	26000	20795	46795				
13	Average	130	103.975	116.9875				
14	Variance	292.4221	156.2356	393.5161				
15								
16	*Total*							
17	Count	400	400					
18	Sum	55600	43992					
19	Average	139	109.98					
20	Variance	359.0877	241.2978					
21								
22								
23	ANOVA							
24	*Source of Variation*	*SS*	*df*	*MS*	*F*	*P-value*	*F crit*	
25	Sample	45030.01	1	45030.01	185.9799	3.29E-38	3.85316	
26	Columns	168432.1	1	168432.1	695.647	1.10E-110	3.85316	
27	Interaction	1794.005	1	1794.005	7.409481	0.00663	3.85316	
28	Within	192729.8	796	242.1229				
29								
30	Total	407985.9	799					

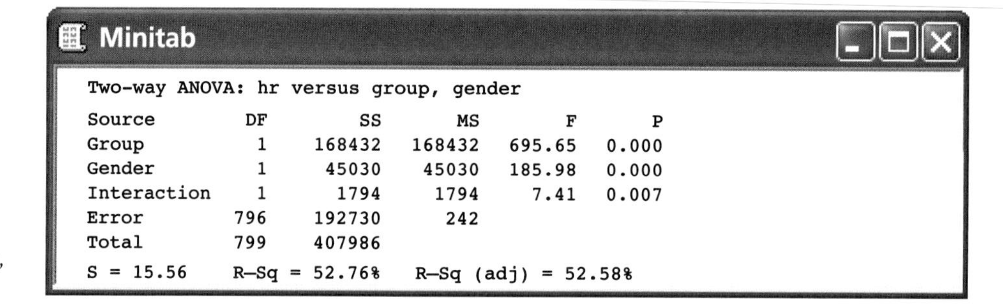
Minitab

Two-way ANOVA: hr versus group, gender

Source	DF	SS	MS	F	P
Group	1	168432	168432	695.65	0.000
Gender	1	45030	45030	185.98	0.000
Interaction	1	1794	1794	7.41	0.007
Error	796	192730	242		
Total	799	407986			

S = 15.56 R-Sq = 52.76% R-Sq (adj) = 52.58%

FIGURE 13.7 Excel and Minitab two-way ANOVA output for the heart rate study, for Example 13.10.

CHAPTER 13 SUMMARY

Two-way analysis of variance is used to compare population means when populations are classified according to two factors.

ANOVA assumes that the populations are Normal with possibly different means and the same standard deviation and that independent SRSs are drawn from each population.

As with one-way ANOVA, preliminary analysis includes examination of means, standard deviations, and Normal quantile plots. **Marginal means** are calculated by taking averages of the cell means across rows and columns. Pooling is used to estimate the within-group variance.

ANOVA separates the total variation into parts for the **model** and **error.** The model variation is separated into parts for each of the **main effects** and the **interaction.**

The calculations are organized into an **ANOVA table.** F statistics and P-values are used to test hypotheses about the main effects and the interaction.

Careful inspection of the means is necessary to interpret significant main effects and interactions. Plots are a useful aid.

CHAPTER 13 EXERCISES

For Exercises 13.1 and 13.2, see pages 603–604.

13.3. What's wrong? For each of the following, explain what is wrong and why.

(a) You should reject the null hypothesis that there is no interaction in a two-way ANOVA when the AB F statistic is small.

(b) Sums of squares are equal to mean squares divided by degrees of freedom.

(c) The test statistics for the main effects in a two-way ANOVA have a chi-square distribution when the null hypothesis is true.

(d) The sums of squares always add in two-way ANOVA.

13.4. Is there an interaction? Each of the following tables gives means for a two-way ANOVA. Make a plot of the means with the levels of Factor A on the x axis. State whether or not there is an interaction, and if there is, describe it.

(a)

	Factor A		
Factor B	1	2	3
1	11	16	21
2	6	16	26

(b)

	Factor A		
Factor B	1	2	3
1	10	15	20
2	40	45	50

(c)

	Factor A		
Factor B	1	2	3
1	10	15	20
2	50	45	40

(d)

	Factor A		
Factor B	1	2	3
1	10	15	20
2	50	55	52

13.5. Describing a two-way ANOVA model. A 3×2 ANOVA was run with 5 observations per cell.

(a) Give the degrees of freedom for the F statistic that is used to test for interaction in this analysis and the entries from Table E that correspond to this distribution.

(b) Sketch a picture of this distribution with the information from the table included.

(c) The calculated value of this F statistic is 3.72. How would you report the P-value?

(d) Would you expect a plot similar to Figure 13.1 (page 601) to have mean profiles that look parallel? Explain your answer.

13.6. Determining the critical value of F. For each of the following situations, state how large the F statistic needs to be for rejection of the null hypothesis at the 5% level. Sketch each distribution and indicate the region where you would reject.

(a) The main effect for the first factor in a 2×4 ANOVA with 3 observations per cell

(b) The interaction in a 4×4 ANOVA with 2 observations per cell

(c) The interaction in a 2×2 ANOVA with 51 observations per cell

13.7. Identifying the factors of a two-way ANOVA model. For each of the following situations, identify both factors and the response variable. Also, state the number of levels for each factor (I and J) and the total number of observations (N).

(a) A child psychologist is interested in studying how a child's percent of pretend play differs with gender and age (4, 8, and 12 months). There are 11 infants assigned to each cell of the experiment.

(b) Brewers malt is produced from germinating barley. A homebrewer wants to determine the best conditions to germinate the barley. A total of 30 lots of barley seed were equally and randomly assigned to 10 germination conditions. The conditions are combinations of the week after harvest (1, 3, 6, 9, or 12 weeks) and the amount of water used in the process (4 or 8 milliliters). The percent of seeds germinating is the outcome variable.

(c) The strength of concrete depends upon the formula used to prepare it. An experiment compares six different mixtures. Nine specimens of concrete are poured from each mixture. Three of these specimens are subjected to 0 cycles of freezing and thawing, three are subjected to 100 cycles, and three are subjected to 500 cycles. The strength of each specimen is then measured.

(d) A company wants to compare four different training programs for its new employees. Each of these programs takes 6 hours to complete. The training can be given for 6 hours on one day or for 3 hours on two consecutive days. The next 80 employees hired by the company will be the subjects for this study.

13.8. Determining the degrees of freedom. For each part in Exercise 13.7, outline the ANOVA table, giving the sources of variation and the degrees of freedom.

13.9. The influences of transaction history and a thank-you statement. A service failure is defined as any service-related problem (real or perceived) that transpires during a customer's experience with a firm. In the hotel industry, there is a high human component, so these sorts of failures commonly occur regardless of extensive training and established policies. As a result, hotel firms must learn to effectively react to these failures. One study investigated the relationship between a consumer's transaction history (levels: long and short) and an employee thank-you statement (levels: yes and no) on a consumer's repurchase intent.[8] Each subject was randomly assigned to one of the four treatment groups and asked to read some service failure/resolution scenarios and respond accordingly. Repurchase intent was measured using a nine-point scale. Here is a summary of the means:

	Thank-you	
History	**No**	**Yes**
Short	5.69	6.80
Long	7.53	7.37

(a) Plot the means. Do you think there is an interaction? If yes, describe the interaction in terms of the two factors.

(b) Find the marginal means. Are they useful for understanding the results of this study? Explain your answer.

13.10. Transaction history and a thank-you statement, continued. Refer to the previous exercise. The number of subjects in each cell was not equal, so the researchers used linear regression to analyze the data. This was done by creating an indicator variable for each factor and the interaction. Below is a partial ANOVA table. Complete it and state your conclusions regarding the main effects and interaction described in the previous exercise.

Source	DF	SS	MS	F	P-value
Transaction history		61.445			
Thank-you statement		21.810			
Interaction		15.404			
Error	160	759.904			

13.11. The effects of proximity and visibility on food intake. A study investigated the influence that proximity and visibility of food have on food intake.[9] A total of 40 secretaries from the University of Illinois participated in the study. A candy dish full of individually wrapped chocolates was placed either at the desk of the participant or at a location two meters from the participant. The candy dish was either a clear (candy visible) or an opaque (candy not visible) covered bowl. After a week, the researchers noted not only the number of candies consumed per day but also the self-reported number of candies consumed by each participant. The following table summarizes the mean difference between these two values (reported minus actual):

	Visibility	
Proximity	**Clear**	**Opaque**
Proximate	−1.2	−0.8
Less proximate	0.5	0.4

(a) Make a plot of the means and describe the patterns that you see. Does the plot suggest an interaction between visibility and proximity?

(b) This study actually took four weeks, with each participant being observed at each treatment combination in a random order. Explain why a "repeated-measures" design like this may be beneficial.

13.12. What can you conclude? A study reported the following results for data analyzed using the methods that we studied in this chapter:

Effect	F	P-value
A	1.50	0.236
B	13.66	0.001
AB	6.14	0.011

(a) What can you conclude from the information given?

(b) What additional information would you need to write a summary of the results for this study?

13.13. The effect of humor. In advertising, humor is often used to overcome sales resistance and stimulate customer purchase behavior. A recent experiment looked at the use of humor as an approach to offset negative feelings often associated with Web site encounters.[10] The setting of the experiment was an online travel agency, and they used a three-factor design, each factor with two levels. The factors were humor (used, not used), process (favorable, unfavorable), and outcome (favorable, unfavorable). For the humor condition, cartoons and jokes of the day about skiing were used on the site. For the no-humor condition, standard pictures of ski sites were used. Two hundred and forty-one business students from a large Dutch university participated in the experiment. Each was randomly assigned to one of the eight treatment conditions. The students were asked to book a skiing holiday and then rate their perceived enjoyment and satisfaction with the process. All responses were measured on a seven-point scale. A summary of the results for satisfaction follows:

Treatment	n	$\bar{x}$	s
No humor - favorable process - unfavorable outcome	27	3.04	0.79
No humor - favorable process - favorable outcome	29	5.36	0.47
No humor - unfavorable process - unfavorable outcome	26	2.84	0.59
No humor - unfavorable process - favorable outcome	31	3.08	0.59
Humor - favorable process - unfavorable outcome	32	5.06	0.59
Humor - favorable process - favorable outcome	30	5.55	0.65
Humor - unfavorable process - unfavorable outcome	36	1.95	0.52
Humor - unfavorable process - favorable outcome	30	3.27	0.71

(a) Plot the means of the four no-humor treatments. Do you think there is an interaction? If yes, describe the interaction in terms of the process and outcome factors.

(b) Plot the means of the four treatments that used humor. Do you think there is an interaction? If yes, describe the interaction in terms of the process and outcome factors.

(c) The three-factor interaction can be assessed by looking at the two interaction plots created in parts (a) and (b). If the relationship between process and outcome is different across the two humor conditions, there is evidence of an interaction among all three factors. Do you think there is a three-factor interaction? Explain your answer.

13.14. Pooling the standard deviations. Refer to the previous exercise. Find the pooled estimate of the standard deviation for these data. What are its degrees of freedom? Using the rule from Chapter 12 (page 560), is it reasonable to use a pooled standard deviation for the analysis? Explain your answer.

13.15. Describing the effects. Refer to Exercise 13.13. The P-values for all main effect and two-factor interactions are significant at the 0.05 level. Using the table, find the marginal means (that is, the mean for the no-humor treatment, the mean for the no-humor and unfavorable process treatment combination, etc.) and use them to describe these effects.

13.16. Acceptance of functional foods. Functional foods are foods that are fortified with health-promoting supplements, like calcium-enriched orange juice or vitamin-enriched cereal. Although the number of functional foods is growing in the marketplace, very little is known about how the next generation of consumers views these foods. Because of this, a questionnaire was given to college students from the United States, Canada, and France.[11] This questionnaire measured both the students' general food and functional food attitudes and beliefs. One of the response variables collected was attitude toward cooking enjoyment. This variable was the average of numerous items, each measured on a 10-point scale, where 1 = most negative value and 10 = most positive value. Here are the means:

	Culture		
Gender	Canada	United States	France
Female	7.70	7.36	6.38
Male	6.39	6.43	5.69

(a) Make a plot of the means and describe the patterns that you see.

(b) Does the plot suggest that there is an interaction between culture and gender? If yes, describe the interaction.

13.17. Estimating the within-group variance. Refer to the previous exercise. Here are the cell standard deviations and sample sizes for cooking enjoyment:

	Culture					
	Canada		United States		France	
Gender	s	n	s	n	s	n
Female	1.668	238	1.736	178	2.024	82
Male	1.909	125	1.601	101	1.875	87

Find the pooled estimate of the standard deviation for these data. Use the rule for examining standard deviations in ANOVA from Chapter 12 (page 560) to determine if it is reasonable to use a pooled standard deviation for the analysis of these data.

13.18. CHALLENGE **Comparing the groups.** Refer to Exercises 13.16 and 13.17. The researchers presented a table of means with different superscripts indicating pairs of means that differed at the 0.05 significance level, using the Bonferroni method.

(a) What denominator degrees of freedom would be used here?

(b) How many pairwise comparisons are there for this problem?

(c) Perform these comparisons using $t^{**} = 2.94$ and summarize your results.

13.19. More on acceptance of functional foods. Refer to Exercise 13.16. The means for four of the response variables associated with functional foods are as follows:

	General Attitude			Product Benefits		
	Culture			Culture		
Gender	**Canada**	**U.S.**	**France**	**Canada**	**U.S.**	**France**
Female	4.93	4.69	4.10	4.59	4.37	3.91
Male	4.50	4.43	4.02	4.20	4.09	3.87

	Credibility of Info.			Purchase Intention		
	Culture			Culture		
Gender	**Canada**	**U.S.**	**France**	**Canada**	**U.S.**	**France**
Female	4.54	4.50	3.76	4.29	4.39	3.30
Male	4.23	3.99	3.83	4.11	3.86	3.41

For each of the four response variables, give a graphical summary of the means. Use this summary to discuss any interactions that are evident. Write a short report summarizing any differences in culture and gender on the response variables measured.

13.20. Interpreting the results. The goal of the study in the previous exercise was to understand cultural and gender differences in functional food attitudes and behaviors among young adults, the next generation of food consumers. The researchers used a sample of undergraduate students and had each participant fill out the survey during class time. How reasonable is it to generalize these results to the young-adult population in these countries? Explain your answer.

13.21. Evaluation of an intervention program. The National Crime Victimization Survey estimates that there were over 400,000 violent crimes committed against women by their intimate partner that resulted in physical injury. An intervention study designed to increase safety behaviors of abused women compared the effectiveness of six telephone intervention sessions with a control group of abused women who received standard care. Fifteen different safety behaviors were examined.[12] One of the variables analyzed was the total number of behaviors (out of 15) that each woman performed. Here is a summary of the means of this variable at baseline (just before the first telephone call) and at follow-up three and six months later:

	Time		
Group	**Baseline**	**3 months**	**6 months**
Intervention	10.4	12.5	11.9
Control	9.6	9.9	10.4

(a) Find the marginal means. Are they useful for understanding the results of this study?

(b) Plot the means. Do you think there is an interaction? Describe the meaning of an interaction for this study. (*Note:* Like Exercise 13.11 [page 610], this exercise is from a repeated-measures design. Also, the data are not particularly Normal, because they are counts with values from 1 to 15. Although we cannot use the methods in this chapter for statistical inference in this setting, the example does illustrate ideas about interactions.)

13.22. CHALLENGE **More on the assessment of an intervention program.** Refer to the previous exercise. Table 13.1 gives the percents of women who responded that they performed each of the 15 safety behaviors studied.

(a) Summarize these data graphically. Do you think that your graphical display is more effective than Table 13.1 for describing the results of this study? Explain why or why not.

(b) Note any particular patterns in the data that would be important to someone who would use these results to design future intervention programs for abused women.

(c) The study was conducted "at a family violence unit of a large urban District Attorney's Office that serves an ethnically diverse population of three million citizens." To what extent do you think that this fact limits the conclusions that can be drawn?

13.23. Conspicuous consumption and men's testosterone levels. It is argued that conspicuous

TABLE 13.1

Safety behaviors of abused women

Behavior	Intervention Group (%) Baseline	3 months	6 months	Control Group (%) Baseline	3 months	6 months
Hide money	68.0	60.0	62.7	60.0	37.8	35.1
Hide extra keys	52.7	76.0	68.9	53.3	33.8	39.2
Abuse code to alert family	30.7	74.7	60.0	22.7	27.0	43.2
Hide extra clothing	37.3	73.6	52.7	42.7	32.9	27.0
Asked neighbors to call police	49.3	73.0	66.2	32.0	45.9	40.5
Know Social Security number	93.2	93.2	100.0	89.3	93.2	98.6
Keep rent, utility receipts	75.3	95.5	89.4	70.3	84.7	80.9
Keep birth certificates	84.0	90.7	93.3	77.3	90.4	93.2
Keep driver's license	93.3	93.3	97.3	94.7	95.9	98.6
Keep telephone numbers	96.0	98.7	100.0	90.7	97.3	100.0
Removed weapons	50.0	70.6	38.5	40.7	23.8	5.9
Keep bank account numbers	81.0	94.3	96.2	76.2	85.5	94.4
Keep insurance policy number	70.9	90.4	89.7	68.3	84.2	94.8
Keep marriage license	71.1	92.3	84.6	63.3	73.2	80.0
Hide valuable jewelry	78.7	84.5	83.9	74.0	75.0	80.3

consumption is a means by which men communicate their social status to prospective mates. One study looked at changes in a male's testosterone level in response to fluctuations in his status created by the consumption of a product.[13] The products considered were a new and luxurious sports car and an old family sedan. Participants were asked to drive either on an isolated highway or a busy urban street. A table of cell means and standard deviations for the change (posttreatment level minus pretreatment level) in testosterone follows:

Car	Location: Highway $\bar{x}$	Highway s	City $\bar{x}$	City s
Old sedan	0.03	0.12	−0.03	0.12
New sports car	0.15	0.14	0.13	0.13

(a) Make a plot of the means and describe the patterns that you see. Does the plot suggest an interaction between location and type of car?

(b) Compute the pooled standard error s_p, assuming equal sample sizes.

(c) The researchers wanted to test the following hypotheses:

1. Testosterone levels will increase more in men who drive the new car.
2. For men driving the new car, testosterone levels will increase more in men who drive in the city.
3. For men driving the old car, testosterone levels will decrease less in men who drive the old car on the highway.

Write out the contrasts for each of these hypotheses.

(d) This study actually involved each male participating in all four combinations. Half of them drove the sedan first and the other half drove the sports car first. Explain why a "repeated-measures" design like this may be beneficial.

13.24. CHALLENGE **The effects of peer pressure on mathematics achievement.** Researchers were interested in comparing the relationship between high achievement in mathematics and peer pressure across several countries.[14] They hypothesized that in countries where high achievement is not valued highly, considerable peer pressure may exist. A questionnaire was distributed to 14-year-olds from three countries (Germany, Canada, and Israel). One of the questions asked students to rate how often they fear being called a nerd or teacher's pet on a four-point scale (1 = never, 4 = frequently). The following table summarizes the response:

Country	Gender	n	$\bar{x}$
Germany	Female	336	1.62
Germany	Male	305	1.39
Israel	Female	205	1.87
Israel	Male	214	1.63
Canada	Female	301	1.91
Canada	Male	304	1.88

(a) The P-values for the interaction and the main effects of country and gender are 0.016, 0.068, and 0.108, respectively. Using the table and P-values, summarize the results both graphically and numerically.

(b) The researchers contend that Germany does not value achievement as highly as Canada and Israel. Do the results from part (a) allow you to address their primary hypothesis? Explain.

(c) The students were also asked to indicate their current grade in mathematics on a six-point scale (1 = excellent, 6 = insufficient). How might both responses be used to address the researchers' primary hypothesis?

13.25. What can you conclude? Analysis of data for a 3×2 ANOVA with 5 observations per cell gave the F statistics in the following table:

Effect	F
A	1.87
B	3.49
AB	2.14

What can you conclude from the information given?

13.26. Changing your major. A study of undergraduate computer science students examined changes in major after the first year.[15] The study examined the fates of 256 students who enrolled as first-year computer science students in the same fall semester. The students were classified according to gender and their declared major at the beginning of the second year. For convenience we use the labels CS for computer science majors, EO for engineering and other science majors, and O for other majors. The explanatory variables included several high school grade summaries coded as 10 = A, 9 = A−, etc. Here are the mean high school mathematics grades for these students:

	Major		
Gender	**CS**	**EO**	**O**
Males	8.68	8.35	7.65
Females	9.11	9.36	8.04

Describe the main effects and interaction using appropriate graphs and calculations.

13.27. More on changing your major. The mean SAT Mathematics scores for the students in the previous exercise are summarized in the following table:

	Major		
Gender	**CS**	**EO**	**O**
Males	628	618	589
Females	582	631	543

Summarize the results of this study using appropriate plots and calculations to describe the main effects and interaction.

13.28. Designing a study. The students studied in the previous two exercises were enrolled at a large midwestern university more than two decades ago. Discuss how you would conduct a similar study at a college or university of your choice today. Include a description of all variables that you would collect for your study.

13.29. Trust of individuals and groups. Trust is an essential element in any exchange of goods or services. The following trust game is often used to study trust experimentally:

> A *sender* starts with \$X and can transfer any amount $x \le X$ to a *responder*. The responder then gets \$$3x$ and can transfer any amount $y \le 3x$ back to the sender. The game ends with final amounts $X - x + y$ and $3x - y$ for the sender and responder respectively.

The value x is taken as a measure of the sender's trust and the value $y/3x$ indicates the responder's trustworthiness. One study used this game to study the dynamics between individuals and groups of three.[16] The following table summarizes the average amount x (in dollars) sent by senders who started with \$100.

Sender	Responder	n	$\bar{x}$	s
Individual	Individual	32	65.5	36.4
Individual	Group	25	76.3	31.2
Group	Individual	25	54.0	41.6
Group	Group	27	43.7	42.4

(a) Find the pooled estimate of the standard deviation for this study and its degrees of freedom.

(b) Is it reasonable to use a pooled standard deviation for the analysis? Explain your answer.

(c) Compute the marginal means.

(d) Plot the means. Do you think there is an interaction? If yes, then describe it.

(e) The F statistics for sender, responder, and interaction are 9.05, 0.001, and 2.08, respectively. Compute the P-values and state your conclusions.

13.30. CHALLENGE **Does the type of cooking pot affect iron content?** Iron-deficiency anemia is the most common form of malnutrition in developing countries, affecting about 50% of children and women and 25% of men. Iron pots for cooking foods had traditionally been used in many of these countries, but they have been largely replaced by aluminum pots, which are cheaper and lighter. Some research has suggested that food cooked in iron pots will contain more iron than food cooked in other types of pots. One study designed to investigate this issue compared the iron content of some Ethiopian foods cooked in aluminum, clay, and iron pots.[17] In Exercise 12.45 (page 589), we analyzed the iron content of *yesiga wet'*, beef cut into small pieces and prepared with several Ethiopian spices. The researchers who conducted this study also examined the iron content of *shiro wet'*, a legume-based mixture of chickpea flour and Ethiopian spiced pepper, and *ye-atkilt allych'a*, a lightly spiced vegetable casserole. Four samples of each food were cooked in each type of pot. The iron in the food was measured in milligrams of iron per 100 grams of cooked food. The data are shown in Table 13.2, where the three foods are labeled meat, legumes, and vegetables.
DATA FILE IRONCONTENT

(a) Make a table giving the sample size, mean, and standard deviation for each type of pot. Is it reasonable to pool the variances? Although the standard deviations vary more than we would like, this is partially due to the small sample sizes, and we will proceed with the analysis of variance.

(b) Plot the means. Give a short summary of how the iron content of foods depends upon the cooking pot.

(c) Run the analysis of variance. Give the ANOVA table, the F statistics with degrees of freedom and P-values, and your conclusions regarding the hypotheses about main effects and interactions.

13.31. Interpreting the results. Refer to the previous exercise. Although there is a statistically significant interaction, do you think that these data support the conclusion that foods cooked in iron pots contain more iron than foods cooked in aluminum or clay pots? Discuss.

13.32. Analysis using a one-way ANOVA. Refer to Exercise 13.30. Rerun the analysis as a one-way ANOVA with 9 groups and 4 observations per group. Report the results of the F test. Examine differences in means using a multiple-comparisons procedure. Summarize your results and compare them with those you obtained in Exercise 13.30.

13.33. Do left-handed people live shorter lives than right-handed people? A study of this question examined a sample of 949 death records and contacted next of kin to determine handedness.[18] Note that there are many possible definitions of "left-handed." The researchers examined the effects of different definitions on the results of their analysis and found that their conclusions were not sensitive to the exact definition used. For the results presented here, people were defined to be right-handed if they wrote, drew, and threw a ball with the right hand. All others were defined to be left-handed. People were classified by gender (female or male) and handedness (left or right), and a 2 × 2 ANOVA was run with the age at death as the response variable. The F statistics were 22.36 (handedness), 37.44 (gender), and 2.10 (interaction). The following marginal mean ages at death (in years) were reported: 77.39 (females), 71.32 (males), 75.00 (right-handed), and 66.03 (left-handed).

(a) For each of the F statistics given, find the degrees of freedom and an approximate P-value. Summarize the results of these tests.

(b) Using the information given, write a short summary of the results of the study.

13.34. A radon exposure study. Scientists believe that exposure to the radioactive gas radon is associated with some types of cancers in the respiratory system. Radon from natural sources is present in many homes in the United States. A group of researchers decided to study the

TABLE 13.2

Iron content (mg/100 g food)

Type of pot	Meat				Legumes				Vegetables			
Aluminum	1.77	2.36	1.96	2.14	2.40	2.17	2.41	2.34	1.03	1.53	1.07	1.30
Clay	2.27	1.28	2.48	2.68	2.41	2.43	2.57	2.48	1.55	0.79	1.68	1.82
Iron	5.27	5.17	4.06	4.22	3.69	3.43	3.84	3.72	2.45	2.99	2.80	2.92

problem in dogs because dogs get similar types of cancers and are exposed to environments similar to those of their owners. Radon detectors are available for home monitoring, but the researchers wanted to obtain actual measures of the exposure of a sample of dogs. To do this, they placed the detectors in holders and attached them to the collars of the dogs. One problem was that the holders might in some way affect the radon readings. The researchers therefore devised a laboratory experiment to study the effects of the holders. Detectors from four series of production were available, so they used a two-way ANOVA design (series with 4 levels and holder with 2, representing the presence or absence of a holder). All detectors were exposed to the same radon source, and the radon measure in picocuries per liter was recorded.[19] The F statistic for the effect of series is 7.02, for the holder it is 1.96, for the interaction it is 1.24, and $N = 69$.

(a) Using Table E or statistical software, find approximate P-values for the three test statistics. Summarize the results of these three significance tests.

(b) The mean radon readings for the four series were 330, 303, 302, and 295. The results of the significance test for series were of great concern to the researchers. Explain why.

13.35. Are insects more attracted to male plants? Some scientists wanted to determine whether there are gender-related differences in the level of herbivory (insects munching on plants) for the jack-in-the-pulpit, a spring-blooming perennial plant common in deciduous forests. A study was conducted in southern Maryland at forests associated with the Smithsonian Environmental Research Center (SERC).[20] To determine the effects of flowering and floral characteristics on herbivory, the researchers altered the floral shape of male and female plants. The three levels of floral characteristics were as follows: (1) the outside part of the flower (spathe) was completely removed; (2) in females, a gap was created in the base of the spathe, and in males, the gap was closed; (3) plants were not altered (control). The percent of leaf area damaged by thrips (an order of insects) between early May and mid-June was recorded for each of 30 plants per combination of sex and floral characteristic. A table of means and standard deviations (in parentheses) follows:

	Floral Characteristic Level		
Gender	**1**	**2**	**3**
Males	0.11 (0.081)	1.28 (0.088)	1.63 (0.382)
Females	0.02 (0.002)	0.58 (0.321)	0.20 (0.035)

(a) Give the degrees of freedom for the F statistics that are used to test for gender, floral characteristic, and the interaction.

(b) Describe the main effects and interaction using appropriate graphs.

(c) The researchers used the natural logarithm of percent area as the response in their analysis. Using the relationship between the means and standard deviations, explain why this was done.

13.36. Change-of-majors study: HSS. Refer to the data given for the change-of-majors study described in Exercise 13.26 (page 614) in the data set MAJORS. Analyze the data for HSS, the high school science grades. Your analysis should include a table of sample sizes, means, and standard deviations; Normal quantile plots; a plot of the means; and a two-way ANOVA using sex and major as the factors. Write a short summary of your conclusions. DATA FILE MAJORS

13.37. Change-of-majors study: HSE. Refer to the data given for the change-of-majors study in the data set MAJORS. Analyze the data for HSE, the high school English grades. Your analysis should include a table of sample sizes, means, and standard deviations; Normal quantile plots; a plot of the means; and a two-way ANOVA using sex and major as the factors. Write a short summary of your conclusions.

13.38. Change-of-majors study: GPA. Refer to the data given for the change-of-majors study in the data set MAJORS. Analyze the data for GPA, the college grade point average. Your analysis should include a table of sample sizes, means, and standard deviations; Normal quantile plots; a plot of the means; and a two-way ANOVA using sex and major as the factors. Write a short summary of your conclusions.

13.39. Change-of-majors study: SATV. Refer to the data given for the change-of-majors study in the data set MAJORS. Analyze the data for SATV, the SAT Verbal score. Your analysis should include a table of sample sizes, means, and standard deviations; Normal quantile plots; a plot of the means; and a two-way ANOVA using sex and major as the factors. Write a short summary of your conclusions.

13.40. Search the Internet. Search the Internet or your library to find a study that is interesting to you and uses a two-way ANOVA to analyze the data. First describe the question or questions of interest and then give the details of how ANOVA was used to provide answers. Be sure to include how the study authors examined the assumptions for the analysis. Evaluate how well the authors used ANOVA in this study. If your evaluation finds the analysis deficient, make suggestions for how it could be improved.

CHAPTER FOURTEEN

LOGISTIC REGRESSION

The simple and multiple linear regression methods we studied in Chapters 10 and 11 are used to model the relationship between a quantitative response variable and one or more explanatory variables. In this chapter we describe similar methods that are used when the response variable has only two possible outcomes, such as a student applicant receives or does not receive financial aid, a patient lives or dies during emergency surgery, or your cell phone coverage is acceptable or not.

In general, we call the two outcomes of the response variable "success" and "failure" and represent them by 1 (for a success) and 0 (for a failure). The mean

binomial setting
p. 285

is then the proportion of ones, $p = P(\text{success})$. If our data are n independent observations, we have the *binomial setting*.

What is *new* in this chapter is that the data now include at least one *explanatory variable* x and the probability p depends on the value of x. For example, suppose that we are studying whether a student applicant receives ($y = 1$) or is denied ($y = 0$) financial aid. Here, p is the probability that an applicant receives aid. Possible explanatory variables include (a) the financial support of the parents, (b) the income and savings of the applicant, and (c) whether or not the applicant has received financial aid before.

Note that the explanatory variables can be either categorical or quantitative. Logistic regression is a statistical method for describing these kinds of relationships.[1]

14.1 The Logistic Regression Model

In Chapter 5 we studied binomial distributions, and in Chapter 6 we learned how to do statistical inference for the proportion p of successes in the binomial setting. We start with a brief review of some of these ideas that we will need in this chapter.

Binomial distributions and odds

EXAMPLE 14.1

Adult video gamers. Example 6.1 (page 317) describes a PEW survey of 2054 U.S. adults aged 18 or older regarding video games. Based on a couple of questions, each respondent was classified as a "gamer" or a "nongamer." The researchers were interested in estimating the proportion of U.S. residents who are gamers. In the notation of Chapter 5, p is the proportion of U.S. adults who are gamers. The number of adults who are gamers in an SRS of size n has the binomial distribution with parameters n and p. The sample size is $n = 2054$ and the number of adult gamers in the sample is 1063. The sample proportion is

$$\hat{p} = \frac{1063}{2054} = 0.5175$$

Logistic regressions work with odds rather than proportions. The odds are simply the ratio of the proportions for the two possible outcomes. If $\hat{p}$ is the proportion for one outcome, then $1 - \hat{p}$ is the proportion for the second outcome:

$$\text{odds} = \frac{\hat{p}}{1 - \hat{p}}$$

A similar formula for the population odds is obtained by substituting p for $\hat{p}$ in this expression.

EXAMPLE 14.2

Odds of being an adult gamer. For the video game data, the proportion of U.S. adults who are gamers in the sample is $\hat{p} = 0.5175$, so the proportion of U.S. adults who are not gamers is

$$1 - \hat{p} = 1 - 0.5175 = 0.4825$$

Therefore, the odds of an adult being a gamer are

$$\begin{aligned} \text{odds} &= \frac{\hat{p}}{1 - \hat{p}} \\ &= \frac{0.5175}{0.4825} \\ &= 1.073 \end{aligned}$$

When people speak about odds, they often round to integers or fractions. Since 1.073 is approximately 14/13, we could say that the odds that an adult is a gamer are 14 to 13. In a similar way, we could describe the odds that an adult is *not* a gamer as 13 to 14.

USE YOUR KNOWLEDGE

14.1 Odds of drawing a black card. If you deal one card from a standard deck, the probability that the card is a black card is 0.50. Find the odds of drawing a black card.

14.2 Given the odds, find the probability. If you know the odds, you can find the probability by solving the preceding equation for odds for the probability. So, $\hat{p} = \text{odds}/(\text{odds} + 1)$. If the odds of an outcome are 3 (or 3 to 1), what is the probability of the outcome?

Odds for two groups

Suppose that this sample of 2054 adults contained 832 men and 1222 women, with 457 men and 606 women classified as gamers. Using the methods of Chapter 6, we could compare the proportions of gamers among men and women using a confidence interval or significance test.

EXAMPLE 14.3

Comparing the proportions of male and female gamers. Figure 14.1 contains output for this comparison. The sample proportion of gamers for men is 0.5493 (54.93%), and the sample proportion of gamers for women is 0.4959 (49.59%). The difference is 0.0534, and the 95% confidence interval is (0.0095, 0.0973). We can summarize this result by saying, "In this sample, the percent of video gamers is 5.3% higher among men than among women. This difference is statistically significant ($z = 2.38$, $P = 0.017$)."

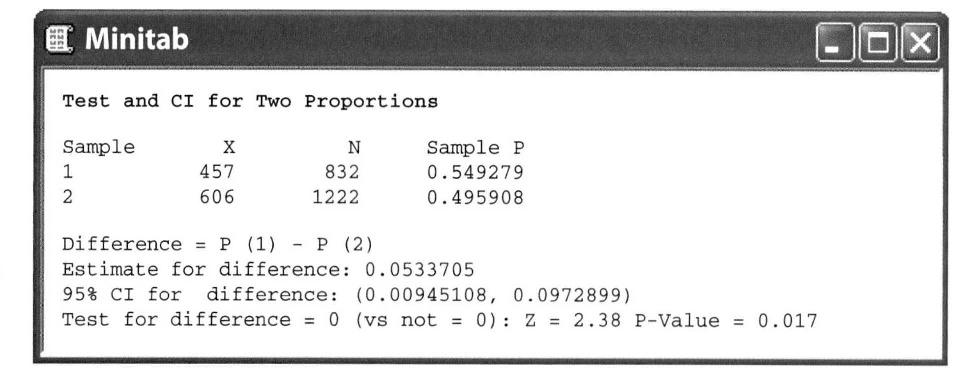

```
Test and CI for Two Proportions

Sample        X          N      Sample P
1           457        832      0.549279
2           606       1222      0.495908

Difference = P (1) - P (2)
Estimate for difference: 0.0533705
95% CI for  difference: (0.00945108, 0.0972899)
Test for difference = 0 (vs not = 0): Z = 2.38 P-Value = 0.017
```

FIGURE 14.1 Minitab output of the comparison between two proportions (adult male gamers versus adult female gamers), for Example 14.3.

Another way to analyze these data is to use logistic regression. The explanatory variable is gender, a categorical variable. To use this in a regression (logistic or otherwise), we need to use a numeric code. The usual way to do this is with an **indicator variable.** For our problem we will use an indicator of whether or not the adult is a male:

indicator variable

$$x = \begin{cases} 1 \text{ if the adult is a male} \\ 0 \text{ if the adult is a female} \end{cases}$$

The response variable is the proportion of video gamers. For use in a logistic regression, we perform two transformations on this variable. First, we convert to odds. For men,

$$\begin{aligned} \text{odds} &= \frac{\hat{p}}{1 - \hat{p}} \\ &= \frac{0.5493}{1 - 0.5493} \\ &= 1.2188 \end{aligned}$$

Similarly, for women we have

$$\begin{aligned} \text{odds} &= \frac{\hat{p}}{1 - \hat{p}} \\ &= \frac{0.4959}{1 - 0.4959} \\ &= 0.9837 \end{aligned}$$

So, for men, the odds of being a gamer are 1.22, and for women, the odds of being a gamer are 0.98.

USE YOUR KNOWLEDGE

14.3 Energy-drink commercials. A study was designed to compare two energy-drink commercials. Each participant was shown the commercials, A and B, in random order and asked to select the better one. There were 130 women and 110 men who participated in the study. Commercial A was selected by 60 women and by 63 men. Find the odds of selecting Commercial A for the men. Do the same for the women.

14.4 Find the odds. Refer to the previous exercise. Find the odds of selecting Commercial B for the men. Do the same for the women.

Model for logistic regression

In simple linear regression we modeled the mean μ of he response variable y as a linear function of the explanatory variable: $\mu = \beta_0 + \beta_1 x$. When y is just 1 or 0 (success or failure), the mean is the probability p of a success. Logistic regression models the mean p in terms of an explanatory variable x. We might try to relate p and x as in simple linear regression: $p = \beta_0 + \beta_1 x$. Unfortunately, this is not a good model. Whenever $\beta_1 \neq 0$, extreme values of x will give values of $\beta_0 + \beta_1 x$ that fall outside the range of possible values of p, $0 \leq p \leq 1$.

log odds
logit

The logistic regression solution to this difficulty is to transform the odds $(p/(1-p))$ using the natural logarithm. We use the term **log odds** or **logit** for this transformation. We model the log odds as a linear function of the explanatory variable:

$$\log\left(\frac{p}{1-p}\right) = \beta_0 + \beta_1 x$$

Figure 14.2 graphs the relationship between p and x for some different values of β_0 and β_1. For logistic regression we use *natural* logarithms. There are tables of natural logarithms, and many calculators have a built-in function for this transformation. As we did with linear regression, we use y for the response variable. So for men,

$$y = \log(\text{odds}) = \log(1.2188) = 0.1979$$

and for women,

$$y = \log(\text{odds}) = \log(0.9837) = -0.0164$$

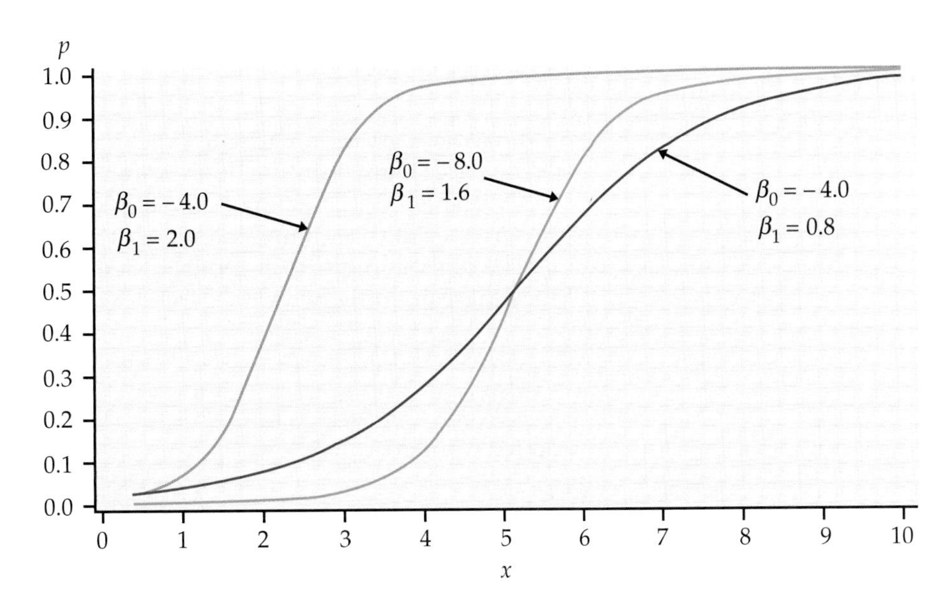

FIGURE 14.2 Plot of p versus x for different logistic regression models.

USE YOUR KNOWLEDGE

14.5 Find the odds. Refer to Exercise 14.3. Find the log odds for the men and the log odds for the women.

14.6 Find the odds. Refer to Exercise 14.4. Find the log odds for the men and the log odds for the women.

In these expressions for the log odds we use y as the observed value of the response variable, the log odds of being a video gamer. We are now ready to build the logistic regression model.

LOGISTIC REGRESSION MODEL

The **statistical model for logistic regression** is

$$\log\left(\frac{p}{1-p}\right) = \beta_0 + \beta_1 x$$

where p is a binomial proportion and x is the explanatory variable. The parameters of the logistic regression model are β_0 and β_1.

EXAMPLE 14.4

Model for adult video gamers. For our video game example, there are $n = 2054$ U.S. adults in the sample. The explanatory variable is gender, which we have coded using an indicator variable with values $x = 1$ for men and $x = 0$ for women. The response variable is also an indicator variable. Thus, the adult is either a gamer or not a gamer. Think of the process of randomly selecting an adult and recording gender (x) and whether or not the adult is a gamer. The model says that the probability (p) that this adult is a gamer depends upon the adult's gender ($x = 1$ or $x = 0$). So there are two possible values for p, which we can call p_{men} and p_{women}.

Logistic regression with an indicator explanatory variable is an important special case. Many multiple logistic regression analyses focus on one or more such variables as the primary explanatory variables of interest. For now, we use this special case to understand a little more about the model.

The logistic regression model specifies the relationship between p and x. Since there are only two values for x, we write both equations. For men,

$$\log\left(\frac{p_{\text{men}}}{1-p_{\text{men}}}\right) = \beta_0 + \beta_1$$

and for women,

$$\log\left(\frac{p_{\text{women}}}{1-p_{\text{women}}}\right) = \beta_0$$

Note that there is a β_1 term in the equation for men because $x = 1$, but it is missing in the equation for women because $x = 0$.

Fitting and interpreting the logistic regression model

In general, the calculations needed to find estimates b_0 and b_1 for the parameters β_0 and β_1 are complex and require the use of software. When the explanatory variable has only two possible values, however, we can easily find the estimates. This simple framework also provides a setting where we can learn what the logistic regression parameters mean.

EXAMPLE 14.5

Log odds for being a video gamer. In the video game example, we found the log odds for men,

$$y = \log\left(\frac{\hat{p}_{\text{men}}}{1 - \hat{p}_{\text{men}}}\right) = 0.1979$$

and for women,

$$y = \log\left(\frac{\hat{p}_{\text{women}}}{1 - \hat{p}_{\text{women}}}\right) = -0.0164$$

The logistic regression model for men is

$$\log\left(\frac{p_{\text{men}}}{1 - p_{\text{men}}}\right) = \beta_0 + \beta_1$$

and for women it is

$$\log\left(\frac{p_{\text{women}}}{1 - p_{\text{women}}}\right) = \beta_0$$

To find the estimates b_0 and b_1, we match the male and female model equations with the corresponding data equations. Thus, we see that the estimate of the intercept b_0 is simply the log(odds) for the women:

$$b_0 = -0.0164$$

and the estimate of the slope is the difference between the log(odds) for the men and the log(odds) for the women:

$$b_1 = 0.1979 - (-0.0164) = 0.214$$

The fitted logistic regression model is

$$\log(\text{odds}) = -0.016 + 0.214x$$

The slope in this logistic regression model is the difference between the log(odds) for men and the log(odds) for women. Most people are not comfortable thinking in the log(odds) scale, so interpretation of the results in terms of the regression slope is difficult. Usually, we apply a transformation to help us. With a little algebra, it can be shown that

$$\frac{\text{odds}_{\text{men}}}{\text{odds}_{\text{women}}} = e^{0.2143} = 1.2390$$

odds ratio

The transformation $e^{0.2143}$ undoes the logarithm and transforms the logistic regression slope into an **odds ratio,** in this case, the ratio of the odds that a man is a video gamer to the odds that a woman is a video gamer. In other words, we can multiply the odds for women by the odds ratio to obtain the odds for men:

$$\text{odds}_{\text{men}} = 1.2390 \times \text{odds}_{\text{women}}$$

In this case, the odds for men are 1.24 times the odds for women.

Notice that we have chosen the coding for the indicator variable so that the regression slope is positive. This will give an odds ratio that is greater than 1. Had we coded women as 1 and men as 0, the sign of the slope would be reversed, the fitted equation would be $\log(\text{odds}) = 0.1979 - 0.2143x$, and the odds ratio would be $e^{-0.2143} = 0.8071$. The odds for women are slightly less than 81% of the odds for men.

Logistic regression with an explanatory variable having two values is a very important special case. Here is an example where the explanatory variable is quantitative.

EXAMPLE 14.6

DATA FILE MOVIES

Will a movie be profitable? The MOVIES data set (described on page 543) includes both the movie's budget and the total U.S. revenue. For this example, we will classify each movie as "profitable" ($y = 1$) if U.S. revenue is larger than the budget and nonprofitable ($y = 0$) otherwise. This is our response variable. The data set contains several explanatory variables, but we will focus here on the natural logarithm of the opening-weekend revenue. Figure 14.3 is a scatterplot of the data with a smoothed function generated by software. The probability that a movie is profitable increases with the log opening-weekend revenue. Because the curve suggested by the smoother is reasonably close

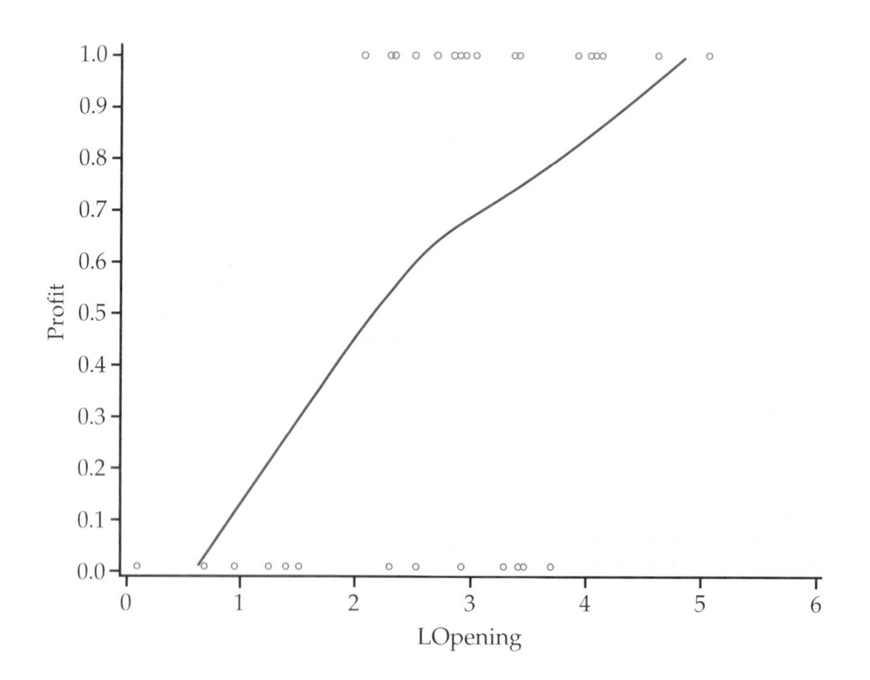

FIGURE 14.3 Scatterplot of profit (Y = 1, N = 0) versus the log opening-weekend revenue (LOpening) with a smooth function, for Example 14.6.

to an S-shaped curve like those in Figure 14.2, we fit the logistic regression model

$$\log\left(\frac{p}{1-p}\right) = \beta_0 + \beta_1 x$$

where p is the probability that the movie is profitable and x is the log opening-weekend revenue. The model for estimated log odds fitted by software is

$$\log(\text{odds}) = b_0 + b_1 x = -3.1658 + 1.3083x$$

The odds ratio is $e^{b_1} = 3.700$. This means that if log opening-weekend revenue x increases by one unit (roughly \$2.71 million), the odds that the movie will be profitable increase by 3.7 times.

USE YOUR KNOWLEDGE

14.7 Find the logistic regression equation and the odds ratio. Refer to Exercises 14.3 and 14.5. Find the logistic regression equation and the odds ratio.

14.8 Find the logistic regression equation and the odds ratio. Refer to Exercises 14.4 and 14.6. Find the logistic regression equation and the odds ratio.

14.2 Inference for Logistic Regression

Statistical inference for logistic regression is very similar to statistical inference for simple linear regression. We calculate estimates of the model parameters and standard errors for these estimates. Confidence intervals are formed in the usual way, but we use standard Normal z^*-values rather than critical values from the t distributions. The ratio of the estimate to the standard error is the basis for hypothesis tests. Often the test statistics are given as the squares of these ratios, and in this case the P-values are obtained from the chi-square distributions with 1 degree of freedom.

Confidence intervals and significance tests

CONFIDENCE INTERVALS AND SIGNIFICANCE TESTS FOR LOGISTIC REGRESSION PARAMETERS

A **level *C* confidence interval for the slope** β_1 is

$$b_1 \pm z^* \text{SE}_{b_1}$$

The ratio of the odds for a value of the explanatory variable equal to $x + 1$ to the odds for a value of the explanatory variable equal to x is the **odds ratio.**

A **level *C* confidence interval for the odds ratio** e^{β_1} is obtained by transforming the confidence interval for the slope

$$\left(e^{b_1 - z^* \text{SE}_{b_1}},\ e^{b_1 + z^* \text{SE}_{b_1}}\right)$$

In these expressions z^* is the value for the standard Normal density curve with area C between $-z^*$ and z^*.

To test the hypothesis H_0: $\beta_1 = 0$, compute the **test statistic**

$$z = \frac{b_1}{\text{SE}_{b_1}}$$

The P-value for the significance test of H_0 against H_a: $\beta_1 \neq 0$ is computed using the fact that, when the null hypothesis is true, z has approximately a standard Normal distribution.

Wald statistic

The statistic z is sometimes called a **Wald statistic.** Output from some statistical software reports the significance test result in terms of the square of the z statistic.

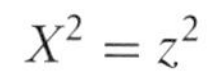

$$X^2 = z^2$$

LOOK BACK
chi-square statistic
p. 471

This statistic is called a chi-square statistic. When the null hypothesis is true, it has a distribution that is approximately a χ^2 distribution with 1 degree of freedom, and the P-value is calculated as $P(\chi^2 \geq X^2)$. Because the square of a standard Normal random variable has a χ^2 distribution with 1 degree of freedom, the z statistic and the chi-square statistic give the same results for statistical inference.

We have expressed the hypothesis-testing framework in terms of the slope β_1 because this form closely resembles what we studied in simple linear regression. In many applications, however, the results are expressed in terms of the odds ratio. A slope of 0 is the same as an odds ratio of 1, so we often express the null hypothesis of interest as "the odds ratio is 1." This means that the two odds are equal and the explanatory variable is not useful for predicting the odds.

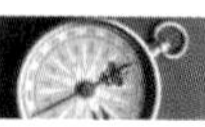

EXAMPLE 14.7

Software output. Figure 14.4 gives the output from SPSS and SAS for the video game data described in Example 14.5. The parameter estimates are given as $b_0 = -0.0164$ and $b_1 = 0.2141$. The standard errors are 0.0572 and 0.0902. A 95% confidence interval for the slope is

$$\begin{aligned} b_1 \pm z^*\text{SE}_{b_1} &= 0.2141 \pm (1.96)(0.0902) \\ &= 0.2141 \pm 0.1768 \end{aligned}$$

We are 95% confident that the slope is between 0.0373 and 0.3909. The SAS output provides the odds ratio 1.239 and a 95% confidence interval. This interval is also easy to compute from the interval for the slope:

$$\begin{aligned} (e^{b_1 - z^*\text{SE}_{b_1}},\ e^{b_1 + z^*\text{SE}_{b_1}}) &= (e^{0.0373},\ e^{0.3909}) \\ &= (1.038,\ 1.478) \end{aligned}$$

For this problem we would report, "Adult males are more likely to be gamers than adult females (odds ratio = 1.24, 95% CI = 1.04 to 1.48)."

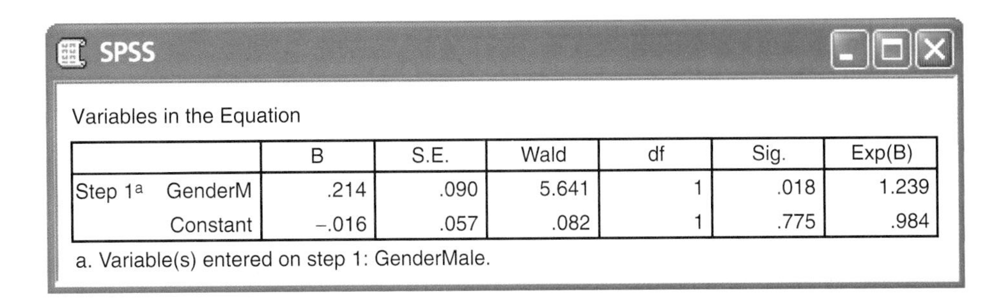
SPSS

Variables in the Equation

		B	S.E.	Wald	df	Sig.	Exp(B)
Step 1[a]	GenderM	.214	.090	5.641	1	.018	1.239
	Constant	-.016	.057	.082	1	.775	.984

a. Variable(s) entered on step 1: GenderMale.

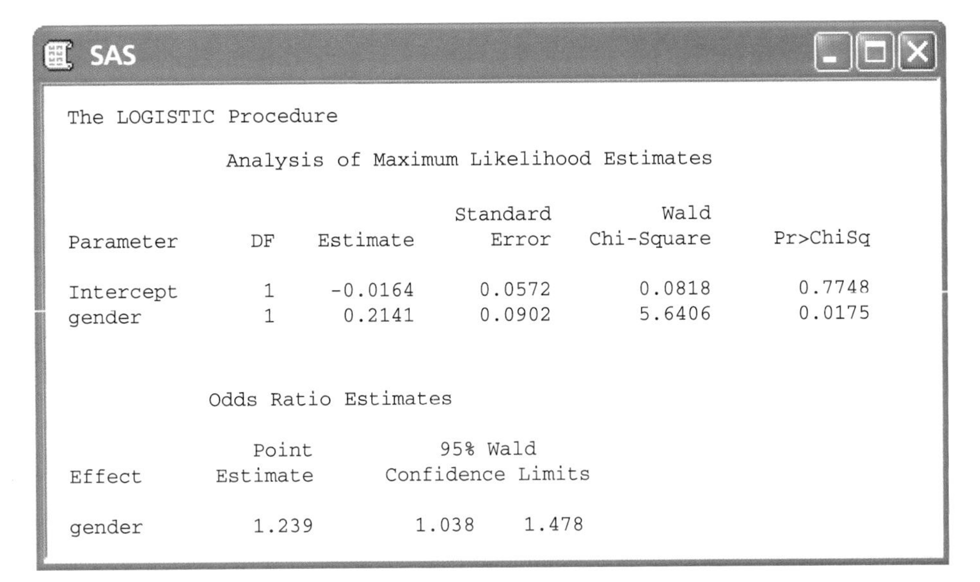
SAS

The LOGISTIC Procedure

Analysis of Maximum Likelihood Estimates

Parameter	DF	Estimate	Standard Error	Wald Chi-Square	Pr>ChiSq
Intercept	1	-0.0164	0.0572	0.0818	0.7748
gender	1	0.2141	0.0902	5.6406	0.0175

Odds Ratio Estimates

Effect	Point Estimate	95% Wald Confidence Limits	
gender	1.239	1.038	1.478

FIGURE 14.4 Logistic regression output from SPSS and SAS for the video game data, for Example 14.7.

In applications such as these, it is standard to use 95% for the confidence coefficient. With this convention, the confidence interval gives us the result of testing the null hypothesis that the odds ratio is 1 for a significance level of 0.05. If the confidence interval does not include 1, we reject H_0 and conclude that the odds for the two groups are different; if the interval does include 1, the data do not provide enough evidence to distinguish the groups in this way.

The following example is typical of many applications of logistic regression. Here there is a designed experiment with five different values for the explanatory variable.

EXAMPLE 14.8

An insecticide for aphids. An experiment was designed to examine how well the insecticide rotenone kills an aphid, called *Macrosiphoniella sanborni*, that feeds on the chrysanthemum plant.[2] The explanatory variable is the concentration (in log of milligrams per liter) of the insecticide. At each concentration, approximately 50 insects were exposed. Each insect was either killed or not killed. We summarize the data using the number killed. The response variable for logistic regression is the log odds of the proportion killed. Here are the data:

Concentration (log)	Number of insects	Number killed
0.96	50	6
1.33	48	16
1.63	46	24
2.04	49	42
2.32	50	44

If we transform the response variable (by taking log odds) and use least squares, we get the fit illustrated in Figure 14.5. The logistic regression fit is given in Figure 14.6. It is a transformed version of Figure 14.5 with the fit calculated using the logistic model.

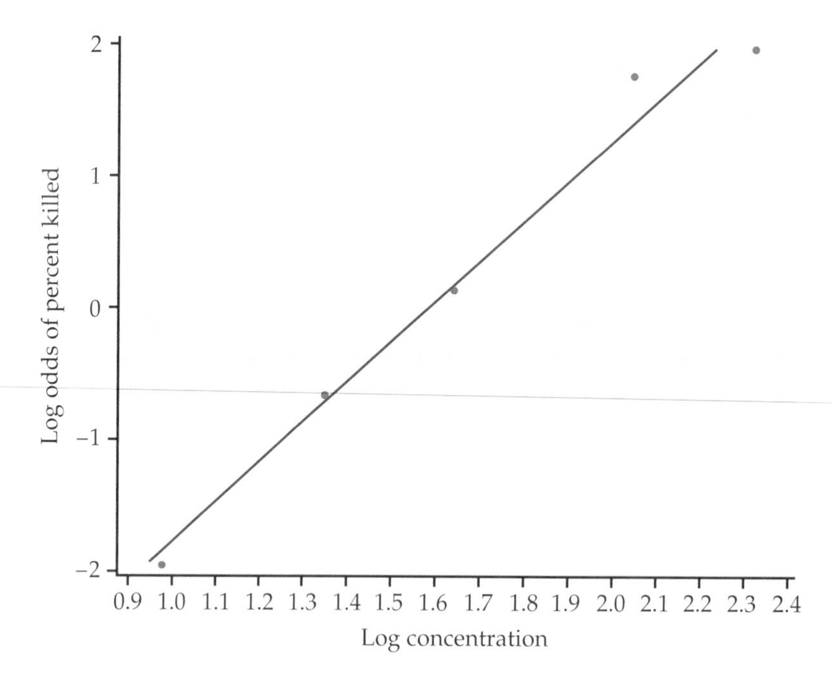

FIGURE 14.5 Plot of log odds of percent killed versus log concentration for the insecticide data, for Example 14.8.

One of the major themes of this text is that we should present the results of a statistical analysis with a graph. For the insecticide example, we have done this with Figure 14.6, and the results appear to be convincing. But suppose that rotenone has no ability to kill *Macrosiphoniella sanborni*. What is the chance that we would observe experimental results at least as convincing as what we observed if this supposition were true? The answer is the P-value for the test of the null hypothesis that the logistic regression slope is zero. If this P-value is not small, our graph may be misleading. Statistical inference provides what we need.

EXAMPLE 14.9

Software output. Figure 14.7 gives the output from SPSS, SAS, and Minitab logistic regression analysis of the insecticide data. The model is

$$\log\left(\frac{p}{1-p}\right) = \beta_0 + \beta_1 x$$

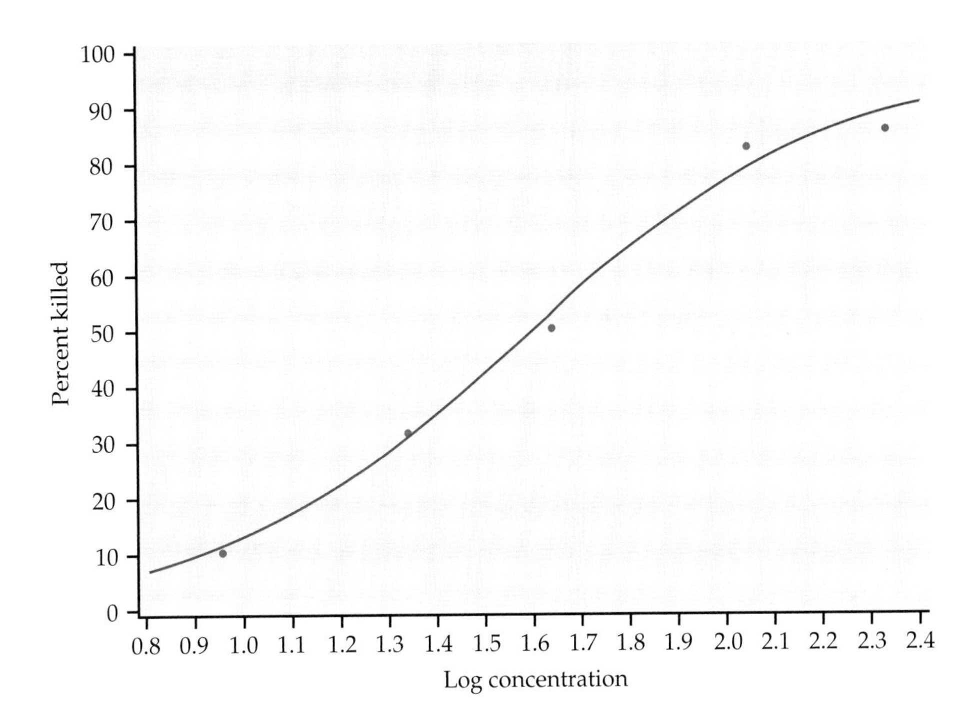

FIGURE 14.6 Plot of the percent killed versus log concentration with the logistic regression fit for the insecticide data, for Example 14.8.

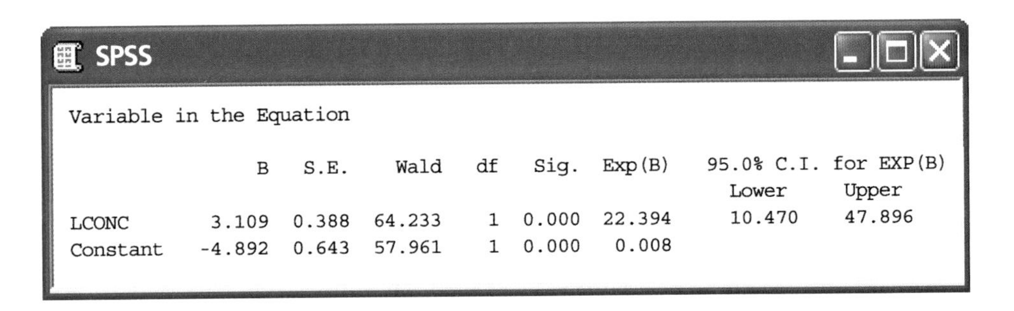

SPSS

Variable in the Equation

	B	S.E.	Wald	df	Sig.	Exp(B)	95.0% C.I. for EXP(B) Lower	Upper
LCONC	3.109	0.388	64.233	1	0.000	22.394	10.470	47.896
Constant	-4.892	0.643	57.961	1	0.000	0.008		

SAS

The LOGISTIC Procedure

Analysis of Maximum Likelihood Estimates

Parameter	DF	Estimate	Standard Error	Wald Chi-Square	Pr>ChiSq
Intercept	1	-4.8923	0.6426	57.9606	<0.001
lconc	1	3.1088	0.3879	64.2332	<0.001

Odds Ratio Estimates

Effect	Point Estimate	95% Wald Confidence Limits	
lconc	22.394	10.470	47.896

FIGURE 14.7 Logistic regression output from SPSS, SAS, and Minitab for the insecticide data, for Example 14.9. (*Continued on next page*)

Minitab

```
Logistic Regression Table

                                                Odds      95% CI
Predictor    Coef     StDev      Z       P     Ratio    Lower   Upper
Constant  -4.8923    0.6426  -7.61   0.000
lconc      3.1088    0.3879   8.01   0.000     22.39    10.47   47.90
```

FIGURE 14.7 *(Continued)*

where the values of the explanatory variable x are 0.96, 1.33, 1.63, 2.04, 2.32. From the output we see that the fitted model is

$$\log(\text{odds}) = b_0 + b_1x = -4.89 + 3.11x$$

This is the fit that we plotted in Figure 14.6. The null hypothesis that $\beta_1 = 0$ is clearly rejected ($X^2 = 64.23$, $P < 0.001$). We calculate a 95% confidence interval for β_1 using the estimate $b_1 = 3.1088$ and its standard error $SE_{b_1} = 0.3879$ given in the output:

$$\begin{aligned} b_1 \pm z^*SE_{b_1} &= 3.1088 \pm (1.96)(0.3879) \\ &= 3.1088 \pm 0.7603 \end{aligned}$$

We are 95% confident that the true value of the slope is between 2.35 and 3.87.

The odds ratio is given on the Minitab output as 22.39. An increase of one unit in the log concentration of insecticide (x) is associated with a 22-fold increase in the odds that an insect will be killed. The confidence interval for the odds is obtained from the interval for the slope:

$$\begin{aligned} (e^{b_1 - z^*SE_{b_1}},\ e^{b_1 + z^*SE_{b_1}}) &= (e^{2.3485},\ e^{3.8691}) \\ &= (10.47,\ 47.90) \end{aligned}$$

Note again that the test of the null hypothesis that the slope is 0 is the same as the test of the null hypothesis that the odds are 1. If we were reporting the results in terms of the odds, we could say, "The odds of killing an insect increase by a factor of 22.4 for each unit increase in the log concentration of insecticide ($X^2 = 64.23$, $P < 0.001$; 95% CI = 10.5 to 47.9)."

In Example 14.6 we studied the problem of predicting whether or not a movie was going to make a profit using the log opening-weekend revenue as the explanatory variable. We now revisit this example and show how statistical inference is an important part of the conclusion.

EXAMPLE 14.10

Software output. Figure 14.8 gives the output from Minitab for a logistic regression analysis using log opening-weekend revenue as the explanatory variable. The fitted model is

$$\log(\text{odds}) = b_0 + b_1x = -3.166 + 1.308x$$

From the output we see that because $P = 0.007$, we can reject the null hypothesis that $\beta_1 = 0$. The value of the test statistic is $X^2 = 7.26$ with 1 degree

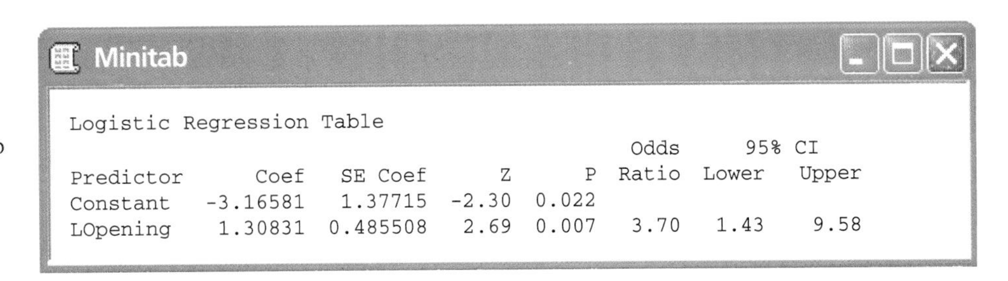

```
Logistic Regression Table
                                                   Odds     95% CI
Predictor      Coef   SE Coef      Z       P  Ratio  Lower   Upper
Constant   -3.16581   1.37715  -2.30   0.022
LOpening    1.30831  0.485508   2.69   0.007   3.70   1.43    9.58
```

FIGURE 14.8 Logistic regression output from Minitab for the movie profitability data with log opening-weekend revenue as the explanatory variable, for Example 14.10.

of freedom. We use the estimate $b_1 = 1.3083$ and its standard error $\text{SE}_{b_1} = 0.4855$ to compute the 95% confidence interval for β_1:

$$\begin{aligned} b_1 \pm z^*\text{SE}_{b_1} &= 1.3083 \pm (1.96)(0.4855) \\ &= 1.3083 \pm 0.9516 \end{aligned}$$

Our estimate of the slope is 1.3083 and we are 95% confident that the true value is between 0.3567 and 2.2599. For the odds ratio, the estimate on the output is 3.70. The 95% confidence interval is

$$\begin{aligned} (e^{b_1 - z^*\text{SE}_{b_1}},\ e^{b_1 + z^*\text{SE}_{b_1}}) &= (e^{0.3567},\ e^{2.2599}) \\ &= (1.43,\ 9.58) \end{aligned}$$

We estimate that an opening-weekend revenue that is one unit larger (roughly \$2.71 million) will increase the odds that a movie is profitable by about 4 times. The data, however, do not give us a very accurate estimate. The odds ratio could be as small as 1.43 or as large as 9.58 with 95% confidence. We have evidence to conclude that movies with higher opening-weekend revenues are more likely to be profitable, but establishing the relationship accurately will require more data.

Multiple logistic regression

The movie example that we just considered naturally leads us to the next topic. The MOVIES data set includes additional explanatory variables. Do these other explanatory variables contain additional information that will give us a better prediction of profitability? We use **multiple logistic regression** to answer this question. Generating the computer output is easy, just as it was when we generalized simple linear regression with one explanatory variable to multiple linear regression with more than one explanatory variable in Chapter 11. The statistical concepts are similar, although the computations are more complex. Here is an example.

multiple logistic regression

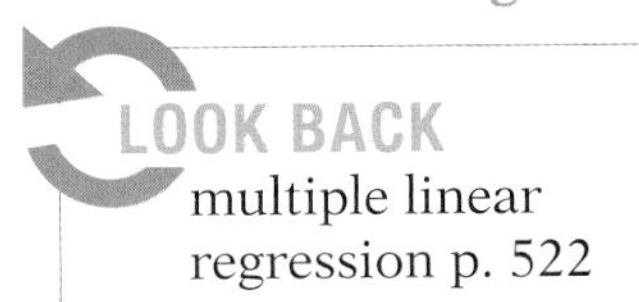

multiple linear regression p. 522

EXAMPLE 14.11

Software output. As in Example 14.10, we will predict the odds that a movie is profitable. The explanatory variables are log opening-weekend revenue (LOpening), number of theaters (Theaters), and the movie's IMDb rating at the end of the first week (Opinion), which is on a 1 to 10 scale (10 being best). Figure 14.9 gives the outputs from SAS, Minitab, and SPSS. The fitted model is

$$\begin{aligned}\log(\text{odds}) &= b_0 + b_1 \text{ LOpening} + b_2 \text{ Theaters} + b_3 \text{ Opinion} \\ &= -2.013 + 2.147 \text{ LOpening} - 0.001 \text{ Theaters} - 0.109 \text{ Opinion}\end{aligned}$$

When analyzing data using multiple linear regression, we first examine the hypothesis that all the regression coefficients for the explanatory variables are zero. We do the same for multiple logistic regression. The hypothesis

$$H_0: \beta_1 = \beta_2 = \beta_3 = 0$$

SAS

Testing Global Null Hypothesis: BETA = 0

Test	Chi-Square	DF	Pr>ChiSq
Likelihood Ratio	12.7157	3	0.0053
Score	10.9325	3	0.0121
Wald	7.1248	3	0.0680

Analysis of Maximum Likelihood Estimates

Parameter	DF	Estimate	Standard Error	Wald Chi-Square	Pr>ChiSq
Intercept	1	-2.0131	3.2320	0.3880	0.5334
LOpening	1	2.1467	0.9749	4.8488	0.0277
Theaters	1	-0.00103	0.000940	1.1924	0.2748
Opinion	1	-0.1095	0.4514	0.0589	0.8083

Odds Ratio Estimates

Effect	Point Estimate	95% Wald Confidence Limits	
LOpening	8.556	1.266	57.823
Theaters	0.999	0.997	1.001
Opinion	0.896	0.370	2.171

Minitab

Logistic Regression Table

Predictor	Coef	SE Coef	Z	P	Odds Ratio	95% CI Lower	Upper
Constant	-2.01319	3.23201	-0.62	0.533			
LOpening	2.14670	0.974874	2.20	0.028	8.56	1.27	57.83
Theaters	-0.0010270	0.0009405	-1.09	0.275	1.00	1.00	1.00
Opinion	-0.109492	0.451356	-0.24	0.808	0.90	0.37	2.17

Log-Likelihood = -17.198
Test that all slopes are zero: G = 12.716, DF = 3, P-Value = 0.005

FIGURE 14.9 Logistic regression output from SAS, Minitab, and SPSS for the movie profitability data with log opening-weekend revenue, number of theaters, and movie rating as the explanatory variables, for Example 14.11. (*Continued on following page*)

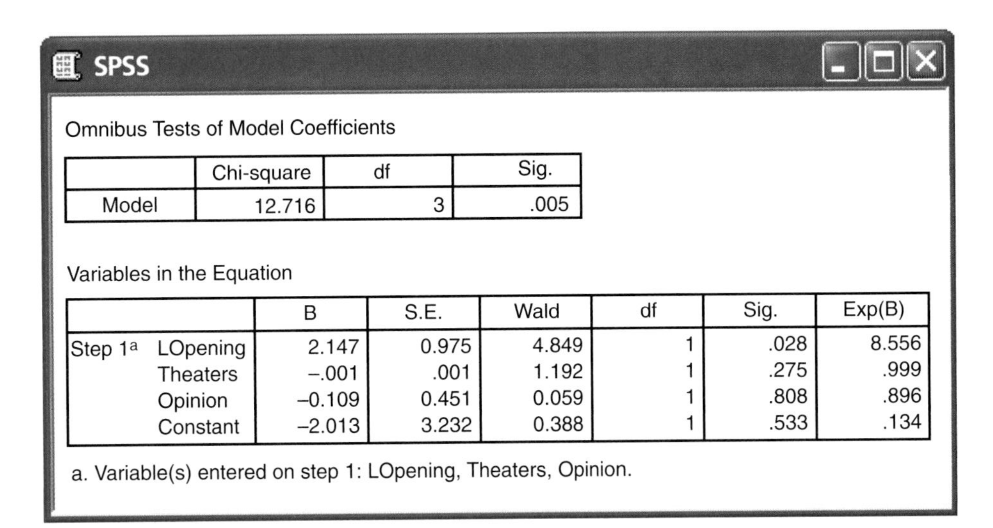
SPSS

Omnibus Tests of Model Coefficients

	Chi-square	df	Sig.
Model	12.716	3	.005

Variables in the Equation

		B	S.E.	Wald	df	Sig.	Exp(B)
Step 1[a]	LOpening	2.147	0.975	4.849	1	.028	8.556
	Theaters	-.001	.001	1.192	1	.275	.999
	Opinion	-0.109	0.451	0.059	1	.808	.896
	Constant	-2.013	3.232	0.388	1	.533	.134

a. Variable(s) entered on step 1: LOpening, Theaters, Opinion.

FIGURE 14.9 *(Continued)*

is tested by a chi-square statistic with 3 degrees of freedom. For Minitab, this is given in the last line of the output and the statistic is called "G." The value is $G = 12.716$ and the P-value is 0.0053 (from the SAS output). We reject H_0 and conclude that one or more of the explanatory variables can be used to predict the odds that a movie is profitable.

We now examine the coefficients for each variable and the tests that each of these is 0 *in a model that contains the other two*. The P-values are 0.028, 0.275, and 0.808. The null hypotheses H_0: $\beta_2 = 0$ and H_0: $\beta_3 = 0$ cannot be rejected. That is, log opening-weekend revenue is the only predictor that adds significant predictive ability once the other two are already in the model.

Our initial multiple logistic regression analysis told us that the explanatory variables contain information that is useful for predicting whether or not the movie is profitable. Because the explanatory variables are correlated, however, we cannot clearly distinguish which variables or combinations of variables are important. Further analysis of these data using subsets of the three explanatory variables is needed to clarify the situation. We leave this work for the exercises.

CHAPTER 14 SUMMARY

If $\hat{p}$ is the sample proportion, then the **odds** are $\hat{p}/(1-\hat{p})$, the ratio of the proportion of times the event happens to the proportion of times the event does not happen.

The **logistic regression model** relates the log of the odds to the explanatory variable:

$$\log\left(\frac{p_i}{1-p_i}\right) = \beta_0 + \beta_1 x_i$$

where the response variables for $i = 1, 2, \ldots, n$ are independent binomial random variables with parameters 1 and p_i; that is, they are independent with distributions $B(1, p_i)$. The explanatory variable is x.

The **parameters** of the logistic regression model are β_0 and β_1.

The **odds ratio** is e^{β_1}, where β_1 is the slope in the logistic regression model.

A **level *C* confidence interval for the intercept** β_0 is

$$b_0 \pm z^* \text{SE}_{b_0}$$

A **level *C* confidence interval for the slope** β_1 is

$$b_1 \pm z^* \text{SE}_{b_1}$$

A **level *C* confidence interval for the odds ratio** e^{β_1} is obtained by transforming the confidence interval for the slope

$$(e^{b_1 - z^* SE_{b_1}}, e^{b_1 + z^* SE_{b_1}})$$

In these expressions z^* is the value for the standard Normal density curve with area C between $-z^*$ and z^*.

To test the hypothesis $H_0: \beta_1 = 0$, compute the **test statistic**

$$z = \frac{b_1}{SE_{b_1}}$$

and use the fact that z has a distribution that is approximately the standard Normal distribution when the null hypothesis is true. This statistic is sometimes called the **Wald statistic.** An alternative equivalent procedure is to report the square of z,

$$X^2 = z^2$$

This statistic has a distribution that is approximately a χ^2 distribution with 1 degree of freedom, and the P-value is calculated as $P(\chi^2 \geq X^2)$. This is the same as testing the null hypothesis that the odds ratio is 1.

In **multiple logistic regression** the response variable has two possible values, as in logistic regression, but there can be several explanatory variables.

CHAPTER 14 EXERCISES

For Exercises 14.1 and 14.2, see page 619; for Exercises 14.3 and 14.4, see pages 620–621; for Exercises 14.5 and 14.6, see page 622; and for Exercises 14.7 and 14.8, see page 625.

14.9. What's wrong? For each of the following, explain what is wrong and why.

(a) For a multiple logistic regression with 5 explanatory variables, the null hypothesis that the regression coefficients of all the explanatory variables are zero is tested with an F test.

(b) For a logistic regression, we assume that the error term in our model has a Normal distribution.

(c) In logistic regression with one explanatory variable, we can use a chi-square statistic to test the null hypothesis H_0: $b_1 = 0$ versus a two-sided alternative.

(d) In multiple logistic regression, we do not have to worry about correlation among explanatory variables when interpreting model coefficient estimates.

14.10. What's wrong? For each of the following, explain what is wrong and why.

(a) If $b_1 = 3$ in a logistic regression analysis with one explanatory variable, we estimate that the probability of an event is multiplied by 3 when the value of the explanatory variable increases by 1 unit.

(b) The intercept β_0 is equal to the odds of an event when $x = 0$.

(c) The odds of an event are 1 minus the probability of the event.

14.11. Will a movie be profitable? In Example 14.6 (page 624), we developed a model to predict whether a movie is profitable based on log opening-weekend revenue. What are the predicted odds of a movie being profitable if the opening-weekend revenue is

(a) \$25 million (LOpening = 3.219)?

(b) \$45 million (LOpening = 3.807)?

(c) \$65 million (LOpening = 4.174)?

14.12. Converting odds to probability. Refer to the previous exercise. For each opening-weekend revenue, compute the estimated probability that the movie is profitable.

14.13. Salt intake and cardiovascular disease. A study was designed to look at the relationship between salt in the diet and the risk of developing cardiovascular disease (CVD).[3] Here are the data:

	Salt in diet		
Developed CVD	**Low**	**High**	**Total**
Yes	88	112	200
No	1081	1134	2215
Total	1169	1246	2415

(a) For each salt level find the probability of developing CVD.

(b) Convert each of the probabilities that you found in part (a) to odds.

(c) Find the log of each of the odds that you found in part (b).

14.14. Salt in the diet and CVD. Refer to Exercise 14.13. Use $x = 1$ for the high-salt diet and $x = 0$ for the low-salt diet.

(a) Find the estimates b_0 and b_1.

(b) Give the fitted logistic regression model.

(c) What is the odds ratio for the high-salt versus the low-salt diet?

14.15. Give a 99% confidence interval for β_1. Refer to Example 14.9 (page 628). Suppose that you wanted to report a 99% confidence interval for β_1. Show how you would use the information provided in the outputs shown in Figure 14.7 to compute this interval.

14.16. Give a 99% confidence interval for the odds ratio. Refer to Example 14.9 (page 628) and the outputs in Figure 14.7. Using the estimate b_1 and its standard error, find the 95% confidence interval for the odds ratio and verify that this agrees with the interval given by the software.

14.17. CHALLENGE **z and the X^2 statistic.** The Minitab output in Figure 14.7 (page 629) does not give the value of X^2. The column labeled "Z" provides similar information.

(a) Find the value under the heading "Z" for the predictor lconc. Verify that Z is simply the estimated coefficient divided by its standard error. This is a z statistic that has approximately the standard Normal distribution if the null hypothesis (slope 0) is true.

(b) Show that the square of z is close to X^2 (with no roundoff error, these two quantities will be equal). The two-sided P-value for z is the same as P for X^2.

(c) Draw sketches of the standard Normal and the chi-square distribution with 1 degree of freedom. (*Hint:* You can use the information in Table F to sketch the chi-square distribution.) Indicate the value of the z and the X^2 statistics on these sketches and use shading to illustrate the P-value.

14.18. Finding the best model? In Example 14.11 (page 631), we looked at a multiple logistic regression for movie profitability based on three explanatory variables. Complete the analysis by looking at the 3 two-explanatory variable models and the 3 single-variable models. Create a table that includes the parameter estimates and their P-values as well as the overall X^2 statistic and degrees of freedom. Based on the results, which model do you feel is the best? Explain your answer. DATA FILE MOVIES

14.19. Tipping behavior in Canada. The Consumer Report on Eating Share Trends (CREST) contains data spanning all provinces of Canada and detailing away-from-home food purchases by roughly 4000 households per quarter. Researchers recently restricted their attention to restaurants at which tips would normally be given.[4] From a total of 73,822 observations, "high" and "low" tipping variables were created based on whether the observed tip rate was above 20% or below 10% respectively. They then used logistic regression to identify explanatory variables associated with either "high" or "low" tips. The following table summarizes what they termed the stereotype-related variables for the low-tip analysis:

Explanatory variable	Odds ratio
Senior adult	1.099
Sunday	1.098
English as second language	1.142
French-speaking Canadian	1.163
Alcoholic drinks	0.713
Lone male	0.858

All coefficients were significant at the 0.01 level. Write a short summary explaining these results in terms of the odds of leaving a low tip.

14.20. Sexual imagery in magazine ads. Exercise 9.28 (page 484) presents some results of a study about how advertisers use sexual imagery to appeal to young people. The clothing worn by the model in each of 1509 ads was classified as "not sexual" or "sexual" based on a standardized criterion. A logistic regression was used to describe the probability that the clothing in the ad was "not sexual" as a function of several explanatory variables. Here are some of the reported results:

Explanatory variable	b	X^2 statistic
Reader age	0.50	13.64
Model gender	1.31	72.15
Men's magazines	−0.05	0.06
Women's magazines	0.45	6.44
Constant	−2.32	135.92

Reader age is coded as 0 for young adult and 1 for mature adult. Therefore, the coefficient of 0.50 for this explanatory variable suggests that the probability that the model clothing is *not* sexual is higher when the target reader age is mature adult. In other words, the model clothing is more likely to be sexual when the target reader age is young adult. Model gender is coded as 0 for female and 1 for male. The explanatory variable men's magazines is 1 if the intended readership is men and is 0 for women's magazines and magazines intended for both men and women. The variable women's magazines is coded similarly.

(a) State the null and alternative hypotheses for each of the explanatory variables.

(b) Perform the significance tests associated with the X^2 statistics.

(c) Interpret the sign of each of the statistically significant coefficients in terms of the probability that the model clothing *is* sexual.

(d) Write an equation for the fitted logistic regression model.

14.21. Interpret the odds ratios. Refer to the previous exercise. The researchers also reported odds ratios with 95% confidence intervals for this logistic regression model. Here is a summary:

		95% Confidence Limits	
Explanatory variable	**Odds ratio**	**Lower**	**Upper**
Reader age	1.65	1.27	2.16
Model gender	3.70	2.74	5.01
Men's magazines	0.96	0.67	1.37
Women's magazines	1.57	1.11	2.23

(a) Explain the relationship between the confidence intervals reported here and the results of the X^2 significance tests that you found in the previous exercise.

(b) Interpret the results in terms of the odds ratios.

(c) Write a short summary explaining the results. Include comments regarding the usefulness of the fitted coefficients versus the odds ratios.

14.22. What purchases will be made? A poll of 1000 adults aged 18 or older asked about purchases they intended to make for the upcoming holiday season.[5] A total of 463 adults listed gift card as a planned purchase.

(a) What proportion of adults plan to purchase a gift card as a present?

(b) What are the odds that an adult will purchase a gift card as a present?

(c) What proportion of adults do not plan to purchase a gift card as a present?

(d) What are the odds that an adult will not buy a gift card as a present?

(e) How are your answers to parts (b) and (d) related?

14.23. High blood pressure and cardiovascular disease. There is much evidence that high blood pressure is associated with increased risk of death from cardiovascular disease. A major study of this association examined 3338 men with high blood pressure and 2676 men with low blood pressure. During the period of the study, 21 men in the low-blood-pressure group and 55 in the high-blood-pressure group died from cardiovascular disease.

(a) Find the proportion of men who died from cardiovascular disease in the high-blood-pressure group. Then calculate the odds.

(b) Do the same for the low-blood-pressure group.

(c) Now calculate the odds ratio with the odds for the high-blood-pressure group in the numerator. Describe the result in words.

14.24. Gender bias in syntax textbooks. To what extent do syntax textbooks, which analyze the structure of sentences, illustrate gender bias? A study of this question sampled sentences from 10 texts.[6] One part of the study examined the use of the words "girl," "boy," "man," and "woman." We will call the first two words juvenile and the last two adult. Here are data from one of the texts:

Gender	n	X(juvenile)
Female	60	48
Male	132	52

(a) Find the proportion of the female references that are juvenile. Then transform this proportion to odds.

(b) Do the same for the male references.

(c) What is the odds ratio for comparing the female references to the male references? (Put the female odds in the numerator.)

14.25. High blood pressure and cardiovascular disease. Refer to the study of cardiovascular disease and blood pressure in Exercise 14.23. Computer output for a logistic regression analysis of these data gives the estimated slope $b_1 = 0.7505$ with standard error $SE_{b_1} = 0.2578$.

(a) Give a 95% confidence interval for the slope.

(b) Calculate the X^2 statistic for testing the null hypothesis that the slope is zero and use Table F to find an approximate P-value.

(c) Write a short summary of the results and conclusions.

14.26. Gender bias in syntax textbooks. The data from the study of gender bias in syntax textbooks given in Exercise 14.24 are analyzed using logistic regression. The estimated slope is $b_1 = 1.8171$ and its standard error is $SE_{b_1} = 0.3686$.

(a) Give a 95% confidence interval for the slope.

(b) Calculate the X^2 statistic for testing the null hypothesis that the slope is zero and use Table F to find an approximate P-value.

(c) Write a short summary of the results and conclusions.

14.27. High blood pressure and cardiovascular disease. The results describing the relationship between blood pressure and cardiovascular disease are given in terms of the change in log odds in Exercise 14.23.

(a) Transform the slope to the odds ratio and the 95% confidence interval for the slope to a 95% confidence interval for the odds ratio.

(b) Write a conclusion using the odds to describe the results.

14.28. Gender bias in syntax textbooks. The gender bias in syntax textbooks is described in the log odds scale in Exercise 14.24.

(a) Transform the slope to the odds ratio and the 95% confidence interval for the slope to a 95% confidence interval for the odds ratio.

(b) Write a conclusion using the odds to describe the results.

14.29. Reducing the number of workers. To be competitive in global markets, many corporations are undertaking major reorganizations. Often these involve "downsizing" or a "reduction in force" (RIF), where substantial numbers of employees are terminated. Federal and various state laws require that employees be treated equally regardless of their age. In particular, employees over the age of 40 years are in a "protected" class, and many allegations of discrimination focus on comparing employees over 40 with their younger coworkers. Here are the data for a recent RIF:

	Over 40	
Terminated	**No**	**Yes**
Yes	7	41
No	504	765

(a) Write the logistic regression model for this problem using the log odds of a RIF as the response variable and an indicator for over and under 40 years of age as the explanatory variable.

(b) Explain the assumptions concerning binomial distributions in terms of the variables in this exercise. To what extent do you think that these assumptions are reasonable?

(c) Software gives the estimated slope $b_1 = 1.3504$ and its standard error $SE_{b_1} = 0.4130$. Transform the results to the odds scale. Summarize the results and write a short conclusion.

(d) If additional explanatory variables were available (for example, a performance evaluation), how would you use this information to study the RIF?

14.30. Internet use in Canada. A study used data from the Canadian Internet Use Survey (CIUS) to explore the relationship between certain demographic variables and Internet use by individuals in Canada.[7] The dependent variable refers to the use of the Internet from any location within the last 12 months. Explanatory variables included age (years), income (thousands of dollars), location (1 = urban, 0 = other), sex (1 = male, 0 = female), education (1 = at least some postsecondary education, 0 = other), language (1 = English, 0 = French), and children (1 = at least one child in household, 0 = no children). The following table summarizes the results:

Explanatory variable	b
Age	−0.063
Income	0.013
Location	0.367
Sex	−0.222
Education	1.080
Language	0.285
Children	0.049
Intercept	2.010

All but Children was significant at the 0.05 level.

(a) Interpret the sign of each of the coefficients (except intercept) in terms of the probability that the individual uses the Internet.

(b) Compute the odds ratio for each of the variables in the table.

(c) What are the odds that a French-speaking, 23-year-old male who lives alone in Montreal and makes $50,000 a year his second year after college is using the Internet?

(d) Convert the odds in part (c) to a probability.

14.31. Predicting physical activity. Participation in physical activities typically declines between high school and young adulthood. This suggests that postsecondary institutions may be an ideal setting to address physical activity. A study looked at the association between physical activity and several behavioral and perceptual characteristics among midwestern college students.[8] Of 663 students who met the vigorous-activity guidelines for the previous week, 169 reported eating fruit two or more times per day. Of the 471 that did not meet the vigorous-activity guidelines in the previous week, 68 reported eating fruit two or more times per day. Model the log odds of vigorous activity using an indicator variable for eating fruit two or more times per day as the explanatory variable. Summarize your findings.

14.32. Online consumer spending. The Consumer Behavior Report is designed to provide insight into online shopping trends.[9] One recent report asked the question "In the past three months, how has the current state of the economy impacted your money spending on online purchasing?" Here are the results from 3156 online consumers:

	Reduced Spending	
Gender	No	Yes
Female	586	708
Male	1074	788

(a) What are the proportions of individuals who reduced spending in each gender?

(b) What is the odds ratio for comparing female individuals to male individuals?

(c) Write the logistic regression model for this problem using the log odds of reducing spending as the response variable and an indicator of gender as the explanatory variable.

(d) Software gives the estimated slope $b_1 = 0.4988$ and its standard error $SE_{b_1} = 0.0729$. Transform this result to the odds scale and compare it with your answer in part (b).

(e) Construct a 95% confidence interval for the odds ratio and write a short conclusion.

14.33. Proximity of fast-food restaurants to schools and adolescent overweight. A California study looked at the relationship between the presence of fast-food restaurants near schools (within a 0.5 mile radius) and being overweight among middle and high school students.[10] Whether a student was overweight was determined based on each student's responses to the California Healthy Kids Survey (CHKS). A database of latitude-longitude coordinates for schools and restaurants was used to determine proximity. Here are the data:

Fast-food nearby	n	X(overweight)
No	238,215	65,080
Yes	291,152	83,143

Use logistic regression to study the question of whether or not being overweight is related to the proximity of fast-food restaurants to schools. Write a short paragraph summarizing your conclusions.

14.34. Overweight and fast-food restaurants, continued. In the article, the researchers state: (1) "CIs were adjusted for clustering at the school level"; and (2) "All models also included controls for the following student characteristics: a female indicator, grade indicator, age indicator, race/ethnicity indicators, and physical exercise indicators. All models also included indicator variables for school location type, including large urban, midsize urban, small urban, large suburban, midsize suburban, small suburban, town, and rural."

(a) What violation of the response distribution is Statement 1 addressing? Explain your answer.

(b) Explain why the researchers controlled for the variables described in Statement 2 when looking at the relationship between overweight and proximity.

The following four exercises use the CSDATA data set. We examine models for relating success, as measured by the GPA, to several explanatory variables. In Chapter 11 we used multiple regression methods for our analysis. Here, we define an indicator variable, say HIGPA, to be 1 if the GPA is 3.0 or better and 0 otherwise. DATA FILE CSDATA

14.35. CHALLENGE **Use high school grades to predict high grade point averages.** Use a logistic regression to predict HIGPA using the three high school grade summaries as explanatory variables.

(a) Summarize the results of the hypothesis test that the coefficients for all three explanatory variables are zero.

(b) Give the coefficient for high school math grades with a 95% confidence interval. Do the same for the two other predictors in this model.

(c) Summarize your conclusions based on parts (a) and (b).

14.36. CHALLENGE **Use SAT scores to predict high grade point averages.** Use a logistic regression to predict HIGPA using the two SAT scores as explanatory variables.

(a) Summarize the results of the hypothesis test that the coefficients for both explanatory variables are zero.

(b) Give the coefficient for the SAT Math score with a 95% confidence interval. Do the same for the SAT Verbal score.

(c) Summarize your conclusions based on parts (a) and (b).

14.37. CHALLENGE **Use high school grades and SAT scores to predict high grade point averages.** Run a logistic regression to predict HIGPA using the three high school grade summaries and the two SAT scores as explanatory variables. We want to produce an analysis that is similar to that done for these data in Chapter 11 (page 519).

(a) Test the null hypothesis that the coefficients of the three high school grade summaries are zero; that is, test H_0: $\beta_{\text{HSM}} = \beta_{\text{HSS}} = \beta_{\text{HSE}} = 0$.

(b) Test the null hypothesis that the coefficients of the two SAT scores are zero; that is, test H_0: $\beta_{\text{SATM}} = \beta_{\text{SATV}} = 0$.

(c) What do you conclude from the tests in (a) and (b)?

14.38. CHALLENGE **Is there an effect of gender?** In this exercise we investigate the effect of gender on the odds of getting a high GPA.

(a) Use gender to predict HIGPA using a logistic regression. Summarize the results.

(b) Perform a logistic regression using gender and the two SAT scores to predict HIGPA. Summarize the results.

(c) Compare the results of parts (a) and (b) with respect to how gender relates to HIGPA. Summarize your conclusions.

14.39. CHALLENGE **An example of Simpson's paradox.** Here is an example of Simpson's paradox, the reversal of the direction of a comparison or an association when data from several groups are combined to form a single group. The data concern two hospitals, A and B, and whether or not patients undergoing surgery died or survived. Here are the data for all patients:

	Hospital A	Hospital B
Died	63	16
Survived	2037	784
Total	2100	800

And here are the more detailed data where the patients are categorized as being in good condition or poor condition before having the surgery:

Good condition

	Hospital A	Hospital B
Died	6	8
Survived	594	592
Total	600	600

Poor condition

	Hospital A	Hospital B
Died	57	8
Survived	1443	192
Total	1500	200

(a) Use a logistic regression to model the odds of death with hospital as the explanatory variable. Summarize the results of your analysis and give a 95% confidence interval for the odds ratio of Hospital A relative to Hospital B.

(b) Rerun your analysis in part (a) using hospital and the condition of the patient as explanatory variables. Summarize the results of your analysis and give a 95% confidence interval for the odds ratio of Hospital A relative to Hospital B.

(c) Explain Simpson's paradox in terms of your results in parts (a) and (b).

14.40. CHALLENGE **Interpret the fitted model.** If we apply the exponential function to the fitted model in Example 14.6 (page 624), we get

$$\text{odds} = e^{-11.0391+3.1709x} = e^{-11.0391} \times e^{3.1709x}$$

Show that for any value of the quantitative explanatory variable x, the odds ratio for increasing x by 1,

$$\frac{\text{odds}_{x+1}}{\text{odds}_x}$$

is $e^{3.1709} = 23.83$. This justifies the interpretation given at the end of Example 14.6.

PART THREE REVIEW

TOPICS IN INFERENCE

In Part III, we were introduced to more advanced methods of statistical inference. These chapters primarily focused on relationships between variables. Chapter 9 focused on the relationship between two categorical variables. Chapter 10 discussed the relationship between a continuous response variable and a single quantitative explanatory variable, and Chapter 11 extended this to more than one explanatory variable. These two chapters introduced the use of least squares and analysis of variance for statistical inference. Chapters 12 and 13 also used analysis of variance and extended the methods of Chapter 7 to handle the comparison of more than two population means. Finally, Chapter 14 considered the relationship between a binary response variable and a set of quantitative explanatory variables.

Chapter 9: Analysis of Two-Way Tables

- The **null hypothesis** for $r \times c$ tables of count data is that there is no relationship between the row variable and the column variable. (page 469)
- **Expected cell counts** under the null hypothesis are computed using the formula

$$\text{expected count} = \frac{\text{row total} \times \text{column total}}{n}$$

(page 471)
- The null hypothesis is tested by the **chi-square statistic,** which compares the observed counts with the expected counts:

$$X^2 = \sum \frac{(\text{observed} - \text{expected})^2}{\text{expected}}$$

(page 471)
- Under the null hypothesis, X^2 has approximately the χ^2 distribution with $(r - 1)(c - 1)$ degrees of freedom. The P-value for the test is

$$P(\chi^2 \geq X^2)$$

where χ^2 is a random variable having the $\chi^2(\text{df})$ distribution with $\text{df} = (r - 1)(c - 1)$. (page 472)
- The chi-square approximation is adequate for practical use when the average expected cell count is 5 or greater and all individual expected counts are 1 or greater, except in the case of 2×2 tables. All four expected counts in a 2×2 table should be 5 or greater. (page 472)
- The **chi-square goodness-of-fit test** is used to compare the sample distribution of a categorical variable from a population with a hypothesized distribution. The data for n observations with k possible outcomes are summarized as observed counts, $n_1, n_2, \ldots, n_k$ in k cells. The **null hypothesis** specifies probabilities $p_1, p_2, \ldots, p_k$ for the possible outcomes. (page 476)
- The analysis of these data is similar to the analyses of two-way tables discussed in Section 9.1. For each cell, the **expected count** is determined by multiplying the total number of observations n by the specified probability p_i. The null hypothesis is tested by the usual **chi-square statistic,** which compares the observed counts, n_i, with the expected counts. Under the null hypothesis, X^2 has approximately the χ^2 distribution with $\text{df} = k - 1$. (page 476)

Chapter 10: Inference for Regression

- The statistical model for **simple linear regression** assumes that the means of the response variable y fall on a line when plotted against x, with the observed y's varying Normally about these means. For n observations, this model can be written

$$y_i = \beta_0 + \beta_1 x_i + \epsilon_i$$

where $i = 1, 2, \ldots, n$, and the ϵ_i are assumed to be independent and Normally distributed with mean 0 and standard deviation σ. Here

$\beta_0 + \beta_1 x_i$ is the mean response when $x = x_i$. The **parameters** of the model are β_0, β_1, and σ. (page 494)

- The **population regression line** intercept and slope, β_0 and β_1, are estimated by the intercept and slope of the **least-squares regression line,** b_0 and b_1. (page 494)
- The parameter σ is estimated by

$$s = \sqrt{\frac{\sum e_i^2}{n-2}}$$

where the e_i are the **residuals**

$$e_i = y_i - \hat{y}_i$$

(page 495)

- Prior to inference, always examine the residuals for Normality, constant variance, and any other remaining patterns in the data. **Plots of the residuals** both against the case number and against the explanatory variable are usually done as part of this examination. (page 497)
- Sometimes a **transformation** of one or both of the variables can make their relationship linear. However, these transformations can harm the assumptions of Normality and constant variance. (page 499)
- A **level *C* confidence interval for β_1** is

$$b_1 \pm t^* \text{SE}_{b_1}$$

where t^* is the value for the $t(n-2)$ density curve with area C between $-t^*$ and t^*. (page 502)

- The **test of the hypothesis** H_0: $\beta_1 = 0$ is based on the ***t* statistic**

$$t = \frac{b_1}{\text{SE}_{b_1}}$$

and the $t(n-2)$ distribution. This tests whether there is a straight-line relationship between y and x. There are similar formulas for confidence intervals and tests for β_0, but these are meaningful only in special cases. (page 502)

- The **estimated mean response** for the subpopulation corresponding to the value x^* of the explanatory variable is

$$\hat{\mu}_y = b_0 + b_1 x^*$$

(page 504)

- A **level *C* confidence interval for the mean response** is

$$\hat{\mu}_y \pm t^* \text{SE}_{\hat{\mu}}$$

where t^* is the value for the $t(n-2)$ density curve with area C between $-t^*$ and t^*. (page 505)

- The **estimated value of the response variable** y for a future observation from the subpopulation corresponding to the value x^* of the explanatory variable is

$$\hat{y} = b_0 + b_1 x^*$$

(page 506)

- A **level *C* prediction interval** for the estimated response is

$$\hat{y} \pm t^* \mathrm{SE}_{\hat{y}}$$

where t^* is the value for the $t(n-2)$ density curve with area C between $-t^*$ and t^*. The standard error for the prediction interval is larger than that for the confidence interval because it also includes the variability of the future observation around its subpopulation mean. (page 507)

Chapter 11: Multiple Regression

- **Data for multiple linear regression** consist of the values of a response variable y and p explanatory variables $x_1, x_2, \ldots, x_p$ for n cases. (page 521) We write the data and enter them into software in the form

	Variables				
Individual	x_1	x_2	...	x_p	y
1	x_{11}	x_{12}	...	x_{1p}	y_1
2	x_{21}	x_{22}	...	x_{2p}	y_2
⋮					
n	x_{n1}	x_{n2}	...	x_{np}	y_n

- The statistical model for **multiple linear regression** with response variable y and p explanatory variables $x_1, x_2, \ldots, x_p$ is

$$y_i = \beta_0 + \beta_1 x_{i1} + \beta_2 x_{i2} + \cdots + \beta_p x_{ip} + \epsilon_i$$

where $i = 1, 2, \ldots, n$. (page 522) The ϵ_i are assumed to be independent and Normally distributed with mean 0 and standard deviation σ. (page 522) The **parameters** of the model are $\beta_0, \beta_1, \beta_2, \ldots, \beta_p$, and σ. (page 522)

- The **multiple regression equation** predicts the response variable by a linear relationship with all the explanatory variables:

$$\hat{y} = b_0 + b_1 x_1 + b_2 x_2 + \cdots + b_p x_p$$

The β's are estimated by $b_0, b_1, b_2, \ldots, b_p$, which are obtained by the **method of least squares.** (page 523)

- The parameter σ is estimated by

$$s = \sqrt{\mathrm{MSE}} = \sqrt{\frac{\sum e_i^2}{n-p-1}}$$

where the e_i are the **residuals,**

$$e_i = y_i - \hat{y}_i$$

Always examine the **distribution of the residuals** and plot them against the explanatory variables prior to inference. (page 524)

- A **level C confidence interval for** β_j is

$$b_j \pm t^* \mathrm{SE}_{b_j}$$

where t^* is the value for the $t(n-p-1)$ density curve with area C between $-t^*$ and t^*. (page 524)

- The test of the hypothesis H_0: $\beta_j = 0$ is based on the ***t* statistic**

$$t = \frac{b_j}{\mathrm{SE}_{b_j}}$$

and the $t(n-p-1)$ distribution. (page 524)

- The estimate b_j of β_j and the test and confidence interval for β_j are all based on a specific multiple linear regression model. The results of all these procedures change if other explanatory variables are added to or deleted from the model. (page 535)
- The **ANOVA table** for a multiple linear regression gives the degrees of freedom, sums of squares, and mean squares for the model, error, and total sources of variation. (page 525)
- The **ANOVA *F* statistic** is the ratio MSM/MSE and is used to test the null hypothesis

$$H_0\text{: } \beta_1 = \beta_2 = \cdots = \beta_p = 0$$

If H_0 is true, this statistic has an $F(p,\ n-p-1)$ distribution. (page 528)

- The **squared multiple correlation** is given by the expression

$$R^2 = \frac{\mathrm{SSM}}{\mathrm{SST}}$$

and is interpreted as the proportion of the variability in the response variable y that is explained by the explanatory variables $x_1, x_2, \ldots, x_p$ in the multiple linear regression. (page 529)

Chapter 12: One-Way Analysis of Variance

- **One-way analysis of variance (ANOVA)** is used to compare several population means based on independent SRSs from each population. The populations are assumed to be Normal with possibly different means and the same standard deviation. (page 553)
- The **null hypothesis** is that the population means are *all* equal. The **alternative hypothesis** is true if there are *any* differences among the population means. (page 562)
- ANOVA is based on separating the total variation observed in the data into two parts: variation **among group means** and variation **within groups.** If the variation among groups is large relative to the variation within groups, we have evidence against the null hypothesis. (page 563)
- An **analysis of variance table** organizes the ANOVA calculations. **Degrees of freedom, sums of squares, and mean squares** appear in the table. (page 565)

- The ***F* statistic** and its ***P*-value** are used to test the null hypothesis. (page 566)
- To do an analysis of variance, first compute sample means and standard deviations for all groups. Side-by-side boxplots and a plot of the means give an overview of the data. Examine histograms or Normal quantile plots (either for each group separately or for the residuals) to detect outliers or extreme deviations from Normality. Compute the ratio of the largest to the smallest sample standard deviation. If this ratio is less than 2 and the data inspection does not find outliers or severe non-Normality, ANOVA can be performed. (page 560)
- The ANOVA F test shares the **robustness** of the two-sample t test. It is relatively insensitive to moderate non-Normality and unequal variances, especially when the sample sizes are similar. (page 550)
- Specific questions formulated before examination of the data can be expressed as **contrasts.** Tests and confidence intervals for contrasts provide answers to these questions. (page 569)
- If no specific questions are formulated before examination of the data and the null hypothesis of equality of population means is rejected, **multiple-comparisons** methods are used to assess the statistical significance of the differences between pairs of means. (page 574)

Chapter 13: Two-Way Analysis of Variance

- **Two-way analysis of variance** is used to compare population means when populations are classified according to two factors. (page 593)
- ANOVA assumes that the populations are Normal with possibly different means and the same standard deviation and that independent SRSs are drawn from each population. (page 598)
- As with one-way ANOVA, preliminary analysis includes examination of means, standard deviations, and Normal quantile plots. **Marginal means** are calculated by taking averages of the cell means across rows and columns. (page 600)
- Pooling is used to estimate the within-group variance. (page 599)
- ANOVA separates the total variation into parts for the **model** and **error.** The model variation is separated into parts for each of the **main effects** and the **interaction.** (page 604)
- The calculations are organized into an **ANOVA table.** F statistics and P-values are used to test hypotheses about the main effects and the interaction. (page 605)
- Careful inspection of the means is necessary to interpret significant main effects and interactions. Plots are a useful aid. (page 599)

Chapter 14: Logistic Regression

- If $\hat{p}$ is the sample proportion, then the **odds** are $\hat{p}/(1 - \hat{p})$, the ratio of the proportion of times the event happens to the proportion of times the event does not happen. (page 618)

- The **logistic regression model** relates the log of the odds to the explanatory variable:

$$\log\left(\frac{p_i}{1-p_i}\right) = \beta_0 + \beta_1 x_i$$

where the response variables for $i = 1, 2, \ldots, n$ are independent binomial random variables with parameters 1 and p_i; that is, they are independent with distributions $B(1, p_i)$. The explanatory variable is x. (page 621)

- The **parameters** of the logistic regression model are β_0 and β_1. (page 622)
- The **odds ratio** is e^{β_1}, where β_1 is the slope in the logistic regression model. (page 624)
- A **level *C* confidence interval for the slope** β_1 is

$$b_1 \pm z^* \mathrm{SE}_{b_1}$$

(page 625)

- A **level *C* confidence interval for the odds ratio** e^{β_1} is obtained by transforming the confidence interval for the slope

$$(e^{b_1 - z^* \mathrm{SE}_{b_1}},\ e^{b_1 + z^* \mathrm{SE}_{b_1}})$$

(page 625)

- In these expressions z^* is the value for the standard Normal density curve with area C between $-z^*$ and z^*. (page 625)
- To test the hypothesis H_0: $\beta_1 = 0$, compute the **test statistic**

$$z = \frac{b_1}{\mathrm{SE}_{b_1}}$$

and use the fact that z has a distribution that is approximately the standard Normal distribution when the null hypothesis is true. This statistic is sometimes called the **Wald statistic.** An alternative equivalent procedure is to report the square of z,

$$X^2 = z^2$$

This statistic has a distribution that is approximately a χ^2 distribution with 1 degree of freedom, and the P-value is calculated as $P(\chi^2 \geq X^2)$. This is the same as testing the null hypothesis that the odds ratio is 1. (page 626)

- In **multiple logistic regression** the response variable has two possible values, as in logistic regression, but there can be several explanatory variables. (page 631)

PART III EXERCISES

III.1. Sexual harassment at middle and high schools. A nationally representative survey of students in grades 7 to 12 asked about the experience of these students with respect to sexual harassment.[1] One question asked how many times the student had witnessed sexual harassment in school. Here are the data, categorized by gender:

	Times Witnessed		
Gender	**Never**	**Once**	**More than once**
Girls	140	192	671
Boys	106	125	732

(a) Discuss different ways to plot the data. Choose one way to make a plot and give reasons for your choice.

(b) Make the plot and describe what it shows.

III.2. Does witnessing sexual harassment depend on gender? Refer to the previous exercise. Does the frequency of witnessing sexual harassment depend on gender?

(a) Formulate this question in terms of appropriate null and alternative hypotheses.

(b) Perform the significance test. Report the test statistic, the degrees of freedom, and the P-value.

(c) Write a short summary explaining the results.

III.3. Sexual harassment online or in person. In the survey described in Exercise III.1, the students were also asked whether they had experienced sexual harassment and whether the harassment was in person or online. Here are the data for the girls:

	Harassed Online	
Harassed in Person	**Yes**	**No**
Yes	321	200
No	40	441

Here are the data for the boys:

	Harassed Online	
Harassed in Person	**Yes**	**No**
Yes	183	154
No	48	578

Analyze the data and prepare a report describing your analytical approach, the results, and your conclusions. Be sure to include graphical and numerical summaries.

III.4. The value of online courses. A Pew survey asked college presidents whether or not they believed that online courses offer an equal educational value when compared with courses taken in the classroom. The presidents were classified by the type of educational institution. Here are the data:[2]

	Institution Type			
Response	**4-year private**	**4-year public**	**2-year private**	**For profit**
Yes	36	50	66	54
No	62	48	34	45

(a) Discuss different ways to plot the data. Choose one way to make a plot and give reasons for your choice.

(b) Make the plot and describe what it shows.

III.5. Do the answers depend upon institution type? Refer to the previous exercise. You want to examine whether or not the data provide evidence that the belief that online courses offer an equal educational value when compared with courses taken in the classroom varies with the type of institution of the president.

(a) Formulate this question in terms of appropriate null and alternative hypotheses.

(b) Perform the significance test. Report the test statistic, the degrees of freedom, and the P-value.

(c) Write a short summary explaining the results.

III.6. Compare the college presidents with the general public. Refer to Exercise III.4. Another Pew survey asked the general public about their opinions of the value of online courses. Of the 2142 people who participated in the survey, 621 responded "Yes" to the question "Do you believe that online courses offer an equal educational value when compared with courses taken in the classroom?"

(a) Use the data given in Exercise III.4 to find the number of college presidents who responded "Yes" to the question.

(b) Construct a two-way table that you can use to compare the responses of the general public with the responses of the college presidents.

(c) Is it meaningful to interpret the marginal totals or percents for this table? Explain your answer.

(d) Analyze the data in your two-way table and summarize the results.

III.7. Pain tolerance among sports teams. Many have argued that sports such as football require the ability to withstand pain from injury for extended periods of time. To see if there is greater pain tolerance among certain sports teams, a group of researchers assessed 183 male Division II athletes from 5 sports.[3] Each athlete was asked to put his dominant hand and forearm in a water bath of 3°C and keep it in there until the pain became intolerable. The total amount of time (in seconds) that each athlete maintained his hand and forearm in the bath was recorded. Following this procedure, each athlete completed a series of surveys on aggression and competitiveness. In their report, the researchers state:

> A univariate between subjects (sports team) ANOVA was performed on the total amount of time athletes were able to keep their hand and forearm in the water bath, and found to be statistically significant, $F(4, 146) = 4.96$, $p < .001$. The lacrosse and soccer players tolerated the pain for a longer period of time than athletes from the other teams. Swimmers tolerated the pain for a significantly shorter period of time than the other teams.

(a) Based on the description of the experiment, what should the degrees of freedom be for this analysis?

(b) Assuming that the degrees of freedom reported are correct, data from how many athletes were used in this analysis?

(c) The researchers do not comment on the missing data in their report. List two reasons why these data may not have been used, and for each, explain how the omission could impact or bias the results.

III.8. College debt versus the percent of students who borrow. Kiplinger's "Best Values in Public Colleges" provides a ranking of U.S. public colleges based on a combination of various measures of academics and affordability.[4] The data set BESTVALUE contains 8 of these measures for a random collection of 40 colleges from Kiplinger's 2011–2012 report. For this exercise, we'll focus on the average debt in dollars at graduation (AvgDebt) and the percent of students who borrow (PercBorrow). DATA FILE BESTVALUE

(a) A scatterplot of these two variables is shown in Figure III.1. Describe the relationship. Are there any possible outliers or unusual values? Does a linear relationship between PercBorrow and AvgDebt seem reasonable?

(b) Based on the scatterplot, approximately how much does the average debt change for a college with 10% more students who borrow?

(c) The State University of New York–Fredonia is a school where 86% of the students borrow. Discuss the appropriateness of using these data to predict average debt for this school.

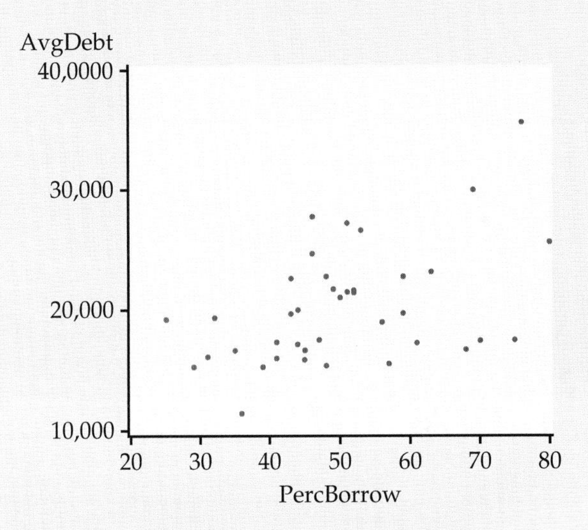

FIGURE III.1 Scatterplot of average debt at graduation (AvgDebt) versus the percent of students who borrow (PercBorrow), for Exercise III.8.

III.9. Can we consider this an SRS? Refer to the previous exercise. The report states that Kiplinger's rankings focus on traditional four-year public colleges with broad-based curricula. Each year, they start with more than 500 schools and then narrow the list down to roughly 120 based on academic quality before ranking them. For our inference purposes, we'd like to consider this SRS of 40 schools from their list of 100 schools to be an SRS from the roughly 500 four-year public colleges. Do you think this is a reasonable consideration? Write a short paragraph explaining your answer.

III.10. Predicting college debt. Refer to Exercise III.8. Figure III.2 contains partial SAS output for the simple linear regression of AvgDebt versus PercBorrow.

(a) State the least-squares regression line.

(b) At Miami University in Oxford, 51% of the students borrow, and the average debt is $27,315. What is the residual?

(c) What percent of the variability in average debt is explained by the percent of students who borrow?

(d) Construct a 95% confidence interval for the slope. What does this interval tell you about the change in average debt for a change in the percent who borrow?

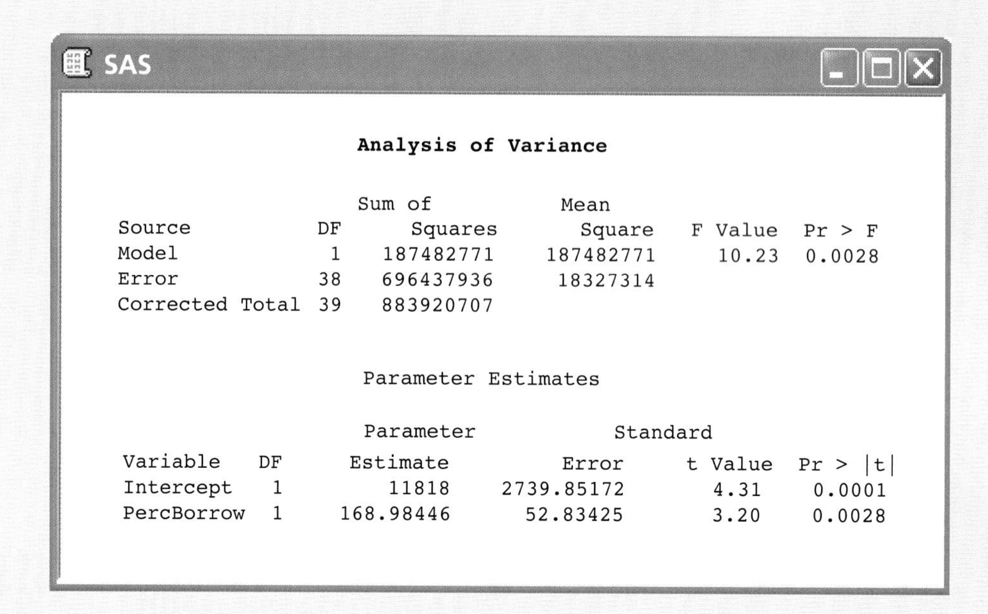
SAS

Analysis of Variance

Source	DF	Sum of Squares	Mean Square	F Value	Pr > F
Model	1	187482771	187482771	10.23	0.0028
Error	38	696437936	18327314		
Corrected Total	39	883920707			

Parameter Estimates

Variable	DF	Parameter Estimate	Standard Error	t Value	Pr > |t|
Intercept	1	11818	2739.85172	4.31	0.0001
PercBorrow	1	168.98446	52.83425	3.20	0.0028

FIGURE III.2 SAS output for Exercise III.10.

III.11. More on predicting college debt. Refer to the previous exercise. The University of Michigan–Ann Arbor is a school where 46% of the students borrow, and the average debt is $27,828. The University of Wisconsin–La Crosse is a school where 69% of the students borrow, and the average debt is $21,420.

(a) Using your answer to part (a) of the previous exercise, what is the predicted average debt for a student at the University of Michigan–Ann Arbor?

(b) What is the predicted average debt for the University of Wisconsin–La Crosse?

(c) Without doing any calculations, would the standard error for the estimated average debt be larger for the University of Michigan or the University of Wisconsin? Explain your answer.

III.12. Predicting college debt: other measures. Refer to Exercise III.8. Let's now look at AvgDebt and its relationship with all seven measures available in BESTVALUE. In addition to the percent of students who borrow (PercBorrow), there is the admittance rate (Admit), the four-year graduation rate (Yr4Grad), in-state tuition after aid (InAfterAid), out-of-state tuition after aid (OutAfterAid), average aid per student (AvgAid), and the number of students per faculty member (StudPerFac).

(a) Generate scatterplots of each explanatory variable and AvgDebt. Do all these relationships look linear? Describe what you see.

(b) Fit each of the predictors separately and create a table that lists the explanatory variable, R^2, and the P-value for the F test. Which variable appears to be the best single predictor of average debt? Explain your answer.

III.13. Multitasking with technology in the classroom. Laptops and other digital technologies with wireless access to the Internet are becoming more and more common in the classroom. While numerous studies have shown that these technologies can be used effectively as part of teaching, there is concern that these technologies can also distract learners if used for off-task behaviors.

Recently, a study looked at the effects of off-task multitasking with digital technologies in the classroom.[5] A total of 145 undergraduates were randomly assigned to one of seven conditions. Each condition involved a task that was conducted simultaneously with a class lecture. The study consisted of three 20-minute lectures, each followed by a 15-item quiz. The score on the quiz is the response variable. The following table summarizes the conditions and results.

Condition	n	Lecture 1	Lecture 2	Lecture 3
Texting	21	0.57	0.75	0.56
Email	20	0.52	0.69	0.50
Facebook	20	0.50	0.68	0.43
MSN Messaging	21	0.48	0.71	0.42
Natural use control	21	0.50	0.78	0.58
Word-processing control	21	0.55	0.75	0.57
Paper-and-pencil control	21	0.60	0.74	0.53

Generate a plot to look at the interaction between lecture and condition. Do you think there is an interaction? If yes, describe the interaction.

III.14. Multitasking, continued. Refer to the previous exercise. Like Example 13.9 (page 602), this study uses a repeated-measures design. Because the paper reports that the interaction is nonsignificant ($P = 0.48$), let's average over the lectures to get a single score per student. Then we can perform a one-factor ANOVA to compare the conditions.

(a) Find the marginal means for condition.

(b) The analysis results in SS(condition) = 0.22178 and SSE = 2.00238. Test the null hypothesis that the mean scores across all conditions are equal.

(c) Using the marginal means from part (a) and the Bonferroni multiple-comparison method, determine which pairs of means differ significantly at the 0.05 significance level. (*Hint:* There are 21 pairwise comparisons, so the critical t value is 3.095.)

(d) Summarize your results from parts (b) and (c) in a short report.

III.15. Contrasts for multitasking. Refer to the previous two exercises. Let $\mu_1, \mu_2, \ldots, \mu_7$ represent the mean scores for the 7 conditions. The first 4 conditions refer to off-task behaviors, while the last 3 conditions represent different sorts of controls.

(a) The researchers hypothesized that the average score for the off-task behaviors would be lower than that for the paper-and-pencil control condition. Write a contrast that expresses this comparison.

(b) For this contrast, give H_0 and an appropriate H_a.

(c) Calculate the test statistic and approximate P-values for the significance test. What do you conclude?

III.16. Predicting college debt: combining measures. Refer to Exercises III.8 and III.12 for a description of the data set BESTVALUE. Now consider fitting a model using all the explanatory variables.

(a) Write out the statistical model for this analysis, making sure to specify all assumptions.

(b) Run the multiple regression model and specify the fitted regression equation.

(c) Obtain the residuals from part (b) and check assumptions. Comment on any unusual residuals or patterns in the residuals.

(d) What percent of the variability in average debt is explained by this model?

III.17. CHALLENGE **Predicting college debt: a simpler model.** Refer to the previous exercise. In the multiple regression analysis using all seven variables, only one variable (StudPerFac) is significant at the 0.05 level. Use the model selection methods described in Chapter 11 to find a multiple regression model that has only significant predictors.

III.18. Comparison of prediction intervals. Refer to the previous two exercises. The Ohio State University has Admit = 68, Yr4Grad = 49, StudPerFac = 19, InAfterAid = 12,680, OutAfterAid = 27,575, AvgAid = 7789, and PercBorrow = 52. Use your software to construct

(a) a 95% prediction interval based on the model with all the predictors.

(b) a 95% prediction interval based on the model using your simpler model.

(c) Compare the two intervals. Do the models give similar predictions and intervals?

III.19. Organic foods and morals? Organic foods are often marketed with moral terms such as "honesty" and "purity." Is this just a marketing strategy or is there a conceptual link between organic food and morality? In one recent experiment, 62 undergraduates were randomly assigned to one of three food conditions (organic, comfort, and control).[6] First, each participant was given a packet of four food types from the assigned condition and told to rate the desirability of each food on a 7-point scale. Then, each was presented with a list of six moral transgressions and asked to rate each on a 7-point scale ranging from 1 = not at all morally wrong to 7 = very morally wrong. The average of these six scores was used as the response. DATA FILE ORGANIC

(a) Make a table giving the sample size, mean, and standard deviation for each group. Is it reasonable to pool the variances?

(b) Generate a histogram for each of the groups. Can we feel confident that the sample means are approximately Normal? Explain your answer.

III.20. Organic foods and morals, continued. Refer to the previous exercise.

(a) Analyze the scores using analysis of variance. Report the test statistic, degrees of freedom, and P-value.

(b) Assess the assumptions necessary for inference by examining the residuals. Summarize your findings.

(c) Compare the groups using the least-significant differences method.

(d) A higher score is associated with a harsher moral judgment. Using the results from parts (a) and (b), write a short summary of your conclusions.

III.21. Organic foods and friendly behavior? Refer to Exercise III.19 for the design of the experiment. After rating the moral transgressions, the participants were told "that another professor from another department is also conducting research and really needs volunteers." They were told that they would not receive compensation or course credit for their help and then were asked to write down the number of minutes (out of 30) that they would be willing to volunteer. This sort of question is often used to measure a person's prosocial behavior.

(a) Figure III.3 contains the Minitab output for the analysis of this response variable. Write a one-paragraph summary of your conclusions.

(b) Figure III.4 contains a residual plot and a Normal quantile plot of the residuals. Are there any concerns regarding the assumptions necessary for inference? Explain your answer.

III.22. Influence of age and gender on motor performance. The slowing of motor performance as humans age is well established. Differences in gender, however, are less so. A recent study assessed the motor performance of 246 healthy adults.[7] One task was to tap the thumb and forefinger of the right hand together 20

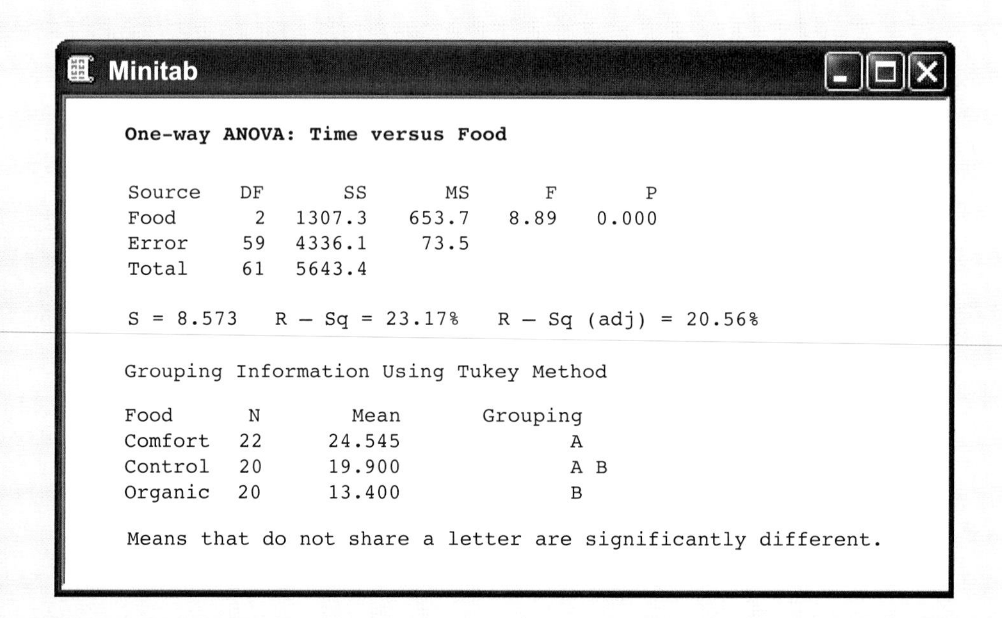

Minitab

One-way ANOVA: Time versus Food

Source	DF	SS	MS	F	P
Food	2	1307.3	653.7	8.89	0.000
Error	59	4336.1	73.5		
Total	61	5643.4			

S = 8.573 R – Sq = 23.17% R – Sq (adj) = 20.56%

Grouping Information Using Tukey Method

Food	N	Mean	Grouping
Comfort	22	24.545	A
Control	20	19.900	A B
Organic	20	13.400	B

Means that do not share a letter are significantly different.

FIGURE III.3 Minitab output for Exercise III.21.

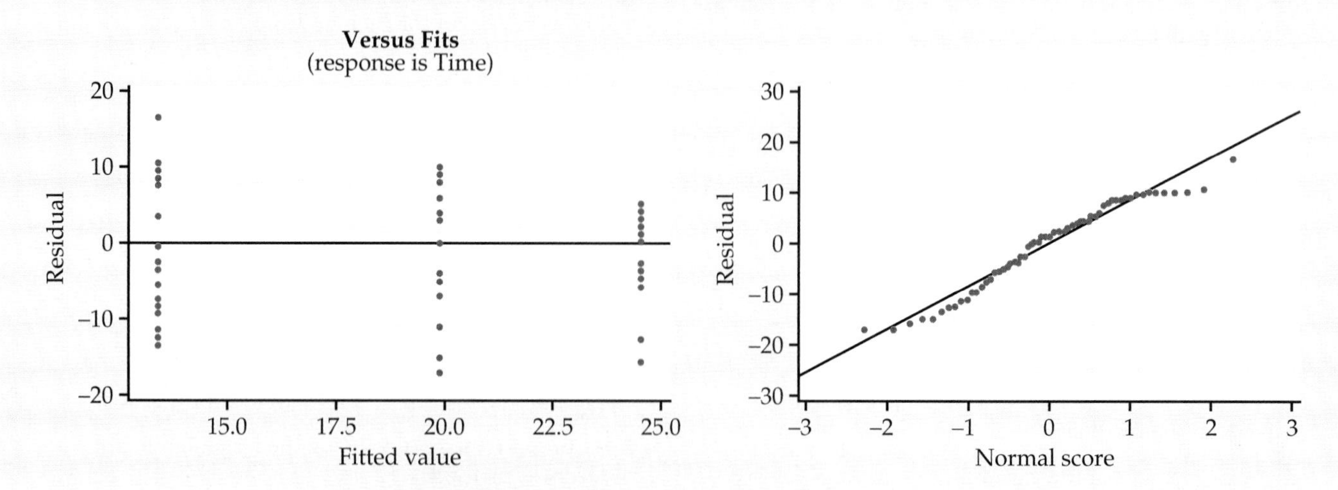

FIGURE III.4 Residual plot and Normal quantile plot for Exercise III.21.

times as quickly as possible. The following table summarizes this time (in seconds) for 7 age classes and 2 genders.

Age class (years)	Males			Females		
	n	$\bar{x}$	s	n	$\bar{x}$	s
41–50	19	4.72	1.31	19	5.88	0.82
51–55	12	4.10	1.62	12	5.93	1.13
56–60	12	4.80	1.04	12	5.85	0.87
61–65	24	5.08	0.98	24	5.81	0.94
66–70	17	5.47	0.85	17	6.50	1.23
71–75	23	5.84	1.44	23	6.12	1.04
> 75	16	5.86	1.00	16	6.19	0.91

Generate a plot to look at changes in the time over age class and across gender. Describe what you see in terms of the main effects for age and gender as well as their interaction.

III.23. Influence of age and gender on motor performance, continued. Refer to the previous exercise.

(a) In their article, the researchers state that each of their response variables was assessed for Normality prior to fitting a two-way ANOVA. Is it necessary for the 246 time measurements to be Normally distributed? Explain your answer.

(b) Is it reasonable to pool the variances?

(c) Suppose for these data that SS(gender) = 44.66, SS(age) = 31.97, SS(interaction) = 13.22, and SSE = 280.95. Construct an ANOVA table and state your conclusions.

III.24. Sexual harassment at middle and high schools. Refer to Exercise III.1, which describes a survey of middle and high school students concerning their experience with sexual harassment. (page 648)

(a) What are the odds that a girl has witnessed sexual harassment?

(b) What are the odds that a boy has witnessed sexual harassment?

(c) Find the odds ratio for girls versus boys.

III.25. Inference for sexual harassment at middle and high schools. Refer to the previous exercise.

(a) Run a logistic regression to predict the log odds of witnessing sexual harassment using gender as a predictor. For the explanatory variable, define X to have values 1 for girls and 0 for boys. Report the estimates of β_0 and β_1.

(b) Give confidence intervals for the estimates.

(c) Report the value of the Wald statistic and its P-value. What do you conclude?

(d) Verify that e^{β_1} is equal to the odds ratio that you calculated in part (c) of the previous exercise.

III.26. The value of online courses. Refer to Exercise III.4, which describes a survey of college presidents with respect to their views on the value of online courses (page 648). Define X_1 to be 1 if the college president is from a four-year private institution and 0 otherwise. Define X_2 to be 1 if the college president is from a four-year public institution and 0 otherwise. Define X_3 to be 1 if the college president is from a two-year private institution and 0 otherwise.

(a) Use the variables X_1, X_2, and X_3 as explanatory variables in a multiple logistic regression to predict the response to the question about whether or not they believe that online courses offer an equal educational value when compared with courses taken in the classroom. Summarize your analysis, results, and conclusions.

(b) Compare what you have found in this exercise with what you found in Exercise III.4.

(c) Which approach do you prefer? Give reasons for your answer.

III.27. Predicting six-year college graduation rates. Many students begin study for a college degree but do not complete it. A major study of degree completion rates used data from 210,056 first-time, full-time students at 356 four-year nonprofit institutions. Logistic regression was the major analytical approach used to examine these data.[8] For one analysis, the response variable Y was 1 if the student graduated in 6 or fewer years, and 0 otherwise. The explanatory variable X was average high school grade, coded as 8 for A+ or A, 7 for A−, 6 for B+, 5 for B, 4 for B−, 3 for C+, 2 for C, and 1 for D. The constant, or intercept, was −2.357, and the coefficient for average high school grade was 0.461.

(a) Write out the fitted logistic regression model for this analysis.

(b) Use the model to find the predicted log odds for a student whose high school grade average is C+.

(c) Use your answer in part (b) to find the probability that the student will graduate in 6 or fewer years.

(d) Repeat parts (b) and (c) for a student whose high school grade average is A.

III.28. Make a plot. Refer to the previous exercise. Compute the probability of graduating in 6 or fewer years for several additional values of high school grade average, and make a plot that shows the relationship between this probability and high school grade average. Write a short summary describing the relationship.

III.29. Predicting six-year college graduation rate with two variables. The study described in Exercise III.27 also examined multiple logistic regression models where more than one explanatory variable was used. One of these analyses used average high school grade and SAT score to predict the six-year graduation rate. SAT was defined to be the sum of the Verbal and Mathematics scores divided by 10. The estimated coefficients were 0.341 for high school grade average, 0.024 for SAT, and −4.206 for the intercept.

(a) Write out the fitted logistic regression model for this analysis.

(b) Use the model to find the predicted log odds for a student whose high school grade average is C+ and whose SAT score is 85 (a total of 850 for the Verbal and Mathematics scores).

(c) Use your answer in part (b) to find the probability that the student will graduate in 6 or fewer years.

(d) Repeat parts (b) and (c) for a student whose high school grade average is A and whose SAT score is 120.

TABLES

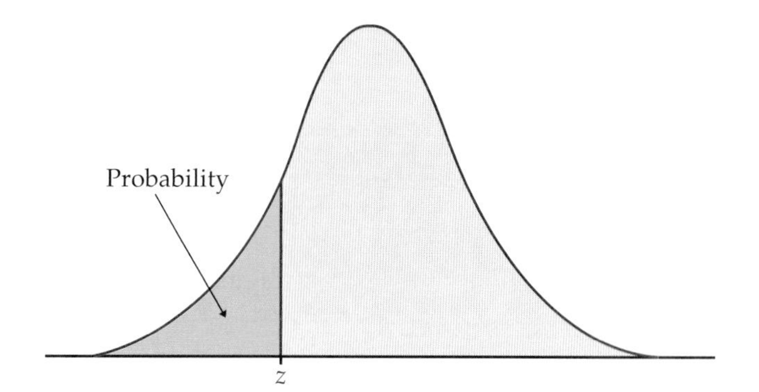

Table entry for z is the area under the standard Normal curve to the left of z.

TABLE A

Standard Normal probabilities

z	.00	.01	.02	.03	.04	.05	.06	.07	.08	.09
−3.4	.0003	.0003	.0003	.0003	.0003	.0003	.0003	.0003	.0003	.0002
−3.3	.0005	.0005	.0005	.0004	.0004	.0004	.0004	.0004	.0004	.0003
−3.2	.0007	.0007	.0006	.0006	.0006	.0006	.0006	.0005	.0005	.0005
−3.1	.0010	.0009	.0009	.0009	.0008	.0008	.0008	.0008	.0007	.0007
−3.0	.0013	.0013	.0013	.0012	.0012	.0011	.0011	.0011	.0010	.0010
−2.9	.0019	.0018	.0018	.0017	.0016	.0016	.0015	.0015	.0014	.0014
−2.8	.0026	.0025	.0024	.0023	.0023	.0022	.0021	.0021	.0020	.0019
−2.7	.0035	.0034	.0033	.0032	.0031	.0030	.0029	.0028	.0027	.0026
−2.6	.0047	.0045	.0044	.0043	.0041	.0040	.0039	.0038	.0037	.0036
−2.5	.0062	.0060	.0059	.0057	.0055	.0054	.0052	.0051	.0049	.0048
−2.4	.0082	.0080	.0078	.0075	.0073	.0071	.0069	.0068	.0066	.0064
−2.3	.0107	.0104	.0102	.0099	.0096	.0094	.0091	.0089	.0087	.0084
−2.2	.0139	.0136	.0132	.0129	.0125	.0122	.0119	.0116	.0113	.0110
−2.1	.0179	.0174	.0170	.0166	.0162	.0158	.0154	.0150	.0146	.0143
−2.0	.0228	.0222	.0217	.0212	.0207	.0202	.0197	.0192	.0188	.0183
−1.9	.0287	.0281	.0274	.0268	.0262	.0256	.0250	.0244	.0239	.0233
−1.8	.0359	.0351	.0344	.0336	.0329	.0322	.0314	.0307	.0301	.0294
−1.7	.0446	.0436	.0427	.0418	.0409	.0401	.0392	.0384	.0375	.0367
−1.6	.0548	.0537	.0526	.0516	.0505	.0495	.0485	.0475	.0465	.0455
−1.5	.0668	.0655	.0643	.0630	.0618	.0606	.0594	.0582	.0571	.0559
−1.4	.0808	.0793	.0778	.0764	.0749	.0735	.0721	.0708	.0694	.0681
−1.3	.0968	.0951	.0934	.0918	.0901	.0885	.0869	.0853	.0838	.0823
−1.2	.1151	.1131	.1112	.1093	.1075	.1056	.1038	.1020	.1003	.0985
−1.1	.1357	.1335	.1314	.1292	.1271	.1251	.1230	.1210	.1190	.1170
−1.0	.1587	.1562	.1539	.1515	.1492	.1469	.1446	.1423	.1401	.1379
−0.9	.1841	.1814	.1788	.1762	.1736	.1711	.1685	.1660	.1635	.1611
−0.8	.2119	.2090	.2061	.2033	.2005	.1977	.1949	.1922	.1894	.1867
−0.7	.2420	.2389	.2358	.2327	.2296	.2266	.2236	.2206	.2177	.2148
−0.6	.2743	.2709	.2676	.2643	.2611	.2578	.2546	.2514	.2483	.2451
−0.5	.3085	.3050	.3015	.2981	.2946	.2912	.2877	.2843	.2810	.2776
−0.4	.3446	.3409	.3372	.3336	.3300	.3264	.3228	.3192	.3156	.3121
−0.3	.3821	.3783	.3745	.3707	.3669	.3632	.3594	.3557	.3520	.3483
−0.2	.4207	.4168	.4129	.4090	.4052	.4013	.3974	.3936	.3897	.3859
−0.1	.4602	.4562	.4522	.4483	.4443	.4404	.4364	.4325	.4286	.4247
−0.0	.5000	.4960	.4920	.4880	.4840	.4801	.4761	.4721	.4681	.4641

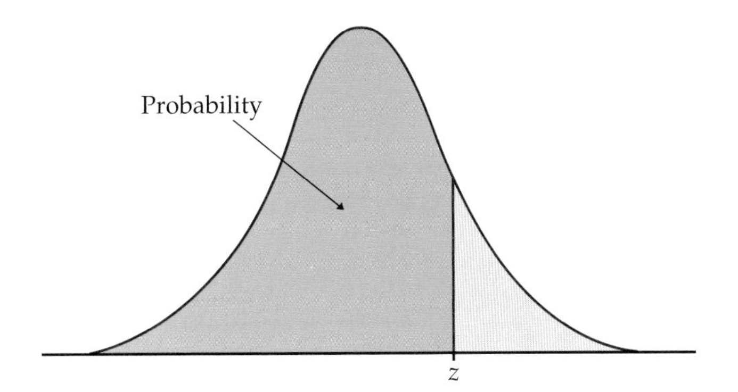

Table entry for z is the area under the standard Normal curve to the left of z.

TABLE A

Standard Normal probabilities (continued)

z	.00	.01	.02	.03	.04	.05	.06	.07	.08	.09
0.0	.5000	.5040	.5080	.5120	.5160	.5199	.5239	.5279	.5319	.5359
0.1	.5398	.5438	.5478	.5517	.5557	.5596	.5636	.5675	.5714	.5753
0.2	.5793	.5832	.5871	.5910	.5948	.5987	.6026	.6064	.6103	.6141
0.3	.6179	.6217	.6255	.6293	.6331	.6368	.6406	.6443	.6480	.6517
0.4	.6554	.6591	.6628	.6664	.6700	.6736	.6772	.6808	.6844	.6879
0.5	.6915	.6950	.6985	.7019	.7054	.7088	.7123	.7157	.7190	.7224
0.6	.7257	.7291	.7324	.7357	.7389	.7422	.7454	.7486	.7517	.7549
0.7	.7580	.7611	.7642	.7673	.7704	.7734	.7764	.7794	.7823	.7852
0.8	.7881	.7910	.7939	.7967	.7995	.8023	.8051	.8078	.8106	.8133
0.9	.8159	.8186	.8212	.8238	.8264	.8289	.8315	.8340	.8365	.8389
1.0	.8413	.8438	.8461	.8485	.8508	.8531	.8554	.8577	.8599	.8621
1.1	.8643	.8665	.8686	.8708	.8729	.8749	.8770	.8790	.8810	.8830
1.2	.8849	.8869	.8888	.8907	.8925	.8944	.8962	.8980	.8997	.9015
1.3	.9032	.9049	.9066	.9082	.9099	.9115	.9131	.9147	.9162	.9177
1.4	.9192	.9207	.9222	.9236	.9251	.9265	.9279	.9292	.9306	.9319
1.5	.9332	.9345	.9357	.9370	.9382	.9394	.9406	.9418	.9429	.9441
1.6	.9452	.9463	.9474	.9484	.9495	.9505	.9515	.9525	.9535	.9545
1.7	.9554	.9564	.9573	.9582	.9591	.9599	.9608	.9616	.9625	.9633
1.8	.9641	.9649	.9656	.9664	.9671	.9678	.9686	.9693	.9699	.9706
1.9	.9713	.9719	.9726	.9732	.9738	.9744	.9750	.9756	.9761	.9767
2.0	.9772	.9778	.9783	.9788	.9793	.9798	.9803	.9808	.9812	.9817
2.1	.9821	.9826	.9830	.9834	.9838	.9842	.9846	.9850	.9854	.9857
2.2	.9861	.9864	.9868	.9871	.9875	.9878	.9881	.9884	.9887	.9890
2.3	.9893	.9896	.9898	.9901	.9904	.9906	.9909	.9911	.9913	.9916
2.4	.9918	.9920	.9922	.9925	.9927	.9929	.9931	.9932	.9934	.9936
2.5	.9938	.9940	.9941	.9943	.9945	.9946	.9948	.9949	.9951	.9952
2.6	.9953	.9955	.9956	.9957	.9959	.9960	.9961	.9962	.9963	.9964
2.7	.9965	.9966	.9967	.9968	.9969	.9970	.9971	.9972	.9973	.9974
2.8	.9974	.9975	.9976	.9977	.9977	.9978	.9979	.9979	.9980	.9981
2.9	.9981	.9982	.9982	.9983	.9984	.9984	.9985	.9985	.9986	.9986
3.0	.9987	.9987	.9987	.9988	.9988	.9989	.9989	.9989	.9990	.9990
3.1	.9990	.9991	.9991	.9991	.9992	.9992	.9992	.9992	.9993	.9993
3.2	.9993	.9993	.9994	.9994	.9994	.9994	.9994	.9995	.9995	.9995
3.3	.9995	.9995	.9995	.9996	.9996	.9996	.9996	.9996	.9996	.9997
3.4	.9997	.9997	.9997	.9997	.9997	.9997	.9997	.9997	.9997	.9998

TABLE B

Random digits

Line								
101	19223	95034	05756	28713	96409	12531	42544	82853
102	73676	47150	99400	01927	27754	42648	82425	36290
103	45467	71709	77558	00095	32863	29485	82226	90056
104	52711	38889	93074	60227	40011	85848	48767	52573
105	95592	94007	69971	91481	60779	53791	17297	59335
106	68417	35013	15529	72765	85089	57067	50211	47487
107	82739	57890	20807	47511	81676	55300	94383	14893
108	60940	72024	17868	24943	61790	90656	87964	18883
109	36009	19365	15412	39638	85453	46816	83485	41979
110	38448	48789	18338	24697	39364	42006	76688	08708
111	81486	69487	60513	09297	00412	71238	27649	39950
112	59636	88804	04634	71197	19352	73089	84898	45785
113	62568	70206	40325	03699	71080	22553	11486	11776
114	45149	32992	75730	66280	03819	56202	02938	70915
115	61041	77684	94322	24709	73698	14526	31893	32592
116	14459	26056	31424	80371	65103	62253	50490	61181
117	38167	98532	62183	70632	23417	26185	41448	75532
118	73190	32533	04470	29669	84407	90785	65956	86382
119	95857	07118	87664	92099	58806	66979	98624	84826
120	35476	55972	39421	65850	04266	35435	43742	11937
121	71487	09984	29077	14863	61683	47052	62224	51025
122	13873	81598	95052	90908	73592	75186	87136	95761
123	54580	81507	27102	56027	55892	33063	41842	81868
124	71035	09001	43367	49497	72719	96758	27611	91596
125	96746	12149	37823	71868	18442	35119	62103	39244
126	96927	19931	36089	74192	77567	88741	48409	41903
127	43909	99477	25330	64359	40085	16925	85117	36071
128	15689	14227	06565	14374	13352	49367	81982	87209
129	36759	58984	68288	22913	18638	54303	00795	08727
130	69051	64817	87174	09517	84534	06489	87201	97245
131	05007	16632	81194	14873	04197	85576	45195	96565
132	68732	55259	84292	08796	43165	93739	31685	97150
133	45740	41807	65561	33302	07051	93623	18132	09547
134	27816	78416	18329	21337	35213	37741	04312	68508
135	66925	55658	39100	78458	11206	19876	87151	31260
136	08421	44753	77377	28744	75592	08563	79140	92454
137	53645	66812	61421	47836	12609	15373	98481	14592
138	66831	68908	40772	21558	47781	33586	79177	06928
139	55588	99404	70708	41098	43563	56934	48394	51719
140	12975	13258	13048	45144	72321	81940	00360	02428
141	96767	35964	23822	96012	94591	65194	50842	53372
142	72829	50232	97892	63408	77919	44575	24870	04178
143	88565	42628	17797	49376	61762	16953	88604	12724
144	62964	88145	83083	69453	46109	59505	69680	00900
145	19687	12633	57857	95806	09931	02150	43163	58636
146	37609	59057	66967	83401	60705	02384	90597	93600
147	54973	86278	88737	74351	47500	84552	19909	67181
148	00694	05977	19664	65441	20903	62371	22725	53340
149	71546	05233	53946	68743	72460	27601	45403	88692
150	07511	88915	41267	16853	84569	79367	32337	03316

TABLE B

Random digits (continued)

Line								
151	03802	29341	29264	80198	12371	13121	54969	43912
152	77320	35030	77519	41109	98296	18984	60869	12349
153	07886	56866	39648	69290	03600	05376	58958	22720
154	87065	74133	21117	70595	22791	67306	28420	52067
155	42090	09628	54035	93879	98441	04606	27381	82637
156	55494	67690	88131	81800	11188	28552	25752	21953
157	16698	30406	96587	65985	07165	50148	16201	86792
158	16297	07626	68683	45335	34377	72941	41764	77038
159	22897	17467	17638	70043	36243	13008	83993	22869
160	98163	45944	34210	64158	76971	27689	82926	75957
161	43400	25831	06283	22138	16043	15706	73345	26238
162	97341	46254	88153	62336	21112	35574	99271	45297
163	64578	67197	28310	90341	37531	63890	52630	76315
164	11022	79124	49525	63078	17229	32165	01343	21394
165	81232	43939	23840	05995	84589	06788	76358	26622
166	36843	84798	51167	44728	20554	55538	27647	32708
167	84329	80081	69516	78934	14293	92478	16479	26974
168	27788	85789	41592	74472	96773	27090	24954	41474
169	99224	00850	43737	75202	44753	63236	14260	73686
170	38075	73239	52555	46342	13365	02182	30443	53229
171	87368	49451	55771	48343	51236	18522	73670	23212
172	40512	00681	44282	47178	08139	78693	34715	75606
173	81636	57578	54286	27216	58758	80358	84115	84568
174	26411	94292	06340	97762	37033	85968	94165	46514
175	80011	09937	57195	33906	94831	10056	42211	65491
176	92813	87503	63494	71379	76550	45984	05481	50830
177	70348	72871	63419	57363	29685	43090	18763	31714
178	24005	52114	26224	39078	80798	15220	43186	00976
179	85063	55810	10470	08029	30025	29734	61181	72090
180	11532	73186	92541	06915	72954	10167	12142	26492
181	59618	03914	05208	84088	20426	39004	84582	87317
182	92965	50837	39921	84661	82514	81899	24565	60874
183	85116	27684	14597	85747	01596	25889	41998	15635
184	15106	10411	90221	49377	44369	28185	80959	76355
185	03638	31589	07871	25792	85823	55400	56026	12193
186	97971	48932	45792	63993	95635	28753	46069	84635
187	49345	18305	76213	82390	77412	97401	50650	71755
188	87370	88099	89695	87633	76987	85503	26257	51736
189	88296	95670	74932	65317	93848	43988	47597	83044
190	79485	92200	99401	54473	34336	82786	05457	60343
191	40830	24979	23333	37619	56227	95941	59494	86539
192	32006	76302	81221	00693	95197	75044	46596	11628
193	37569	85187	44692	50706	53161	69027	88389	60313
194	56680	79003	23361	67094	15019	63261	24543	52884
195	05172	08100	22316	54495	60005	29532	18433	18057
196	74782	27005	03894	98038	20627	40307	47317	92759
197	85288	93264	61409	03404	09649	55937	60843	66167
198	68309	12060	14762	58002	03716	81968	57934	32624
199	26461	88346	52430	60906	74216	96263	69296	90107
200	42672	67680	42376	95023	82744	03971	96560	55148

TABLE C

Binomial probabilities

Entry is $P(X = k) = \binom{n}{k} p^k (1 - p)^{n-k}$

		p								
n	*k*	.01	.02	.03	.04	.05	.06	.07	.08	.09
2	0	.9801	.9604	.9409	.9216	.9025	.8836	.8649	.8464	.8281
	1	.0198	.0392	.0582	.0768	.0950	.1128	.1302	.1472	.1638
	2	.0001	.0004	.0009	.0016	.0025	.0036	.0049	.0064	.0081
3	0	.9703	.9412	.9127	.8847	.8574	.8306	.8044	.7787	.7536
	1	.0294	.0576	.0847	.1106	.1354	.1590	.1816	.2031	.2236
	2	.0003	.0012	.0026	.0046	.0071	.0102	.0137	.0177	.0221
	3				.0001	.0001	.0002	.0003	.0005	.0007
4	0	.9606	.9224	.8853	.8493	.8145	.7807	.7481	.7164	.6857
	1	.0388	.0753	.1095	.1416	.1715	.1993	.2252	.2492	.2713
	2	.0006	.0023	.0051	.0088	.0135	.0191	.0254	.0325	.0402
	3			.0001	.0002	.0005	.0008	.0013	.0019	.0027
	4									.0001
5	0	.9510	.9039	.8587	.8154	.7738	.7339	.6957	.6591	.6240
	1	.0480	.0922	.1328	.1699	.2036	.2342	.2618	.2866	.3086
	2	.0010	.0038	.0082	.0142	.0214	.0299	.0394	.0498	.0610
	3		.0001	.0003	.0006	.0011	.0019	.0030	.0043	.0060
	4						.0001	.0001	.0002	.0003
	5									
6	0	.9415	.8858	.8330	.7828	.7351	.6899	.6470	.6064	.5679
	1	.0571	.1085	.1546	.1957	.2321	.2642	.2922	.3164	.3370
	2	.0014	.0055	.0120	.0204	.0305	.0422	.0550	.0688	.0833
	3		.0002	.0005	.0011	.0021	.0036	.0055	.0080	.0110
	4					.0001	.0002	.0003	.0005	.0008
	5									
	6									
7	0	.9321	.8681	.8080	.7514	.6983	.6485	.6017	.5578	.5168
	1	.0659	.1240	.1749	.2192	.2573	.2897	.3170	.3396	.3578
	2	.0020	.0076	.0162	.0274	.0406	.0555	.0716	.0886	.1061
	3		.0003	.0008	.0019	.0036	.0059	.0090	.0128	.0175
	4				.0001	.0002	.0004	.0007	.0011	.0017
	5								.0001	.0001
	6									
	7									
8	0	.9227	.8508	.7837	.7214	.6634	.6096	.5596	.5132	.4703
	1	.0746	.1389	.1939	.2405	.2793	.3113	.3370	.3570	.3721
	2	.0026	.0099	.0210	.0351	.0515	.0695	.0888	.1087	.1288
	3	.0001	.0004	.0013	.0029	.0054	.0089	.0134	.0189	.0255
	4			.0001	.0002	.0004	.0007	.0013	.0021	.0031
	5							.0001	.0001	.0002
	6									
	7									
	8									

TABLE C

Binomial probabilities (continued)

Entry is $P(X = k) = \binom{n}{k} p^k (1-p)^{n-k}$

n	k	.10	.15	.20	.25	.30	.35	.40	.45	.50
2	0	.8100	.7225	.6400	.5625	.4900	.4225	.3600	.3025	.2500
	1	.1800	.2550	.3200	.3750	.4200	.4550	.4800	.4950	.5000
	2	.0100	.0225	.0400	.0625	.0900	.1225	.1600	.2025	.2500
3	0	.7290	.6141	.5120	.4219	.3430	.2746	.2160	.1664	.1250
	1	.2430	.3251	.3840	.4219	.4410	.4436	.4320	.4084	.3750
	2	.0270	.0574	.0960	.1406	.1890	.2389	.2880	.3341	.3750
	3	.0010	.0034	.0080	.0156	.0270	.0429	.0640	.0911	.1250
4	0	.6561	.5220	.4096	.3164	.2401	.1785	.1296	.0915	.0625
	1	.2916	.3685	.4096	.4219	.4116	.3845	.3456	.2995	.2500
	2	.0486	.0975	.1536	.2109	.2646	.3105	.3456	.3675	.3750
	3	.0036	.0115	.0256	.0469	.0756	.1115	.1536	.2005	.2500
	4	.0001	.0005	.0016	.0039	.0081	.0150	.0256	.0410	.0625
5	0	.5905	.4437	.3277	.2373	.1681	.1160	.0778	.0503	.0313
	1	.3280	.3915	.4096	.3955	.3602	.3124	.2592	.2059	.1563
	2	.0729	.1382	.2048	.2637	.3087	.3364	.3456	.3369	.3125
	3	.0081	.0244	.0512	.0879	.1323	.1811	.2304	.2757	.3125
	4	.0004	.0022	.0064	.0146	.0284	.0488	.0768	.1128	.1562
	5		.0001	.0003	.0010	.0024	.0053	.0102	.0185	.0312
6	0	.5314	.3771	.2621	.1780	.1176	.0754	.0467	.0277	.0156
	1	.3543	.3993	.3932	.3560	.3025	.2437	.1866	.1359	.0938
	2	.0984	.1762	.2458	.2966	.3241	.3280	.3110	.2780	.2344
	3	.0146	.0415	.0819	.1318	.1852	.2355	.2765	.3032	.3125
	4	.0012	.0055	.0154	.0330	.0595	.0951	.1382	.1861	.2344
	5	.0001	.0004	.0015	.0044	.0102	.0205	.0369	.0609	.0937
	6			.0001	.0002	.0007	.0018	.0041	.0083	.0156
7	0	.4783	.3206	.2097	.1335	.0824	.0490	.0280	.0152	.0078
	1	.3720	.3960	.3670	.3115	.2471	.1848	.1306	.0872	.0547
	2	.1240	.2097	.2753	.3115	.3177	.2985	.2613	.2140	.1641
	3	.0230	.0617	.1147	.1730	.2269	.2679	.2903	.2918	.2734
	4	.0026	.0109	.0287	.0577	.0972	.1442	.1935	.2388	.2734
	5	.0002	.0012	.0043	.0115	.0250	.0466	.0774	.1172	.1641
	6		.0001	.0004	.0013	.0036	.0084	.0172	.0320	.0547
	7				.0001	.0002	.0006	.0016	.0037	.0078
8	0	.4305	.2725	.1678	.1001	.0576	.0319	.0168	.0084	.0039
	1	.3826	.3847	.3355	.2670	.1977	.1373	.0896	.0548	.0313
	2	.1488	.2376	.2936	.3115	.2965	.2587	.2090	.1569	.1094
	3	.0331	.0839	.1468	.2076	.2541	.2786	.2787	.2568	.2188
	4	.0046	.0185	.0459	.0865	.1361	.1875	.2322	.2627	.2734
	5	.0004	.0026	.0092	.0231	.0467	.0808	.1239	.1719	.2188
	6		.0002	.0011	.0038	.0100	.0217	.0413	.0703	.1094
	7			.0001	.0004	.0012	.0033	.0079	.0164	.0312
	8					.0001	.0002	.0007	.0017	.0039

(Continued)

TABLE C

Binomial probabilities (continued)

Entry is $P(X = k) = \binom{n}{k} p^k (1-p)^{n-k}$

n	k	.01	.02	.03	.04	.05	.06	.07	.08	.09
9	0	.9135	.8337	.7602	.6925	.6302	.5730	.5204	.4722	.4279
	1	.0830	.1531	.2116	.2597	.2985	.3292	.3525	.3695	.3809
	2	.0034	.0125	.0262	.0433	.0629	.0840	.1061	.1285	.1507
	3	.0001	.0006	.0019	.0042	.0077	.0125	.0186	.0261	.0348
	4			.0001	.0003	.0006	.0012	.0021	.0034	.0052
	5						.0001	.0002	.0003	.0005
	6									
	7									
	8									
	9									
10	0	.9044	.8171	.7374	.6648	.5987	.5386	.4840	.4344	.3894
	1	.0914	.1667	.2281	.2770	.3151	.3438	.3643	.3777	.3851
	2	.0042	.0153	.0317	.0519	.0746	.0988	.1234	.1478	.1714
	3	.0001	.0008	.0026	.0058	.0105	.0168	.0248	.0343	.0452
	4			.0001	.0004	.0010	.0019	.0033	.0052	.0078
	5					.0001	.0001	.0003	.0005	.0009
	6									.0001
	7									
	8									
	9									
	10									
12	0	.8864	.7847	.6938	.6127	.5404	.4759	.4186	.3677	.3225
	1	.1074	.1922	.2575	.3064	.3413	.3645	.3781	.3837	.3827
	2	.0060	.0216	.0438	.0702	.0988	.1280	.1565	.1835	.2082
	3	.0002	.0015	.0045	.0098	.0173	.0272	.0393	.0532	.0686
	4		.0001	.0003	.0009	.0021	.0039	.0067	.0104	.0153
	5				.0001	.0002	.0004	.0008	.0014	.0024
	6							.0001	.0001	.0003
	7									
	8									
	9									
	10									
	11									
	12									
15	0	.8601	.7386	.6333	.5421	.4633	.3953	.3367	.2863	.2430
	1	.1303	.2261	.2938	.3388	.3658	.3785	.3801	.3734	.3605
	2	.0092	.0323	.0636	.0988	.1348	.1691	.2003	.2273	.2496
	3	.0004	.0029	.0085	.0178	.0307	.0468	.0653	.0857	.1070
	4		.0002	.0008	.0022	.0049	.0090	.0148	.0223	.0317
	5			.0001	.0002	.0006	.0013	.0024	.0043	.0069
	6						.0001	.0003	.0006	.0011
	7								.0001	.0001
	8									
	9									
	10									
	11									
	12									
	13									
	14									
	15									

TABLE C

Binomial probabilities (continued)

Entry is $P(X = k) = \binom{n}{k} p^k (1-p)^{n-k}$

		p								
n	*k*	.10	.15	.20	.25	.30	.35	.40	.45	.50
9	0	.3874	.2316	.1342	.0751	.0404	.0207	.0101	.0046	.0020
	1	.3874	.3679	.3020	.2253	.1556	.1004	.0605	.0339	.0176
	2	.1722	.2597	.3020	.3003	.2668	.2162	.1612	.1110	.0703
	3	.0446	.1069	.1762	.2336	.2668	.2716	.2508	.2119	.1641
	4	.0074	.0283	.0661	.1168	.1715	.2194	.2508	.2600	.2461
	5	.0008	.0050	.0165	.0389	.0735	.1181	.1672	.2128	.2461
	6	.0001	.0006	.0028	.0087	.0210	.0424	.0743	.1160	.1641
	7			.0003	.0012	.0039	.0098	.0212	.0407	.0703
	8				.0001	.0004	.0013	.0035	.0083	.0176
	9						.0001	.0003	.0008	.0020
10	0	.3487	.1969	.1074	.0563	.0282	.0135	.0060	.0025	.0010
	1	.3874	.3474	.2684	.1877	.1211	.0725	.0403	.0207	.0098
	2	.1937	.2759	.3020	.2816	.2335	.1757	.1209	.0763	.0439
	3	.0574	.1298	.2013	.2503	.2668	.2522	.2150	.1665	.1172
	4	.0112	.0401	.0881	.1460	.2001	.2377	.2508	.2384	.2051
	5	.0015	.0085	.0264	.0584	.1029	.1536	.2007	.2340	.2461
	6	.0001	.0012	.0055	.0162	.0368	.0689	.1115	.1596	.2051
	7		.0001	.0008	.0031	.0090	.0212	.0425	.0746	.1172
	8			.0001	.0004	.0014	.0043	.0106	.0229	.0439
	9					.0001	.0005	.0016	.0042	.0098
	10							.0001	.0003	.0010
12	0	.2824	.1422	.0687	.0317	.0138	.0057	.0022	.0008	.0002
	1	.3766	.3012	.2062	.1267	.0712	.0368	.0174	.0075	.0029
	2	.2301	.2924	.2835	.2323	.1678	.1088	.0639	.0339	.0161
	3	.0852	.1720	.2362	.2581	.2397	.1954	.1419	.0923	.0537
	4	.0213	.0683	.1329	.1936	.2311	.2367	.2128	.1700	.1208
	5	.0038	.0193	.0532	.1032	.1585	.2039	.2270	.2225	.1934
	6	.0005	.0040	.0155	.0401	.0792	.1281	.1766	.2124	.2256
	7		.0006	.0033	.0115	.0291	.0591	.1009	.1489	.1934
	8		.0001	.0005	.0024	.0078	.0199	.0420	.0762	.1208
	9			.0001	.0004	.0015	.0048	.0125	.0277	.0537
	10					.0002	.0008	.0025	.0068	.0161
	11						.0001	.0003	.0010	.0029
	12								.0001	.0002
15	0	.2059	.0874	.0352	.0134	.0047	.0016	.0005	.0001	.0000
	1	.3432	.2312	.1319	.0668	.0305	.0126	.0047	.0016	.0005
	2	.2669	.2856	.2309	.1559	.0916	.0476	.0219	.0090	.0032
	3	.1285	.2184	.2501	.2252	.1700	.1110	.0634	.0318	.0139
	4	.0428	.1156	.1876	.2252	.2186	.1792	.1268	.0780	.0417
	5	.0105	.0449	.1032	.1651	.2061	.2123	.1859	.1404	.0916
	6	.0019	.0132	.0430	.0917	.1472	.1906	.2066	.1914	.1527
	7	.0003	.0030	.0138	.0393	.0811	.1319	.1771	.2013	.1964
	8		.0005	.0035	.0131	.0348	.0710	.1181	.1647	.1964
	9		.0001	.0007	.0034	.0116	.0298	.0612	.1048	.1527
	10			.0001	.0007	.0030	.0096	.0245	.0515	.0916
	11				.0001	.0006	.0024	.0074	.0191	.0417
	12					.0001	.0004	.0016	.0052	.0139
	13						.0001	.0003	.0010	.0032
	14								.0001	.0005
	15									

(Continued)

TABLE C

Binomial probabilities (continued)

		p								
n	*k*	.01	.02	.03	.04	.05	.06	.07	.08	.09
20	0	.8179	.6676	.5438	.4420	.3585	.2901	.2342	.1887	.1516
	1	.1652	.2725	.3364	.3683	.3774	.3703	.3526	.3282	.3000
	2	.0159	.0528	.0988	.1458	.1887	.2246	.2521	.2711	.2818
	3	.0010	.0065	.0183	.0364	.0596	.0860	.1139	.1414	.1672
	4		.0006	.0024	.0065	.0133	.0233	.0364	.0523	.0703
	5			.0002	.0009	.0022	.0048	.0088	.0145	.0222
	6				.0001	.0003	.0008	.0017	.0032	.0055
	7						.0001	.0002	.0005	.0011
	8								.0001	.0002
	9									
	10									
	11									
	12									
	13									
	14									
	15									
	16									
	17									
	18									
	19									
	20									

		p								
n	*k*	.10	.15	.20	.25	.30	.35	.40	.45	.50
20	0	.1216	.0388	.0115	.0032	.0008	.0002	.0000	.0000	.0000
	1	.2702	.1368	.0576	.0211	.0068	.0020	.0005	.0001	.0000
	2	.2852	.2293	.1369	.0669	.0278	.0100	.0031	.0008	.0002
	3	.1901	.2428	.2054	.1339	.0716	.0323	.0123	.0040	.0011
	4	.0898	.1821	.2182	.1897	.1304	.0738	.0350	.0139	.0046
	5	.0319	.1028	.1746	.2023	.1789	.1272	.0746	.0365	.0148
	6	.0089	.0454	.1091	.1686	.1916	.1712	.1244	.0746	.0370
	7	.0020	.0160	.0545	.1124	.1643	.1844	.1659	.1221	.0739
	8	.0004	.0046	.0222	.0609	.1144	.1614	.1797	.1623	.1201
	9	.0001	.0011	.0074	.0271	.0654	.1158	.1597	.1771	.1602
	10		.0002	.0020	.0099	.0308	.0686	.1171	.1593	.1762
	11			.0005	.0030	.0120	.0336	.0710	.1185	.1602
	12			.0001	.0008	.0039	.0136	.0355	.0727	.1201
	13				.0002	.0010	.0045	.0146	.0366	.0739
	14					.0002	.0012	.0049	.0150	.0370
	15						.0003	.0013	.0049	.0148
	16							.0003	.0013	.0046
	17								.0002	.0011
	18									.0002
	19									
	20									

Table entry for p and C is the critical value t^* with probability p lying to its right and probability C lying between $-t^*$ and t^*.

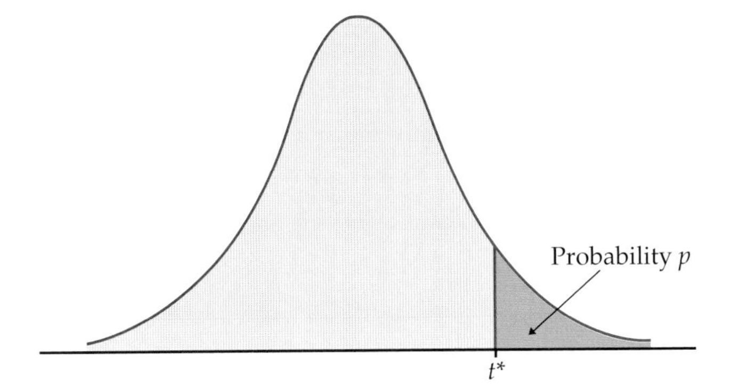

TABLE D

t distribution critical values

df	Upper-tail probability p											
df	.25	.20	.15	.10	.05	.025	.02	.01	.005	.0025	.001	.0005
1	1.000	1.376	1.963	3.078	6.314	12.71	15.89	31.82	63.66	127.3	318.3	636.6
2	0.816	1.061	1.386	1.886	2.920	4.303	4.849	6.965	9.925	14.09	22.33	31.60
3	0.765	0.978	1.250	1.638	2.353	3.182	3.482	4.541	5.841	7.453	10.21	12.92
4	0.741	0.941	1.190	1.533	2.132	2.776	2.999	3.747	4.604	5.598	7.173	8.610
5	0.727	0.920	1.156	1.476	2.015	2.571	2.757	3.365	4.032	4.773	5.893	6.869
6	0.718	0.906	1.134	1.440	1.943	2.447	2.612	3.143	3.707	4.317	5.208	5.959
7	0.711	0.896	1.119	1.415	1.895	2.365	2.517	2.998	3.499	4.029	4.785	5.408
8	0.706	0.889	1.108	1.397	1.860	2.306	2.449	2.896	3.355	3.833	4.501	5.041
9	0.703	0.883	1.100	1.383	1.833	2.262	2.398	2.821	3.250	3.690	4.297	4.781
10	0.700	0.879	1.093	1.372	1.812	2.228	2.359	2.764	3.169	3.581	4.144	4.587
11	0.697	0.876	1.088	1.363	1.796	2.201	2.328	2.718	3.106	3.497	4.025	4.437
12	0.695	0.873	1.083	1.356	1.782	2.179	2.303	2.681	3.055	3.428	3.930	4.318
13	0.694	0.870	1.079	1.350	1.771	2.160	2.282	2.650	3.012	3.372	3.852	4.221
14	0.692	0.868	1.076	1.345	1.761	2.145	2.264	2.624	2.977	3.326	3.787	4.140
15	0.691	0.866	1.074	1.341	1.753	2.131	2.249	2.602	2.947	3.286	3.733	4.073
16	0.690	0.865	1.071	1.337	1.746	2.120	2.235	2.583	2.921	3.252	3.686	4.015
17	0.689	0.863	1.069	1.333	1.740	2.110	2.224	2.567	2.898	3.222	3.646	3.965
18	0.688	0.862	1.067	1.330	1.734	2.101	2.214	2.552	2.878	3.197	3.611	3.922
19	0.688	0.861	1.066	1.328	1.729	2.093	2.205	2.539	2.861	3.174	3.579	3.883
20	0.687	0.860	1.064	1.325	1.725	2.086	2.197	2.528	2.845	3.153	3.552	3.850
21	0.686	0.859	1.063	1.323	1.721	2.080	2.189	2.518	2.831	3.135	3.527	3.819
22	0.686	0.858	1.061	1.321	1.717	2.074	2.183	2.508	2.819	3.119	3.505	3.792
23	0.685	0.858	1.060	1.319	1.714	2.069	2.177	2.500	2.807	3.104	3.485	3.768
24	0.685	0.857	1.059	1.318	1.711	2.064	2.172	2.492	2.797	3.091	3.467	3.745
25	0.684	0.856	1.058	1.316	1.708	2.060	2.167	2.485	2.787	3.078	3.450	3.725
26	0.684	0.856	1.058	1.315	1.706	2.056	2.162	2.479	2.779	3.067	3.435	3.707
27	0.684	0.855	1.057	1.314	1.703	2.052	2.158	2.473	2.771	3.057	3.421	3.690
28	0.683	0.855	1.056	1.313	1.701	2.048	2.154	2.467	2.763	3.047	3.408	3.674
29	0.683	0.854	1.055	1.311	1.699	2.045	2.150	2.462	2.756	3.038	3.396	3.659
30	0.683	0.854	1.055	1.310	1.697	2.042	2.147	2.457	2.750	3.030	3.385	3.646
40	0.681	0.851	1.050	1.303	1.684	2.021	2.123	2.423	2.704	2.971	3.307	3.551
50	0.679	0.849	1.047	1.299	1.676	2.009	2.109	2.403	2.678	2.937	3.261	3.496
60	0.679	0.848	1.045	1.296	1.671	2.000	2.099	2.390	2.660	2.915	3.232	3.460
80	0.678	0.846	1.043	1.292	1.664	1.990	2.088	2.374	2.639	2.887	3.195	3.416
100	0.677	0.845	1.042	1.290	1.660	1.984	2.081	2.364	2.626	2.871	3.174	3.390
1000	0.675	0.842	1.037	1.282	1.646	1.962	2.056	2.330	2.581	2.813	3.098	3.300
z^*	0.674	0.841	1.036	1.282	1.645	1.960	2.054	2.326	2.576	2.807	3.091	3.291
	50%	60%	70%	80%	90%	95%	96%	98%	99%	99.5%	99.8%	99.9%
	Confidence level C											

Table entry for p is the critical value F^* with probability p lying to its right.

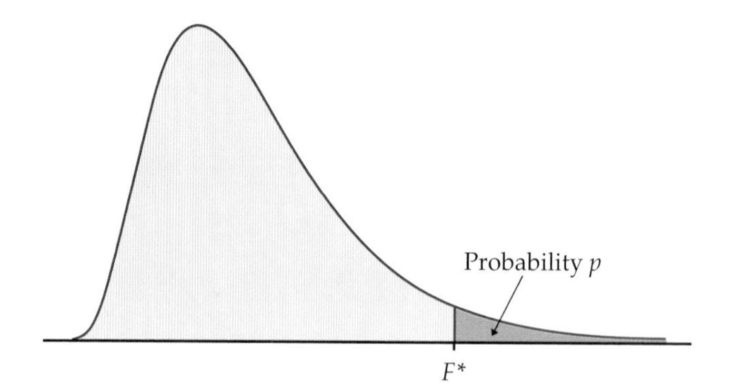

TABLE E

F distribution critical values

Degrees of freedom in the denominator	p	Degrees of freedom in the numerator: 1	2	3	4	5	6	7	8	9
	.100	39.86	49.50	53.59	55.83	57.24	58.20	58.91	59.44	59.86
	.050	161.45	199.50	215.71	224.58	230.16	233.99	236.77	238.88	240.54
1	.025	647.79	799.50	864.16	899.58	921.85	937.11	948.22	956.66	963.28
	.010	4052.2	4999.5	5403.4	5624.6	5763.6	5859.0	5928.4	5981.1	6022.5
	.001	405284	500000	540379	562500	576405	585937	592873	598144	602284
	.100	8.53	9.00	9.16	9.24	9.29	9.33	9.35	9.37	9.38
	.050	18.51	19.00	19.16	19.25	19.30	19.33	19.35	19.37	19.38
2	.025	38.51	39.00	39.17	39.25	39.30	39.33	39.36	39.37	39.39
	.010	98.50	99.00	99.17	99.25	99.30	99.33	99.36	99.37	99.39
	.001	998.50	999.00	999.17	999.25	999.30	999.33	999.36	999.37	999.39
	.100	5.54	5.46	5.39	5.34	5.31	5.28	5.27	5.25	5.24
	.050	10.13	9.55	9.28	9.12	9.01	8.94	8.89	8.85	8.81
3	.025	17.44	16.04	15.44	15.10	14.88	14.73	14.62	14.54	14.47
	.010	34.12	30.82	29.46	28.71	28.24	27.91	27.67	27.49	27.35
	.001	167.03	148.50	141.11	137.10	134.58	132.85	131.58	130.62	129.86
	.100	4.54	4.32	4.19	4.11	4.05	4.01	3.98	3.95	3.94
	.050	7.71	6.94	6.59	6.39	6.26	6.16	6.09	6.04	6.00
4	.025	12.22	10.65	9.98	9.60	9.36	9.20	9.07	8.98	8.90
	.010	21.20	18.00	16.69	15.98	15.52	15.21	14.98	14.80	14.66
	.001	74.14	61.25	56.18	53.44	51.71	50.53	49.66	49.00	48.47
	.100	4.06	3.78	3.62	3.52	3.45	3.40	3.37	3.34	3.32
	.050	6.61	5.79	5.41	5.19	5.05	4.95	4.88	4.82	4.77
5	.025	10.01	8.43	7.76	7.39	7.15	6.98	6.85	6.76	6.68
	.010	16.26	13.27	12.06	11.39	10.97	10.67	10.46	10.29	10.16
	.001	47.18	37.12	33.20	31.09	29.75	28.83	28.16	27.65	27.24
	.100	3.78	3.46	3.29	3.18	3.11	3.05	3.01	2.98	2.96
	.050	5.99	5.14	4.76	4.53	4.39	4.28	4.21	4.15	4.10
6	.025	8.81	7.26	6.60	6.23	5.99	5.82	5.70	5.60	5.52
	.010	13.75	10.92	9.78	9.15	8.75	8.47	8.26	8.10	7.98
	.001	35.51	27.00	23.70	21.92	20.80	20.03	19.46	19.03	18.69
	.100	3.59	3.26	3.07	2.96	2.88	2.83	2.78	2.75	2.72
	.050	5.59	4.74	4.35	4.12	3.97	3.87	3.79	3.73	3.68
7	.025	8.07	6.54	5.89	5.52	5.29	5.12	4.99	4.90	4.82
	.010	12.25	9.55	8.45	7.85	7.46	7.19	6.99	6.84	6.72
	.001	29.25	21.69	18.77	17.20	16.21	15.52	15.02	14.63	14.33

Table entry for p is the critical value F^* with probability p lying to its right.

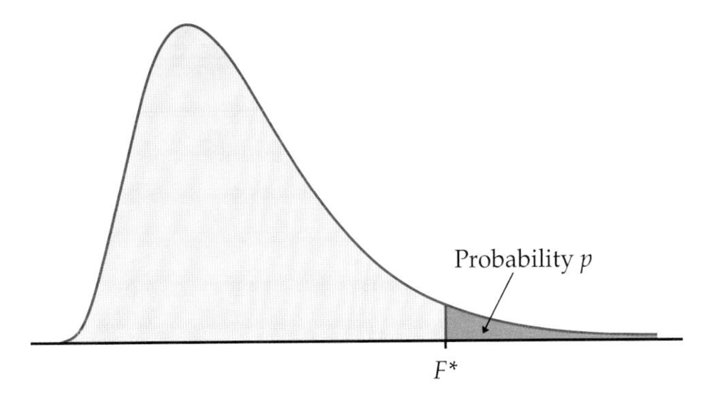

TABLE E

F distribution critical values (continued)

Degrees of freedom in the numerator										
10	12	15	20	25	30	40	50	60	120	1000
60.19	60.71	61.22	61.74	62.05	62.26	62.53	62.69	62.79	63.06	63.30
241.88	243.91	245.95	248.01	249.26	250.10	251.14	251.77	252.20	253.25	254.19
968.63	976.71	984.87	993.10	998.08	1001.4	1005.6	1008.1	1009.8	1014.0	1017.7
6055.8	6106.3	6157.3	6208.7	6239.8	6260.6	6286.8	6302.5	6313.0	6339.4	6362.7
605621	610668	615764	620908	624017	626099	628712	630285	631337	633972	636301
9.39	9.41	9.42	9.44	9.45	9.46	9.47	9.47	9.47	9.48	9.49
19.40	19.41	19.43	19.45	19.46	19.46	19.47	19.48	19.48	19.49	19.49
39.40	39.41	39.43	39.45	39.46	39.46	39.47	39.48	39.48	39.49	39.50
99.40	99.42	99.43	99.45	99.46	99.47	99.47	99.48	99.48	99.49	99.50
999.40	999.42	999.43	999.45	999.46	999.47	999.47	999.48	999.48	999.49	999.50
5.23	5.22	5.20	5.18	5.17	5.17	5.16	5.15	5.15	5.14	5.13
8.79	8.74	8.70	8.66	8.63	8.62	8.59	8.58	8.57	8.55	8.53
14.42	14.34	14.25	14.17	14.12	14.08	14.04	14.01	13.99	13.95	13.91
27.23	27.05	26.87	26.69	26.58	26.50	26.41	26.35	26.32	26.22	26.14
129.25	128.32	127.37	126.42	125.84	125.45	124.96	124.66	124.47	123.97	123.53
3.92	3.90	3.87	3.84	3.83	3.82	3.80	3.80	3.79	3.78	3.76
5.96	5.91	5.86	5.80	5.77	5.75	5.72	5.70	5.69	5.66	5.63
8.84	8.75	8.66	8.56	8.50	8.46	8.41	8.38	8.36	8.31	8.26
14.55	14.37	14.20	14.02	13.91	13.84	13.75	13.69	13.65	13.56	13.47
48.05	47.41	46.76	46.10	45.70	45.43	45.09	44.88	44.75	44.40	44.09
3.30	3.27	3.24	3.21	3.19	3.17	3.16	3.15	3.14	3.12	3.11
4.74	4.68	4.62	4.56	4.52	4.50	4.46	4.44	4.43	4.40	4.37
6.62	6.52	6.43	6.33	6.27	6.23	6.18	6.14	6.12	6.07	6.02
10.05	9.89	9.72	9.55	9.45	9.38	9.29	9.24	9.20	9.11	9.03
26.92	26.42	25.91	25.39	25.08	24.87	24.60	24.44	24.33	24.06	23.82
2.94	2.90	2.87	2.84	2.81	2.80	2.78	2.77	2.76	2.74	2.72
4.06	4.00	3.94	3.87	3.83	3.81	3.77	3.75	3.74	3.70	3.67
5.46	5.37	5.27	5.17	5.11	5.07	5.01	4.98	4.96	4.90	4.86
7.87	7.72	7.56	7.40	7.30	7.23	7.14	7.09	7.06	6.97	6.89
18.41	17.99	17.56	17.12	16.85	16.67	16.44	16.31	16.21	15.98	15.77
2.70	2.67	2.63	2.59	2.57	2.56	2.54	2.52	2.51	2.49	2.47
3.64	3.57	3.51	3.44	3.40	3.38	3.34	3.32	3.30	3.27	3.23
4.76	4.67	4.57	4.47	4.40	4.36	4.31	4.28	4.25	4.20	4.15
6.62	6.47	6.31	6.16	6.06	5.99	5.91	5.86	5.82	5.74	5.66
14.08	13.71	13.32	12.93	12.69	12.53	12.33	12.20	12.12	11.91	11.72

(Continued)

TABLE E

F distribution critical values (continued)

Degrees of freedom in the denominator	p	Degrees of freedom in the numerator 1	2	3	4	5	6	7	8	9
8	.100	3.46	3.11	2.92	2.81	2.73	2.67	2.62	2.59	2.56
	.050	5.32	4.46	4.07	3.84	3.69	3.58	3.50	3.44	3.39
	.025	7.57	6.06	5.42	5.05	4.82	4.65	4.53	4.43	4.36
	.010	11.26	8.65	7.59	7.01	6.63	6.37	6.18	6.03	5.91
	.001	25.41	18.49	15.83	14.39	13.48	12.86	12.40	12.05	11.77
9	.100	3.36	3.01	2.81	2.69	2.61	2.55	2.51	2.47	2.44
	.050	5.12	4.26	3.86	3.63	3.48	3.37	3.29	3.23	3.18
	.025	7.21	5.71	5.08	4.72	4.48	4.32	4.20	4.10	4.03
	.010	10.56	8.02	6.99	6.42	6.06	5.80	5.61	5.47	5.35
	.001	22.86	16.39	13.90	12.56	11.71	11.13	10.70	10.37	10.11
10	.100	3.29	2.92	2.73	2.61	2.52	2.46	2.41	2.38	2.35
	.050	4.96	4.10	3.71	3.48	3.33	3.22	3.14	3.07	3.02
	.025	6.94	5.46	4.83	4.47	4.24	4.07	3.95	3.85	3.78
	.010	10.04	7.56	6.55	5.99	5.64	5.39	5.20	5.06	4.94
	.001	21.04	14.91	12.55	11.28	10.48	9.93	9.52	9.20	8.96
11	.100	3.23	2.86	2.66	2.54	2.45	2.39	2.34	2.30	2.27
	.050	4.84	3.98	3.59	3.36	3.20	3.09	3.01	2.95	2.90
	.025	6.72	5.26	4.63	4.28	4.04	3.88	3.76	3.66	3.59
	.010	9.65	7.21	6.22	5.67	5.32	5.07	4.89	4.74	4.63
	.001	19.69	13.81	11.56	10.35	9.58	9.05	8.66	8.35	8.12
12	.100	3.18	2.81	2.61	2.48	2.39	2.33	2.28	2.24	2.21
	.050	4.75	3.89	3.49	3.26	3.11	3.00	2.91	2.85	2.80
	.025	6.55	5.10	4.47	4.12	3.89	3.73	3.61	3.51	3.44
	.010	9.33	6.93	5.95	5.41	5.06	4.82	4.64	4.50	4.39
	.001	18.64	12.97	10.80	9.63	8.89	8.38	8.00	7.71	7.48
13	.100	3.14	2.76	2.56	2.43	2.35	2.28	2.23	2.20	2.16
	.050	4.67	3.81	3.41	3.18	3.03	2.92	2.83	2.77	2.71
	.025	6.41	4.97	4.35	4.00	3.77	3.60	3.48	3.39	3.31
	.010	9.07	6.70	5.74	5.21	4.86	4.62	4.44	4.30	4.19
	.001	17.82	12.31	10.21	9.07	8.35	7.86	7.49	7.21	6.98
14	.100	3.10	2.73	2.52	2.39	2.31	2.24	2.19	2.15	2.12
	.050	4.60	3.74	3.34	3.11	2.96	2.85	2.76	2.70	2.65
	.025	6.30	4.86	4.24	3.89	3.66	3.50	3.38	3.29	3.21
	.010	8.86	6.51	5.56	5.04	4.69	4.46	4.28	4.14	4.03
	.001	17.14	11.78	9.73	8.62	7.92	7.44	7.08	6.80	6.58
15	.100	3.07	2.70	2.49	2.36	2.27	2.21	2.16	2.12	2.09
	.050	4.54	3.68	3.29	3.06	2.90	2.79	2.71	2.64	2.59
	.025	6.20	4.77	4.15	3.80	3.58	3.41	3.29	3.20	3.12
	.010	8.68	6.36	5.42	4.89	4.56	4.32	4.14	4.00	3.89
	.001	16.59	11.34	9.34	8.25	7.57	7.09	6.74	6.47	6.26
16	.100	3.05	2.67	2.46	2.33	2.24	2.18	2.13	2.09	2.06
	.050	4.49	3.63	3.24	3.01	2.85	2.74	2.66	2.59	2.54
	.025	6.12	4.69	4.08	3.73	3.50	3.34	3.22	3.12	3.05
	.010	8.53	6.23	5.29	4.77	4.44	4.20	4.03	3.89	3.78
	.001	16.12	10.97	9.01	7.94	7.27	6.80	6.46	6.19	5.98
17	.100	3.03	2.64	2.44	2.31	2.22	2.15	2.10	2.06	2.03
	.050	4.45	3.59	3.20	2.96	2.81	2.70	2.61	2.55	2.49
	.025	6.04	4.62	4.01	3.66	3.44	3.28	3.16	3.06	2.98
	.010	8.40	6.11	5.19	4.67	4.34	4.10	3.93	3.79	3.68
	.001	15.72	10.66	8.73	7.68	7.02	6.56	6.22	5.96	5.75

TABLE E
F distribution critical values (continued)

Degrees of freedom in the numerator										
10	12	15	20	25	30	40	50	60	120	1000
2.54	2.50	2.46	2.42	2.40	2.38	2.36	2.35	2.34	2.32	2.30
3.35	3.28	3.22	3.15	3.11	3.08	3.04	3.02	3.01	2.97	2.93
4.30	4.20	4.10	4.00	3.94	3.89	3.84	3.81	3.78	3.73	3.68
5.81	5.67	5.52	5.36	5.26	5.20	5.12	5.07	5.03	4.95	4.87
11.54	11.19	10.84	10.48	10.26	10.11	9.92	9.80	9.73	9.53	9.36
2.42	2.38	2.34	2.30	2.27	2.25	2.23	2.22	2.21	2.18	2.16
3.14	3.07	3.01	2.94	2.89	2.86	2.83	2.80	2.79	2.75	2.71
3.96	3.87	3.77	3.67	3.60	3.56	3.51	3.47	3.45	3.39	3.34
5.26	5.11	4.96	4.81	4.71	4.65	4.57	4.52	4.48	4.40	4.32
9.89	9.57	9.24	8.90	8.69	8.55	8.37	8.26	8.19	8.00	7.84
2.32	2.28	2.24	2.20	2.17	2.16	2.13	2.12	2.11	2.08	2.06
2.98	2.91	2.85	2.77	2.73	2.70	2.66	2.64	2.62	2.58	2.54
3.72	3.62	3.52	3.42	3.35	3.31	3.26	3.22	3.20	3.14	3.09
4.85	4.71	4.56	4.41	4.31	4.25	4.17	4.12	4.08	4.00	3.92
8.75	8.45	8.13	7.80	7.60	7.47	7.30	7.19	7.12	6.94	6.78
2.25	2.21	2.17	2.12	2.10	2.08	2.05	2.04	2.03	2.00	1.98
2.85	2.79	2.72	2.65	2.60	2.57	2.53	2.51	2.49	2.45	2.41
3.53	3.43	3.33	3.23	3.16	3.12	3.06	3.03	3.00	2.94	2.89
4.54	4.40	4.25	4.10	4.01	3.94	3.86	3.81	3.78	3.69	3.61
7.92	7.63	7.32	7.01	6.81	6.68	6.52	6.42	6.35	6.18	6.02
2.19	2.15	2.10	2.06	2.03	2.01	1.99	1.97	1.96	1.93	1.91
2.75	2.69	2.62	2.54	2.50	2.47	2.43	2.40	2.38	2.34	2.30
3.37	3.28	3.18	3.07	3.01	2.96	2.91	2.87	2.85	2.79	2.73
4.30	4.16	4.01	3.86	3.76	3.70	3.62	3.57	3.54	3.45	3.37
7.29	7.00	6.71	6.40	6.22	6.09	5.93	5.83	5.76	5.59	5.44
2.14	2.10	2.05	2.01	1.98	1.96	1.93	1.92	1.90	1.88	1.85
2.67	2.60	2.53	2.46	2.41	2.38	2.34	2.31	2.30	2.25	2.21
3.25	3.15	3.05	2.95	2.88	2.84	2.78	2.74	2.72	2.66	2.60
4.10	3.96	3.82	3.66	3.57	3.51	3.43	3.38	3.34	3.25	3.18
6.80	6.52	6.23	5.93	5.75	5.63	5.47	5.37	5.30	5.14	4.99
2.10	2.05	2.01	1.96	1.93	1.91	1.89	1.87	1.86	1.83	1.80
2.60	2.53	2.46	2.39	2.34	2.31	2.27	2.24	2.22	2.18	2.14
3.15	3.05	2.95	2.84	2.78	2.73	2.67	2.64	2.61	2.55	2.50
3.94	3.80	3.66	3.51	3.41	3.35	3.27	3.22	3.18	3.09	3.02
6.40	6.13	5.85	5.56	5.38	5.25	5.10	5.00	4.94	4.77	4.62
2.06	2.02	1.97	1.92	1.89	1.87	1.85	1.83	1.82	1.79	1.76
2.54	2.48	2.40	2.33	2.28	2.25	2.20	2.18	2.16	2.11	2.07
3.06	2.96	2.86	2.76	2.69	2.64	2.59	2.55	2.52	2.46	2.40
3.80	3.67	3.52	3.37	3.28	3.21	3.13	3.08	3.05	2.96	2.88
6.08	5.81	5.54	5.25	5.07	4.95	4.80	4.70	4.64	4.47	4.33
2.03	1.99	1.94	1.89	1.86	1.84	1.81	1.79	1.78	1.75	1.72
2.49	2.42	2.35	2.28	2.23	2.19	2.15	2.12	2.11	2.06	2.02
2.99	2.89	2.79	2.68	2.61	2.57	2.51	2.47	2.45	2.38	2.32
3.69	3.55	3.41	3.26	3.16	3.10	3.02	2.97	2.93	2.84	2.76
5.81	5.55	5.27	4.99	4.82	4.70	4.54	4.45	4.39	4.23	4.08
2.00	1.96	1.91	1.86	1.83	1.81	1.78	1.76	1.75	1.72	1.69
2.45	2.38	2.31	2.23	2.18	2.15	2.10	2.08	2.06	2.01	1.97
2.92	2.82	2.72	2.62	2.55	2.50	2.44	2.41	2.38	2.32	2.26
3.59	3.46	3.31	3.16	3.07	3.00	2.92	2.87	2.83	2.75	2.66
5.58	5.32	5.05	4.78	4.60	4.48	4.33	4.24	4.18	4.02	3.87

(Continued)

TABLE E

F distribution critical values (continued)

Degrees of freedom in the denominator	p	1	2	3	4	5	6	7	8	9
		Degrees of freedom in the numerator								
18	.100	3.01	2.62	2.42	2.29	2.20	2.13	2.08	2.04	2.00
	.050	4.41	3.55	3.16	2.93	2.77	2.66	2.58	2.51	2.46
	.025	5.98	4.56	3.95	3.61	3.38	3.22	3.10	3.01	2.93
	.010	8.29	6.01	5.09	4.58	4.25	4.01	3.84	3.71	3.60
	.001	15.38	10.39	8.49	7.46	6.81	6.35	6.02	5.76	5.56
19	.100	2.99	2.61	2.40	2.27	2.18	2.11	2.06	2.02	1.98
	.050	4.38	3.52	3.13	2.90	2.74	2.63	2.54	2.48	2.42
	.025	5.92	4.51	3.90	3.56	3.33	3.17	3.05	2.96	2.88
	.010	8.18	5.93	5.01	4.50	4.17	3.94	3.77	3.63	3.52
	.001	15.08	10.16	8.28	7.27	6.62	6.18	5.85	5.59	5.39
20	.100	2.97	2.59	2.38	2.25	2.16	2.09	2.04	2.00	1.96
	.050	4.35	3.49	3.10	2.87	2.71	2.60	2.51	2.45	2.39
	.025	5.87	4.46	3.86	3.51	3.29	3.13	3.01	2.91	2.84
	.010	8.10	5.85	4.94	4.43	4.10	3.87	3.70	3.56	3.46
	.001	14.82	9.95	8.10	7.10	6.46	6.02	5.69	5.44	5.24
21	.100	2.96	2.57	2.36	2.23	2.14	2.08	2.02	1.98	1.95
	.050	4.32	3.47	3.07	2.84	2.68	2.57	2.49	2.42	2.37
	.025	5.83	4.42	3.82	3.48	3.25	3.09	2.97	2.87	2.80
	.010	8.02	5.78	4.87	4.37	4.04	3.81	3.64	3.51	3.40
	.001	14.59	9.77	7.94	6.95	6.32	5.88	5.56	5.31	5.11
22	.100	2.95	2.56	2.35	2.22	2.13	2.06	2.01	1.97	1.93
	.050	4.30	3.44	3.05	2.82	2.66	2.55	2.46	2.40	2.34
	.025	5.79	4.38	3.78	3.44	3.22	3.05	2.93	2.84	2.76
	.010	7.95	5.72	4.82	4.31	3.99	3.76	3.59	3.45	3.35
	.001	14.38	9.61	7.80	6.81	6.19	5.76	5.44	5.19	4.99
23	.100	2.94	2.55	2.34	2.21	2.11	2.05	1.99	1.95	1.92
	.050	4.28	3.42	3.03	2.80	2.64	2.53	2.44	2.37	2.32
	.025	5.75	4.35	3.75	3.41	3.18	3.02	2.90	2.81	2.73
	.010	7.88	5.66	4.76	4.26	3.94	3.71	3.54	3.41	3.30
	.001	14.20	9.47	7.67	6.70	6.08	5.65	5.33	5.09	4.89
24	.100	2.93	2.54	2.33	2.19	2.10	2.04	1.98	1.94	1.91
	.050	4.26	3.40	3.01	2.78	2.62	2.51	2.42	2.36	2.30
	.025	5.72	4.32	3.72	3.38	3.15	2.99	2.87	2.78	2.70
	.010	7.82	5.61	4.72	4.22	3.90	3.67	3.50	3.36	3.26
	.001	14.03	9.34	7.55	6.59	5.98	5.55	5.23	4.99	4.80
25	.100	2.92	2.53	2.32	2.18	2.09	2.02	1.97	1.93	1.89
	.050	4.24	3.39	2.99	2.76	2.60	2.49	2.40	2.34	2.28
	.025	5.69	4.29	3.69	3.35	3.13	2.97	2.85	2.75	2.68
	.010	7.77	5.57	4.68	4.18	3.85	3.63	3.46	3.32	3.22
	.001	13.88	9.22	7.45	6.49	5.89	5.46	5.15	4.91	4.71
26	.100	2.91	2.52	2.31	2.17	2.08	2.01	1.96	1.92	1.88
	.050	4.23	3.37	2.98	2.74	2.59	2.47	2.39	2.32	2.27
	.025	5.66	4.27	3.67	3.33	3.10	2.94	2.82	2.73	2.65
	.010	7.72	5.53	4.64	4.14	3.82	3.59	3.42	3.29	3.18
	.001	13.74	9.12	7.36	6.41	5.80	5.38	5.07	4.83	4.64
27	.100	2.90	2.51	2.30	2.17	2.07	2.00	1.95	1.91	1.87
	.050	4.21	3.35	2.96	2.73	2.57	2.46	2.37	2.31	2.25
	.025	5.63	4.24	3.65	3.31	3.08	2.92	2.80	2.71	2.63
	.010	7.68	5.49	4.60	4.11	3.78	3.56	3.39	3.26	3.15
	.001	13.61	9.02	7.27	6.33	5.73	5.31	5.00	4.76	4.57

TABLE E

F distribution critical values (continued)

Degrees of freedom in the numerator										
10	12	15	20	25	30	40	50	60	120	1000
1.98	1.93	1.89	1.84	1.80	1.78	1.75	1.74	1.72	1.69	1.66
2.41	2.34	2.27	2.19	2.14	2.11	2.06	2.04	2.02	1.97	1.92
2.87	2.77	2.67	2.56	2.49	2.44	2.38	2.35	2.32	2.26	2.20
3.51	3.37	3.23	3.08	2.98	2.92	2.84	2.78	2.75	2.66	2.58
5.39	5.13	4.87	4.59	4.42	4.30	4.15	4.06	4.00	3.84	3.69
1.96	1.91	1.86	1.81	1.78	1.76	1.73	1.71	1.70	1.67	1.64
2.38	2.31	2.23	2.16	2.11	2.07	2.03	2.00	1.98	1.93	1.88
2.82	2.72	2.62	2.51	2.44	2.39	2.33	2.30	2.27	2.20	2.14
3.43	3.30	3.15	3.00	2.91	2.84	2.76	2.71	2.67	2.58	2.50
5.22	4.97	4.70	4.43	4.26	4.14	3.99	3.90	3.84	3.68	3.53
1.94	1.89	1.84	1.79	1.76	1.74	1.71	1.69	1.68	1.64	1.61
2.35	2.28	2.20	2.12	2.07	2.04	1.99	1.97	1.95	1.90	1.85
2.77	2.68	2.57	2.46	2.40	2.35	2.29	2.25	2.22	2.16	2.09
3.37	3.23	3.09	2.94	2.84	2.78	2.69	2.64	2.61	2.52	2.43
5.08	4.82	4.56	4.29	4.12	4.00	3.86	3.77	3.70	3.54	3.40
1.92	1.87	1.83	1.78	1.74	1.72	1.69	1.67	1.66	1.62	1.59
2.32	2.25	2.18	2.10	2.05	2.01	1.96	1.94	1.92	1.87	1.82
2.73	2.64	2.53	2.42	2.36	2.31	2.25	2.21	2.18	2.11	2.05
3.31	3.17	3.03	2.88	2.79	2.72	2.64	2.58	2.55	2.46	2.37
4.95	4.70	4.44	4.17	4.00	3.88	3.74	3.64	3.58	3.42	3.28
1.90	1.86	1.81	1.76	1.73	1.70	1.67	1.65	1.64	1.60	1.57
2.30	2.23	2.15	2.07	2.02	1.98	1.94	1.91	1.89	1.84	1.79
2.70	2.60	2.50	2.39	2.32	2.27	2.21	2.17	2.14	2.08	2.01
3.26	3.12	2.98	2.83	2.73	2.67	2.58	2.53	2.50	2.40	2.32
4.83	4.58	4.33	4.06	3.89	3.78	3.63	3.54	3.48	3.32	3.17
1.89	1.84	1.80	1.74	1.71	1.69	1.66	1.64	1.62	1.59	1.55
2.27	2.20	2.13	2.05	2.00	1.96	1.91	1.88	1.86	1.81	1.76
2.67	2.57	2.47	2.36	2.29	2.24	2.18	2.14	2.11	2.04	1.98
3.21	3.07	2.93	2.78	2.69	2.62	2.54	2.48	2.45	2.35	2.27
4.73	4.48	4.23	3.96	3.79	3.68	3.53	3.44	3.38	3.22	3.08
1.88	1.83	1.78	1.73	1.70	1.67	1.64	1.62	1.61	1.57	1.54
2.25	2.18	2.11	2.03	1.97	1.94	1.89	1.86	1.84	1.79	1.74
2.64	2.54	2.44	2.33	2.26	2.21	2.15	2.11	2.08	2.01	1.94
3.17	3.03	2.89	2.74	2.64	2.58	2.49	2.44	2.40	2.31	2.22
4.64	4.39	4.14	3.87	3.71	3.59	3.45	3.36	3.29	3.14	2.99
1.87	1.82	1.77	1.72	1.68	1.66	1.63	1.61	1.59	1.56	1.52
2.24	2.16	2.09	2.01	1.96	1.92	1.87	1.84	1.82	1.77	1.72
2.61	2.51	2.41	2.30	2.23	2.18	2.12	2.08	2.05	1.98	1.91
3.13	2.99	2.85	2.70	2.60	2.54	2.45	2.40	2.36	2.27	2.18
4.56	4.31	4.06	3.79	3.63	3.52	3.37	3.28	3.22	3.06	2.91
1.86	1.81	1.76	1.71	1.67	1.65	1.61	1.59	1.58	1.54	1.51
2.22	2.15	2.07	1.99	1.94	1.90	1.85	1.82	1.80	1.75	1.70
2.59	2.49	2.39	2.28	2.21	2.16	2.09	2.05	2.03	1.95	1.89
3.09	2.96	2.81	2.66	2.57	2.50	2.42	2.36	2.33	2.23	2.14
4.48	4.24	3.99	3.72	3.56	3.44	3.30	3.21	3.15	2.99	2.84
1.85	1.80	1.75	1.70	1.66	1.64	1.60	1.58	1.57	1.53	1.50
2.20	2.13	2.06	1.97	1.92	1.88	1.84	1.81	1.79	1.73	1.68
2.57	2.47	2.36	2.25	2.18	2.13	2.07	2.03	2.00	1.93	1.86
3.06	2.93	2.78	2.63	2.54	2.47	2.38	2.33	2.29	2.20	2.11
4.41	4.17	3.92	3.66	3.49	3.38	3.23	3.14	3.08	2.92	2.78

(Continued)

TABLE E

F distribution critical values (continued)

Degrees of freedom in the denominator	p	1	2	3	4	5	6	7	8	9
		Degrees of freedom in the numerator								
28	.100	2.89	2.50	2.29	2.16	2.06	2.00	1.94	1.90	1.87
	.050	4.20	3.34	2.95	2.71	2.56	2.45	2.36	2.29	2.24
	.025	5.61	4.22	3.63	3.29	3.06	2.90	2.78	2.69	2.61
	.010	7.64	5.45	4.57	4.07	3.75	3.53	3.36	3.23	3.12
	.001	13.50	8.93	7.19	6.25	5.66	5.24	4.93	4.69	4.50
29	.100	2.89	2.50	2.28	2.15	2.06	1.99	1.93	1.89	1.86
	.050	4.18	3.33	2.93	2.70	2.55	2.43	2.35	2.28	2.22
	.025	5.59	4.20	3.61	3.27	3.04	2.88	2.76	2.67	2.59
	.010	7.60	5.42	4.54	4.04	3.73	3.50	3.33	3.20	3.09
	.001	13.39	8.85	7.12	6.19	5.59	5.18	4.87	4.64	4.45
30	.100	2.88	2.49	2.28	2.14	2.05	1.98	1.93	1.88	1.85
	.050	4.17	3.32	2.92	2.69	2.53	2.42	2.33	2.27	2.21
	.025	5.57	4.18	3.59	3.25	3.03	2.87	2.75	2.65	2.57
	.010	7.56	5.39	4.51	4.02	3.70	3.47	3.30	3.17	3.07
	.001	13.29	8.77	7.05	6.12	5.53	5.12	4.82	4.58	4.39
40	.100	2.84	2.44	2.23	2.09	2.00	1.93	1.87	1.83	1.79
	.050	4.08	3.23	2.84	2.61	2.45	2.34	2.25	2.18	2.12
	.025	5.42	4.05	3.46	3.13	2.90	2.74	2.62	2.53	2.45
	.010	7.31	5.18	4.31	3.83	3.51	3.29	3.12	2.99	2.89
	.001	12.61	8.25	6.59	5.70	5.13	4.73	4.44	4.21	4.02
50	.100	2.81	2.41	2.20	2.06	1.97	1.90	1.84	1.80	1.76
	.050	4.03	3.18	2.79	2.56	2.40	2.29	2.20	2.13	2.07
	.025	5.34	3.97	3.39	3.05	2.83	2.67	2.55	2.46	2.38
	.010	7.17	5.06	4.20	3.72	3.41	3.19	3.02	2.89	2.78
	.001	12.22	7.96	6.34	5.46	4.90	4.51	4.22	4.00	3.82
60	.100	2.79	2.39	2.18	2.04	1.95	1.87	1.82	1.77	1.74
	.050	4.00	3.15	2.76	2.53	2.37	2.25	2.17	2.10	2.04
	.025	5.29	3.93	3.34	3.01	2.79	2.63	2.51	2.41	2.33
	.010	7.08	4.98	4.13	3.65	3.34	3.12	2.95	2.82	2.72
	.001	11.97	7.77	6.17	5.31	4.76	4.37	4.09	3.86	3.69
100	.100	2.76	2.36	2.14	2.00	1.91	1.83	1.78	1.73	1.69
	.050	3.94	3.09	2.70	2.46	2.31	2.19	2.10	2.03	1.97
	.025	5.18	3.83	3.25	2.92	2.70	2.54	2.42	2.32	2.24
	.010	6.90	4.82	3.98	3.51	3.21	2.99	2.82	2.69	2.59
	.001	11.50	7.41	5.86	5.02	4.48	4.11	3.83	3.61	3.44
200	.100	2.73	2.33	2.11	1.97	1.88	1.80	1.75	1.70	1.66
	.050	3.89	3.04	2.65	2.42	2.26	2.14	2.06	1.98	1.93
	.025	5.10	3.76	3.18	2.85	2.63	2.47	2.35	2.26	2.18
	.010	6.76	4.71	3.88	3.41	3.11	2.89	2.73	2.60	2.50
	.001	11.15	7.15	5.63	4.81	4.29	3.92	3.65	3.43	3.26
1000	.100	2.71	2.31	2.09	1.95	1.85	1.78	1.72	1.68	1.64
	.050	3.85	3.00	2.61	2.38	2.22	2.11	2.02	1.95	1.89
	.025	5.04	3.70	3.13	2.80	2.58	2.42	2.30	2.20	2.13
	.010	6.66	4.63	3.80	3.34	3.04	2.82	2.66	2.53	2.43
	.001	10.89	6.96	5.46	4.65	4.14	3.78	3.51	3.30	3.13

TABLE E

***F* distribution critical values (continued)**

Degrees of freedom in the numerator										
10	12	15	20	25	30	40	50	60	120	1000
1.84	1.79	1.74	1.69	1.65	1.63	1.59	1.57	1.56	1.52	1.48
2.19	2.12	2.04	1.96	1.91	1.87	1.82	1.79	1.77	1.71	1.66
2.55	2.45	2.34	2.23	2.16	2.11	2.05	2.01	1.98	1.91	1.84
3.03	2.90	2.75	2.60	2.51	2.44	2.35	2.30	2.26	2.17	2.08
4.35	4.11	3.86	3.60	3.43	3.32	3.18	3.09	3.02	2.86	2.72
1.83	1.78	1.73	1.68	1.64	1.62	1.58	1.56	1.55	1.51	1.47
2.18	2.10	2.03	1.94	1.89	1.85	1.81	1.77	1.75	1.70	1.65
2.53	2.43	2.32	2.21	2.14	2.09	2.03	1.99	1.96	1.89	1.82
3.00	2.87	2.73	2.57	2.48	2.41	2.33	2.27	2.23	2.14	2.05
4.29	4.05	3.80	3.54	3.38	3.27	3.12	3.03	2.97	2.81	2.66
1.82	1.77	1.72	1.67	1.63	1.61	1.57	1.55	1.54	1.50	1.46
2.16	2.09	2.01	1.93	1.88	1.84	1.79	1.76	1.74	1.68	1.63
2.51	2.41	2.31	2.20	2.12	2.07	2.01	1.97	1.94	1.87	1.80
2.98	2.84	2.70	2.55	2.45	2.39	2.30	2.25	2.21	2.11	2.02
4.24	4.00	3.75	3.49	3.33	3.22	3.07	2.98	2.92	2.76	2.61
1.76	1.71	1.66	1.61	1.57	1.54	1.51	1.48	1.47	1.42	1.38
2.08	2.00	1.92	1.84	1.78	1.74	1.69	1.66	1.64	1.58	1.52
2.39	2.29	2.18	2.07	1.99	1.94	1.88	1.83	1.80	1.72	1.65
2.80	2.66	2.52	2.37	2.27	2.20	2.11	2.06	2.02	1.92	1.82
3.87	3.64	3.40	3.14	2.98	2.87	2.73	2.64	2.57	2.41	2.25
1.73	1.68	1.63	1.57	1.53	1.50	1.46	1.44	1.42	1.38	1.33
2.03	1.95	1.87	1.78	1.73	1.69	1.63	1.60	1.58	1.51	1.45
2.32	2.22	2.11	1.99	1.92	1.87	1.80	1.75	1.72	1.64	1.56
2.70	2.56	2.42	2.27	2.17	2.10	2.01	1.95	1.91	1.80	1.70
3.67	3.44	3.20	2.95	2.79	2.68	2.53	2.44	2.38	2.21	2.05
1.71	1.66	1.60	1.54	1.50	1.48	1.44	1.41	1.40	1.35	1.30
1.99	1.92	1.84	1.75	1.69	1.65	1.59	1.56	1.53	1.47	1.40
2.27	2.17	2.06	1.94	1.87	1.82	1.74	1.70	1.67	1.58	1.49
2.63	2.50	2.35	2.20	2.10	2.03	1.94	1.88	1.84	1.73	1.62
3.54	3.32	3.08	2.83	2.67	2.55	2.41	2.32	2.25	2.08	1.92
1.66	1.61	1.56	1.49	1.45	1.42	1.38	1.35	1.34	1.28	1.22
1.93	1.85	1.77	1.68	1.62	1.57	1.52	1.48	1.45	1.38	1.30
2.18	2.08	1.97	1.85	1.77	1.71	1.64	1.59	1.56	1.46	1.36
2.50	2.37	2.22	2.07	1.97	1.89	1.80	1.74	1.69	1.57	1.45
3.30	3.07	2.84	2.59	2.43	2.32	2.17	2.08	2.01	1.83	1.64
1.63	1.58	1.52	1.46	1.41	1.38	1.34	1.31	1.29	1.23	1.16
1.88	1.80	1.72	1.62	1.56	1.52	1.46	1.41	1.39	1.30	1.21
2.11	2.01	1.90	1.78	1.70	1.64	1.56	1.51	1.47	1.37	1.25
2.41	2.27	2.13	1.97	1.87	1.79	1.69	1.63	1.58	1.45	1.30
3.12	2.90	2.67	2.42	2.26	2.15	2.00	1.90	1.83	1.64	1.43
1.61	1.55	1.49	1.43	1.38	1.35	1.30	1.27	1.25	1.18	1.08
1.84	1.76	1.68	1.58	1.52	1.47	1.41	1.36	1.33	1.24	1.11
2.06	1.96	1.85	1.72	1.64	1.58	1.50	1.45	1.41	1.29	1.13
2.34	2.20	2.06	1.90	1.79	1.72	1.61	1.54	1.50	1.35	1.16
2.99	2.77	2.54	2.30	2.14	2.02	1.87	1.77	1.69	1.49	1.22

Table entry for p is the critical value $(\chi^2)^*$ with probability p lying to its right.

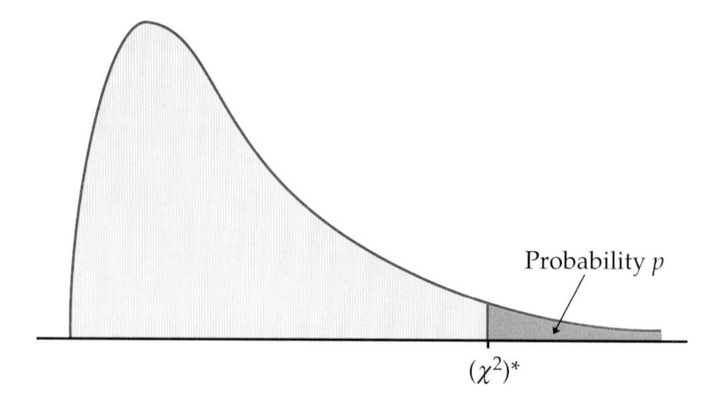

TABLE F

χ^2 distribution critical values

	Tail probability p											
df	.25	.20	.15	.10	.05	.025	.02	.01	.005	.0025	.001	.0005
1	1.32	1.64	2.07	2.71	3.84	5.02	5.41	6.63	7.88	9.14	10.83	12.12
2	2.77	3.22	3.79	4.61	5.99	7.38	7.82	9.21	10.60	11.98	13.82	15.20
3	4.11	4.64	5.32	6.25	7.81	9.35	9.84	11.34	12.84	14.32	16.27	17.73
4	5.39	5.99	6.74	7.78	9.49	11.14	11.67	13.28	14.86	16.42	18.47	20.00
5	6.63	7.29	8.12	9.24	11.07	12.83	13.39	15.09	16.75	18.39	20.51	22.11
6	7.84	8.56	9.45	10.64	12.59	14.45	15.03	16.81	18.55	20.25	22.46	24.10
7	9.04	9.80	10.75	12.02	14.07	16.01	16.62	18.48	20.28	22.04	24.32	26.02
8	10.22	11.03	12.03	13.36	15.51	17.53	18.17	20.09	21.95	23.77	26.12	27.87
9	11.39	12.24	13.29	14.68	16.92	19.02	19.68	21.67	23.59	25.46	27.88	29.67
10	12.55	13.44	14.53	15.99	18.31	20.48	21.16	23.21	25.19	27.11	29.59	31.42
11	13.70	14.63	15.77	17.28	19.68	21.92	22.62	24.72	26.76	28.73	31.26	33.14
12	14.85	15.81	16.99	18.55	21.03	23.34	24.05	26.22	28.30	30.32	32.91	34.82
13	15.98	16.98	18.20	19.81	22.36	24.74	25.47	27.69	29.82	31.88	34.53	36.48
14	17.12	18.15	19.41	21.06	23.68	26.12	26.87	29.14	31.32	33.43	36.12	38.11
15	18.25	19.31	20.60	22.31	25.00	27.49	28.26	30.58	32.80	34.95	37.70	39.72
16	19.37	20.47	21.79	23.54	26.30	28.85	29.63	32.00	34.27	36.46	39.25	41.31
17	20.49	21.61	22.98	24.77	27.59	30.19	31.00	33.41	35.72	37.95	40.79	42.88
18	21.60	22.76	24.16	25.99	28.87	31.53	32.35	34.81	37.16	39.42	42.31	44.43
19	22.72	23.90	25.33	27.20	30.14	32.85	33.69	36.19	38.58	40.88	43.82	45.97
20	23.83	25.04	26.50	28.41	31.41	34.17	35.02	37.57	40.00	42.34	45.31	47.50
21	24.93	26.17	27.66	29.62	32.67	35.48	36.34	38.93	41.40	43.78	46.80	49.01
22	26.04	27.30	28.82	30.81	33.92	36.78	37.66	40.29	42.80	45.20	48.27	50.51
23	27.14	28.43	29.98	32.01	35.17	38.08	38.97	41.64	44.18	46.62	49.73	52.00
24	28.24	29.55	31.13	33.20	36.42	39.36	40.27	42.98	45.56	48.03	51.18	53.48
25	29.34	30.68	32.28	34.38	37.65	40.65	41.57	44.31	46.93	49.44	52.62	54.95
26	30.43	31.79	33.43	35.56	38.89	41.92	42.86	45.64	48.29	50.83	54.05	56.41
27	31.53	32.91	34.57	36.74	40.11	43.19	44.14	46.96	49.64	52.22	55.48	57.86
28	32.62	34.03	35.71	37.92	41.34	44.46	45.42	48.28	50.99	53.59	56.89	59.30
29	33.71	35.14	36.85	39.09	42.56	45.72	46.69	49.59	52.34	54.97	58.30	60.73
30	34.80	36.25	37.99	40.26	43.77	46.98	47.96	50.89	53.67	56.33	59.70	62.16
40	45.62	47.27	49.24	51.81	55.76	59.34	60.44	63.69	66.77	69.70	73.40	76.09
50	56.33	58.16	60.35	63.17	67.50	71.42	72.61	76.15	79.49	82.66	86.66	89.56
60	66.98	68.97	71.34	74.40	79.08	83.30	84.58	88.38	91.95	95.34	99.61	102.7
80	88.13	90.41	93.11	96.58	101.9	106.6	108.1	112.3	116.3	120.1	124.8	128.3
100	109.1	111.7	114.7	118.5	124.3	129.6	131.1	135.8	140.2	144.3	149.4	153.2

Chapter 1

1.1. For example: "Do you think you will use statistics in your major?" "Do you think you will use statistics in your career?" "How often do you see statistics in your daily life?"

1.3. No; math majors in their fourth year might think statistics is more important than other students do.

1.7. (a) Data. (b) Research question. (c) Research question. (d) Conclusion.

1.9. For example: "How much do you have in student loans right now?" "Are you concerned about an increase in tuition?" "If tuition increases, do you think your textbook-buying habits will change?"

1.13. (a) 10 times.

1.17. The mean is 114.9, and the standard deviation is 14.8. The minimum is 81, and the maximum is 145. The median is 114.

1.19. For example, we could plot the smoking rate per state against the fruit and vegetable consumption by state.

1.21. (a) For example, below top. (b) For example, below middle. (c) For example, below bottom.

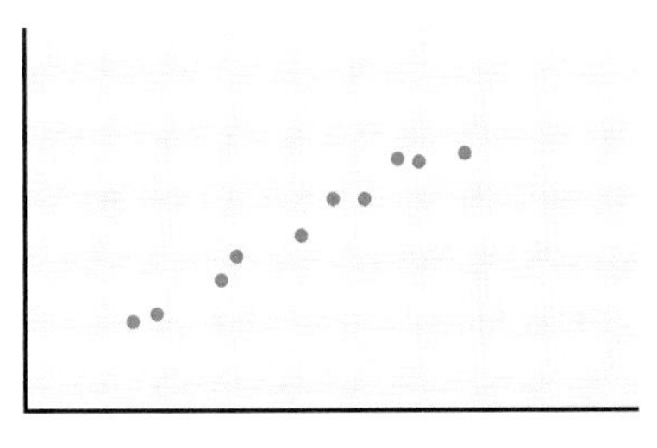

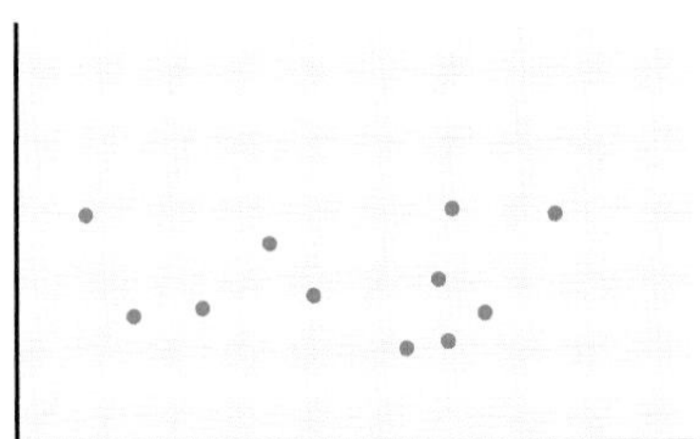

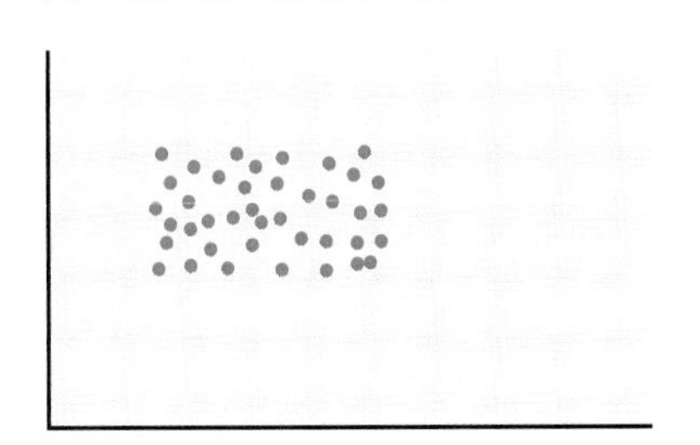

1.23. The mean is 10.0, and the standard deviation is 1.58.

1.25. (a) The standard deviation should be larger. (b) The standard deviation is 3.16.

1.27. The median is equal to the value of the middle data point when the data are arranged in increasing order.

1.29. (a) The mean and the median are (about) equal. (b) The mean increased and then decreased. The median remained the same.

1.33. The variation decreases as the number of tosses increases. The proportion of heads gets closer to 0.9.

1.35. For example, select a card from a standard deck of cards. Record the color (black or red). Select another card without replacing the first card. Record the color.

1.37. The percent hit should be close to 95.

1.39. (a) False. (b) True.

1.41. The percent hit should be close to 95.

1.43. The percent hit should be close to 80 and 99, respectively.

1.49. (a) 1, 1, 1, 5, 5, 5. (b) For example, 2, 2, 2, 2, 2, 2.

1.51. (a) The sample proportion should settle down to 0.5.

Chapter 2

2.1. Any group of friends is unlikely to include a representative cross section of all students.

2.3. A hard-core runner and her friends are not representative of all young people.

2.5. For example, who owns the Web site? Do they have data to back up this statement, and if so, what was the source of the data?

2.7. An experiment: each subject is (presumably randomly) assigned to a treatment group.

2.9. An experiment: each subject is (presumably randomly) assigned to a treatment group.

2.11. Experimental units: food samples. Factor: radiation exposure. Levels: nine different levels of radiation. Response variable: lipid oxidation. It is likely that different lipids react to radiation differently.

2.13. Those who volunteer to use the software may be better students (or they may be worse students).

2.15.

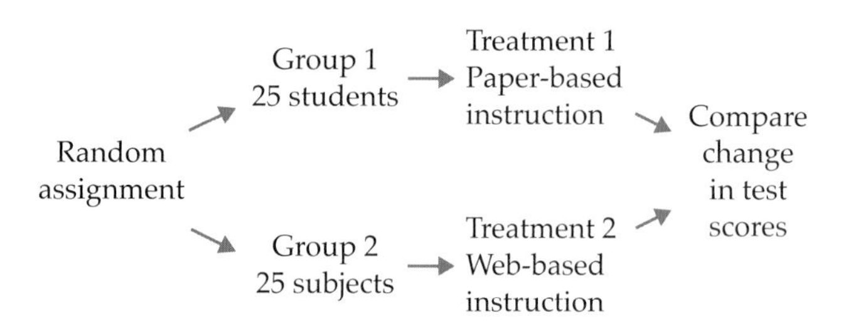

2.17. **(a)** Shopping patterns may differ on Friday and Saturday. **(b)** Responses may vary in different states. **(c)** A control is needed for comparison.

2.19. Students will know which method was used to teach them.

2.21. For example, new employees should be randomly assigned to either the current program or the new one.

2.23. (a) Factors: calcium dose and vitamin D dose. There are 9 treatments (each calcium/vitamin D combination). (b) Assign 20 students to each group, with 10 of each gender.

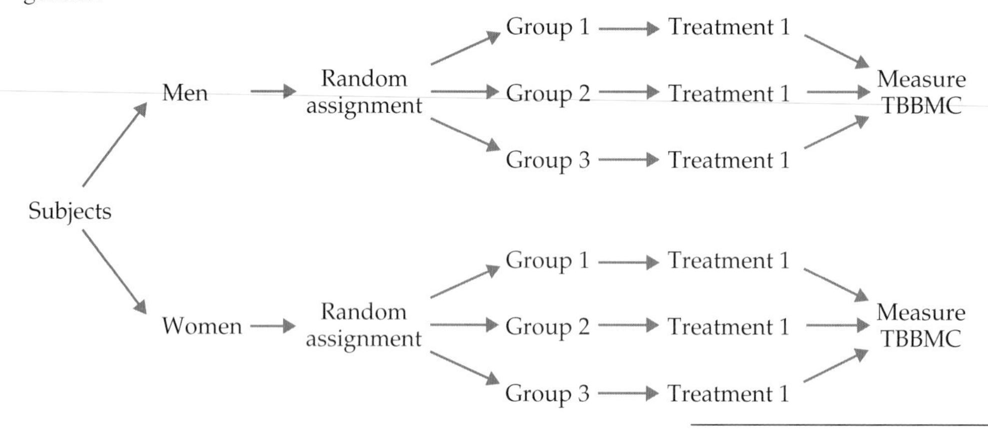

2.25. (a) For example, flip a coin for each customer to choose which variety he or she will taste. (b) For example, flip a coin for each customer to choose which variety he or she will taste first. (c) If each customer tastes both varieties, we only need to ask which was preferred. This design also controls for person-to-person variation.

2.27. Experimental units: pine tree seedlings. Factor: amount of light. Treatments: full light or shaded to 5% of normal. Response variable: dry weight at end of study.

2.29. Subjects: adults from selected households. Factors: level of identification and offer of survey results. Six treatments: interviewer's name/university name/both names, with or without results. Response variable: whether or not the interview is completed.

2.31. Assign 9 subjects to each treatment. The first three groups are

03, 22, 29, 26, 01, 12, 11, 31, 21;
32, 30, 09, 23, 07, 27, 20, 06, 33;
05, 16, 28, 10, 18, 13, 25, 19, 04.

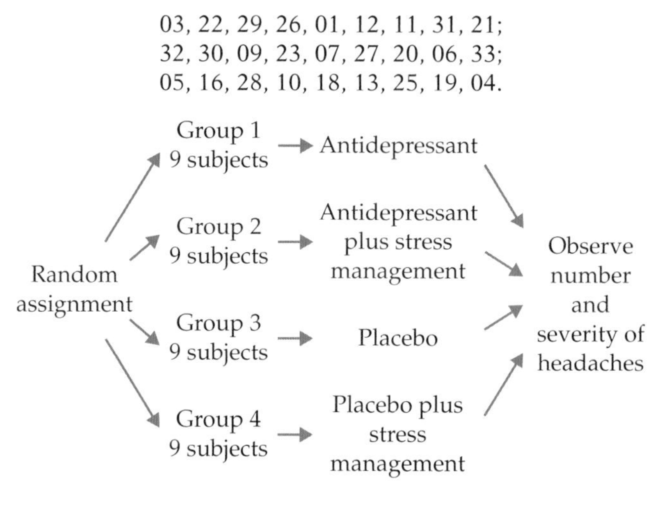

2.33. Assign 6 schools to each treatment group. Choose 16, 21, 06, 12, 02, 04 for Group 1; 14, 15, 23, 11, 09, 03 for Group 2; 07, 24, 17, 22, 01, 13 for Group 3; and the rest for Group 4.

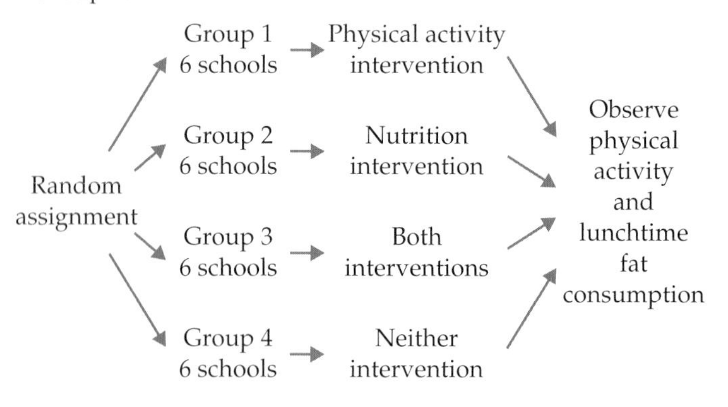

2.35. (a) Population = 1 to 150, sample size 25, then click "Reset" and "Sample." (b) Without resetting, click "Sample" again.

2.37. Design (a) is an experiment, while (b) is an observational study; with the first, any difference in colon health between the two groups could be attributed to the treatment (bee pollen or not).

2.39. (a) Randomly select half the girls to get high-calcium punch; the other half will get low-calcium punch.

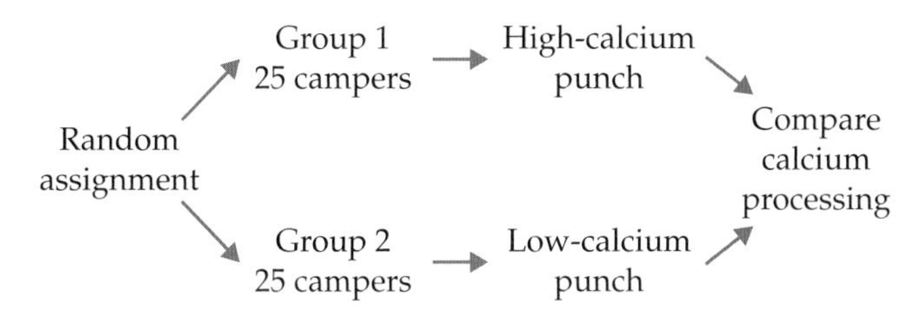

(b) Randomly select half of the girls to receive high-calcium punch first (low-calcium punch later), while the other half get low-calcium punch first (followed by high-calcium punch). This is a better design because it deals with person-to-person variation. (c) The first five subjects selected are 38, 44, 18, 33, and 46. In the completely randomized design, the first group receives high-calcium punch all summer; in the matched pairs design, they receive high-calcium punch for the first half of the summer (and low-calcium punch for the second half).

2.41. The population is faculty members at Mongolian public universities, and the sample is the 300 faculty members to whom the survey was sent. Because we do not know how many responses were received, we cannot determine the response rate.

2.43. Number the list from 0 to 9 and choose two single digits from a line in Table B.

2.45. (a) The content of a single chapter is not random; choose random words from random pages. (b) Students who are registered for an early-morning class might have different characteristics from those who avoid such classes. (c) Alphabetic order is not random; for example, some last names occur more often in some ethnic groups.

2.47. Population: local businesses. Sample: the 150 businesses surveyed (or the 73 businesses that responded). Nonresponse rate: 51.3%.

2.49. Note that the numbers add to 100% across the rows; that is, 34% is the percent of Republicans who are Fox viewers, not the percent of Fox viewers who are Republicans.

2.51. 12 (Country View), 14 (Crestview), 11 (Country Squire), 16 (Fairington), and 08 (Burberry).

2.53. Population = 1 to 200, sample size 25, then click "Reset" and "Sample."

2.55. The results would not be random.

2.57. Each student has chance 1/45 of being selected, but the sample is not an SRS, because the only possible samples have exactly 1 name from the first 45, 1 name from the second 45, and so on.

2.59. Assign labels 01 to 36 for the Climax 1 group, 01 to 72 for the Climax 2 group, and so on; then choose (from Table B) 12, 32, 13, 04; 51, 44, 72, 32, 18, 19, 40; 24, 28, 23; and 29, 12, 16, 25.

2.61. Each student has a 10% chance, but the only possible samples are those with 3 students over age 21 and 2 students under age 21.

2.63. (a) Households without telephones or with unlisted numbers. Such households would likely be made up of poor individuals, those who choose not to have landlines, and those who do not wish to have their phone number published. (b) Those with unlisted numbers.

2.65. Questions slanted in opposing ways can produce opposing results.

2.67. (a) A nonscientist might have different viewpoints and raise different concerns from those considered by scientists.

2.73. Alternate control and cancer samples.

2.77. Interviews conducted in person cannot be anonymous, but they can be confidential.

2.85. (a) Your friend may be misremembering or stretching the truth. Have an unbiased observer report whether your friend has won. (b) Randomize students to seats. Students who voluntarily sit in the first two rows may be better students.

2.87. This is an experiment because treatments are assigned. Explanatory variable: price history (steady or fluctuating). Response variable: price the subject expects to pay.

2.91. Randomly choose the order in which the treatments (gear and steepness combination) are tried.

2.93. (a) One possibility: full-time undergraduate students in the fall term on a list provided by the registrar. (b) One possibility: a stratified sample with 125 students from each year. (c) Nonresponse might be higher with mailed (or emailed) questionnaires; telephone interviews exclude some students and may require repeated calling for those who are not home; face-to-face interviews might be too costly. The topic might also be subject to response bias.

2.95. Use a block design: separate men and women, and randomly allocate each gender among the six treatments.

2.97. CASI will typically produce more honest responses to embarrassing questions.

Chapter 3

3.1. Working in seconds means avoiding decimals and fractions.

3.3. Exam1 = 79, Exam2 = 88, Final = 88.

3.5. Cases: apartments. Five variables: rent (quantitative), cable (categorical), pets (categorical), bedrooms (quantitative), distance to campus (quantitative).

3.7. Scores are slightly left-skewed; most range from 70 to the low 90s.

Stem	Leaves
5	58
6	0
6	58
7	0023
7	5558
8	00003
8	5557
9	0002233
9	8

3.9. (a) Stemplot below

Stem	Leaves
1	6
2	3
2	568
3	35
3	55678
4	012233
4	8
5	1

(b) Use two stems, since the stemplot shows the left-skew in the data.

3.11. The larger classes hide a lot of detail. For example,

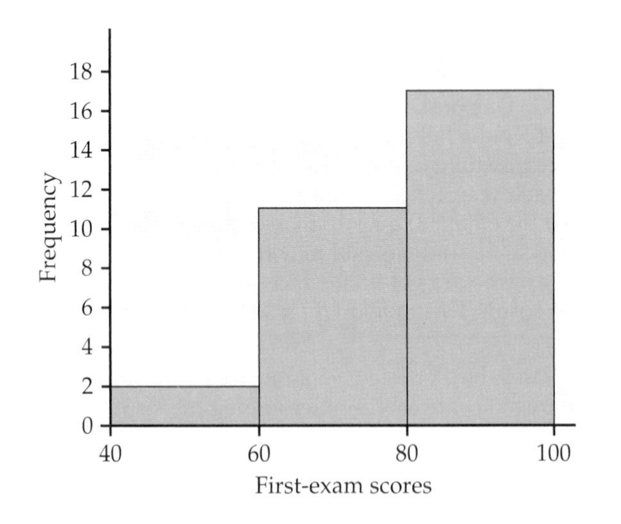

3.13. A stemplot or histogram can be used; the distribution is left-skewed, centered near 80, and spread from 55 to 98.

3.15. For example, crime rates, income, cost of living, entertainment and cultural activities, or taxes.

3.19. For example, blue is by far the most popular choice; 70% of respondents chose 3 of the 10 options.

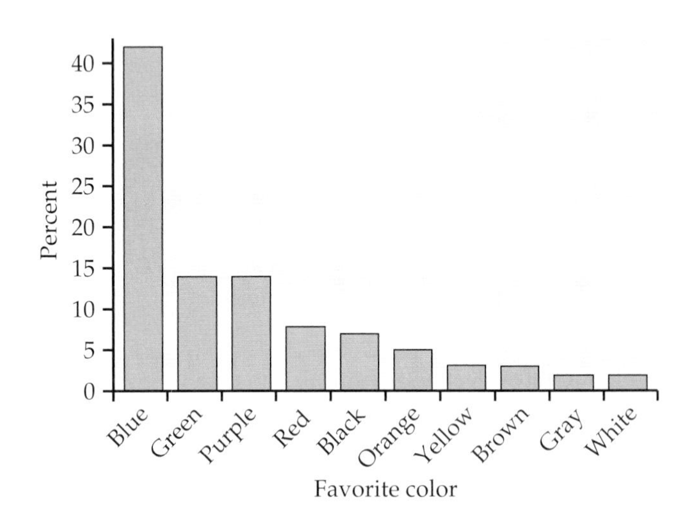

3.21. (a) 232 total respondents; 4.31%, 41.81%, 30.17%, 15.52%, 6.03%, 2.16%. (b) Bar graph below.

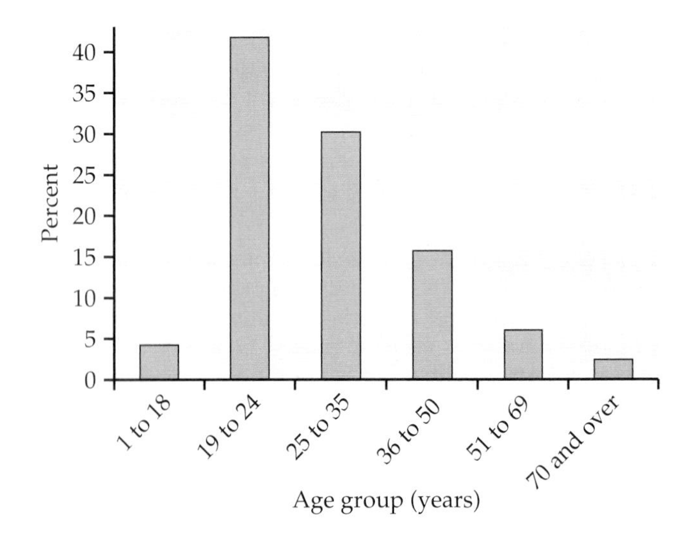

(c) For example, 87.5% of the group were between 19 and 50. (d) The age group classes do not have equal width.

3.23. Ordering bars by decreasing height shows the phones most affected by iPhone sales.

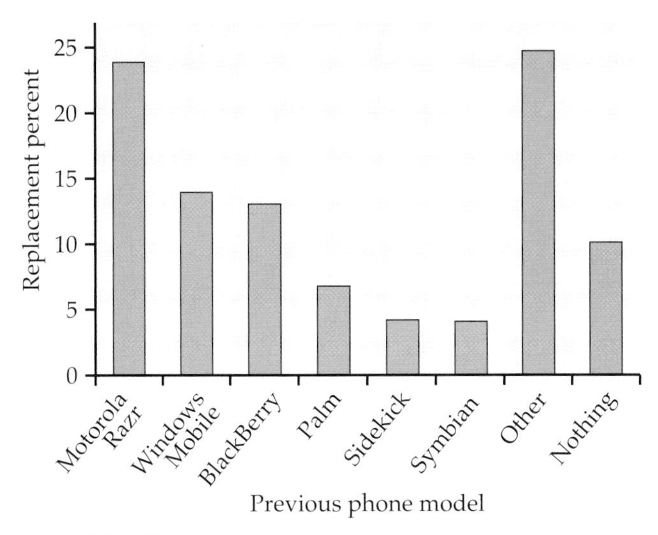

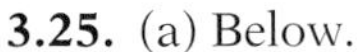
3.25. (a) Below.

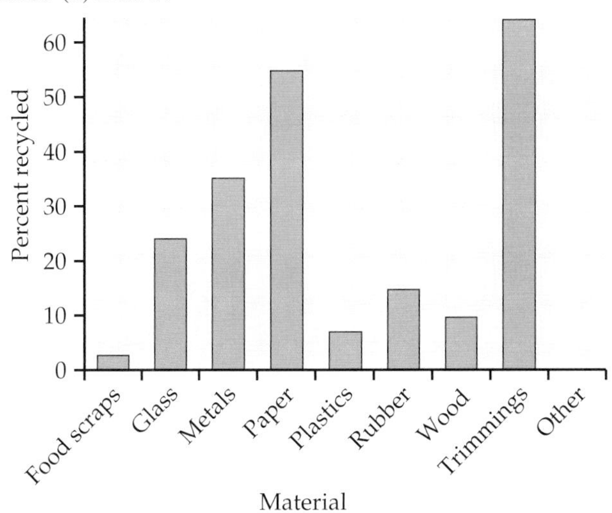

(b) Below.

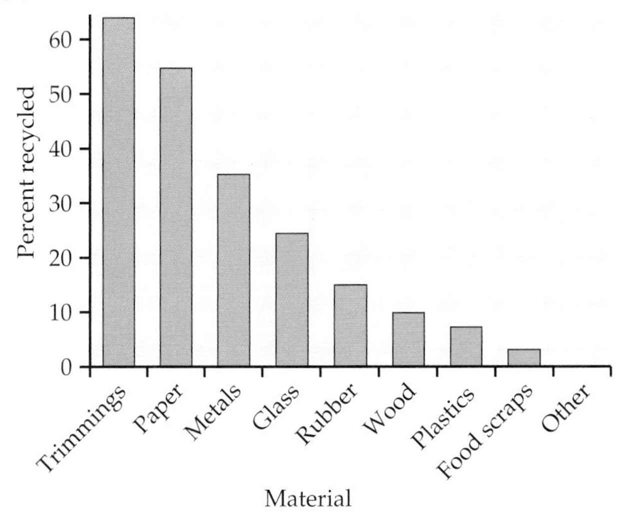

(c) When ordered by height, it is easier to compare categories. (d) Each percent represents part of a different whole.

3.27.

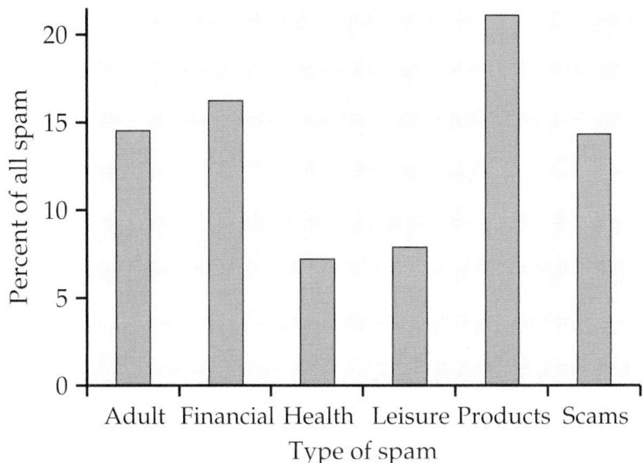

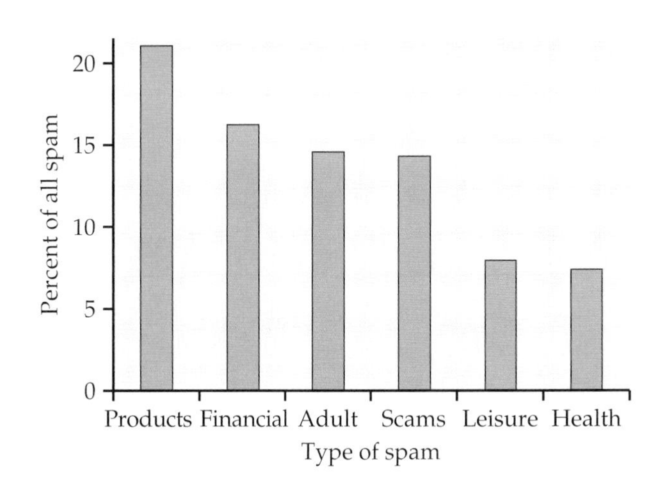

3.29. (a) Table below

Region	Percent of Facebook users
Asia	4.7
Africa	3.7
Europe	27.3
Latin America	25.5
North America	50.4
Middle East	8.3
The Caribbean	14.6
Oceania/Australia	37.1

(b) North America is a high outlier (almost double Europe). Africa, Asia, and the Middle East may be low outliers. (c) In the stemplot below, stems are 10s, and leaves are 1s.

0	34
0	8
1	
1	5
2	
2	57
3	
3	7
4	
4	
5	0

(c) The percents of Facebook users are quite spread out. The outliers are visibly different from the middle of the data. (e) The stemplot is not effective, since there are very few values in the data set. (f) Here we are looking at the percent of the population in each region who are users (instead of the percent of users who are from each region).

3.31. (a) Below left. (b) Below right.

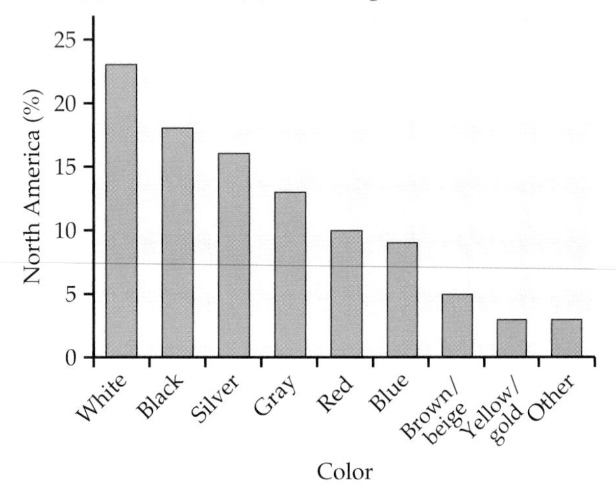

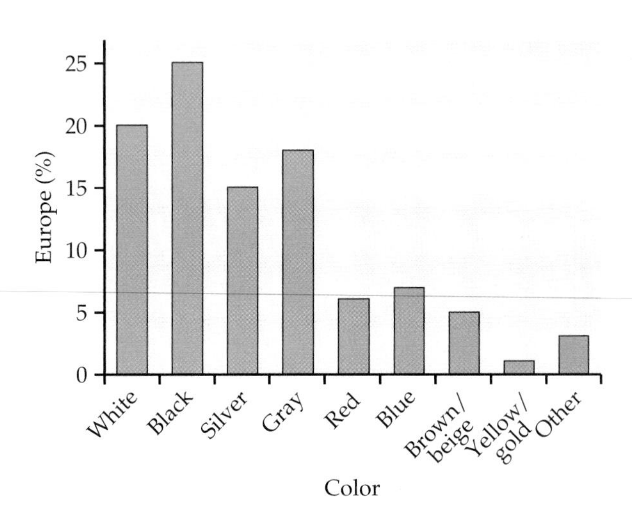

(c) Below.

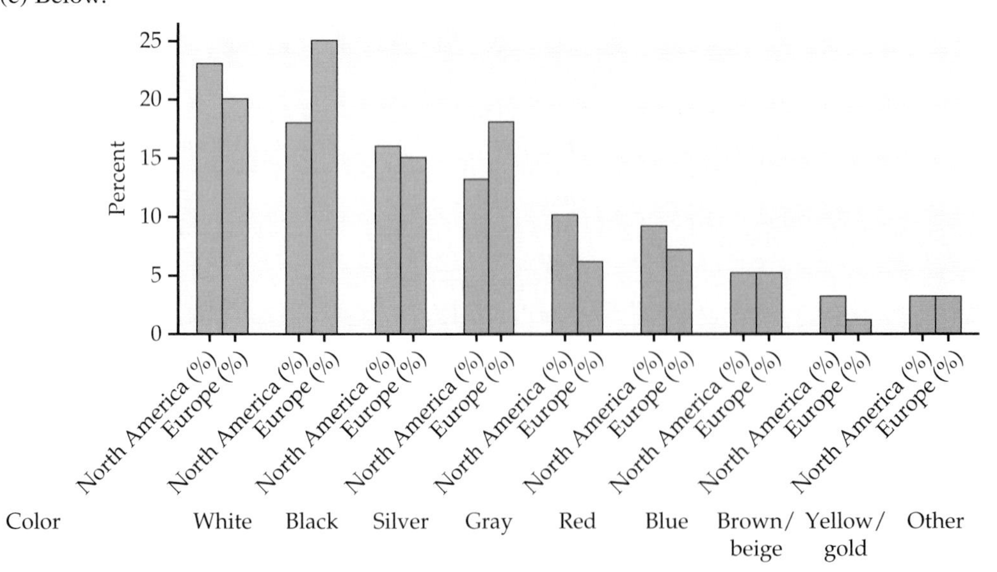

3.33. 359 mg/dl is an outlier; only four are in the desired range.

0	799
1	0134444
1	5577
2	0
2	57
3	
3	5

3.35. Roughly symmetric, centered near 7, spread from 2 to 13.

3.37. Right-skewed, centered near 5 or 6, spread from 0 to 18, no outliers.

```
0 | 0000000000000000000000000000000000011111111111111111111
0 | 22222222222222222233333333333333333333333
0 | 4444444444444444444445555555555555555555
0 | 66666666666666666666677777777777777
0 | 8888888888888889999999999999999999
1 | 00000000000011111111
1 | 222222222222333333333333
1 | 44444455
1 | 66666777
1 | 8
```

3.39. Top-left histogram, 4; top right, 2; bottom left, 1; bottom right, 3.

3.41. $\bar{x} = 82.1$.

3.43. The median of the service times is 103.5 seconds.

3.45. $Q_1 = 75$, $Q_3 = 92$.

3.47. 55, 75, 82.5, 92, 98.

3.49. $IQR = Q_3 - Q_1 = 92 - 75 = 17$, so outliers are below $Q_1 - 1.5 \times IQR = 75 - 1.5(17) = 49.5$ or above $Q_3 + 1.5 \times IQR = 92 + 1.5(17) = 117.5$. There are no outliers.

3.51. All five cases must be equal. For example: 1, 1, 1, 1, 1 or 16, 16, 16, 16, 16.

3.53. $950/4 = 237.5$ points.

3.55. (a) In the stemplot below, leaves are 1000s.

```
0 | 33333333333333333444444444444444
0 | 55555555666666666667777778888888889999
1 | 1111222224444
1 | 679
2 | 3334
2 | 557789
3 | 3
3 | 55
4 | 2
4 |
5 |
5 | 59
6 |
6 | 9
7 | 1
```

(b) In millions of dollars, the five-number summary is 3512, 4429, 7305, 14,086, and 71,861. (c) For example, the distribution is sharply right-skewed; the top four could be considered outliers.

3.57. (a) $\bar{x}$ changes from 4.76% (with) to 4.81% (without); the median (4.7%) does not change. (b) s changes from 0.7523% to 0.5864%; Q_1 changes from 4.3% to 4.35%, while $Q_3 = 5\%$ does not change. (c) A low outlier decreases $\bar{x}$; any kind of outlier increases s. Outliers have little or no effect on the median and quartiles.

3.59. (a) The distribution is left-skewed; the five-number summary (in ounces) is 3.7, 4.95, 6.7, 7.85, and 8.2. (b) The numerical summary does not reveal the two weight clusters (visible in a stemplot or histogram). (c) For the small potatoes, $\bar{x} = 4.662$ and $s = 0.501$ ounces; for the large potatoes, $\bar{x} = 7.300$ and $s = 0.755$ ounces. Because there are clearly two groups, it seems appropriate to treat them separately.

3.61. (a) Min $= Q_1 = 0$, $M = 5.085$, $Q_3 = 9.47$, Max $= 73.2$. (b) Below left. (c) Below right.

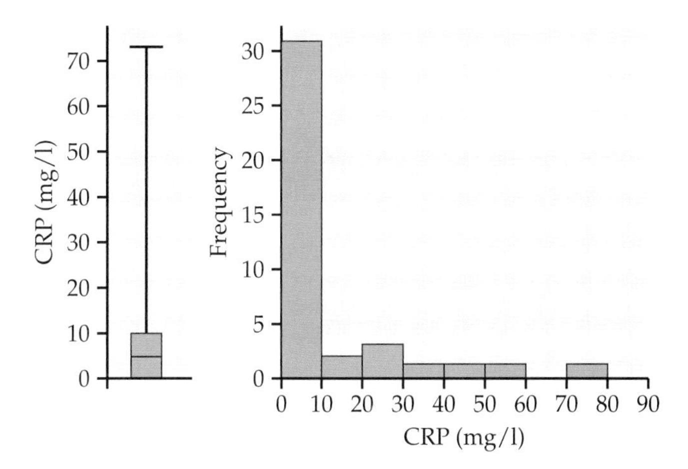

(d) The distribution is sharply right-skewed. The histogram seems to convey the distribution better.

3.63. Min $= 0.24$, $Q_1 = 0.355$, $M = 0.76$, $Q_3 = 1.03$, Max $= 1.9$. The distribution is right-skewed. A histogram or stemplot reveals an important feature not evident from a boxplot: this distribution has two peaks.

3.65. This distribution will almost always be strongly skewed to the right.

3.67. $\bar{x} = \$76,667$; six of the nine employees earn less than the mean. $M = \$35,000$.

3.69. The mean rises to $91,667, while the median is unchanged.

3.71. The mean and median always agree for two observations.

3.73. (a) Place the new point at the current median.

3.75. (a) *Bihai*: $\bar{x} \doteq 47.5975$, $s \doteq 1.2129$. Red: $\bar{x} \doteq 39.7113$, $s \doteq 1.7988$. Yellow: $\bar{x} \doteq 36.1800$, $s \doteq 0.9753$ (all in mm). (b) *Bihai* and red appear to be right-skewed (although it is difficult to tell with such small samples).

bihai		red		yellow	
46	3466789	37	4789	34	56
47	114	38	001 2278	35	146
48	01 33	39	167	36	001 5678
49		40	56	37	01
50	12	41	4699	38	1
		42	01		
		43	0		

3.77. The five-number summary is 1, 3, 4, 5, and 12 letters.

3.79. Take six or more numbers, with the largest number much larger than Q_3.

3.81. (a) Any set of four identical numbers works. (b) 0, 0, 20, 20 is the only possible answer.

3.83. Multiply each value by 0.03937.

3.85. $\bar{x} = 5.32$ pounds and $s = 2.60$ pounds.

3.87. Take the mean plus or minus three standard deviations: $572 \pm 3(51) = 419$ to 725.

3.89. $z = \frac{510-572}{51} \doteq -1.22$. This is negative because an ISTEP score of 510 is below average; specifically, it is 1.22 standard deviations below the mean.

3.91. Using Table A, the proportion below 620 ($z \doteq 0.94$) is 0.8264, and the proportion below 660 ($z \doteq 1.73$) is 0.9582. The area between 620 and 660 is $0.9582 - 0.8264 = 0.1318$.

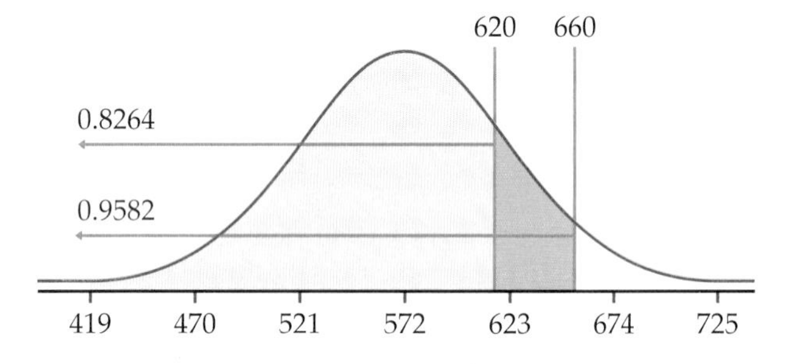

3.93. Using Table A, x should correspond to a standard score of $z \doteq -0.84$ (software gives -0.8416), so the ISTEP score (unstandardized) is $x = 572 - 0.84(51) \doteq 529.2$ (software: 529.1).

3.97. (a) The 68–95–99.7 ranges for women are 8489 to 20,919, 2274 to 27,134, and −3941 to 33,349. For men, they are 7158 to 22,886, −706 to 30,750, and −8570 to 38,614. In both cases, some of the lower limits are negative, which does not make sense; this happens because the women's distribution is skewed and because the men's distribution has an outlier. Contrary to the conventional wisdom, the men's mean is slightly higher. (b) The means suggest that Mexican men and women tend to speak more than those from the United States.

3.99. (a) −1.64, −1.04, 0.13, and 1.04. (b) 53.6, 59.6, 71.3, and 80.4.

3.101. (a) 1/4. (b) 0.25. (c) 0.5.

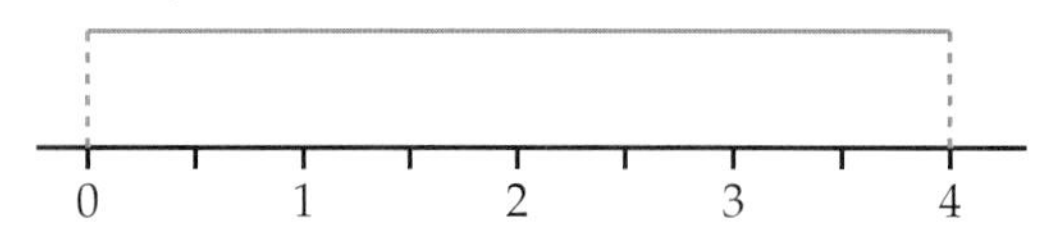

3.103. (a) Mean C, median B. (b) Mean A, median A. (c) Mean A, median B.

3.105. (a) 0.6826; the 68–95–99.7 rule gives 0.68. (b) 0.9544 (compared with 0.95); 0.9974 (compared with 0.997).

3.107. (a) 327 to 345 days. (b) 16%.

3.109. $\bar{x} = 5.4256$ and $s = 0.5379$; 67.62% within $\bar{x} - s$ and $\bar{x} + s$, 95.24% within $\bar{x} \pm 2s$, all within $\bar{x} \pm 3s$.

3.111. (a) 0.0359. (b) 0.9641. (c) 0.0548. (d) 0.9093.

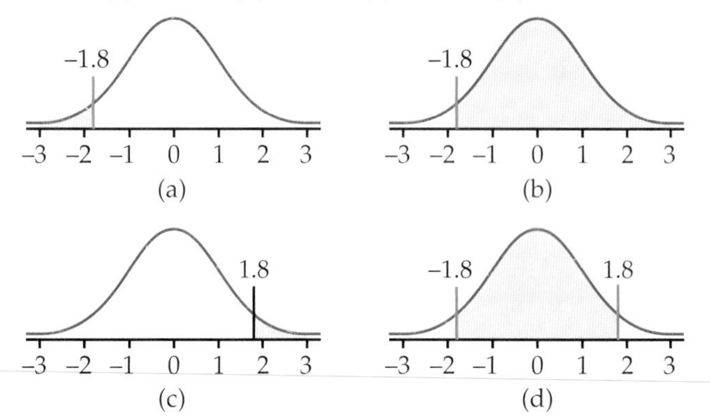

3.113. About 2.5%.

3.115. Jayden's score ($z = -1.02$) is higher than Emily's ($z = -1.52$).

3.117. About 2014.

3.119. 32.3%.

3.121. 1239 and below.

3.123. 1239, 1428, 1590, and 1779.

3.125. (a) 33%. (b) 15%. (c) 52%.

3.127. (a) About 1.35. (b) 1.35.

3.129. (a) The yellow variety is the nearest to a straight line. See below and on the next page.

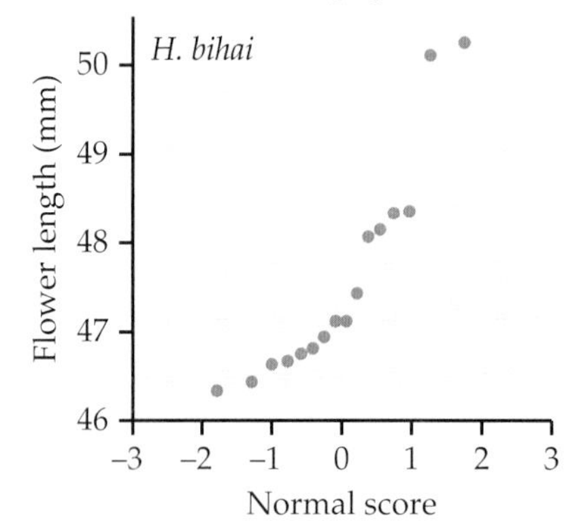

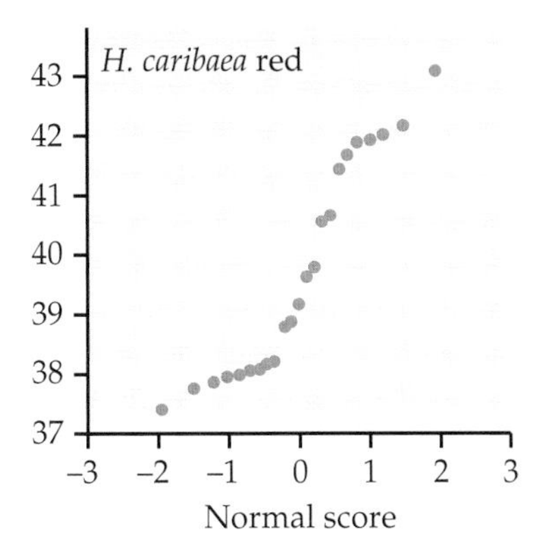

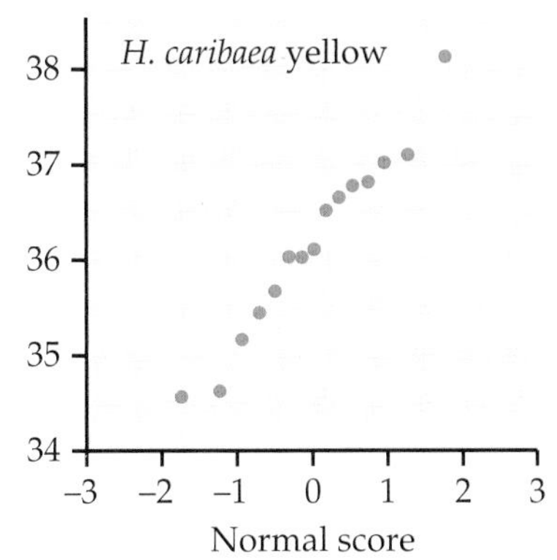

(b) The other two distributions are both slightly right-skewed, and the *bihai* variety appears to have a couple of high outliers.

3.131. Histograms will suggest (but not exactly match) Figure 3.33. The uniform distribution does not extend as low or as high as a Normal distribution.

3.133. (a) The distribution appears to be roughly Normal.

```
 8 | 28
 9 |
10 | 58
11 | 34
12 | 023689
13 | 015788
14 | 0077
15 | 13466889
16 | 01567
17 | 45677789
18 | 8
19 | 148
20 | 2
21 | 6
22 | 8
```

(b) One could justify either $\bar{x} = 15.27\%$ and $s = 3.118\%$, or the five-number summary (8.2%, 13%, 15.5%, 17.6%, and 22.8%). (c) For example, binge-drinking rates are typically 10% to 20%. Which states are high, and which are low?

3.135. For example, white is considerably less popular in Europe, and gray is less common in China.

3.139. The distribution is somewhat right-skewed, with only one country (Bosnia and Herzegovina) in the20s. The mean and standard deviation are 39.85 and 22.05 users per 100 people.

3.141. The given description is true on the average, but the curves (and a few calculations) give a more complete picture. For example, a score of about 675 is about the 97.5th percentile for both genders, so the top boys and top girls have very similar scores.

3.143. Slightly right-skewed, with one (or more) high outliers. Five-number summary: 22, 23.735, 24.31, 24.845, and 28.55 hours.

```
22 | 013
22 | 7899
23 | 00001122223334444 4
23 | 5556666666777777888888899 9
24 | 00000011111112222222223333333333444444
24 | 555555666666666777777888888999999
25 | 000011112333 44
25 | 56666889
26 | 2
26 | 56
27 | 2
27 |
28 |
28 | 5
```

3.145. The many smaller providers have $100 - 70.9 = 29.1\%$.

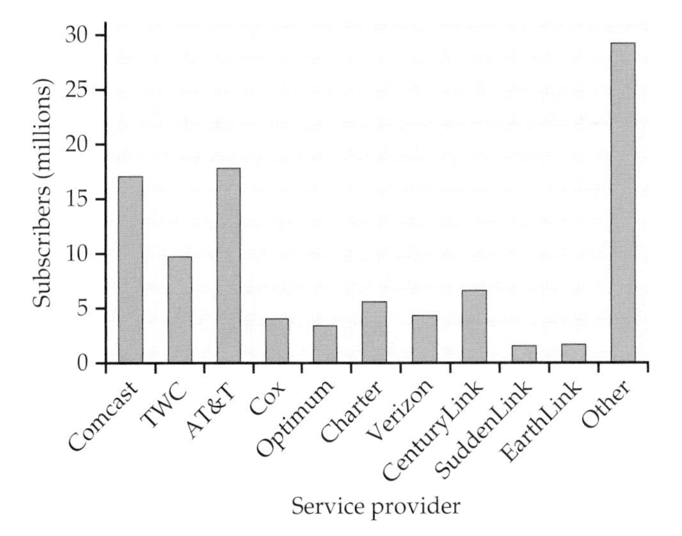

3.147. (a) Makes: bar graph or pie chart. Age: histogram, stemplot, or boxplot. (b) Study time: histogram, stemplot, or boxplot. To show change over time, use a time plot. (c) Bar graph or pie chart. (d) Normal quantile plot.

3.149. No to both questions; no summary can exactly describe a distribution that can include any number of values.

Chapter 4

4.1. Students.

4.3. Cases: cups of Mocha Frappuccino. Variables: size and price (both quantitative).

4.5. (b) 10 cases. (c) Choices will vary. (d) The second and third columns are quantitative.

4.7. (a) and (b)

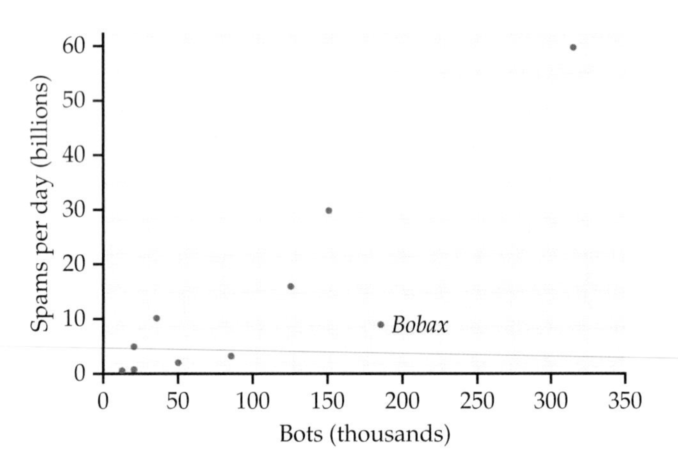

4.9. Size should be explanatory. The scatterplot shows a positive association between size and price.

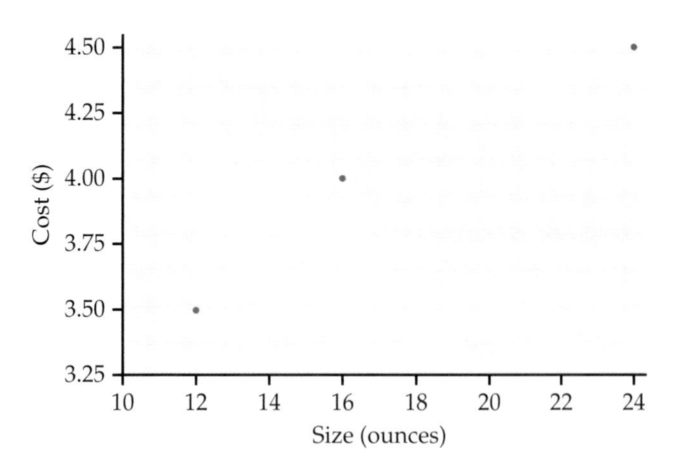

4.11. There is still a strong positive relationship, but the new points are far away from the others.

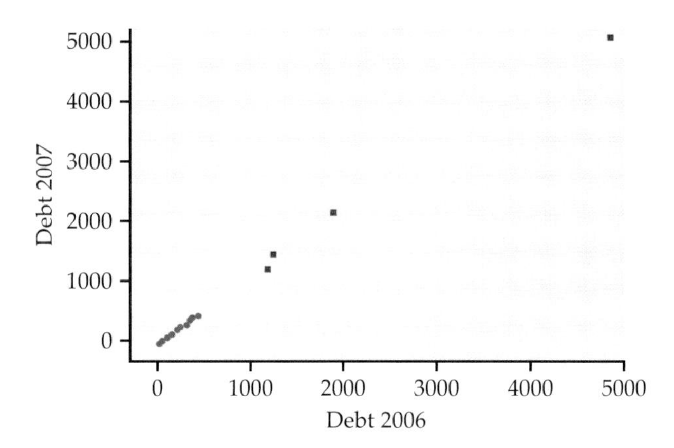

4.13. (a) A boxplot summarizes the distribution of one variable. (b) This is correct only if there is an explanatory-response relationship. (c) High values go with high values.

4.15. (a) Below top. (b) There are 49,904 thousand uninsured; divide each number in the second column by this amount. (c) Below bottom. (d) The plots differ only in the vertical scale. (e) The uninsured are found in similar numbers for the five lowest age groups (with slightly more in those aged 25–34 and 45–64), and fewer among those over 65.

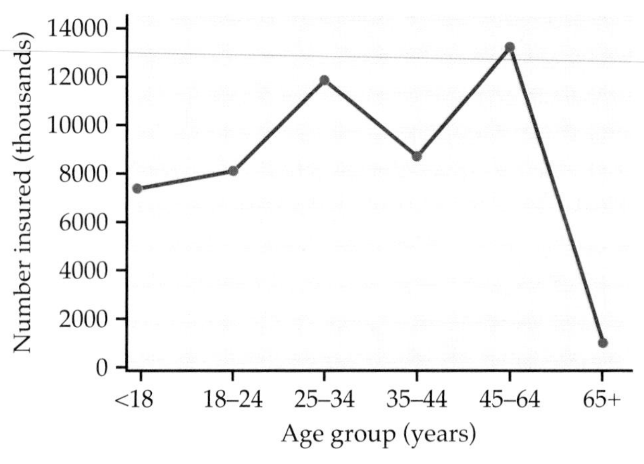

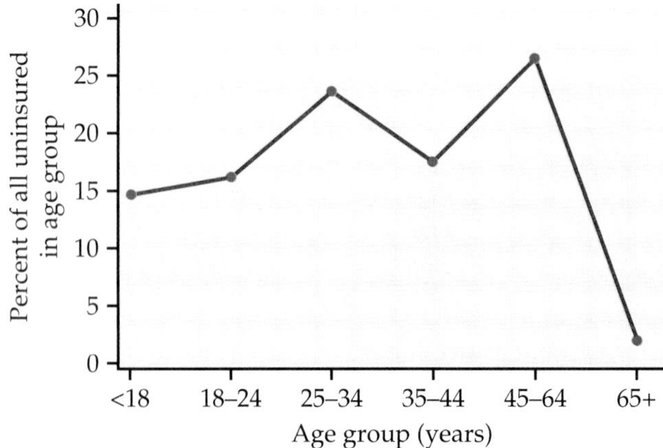

4.17. The percents in Exercise 4.15 show what fraction of the uninsured fall in each age group. The percents in Exercise 4.16 show what fraction of each age group is uninsured.

4.19. (a) Scatterplot below.

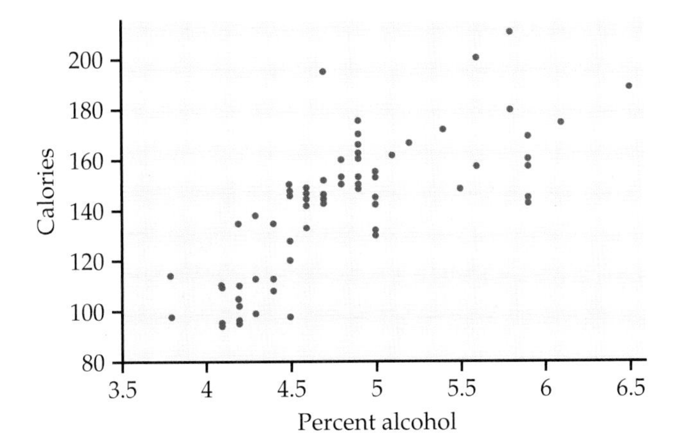

(b) There is a moderate positive linear relationship.

4.21. There is a moderate positive linear relationship; the relationship for all countries is less linear because of the wide range in life expectancy among countries with low Internet use.

4.23. (a) "Month" (the passage of time) explains changes in temperature (not vice versa). (b) Temperature increases linearly with time (about 10 degrees per month); the relationship is strong.

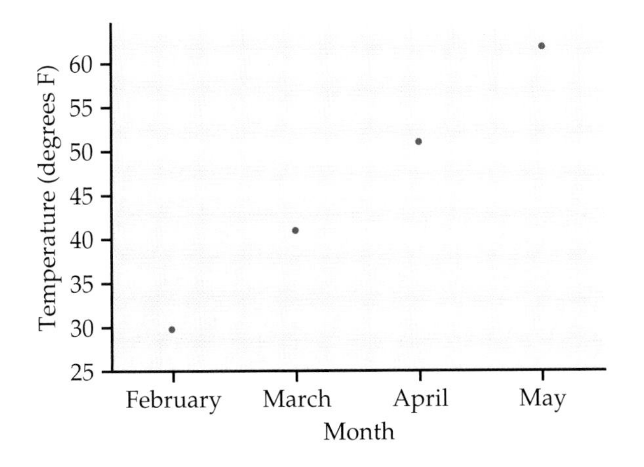

4.25. (a) The second test happens before the final exam. (b) The plot shows a weak positive association.

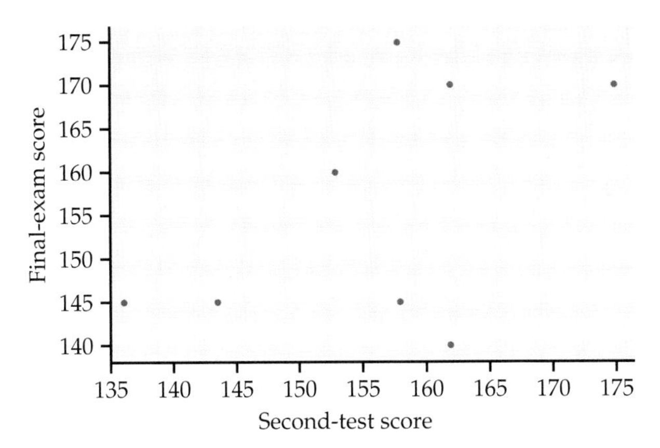

(c) Students' study habits are more established by the middle of the term.

4.27. (a) Explanatory: age. Response: weight. (b) Explore the relationship. (c) Explanatory: number of bedrooms. Response: price. (d) Explanatory: amount of sugar. Response: sweetness. (e) Explore the relationship.

4.29. (a) Areas with many breeding pairs would correspondingly have more males that might potentially return. (c) The theory suggests a negative association; the scatterplot shows this.

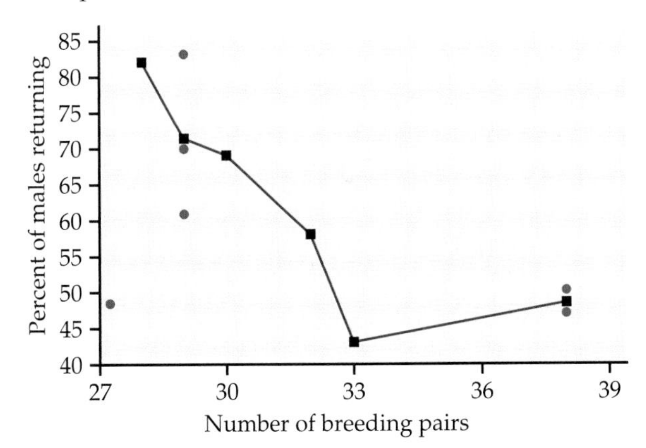

4.31. (a) Scatterplot below.

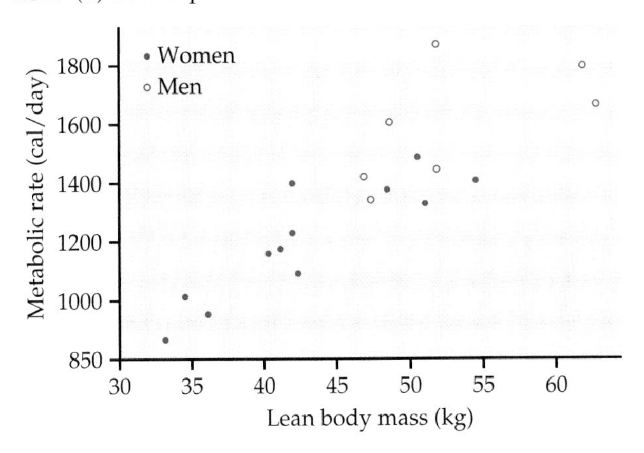

(b) The association is linear and positive and is stronger for women. Males typically have larger values for both variables.

4.33. (a) $r = 0.8839$. (b) They are equal. (c) Units do not affect correlation.

4.35. (a) $r = 1$. (b) $r = 1$.

4.37. $r = 0.2873$.

4.39. $r = 0.6701$.

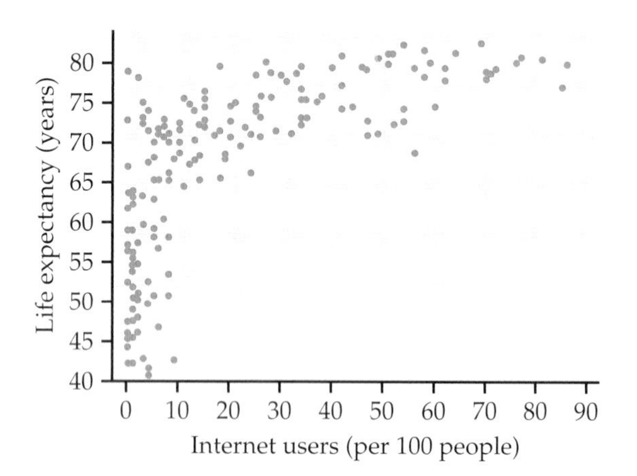

4.41. (a) $r = 0.5194$. (b) The first-test/final-exam correlation will be lower.

4.43. The correlation increases.

4.45. (a) Positive, but not close to 1.

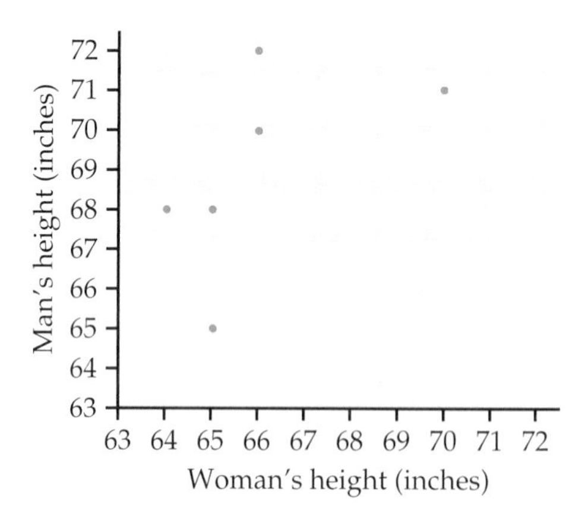

(b) $r = 0.5653$. (c) r would not change; it does not tell us that the men were generally taller. (d) r would not change. (e) 1.

4.47. (a) $r = \pm 1$ for a line. (c) Leave some space above your vertical stack. (d) The curve must be higher at the right than at the left.

4.49. (a) Occupation is not quantitative. (b) r cannot exceed 1. (c) Neither variable is quantitative.

4.51. There is little linear association between research and teaching—for example, knowing that a professor is a good researcher gives little information about whether she is a good or bad teacher.

4.53. 1.441 kg.

4.55. Expressed as percentages, these fractions are 81%, 25%, 9%, 0%, 9%, 25%, and 81%.

4.57. Los Angeles (5995) is best, and Chicago (−7283) is worst.

Los Angeles	5994.85
Washington, DC	2763.75
Minneapolis	2107.59
Philadelphia	169.42
Oakland	27.91
Boston	20.96
San Francisco	−75.78
Baltimore	−131.55
New York	−282.99
Long Beach	−1181.70
Miami	−2129.21
Chicago	−7283.26

4.59. For Baltimore, for example, this rate is $\frac{5091}{651} \doteq 7.82$ acres per thousand people. This scatterplot is much less linear; the regression equation $\hat{y} = 8.739 - 0.000424x$ explains only 8.7% of the variation in open space.

4.61. $\hat{y} = 3.379 + 1.6155x$.

4.63. (a) All correlations are approximately 0.816 or 0.817, and the regression lines are $\hat{y} = 3.000 + 0.500x$. We predict $\hat{y} \doteq 8$ when $x = 10$. (b) Scatterplots below.

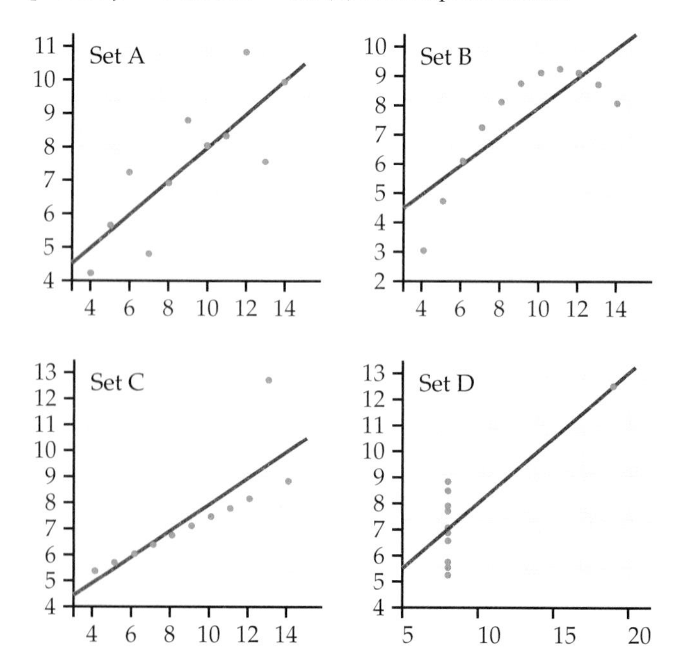

(c) This regression formula is appropriate only for Set A.

4.65. (a) The relationship is linear, positive, and fairly strong. (b) $\hat{y} = 1.470 + 1.443x$. (c) 71.1%. (d) The new point makes the relationship stronger, but its location has a large impact on the regression equation.

4.67. (a) $y = 42$. (b) y increases by 6. (c) 12.

4.69. For metabolic rate on body mass, the slope is 26.9 cal/day per kg. For body mass on metabolic rate, the slope is 0.0278 kg per cal/day.

4.71. When $x = \bar{x}$, $\hat{y} = a + b\bar{x} = (\bar{y} - b\bar{x}) + b\bar{x} = \bar{y}$.

4.73. $r = 0.40$.

4.75. The sum is 0.01.

4.77. (a) A high correlation means strong association, not causation. (b) Outliers in the y direction (and some other data points) will have large residuals. (c) It is not extrapolation if $1 \leq x \leq 5$.

4.79. For example, it may be that doing well makes students or workers feel good about themselves rather than vice versa.

4.81. The explanatory and response variables were "consumption of herbal tea" and "cheerfulness/health." One lurking variable is social interaction; many of the nursing-home residents may have been lonely before the students started visiting.

4.83. (a) The plot (below) is curved (low at the beginning and end of the year, high in the middle).
(b) $\hat{y} = 39.392 + 1.4832x$; it does not fit well.
(c) Residuals (top of next column) are negative for January through March and October through December, and positive from April to September. (d) A similar pattern would be expected in any city that is subject to seasonal temperature variation. (e) Seasons in the Southern Hemisphere are reversed.

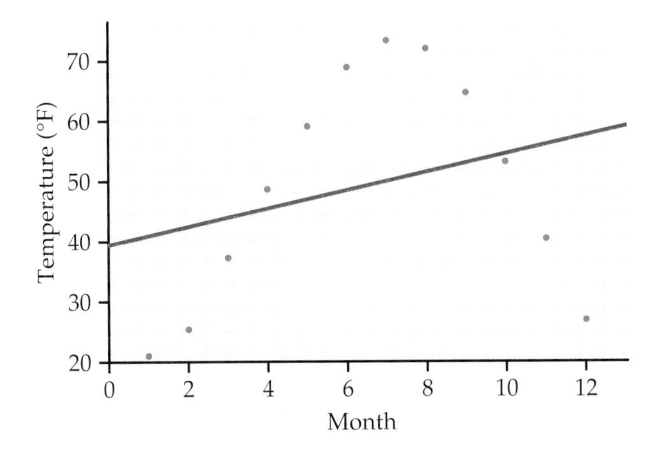

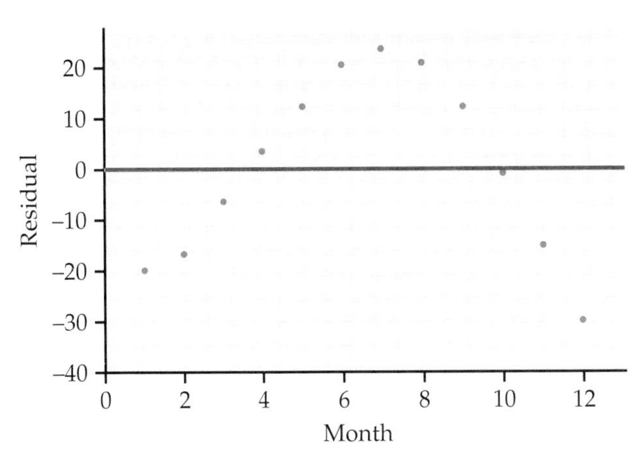

4.85. For example, a student who in the past might have received a grade of B (and a lower SAT score) now receives an A (but has a lower SAT score than an A student in the past).

4.87. $r = 0.08795$ and $b = 0.00811$ kg/cal.

4.89. (a)

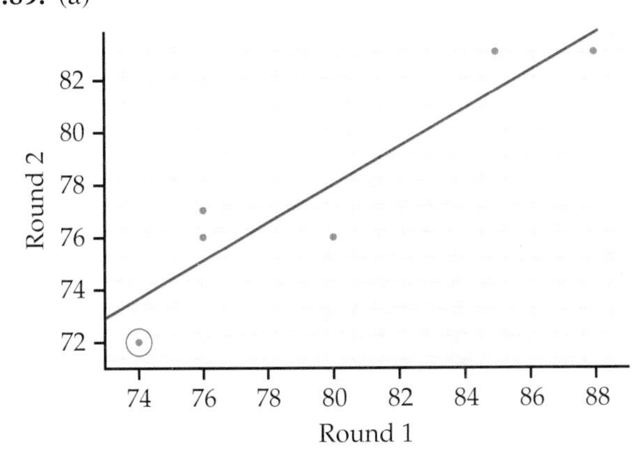

(b) The plot shows a strong positive linear relationship. (c) $\hat{y} = 20.40 + 0.7194x$. (d) Hernandez's point is in the lower left.

4.91. (a) Drawing the "best line" by eye is a very inaccurate process.

4.93. The plot should show a positive association when either group of points is viewed separately and should show a large number of bachelor's degree economists in business and graduate degree economists in academia.

4.95. 1684 are binge drinkers; 8232 are not.

4.97. $\dfrac{8232}{17{,}096} \doteq 0.482$.

4.99. $\dfrac{1630}{7180} \doteq 0.227$.

4.101. (a) 50.5% get enough sleep; 49.5% do not.
(b) 32.2% get enough sleep; 67.8% do not. (c) Those who exercise more than the median are more likely to get enough sleep.

4.103. 3.0% of Hospital A's patients died, compared with 2.0% at Hospital B.

4.105. (a) About 3,388,000. (b) 0.207, 0.024; 0.320, 0.071; 0.104, 0.104; 0.046, 0.125. (c) 0.230, 0.391, 0.208, 0.171. (d) 0.677, 0.323.

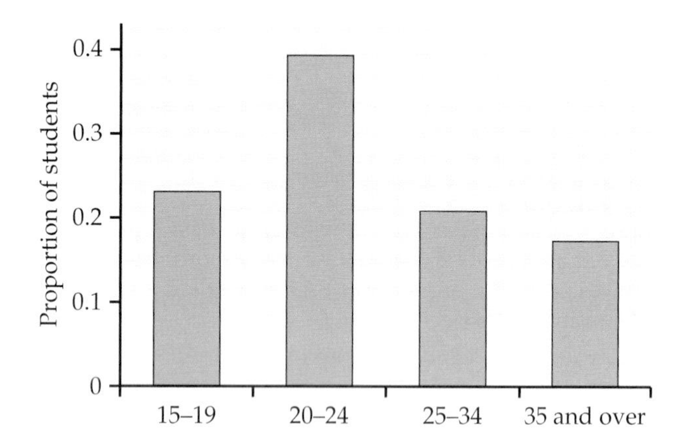

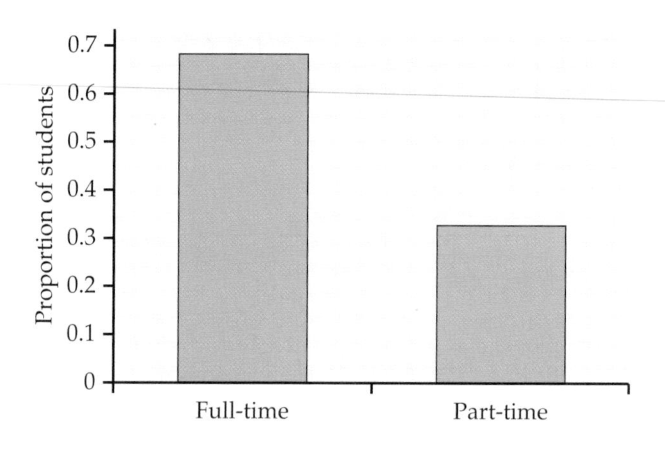

4.107. Full-time: 0.305, 0.472, 0.154, 0.069. Part-time: 0.073, 0.220, 0.321, 0.386. Full-time students are dominated by younger ages, while part-time students are more likely to be older.

4.109. There are many possibilities, but the simplest such table would have all nine counts equal to one another.

4.111. (a) Plot below (with regression line). (b) $\hat{y} = 112{,}323 - 55.6x$. (c) The plot shows a clear negative association, and the slope of the regression line says that the rank is decreasing at an average rate of about 56 per year. Because a *lower* rank means *higher* popularity, this means that "Atticus" is getting more popular.

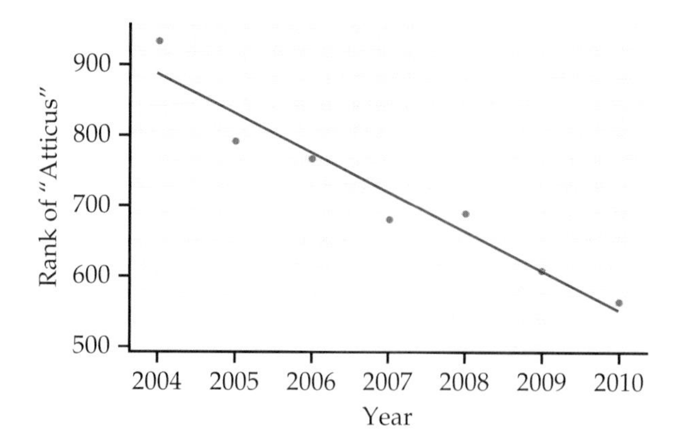

4.113. (a) The scatterplot shows a positive curved relationship. (b) The regression explains about $R^2 \doteq 98.3\%$ of the variation in salary. While this indicates that the relationship is strong, and *close* to linear, we can see from the scatterplot that the actual relationship is curved.

4.115. (a) Both plots show a positive association, but the log-salary plot is linear rather than curved. (b) The residuals in Figure 4.26 were positive at the beginning and end and negative in the middle. The log-salary residuals in Figure 4.27 show no particular pattern.

4.117. (a) Plot below. (b) The scatterplot shows a very strong positive linear association. (c) 69.6% of the variation in 2011–2012 salaries is explained by 2010–2011 salaries.

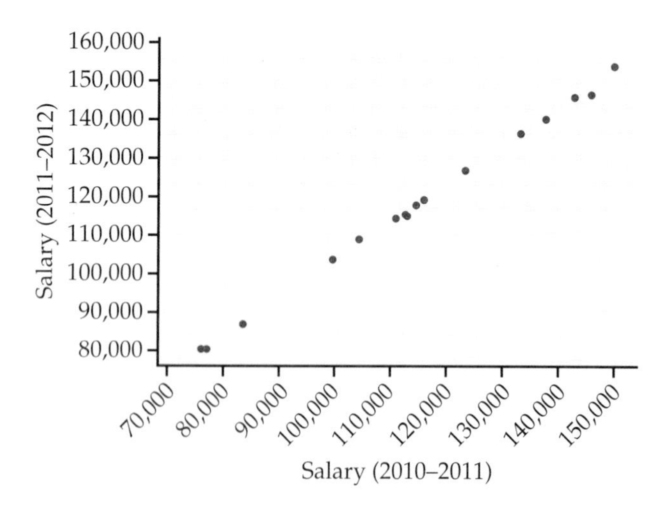

4.119. (a) Plot below (with regression line). (b) $\hat{y} = 8.422 - 0.00005x$. (c) A plot of residuals versus 2010–2011 salaries reveals no outliers or other causes for concern. (d) The data show that, in general, those with lower salaries are given greater raise percents.

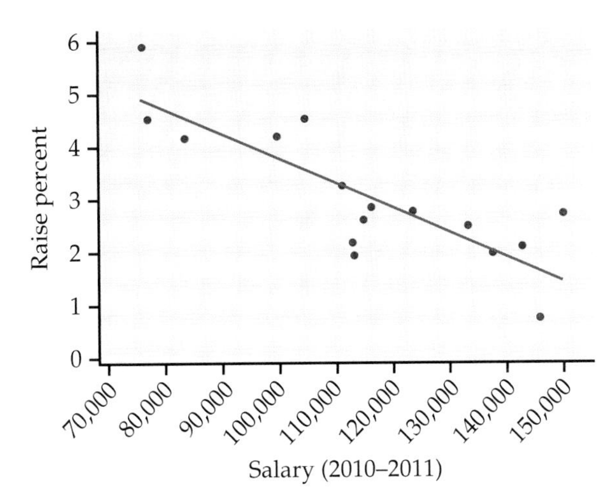

4.121. (a) The association is negative and roughly linear. (b) The correlation is $r \doteq -0.5503$. (c) Utah is the farthest point to the left (that is, it has the lowest smoking rate) and lies well below the line (i.e., the proportion of adults who eat fruits and vegetables is lower than we would expect). (d) California has the second-lowest smoking rate and one of the highest fruit/vegetable rates. This point lies above the line, meaning that the proportion of California adults who eat fruits and vegetables is higher than we would expect.

4.123. (a) The plot (below) shows a fairly strong positive linear association. (b) The regression equation is $\hat{y} = -6.202 + 1.2081x$. (c) If $x = 62$ pages, we predict $\hat{y} \doteq 68.7$ pages.

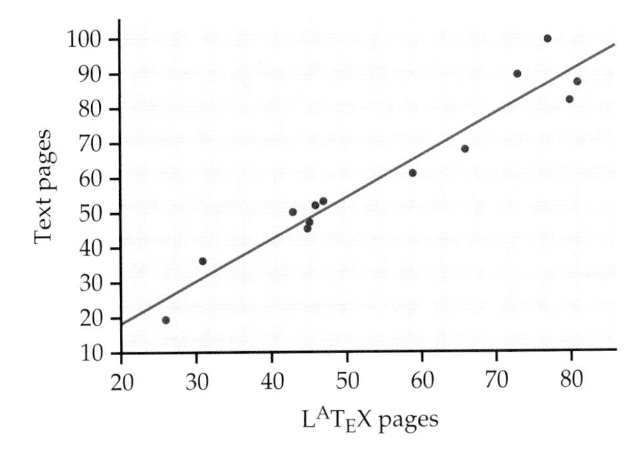

4.127. Based on the quantile plot (below), the distribution is close to Normal.

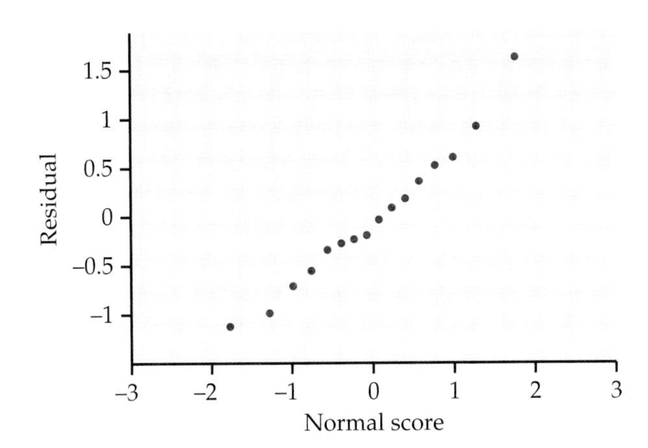

4.129. About 4.1 points above the final-exam mean.

4.131. There are considerably more social science degrees and fewer engineering degrees in the United States. The Western Europe and Asia distributions are similar.

	United States	Western Europe	Asia	Overall
Engineering	17.44	38.26	36.96	32.78
Natural science	31.29	33.73	31.97	32.29
Social science	51.28	28.01	31.07	34.93

4.133. Women are more likely to volunteer.

Gender	Strictly voluntary	Court-ordered	Other
Men	79.16	5.21	15.63
Women	85.19	2.14	12.67

4.135. For example, if the four schools admit 50%, 60%, 70%, and 80% of applicants, and men are more likely to apply to the first two, while women are more likely to apply to the latter two, women will be admitted more often.

4.137. (a)

	Students	Nonstudents	Total
Agree	22	30	52
Disagree	39	29	68
Total	61	59	120

(b) We should not generalize our conclusions too far beyond the populations "upper-level northeastern college students taking a course in Internet marketing" and "West Coast residents willing to participate in commercial focus groups."

Part I Review

I.1. (a) Factor: smartphone app. Levels: three versions. (b) Experimental units: 30 college-aged women, 30 college-aged men, 30 high-school-aged women, 30 high-school-aged men. (c) Partial diagram below (this is the part of the diagram for college-aged women only).

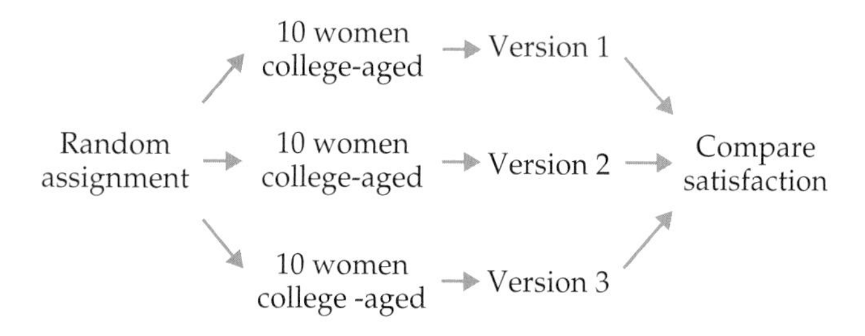

I.3. For example, "How likely are you to use the app regularly?" or "How likely are you to suggest the app to your friends?"

I.5. For example, a histogram or stemplot could be used for the number of hours spent per week in cocurricular activities, and a bar graph could be used for satisfaction with educational experience.

I.7. Since this is not an experiment, we cannot conclude a causal relationship.

I.9. (a) A control group will allow for comparison. (b) No; students may respond to the questions differently if they knew the purpose of the experiment.

I.11. (a) Explanatory: year. Response: dollars. (b) The relationship is strong, positive, and somewhat linear.

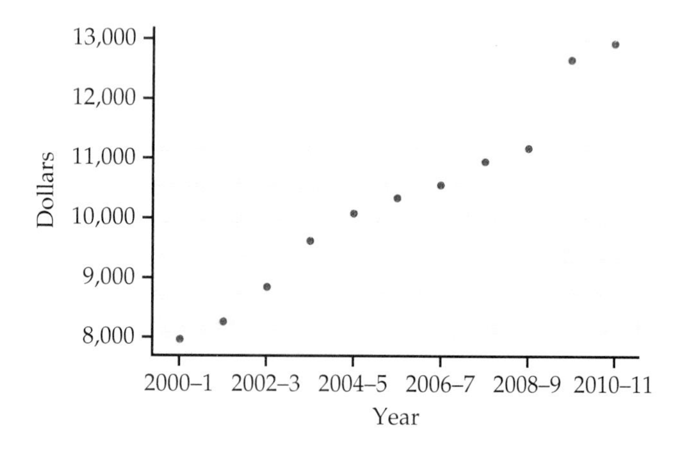

(c) Some might consider the years 2009–2010 and 2010–2011 to be outliers. (d) The years mentioned in part (c) might be influential.

I.13. (a) The residuals are 51.091, −128.073, −32.236, 264.600, 244.436, 23.273, −238.891, −296.055, −568.218, 444.618, 235.455 (b) Plot below.

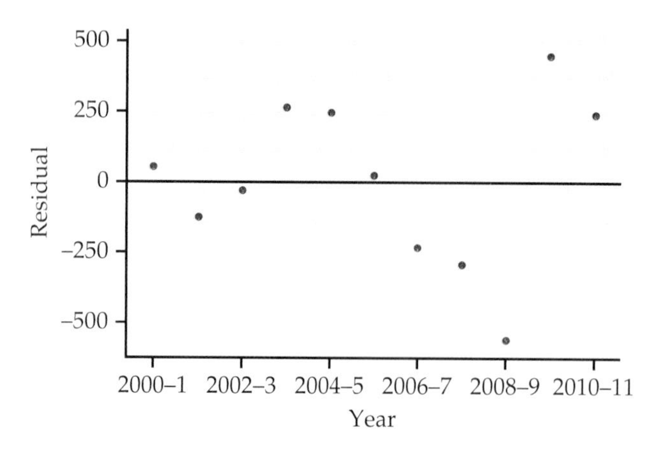

(c) There appears to be an irregular curving pattern in the residual plot.

I.15. (a) and (b) Histogram and stemplot below.

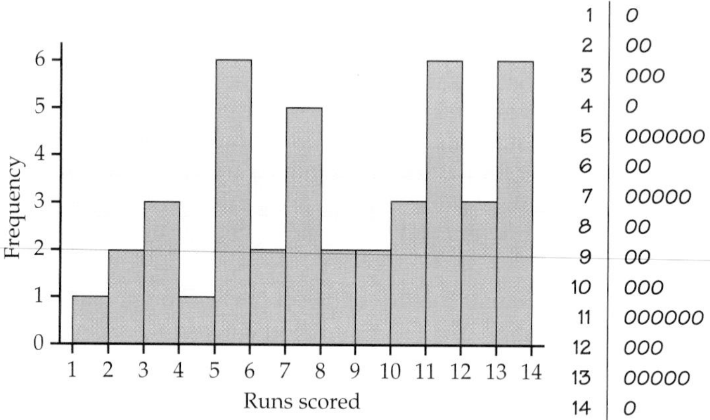

I.17. (a) $IQR = 11 - 8 = 3$ runs. (b) $s = 3.7$ runs.

I.19. (a) Plot below.

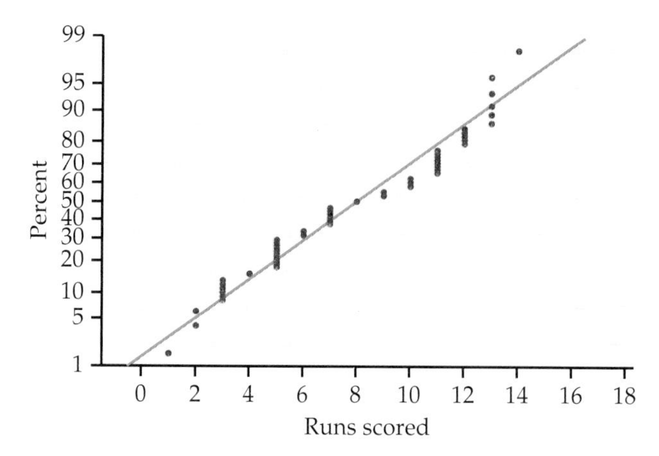

(b) The numbers of runs scored do not appear Normal, since there are curves in the plot.

I.21. The mean and the median are similar between the National League and the American League. The big difference between the two is that the American League is somewhat right-skewed, and the maximum is almost double that of the American League.

I.23. (a) 0.1587. (b) For example, if I want to be in the top 5% or 10% of adults, I should eat more foods rich in calcium. (c) 996 mg/day.

I.25. (a) Unimodal and left-skewed with some potential outliers. The median is 100 calories, and the interquartile range is 30 calories. (b) Unimodal and left-skewed with some potential outliers. The median is 100 calories, and the interquartile range is 13 calories. (c) Back-to-back stemplot below. The cookie calories with the meal vary more than those after the meal.

Cookie Calories With		Cookie Calories After
4	1	
	2	
76	3	
2	4	3
	5	347
0	6	
62	7	2
0	8	
963	9	05799
9752211000	10	00111223349
32	11	04

I.27. (a) Scatterplot (with regression line) below.

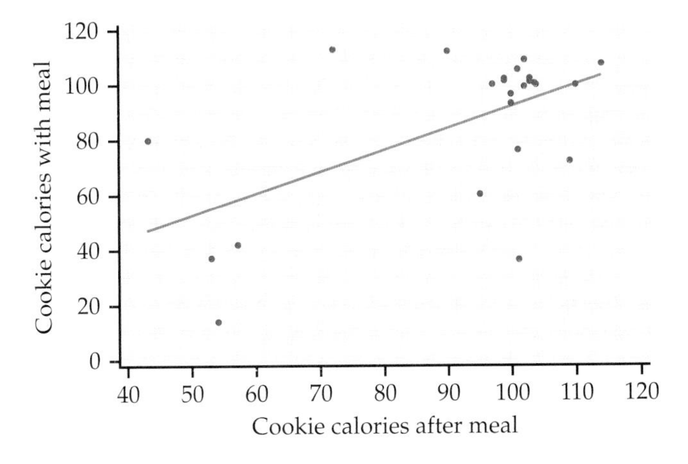

(b) There is a positive relationship; however, there is no clear form, and the relationship is not very strong. (c) There is a group of four points toward the left of the scatterplot that could be considered outliers. Because of the weak relationship, it is difficult to tell. (d) $\hat{y} = 13.12 + 0.785x$. (e) $r = 0.567$. (f) Scatterplot (with regression line) below.

There is a strong positive linear relationship. The point at (176, 113) might be an outlier. $\hat{y} = 13.38 + 0.617x$, $r = 0.860$. The regression line for meal calories has approximately the same intercept, but its slope is less steep. (Without the outlier, $\hat{y} = 9.72 + 0.698x$, $r = 0.803$.)

I.31. (a) Comparative histograms below.

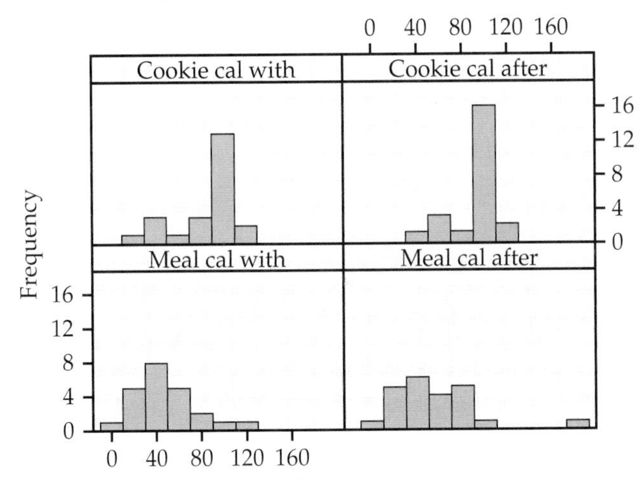

(b) Numerical summaries below.

Variable	N	Mean	StDev	Minimum	Q1	Median	Q3	Maximum
CookieCalWith	23	85.0	28.30	14.00	72.00	100.00	102.00	113.00
CookieCalAfter	23	91.70	20.43	43.00	90.00	100.00	103.00	114.00
MealCalWith	23	47.70	26.10	9.00	28.00	44.00	64.00	113.00
MealCalAfter	23	55.61	36.37	9.00	28.00	49.00	79.00	176.00

(c) For example, histograms allow us to see skew and modality. (d) All distributions are unimodal. Cookie calories tend to be left-skewed, while meal calories tend to be right-skewed.

Chapter 5

5.1. The proportion of heads is 0.3.

5.5. If you hear music (or talking) one time, you will almost certainly hear the same thing for several more checks after that.

5.7. Out of a very large number of patients taking this medication, the fraction who experience this side effect is about 0.00001.

5.9. For example, {blonde, brunette (or brown), black, red, gray, bald, other}.

5.11. $P(\text{Black or White}) = 0.07 + 0.02 = 0.09$.

5.13. $P(\text{not } 1) = 1 - 0.301 = 0.699$.

5.15. For each possible value (1, 2, ..., 6), the probability is 1/6.

5.17. If A_k is the event "the kth card drawn is an ace," then A_1 and A_2 are not independent; in particular, if we know that A_1 occurred, then the probability of A_2 is only 3/51.

5.19. There are 6 possible outcomes: {link1, link2, link3, link4, link5, leave}.

5.21. (a) $0.180 + 0.068 = 0.248$. (b) $1 - 0.248 = 0.752$.

5.23. (a) 0.03, so the sum equals 1. (b) 0.55.

5.25. (a) The probabilities sum to 2. (b) Legitimate (for a nonstandard deck). (c) Legitimate (for a nonstandard die).

5.27. (a) 0.28. (b) 0.88.

5.29. Take each blood type probability and multiply by 0.84 and by 0.16. For example, the probability for A-positive blood is $(0.42)(0.84) = 0.3528$.

5.31. (a) 0.006. (b) 0.001.

5.33. 0.5160.

5.35. Observe that
$P(A \text{ and } B^c) = P(A) - P(A \text{ and } B) = P(A) - P(A)P(B)$.

5.37. (a) Either B or O. (b) $P(B) = 0.75$, $P(O) = 0.25$.

5.39. (a) 0.25. (b) 0.015625; 0.140625.

5.41. Possible values: 0, 1, 2. Probabilities: 1/4, 1/2, 1/4.

5.43. (a) Discrete *random variable*. (b) Continuous random variables can take values from *any interval*. (c) Normal random variables are *continuous*.

5.45. (a) $P(T) = 0.19$. (b) $P(TTT) = 0.0069$, $P(TTT^c) = P(TT^cT) = P(T^cTT) = 0.0292$, $P(TT^cT^c) = P(T^cTT^c) = P(T^cT^cT) = 0.1247$, and $P(T^cT^cT^c) = 0.5314$. (c) $P(X = 3) = 0.0069$, $P(X = 2) = 0.0877$, $P(X = 1) = 0.3740$, and $P(X = 0) = 0.5314$.

5.47. (a) Continuous. (b) Discrete. (c) Discrete.

5.49. (b) $P(X \geq 1) = 0.9$. (c) "No more than two nonword errors."

5.51. (a) Note that, for example, "(1, 2)" and "(2, 1)" are distinct outcomes. (b) 1/36. (c) For example, four pairs add to 5, so $P(X = 5) = 4/36 = 1/9$. (d) 2/9. (e) 5/6.

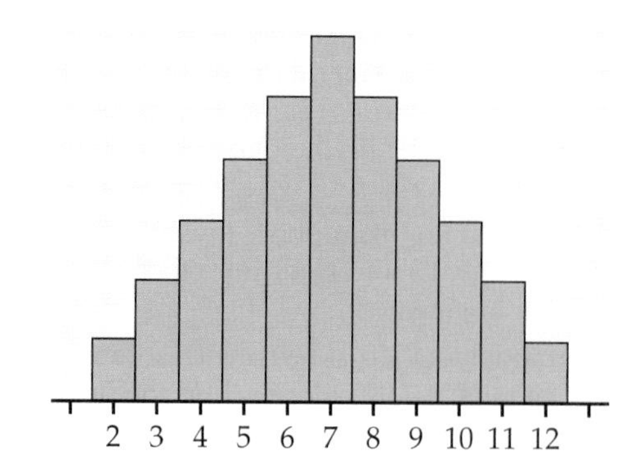

5.53. (a) 0.6. (b) 0.6. (c) "Equal to" has no effect on the probability.

5.55. (a) The height should be 0.5. (b) 0.8. (c) 0.6. (d) 0.525.

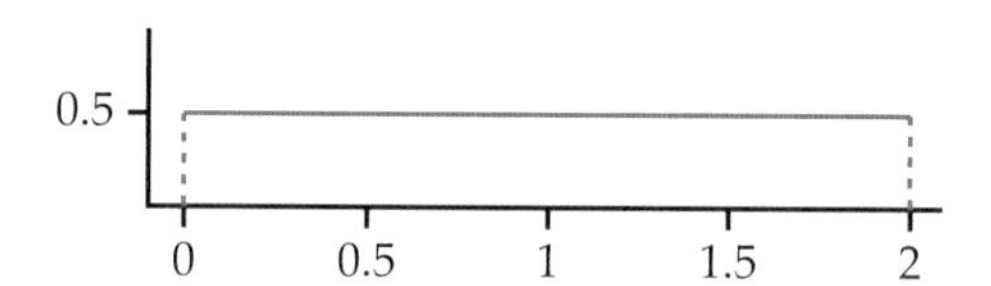

5.57. Very close to 1.

5.59. Possible values: 0 and 1. Probabilities: 0.5 and 0.5. Mean: 0.50.

5.61. $\mu_Y = 95$.

5.63. $\sigma_X^2 = 1$ and $\sigma_X = 1$.

5.65. Mean: 2.3 servings; standard deviation: 1.9 servings.

5.67. 2.88.

5.69. 0.3730 aces.

5.71. (a) \$85.48. (b) This is larger; the negative correlation decreased the variance.

5.73. The exercise describes a positive correlation between calcium intake and compliance. Because of this, the variance of total calcium intake is greater than the variance we would see if there were no correlation.

5.75. (a) $\mu_1 = \sigma_1 = 0.5$. (c) $\mu_4 = 2$ and $\sigma_4 = 1$.

5.77. Mean: 16 cm. Standard deviation: 0.0042 mm.

5.79. (a) Not independent. (b) Independent.

5.81. Show that $\sigma_{X+Y}^2 = (\sigma_X + \sigma_Y)^2$.

5.83. $n = 250$, $X = 100$.

5.85. Assuming no multiple births (twins, triplets, quadruplets), the number of children with type O blood has the $B(4, 0.25)$ distribution.

5.87. 0.0064.

5.89. (a) 0.3588. (b) If the coin were fair: 0.3125.

5.91. (a) A $B(200, p)$ distribution seems reasonable (p not known). (b) Not binomial (no fixed n). (c) A $B(500, 1/12)$ distribution. (d) Not binomial (dependent trials).

5.93. (a) Caught: $B(10, 0.7)$. Missed: $B(10, 0.3)$. (b) 0.3503.

5.95. (a) 7 errors caught, 3 missed. (b) 1.4491 errors. (c) With $p = 0.9$, 0.9487 errors; with $p = 0.99$, 0.3146 errors. σ decreases toward 0 as p approaches 1.

5.97. $m = 6$.

5.99. (a) $n = 4$ and $p = 0.25$. (b) The probabilities are 0.3164, 0.4219, 0.2109, 0.0469, 0.0039. (c) $\mu = 1$ child.

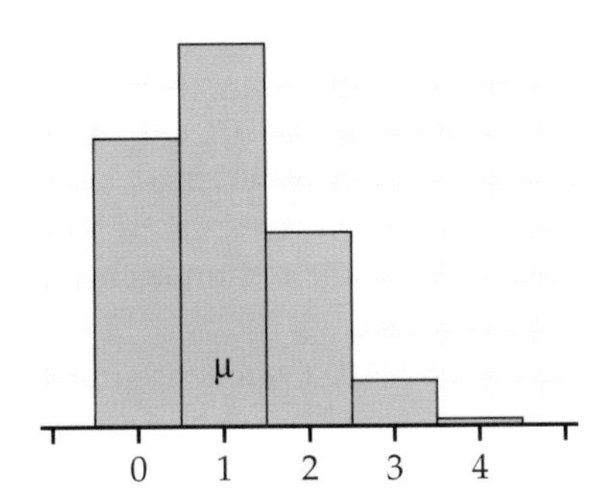

5.101. (a) The only way to distribute n successes among n observations is for all observations to be successes. (b) To distribute $n - 1$ successes among n observations, the one failure must be either observation $1, 2, 3, \ldots, n - 1$, or n. (c) Distributing k successes is equivalent to distributing $n - k$ failures.

5.103. Possible values: $1, 2, 3, \ldots . P(Y = k) = pq^{k-1}$.

5.105. $4/6 = 2/3$.

5.107. $P(A_1 \text{ and } A_2) = \frac{10}{27} \cdot \frac{9}{26} = \frac{5}{39} \doteq 0.1282$.

5.109. $\frac{0.21}{0.21+0.02} \doteq 0.9130$.

5.111. (a) 0.40. (b) 0.10. (c) 0.20. (d) For (a) and (b), use the addition rule; for (c), use the addition rule, and note that D^c and $S^c = (D \text{ or } S)^c$.

5.113. 0.73; use the addition rule.

5.115. (a) The four probabilities sum to 1. (b) 0.77. (c) 0.7442. (d) The events are not independent.

5.117. (a) The four entries are 0.2684, 0.3416, 0.1599, 0.2301. (b) 0.5975.

5.119. For example, the probability of selecting a female student is 0.5717; the probability that she comes from a four-year institution is 0.5975.

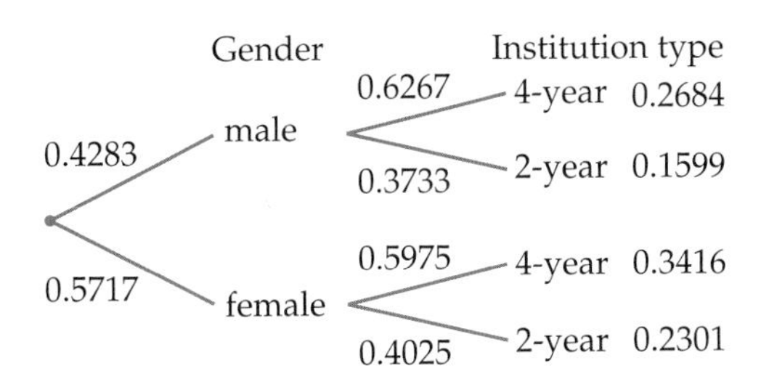

5.121. $P(A|B) \doteq 0.3142$.

5.123. (a) $P(A^c) = 0.69$. (b) $P(A \text{ and } B) = 0.08$.

5.125. (a) 0.6364. (b) Not independent.

5.127. 1.

5.129. $P(B|C) = 1/3$. $P(C|B) = 0.2$.

5.131. (a) $P(M) \doteq 0.4124$. (b) $P(B|M) \doteq 0.6670$. (c) $P(M \text{ and } B) \doteq 0.2751$.

5.133. Close to $\mu_X = 1.1$.

5.135. (a) Possible values 1 and 10, with probabilities 0.4 and 0.6. (b) $\mu_Y = 6.4$ and $\sigma_Y \doteq 4.4091$. (c) Those rules are for transformations of the form $aX + b$.

5.137. (a) $P(\text{A}) = 5/36$ and $P(\text{B}) = 5/18$. (b) $P(\text{A}) = 5/36$ and $P(\text{B}) = 7/12$. (c) $P(\text{A}) = 5/12$ and $P(\text{B}) = 5/18$. (d) $P(\text{A}) = 5/12$ and $P(\text{B}) = 5/18$.

5.139. For example, if the point is 4 or 10, the expected gain is $(1/3)(+20) + (2/3)(-10) = 0$.

5.141. (a) All probabilities are greater than or equal to 0, and their sum is 1. (b) 0.61. (c) Both probabilities are 0.39.

5.143. 0.005352.

5.145. 0.7356.

5.147. $P(\text{no point}) = 1/3$. The probability of winning (losing) an odds bet is 1/36 (1/18) on 4 or 10; 2/45 (1/15) on 5 or 9; 25/396 (5/66) on 6 or 8.

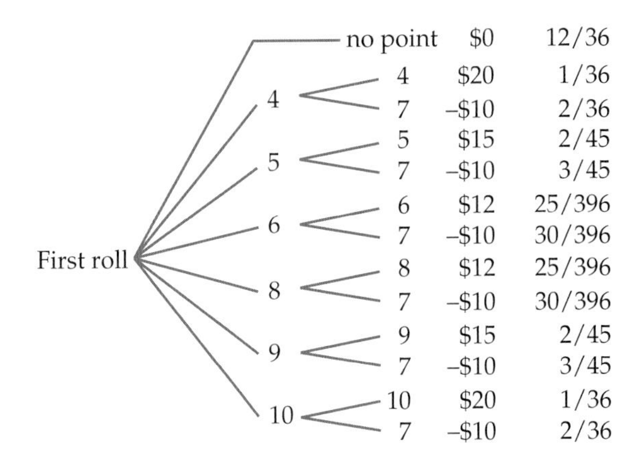

Chapter 6

6.1. The sample is 2036 undergraduate college students aged 18 to 24. The population is college students.

6.3. (a) $n = 760$ banks. (b) $X = 283$ banks expected to acquire another bank. (c) $\hat{p} \doteq 0.3724$.

6.5. $\mu_{\hat{p}} = 0.60$ and $\sigma_{\hat{p}} = 0.0139$. The mean is the same, but the standard deviation is larger.

6.7. The mean would remain the same, but the standard deviation would be smaller for the SRS of size 400.

6.9. (a) 0.01754. (b) $\$0.3724 \pm \$0.0344$. (c) 33.8% to 40.7%.

6.11. $H_0 : \mu = 4$ versus $H_a : \mu > 4$.

6.13. 0.0156.

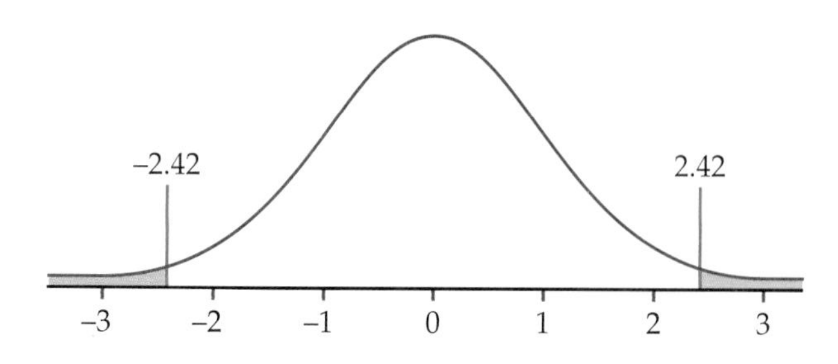

6.15. Shade below −1.34 and above 1.34.

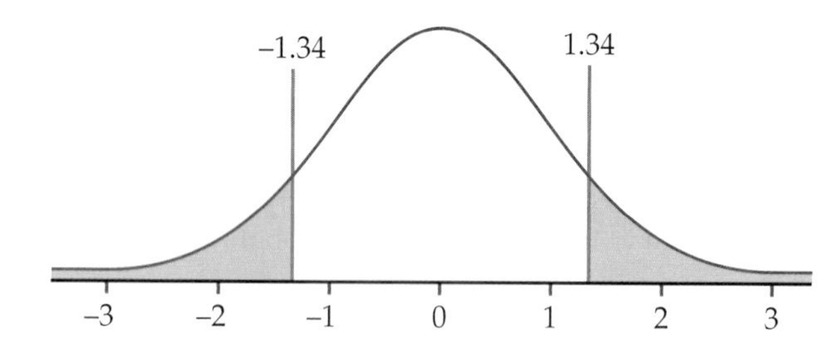

6.17. $\hat{p} = 0.75$, $z = 2.24$, $P = 0.250$.

6.19. (a) $z = -1.34$, $P = 0.1802$. (b) 0.1410 to 0.5590—the complement of the interval shown in Figure 6.10.

6.21. (a) The margin of error accounts for only random sampling error. (b) P-values measure the strength of the evidence against H_0, not the probability of it being true. (c) The confidence level cannot exceed 100%.

6.23. $\hat{p} = 0.6548$, 0.6416 to 0.6680.

6.25. (a) Without that number, we do not know the margin of error for the statistic. (b) About 2365. (c) 0.65 to 0.69. (d) We do not know the sampling methods used, which might make these methods unreliable.

6.27. (a) 0.0021. (b) Other sources of error are much more significant than sampling error.

6.29. (a) No. (b) Yes. (c) Yes. (d) No.

6.31. (a) $\hat{p} = 0.3275$; 0.3008 to 0.3541. (b) Speakers and listeners probably perceive sermon length differently.

6.33. (a) 0.3506 to 0.4094. (b) Yes; some respondents might not admit to such behavior.

6.35. 0.2180 to 0.2510.

6.37. 0.8230 to 0.9370.

6.39. Mean: −0.1. Standard deviation: 0.1339.

6.41. (a) Means: p_1 and p_2. Standard deviations: $\sqrt{\frac{p_1(1-p_1)}{n_1}}$ and $\sqrt{\frac{p_2(1-p_2)}{n_2}}$. (b) $p_1 - p_2$. (c) $\frac{p_1(1-p_1)}{n_1} + \frac{p_2(1-p_2)}{n_2}$.

6.43. The interval for $q_w - q_m$ is −0.0030 to 0.2516.

6.45. The sample proportions support the alternative hypothesis $p_m > p_w$; $P = 0.0288$.

6.47. (a) H_0 should refer to p_1 and p_2. (b) Only if $n_1 = n_2$. (c) Confidence intervals account for only sampling error.

6.49. $z = 11.99$, for which P is tiny.

6.51. (a) $n_1 = 1063$, $\hat{p}_1 \doteq 0.54$, $n_2 = 1064$, $\hat{p}_2 \doteq 0.89$. (We can estimate $X_1 \doteq 574$ and $X_2 \doteq 947$.) (b) 0.35. (c) Yes; large, independent samples from two populations. (d) 0.3146 to 0.3854. (e) 35%; 31.5% to 38.5%. (f) A possible concern: adults were surveyed before Christmas.

6.53. (a) $n_1 = 1063$, $\hat{p}_1 \doteq 0.73$, $n_2 = 1064$, $\hat{p}_2 \doteq 0.76$. (We can estimate $X_1 \doteq 776$ and $X_2 \doteq 809$.) (b) 0.03. (c) Yes; large, independent samples from two populations. (d) −0.0070 to 0.0670. (e) 3%; −0.7% to 6.7%. (f) A possible concern: adults were surveyed before Christmas.

6.55. No; we need independent samples from different populations.

6.57. $z = 11.15$, $P < 0.0001$. Confidence interval: 0.0578 to 0.0822.

6.59. (a) $\hat{p}_f = 0.8$, $SE \doteq 0.05164$; $\hat{p}_m = 0.3939$, $SE \doteq 0.04253$. (b) 0.2960 to 0.5161.

6.61. (a) $\hat{p}_1 \doteq 0.3388$ for men, $\hat{p}_2 \doteq 0.1414$ for women. $SE_D \doteq 0.02798$. Confidence interval: 0.1426 to 0.2523. (b) The female contribution is larger because the sample size for women is much smaller.

6.63. (a) We have six chances to make an error. (b) and (c) Use $z^* = 2.65$. (d) See table below.

Genre	Interval
Racing	0.705 to 0.775
Puzzle	0.684 to 0.756
Sports	0.643 to 0.717
Action	0.632 to 0.708
Adventure	0.622 to 0.698
Rhythm	0.571 to 0.649

6.65. (a) $\hat{p} = 0.316$ and $X = 4740$. (b) 0.3085 to 0.3234. (c) 31.6%; 30.9% to 32.3%. (d) 0.211. (e) 0.0092.

6.67. As the sample size increases, the margin of error decreases.

6.69. −0.0298 to 0.0898.

6.71. The margin of error is ±2.7%.

6.73. Answers will vary.

6.75. $z = 8.95$, $P < 0.0001$; 0.3720 to 0.5613.

6.77. All $\hat{p}$ values are greater than 0.5. Texts 3, 7, and 8 have (respectively) $z = 0.82$, $P = 0.4122$; $z = 3.02$, $P = 0.0025$; $z = 2.10$, $P = 0.0357$. For the other texts, $z \geq 4.64$ and $P < 0.00005$.

6.79. z: 0.90, 1.01, 1.27, 1.42, 2.84, 3.18, 4.49. P: 0.3681, 0.3125, 0.2041, 0.1556, 0.0045, 0.0015, 0.0000.

6.81. (a) 0.5278 to 0.5822. (b) 0.5167 to 0.5713. (c) 0.3170 to 0.3690. (d) 0.5620 to 0.6160. (e) 0.5620 to 0.6160. (f) 0.6903 to 0.7397.

Chapter 7

7.1. Population: iPhone AppsFire users. Statistic: a median of 88 downloaded apps per device. Likely values will vary.

7.3. $\mu_{\bar{x}} = 320$ and $\sigma_{\bar{x}} = 1.5$.

7.5. About 95% of the time $\bar{x}$ is between 319 and 321.

7.7. (a) \$27.75. (b) 15.

7.9. \$561.86 to \$664.14.

7.11. (a) Yes. (b) No. (c) Graphs below.

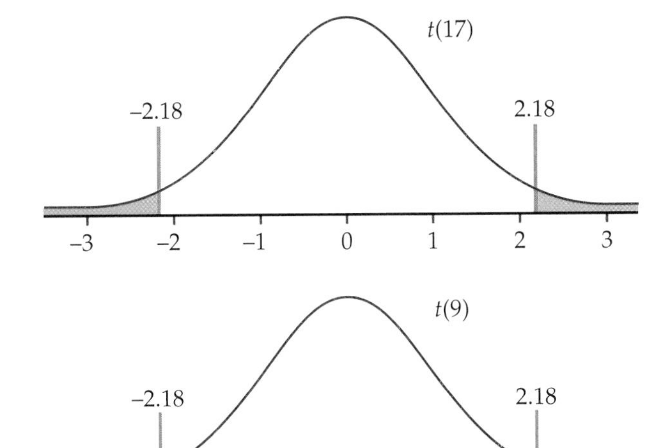

7.15. −1.33 to 10.13.

7.17. (a) "Variance" should be "standard deviation." (b) Standard deviation decreases with increasing sample size. (c) $\mu_{\bar{x}}$ always equals μ.

7.19. (a) We need a larger sample size, to decrease the 95% range. (b) $\sigma_{\bar{x}} \leq 0.0255$. (c) 2033 students.

7.21. (a) $t^* = 2.228$. (b) $t^* = 2.080$. (c) $t^* = 1.721$. (d) t^* decreases with increasing sample size, and it increases with increasing confidence.

7.23. $t^* = 1.740$ (or $t^* = -1.740$).

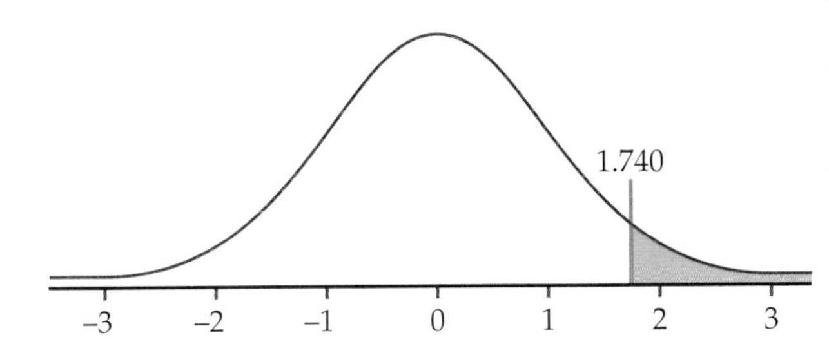

7.25. For the alternative $\mu < 0$, the answer would be the same ($P = 0.037$). For the alternative $\mu > 0$, the answer would be $P = 0.963$.

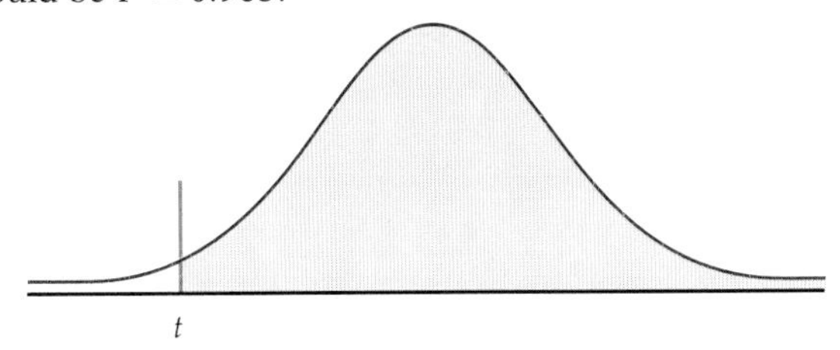

7.27. (a) $H_0 : \mu = 10$; $H_a : \mu < 10$. (b) $t \doteq -5.26$, df = 33, $P < 0.0001$.

7.29. (a) Not Normal (lots of 1s and 10s), but no outliers. (b) 4.92 to 6.88. (c) If there are two distinct groups, the assumption that the sample comes from one population may not be valid. (d) Because this is not a random sample, it may not represent other children well.

7.31. (a) $H_0 : \mu_c = \mu_d$; $H_a : \mu_c \neq \mu_d$. (b) $t \doteq 4.358$, $P \doteq 0.0003$; we reject H_0.

7.33. (a) $t = 5.13$, df = 15, $P < 0.001$. (b) With 95% confidence, the mean NEAT increase is between 192 and 464 calories.

7.35. (a) The differences are spread from −0.018 to 0.020 g. (b) $t = -0.347$, df = 7, $P = 0.7388$. (c) −0.0117 to 0.0087 g. (d) They may be representative of future subjects, but the results are suspect because this is not a random sample.

7.37. (a) $H_0 : \mu = 0$; $H_a : \mu > 0$. (b) Slightly left-skewed; $\bar{x} = 2.5$ and $s = 2.893$. (c) $t = 3.865$, df = 19, $P < 0.0005$. (d) 1.146 to 3.854.

7.39. $\bar{x} = 114.98$ and $s = 14.801$; 111.16 to 118.81.

7.41. $0.02 < P < 0.04$. That is sufficient to reject H_0 at $\alpha = 0.05$.

7.43. −20.3163 to 0.3163; do not reject H_0.

7.45. (a) Hypotheses should involve μ_1 and μ_2. (b) The samples are not independent. (c) We need P to be small (for example, less than 0.10) to reject H_0. (d) t should be negative.

7.47. (a) No (in fact, $P \doteq 0.0542$). (b) Yes ($P \doteq 0.0271$).

7.49. (a) While the distributions do not look particularly Normal, they have no extreme outliers or skewness.
(b)

Group	n	$\bar{x}$	s	df	t^*	Confidence interval
Neutral	14	\$0.5714	\$0.7300	26.5	2.0538	−2.2842 to −0.8082
Sad	17	\$2.1176	\$1.2441	13	2.160	−2.3224 to −0.7701

(c) $H_0 : \mu_N = \mu_S$; $H_a : \mu_N < \mu_S$. (d) $t = -4.303$, so $P \doteq 0.0001$ (df $\doteq$ 26.5) or $P < 0.0005$ (df = 13). (e) −2.2842 to −0.8082 (df $\doteq$ 26.5) or −2.3225 to −0.7699 (df = 13).

7.51. (a) $\bar{x}_F \doteq 4.0791$, $s_F \doteq 0.9861$; $\bar{x}_M \doteq 3.8326$, $s_M \doteq 1.0677$. (b) Both distributions are somewhat skewed, but because the ratings range from 1 to 5, there are no outliers. (c) $t \doteq 2.898$, $P \doteq 0.0040$ (df $\doteq$ 402.2) or $0.002 < P < 0.005$ (df = 220). (d) 0.0793 to 0.4137 (df $\doteq$ 402.2) or 0.0777 to 0.4153 (df = 220). (e) The difference in means might not be as large as 0.25.

7.53. (a) This may be close enough to an SRS if this company's working conditions are similar to those of other workers. (b) 9.99 to 13.01 mg.y/m^3. (c) $t = 15.08$, $P < 0.0001$ with either df = 137 or 114. (d) The sample sizes are large enough that skewness should not matter.

7.55. You need either sample sizes and standard deviations or degrees of freedom and a more accurate value for the P-value. The confidence interval will give us useful information about the magnitude of the difference.

7.57. This is a matched pairs design.

7.59. The next 10 employees who need screens might not be an independent group—perhaps they all come from the same department, for example.

7.61. (a) Either −0.90 to 6.90 units (df = 122.5) or −0.95 to 6.95 units (df = 54). (b) Random fluctuation may account for the difference in the two averages.

7.63. (a) $H_0 : \mu_B = \mu_F$; $H_a : \mu_B > \mu_F$; $t = 1.654$, $P = 0.053$ (df = 37.6) or $P = 0.058$ (df = 18). (b) −0.2 to 2.0. (c) We need two independent SRSs from Normal populations.

7.65. $s_p = 0.9347$; $t \doteq 3.636$, df = 40, $P \doteq 0.0008$; 0.4663 to 1.6337. Both results are similar to those for Exercise 7.50.

7.67. We have pooled standard deviation $s_p \doteq 8.1772$, so $SE \doteq 1.2141$ and $t \doteq 4.777$. With either df = 236 or 100, we find that $P < 0.001$.

7.69. $\bar{x} = 141.25$, $s \doteq 10.81$, $s_{\bar{x}} \doteq 5.41$. We cannot consider these four scores to be an SRS.

7.71. As df increases, t^* approaches 1.96.

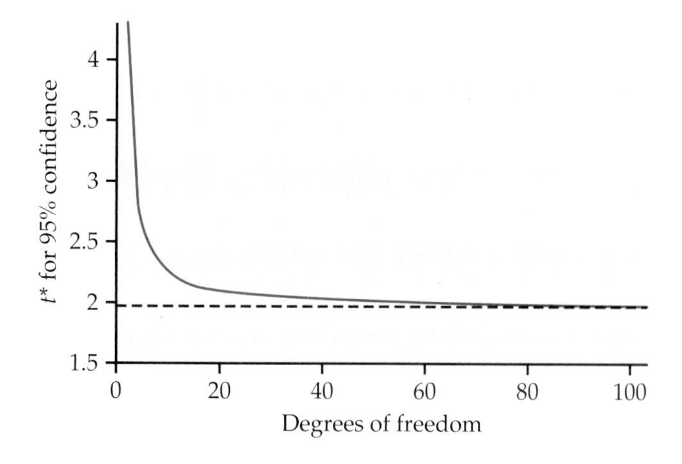

7.73. Margins of error decrease with increasing sample size.

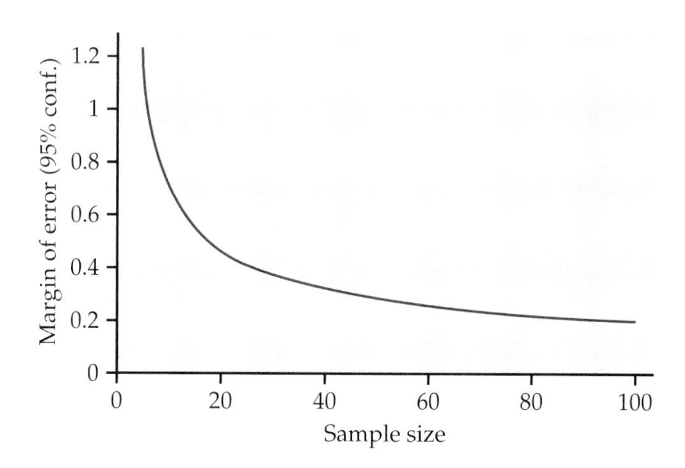

7.75. (a) Two independent samples. (b) Matched pairs. (c) Single sample.

7.77. (a) $H_0 : \mu = 0.75$; $H_a : \mu < 0.75$. $t \doteq -1.607$, $P \doteq 0.0548$. (b) 0.5273 to 0.7727. (d) The sample size should be large enough to make t procedures safe.

7.79. (a) Body weight: mean = −0.7 kg, SE = 2.298 kg. Caloric intake: mean = 14 cal, SE = 56.125 cal. (b) $t_1 = -0.305$ (body weight) and $t_2 = 0.249$ (caloric intake), both with df = 13; both P-values are about 0.8. (c) −5.66 to 4.26 kg and −107.23 to 135.23 cal.

7.81. How much a person eats and drinks may depend on how many people he or she is sitting with.

7.83. (a) At each nest, the same mockingbird responded on each day. (b) 6.9774 m. (c) $t \doteq 6.32$, $P < 0.0001$. (d) 5.5968 m; $t \doteq -1.05$, $P \doteq 0.3045$. (e) The significant difference between Day 1 and Day 4 suggests that the mockingbirds altered their behavior when approached by the same person for four consecutive days; seemingly, the birds perceived an escalating threat. When approached by a new person on Day 5, the response was not significantly different from Day 1; this suggests that the birds saw the new person as less threatening than a return visit from the first person.

7.85. 78.3% ± 13.8%, or 64.5% to 92.1%.

7.87. $t = 3.65$, df = 237.0 or 115, $P < 0.0005$. 95% confidence interval for the difference: 0.78 to 2.60.

7.89. $t = -0.3533$, df = 179, $P = 0.3621$.

7.91. No; what we have is nothing like an SRS.

7.93. (a) Three-bedroom houses are slightly right-skewed; four-bedroom houses are generally more expensive. The top price from the three-bedroom distribution qualifies as an outlier.

3BR		4BR
87	0	
433	1	3
9888776	1	78
43321	2	224
	2	579
	3	01
5	3	9
	4	2
	4	
	5	2

(b) $t \doteq -3.06$ with either df $= 20.23$ ($P \doteq 0.0061$) or df $= 20$ ($P = 0.006$); we reject H_0. (c) It would be reasonable to guess that $\mu_3 < \mu_4$. (d) \$31,339 to \$165,109 (df $= 20.23$) or \$31,357 to \$165,158 (df $= 20$). (e) It seems that these houses should be a fair representation of three- and four-bedroom houses in West Lafayette.

Chapter 8

8.1. The plot is symmetric about 0.5, where it has its maximum.

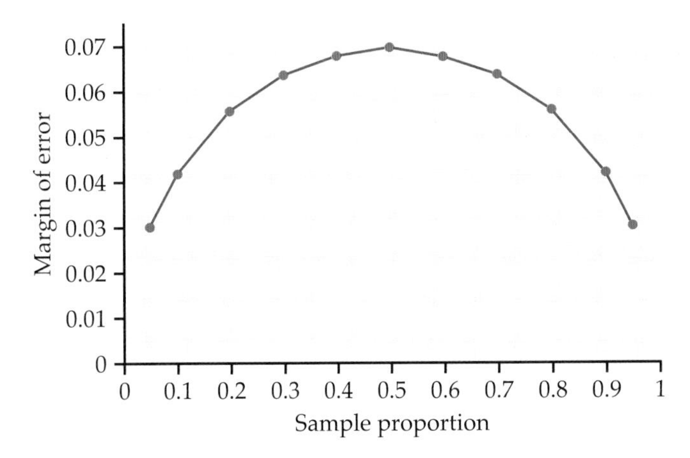

8.3. $n = 579$.

8.5. (a) No. (b) Yes.

8.7. No; the students who did not respond are (obviously) not represented in the results.

8.9. (a) She did not divide the standard deviation by $\sqrt{n} = 20$. (b) Confidence intervals concern the *population* mean. (c) 0.95 is a confidence level, not a probability. (d) The large sample size affects the distribution of the sample mean, not the individual ratings.

8.11. Margins of error: 0.1550, 0.1096, 0.0895, 0.0775; interval width decreases with increasing sample size.

8.13. Scenario B has a smaller margin of error; less variability at a single school.

8.15. (a) Smaller. (b) Equal. (c) 1399.

8.17. We need 1225 responses, so the registrar should email 2228 students.

8.19. No; this is a range of values for the mean rent, not for individual rents.

8.21. (a) Not certain (only 95% confident). (b) We obtained the interval 86.5% to 88.5% by a method that gives a correct result 95% of the time. (c) About 0.5%. (d) No (only random sampling error).

8.23. No; confidence interval methods can be applied only to an SRS.

8.25. To determine the effectiveness of alarm systems, we need to know the percent of all homes with alarm systems and the percent of burglarized homes with alarm systems.

8.27. The first test was barely significant at $\alpha = 0.05$, while the second was significant at any reasonable α.

8.29. A significance test answers only Question b.

8.31. (a) The differences observed might occur by chance even if SES had no effect. (b) This tells us that the statistically insignificant test result did not occur merely because of a small sample size.

8.33. (a) $P = 0.2843$. (b) $P = 0.1020$. (c) $P = 0.0023$.

8.35. With a larger sample, we might have significant results.

8.37. (a) For example, collect data from all people who purchased the iPhone 5 on the day it was available for preorder. Ask each person how long it took to place his or her order. (b) For example, collect data from a random sample of 500 adult Americans who purchased the iPhone 5 on the day it was available for preorder. Ask each person how long it took to place his or her order.

8.39. For example, perform 100 hypothesis tests at the $\alpha = 0.05$ level and report only the statistically significant results.

8.41. n should be about 100,000.

8.43. It is essentially correct.

8.45. Larger samples give more power.

8.47. (a) For example, if μ is the mean difference in scores, $H_0 : \mu = 0$ versus $H_a : \mu \neq 0$. (b) No. (c) For example: Was this an experiment? What was the design? How big were the samples?

8.49. (a) 4.67 to 5.99 mg/dl. (b) $n = 20$.

8.51. (a) 0.7738. (b) 0.9774.

8.53. $n = 171$ or 172.

8.55. The sample sizes are 35, 62, 81, 93, 97, 93, 81, 62, and 35; take $n = 97$.

8.57. $n = 73$.

8.59. (a) $n = 342$. (b) $n = (z^*/m)^2/2$.

8.61. $n = 247$.

8.63. Look back at Figure 8.6 (page 445). We have a Normal distribution centered at 0.40 and then two other distributions centered at 0.55 and 0.5. The blue shaded area (probability of Type II error) will be larger for the Normal distribution with mean 0.5. This means that the power will be lower.

Part II Review

II.1. (a) S = {Ontario, Quebec, British Columbia, Alberta, Manitoba, Saskatchewan, Nova Scotia, New Brunswick, Newfoundland and Labrador, Prince Edward Island, Northwest Territories, Yukon, Nunavut}. (b) 1/13. (c) 12/13. (d) 3/13. (e) Complement; disjoint events.

II.3. (a) Mean: 20; standard deviation: 50. (b) Mean: 20; standard deviation: 50.

II.5. X = number of Facebook friendships; n = number of requests for friendship; p = underlying probability of accepting a friendship.

II.7. (a) 1/3. (b) No. (c) Yes.

II.9. (a) 682 to 730 minutes. (b) 712 to 772 minutes. (c) There is a lower limit for the number of minutes but no defined upper limit. Also, the standard deviations are large compared with the means. The sample size is large enough to overcome the skew.

II.11. (a) 0.839 to 0.902. (b) $H_0: p = 0.85$ versus $H_a: p > 0.85$; $z = 1.20$, $P = 0.1148$.

II.13. For example, the athletes might answer differently because they would be working with the dietitians.

II.15. No; we are looking at the complement of the previous event.

II.17. (a) The sample sizes are large enough. (b) $H_0: \mu_B = \mu_W$ versus $H_a: \mu_B \neq \mu_W$. (c) First year: $t = 0.98$, $P = 0.3305$. Third year: $t = 2.126$, $P = 0.0388$. (d) Those white students who were randomly assigned a black roommate their first year tend to have a higher proportion of black friends by their third year.

II.19. (a) Disagree; bias has to do with data collection methods, not sample size. (b) Disagree; the shape of the histogram of the response variable does not depend on sample size. (c) Agree.

II.21. (a) True. (b) False; the P-value tells us the chance of observing data as extreme as ours given that the null hypothesis is true. (c) False; we cannot guarantee that a Type I error occurred. (d) True.

II.23. (a) Marijuana use: $z = -10.19$, $P < 0.0001$. Unprotected sex: $z = 4.79$, $P < 0.0001$. Sex while drunk/high: $z = -1.27$, $P = 0.2030$. Driving while drunk/high: $z = -7.39$, $P < 0.0001$. (b) No; all P-values are clearly under or over 0.005. (c) Marijuana use, unprotected sex, driving while drunk/high.

II.25. (a) −2.859 to 1.153. (b) −1.761 to 0.055. (c) The centers of the intervals are the same, but the margin of error is smaller for the matched pairs interval.

II.27. (a) Adolescents: 0.72 to 3.08. Middle-aged adults: 1.44 to 3.76. (b) Middle-aged adults seem to have a larger difference in group means. (c) The samples are not independent.

II.29. (a) 0.57 to 0.61. (b) $H_0: p = 0.48$ versus $H_a: p > 0.48$; $z = 10.55$, $P < 0.0001$.

Chapter 9

9.1. (a) Given Explanatory = 1: 35% "Yes," 65% "No." Given Explanatory = 2: 45% "Yes," 55% "No." (b) One possible graph is below.

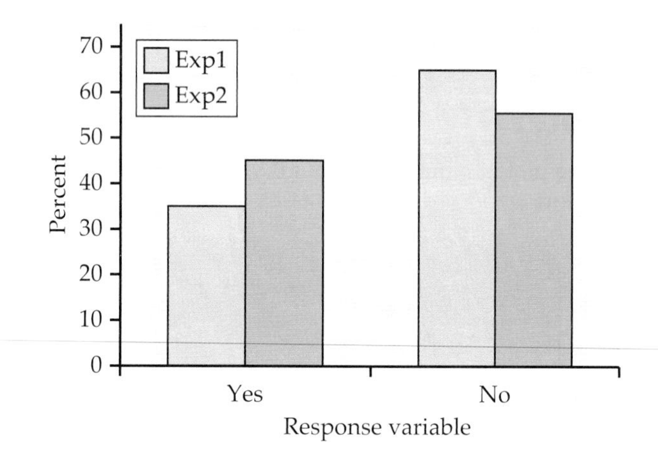

(c) When Explanatory = 2, "Yes" and "No" are closer to being evenly split.

9.3. The expected counts were rounded.

9.5. The other five chi-square components are 0.5820, 0.0000, 0.0196, 0.0660, and 0.2264; the six values add up to 0.93.

9.7. (a) The 3 × 2 table is below.

	Allowed?		
Stratum	**Yes**	**No**	**Total**
Small	45	12	57
Medium	7	10	17
Large	3	2	5
Total	55	24	79

(b) 21.1% (small claims), 58.8% (medium), and 40.0% (large) were not allowed. (c) In the 3 × 2 table, the expected count for large/not allowed is too small.

(d) There is no relationship between claim size and whether a claim is allowed. (e) $X^2 = 8.419$, df = 1, $P = 0.0037$.

9.9. There is strong evidence of a change. $X^2 = 308.3$, df = 2, $P < 0.0001$.

9.11. (a) For example, among those students in trades, 320 enrolled right after high school, and 622 enrolled later. Table below.

Field of study	Time of Entry: Right after high school	Later	Total
Trades	320.28	621.72	942
Design	274.48	309.52	584
Health	2034	3051	5085
Media/IT	975.88	2172.12	3148
Service	486	864	1350
Other	1172.60	1082.40	2255
Total	5263.24	8100.76	13,364

(b) For example, in addition to the given percents, 39.4% of these students enrolled right after high school. (c) $X^2 = 276.1$, df = 5, $P < 0.0001$.

9.13. (a) For example, among those students in trades, 188 relied on parents, family, or spouse, and 754 did not. Table below.

Field of study	Time of Entry: Right after high school	Later	Total
Trades	188.4	753.6	942
Design	221.63	377.37	599
Health	1360.84	3873.16	5234
Media/IT	518.08	2719.92	3238
Service	248.04	1129.96	1378
Other	943	1357	2300
Total	3479.99	10,211.01	13,691

(b) $X^2 = 544.8$, df = 5, $P < 0.0001$. (c) In addition to the given percents, 25.4% of all students relied on family support.

9.15. (a) 50.5% get enough sleep; 49.5% do not. (b) 32.2% get enough sleep; 67.8% do not. (c) Those who exercise more than the median are more likely to get enough sleep. (d) $X^2 = 22.577$, df = 1, $P < 0.0001$.

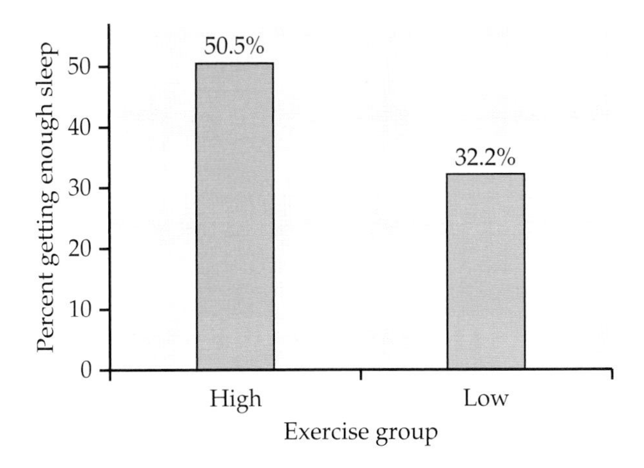

9.17. (a) Table below.

Lied?	Male	Female	Total
Yes	11,724	14,169	25,893
No	1,163	746	1,909
Total	12,887	14,915	27,801

(b) 91% of males and 95% of females agreed that trust and honesty are essential. (c) Females are slightly more likely to view trust and honesty as essential. (d) $X^2 = 175.0$, df = 1, $P < 0.0001$.

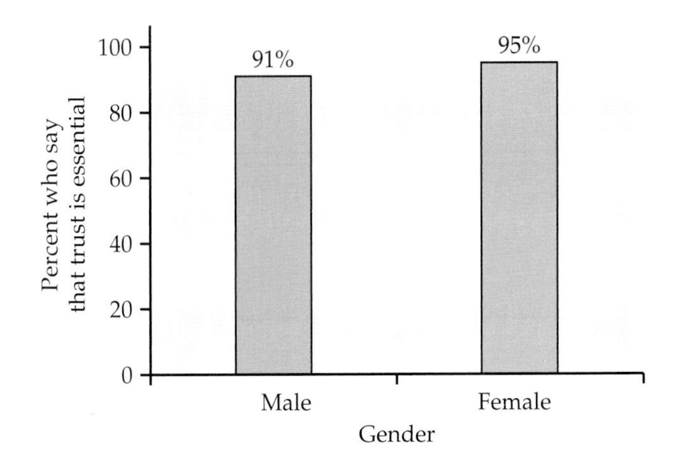

9.19. (a) The proportions for the joint distribution are the italicized entries in the table below. (b) The marginal distribution of age is in the right margin of the table below. (c) The marginal distribution of status is in the bottom margin of the table below.

Age	Full-time	Part-time	
15–19	*0.2044*	*0.0189*	0.2234
20–24	*0.3285*	*0.0699*	0.3984
25–34	*0.1050*	*0.1072*	0.2123
35+	*0.0518*	*0.1141*	0.1659
	0.6898	0.3102	

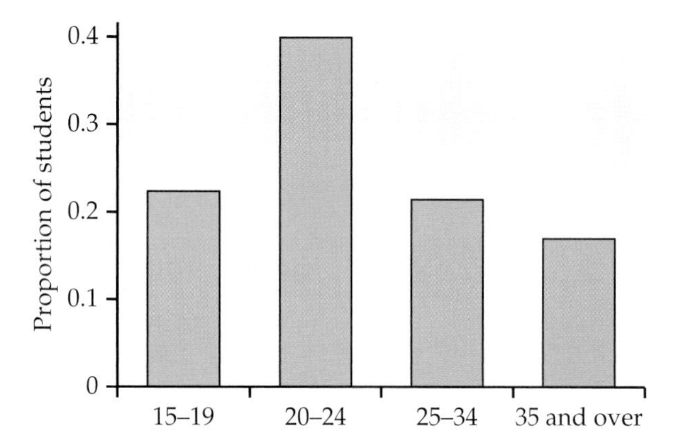

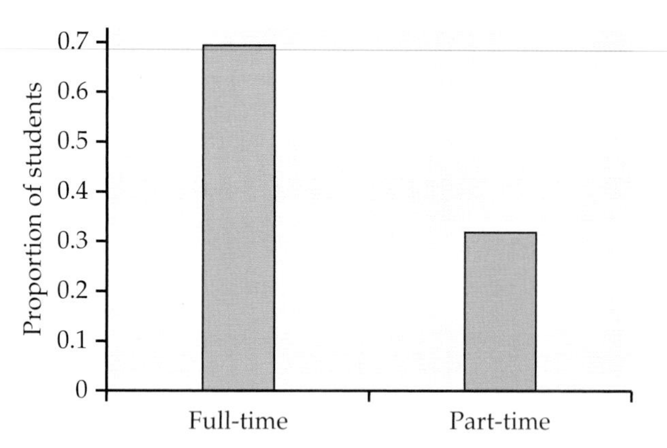

(d) The conditional distributions are given in the table below.

Age	Full-time	Part-time
15–19	0.2964	0.0610
20–24	0.4763	0.2254
25–34	0.1522	0.3458
35+	0.0752	0.3678

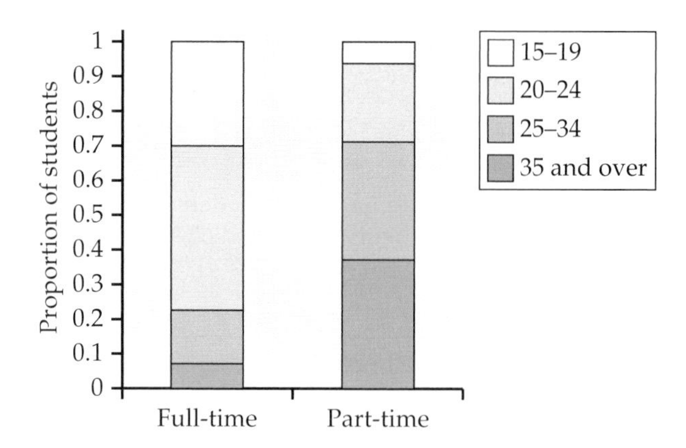

9.21. (a) 57.98%. (b) 30.25%. (c) To test "There is no relationship between waking and bedtime symptoms" versus "There is a relationship," we find $X^2 \doteq 2.275$, df = 1, $P \doteq 0.132$.

9.23. Start by setting a equal to any number from 0 to 50.

9.25. (a) A notably higher percent of women are "strictly voluntary" participants. (b) 40.3% of men and 51.3% of women are participants; the relative risk is 1.27.

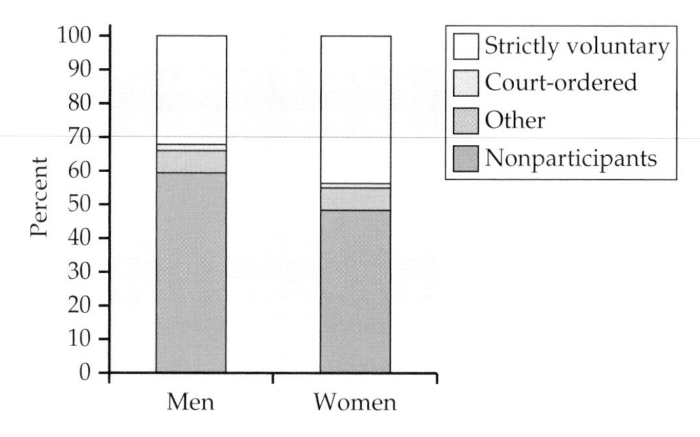

9.27. (a) 146 women/No, 97 men/No. (b) For example, 19.34% of women, versus 7.62% of men, have tried low-fat diets. (c) $X^2 = 7.143$, df = 1, $P = 0.008$.

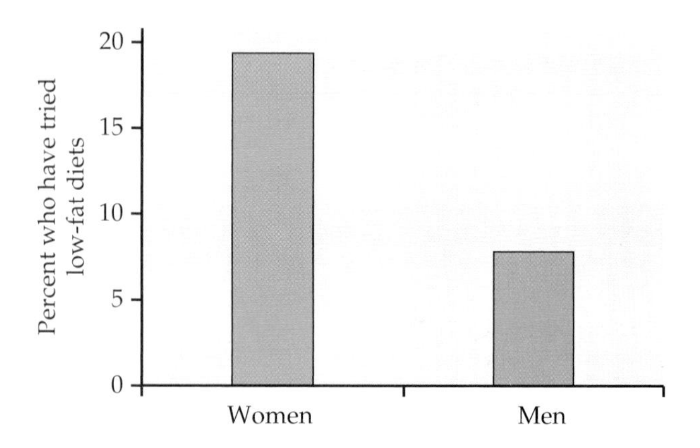

9.29. (a) For example, see graph below.

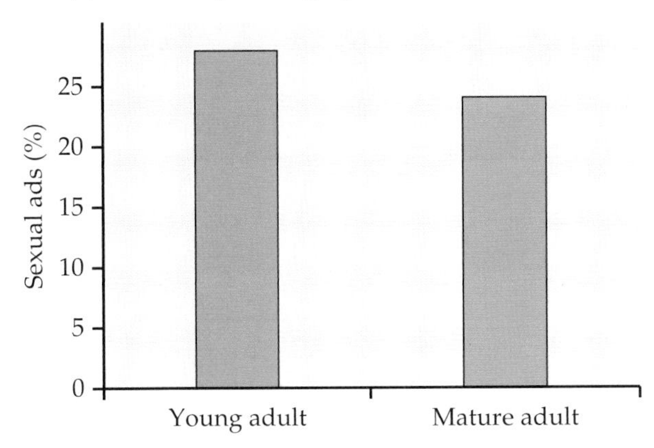

(b) $X^2 = 2.591$, df = 1, $P = 0.108$.

9.31. (a) $X^2 = 76.7$, df = 2, $P < 0.0001$. (b) Even with much smaller numbers of students, P is still very small. (c) Our conclusion might not hold for the true percents. (d) Lack of independence could cause the estimated percents to be too large or too small.

9.33. The missing entries are 202, 64, 38, 33. $X^2 = 50.5$, df = 9, $P < 0.0005$. The largest contributions to X^2 come from chemistry/engineering, physics/engineering, and biology/liberal arts (more than expected), and from biology/engineering and chemistry/liberal arts (less than expected).

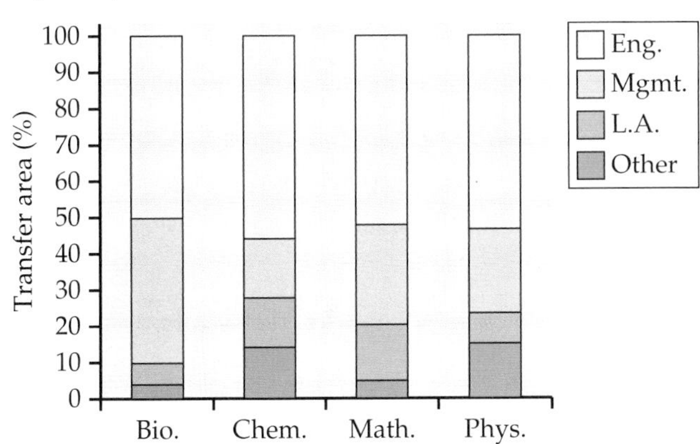

9.35. $X^2 = 852.433$, df = 1, $P < 0.0005$.

9.37. $X^2 = 3.781$, df = 3, $P = 0.2861$.

9.43. (a) We expect each quadrant to contain one-fourth of the 100 trees. (b) *Some* random variation would not surprise us. (c) $X^2 = 10.8$, df = 3, $P = 0.0129$.

Chapter 10

10.1. (a) 1.7. (b) When x increases by 1, μ_y increases by 1.7. (c) 62.2. (d) 53.8 to 70.6.

10.3. (a) $t = 1.90$, $P = 0.0705$. (b) $t = 3.62$, $P = 0.0014$. (a) $t = 1.90$, $P = 0.0608$.

10.5. (a) 0.7 kg/m^2. (b) Smaller.

10.7. (a) A decrease of 0.341 to 0.969 kg/m^2. (b) An increase of 0.341 to 0.969 kg/m^2. (c) A decrease of 0.1705 to 0.4845 kg/m^2.

10.9. (a) β_0, β_1, and σ. (b) H_0 should refer to β_1. (c) The confidence interval will be narrower than the prediction interval.

10.11. (a) 1.3649 to 2.0199; a \$1 difference in tuition in 2000 changes 2008 tuition by between \$1.36 and \$2.02. (b) 78.2%. (c) \$9764. (d) \$15,856. (e) The Moneypit U prediction requires extrapolation.

10.13. (a) $\hat{y} = -0.0127 + 0.0180x$, $r^2 \doteq 80.0\%$.

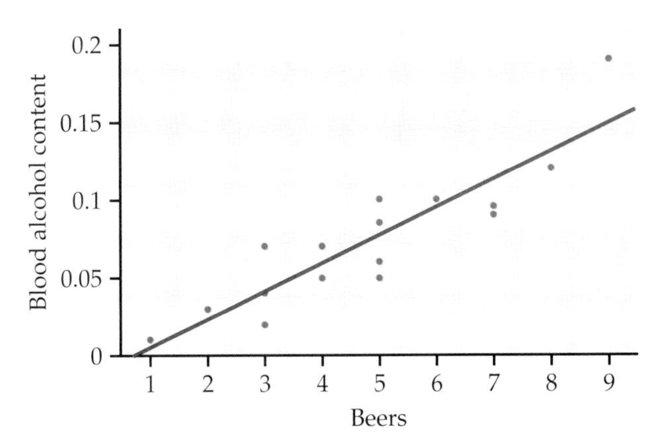

(b) $H_0: \beta_1 = 0$ versus $H_a: \beta_1 > 0$; $t = 7.48$, $P < 0.0001$. (c) The predicted mean is 0.07712; the interval is 0.040 to 0.114.

10.15. (a) Both distributions are sharply right-skewed. The five-number summaries are

	Min	Q_1	M	Q_3	Max
Incentives as percent of salary	0	0.306	1.43	17.65	85.01
Overall player rating	0	2.25	6.31	12.69	27.88

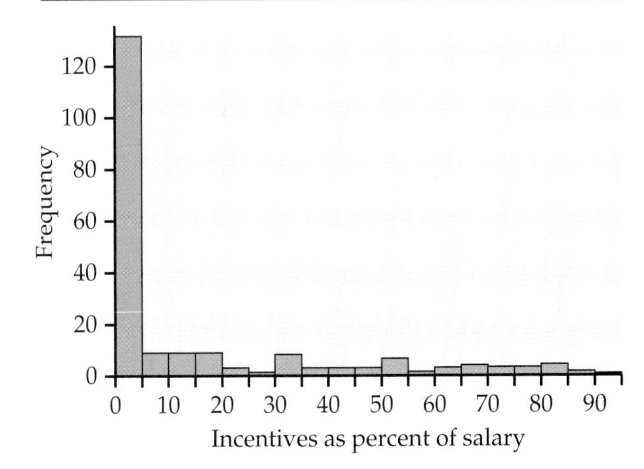

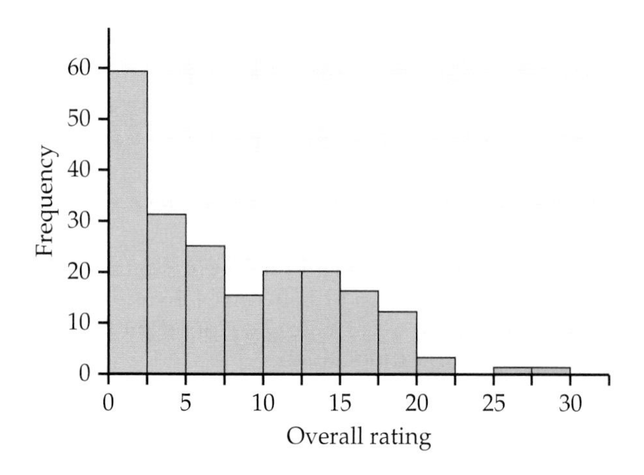

(b) No; x and y do not need to be Normal. (c) There is a weak positive linear relationship.

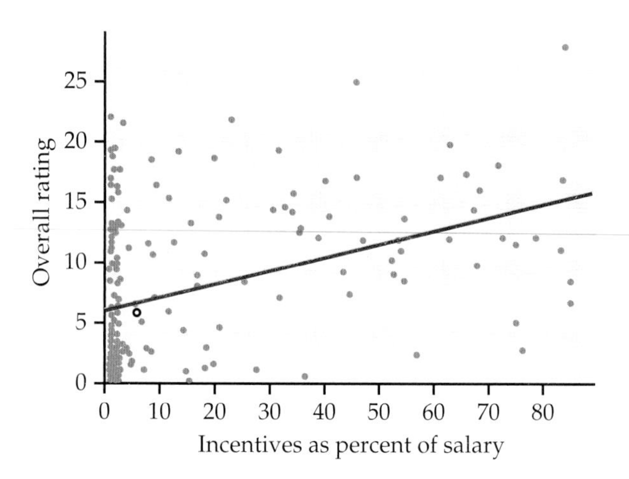

(d) $\hat{y} = 6.247 + 0.1063x$. (e) The residuals are slightly right-skewed.

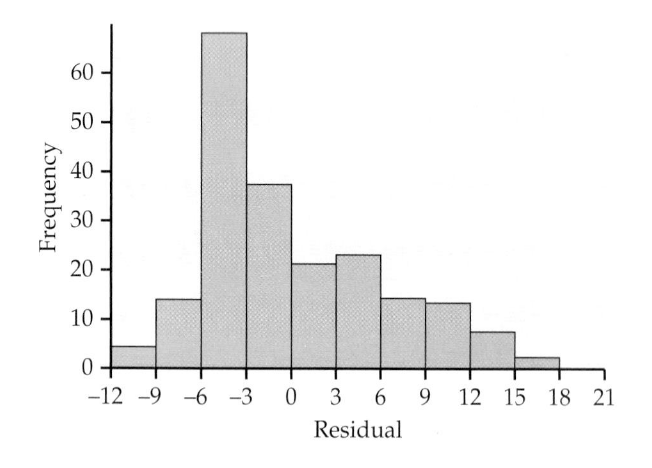

10.17. (a) 22 of these 30 homes sold for more than their assessed values. (b) A moderately strong linear association. (c) $\hat{y} = 21.50 + 0.9468x$. (d) There are no obvious unusual features. (e) A stemplot or histogram looks reasonably Normal.

−3	3
−2	6543
−1	9422
−0	984430
0	1134579
1	1236
2	26
3	
4	17

(f) There are no clear violations of the assumptions.

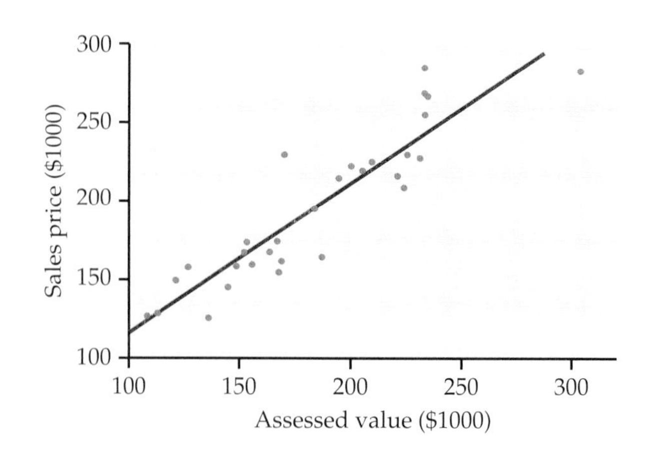

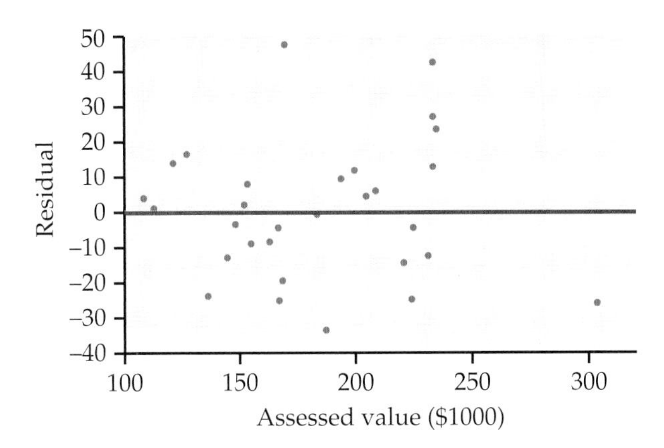

10.19. (a) The plot is roughly linear and increasing. The 2004 and 2008 counts are noticeably high. (b) $\hat{y} \doteq -28{,}438 + 14.82x$; the confidence interval for the slope is 11.89 to 17.76 tornadoes per year. (c) The scatter might be greater in recent years, and the 2004 residual is particularly high. (d) The 2004 residual is an outlier; the other residuals appear to be roughly Normal.

Stem	Leaves
−3	520
−2	931
−1	99843310
−0	9887654443211110
0	001223556778
1	001224
2	011789
3	6
4	
5	5

(e) $\hat{y} \doteq -26{,}584 + 13.88x$.

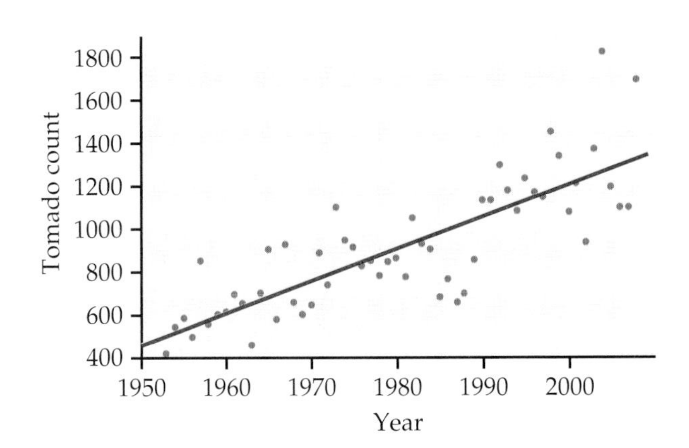

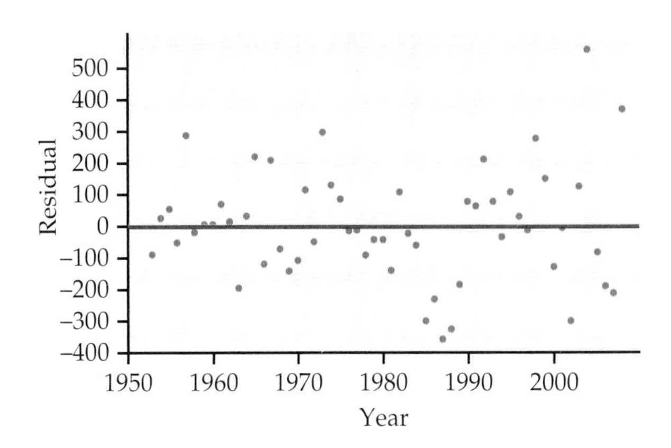

10.21. (a) Both distributions are fairly symmetric. For x (MOE), $\bar{x} \doteq 1{,}799{,}180$ and $s_x \doteq 329{,}253$; for y (MOR), $\bar{y} \doteq 11{,}185$ and $s_y \doteq 1980$. (b) Put MOE on the x axis. (c) $y_i = \beta_0 + \beta_1 x_i + \varepsilon_i$, where $i = 1, 2, \ldots, 32$; ε_i are independent $N(0, \sigma)$ variables. $\widehat{\text{MOR}} = 2653 + 0.00474\,\text{MOE}$, $s = 1238$, $t = 7.02$, $P < 0.0001$. (d) Assumptions appear to be met.

MOE

Stem	Leaves
11	6
12	
13	55
14	1578
15	5589
16	14
17	2479
18	447
19	358
20	0348
21	8
22	1
23	47
24	
25	3

MOR

Stem	Leaves
6	3
7	
8	3588
9	222
10	22356
11	223455799
12	00777
13	469
14	5
15	3

Residuals

Stem	Leaves
−3	3
−2	
−2	
−1	6
−1	31110
−0	76555
−0	43221
0	00223
0	78
1	1334
1	599
2	1

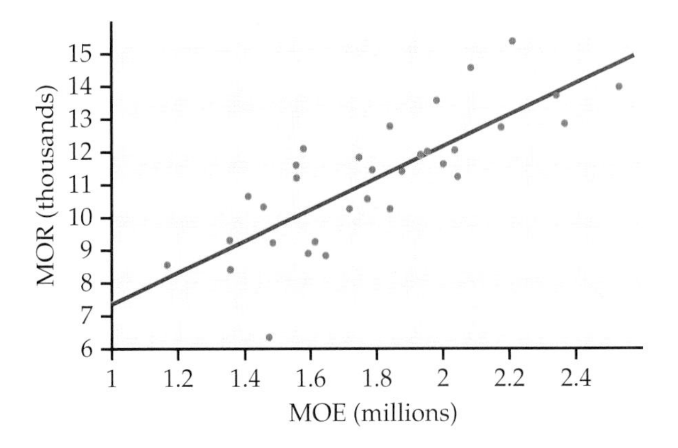

10.23. (a) x (percent forested) is right-skewed; $\overline{x} = 39.3878\%$, $s_x = 32.2043\%$. y (IBI) is left-skewed; $\overline{y} = 65.9388$, $s_y = 18.2796$.

Percent forested

```
 0 | 00000033789
 1 | 0014778
 2 | 125
 3 | 123339
 4 | 133799
 5 | 229
 6 | 38
 7 | 599
 8 | 069
 9 | 055
10 | 00
```

(b) A weak positive association, with more scatter in y for small x.

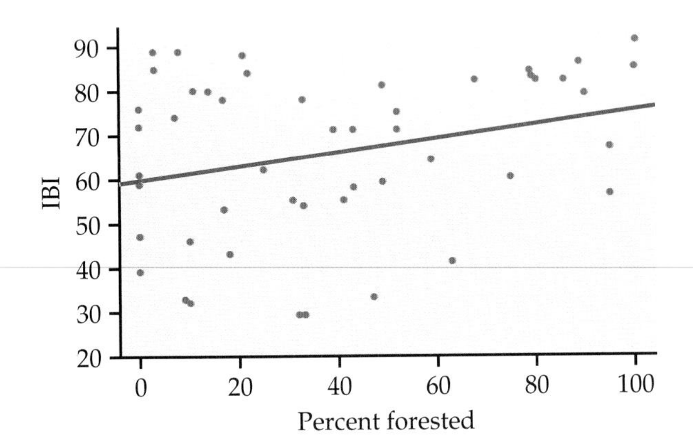

(c) $y_i = \beta_0 + \beta_1 x_i + \varepsilon_i$, where $i = 1, 2, \ldots, 49$; ε_i are independent $N(0, \sigma)$ variables. (d) $H_0 : \beta_1 = 0$ versus $H_a : \beta_1 \neq 0$. (e) $\widehat{\text{IBI}} = 59.9 + 0.153\,\text{Area}$; $s = 17.79$. For testing the hypotheses in (d), $t = 1.92$ and $P = 0.061$. (f) Residual plot (below) shows a slight curve.

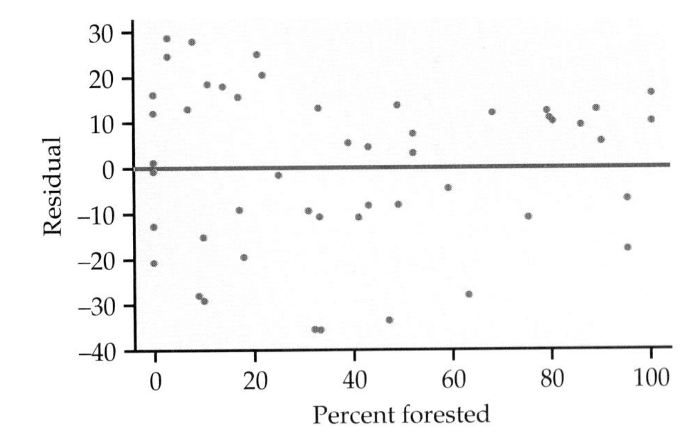

(g) Residuals (stemplot and Normal quantile plot below) are left-skewed.

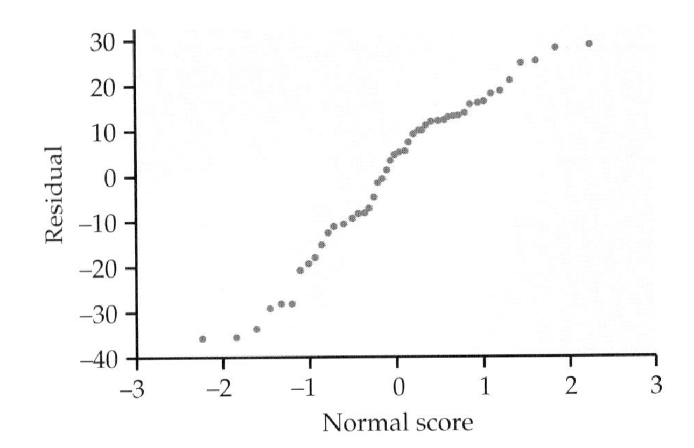

10.25. The first change decreases P (that is, the relationship is more significant) because it accentuates the positive association. The second change weakens the association, so P increases (the relationship is less significant).

10.27. Using area = 10, $\hat{y} \doteq 57.52$; using percent forest = 63, $\hat{y} \doteq 69.55$. Both predictions have a lot of uncertainty (the prediction intervals are about 70 units wide).

10.29. (a) Scatterplot shows a weak negative association. $\widehat{\text{Bonds}} = 55.58 - 0.1769\,\text{Stocks}$, $s = 54.55$. (Solid line in plot.) (b) $H_0 : \beta_1 = 0$ versus $H_a : \beta_1 \neq 0$; $t = -1.66$, $P = 0.111$. (c) $\widehat{\text{Bonds}} = 69.46 - 0.2814\,\text{Stocks}$, $s = 53.12$. (Dotted line in plot.) The slope is now significantly different from 0 ($t = -2.24$, $P = 0.036$). (d) We should explore whether something happened in 2008 that might explain why that point strayed from the line.

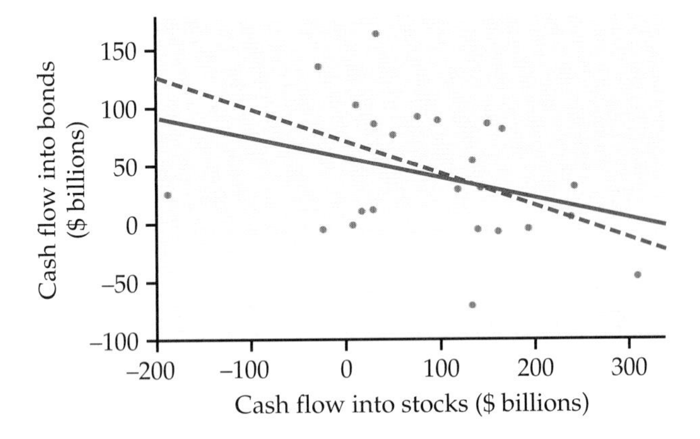

10.31. (a) The means are identical (21.133). (b) For the observed ACT scores, $s_y = 4.714$; for the fitted values, $s_{\hat{y}} = 3.850$. (c) For $z = 1$, the SAT score is 1092.8. The predicted ACT score is $\hat{y} \doteq 25$. (d) For $z = -1$, the SAT score is 732.6. The predicted ACT score is $\hat{y} \doteq 17.3$. (e) It appears that the standard score of the predicted value is the same as the explanatory variable's standard score.

10.33. (a) It appears to be quite linear. (b) $\widehat{\text{Lean}} = -61.12 + 9.3187\,\text{Year}$; $r^2 = 98.8\%$. (c) 8.36 to 10.28 tenths of a millimeter/year.

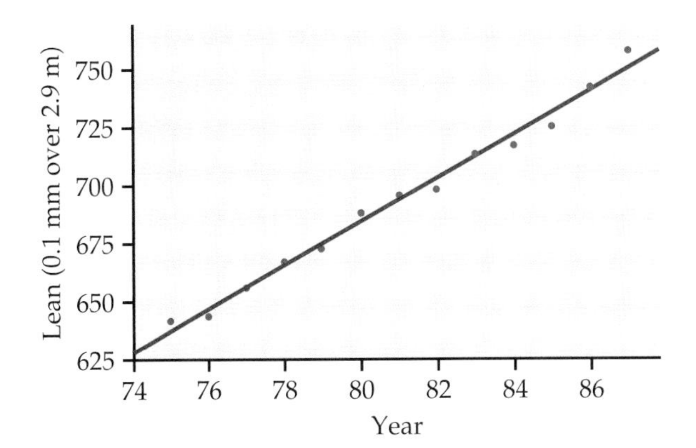

10.35. (a) $x = 112$. (b) 2.9983 m. (c) Use a prediction interval.

10.37. (a) For squared length: $\widehat{\text{Weight}} = -118 + 0.497\,\text{SQLEN}$; $r^2 = 97.7\%$. (b) For squared width: $\widehat{\text{Weight}} = -99 + 18.7\,\text{SQWID}$; $r^2 = 96.5\%$.

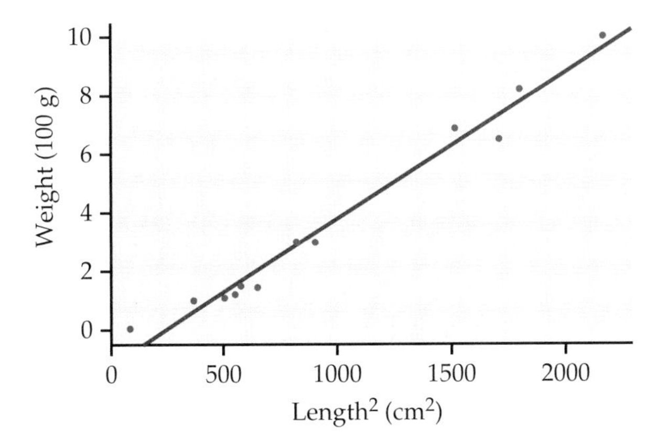

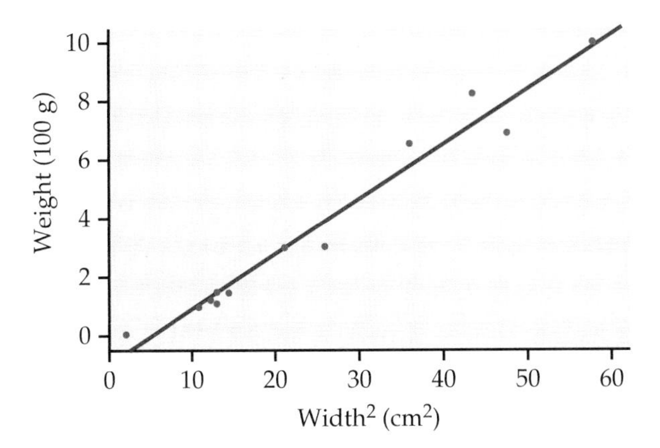

10.39. (a) 95% confidence interval for women: 14.73 to 33.33. For men: −9.47 to 42.97. These intervals overlap quite a bit. (b) For women: 22.78. For men: 16.38. The women's slope standard error is smaller in part because it is divided by a large number. (c) Choose men with a wider variety of lean body masses.

Chapter 11

11.1. (a) Math GPA. (b) $n = 106$. (c) $p = 4$. (d) SAT Math, SAT Verbal, class rank, and mathematics placement score.

11.3. (a) Math GPA should increase when any explanatory variable increases. (b) DFM = 4, DFE = 81. (c) All four coefficients are significantly different from 0 (although the intercept is not).

11.5. The correlations are found in Figure 11.5. The scatterplots for the pairs with the largest correlations are easy to pick out. The whole-number scale for high school grades causes point clusters in those scatterplots.

11.7. (a) H_0 should refer to β_2. (b) Squared multiple correlation. (c) Small P implies that at least one coefficient is different from 0.

11.9. (a) −0.1112 to 13.5112. (b) 0.0703 to 13.3297. (c) −0.1277 to 13.5277. (d) 0.1660 to 13.2340.

11.11. (a) $y_i = \beta_0 + \beta_1 x_{i1} + \beta_2 x_{i2} + \cdots + \beta_8 x_{i8} + \varepsilon_i$, where $i = 1, 2, \ldots, 135$; ε_I are independent $N(0, \sigma)$ random variables. (b) Model (df = 8), error (df = 126), and total (df = 134).

11.13. (a) 8 and 786. (b) 7.84%; this model is not very predictive. (c) Males and Hispanics consume energy drinks more frequently. Consumption increases with risk-taking scores. (d) Within a group of students with identical (or similar) values of those other variables, energy-drink consumption increases with increasing jock identity and increasing risk taking.

11.15. (a) $\widehat{\text{BMI}} = 23.4 - 0.682x_1 + 0.102x_2$. (b) $R^2 = 17.7\%$. (c) The residuals look roughly Normal and show no obvious remaining patterns. (d) $t = 1.83$, df = 97, $P = 0.070$.

11.17. (a) Budget and Opening are right-skewed; Theaters and Opinion are roughly symmetric (slightly left-skewed). Five-number summaries for Budget and Opening are appropriate; mean and standard deviation could be used for the other two variables. (b) All relationships are positive. The Budget/Theaters and Opening/Theaters relationships appear to be curved; the others are reasonably linear. The correlations between Budget, Opening, and Theaters are all greater than 0.7. Opinion is less correlated with the other three variables—about 0.4 with Budget and Opening and only 0.156 with Theaters.

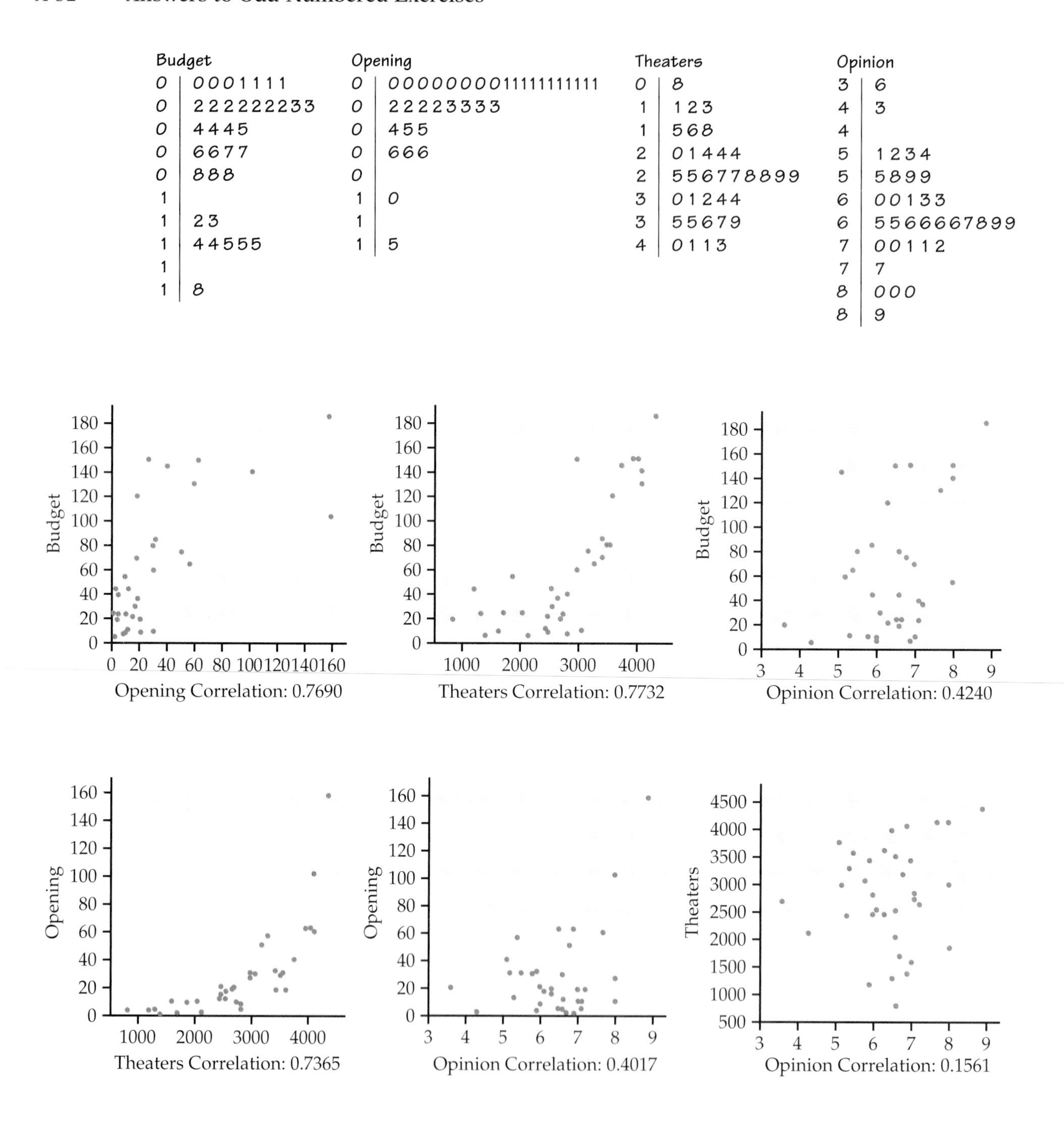

11.19. (a) USRevenue $= \beta_0 + \beta_1$Budget $+ \beta_2$Opening $+ \beta_3$Theaters $+ \beta_4$Opinion $+ \varepsilon_i$, where $i = 1, 2, \ldots, 35$; ε_I are independent $N(0, \sigma)$ random variables. (b) $\widehat{\text{USRevenue}} = -67.72 + 0.1351$ Budget $+ 3.0165$ Opening $-$ 0.00223 Theaters $+ 10.262$ Opinion. (c) *The Dark Knight* may be influential. The spread of the residuals appears to increase with Theaters. (d) 98.1%.

11.21. (a) \$86.86 to \$154.91 million. (b) \$89.93 to \$155.00 million. (c) The intervals are very similar.

11.23. (a) TEACH is bimodal (or irregular) and right-skewed; RESEARCH is unimodal and right-skewed; CITATIONS is irregular. (b) All three variables are positively correlated. The strongest correlation is between TEACH and RESEARCH ($r = 0.905$).

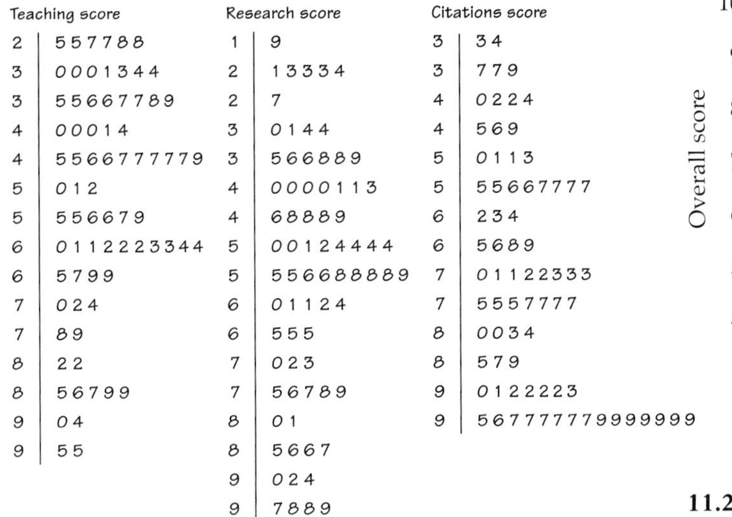

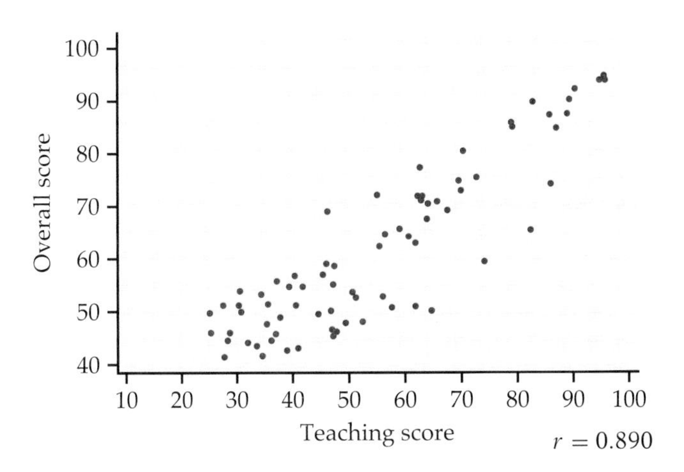

$r = 0.890$

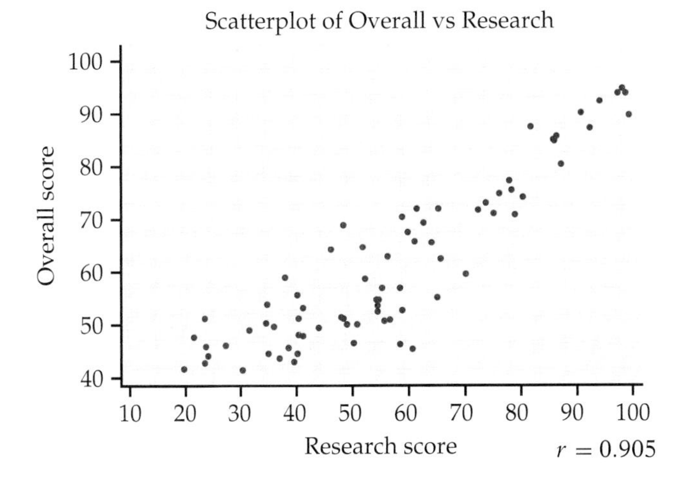

$r = 0.905$

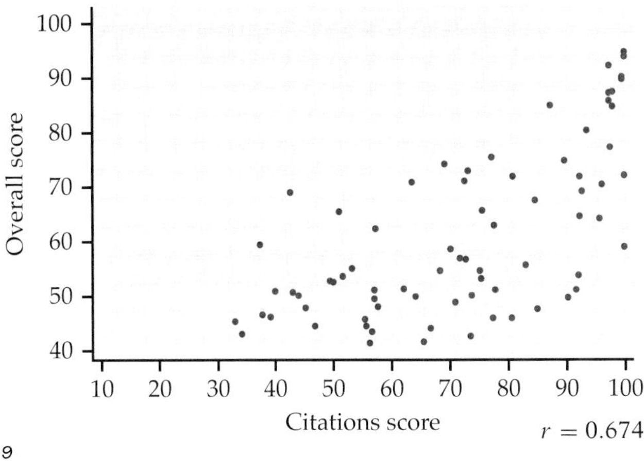

$r = 0.674$

11.25. (a) $y_i = \beta_0 + \beta_1 x_{i1} + \beta_2 x_{i2} + \beta_3 x_{i3} + \varepsilon_i$, where $i = 1, 2, \ldots, 75$; ε_I are independent $N(0, \sigma)$ random variables. (b) $\widehat{\text{OVERALL}} = 7.722 + 0.2307x_1 + 0.3618x_2 + 0.2822x_3$. All coefficients are significantly different from 0. (c) TEACH: 0.1456 to 0.3158; RESEARCH: 0.2832 to 0.4404; CITATIONS: 0.2439 to 0.3205. None of these intervals should contain 0, since all coefficients are significantly different from 0. (d) $R^2 = 0.962$, $s = 3.065$.

11.27. (a) For example: All distributions are skewed to varying degrees—GINI and CORRUPT to the right, the other three to the left. CORRUPT and DEMOCRACY have the most skewness. (b) GINI is negatively correlated to the other four variables (ranging from −0.384 to −0.139), while all other correlations are positive and more substantial (0.533 or more).

11.29. (a) Refer to your regression output (see following page). (b) For example, the t statistic for the GINI coefficient grows from $t = -1.18$ ($P = 0.243$) to $t = 3.92$ ($P < 0.0005$). The DEMOCRACY t is 3.27 in the third model ($P = 0.002$) but drops to 0.60 ($P = 0.552$) in the fourth model. (c) A good choice is to use GINI, LIFE, and CORRUPT: all three coefficients are significant, and $R^2 = 77.3\%$ is nearly the same as in the fourth model from Exercise 11.28.

Minitab output: Regression of LSI on GINI (Model 1)

LSI = 7.02 – 0.0201 GINI

Predictor	Coef	Stdev	t-ratio	p
Constant	7.0238	0.6660	10.55	0.000
GINI	–0.02014	0.01710	–1.18	0.243

s = 1.274 R-sq = 1.9% R-sq (adj) = 0.5%

Regression of LSI on GINI and LIFE (Model 2)

LSI = – 3.83 + 0.0287 GINI + 0.125 LIFE

Predictor	Coef	Stdev	t-ratio	p
Constant	–3.8257	0.9746	–3.93	0.000
GINI	0.02873	0.01056	2.72	0.008
LIFE	0.12503	0.01034	12.09	0.000

s = 0.7266 R-sq = 68.6% R-sq (adj) = 67.6%

Regression of LSI on GINI, LIFE , and DEMOCRACY (Model 3)

LSI = – 3.25 + 0.0280 GINI + 0.106 LIFE + 0.186 DEMOCRACY

Predictor	Coef	Stdev	t-ratio	p
Constant	–3.2524	0.9293	–3.50	0.001
GINI	0.028049	0.009891	2.84	0.006
LIFE	0.10634	0.01125	9.46	0.000
DEMOCRACY	0.18575	0.05682	3.27	0.002

s = 0.6804 R-sq = 72.8% R-sq (adj) = 71.6%

Regression of LSI on all four variables (Model 4)

LSI = – 2.72 + 0.0368 GINI + 0.0905 LIFE + 0.0392 DEMOCRACY + 0.186 CORRUPT

Predictor	Coef	Stdev	t-ratio	p
Constant	–2.7201	0.8661	–3.14	0.003
GINI	0.036782	0.009393	3.92	0.000
LIFE	0.09048	0.01120	8.08	0.000
DEMOCRACY	0.03925	0.06566	0.60	0.552
CORRUPT	0.18554	0.05042	3.68	0.000

s = 0.6252 R–sq = 77.4% R–sq (adj) = 76.0%

11.31. (a) $y_i = \beta_0 + \beta_1 x_{i1} + \beta_2 x_{i2} + \beta_3 x_{i3} + \beta_4 x_{i4} + \varepsilon_i$, where $i = 1, 2, \ldots, 69$; ε_I are independent $N(0, \sigma)$ random variables. (b) $\widehat{\text{PCB}} = 0.94 + 11.9x_1 + 3.76x_2 + 3.88x_3 + 4.18x_4$. All coefficients are significantly different from 0, although the constant 0.937 is not ($t = 0.76$, $P = 0.449$). $R^2 = 0.989$, $s = 6.382$. (c) The residuals appear to be roughly Normal, but with two outliers. There are no clear patterns when plotted against the explanatory variables.

11.33. (a) $\widehat{\text{PCB}} = -1.02 + 12.6x_1 + 0.313x_2 + 8.25x_3$, $R^2 = 0.973$, $s = 9.945$. (b) $b_2 = 0.313$, $P = 0.708$. (c) In Exercise 11.31, $b_2 = 3.76$, $P < 0.0005$.

11.35. The model is $y_i = \beta_0 + \beta_1 x_{i1} + \beta_2 x_{i2} + \beta_3 x_{i3} + \beta_4 x_{i4} + \varepsilon_i$, where $i = 1, 2, \ldots, 69$; ε_I are independent $N(0, \sigma)$ random variables. Regression gives $\widehat{\text{TEQ}} = 1.06 - 0.097x_1 + 0.306x_2 + 0.106x_3 - 0.0039x_4$, $R^2 = 0.677$. Only the constant (1.06) and the PCB118 coefficient (0.306) are significantly different from 0. Residuals are slightly right-skewed and show no clear patterns when plotted with the explanatory variables.

11.37. (a) The correlations are all positive, ranging from 0.227 (LPCB28 and LPCB180) to 0.956 (LPCB and LPCB138). LPCB28 has one outlier (Specimen 39) when plotted with the other variables; except for that point, all scatterplots appear fairly linear. (b) All correlations are higher with the transformed data.

11.39. It appears that a good model is LPCB126 and LPCB28 ($R^2 = 0.768$).

Chapter 12

12.1. (a) H_0 says the population means are all equal. (b) Experiments are best for establishing causation. (c) ANOVA is used to compare means. (d) Multiple-comparisons procedures are used when we wish to determine which means are significantly different, but when we have no specific relations in mind before looking at the data.

12.3. (a) Yes; $6/5 = 1.2 < 2$. (b) 25, 25, and 36. (c) 29.3676. (d) 5.4192.

12.5. (a) This is the description of *between*-group variation. (b) The *sums* of squares will add. (c) σ is a parameter. (d) A small P means that the means are not all the same, but the distributions may still overlap.

12.7. Assuming the t (ANOVA) test establishes that the means are different, contrasts and multiple comparisons provide no further useful information.

12.9. (a) df 4 and 30; $2.69 < F < 3.25$. (b)

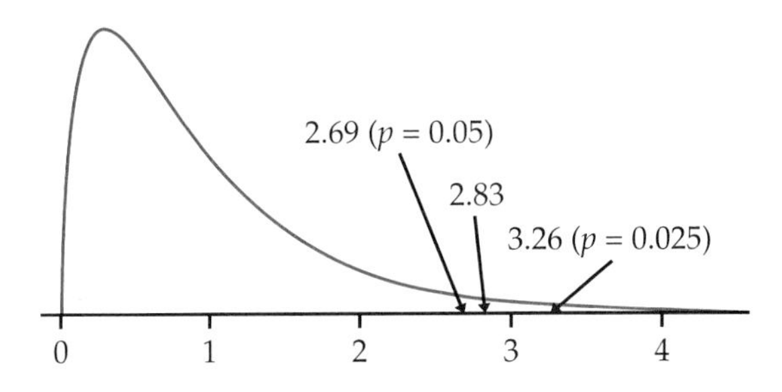

(c) $0.025 < P < 0.050$. (d) We can conclude that at least one mean is different.

12.11. (a) df = 2 and 30; $F = 2.54$, $0.05 < P < 0.10$. (b) df = 3 and 28; $F \doteq 2.97$, $0.025 < P < 0.05$.

12.13. (a) Response: egg cholesterol level. Populations: chickens with different diets or drugs. $I = 3$, $n_1 = n_2 = n_3 = 25$, $N = 75$. (b) Response: rating on five-point scale. Populations: the three groups of students. $I = 3$, $n_1 = 31$, $n_2 = 18$, $n_3 = 45$, $N = 94$. (c) Response: quiz score. Populations: students in each TA group. $I = 3$, $n_1 = n_2 = n_3 = 14$, $N = 42$.

12.15. For all three situations, we test $H_0: \mu_1 = \mu_2 = \mu_3$; H_a: at least one mean is different. (a) DFM = 2, DFE = 72, DFT = 74. $F(2, 72)$. (b) DFM = 2, DFE = 91, DFT = 93. $F(2, 91)$. (c) DFM = 2, DFE = 39, DFT = 41. $F(2, 39)$.

12.17. (a) This sounds like a fairly well designed experiment, so the results should apply at least to this farmer's breed of chicken. (b) It would be good to know what proportion of the total student body falls in each of these groups—that is, is anyone overrepresented in this sample? (c) Effectiveness in teaching one topic (power calculations) might not reflect overall effectiveness.

12.19. (a) 2 and 117. (b) $P < 0.001$, or $P = 0.0003$. (c) We should hesitate to generalize these results beyond similar informal shops in Mexico.

12.21. (a) F can be made very small (close to 0), and P close to 1. (b) F increases, and P decreases.

12.23. (a) Based on the sample means, fiber is cheapest and cable is most expensive.

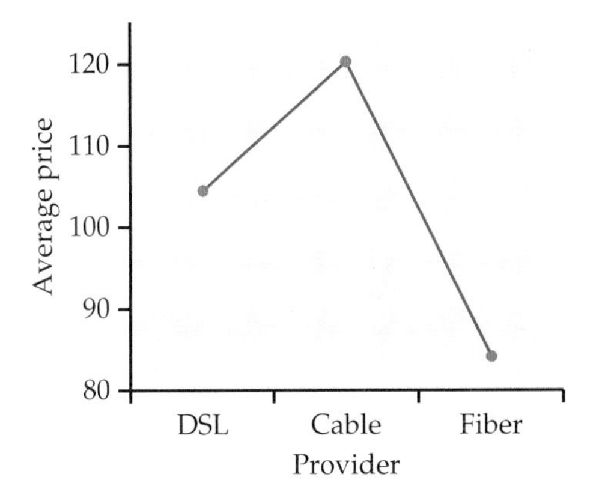

(b) Yes; the ratio is 1.55. (c) df = 2 and 44; $0.025 < P < 0.05$, or $P = 0.0427$.

12.25. (a) Matched pairs t methods; we examine the change in reaction time for each subject. (b) No; we do not have 4 independent samples.

12.27. (a) Activity seems to increase with both drugs, and Drug B appears to have a greater effect.

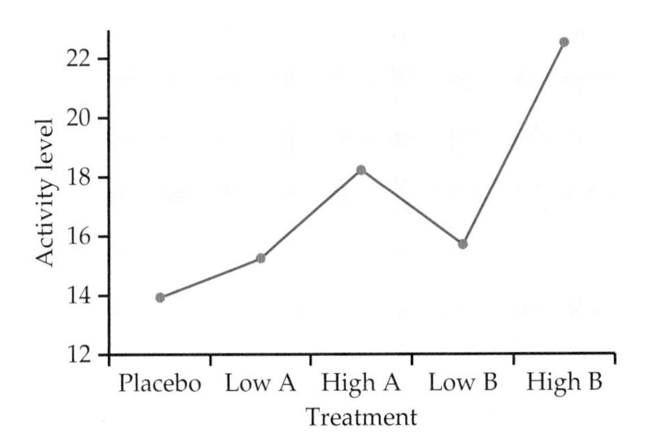

(b) Yes; the standard deviation ratio is 1.44. $s_p \doteq 3.256$. (c) df = 4 and 15. (d) $0.01 < P < 0.025$ ($P = 0.0165$).

12.29. (a) ANOVA is not appropriate for these data. (b) There may be a number of local factors (for example, student demographics or teachers' attitudes toward accommodations) that affected grades; these effects might not be the same elsewhere. (c) It suggests neither. Recall that ANOVA is not appropriate for these data.

12.31. (a) The variation in sample size is some cause for concern, but there can be no extreme outliers in a 1-to-7 scale, so ANOVA is probably reliable. (b) Yes: $1.26/1.03 = 1.22 < 2$. (c) $F(4, 405)$, $P = 0.0002$.

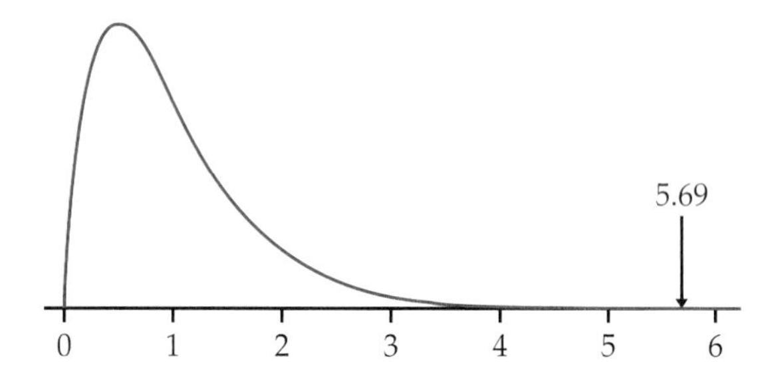

(d) Hispanic Americans are highest; Japanese are in the middle; the other three are lowest.

12.33. (a) $\psi_1 = \mu_2 - (\mu_1 + \mu_4)/2$.
(b) $\psi_2 = (\mu_1 + \mu_2 + \mu_4)/3 - \mu_3$.

12.35. *H. bihai* and *H. caribaea* red distributions are slightly skewed. *H. bihai*: $n = 16$, $\bar{x} = 47.597$, $s = 1.213$ mm. *H. caribaea* red: $n = 23$, $\bar{x} = 39.711$, $s = 1.799$ mm. *H. caribaea* yellow: $n = 15$, $\bar{x} = 36.180$, $s = 0.975$ mm. This just meets our rule for standard deviations. ANOVA gives $F = 259.12$, df = 2 and 51, $P < 0.0005$, so we conclude that the means differ.

12.37. *H. bihai*: $n = 16$, $\bar{x} = 3.8625$, $s = 0.0251$. *H. caribaea* red: $n = 23$, $\bar{x} = 3.6807$, $s = 0.0450$. *H. caribaea* yellow: $n = 15$, $\bar{x} = 3.5882$, $s = 0.0270$. ANOVA gives $F = 244.27$, df = 2 and 51, $P < 0.0005$, so we conclude that the means differ.

12.39. (a) Yes; the ratio is 1.25. $s_p \doteq 0.7683$. (b) df = 2 and 767; $P < 0.001$.

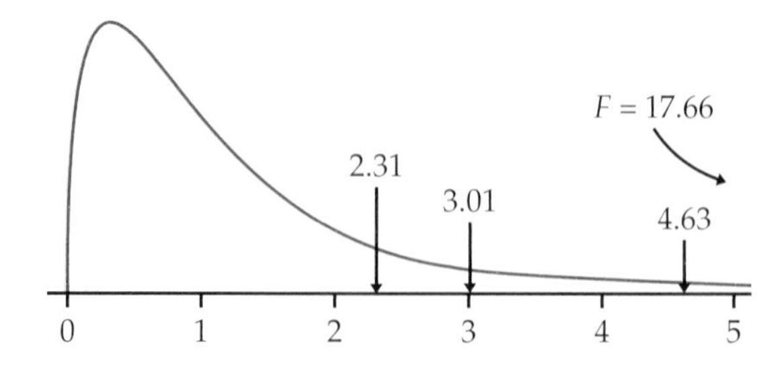

(c) With $\psi = \mu_2 - (\mu_1 + \mu_3)/2$, we test $H_0 : \psi = 0$; $H_a : \psi > 0$. We find $c = 0.585$, $t = 5.99$, and $P < 0.0001$.

12.41. (a) The mean responses were not significantly different. (b) For Bonferroni, $t^{**} = 2.827$. Only the largest difference within each set of means is significant: $t_{14} = -2.89$ (experience culture), $t_{23} = 3.55$ (group tour), and $t_{45} = 2.86$ (ocean sports).

12.43. (a)

Group	n	$\bar{x}$	s
Control	10	601.1	27.36
Low jump	10	612.5	19.33
High jump	10	638.7	16.59

(b) $F = 7.98$, df = 2 and 27, $P = 0.002$. We conclude that not all means are equal.

12.45. (a) Pooling is risky because $0.6283/0.2520 = 2.49 > 2$.

Pot	n	$\bar{x}$	s
Aluminum	4	2.0575	0.2520
Clay	4	2.1775	0.6213
Iron	4	4.6800	0.6283

(b) $F = 31.16$, df = 2 and 9, $P < 0.0005$. We cautiously conclude that the means are not the same.

12.47. (a) Pooling is risky because $8.66/2.89 > 2$.

Material	n	$\bar{x}$	s
ECM1	3	65.0%	8.6603%
ECM2	3	63.3%	2.8868%
ECM3	3	73.3%	2.8868%
MAT1	3	23.3%	2.8868%
MAT2	3	6.6%	2.8868%
MAT3	3	11.6%	2.8868%

(b) $F = 137.94$, df = 5 and 12, $P < 0.0005$. We cautiously conclude that the means are not the same.

12.49. (a) $\psi_1 = \mu_1 - (\mu_2 + \mu_4)/2$ and $\psi_2 = \mu_3 - \mu_2 - (\mu_5 - \mu_4)$. (b) $c_1 = -1.5$, $SE_{c_1} \doteq 1.9337$, $c_2 = -3.75$, $SE_{c_2} \doteq 3.2558$. (c) Neither contrast is significant ($t_1 \doteq -0.752$ and $t_2 \doteq -1.152$).

12.51. The means all increase by 5%, but everything else (standard deviations, standard errors, and the ANOVA table) is unchanged.

12.53. All distributions are reasonably Normal, and standard deviations are close enough to justify pooling. For PRE1, $F = 1.13$, df = 2 and 63, $P = 0.329$. For PRE2, $F = 0.11$, df = 2 and 63, $P = 0.895$. Neither set of pretest scores suggests a difference in means.

12.55. $\widehat{\text{Score}} = 4.432 - 0.000102$ Friends. The slope is not significantly different from 0 ($t = -0.28$, $P = 0.782$), and the regression explains only 0.1% of the variation in score. Residuals suggest a possible curved relationship.

Chapter 13

13.1. (a) Two-way ANOVA is used when there are two factors. (b) Each level of A should occur with all three levels of B. (c) The RESIDUAL part of the model represents the error. (d) DFAB = $(I - 1)(J - 1)$.

13.3. (a) Reject H_0 when F is large. (b) Mean squares equal sum of squares divided by degrees of freedom. (c) The test statistics have an F distribution. (d) If the sample sizes are not the same, the sums of squares may not add.

13.5. (a) df = 2 and 24. (b) and (c) $0.025 < P < 0.05$.

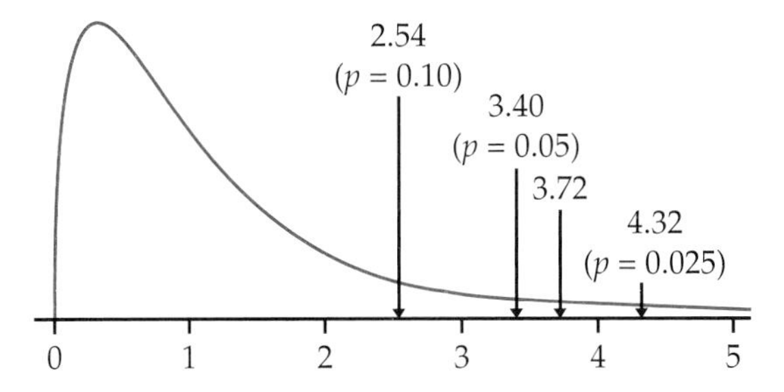

(d) The interaction term is significantly different from 0, so the mean profiles should not be parallel.

13.7. (a) Factors: gender ($I = 2$) and age ($J = 3$). Response: percent of pretend play. $N = 66$. (b) Factors: time after harvest ($I = 5$) and amount of water ($J = 2$). Response: percent of seeds germinating. $N = 30$. (c) Factors: mixture ($I = 6$) and freezing/thawing cycles ($J = 3$). Response: Strength. $N = 54$. (d) Factors: training programs ($I = 4$) and number of days to give the training ($J = 2$). Response: some (unspecified) measure of the training's effectiveness. $N = 80$.

13.9. (a) There appears to be an interaction; a thank-you note increases repurchase intent for those with short history and decreases it for customers with long history.

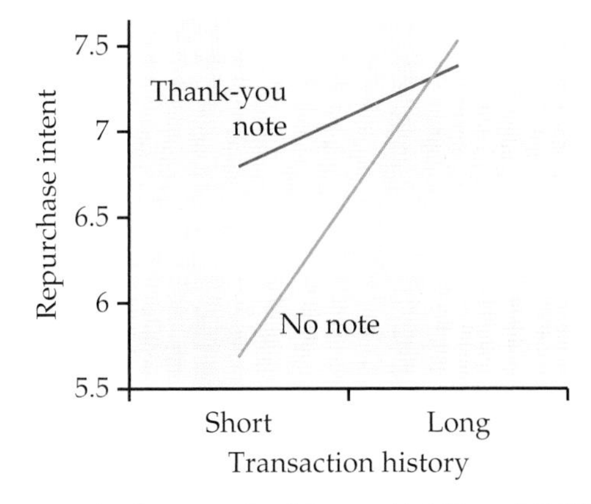

(b) The marginal means for history (6.245 and 7.45) convey the fact that repurchase intent is higher for customers with long history. The thank-you note marginal means (6.61 and 7.085) are less useful because of the interaction.

13.11. (a) The plot suggests a possible interaction.

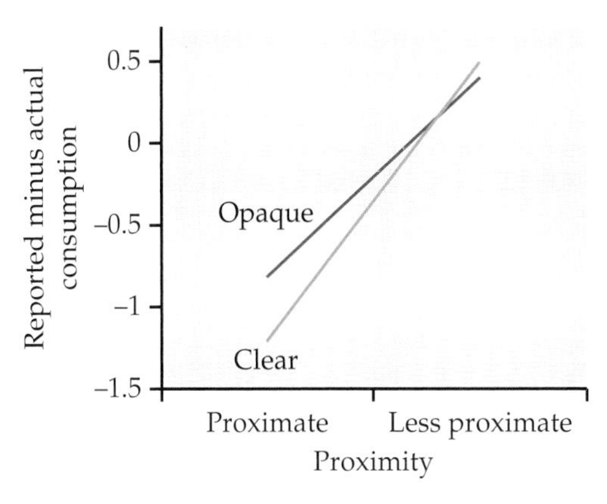

(b) By subjecting the same individual to all four treatments, rather than four individuals to one treatment each, we reduce the within-groups variability.

13.13. (a) There may be an interaction; for a favorable process, a favorable outcome increases satisfaction quite a bit more than for an unfavorable process (+2.32 compared with +0.24). (b) This time, the increase in satisfaction from a favorable outcome is less for a favorable process (+0.49 compared with +1.32).

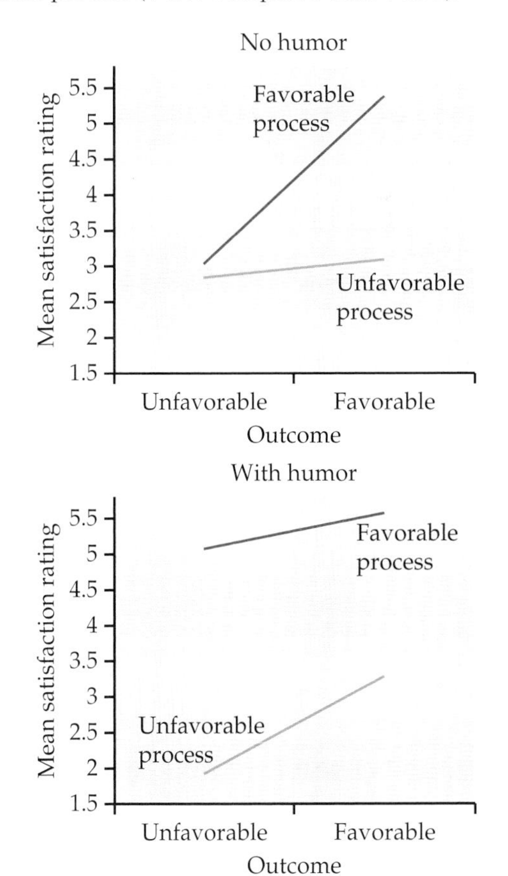

(c) There seems to be a three-factor interaction, because the interactions in parts (a) and (b) are different.

13.15. Humor slightly increases satisfaction (3.58 with no humor, 3.96 with humor). The process and outcome effects are greater: favorable process 4.75, unfavorable process 2.79; favorable outcome 4.32, unfavorable outcome 3.22.

13.17. The largest-to-smallest ratio is 1.26, and the pooled standard deviation is 1.7746.

13.19. Except for female responses to purchase intention, means decreased from Canada to the United States to France. Females had higher means than men in almost every case, except for French responses to credibility and purchase intention (a modest interaction).

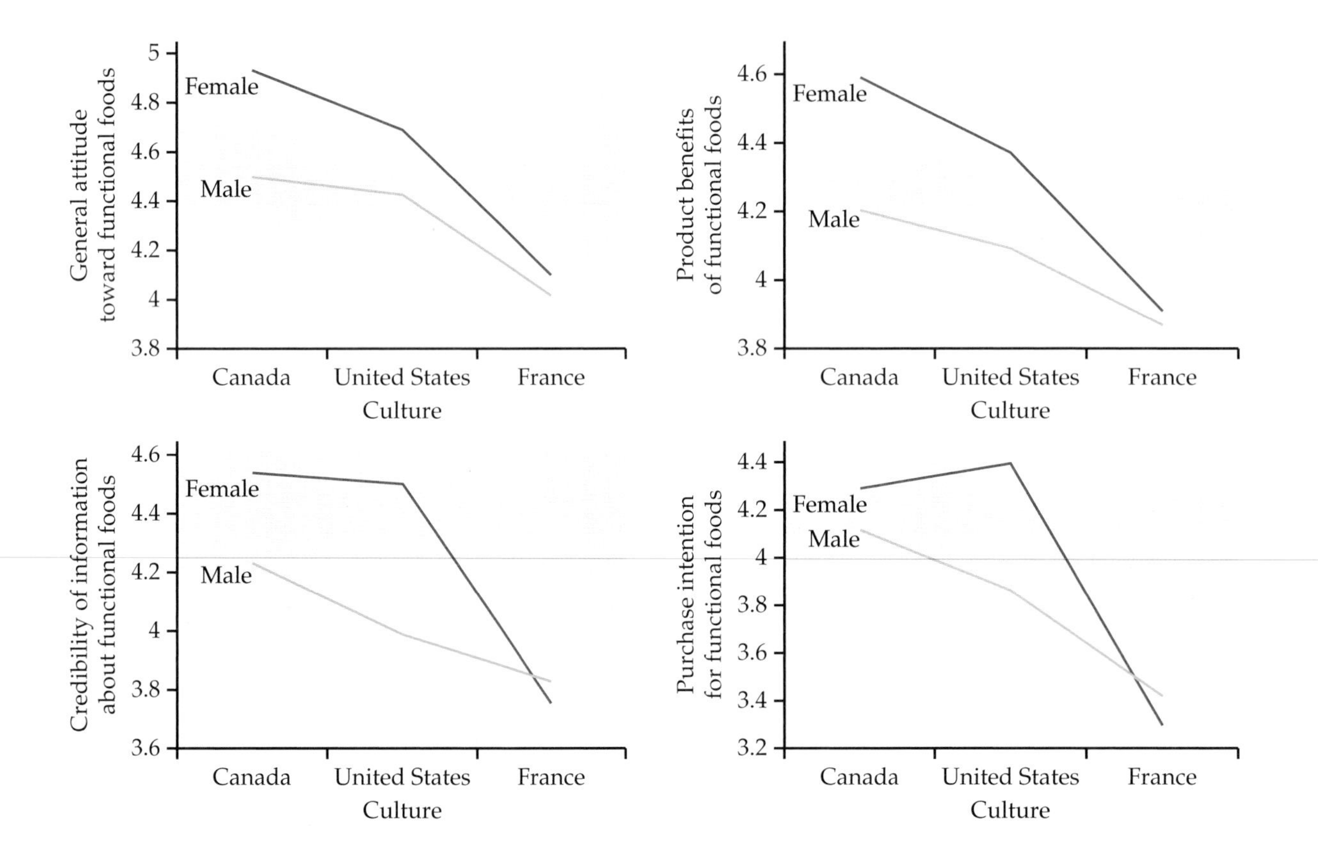

13.21. (a) The row means suggest that the intervention group showed more improvement than the control group.

Group	Time			Mean
	Baseline	**3 mo.**	**6 mo.**	
Intervention	10.4	12.5	11.9	11.6
Control	9.6	9.9	10.4	9.967
Mean	10.0	11.2	11.15	10.783

(b) Interaction means that the mean number of actions changes differently over time for the two groups.

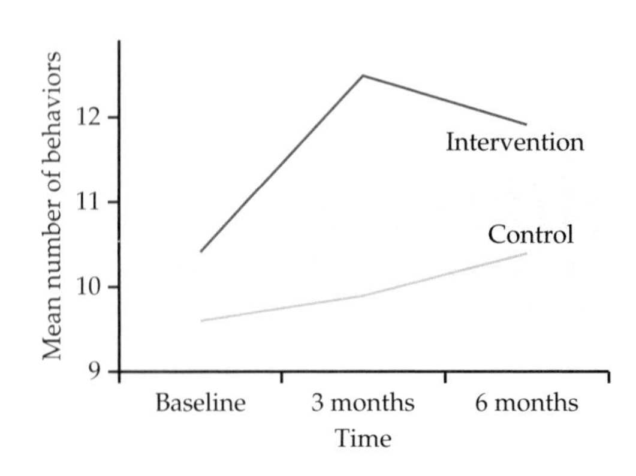

13.23. (a) There is little evidence of an interaction.

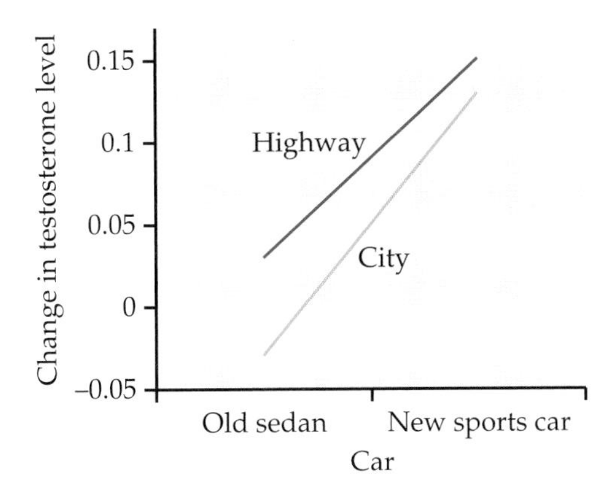

(b) $s_p \doteq 0.1278$. (c) $\psi_1 = (\mu_{\text{new, city}} + \mu_{\text{new, highway}})/2 - (\mu_{\text{old, city}} + \mu_{\text{old, highway}})/2$. $\psi_2 = \mu_{\text{new, city}} - \mu_{\text{new, highway}}$. $\psi_3 = \mu_{\text{old, highway}} - \mu_{\text{old, city}}$.

13.25. There are no significant effects (although B is close): F_A(df = 2 and 24) has $P = 0.1759$. F_B(df = 1 and 24) has $P = 0.0740$. F_{AB}(df = 2 and 24) has $P = 0.1396$.

13.27. (a) Plot below

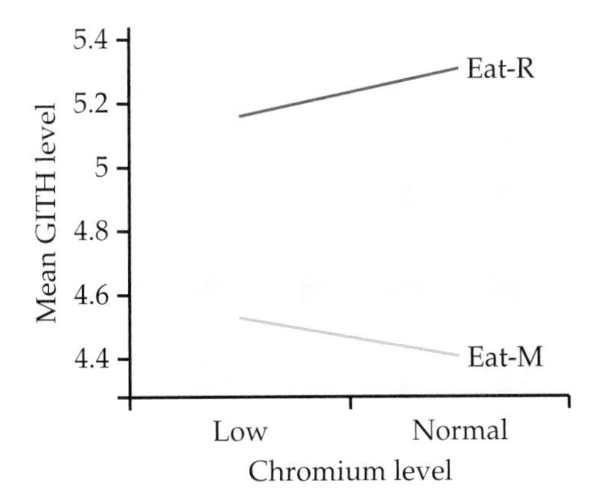

(b) There seems to be a fairly large difference between the means based on how much the rats were allowed to eat but not very much difference based on the chromium level. There may be an interaction: The NM mean is lower than the LM mean, while the NR mean is higher than the LR mean. (c) L mean: 4.86. N mean: 4.871. M mean: 4.485. R mean: 5.246. LR minus LM: 0.63. NR minus NM: 0.892. Mean GITH levels are lower for M than for R; there is not much difference between L and N. The difference between M and R is greater among rats who had normal chromium levels in their diets (N).

13.29. (a) $s_p \doteq 38.14$, df = 105. (b) Yes; the largest-to-smallest ratio is 1.36. (c) Individual sender, 70.90; group sender, 48.85; individual responder, 59.75; group responder, 60.00. (d) There appears to be an interaction (see plot below); individuals send more money to groups, while groups send more money to individuals. (e) $P = 0.0033$, $P = 0.9748$, and $P = 0.1522$. Only the main effect of sender is significant.

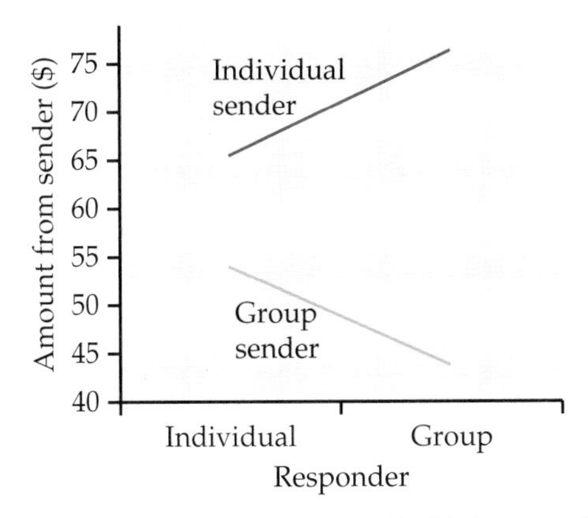

13.31. Yes; the iron-pot means are the highest, and F for testing the effect of the pot type is very large.

13.33. (a) All three F-values have df = 1 and 945; the P-values are <0.001, <0.001, and 0.1477. Gender and handedness both have significant effects on mean lifetime, but there is no interaction. (b) Women live about 6 years longer than men (on the average), while right-handed people average 9 more years of life than left-handed people. Handedness affects both genders in the same way, and vice versa.

13.35. (a) Gender: df = 1 and 174. Floral characteristic: df = 2 and 174. Interaction: df = 2 and 174. (b) Damage to males was higher for all characteristics. For males, damage was highest under characteristic level 3, while for females, the highest damage occurred at level 2. (c) Three of the standard deviations are at least half as large as the means. Because the response variable (leaf damage) had to be nonnegative, this suggests that these distributions are right-skewed.

13.37. The means suggest that females have higher HSE grades than males. For a given gender, there is not too much difference among majors. Normal quantile plots show no great deviations from Normality, apart from the granularity of the grades (most evident among women in EO). In the ANOVA, only the effect of gender is significant ($F = 50.32$, df = 1 and 228, $P < 0.0005$).

		Major		
Gender		**CS**	**EO**	**Other**
Male	$n =$	39	39	39
	$\bar{x} =$	7.79487	7.48718	7.41023
	$s =$	1.50752	2.15054	1.56807
Female	$n =$	39	39	39
	$\bar{x} =$	8.84615	9.25641	8.61539
	$s =$	1.13644	0.75107	1.16111

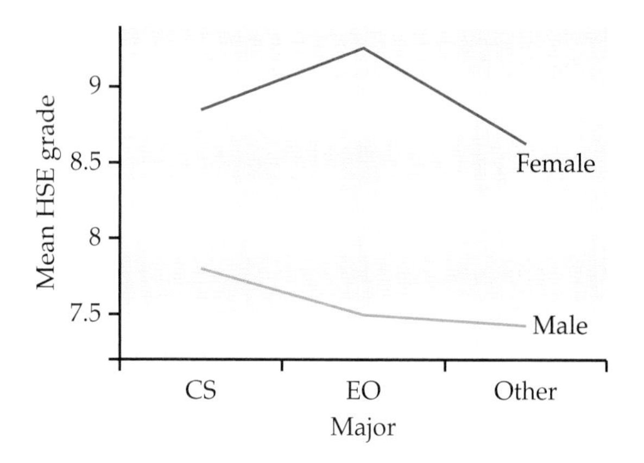

13.39. The means suggest that students who stay in the sciences have higher mean SATV scores than those who end up in the "Other" group. Female CS and EO students have higher scores than males in those majors, but males have the higher mean in the Other group. Normal quantile plots suggest some right-skewness in the "Women in CS" group and also some non-Normality in the tails of the "Women in EO" group. Other groups look reasonably Normal. In the ANOVA, only the effect of major is significant ($F = 9.32$, df = 2 and 228, $P < 0.0005$).

Gender		Major		
		CS	**EO**	**Other**
Male	$n =$	39	39	39
	$\bar{x} =$	526.949	507.846	487.564
	$s =$	100.937	57.213	108.779
Female	$n =$	39	39	39
	$\bar{x} =$	543.385	538.205	465.026
	$s =$	77.654	102.209	82.184

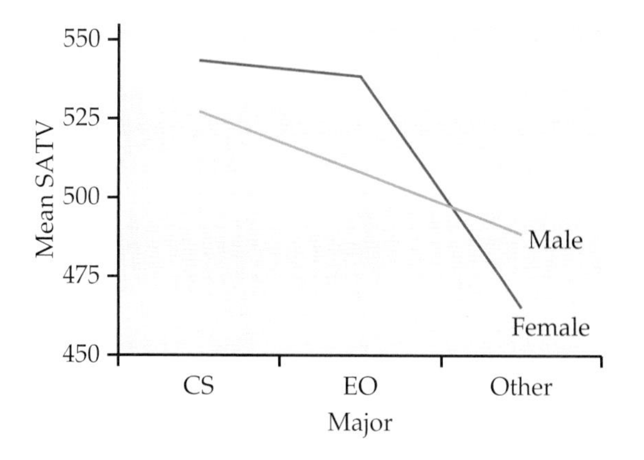

Chapter 14

14.1. 1 to 1.

14.3. For men: 1.3404. For women: 0.8571 (or 6 to 7).

14.5. For men: 0.2930. For women: −0.1542.

14.7. If $x = 1$ for men and 0 for women, $\log(\text{odds}) = -0.1542 + 0.4471x$. (If vice versa, $\log(\text{odds}) = 0.2930 - 0.1542x$.) The odds ratio is 1.5638 (or 0.6395).

14.9. (a) Use a chi-square test with df = 5. (b) The logistic regression model has no error term. (c) H_0 should refer to b_1. (d) The interpretation of coefficients is affected by correlations among explanatory variables.

14.11. (a) 2.85. (b) 6.14. (c) 9.92.

14.13. (a) $\hat{p}_{\text{low}} \doteq 0.0753$ and $\hat{p}_{\text{high}} \doteq 0.0899$. (b) 0.0814 and 0.0988. (c) −2.5083 and −2.3150.

14.15. 2.1096 to 4.1080.

14.17. (a) $z \doteq 8.01$. (b) $z^2 \doteq 64.23$. (c) Graphs below.

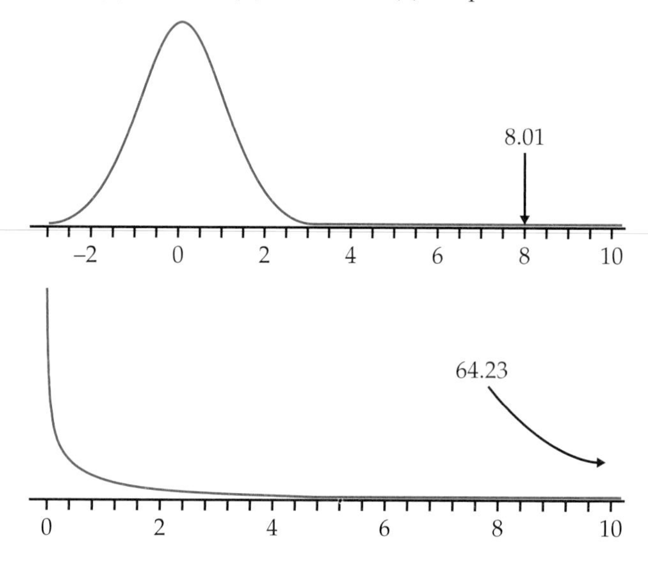

14.19. The odds favor a low tip from senior adults, those dining on Sunday, those who speak English as a second language, and French-speaking Canadians. Diners who drink alcohol and lone males are less likely to leave low tips. For example, for a senior adult, the odds of leaving a low tip were 1.099 (for a probability of 0.5236).

14.21. (a) The men's magazines confidence interval includes 1, consistent with failing to reject H_0. For other explanatory variables, the interval does not include 1. (b) The odds that the model's clothing is not sexual are 1.27 to 2.16 times higher for magazines targeted at mature adults, 2.74 to 5.01 times higher when the model is male, and 1.11 to 2.23 times higher for magazines aimed at women. (c) For example, it is easier to interpret the odds ratio than the regression coefficients because it is hard to think in terms of a log-odds scale.

14.23. (a) $\hat{p}_{\text{hi}} \doteq 0.01648$ and $\text{odds}_{\text{hi}} \doteq 0.01675$, or about 1 to 60. (b) $\hat{p}_{\text{lo}} \doteq 0.00785$ and $\text{odds}_{\text{lo}} \doteq 0.00791$, or about 1 to 126. (c) The odds ratio is 2.1181.

14.25. (a) 0.2452 to 1.2558. (b) $X^2 \doteq 8.47$, $0.0025 < P < 0.005$. (c) We have strong evidence of a difference in risk between the two groups.

14.27. (a) The estimated odds ratio is 2.1181; the odds-ratio interval is 1.28 to 3.51. (b) We are 95% confident that the odds of death from cardiovascular disease are about 1.3 to 3.5 times greater in the high-blood-pressure group.

14.29. (a) $\log(\text{odds}) = b_0 + b_1 x$, where $x = 1$ if the person is over 40, and 0 if the person is under 40. (b) p_i is the probability that the ith person is terminated; this model assumes that the probability of termination depends on age (over/under 40). (c) The estimated odds ratio is 3.859. A 95% confidence interval for b_1 is 0.5409 to 2.1599. The odds of being terminated are 1.7 to 8.7 times greater for those over 40. (d) Use a multiple logistic regression model.

14.31. The model is $\log(\text{odds}) = 0.2036 + 0.7068x$.

14.33. $\log(\text{odds}) = -0.9785 + 0.0614x$; the slope is significantly different from 0.

14.35. (a) $X^2 \doteq 33.65$, df = 3, $P = 0.0001$. (b) $\log(\text{odds}) = -6.053 + 0.3710\text{HSM} + 0.2489\text{HSS} + 0.03605\text{HSE}$. 95% confidence intervals: 0.1158 to 0.6262, −0.0010 to 0.4988, and −0.2095 to 0.2816. (c) Only the coefficient of HSM is significantly different from 0, though HSS may also be useful.

14.37. (a) $X^2 \doteq 19.2256$, df = 3, $P = 0.0002$. (b) $X^2 \doteq 3.4635$, df = 2, $P = 0.1770$. (c) High school grades (especially HSM and, to a lesser extent, HSS) are useful, while SAT scores are not.

14.39. (a) $\log(\text{odds}) = -3.892 + 0.4157$ Hospital, using 1 for Hospital A and 0 for Hospital B. The slope is not significantly different from 0 ($z = -1.47$ or $X^2 = 2.16$, $P = 0.1420$). A 95% confidence interval for b_1 is −0.1392 to 0.9706. The odds ratio is 1.515, with confidence interval 0.87 to 2.64. (b) $\log(\text{odds}) = -3.109 - 0.1320$ Hospital -1.266 Condition, using 1 for Hospital A and 0 for Hospital B, and 1 for good condition and 0 for poor. The odds ratio is 0.8764, with confidence interval 0.48 to 1.60. (c) In the model with Hospital alone, the slope was positive and the odds ratio was greater than 1. When Condition is added to the model, the Hospital coefficient is negative and the odds ratio is less than 1.

Part III Review

III.1. (a) For example, a comparative bar graph might be good. (b) Plot below.

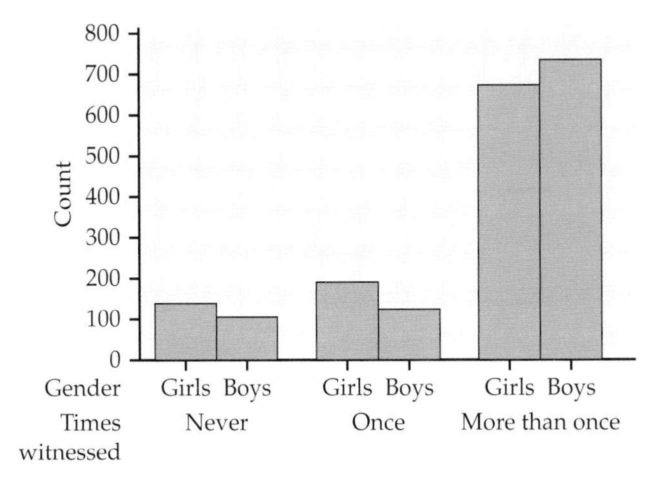

III.3. Girls: $X^2 = 308.23$, df = 1, $P < 0.0001$. Boys: $X^2 = 261.29$, df = 1, $P < 0.0001$.

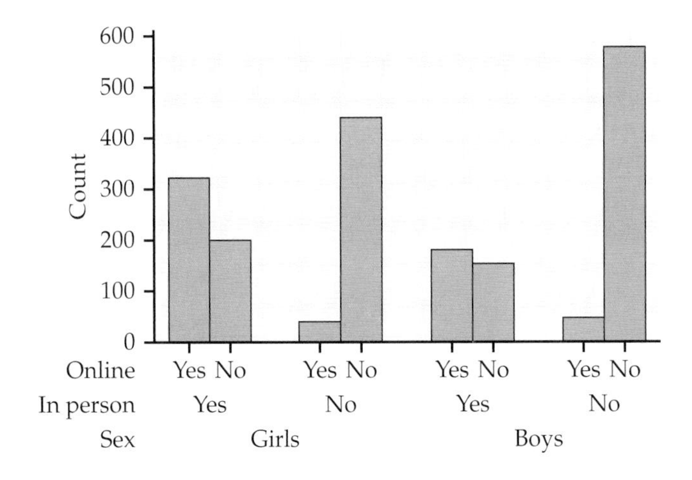

III.5. (a) H_0: Response and Institution Type are independent versus H_a: Response and Institution Type are not independent. (b) $X^2 = 17.30$, df = 3, $P = 0.0006$. (c) There appears to be a relationship between Response and Institution Type.

III.7. (a) df = 4 and 178. (b) Data from 151 athletes were used. (c) For example, the athletes refuse to answer the surveys.

III.9. Because Kiplinger narrows down the number of colleges, the 40 colleges selected are an SRS from this list, not from the original 500 four-year public colleges.

III.11. (a) \$19,591. (b) \$23,477. (c) The standard error would be larger for the University of Wisconsin since percent borrowing is farther from 50%.

III.13. It appears that Lecture 2 has higher means for all conditions, but that there is an interaction between lecture and condition for Lectures 1 and 3.

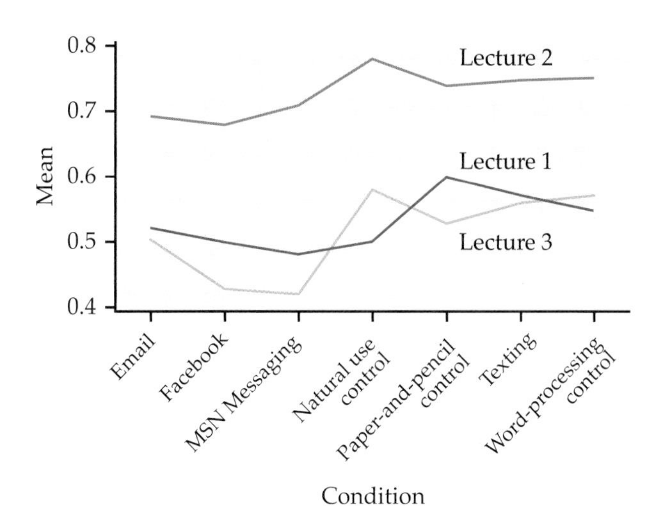

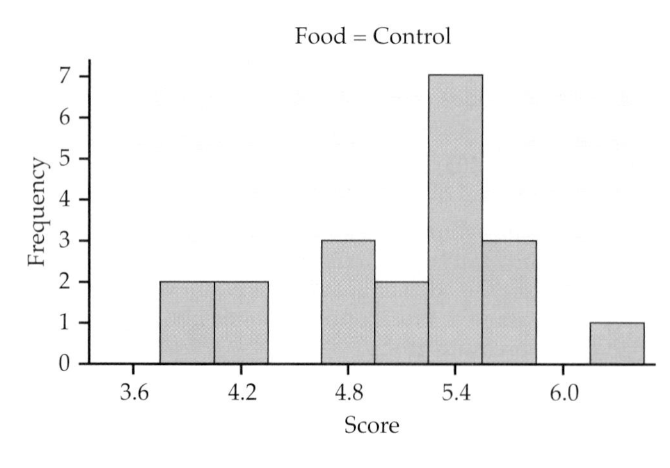

III.15. (a) $\psi = (\mu_1 + \mu_2 + \mu_3 + \mu_4)/4 - \mu_7$. (b) $H_0 : \psi = 0$ versus $H_a : \psi < 0$. (c) $c = -0.0558$, $\text{SE}_c = 0.0082$; $t = -6.805$, $P < 0.0001$. We have sufficient evidence to conclude that the average score of the off-task behaviors is lower than that for the paper-and-pencil control condition.

III.17. $\hat{y} = 381.4 + 0.36\,\text{InAfterAid} + 122\,\text{PercBorrow} + 57\,\text{Admit} + 62\,\text{Yr4Grad} + 194\,\text{StudPerFac}$.

III.19. (a) Table below. It is reasonable to pool the variances.

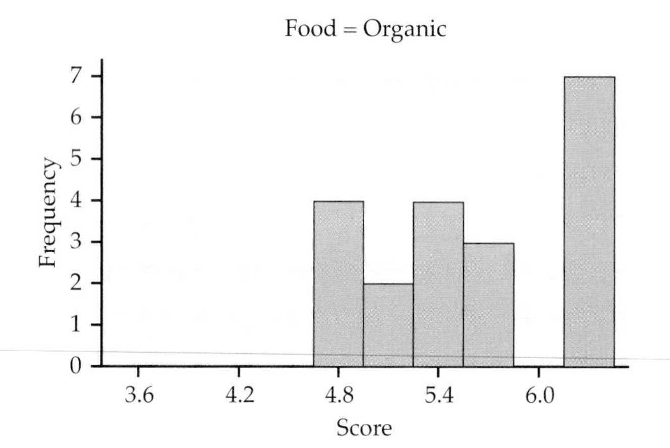

Group	n	$\bar{x}$	s
Comfort	22	4.887	0.573
Control	20	5.082	0.622
Organic	20	5.584	0.594

(b) The sample means for Comfort appear relatively Normal. We might question the Normality of the means for the other two groups.

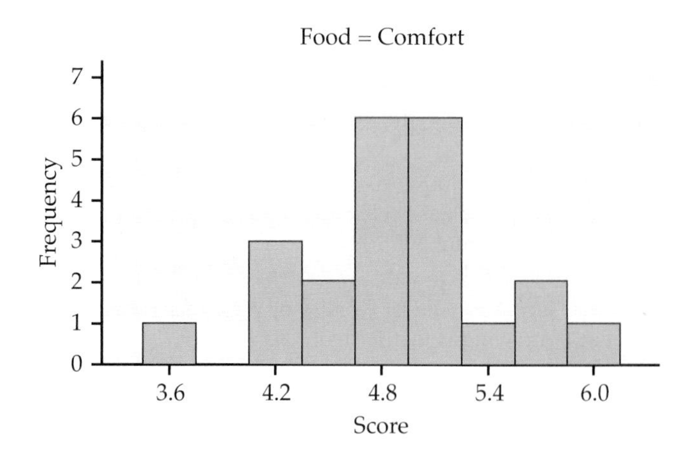

III.21. (a) The means are not all equal for the three groups. Comfort and Organic appear to differ; Control is not significantly different from either Comfort or Organic. (b) The decrease in variability for the three groups and the curve in the Normal quantile plot might make us question Normality.

III.23. (a) The individual time measurements do not need to be Normally distributed. (b) It is (barely) reasonable to pool the variances. (c) The interaction between Gender and Age is not statistically significant. Both Gender and Age main effects are statistically significant.

Source	df	SS	MS	F	P
Gender	1	44.66	44.66	36.88	< 0.0001
Age	6	31.97	5.328	4.40	0.0003
Interaction	6	13.22	2.203	1.82	0.0962
Error	232	280.95	1.211		
Total	245	370.80			

III.25. (a) $b_0 = 2.0900$, $b_1 = -0.2712$. (b) $SE_{b_0} = 0.1030$, $SE_{b_1} = 0.1375$; $b_0 \pm 1.96SE_{b_0} = (1.8881, 2.2919)$; $b_1 \pm 1.96SE_{b_1} = (-0.5407, -0.0017)$. (c) $z = \frac{-0.2713}{0.1375} = -1.973$, $P = 0.0485$. (d) odds ratio $= \frac{(863)(106)}{(857)(140)} = 0.76244$; $e^{b_1} = e^{-0.2712} = 0.76246$. These are approximately equal.

III.27. (a) $\log(\text{odds}) = -2.357 + 0.461x$. (b) -0.974. (c) $\frac{e^{-0.974}}{1+e^{-0.974}} = 0.2741$. (d) 1.331; $\frac{e^{1.331}}{1+e^{1.331}} = 0.7910$.

III.29. (a) $\log(\text{odds}) = -4.206 + 0.341\,\text{HSgrade} + 0.024\,\text{SAT}$. (b) -1.143. (c) $\frac{e^{-1.143}}{1+e^{-1.143}} = 0.2418$. (d) 1.402; $\frac{e^{1.402}}{1+e^{1.402}} = 0.8025$.

NOTES AND DATA SOURCES

Chapter 1

1. For more information and references about Discovery Learning, see en.wikipedia.org/wiki/Discovery_learning.
2. See www.cdc.gov/brfss/.
3. Amanda Lenhart, et al. "Teens, kindness and cruelty on social network sites," November 9, 2011. The report is available from the Pew Internet and American Life Project Web site, pewinternet.org/Reports/2011/Teens-and-social-media.aspx.

Chapter 2

1. See the NORC Web site, at norc.uchicago.edu.
2. See www.cdc.gov/mmwr/preview/mmwrhtml/mm5839a3.htm.
3. See Jennifer A. Manganello and Catherine A. Taylor, "Television behavior as a risk factor for aggressive behavior among 3-year-old children," *Archives of Pediatric and Adolescent Medicine,* 163:11 (2009), pp. 1037–1045.
4. National Institute of Child Health and Human Development Early Child Care Research Network, "Does amount of time spent in child care predict socioemotional adjustment during the transition to kindergarten?" *Child Development,* 74:4 (2003), pp. 976–1005.
5. The quotation is from the summary on the NICHD Web site, nichd.nih.gov.
6. For a full description of the STAR program and its follow-up studies, go to heros-inc.org/star.htm.
7. Simplified from Arno J. Rethans, John L. Swasy, and Lawrence J. Marks, "Effects of television commercial repetition, receiver knowledge, and commercial length: A test of the two-factor model," *Journal of Marketing Research,* 23 (February 1986), pp. 50–61.
8. Based on an experiment performed by Jake Gandolph under the direction of Professor Lisa Mauer in the Purdue University Department of Food Science.
9. Based on an experiment performed by Evan Whalen under the direction of Professor Patrick Connolly in the Purdue University Department of Computer Graphics Technology.
10. John H. Kagel, Raymond C. Battalio, and C. G. Miles, "Marijuana and work performance: Results from an experiment," *Journal of Human Resources,* 15 (1980), pp. 373–395. A general discussion of failures of blinding is Dean Ferguson et al., "Turning a blind eye: the success of blinding reported in a random sample of randomised, placebo controlled trials," *British Medical Journal,* 328 (2004), p. 432.
11. Joel Brockner et al., "Layoffs, equity theory, and work performance: Further evidence of the impact of survivor guilt," *Academy of Management Journal,* 29 (1986), pp. 373–384.
12. Simplified from Sanjay K. Dhar, Claudia González-Vallejo, and Dilip Soman, "Modeling the effects of advertised price claims: Tensile versus precise pricing," *Marketing Science,* 18 (1999), pp. 154–177.
13. Based on a study conducted by Sandra Simonis under the direction of Professor Jon Harbor in the Purdue University Earth and Atmospheric Sciences Department.
14. Based on a study conducted by Tammy Younts under the direction of Professor Deb Bennett of the Purdue University Department of Educational Studies. For more information about Reading Recovery, see readingrecovery.org.
15. Based on a study conducted by Rajendra Chaini under the direction of Professor Bill Hoover in the Purdue University Department of Forestry and Natural Resources.
16. Before the sort, the random numbers in column B need to be designated as values rather than formulas. An easy way to do this is to copy the column and then do a "paste special" into a new column.

17. From the Hot Ringtones list at billboard.com/ on February 19, 2012.

18. From the Rock Songs list at billboard.com/ on February 19, 2012.

19. From the online version of the Bureau of Labor Statistics, *Handbook of Methods,* modified April 17, 2003, at bls.gov. The details of the design are more complicated than the text describes.

20. For more detail on the material of this section and complete references, see P. E. Converse and M. W. Traugott, "Assessing the accuracy of polls and surveys," *Science,* 234 (1986), pp. 1094–1098.

21. The nonresponse rate for the CPS comes from the source cited in Note 19. The GSS reports its response rate on its Web site, norc.org/projects/gensoc.asp. The Pew study is described in Gregory Flemming and Kimberly Parker, "Possible consequences of non-response for pre-election surveys: Race and reluctant respondents," May 16, 1998, found at people-press.org/1998/05/16.

22. For more detail on the limits of memory in surveys, see N. M. Bradburn, L. J. Rips, and S. K. Shevell, "Answering autobiographical questions: The impact of memory and inference on surveys," *Science,* 236 (1987), pp. 157–161.

23. Robert F. Belli et al., "Reducing vote overreporting in surveys: Social desirability, memory failure, and source monitoring," *Public Opinion Quarterly,* 63 (1999), pp. 90–108.

24. Sex: Tom W. Smith, "The *JAMA* controversy and the meaning of sex," *Public Opinion Quarterly,* 63 (1999), pp. 385–400. Welfare: from a *New York Times*/CBS News Poll reported in the *New York Times,* July 5, 1992. Scotland: "All set for independence?" *Economist,* September 12, 1998. Many other examples appear in T. W. Smith, "That which we call welfare by any other name would smell sweeter," *Public Opinion Quarterly,* 51 (1987), pp. 75–83.

25. From gallup.com on November 10, 2009.

26. From pewresearch.org on February 19, 2012.

27. From *CIS Boletin 9, Spaniards' Economic Awareness,* found online at cis.es/ingles/opinion/economia.htm.

28. John C. Bailar III, "The real threats to the integrity of science," *Chronicle of Higher Education,* April 21, 1995, pp. B1–B2.

29. The difficulties of interpreting guidelines for informed consent and for the work of institutional review boards in medical research are a main theme of Beverly Woodward, "Challenges to human subject protections in U.S. medical research," *Journal of the American Medical Association,* 282 (1999), pp. 1947–1952. The references in this paper point to other discussions.

30. Quotation from the *Report of the Tuskegee Syphilis Study Legacy Committee,* May 20, 1996. A detailed history is James H. Jones, *Bad Blood: The Tuskegee Syphilis Experiment,* Free Press, 1993.

31. Dr. Hennekens's words are from an interview in the Annenberg/Corporation for Public Broadcasting video series *Against All Odds: Inside Statistics.*

32. See ftc.gov/opa/2009/04/kellogg.shtm.

33. See findarticles.com/p/articles/mi_m0CYD/is_8_40/ai_n13675065/.

34. R. D. Middlemist, E. S. Knowles, and C. F. Matter, "Personal space invasions in the lavatory: Suggestive evidence for arousal," *Journal of Personality and Social Psychology,* 33 (1976), pp. 541–546.

35. The report was issued in February 2009 and is available at ftc.gov/os/2009/02/P085400behavadreport.pdf.

Chapter 3

1. Data from the Bureau of Labor Statistics. See www.bls.gov/iif/oshsum.htm.

2. Reported in the May 8, 2008, edition of *GPS Magazine.* See www.gpsmagazine.com/print/000443.php.

3. Data collected in the lab of Connie Weaver, Department of Foods and Nutrition, Purdue University, and provided by Linda McCabe.

4. Haipeng Shen, "Nonparametric regression for problems involving lognormal distributions," PhD thesis, University of Pennsylvania, 2003. Thanks to Haipeng Shen and Larry Brown for sharing the data.

5. From the Digest of Education Statistics at the Web site of the National Center for Education Statistics, www.nces.ed.gov/programs/digest.

6. See Note 3.

7. Based on Barbara Ernst et al., "Seasonal variation in the deficiency of 25–hydroxyvitamin D_3 in mildly to extremely obese subjects," *Obesity Surgery*, 19 (2009), pp. 180–183.

8. From the Color Assignment Web site of Joe Hallock, www.joehallock.com/edu/COM498/index.html.

9. "The Apple iPhone: Successes and challenges for the mobile industry," March 31, 2008, Rubicon Win Markets. See rubiconconsulting.com.

10. U.S. Environmental Protection Agency, *Municipal Solid Waste in the United States: 2007 Facts and Figures.*

11. February 2012 data from marketshare.hitslink.com.

12. Robyn Greenspan, "The deadly duo: Spam and viruses, October 2003," at cyberatlas.internet.com.

13. See www.internetworldstats.com/facebook.htm.

14. See previous note.

15. National Center for Education Statistics, NEDRC Table Library, nces.ed.gov/surveys/npsas/table_library.

16. Color popularity for 2011 from the Dupont Automotive Color report; see www2.dupont.com/Media_Center/en_US/color_popularity.

17. Debora L. Arsenau, "Comparison of diet management instruction for patients with non-insulin dependent diabetes mellitus: Learning activity package vs. group instruction," MS thesis, Purdue University, 1993.

18. Data from Gary Community School Corporation, courtesy of Celeste Foster, Department of Education, Purdue University.

19. Found online at earthtrends.wri.org.

20. We thank Heeseung Roh Ryu for supplying the data, from her dissertation in the Department of Health and Kinesiology, Purdue University.

21. This exercise was provided by Nicolas Fisher.

22. Data from the World Bank, Quick Query option from Key Development Data and Statistics page, at worldbank.org.

23. *Extreme Weather Sourcebook, 2001,* found online at sciencepolicy.colorado.edu/sourcebook.

24. From the Interbrand Web site; see interbrand.com/best_global_brands.aspx?langid=1000.

25. From beer100.com/beercalories.htm.

26. See Noel Cressie, *Statistics for Spatial Data*, Wiley, 1993.

27. Data provided by Francisco Rosales, Department of Nutritional Sciences, Pennsylvania State University.

28. Data provided by Betsy Hoza, Department of Psychological Sciences, University of Vermont.

29. "Changes in U.S. family finances from 2004 to 2007: Evidence from the survey of consumer finances," Federal Reserve Bulletin, 2009.

30. We thank Ethan J. Temeles of Amherst College for providing the data. His work is described in Ethan J. Temeles and W. John Kress, "Adaptation in a plant-hummingbird association," *Science*, 300 (2003), pp. 630–633.

31. Information about the Indiana Statewide Testing for Educational Progress program can be found at doe.state.in.us/istep/.

32. Matthias R. Mehl et al., "Are women really more talkative than men?" *Science*, 317, (2007), p. 82. The raw data were provided by Matthias Mehl.

33. From the American Heart Association Web site, americanheart.org.

34. From fueleconomy.gov/.

35. From cdc.gov/brfss.

36. See Note 16.

37. See worldbank.org.

38. Data are from the Open Accessible Space Information System for New York City. See oasisnyc.net.

39. We thank C. Robertson McClung of Dartmouth College for supplying the data. The study is reported in Todd P. Michael et al., "Enhanced fitness conferred by naturally occurring variation in the circadian clock," *Science*, 302 (2003), pp. 1049–1053.

40. From isp-review.toptenreviews.com on February 26, 2012.

Chapter 4

1. Hannah G. Lund et al., "Sleep patterns and predictors of disturbed sleep in a large population of college students," *Adolescent Health,* 46, No. 2 (2010), pp. 97–99.
2. See previous note.
3. See cfs.purdue.edu/FN/campcalcium/public.htm for information about the 2010 camp.
4. See "Happy spamiversary! Spam reaches 30," *NewScientistTech,* April 25, 2008; available at NewScientist.com.
5. Thanks to Doug Crabill, manager of Computer Systems for the Purdue University Department of Statistics, for providing this background information about spam botnets.
6. See symantec.com/security_response/writeup.jsp?docid=2007-062007-0946-99.
7. OECD StatExtracts, Organisation for Economic Co-operation and Development, downloaded from stats.oecd.org/wbos on June 29, 2008.
8. See forbes.com/lists/2008/6/biz_bizcountries08_Best-Countries-for-Business_Rank.html.
9. A sophisticated treatment of improvements and additions to scatterplots is W. S. Cleveland and R. McGill, "The many faces of a scatterplot," *Journal of the American Statistical Association,* 79 (1984), pp. 807–822.
10. Data for 2010, from the U.S. Census Bureau Web site, census.gov/hhes/www/hlthins/data/incpovhlth/2010/table8.pdf.
11. See beer100.com/beercalories.htm.
12. See worldbank.org.
13. Christer G. Wiklund, "Food as a mechanism of density-dependent regulation of breeding numbers in the merlin *Falco columbarius,*" *Ecology,* 82 (2001), pp. 860–867.
14. We thank C. Robertson McClung of Dartmouth College for supplying the data. The study is reported in Todd P. Michael et al., "Enhanced fitness conferred by naturally occurring variation in the circadian clock," *Science,* 302 (2003), pp. 1049–1053.
15. A careful study of this phenomenon is W. S. Cleveland, P. Diaconis, and R. McGill, "Variables on scatterplots look more highly correlated when the scales are increased," *Science,* 216 (1982), pp. 1138–1141.
16. Data from a plot in James A. Levine, Norman L. Eberhardt, and Michael D. Jensen, "Role of nonexercise activity thermogenesis in resistance to fat gain in humans," *Science,* 283 (1999), pp. 212–214.
17. From the Web site oasisnyc.net.
18. Frank J. Anscombe, "Graphs in statistical analysis," *The American Statistician,* 27 (1973), pp. 17–21.
19. From the Web site of the National Center for Education Statistics, nces.ed.gov.
20. See Chapter 3, Note 17.
21. Zeinab E. M. Afifi,"Principal components analysis of growth of Nahya infants: Size, velocity and two physique factors," *Human Biology,* 57 (1985), pp. 659–669.
22. D. A. Kurtz (ed.), *Trace Residue Analysis,* American Chemical Society Symposium Series No. 284, 1985, Appendix.
23. The scores are from the Purdue University 2008–2009 women's golf team in the NCAA tournament.
24. Data from a plot in Feng Sheng Hu et al., "Cyclic variation and solar forcing of Holocene climate in the Alaskan subarctic," *Science,* 301 (2003), pp. 1890–1893.
25. See Note 18.
26. Results of this survey are reported in Henry Wechsler et al., "Health and behavioral consequences of binge drinking in college," *Journal of the American Medical Association,* 272 (1994), pp. 1672–1677.
27. You can find a clear and comprehensive discussion of numerical measures of association for categorical data in Chapter 2 of Alan Agresti, *Categorical Data Analysis,* 2nd ed., Wiley, 2002.
28. From M.-Y. Chen et al., "Adequate sleep among adolescents is positively associated with health status and health-related behaviors," *BMC Public Health,* 6, No. 59 (2006); available at biomedicalcentral.com/1471-2458/6/59.
29. See the U.S. Census Bureau Web site at census.gov/population/socdemo/school for these and similar data.

30. From the Social Security Web site, ssa.gov/OACT/babynames.

31. Information about this procedure was provided by Samuel Flanigan of *U.S. News & World Report*. See usnews.com/usnews/rankguide/rghome.htm for a description of the variables used to construct the ranks and for the most recent ranks.

32. See cdc.gov/brfss/. The data file BRFSS contains several variables from this source.

33. We thank Zhiyong Cai of Texas A&M University for providing the data. The data are from work performed in connection with his PhD dissertation in the Department of Forestry and Natural Resources, Purdue University.

34. Gary Smith, "Do statistics test scores regress toward the mean?" *Chance*, 10, No. 4 (1997), pp. 42–45.

35. Data from National Science Foundation Science Resources Studies Division, *Data Brief*, 12 (1996), p. 1.

36. Based on data in Mike Planty et al., "Volunteer service by young people from high school through early adulthood," National Center for Education Statistics Report, NCES 2004–365.

37. Although these data are fictitious, similar though less simple situations occur. See P. J. Bickel and J. W. O'Connell, "Is there a sex bias in graduate admissions?" *Science*, 187 (1975), pp. 398–404.

38. See George R. Milne, "How well do consumers protect themselves from identity theft?" *Journal of Consumer Affairs*, 37 (2003), pp. 388–402.

39. Based on information in "NCAA 2003 national study of collegiate sports wagering and associated health risks," which can be found at the NCAA Web site, ncaa.org.

PART I Review

1. See the National Survey of Student Engagement Web site at nsse.iub.edu/.

2. See the College Board Advocacy and Policy Center Web site at trends.collegeboard.org/.

3. From the report "Dietary reference intakes for calcium and vitamin D," 2010, available at iom.edu/Reports.

4. Data from Professor Sibylle Kranz and Lyndsey Huss, Department of Nutrition Science, Purdue University.

Chapter 5

1. An informative and entertaining account of the origins of probability theory is Florence N. David, *Games, Gods and Gambling*, Charles Griffin, 1962.

2. Based on information from the Color Assignment Web site of Joe Hallock, joehallock.com/edu/COM498/index.html.

3. You can find a mathematical explanation of Benford's law in Ted Hill, "The first-digit phenomenon," *American Scientist*, 86 (1996), pp. 358–363; and Ted Hill, "The difficulty of faking data," *Chance*, 12, No. 3 (1999), pp. 27–31. Applications in fraud detection are discussed in the second paper by Hill and in Mark A. Nigrini, "I've got your number," *Journal of Accountancy*, May 1999, available online at aicpa.org/pubs/jofa/joaiss.htm.

4. Royal Statistical Society news release, "Royal Statistical Society concerned by issues raised in Sally Clark case," October 23, 2001, at www.rss.org.uk. For background, see an editorial and article in the *Economist*, January 22, 2004. The editorial is entitled "The probability of injustice."

5. See cdc.gov/mmwr/preview/mmwrhtml/mm57e618a1.htm.

6. See the previous note.

7. From funtonia.com/top_ringtones_chart.asp. This Web site gives popularity scores based on download activity on the Internet. These scores were converted to probabilities for this exercise by dividing each popularity score by the sum of the scores for the top 10 ringtones.

8. See bloodbook.com/world-abo.html for the distribution of blood types for various groups of people.

9. From Statistics Canada, www.statcan.ca.

10. We use $\bar{x}$ both for the random variable, which takes different values in repeated sampling, and for the numerical value of the random variable in a particular sample. Similarly, s and $\hat{p}$ stand both for random variables and for specific

values. This notation is mathematically imprecise but statistically convenient.

11. We will consider only the case in which X takes a finite number of possible values. The same ideas, implemented with more advanced mathematics, apply to random variables with an infinite but still countable collection of values.

12. Based on a Pew Internet report, "Teens and distracted driving," available at `pewinternet.org/Reports/2009/Teens-and-Distracted-Driving.aspx`.

13. See `pewinternet.org/Reports/2009/17-Twitter-and-Status-Updating-Fall-2009.aspx`.

14. The mean of a continuous random variable X with density function $f(x)$ can be found by integration:

$$\mu_X = \int x f(x) dx$$

This integral is a kind of weighted average, analogous to the discrete-case mean

$$\mu_X = \sum x P(X\overline{x})$$

The variance of a continuous random variable X is the average squared deviation of the values of X from their mean, found by the integral

$$\sigma_X^2 = \int (x - \mu)^2 f(x) dx$$

15. See A. Tversky and D. Kahneman, "Belief in the law of small numbers," *Psychological Bulletin*, 76 (1971), pp. 105–110, and other writings of these authors for a full account of our misperception of randomness.

16. Probabilities involving runs can be quite difficult to compute. That the probability of a run of three or more heads in 10 independent tosses of a fair coin is $(1/2) + (1/128) = 0.508$ can be found by clever counting. A general treatment using advanced methods appears in Section XIII.7 of William Feller, *An Introduction to Probability Theory and Its Applications*, Vol. 1, 3rd ed., Wiley, 1968.

17. R. Vallone and A. Tversky, "The hot hand in basketball: On the misperception of random sequences," *Cognitive Psychology*, 17 (1985), pp. 295–314. A later series of articles that debate the independence question is A. Tversky and T. Gilovich, "The cold facts about the 'hot hand' in basketball," *Chance*, 2, No. 1 (1989), pp. 16–21; P. D. Larkey, R. A. Smith, and J. B. Kadane, "It's OK to believe in the 'hot hand,'" *Chance*, 2, No. 4 (1989), pp. 22–30; and A. Tversky and T. Gilovich, "The 'hot hand': statistical reality or cognitive illusion?" *Chance*, 2, No. 4 (1989), pp. 31–34.

18. Based on a study discussed in S. Atkinson, G. McCabe, C. Weaver, S. Abrams, and K. O'Brien, "Are current calcium recommendations for adolescents higher than needed to achieve optimal peak bone mass? The controversy," *Journal of Nutrition*, 138, No. 6 (2008), pp. 1182–1186.

19. S. A. Rahimtoola, "Outcomes 15 years after valve replacement with a mechanical vs. a prosthetic valve: Final report of the Veterans Administration randomized trial," American College of Cardiology, found online at `acc.org/education/online/trials/acc2000/15yr.htm`.

20. A description and summary of this 2009 survey can be found at `musically.com/theleadingquestion/downloads/090713-filesharing.pdf`.

21. Information about Internet users comes from sample surveys carried out by the Pew Internet and American Life Project, found online at `pewinternet.org`. The music-downloading data were collected in 2003.

22. These probabilities come from studies by the sociologist Harry Edwards, reported in the *New York Times*, February 25, 1986.

23. See Note 21.

24. Based on *The Ethics of American Youth—2008*, available from the Josephson Institute at `charactercounts.org/programs/reportcard`.

25. From `irs.gov/taxstats`.

26. See `nces.ed.gov/programs/digest`. Data are from Table 311 of the 2007 *Digest of Education Statistics*.

27. From the 2006 *Statistical Abstract of the United States*, Table 474.

28. From the Bureau of Labor Statistics, found online at `bls.gov/data`

Chapter 6

1. *Drawing the Line: Sexual Harassment on Campus*, a report from the American Association of

University Women (AAUW) Educational Foundation published in 2006. See www.aauw.org/.

2. From the DFC Intelligence Web site dfcint.com/.

3. Pew Internet Project Memo by Amanda Lenhart et al., dated December 7, 2008; available at pewinternet.org/pdfs/PIP_Adult_gaming_memo.pdf.

4. "Community Bank Competitiveness Survey," *ABA Banking Journal,* 2008. The survey is available at nxtbook.com/nxtbooks/sb/ababj-compsurv08/index.php.

5. The actual distribution of X based on an SRS from a finite population is the *hypergeometric distribution*. Details regarding this distribution can be found in Sheldon Ross, *A First Course in Probability,* 8th ed., Prentice Hall, 2010.

6. The full online clothing store ratings are featured in the December 2008 issue of *Consumer Reports* and are available online at www.ConsumerReports.org/.

7. Details of exact binomial procedures can be found in Myles Hollander and Douglas Wolfe, *Nonparametric Statistical Methods,* 2nd ed., Wiley, 1999.

8. J. H. Pryor et al., *The American Freshman: National Norms, Fall 2011,* Higher Education Research Institute, UCLA, 2011.

9. Heather Tait, *Aboriginal Peoples Survey, 2006: Inuit Health and Social Conditions,* Social and Aboriginal Statistics Division, Statistics Canada, 2008; available at statcan.gc.ca/pub.

10. From a story posted on December 20, 2008, at stuff.co.nz. A question based on this survey was used in Michael Feldman's "Whad' Ya Know Quiz" in December 2008. See notmuch.com.

11. See news.teamxbox.com/xbox/18254.

12. See the "National Survey of Student Engagement: The College Student Report 2011", available at nsse.iub.edu/index.cfm.

13. Information about the survey can be found online at www.soc.duke.edu/natcong.

14. Results from the 2006 Pew Research Center report titled "Americans and their cars: Is the romance on the skids?" The report is available online at pewresearch.org.

15. "Report card 2010: The ethics of American youth," 2010; available at josephsoninstitute.org/reportcard.

16. Data from Roland J. Thorpe, Jr., and based on his analysis of "Health ABC," a 10-year longitudinal study of older adults supported by the Laboratory of Epidemiology, Demography, and Biometry of the National Institute on Aging. Additional analyses are given in his PhD dissertation, "Relationship between pet ownership, physical activity, and human health in an elderly population," Purdue University, 2004.

17. Michael McCulloch et al., "Diagnostic accuracy of canine scent detection in early- and late-stage lung and breast cancers," *Integrative Cancer Therapies,* 5, No. 1 (2006), pp. 30–39.

18. Data from Guohua Li and Susan P. Baker, "Alcohol in fatally injured bicyclists," *Accident Analysis and Prevention,* 26 (1994), pp. 543–548.

19. See, for example, sciencedaily.com/releases/2009/11/091117094833.htm for a collection of links concerning this problem.

20. This information was posted on the Podcast Alley Web site, podcastalley.com, on March 27, 2012.

21. Pew Internet Project Data Memo by Mary Madden, dated August 2008; available at pewinternet.org.

22. Pew Internet Project Data Memo by Amanda Lenhart et al., dated December 7, 2008; available at pewinternet.org.

23. This 2011 report can be found at www.npd.com.

24. Monica Macaulay and Colleen Brice, "Don't touch my projectile: Gender bias and stereotyping in syntactic examples," *Language,* 73, No. 4 (1997), pp. 798–825. The first part of the title is a direct quote from one of the texts.

25. From the Entertainment Software Association Web site at theesa.com/facts/.

26. From a Pew Internet report "Teens, video games, and civics," by Amanda Lehnart et al., September 16, 2008; available at pewinternet.org.

27. Data can be found at the International Association for the Wireless Telecommunications Industry Web site www.ctia.org.

28. Lee Rainie et al., "The state of music downloading and file-sharing online," Pew Internet Project and Comscore Media Metrix Data Memo, 2004; available at pewinternet.org.

29. Based on a Pew Research Poll conducted in March-May 2011; availabgle at pewresearch.org.

30. Based on Robert T. Driescher, "A quality swing with Ping," *Quality Progress,* August 2001, pp. 37–41.

31. Dennis N. Bristow and Richard J. Sebastian, "Holy cow! Wait till next year! A closer look at the brand loyalty of Chicago Cubs baseball fans," *Journal of Consumer Marketing,* 18 (2001), pp. 256–275.

32. S. W. Lagakos, B. J. Wessen, and M. Zelen, "An analysis of contaminated well water and health effects in Woburn, Massachusetts," *Journal of the American Statistical Association,* 81 (1986), pp. 583–596, and the following discussion. This case is the basis for the movie *A Civil Action*.

33. This survey and others that study issues related to college students can be found at nelliemae.com/.

Chapter 7

1. H. G. Lund et al., "Sleep patterns and predictors of disturbed sleep in a large population of college students," *Journal of Adolescent Health,* 46 (2010), pp. 124–132.

2. These results were published on January 27, 2011 at blog.appsfire.com.

3. Haipeng Shen, "Nonparametric regression for problems involving lognormal distributions," PhD diss., University of Pennsylvania, 2003. Thanks to Haipeng Shen and Larry Brown for sharing the data.

4. Average hours per month obtained from "The cross-platform report, quarter 3, 2011," Nielsen Company, 2012.

5. C. Don Wiggins, "The legal perils of 'underdiversification'—a case study," *Personal Financial Planning,* 1, No. 6 (1999), pp. 16–18.

6. These data were collected as part of a larger study of dementia patients conducted by Nancy Edwards, School of Nursing, and Alan Beck, School of Veterinary Medicine, Purdue University.

7. These recommendations are based on extensive computer work. See, for example, Harry O. Posten, "The robustness of the one-sample *t*-test over the Pearson system," *Journal of Statistical Computation and Simulation,* 9 (1979), pp. 133–149; and E. S. Pearson and N. W. Please, "Relation between the shape of population distribution and the robustness of four simple test statistics," *Biometrika,* 62 (1975), pp. 223–241.

8. K. Hampton et al., "Why most Facebook users get more than they give," Pew Internet and American Life Project, 2012.

9. Christine L. Porath and Amir Erez, "Overlooked but not untouched: How rudeness reduces onlookers' performance on routine and creative tasks," *Organizational Behavior and Human Decision Processes,* 109 (2009), pp. 29–44.

10. The vehicle is a 2002 Toyota Prius.

11. Data provided by Betsy Hoza, Department of Psychological Sciences, Purdue University.

12. Victor Lun et al., "Evaluation of nutritional intake in Canadian high-performance athletes," *Clinical Journal of Sports Medicine,* 19, No. 5 (2009), pp. 405–411.

13. James A. Levine, Norman L. Eberhardt, and Michael D. Jensen, "Role of nonexercise activity thermogenesis in resistance to fat gain in humans," *Science,* 283 (1999), pp. 212–214. Data for this study are available from the *Science* Web site, www.sciencemag.org.

14. These data were collected in connection with a bone health study at Purdue University and were provided by Linda McCabe.

15. Data provided by Joseph A. Wipf, Department of Foreign Languages and Literatures, Purdue University.

16. Summary information can be found at the National Center for Health Statistics Web site, www.cdc.gov/nchs/nhanes.htm.

17. Detailed information about the conservative *t* procedures can be found in Paul Leaverton and John J. Birch, "Small sample power curves for the two sample location problem," *Technometrics,* 11 (1969), pp. 299–307; in Henry Scheffé, "Practical solutions of the Behrens-Fisher

problem," *Journal of the American Statistical Association,* 65 (1970), pp. 1501–1508; and in D. J. Best and J. C. W. Rayner, "Welch's approximate solution for the Behrens-Fisher problem," *Technometrics,* 29 (1987), pp. 205–210.

18. This example is adapted from Maribeth C. Schmitt, "The effects of an elaborated directed reading activity on the metacomprehension skills of third graders," PhD diss., Purdue University, 1987.

19. See the extensive simulation studies in Harry O. Posten, "The robustness of the two-sample *t* test over the Pearson system," *Journal of Statistical Computation and Simulation,* 6 (1978), pp. 295–311.

20. Sogol Javaheri et al., "Sleep quality and elevated blood pressure in adolescents," *Circulation,* 118 (2008), pp. 1034–1040.

21. This study is reported in Roseann M. Lyle et al., "Blood pressure and metabolic effects of calcium supplementation in normotensive white and black men," *Journal of the American Medical Association,* 257 (1987), pp. 1772–1776. The individual measurements in Table 7.4 were provided by Dr. Lyle.

22. C. E. Cryfer et al., "Misery is not miserly: Sad and self-focused individuals spend more," *Psychological Science,* 19 (2008), pp. 525–530.

23. A. A. Labroo et al., "Of frog wines and frowning watches: semantic priming, perceptual fluency, and brand evaluation," *Journal of Consumer Research,* 34 (2008), pp. 819–831.

24. The 2008 study can be found at `www.qsrmagazine.com/reports/drive-thru_time_study/2008/consumers-1.phtml`.

25. Grant D. Brinkworth et al., "Long-term effects of a very low-carbohydrate diet and a low-fat diet on mood and cognitive function," *Archives of Internal Medicine,* 169 (2009), pp. 1873–1880.

26. B. Bakke et al., "Cumulative exposure to dust and gases as determinants of lung function decline in tunnel construction workers," *Occupational Environmental Medicine,* 61 (2004), pp. 262–269.

27. Samara Joy Nielsen and Barry M. Popkin, "Patterns and trends in food portion sizes, 1977–1998," *Journal of the American Medical Association,* 289 (2003), pp. 450–453.

28. Gordana Mrdjenovic and David A. Levitsky, "Nutritional and energetic consequences of sweetened drink consumption in 6- to 13-year old children," *Journal of Pediatrics,* 142 (2003), pp. 604–610.

29. David Han-Kuen Chu, "A test of corporate advertising using the elaboration likelihood model," MS thesis, Purdue University, 1993.

30. M. F. Picciano and R. H. Deering, "The influence of feeding regimens on iron status during infancy," *American Journal of Clinical Nutrition,* 33 (1980), pp. 746–753.

31. This city's restaurant inspection data can be found at `www.jsonline.com/watchdog/dataondemand`.

32. G. E. Smith et al., "A cognitive training program based on principles of brain plasticity: Results from the improvement in memory with plasticity-based adaptive cognitive training (IMPACT) study," *Journal of the American Geriatrics Society,* 57, No. 4 (2009), pp. 594–603.

33. Based on Loren Cordain et al., "Influence of moderate daily wine consumption on body weight regulation and metabolism in healthy free-living males," *Journal of the American College of Nutrition,* 16 (1997), pp. 134–139.

34. B. Wansink et al., "Fine as North Dakota wine: Sensory expectations and the intake of companion foods," *Physiology & Behavior,* 90 (2007), pp. 712–716.

35. Douglas J. Levey et al., "Urban mockingbirds quickly learn to identify individual humans," *Proceedings of the National Academy of Sciences,* 106 (2009), pp. 8959–8962.

36. Anne Z. Hoch et al., "Prevalence of the female athlete triad in high school athletes and sedentary students," *Clinical Journal of Sports Medicine,* 19 (2009), pp. 421-428.

37. This exercise is based on events that are real. The data and details have been altered to protect the privacy of the individuals involved.

38. Based loosely on D. R. Black et al., "Minimal interventions for weight control: A cost-effective alternative," *Addictive Behaviors,* 9 (1984), pp. 279–285.

39. These data were provided by Professor Sebastian Heath, School of Veterinary Medicine, Purdue University.

40. J. W. Marr and J. A. Heady, "Within- and between-person variation in dietary surveys: Number of days needed to classify individuals," *Human Nutrition: Applied Nutrition,* 40A (1986), pp. 347–364.

41. Data taken from the Web site www.realtor.com on April 5, 2012.

Chapter 8

1. See, for example, www.pilatesmethodalliance.org/.

2. The average starting salary is taken from a 2011 summer salary survey by the National Association of Colleges and Employers (NACE).

3. Average hours per month obtained from "The cross-platform report, quarter 3, 2011," Nielsen Company, 2012.

4. Based on information reported in "How undergraduate students use credit cards: Sallie Mae's National Study of Usage Rates and Trends 2009," found online at www.salliemae.com/about/news_info/research/credit_card_study.

5. The site can be found at thekaraokechannel.com.

6. Results found in the analysis presented in Mintel's "Mobile and casual gaming, U.S.," December 2010.

7. 2011 press release from the *Student Monitor* ; available at www.studentmonitor.com.

8. Elizabeth Mendes, "U.S. job satisfaction struggles to recover to 2008 levels," Gallup News Service, May 31, 2011; available at www.gallup.com/poll/.

9. R. A. Fisher, "The arrangement of field experiments," *Journal of the Ministry of Agriculture of Great Britain,* 33 (1926), p. 504, quoted in Leonard J. Savage, "On rereading R. A. Fisher," *Annals of Statistics,* 4 (1976), p. 471. Fisher's work is described in a biography by his daughter: Joan Fisher Box, *R. A. Fisher: The Life of a Scientist,* Wiley, 1978.

10. The editorial was written by Phil Anderson. See *British Medical Journal,* 328 (2004), pp. 476–477. A letter to the editor on this topic by Doug Altman and J. Martin Bland appeared shortly after. See "Confidence intervals illuminate absence of evidence," *British Medical Journal,* 328 (2004), pp. 1016–1017.

11. A. Kamali et al., "Syndromic management of sexually-transmitted infections and behavior change interventions on transmission of HIV-1 in rural Uganda: A community randomised trial," *Lancet,* 361 (2003), pp. 645–652.

12. T. D. Sterling, "Publication decisions and their possible effects on inferences drawn from tests of significance—or vice versa," *Journal of the American Statistical Association,* 54 (1959), pp. 30–34. Related comments appear in J. K. Skipper, A. L. Guenther, and G. Nass, "The sacredness of 0.05: A note concerning the uses of statistical levels of significance in social science," *American Sociologist,* 1 (1967), pp. 16–18.

13. For a good overview of these issues, see Bruce A. Craig, Michael A. Black, and Rebecca W. Doerge, "Gene expresssion data: The technology and statistical analysis," *Journal of Agricultural, Biological, and Environmental Statistics,* 8 (2003), pp. 1–28.

14. Erick H. Turner et al., "Selective publication of antidepressant trials and its influence on apparent efficacy," *New England Journal of Medicine,* 358 (2008), pp. 252–260.

15. See Chapter 7, Note 12.

16. Padmaja Ayyagari and Jody L. Sindelar, "The impact of job stress on smoking and quitting: Evidence from the HRS," National Bureau of Economic Research, Working Paper 15232 (2009).

17. Data from Joan M. Susic, "Dietary phosphorus intakes, urinary and peritoneal phosphate excretion and clearance in continuous ambulatory peritoneal dialysis patients," MS thesis, Purdue University, 1985.

18. Mugdha Gore and Joseph Thomas, "Store image as a predictor of store patronage for nonprescription medication purchases: A multiattribute model approach," *Journal of Pharmaceutical Marketing & Management,* 10 (1996), pp. 45–68.

19. C. M. Weaver et al., "Quantification of biochemical markers of bone turnover by kinetic measures of bone formation and resorption in young healthy females," *Journal of Bone and Mineral Research,* 12 (1997), pp. 1714–1720.

PART II Review

1. From statcan.gc.ca/tables-tableaux/sum-som/l01/cst01/demo02a-eng.htm.
2. Reynol Junco, "Too much face and not enough books: The relationship between multiple indices of Facebook use and academic performance," *Computers in Human Behavior,* 2011, doi:10.1016/j.chb.2011.08.026.
3. Victor Lun et al., "Dietary supplementation practices in Canadian high-performance athletes," *International Journal of Sport Nutrition and Exercise Metabolism,* 22 (2012), pp. 31–37.
4. Braz Camargo et al., "Interracial friendships in college," *Journal of Labor Economics,* 28 (2010), pp. 861–892.
5. Seth J. Schwartz et al., "The association of well-being with health risk behaviors in college-attending young adults," *Applied Developmental Science,* 15, No. 1 (2011), pp. 20–36.
6. Jihui Zhang et al., "Insomnia, sleep quality, pain, and somatic symptoms: Sex differences and shared genetic components," *Pain,* 153 (2012), pp. 666–673.
7. Find this report and additional resources on helping teens become responsible drivers at www.libertymutual.com/teendriving.

Chapter 9

1. J. Cantor, "Long-term memories of frightening media often include lingering trauma symptoms," poster paper presented at the Association for Psychological Science Convention, New York, May 26, 2006.
2. When the expected cell counts are small, it is best to use a test based on the exact distribution rather than the chi-square approximation, particularly for 2×2 tables. Many statistical software systems offer an "exact" test as well as the chi-square test for 2×2 tables.
3. E. Y. Peck, "Gender differences in film-induced fear as a function of type of emotion measure and stimulus content: A meta-analysis and laboratory study," PhD dissertation. University of Wisconsin-Madison.
4. The sampling procedure was designed by George McCabe. It was carried out by Amy Conklin, an undergraduate Honors student in the Department of Foods and Nutrition at Purdue University.
5. The analysis could also be performed using a two-way table to compare the states of the selected and not-selected students. Since the selected students are a relatively small percent of the total sample, the results will be approximately the same.
6. See the M&M Mars Web site at us.mms.com/us/about/products for this and other information.
7. See nhcaa.org.
8. These data are a composite based on several actual audits of this type.
9. See, for example, R. Benford and J. Gess-Newsome, "Factors affecting student academic success in gateway courses at Northern Arizona University," 2006; available at eric.ed.gov.
10. Data provided by Professor Marcy Towns, Department of Chemistry, Purdue University.
11. See Note 10.
12. From the "Survey of Canadian career college students phase II: In-school student survey," 2008; available at hrsdc.gc.ca/eng/publications_resources.
13. For an overview of remote deposit capture, see remotedepositcapture.com/overview/rdc.overview.aspx.
14. From the "Community Bank Competitiveness Survey," 2008, *ABA Banking Journal;* available at nxtbook.com/nxtbooks/sb/ababj-compsurv08/index.php.
15. See Chapter 4, Note 28.
16. Based on *The Ethics of American Youth—2008,* available from the Josephson Institute at charactercounts.org/programs/reportcard.
17. See Note 1.
18. See the U.S. Bureau of the Census Web site at census.gov/population/socdemo/school for these and similar data.
19. Based on data in Mike Planty et al., "Volunteer service by young people from high school through early adulthood," National Center for Educational Statistics Report NCES 2004–365.

20. S. R. Davy, B. A. Benes, and J. A. Driskell, "Sex differences in dieting trends, eating habits, and nutrition beliefs of a group of midwestern college students," *Journal of the American Dietetic Association,* 10 (2006), pp. 1673–1677.

21. Tom Reichert, "The prevalence of sexual imagery in ads targeted to young adults," *Journal of Consumer Affairs,* 37 (2003), pp. 403–412.

22. George R. Milne, "How well do consumers protect themselves from identity theft?" *Journal of Consumer Affairs,* 37 (2003), pp. 388–402.

23. Based on information in "NCAA 2003 national study of collegiate sports wagering and associated health risks"; available at the NCAA Web site, `ncaa.org`.

24. Ethan J. Temeles and W. John Kress, "Adaption in a plant-hummingbird association," *Science,* 300 (2003), pp. 630–633.

25. These data are from an Undergraduate Task Force study of student retention in the Purdue University College of Science.

26. Daniel Cassens et al., "Face check development in veneered furniture panels," *Forest Products Journal,* 53 (2003), pp. 79–86.

27. James A. Taylor et al., "Efficacy and safety of echinacea in treating upper respiratory tract infections in children," *Journal of the American Medical Association,* 290 (2003), pp. 2824–2830.

28. Karen L. Dean, "Herbalists respond to *JAMA* echinacea study," *Alternative and Complementary Therapies,* 10 (2004), pp. 11–12.

29. The report can be found at `nsse.iub.edu/pdf/NSSE2005\_annual\_report.pdf`.

30. Noel Cressie, *Statistics for Spatial Data,* Wiley, 1993. The significance test result that we report is one of several that could be used to address this question. See pp. 607–609 of the Cressie book for more details.

Chapter 10

1. Data based on Michael L. Mestek et al., "The relationship between pedometer-determined and self-reported physical activity and body composition variables in college-aged men and women," *Journal of American College Health,* 57 (2008), pp. 39–44.

2. The vehicle is a 1997 Pontiac transport van.

3. Tuition rates for 2000 from the "2000–2001 Tuition and required fees report," University of Missouri. Tuition rates for 2008 are available at `colleges.collegetoolkit.com/college/main.aspx`.

4. These are some of the data from the EESEE story "Blood Alcohol Content," found on the text Web site.

5. M. Mondello and J. Maxcy, "The impact of salary dispersion and performance bonuses in NFL organizations," *Management Decision,* 47 (2009), pp. 110–123.

6. These data were collected from `www.cbssports.com/nfl/playerrankings/regularseason/` and `content.usatoday.com/sports/football/nfl/salaries/`.

7. Selling price and assessment value available at `php.jconline.com/propertysales/propertysales.php`.

8. Data available at `www.ncdc.noaa.gov/oa/climate/sd`.

9. Based on Dan Dauwalter's master's thesis in the Department of Forestry and Natural Resources at Purdue University. More information is available in Daniel C. Dauwalter et al., "An index of biotic integrity for fish assemblages in Ozark Highland streams of Arkansas," *Southeastern Naturalist,* 2 (2003), pp. 447–468. These data were provided by Emmanuel Frimpong.

10. Net cash flow data from Sean Collins, *Mutual Fund Assets and Flows in 2000,* Investment Company Institute, 2001; and "Trends in mutual fund investing" at `www.ici.org/stats/mf/arctrends/index.html`. The raw data were converted to real dollars using annual average values of the Consumer Price Index.

11. These data are from G. Geri and B. Palla, "Considerazioni sulle più recenti osservazioni ottiche alla Torre Pendente di Pisa," *Estratto dal Bollettino della Società Italiana di Topografia e Fotogrammetria,* 2 (1988), pp. 121–135. Professor Julia Mortera of the University of Rome provided valuable assistance with the translation.

12. Data on a sample of 12 of 56 perch in a data set contributed to the *Journal of Statistics Education* data archive `www.amstat.org/`

publications/jse/ by Juha Puranen of the University of Helsinki.

Chapter 11

1. Results of the study are reported in P. F. Campbell and G. P. McCabe, "Predicting the success of freshmen in a computer science major," *Communications of the ACM,* 27 (1984), pp. 1108–1113.
2. R. M. Smith and P. A. Schumacher, "Predicting success for actuarial students in undergraduate mathematics courses," *College Student Journal,* 39, No. 1 (2005), pp. 165–177.
3. Kathleen E. Miller, "Wired: Energy drinks, jock identity, masculine norms, and risk taking," *Journal of American College Health,* 56 (2008), pp. 481–489.
4. From a table entitled "Largest Indianapolis-area architectural firms," *Indianapolis Business Journal,* December 16, 2003.
5. The data were obtained from the Internet Movie Database (IMDb), available at www.imdb.com, on April 20, 2010.
6. The 2011–2012 table of 200 top universities can be found at www.timeshighereducation.co.uk.
7. This data set was provided by Joanne Lasrado, Department of Foods and Nutrition, Purdue University.

Chapter 12

1. R. Kanai et al., "Online social network size is reflected in human brain structure," *Proceedings of the Royal Society—Biological Sciences,* published online October 19, 2011, in advance of the print journal.
2. Based on Stephanie T. Tong et al., "Too much of a good thing? The relationship between number of friends and interpersonal impressions on Facebook," *Journal of Computer-Mediated Communication,* 13 (2008), pp. 531–549.
3. This rule is intended to provide a general guideline for deciding when serious errors may result by applying ANOVA procedures. When the sample sizes in each group are very small, this rule may be a little too conservative. For unequal sample sizes, particular difficulties can arise when a relatively small sample size is associated with a population having a relatively large standard deviation.
4. Penny M. Simpson et al., "The eyes have it, or do they? The effects of model eye color and eye gaze on consumer ad response," *Journal of Applied Business and Economics,* 8 (2008), pp. 60–71.
5. Jesus Tanguma et al., "Shopping and bargaining in Mexico: The role of women," *Journal of Applied Business and Economics,* 9 (2009), pp. 34–40.
6. A. A. Labroo et al., "Of frog wines and frowning watches: semantic priming, perceptual fluency, and brand evaluation," *Journal of Consumer Research,* 34 (2008), pp. 819–831.
7. Sangwon Lee and Seonmi Lee, "Multiple play strategy in global telecommunication markets: An empirical analysis," *International Journal of Mobile Marketing,* 3 (2008), pp. 44–53.
8. P. Bartel et al., "Attention and working memory in resident anaesthetists after night duty: Group and individual effects," *Occupational and Environmental Medicine,* 61 (2004), pp. 167–170.
9. Adrian C. North et al., "The effect of musical style on restaurant consumers' spending," *Environment and Behavior,* 35 (2003), pp. 712–718.
10. Based on Jack K. Trammel, "The impact of academic accommodations on final grades in a post secondary setting," *Journal of College Reading and Learning,* 34 (2003), pp. 76–90.
11. Christie N. Scollon et al., "Emotions across cultures and methods," *Journal of Cross-cultural Psychology,* 35 (2004), pp. 304–326.
12. We thank Ethan J. Temeles of Amherst College for providing the data. His work is described in Ethan J. Temeles and W. John Kress, "Adaptation in a plant-hummingbird association," *Science,* 300 (2003), pp. 630–633.
13. The data were provided by James Kaufman. The study is described in James C. Kaufman, "The cost of the muse: Poets die young," *Death Studies,* 27 (2003), pp. 813–821. The quotation from Yeats appears in this article.
14. Woo Gon Kim et al., "Influence of institutional DINESERV on customer satisfaction, return intention, and word-of-mouth," *International*

Journal of Hospitality Management, 28 (2009), pp. 10–17.

15. Samuel S. Kim and Jerome Agrusa, "Segmenting Japanese tourists to Hawaii according to tour purpose," *Journal of Travel and Tourism Marketing,* 24 (2008), pp. 63–80.
16. Data provided by Jo Welch, Department of Foods and Nutrition, Purdue University.
17. Based on A. A. Adish et al., "Effect of consumption of food cooked in iron pots on iron status and growth of young children: A randomised trial," *Lancet,* 353 (1999), pp. 712–716.
18. Steve Badylak et al., "Marrow-derived cells populate scaffolds composed of xenogeneic extracellular matrix," *Experimental Hematology,* 29 (2001), pp. 1310–1318.
19. This exercise is based on data provided from a study conducted by Jim Baumann and Leah Jones, School of Education, Purdue University.

Chapter 13

1. See `www.who.int/topics/malaria/en/` for more information about malaria.
2. This example is based on a study described at `clinicaltrials.gov/ct2/show/NCT00623857`.
3. We present the two-way ANOVA model and analysis for the general case in which the sample sizes may be unequal. If the sample sizes vary a great deal, serious complications can arise. There is no longer a single standard ANOVA analysis. Most computer packages offer several options for the computation of the ANOVA table when cell counts are unequal. When the counts are approximately equal, all methods give essentially the same results.
4. Sara N. Bleich et al., "Increasing consumption of sugar-sweetened beverages among US adults: 1988–1994 to 1999–2004," *American Journal of Clinical Nutrition,* 89 (2009), pp. 372–381.
5. Rick Bell and Patricia L. Pliner, "Time to eat: The relationship between the number of people eating and meal duration in three lunch settings," *Appetite,* 41 (2003), pp. 215–218.
6. Carolyn Drake and Jamel Ben El Heni, "Synchronizing with music: Intercultural differences," *Annals of the New York Academy of Sciences,* 99 (2003), pp. 429–437.
7. Example 13.10 is based on a study described in P. D. Wood et al., "Plasma lipoprotein distributions in male and female runners," in P. Milvey (ed.), *The Marathon: Physiological, Medical, Epidemiological, and Psychological Studies,* New York Academy of Sciences, 1977.
8. Vincent P. Magnini and Kiran Karande, "The influences of transaction history and thank you statements in service recovery," *International Journal of Hospitality Management,* 28 (2009), pp. 540–546.
9. Brian Wansink et al., "The office candy dish: Proximity's influence on estimated and actual consumption," *International Journal of Obesity,* 30 (2006), pp. 871–875.
10. Willemijn M. van Dolen, Ko de Ruyter, and Sandra Streukens, "The effect of humor in electronic service encounters," *Journal of Economic Psychology,* 29 (2008), pp. 160–179.
11. Jane Kolodinsky et al., "Sex and cultural differences in the acceptance of functional foods: A comparison of American, Canadian, and French college students," *Journal of American College Health,* 57 (2008), pp. 143–149.
12. Judith McFarlane et al., "An intervention to increase safety behaviors of abused women," *Nursing Research,* 51 (2002), pp. 347–354.
13. Gad Saad and John G. Vongas, "The effect of conspicuous consumption on men's testosterone levels," *Organizational Behavior and Human Decision Processes,* 110 (2009), pp. 80–92.
14. Klaus Boehnke et al., "On the interrelation of peer climate and school performance in mathematics: A German-Canadian-Israeli comparison of 14-year-old school students," in B. N. Setiadi, A. Supratiknya, W. J. Lonner, and Y. H. Poortinga (eds.), *Ongoing Themes in Psychology and Culture* (online ed.), International Association for Cross-Cultural Psychology, 2004.
15. Results of the study are reported in P. F. Campbell and G. P. McCabe, "Predicting the success of freshmen in a computer science major," *Communications of the ACM,* 27 (1984), pp. 1108–1113.

16. Tamar Kugler et al., "Trust between individuals and groups: Groups are less trusting than individuals but just as trustworthy," *Journal of Economic Psychology,* 28 (2007), pp. 646–657.

17. Based on A. A. Adish et al., "Effect of consumption of food cooked in iron pots on iron status and growth of young children: A randomised trial," *Lancet,* 353 (1999), pp. 712–716.

18. For a summary of this study and other research in this area, see Stanley Coren and Diane F. Halpern, "Left-handedness: A marker for decreased survival fitness," *Psychological Bulletin,* 109 (1991), pp. 90–106.

19. Data provided by Neil Zimmerman of the Purdue University School of Health Sciences.

20. I. C. Feller et al., "Sex-biased herbivory in jack-in-the-pulpit (*Arisaema triphyllum*) by a specialist thrips (*Heterothrips arisaemae*)," in *Proceedings of the 7th International Thysanoptera Conference,* Reggio Calabria, Italy, pp. 163–172.

Chapter 14

1. Logistic regression models for the general case where there are more than two possible values for the response variable have been developed. These are considerably more complicated and are beyond the scope of our present study. For more information on logistic regression, see A. Agresti, *An Introduction to Categorical Data Analysis,* 2nd ed., Wiley, 2002; and D. W. Hosmer and S. Lemeshow, *Applied Logistic Regression,* 2nd ed., Wiley, 2000.

2. This example is taken from a classic text written by a contemporary of R. A. Fisher, the person who developed many of the fundamental ideas of statistical inference that we use today. The reference is D. J. Finney, *Probit Analysis,* Cambridge University Press, 1947. Although not included in the analysis, it is important to note that the experiment included a control group that received no insecticide. No aphids died in this group. Also, although we have chosen to call the response "dead," in Finney's book the category is described as "apparently dead, moribund, or so badly affected as to be unable to walk more than a few steps." This is an early example of the need to make careful judgments when defining variables to be used in a statistical analysis. An insect that is "unable to walk more than a few steps" is unlikely to eat very much of a chrysanthemum plant!

3. P. Strazzullo et al., "Salt intake, stroke, and cardiovascular disease," *BMJ,* 339 (2009). The meta-analysis combined data from 14 study cohorts taken from 10 different studies.

4. Based on Leigh J. Maynard and Malvern Mupandawana, "Tipping behavior in Canadian restaurants," *International Journal of Hospitality Management,* 28 (2009), pp. 597–603.

5. The poll is part of the American Express Retail Index Project and is reported in *Stores*, December 2000, pp. 38–40.

6. Monica Macaulay and Colleen Brice, "Don't touch my projectile: Gender bias and stereotyping in syntactic examples," *Language,* 73, No. 4 (1997), pp. 798–825.

7. Anthony A. Noce and Larry McKeown, "A new benchmark for Internet use: A logistic modeling of factors influencing Internet use in Canada, 2005," *Government Information Quarterly,* 25 (2008), pp. 462–476.

8. Dong-Chul Seo et al., "Relations between physical activity and behavioral and perceptual correlates among midwestern college students," *Journal of American College Health,* 56 (2007), pp. 187–197.

9. These economic trend reports can be found at `mr.pricegrabber.com`. These results are based on the June 2009 report.

10. Brennan Davis and Christopher Carpenter, "Proximity of fast-food restaurants to schools and adolescent obesity," *American Journal of Public Health,* 99 (2009), pp. 505–510.

PART III Review

1. Catherine Hill and Holly Kearl, *Crossing the Line: Sexual Harassment at School*, American Association of University Women, 2011.

2. Based on `pewsocialtrends.org/files/2011/08/online-learning.pdf`.

3. Bryan Raudenbush et al., "Pain threshold and tolerance differences among intercollegiate athletes: Implication of past sports injuries and willingness to compete among sports teams,"

North American Journal of Psychology, 14 (2012), pp. 85–94.

4. This annual report can be found at `www.kiplinger.com`.

5. Eileen Wood et al., "Examining the impact of off-task multi-tasking with technology on real-time classroom learning," *Computers & Education*, 58 (2012), pp. 365–374.

6. Kendall J. Eskine, "Wholesome foods and wholesome morals? Organic foods reduce prosocial behavior and harshen moral judgments," *Social Psychological and Personality Science*, 2012, doi: 10.1177/1948550612447114.

7. Felix Javier Jimenez-Jimenez et al., "Influence of age and gender in motor performance in healthy subjects," *Journal of the Neurological Sciences*, 302 (2011), pp. 72–80.

8. Linda DeAngelo et al., *Completing College: Assessing Graduation Rates at Four-Year Institutions*, Higher Education Research Institute, 2011.

PHOTO CREDITS

Cover	Christine Balderas/Getty Images
p. 1	Stephen Voss/Alamy
p. 3	Design provided by: Lucille S. Chang/Words & Unwords Fulfillment vendor: Zazzle.com Photo by: Ted Szczepanski
p. 5	Steve Debenport/iStockphoto
p. 15	Igor Mojzes/iStockphoto
p. 35	Robert Daly/Alamy
p. 45	David Paul Morris/Bloomberg via Getty Images
p. 59	Mark Bowden/iStockphoto
p. 75	National Archives and Records Administration NARA
p. 77	Christo Doherty
p. 83	D. Hurst/Alamy
p. 90	Thomas Northcut/Getty Images
p. 102	Nucleus Medical Art, Inc/Phototake351
p. 104	Ken Chernus/Getty Images
p. 129	Art Wolfe/Getty Images
p. 137	Novastock/Alamy
p. 153	PhotoAlto/Alamy
p. 154	Alamy
p. 170	Brian Kushner/Dreamstime.com
p. 178	Gene Chutka/iStockphoto
p. 203	INSADCO Photography/Alamy
p. 210	Jacob Wackerhausen/iStockphoto
p. 235	Andrew Davidhazy
p. 250	Brand X/Getty Images
p. 251	Profimedia.CZ a.s./Alamy
p. 273	Mitchell Layton/Getty Images
p. 312	David Moore
p. 315	Mark Andersen/Getty Images
p. 319	Andrew Bret Wallis/Getty Images
p. 332	Eric Audras/Getty Images
p. 341	Marilyn Angel Wynn/Getty Images
p. 356	Chuck Eckert/Alamy
p. 363	Paul Simcock/Getty Images
p. 384	Richard Kail/Photo Researchers, Inc.
p. 396	Robert Warren/Getty Images
p. 401	AP/Wide World Photos
p. 403	Olivier Lantzendörffer/iStockphoto
p. 420	Mary Kempel, Maple River Winery
p. 420	Millard H. Sharp/Photo Researchers, Inc.
p. 423	Getty Images
p. 443	Martin Shields
p. 444	Photo by The Photo Works
p. 469	moodboard/Alamy
p. 470	Jame Shaffe/PhotoEdit
p. 486	Design Pics Inc./Alamy
p. 489	Charles J. Sharp/Wikimedia Commons
p. 493	Ruth Jenkinson/Getty Images
p. 496	Bubbles Photolibrary/Alamy
p. 503	© General Motors Corp. Used with permission. GM Media Archives
p. 516	Chuck Doswell/Visuals Unlimited
p. 519	FABIO MUZZI/AFP/Getty Images via Newscom
p. 522	Barry Austin Photography/Getty Images
p. 549	John Lund/Getty Images
p. 551	Radius Images/Jupiter Images
p. 579	The Advertising Archives
p. 582	Ocean/Corbis
p. 590	Photononstop/SuperStock
p. 593	Allan Munsie/Alamy
p. 594	Chris Cooper-Smith/Alamy
p. 597	Custom Medical Stock Photo
p. 602	Robert Ghement/epa/Corbis
p. 610	Jason Koski/Copyright Cornell University
p. 619	Photograph by Brad Fulton. Copyright 2011 by Dan and Dave Industries
p. 630	Nigel Cattlin/Alamy
p. 636	i love images/Food & Drink/Alamy

INDEX

Table entry for p and C is the critical value t^* with probability p lying to its right and probability C lying between $-t^*$ and t^*.

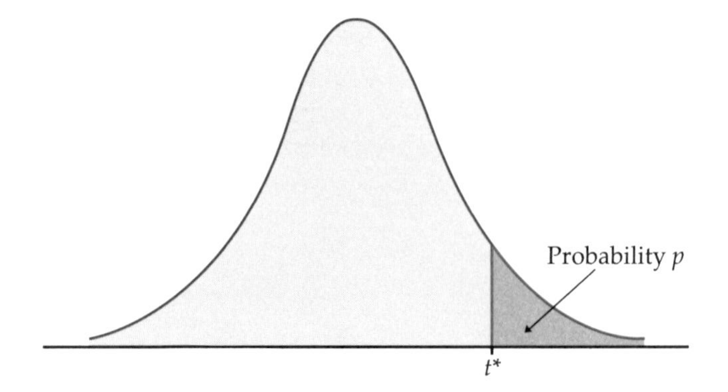

TABLE D

t distribution critical values

	Upper-tail probability p											
df	.25	.20	.15	.10	.05	.025	.02	.01	.005	.0025	.001	.0005
1	1.000	1.376	1.963	3.078	6.314	12.71	15.89	31.82	63.66	127.3	318.3	636.6
2	0.816	1.061	1.386	1.886	2.920	4.303	4.849	6.965	9.925	14.09	22.33	31.60
3	0.765	0.978	1.250	1.638	2.353	3.182	3.482	4.541	5.841	7.453	10.21	12.92
4	0.741	0.941	1.190	1.533	2.132	2.776	2.999	3.747	4.604	5.598	7.173	8.610
5	0.727	0.920	1.156	1.476	2.015	2.571	2.757	3.365	4.032	4.773	5.893	6.869
6	0.718	0.906	1.134	1.440	1.943	2.447	2.612	3.143	3.707	4.317	5.208	5.959
7	0.711	0.896	1.119	1.415	1.895	2.365	2.517	2.998	3.499	4.029	4.785	5.408
8	0.706	0.889	1.108	1.397	1.860	2.306	2.449	2.896	3.355	3.833	4.501	5.041
9	0.703	0.883	1.100	1.383	1.833	2.262	2.398	2.821	3.250	3.690	4.297	4.781
10	0.700	0.879	1.093	1.372	1.812	2.228	2.359	2.764	3.169	3.581	4.144	4.587
11	0.697	0.876	1.088	1.363	1.796	2.201	2.328	2.718	3.106	3.497	4.025	4.437
12	0.695	0.873	1.083	1.356	1.782	2.179	2.303	2.681	3.055	3.428	3.930	4.318
13	0.694	0.870	1.079	1.350	1.771	2.160	2.282	2.650	3.012	3.372	3.852	4.221
14	0.692	0.868	1.076	1.345	1.761	2.145	2.264	2.624	2.977	3.326	3.787	4.140
15	0.691	0.866	1.074	1.341	1.753	2.131	2.249	2.602	2.947	3.286	3.733	4.073
16	0.690	0.865	1.071	1.337	1.746	2.120	2.235	2.583	2.921	3.252	3.686	4.015
17	0.689	0.863	1.069	1.333	1.740	2.110	2.224	2.567	2.898	3.222	3.646	3.965
18	0.688	0.862	1.067	1.330	1.734	2.101	2.214	2.552	2.878	3.197	3.611	3.922
19	0.688	0.861	1.066	1.328	1.729	2.093	2.205	2.539	2.861	3.174	3.579	3.883
20	0.687	0.860	1.064	1.325	1.725	2.086	2.197	2.528	2.845	3.153	3.552	3.850
21	0.686	0.859	1.063	1.323	1.721	2.080	2.189	2.518	2.831	3.135	3.527	3.819
22	0.686	0.858	1.061	1.321	1.717	2.074	2.183	2.508	2.819	3.119	3.505	3.792
23	0.685	0.858	1.060	1.319	1.714	2.069	2.177	2.500	2.807	3.104	3.485	3.768
24	0.685	0.857	1.059	1.318	1.711	2.064	2.172	2.492	2.797	3.091	3.467	3.745
25	0.684	0.856	1.058	1.316	1.708	2.060	2.167	2.485	2.787	3.078	3.450	3.725
26	0.684	0.856	1.058	1.315	1.706	2.056	2.162	2.479	2.779	3.067	3.435	3.707
27	0.684	0.855	1.057	1.314	1.703	2.052	2.158	2.473	2.771	3.057	3.421	3.690
28	0.683	0.855	1.056	1.313	1.701	2.048	2.154	2.467	2.763	3.047	3.408	3.674
29	0.683	0.854	1.055	1.311	1.699	2.045	2.150	2.462	2.756	3.038	3.396	3.659
30	0.683	0.854	1.055	1.310	1.697	2.042	2.147	2.457	2.750	3.030	3.385	3.646
40	0.681	0.851	1.050	1.303	1.684	2.021	2.123	2.423	2.704	2.971	3.307	3.551
50	0.679	0.849	1.047	1.299	1.676	2.009	2.109	2.403	2.678	2.937	3.261	3.496
60	0.679	0.848	1.045	1.296	1.671	2.000	2.099	2.390	2.660	2.915	3.232	3.460
80	0.678	0.846	1.043	1.292	1.664	1.990	2.088	2.374	2.639	2.887	3.195	3.416
100	0.677	0.845	1.042	1.290	1.660	1.984	2.081	2.364	2.626	2.871	3.174	3.390
1000	0.675	0.842	1.037	1.282	1.646	1.962	2.056	2.330	2.581	2.813	3.098	3.300
z^*	0.674	0.841	1.036	1.282	1.645	1.960	2.054	2.326	2.576	2.807	3.091	3.291
	50%	60%	70%	80%	90%	95%	96%	98%	99%	99.5%	99.8%	99.9%
	Confidence level C											